Advances in Science, Technology & Innovation

IEREK Interdisciplinary Series for Sustainable Development

Advances in Science, Technology & Innovation (ASTI) is a series of peer-reviewed books based on important emerging research that redefines the current disciplinary boundaries in science, technology, and innovation (ASTI) in order to develop integrated concepts for sustainable development. It not only discusses the progress made towards securing more resources, allocating smarter solutions, and rebalancing the relationship between nature and people, but also provides in-depth insights from comprehensive research that addresses the **17 sustainable development goals (SDGs)** as set out by the UN for 2030.

The series promotes the creation and development of viable solutions for a sustainable future and a positive societal transformation with the help of integrated and innovative science-based approaches. Including interdisciplinary contributions, it presents innovative approaches and highlights how they can best support both economic and sustainable development, through better use of data, more effective institutions, and global, local, and individual action, for the welfare of all societies.

The series particularly features conceptual and empirical contributions from various interrelated fields of science, technology, and innovation, with an emphasis on digital transformation, that focus on providing practical solutions to ensure food, water, and energy security to achieve the SDGs. It also presents new case studies offering concrete examples of how to resolve sustainable urbanization and environmental issues in different regions of the world. The series is intended for professionals in research and teaching, consultancies and industry, and government and international organizations. Published in collaboration with IEREK.

The series draws on the best research papers from various IEREK's international conferences, while external conference proceedings proposals are no longer accepted as direct publications. The series does invite significant expansions of selected papers previously submitted to conferences. The Editors' role should be evident in the reviews through an intensive **double-blind peer review** process, helping authors with edits, and collaborating with them to demonstrate **great academic significance** through clear research findings and innovative solutions that go well beyond a typical conference submission.

The ASTI series welcomes external proposals and requests to publish **edited books, contributed volumes, monographs,** and **textbooks**. With the commitment to the aforementioned instructions and policy.

The ASTI series has now been accepted for Scopus (September 2020). All content published in this series will start appearing on the Scopus site in early 2021.

ASTI's editorial office helps authors and editors in producing work of the highest possible quality. For any assistance for those interested in submitting proposals please contact the editorial office at asti.serties@ierek.com

Francesco Alberti • Paola Gallo • Hasim Altan
Antonio Labalestra • Eric J. Strauss
Editors

Towards Sustainable Urban Regeneration: Utilizing Environmental, Cultural, and Social Assets

Springer

Editors
Francesco Alberti
Department of Architecture (DIDA)
University of Florence
Florence, Italy

Paola Gallo
Department of Architecture (DIDA)
University of Florence
Florence, Italy

Hasim Altan
Department of Architectural Engineering, College of Engineering
United Arab Emirates University
Al Ain, United Arab Emirates

Antonio Labalestra
Department of Architecture Constructions and Design (ArCoD)
Politecnico di Bari
Bari, Italy

Eric J. Strauss
Urban and Regional Planning, School of Planning Design and Construction Michigan State University
Michigan, MI, USA

ISSN 2522-8714 ISSN 2522-8722 (electronic)
Advances in Science, Technology & Innovation
IEREK Interdisciplinary Series for Sustainable Devel
ISBN 978-3-032-02090-1 ISBN 978-3-032-02091-8 (eBook)
https://doi.org/10.1007/978-3-032-02091-8

This Springer imprint is published by the registered company Springer Nature Switzerland AG
The registered company address is: Gewerbestrasse 11, 6330 Cham, Switzerland

Editorial Board

Preface

This volume combines a collection of research and case studies that collectively address the complex challenges and opportunities facing contemporary cities. It explores a wide spectrum of themes, ranging from green infrastructure and urban climate resilience to mobility strategies, cultural heritage, social sustainability, innovative construction, energy efficiency, analytical tools, and the digital transformation of cities. With contributions from various global contexts, the book highlights interdisciplinary approaches to urban planning and design that prioritize ecological integrity, cultural identity, and community well-being. Emphasizing both theoretical frameworks and practical applications, the chapters offer valuable insights into how cities can become more resilient, inclusive, and adaptable in response to climate change, population growth, and technological advancement. Through the integration of nature-based solutions, participatory design, sustainable building practices, and smart technologies, this volume visualizes an urban future that is not only more sustainable but also socially and environmentally responsible.

Florence, Italy	Francesco Alberti
Florence, Italy	Paola Gallo
Al Ain, United Arab Emirates	Hasim Altan
Bari, Italy	Antonio Labalestra
Michigan, MI, USA	Eric J. Strauss

Introduction

In the 30 years since the launch of the Aalborg Charter in 1994, commitment to sustainable urban development has profoundly shaped the international discourse on cities and the built environment. The Charter, emerging from the first European Conference on Sustainable Cities and Towns, laid the foundations for a shared understanding of sustainability at the urban level—an understanding that transcended national boundaries and inspired action far beyond Europe. It was the beginning of a process that saw thousands of cities across continents adopting Local Agenda 21 plans, striving to integrate environmental concerns with social and economic priorities.

Yet, as we approach the midpoint of the 2030 Agenda for Sustainable Development—adopted by all UN member states some 20 years after the Aalborg Charter—and reflect on the ambitious trajectory set by that foundational document, there is cause for deep concern. The transformative path toward sustainable development, once widely embraced, now appears increasingly fragile. A growing number of national governments are wavering in their commitments to the Sustainable Development Goals (SDGs) and the Paris Agreement on Climate Change, raising the alarming prospect that the vision of sustainability—of inclusive, resilient, and climate-conscious urban futures—could falter under the weight of political inertia, short-term interests, and global instability.

The SDGs were designed not only as a roadmap but as a pact—a promise among nations to collaborate for a better world. Within this framework, Goal 11, "Sustainable Cities and Communities," holds a pivotal role, recognizing the centrality of urban areas in the quest for a livable planet. But cities cannot thrive in isolation. They depend on the integrity of the entire system envisioned by the Agenda, a system built on the five interconnected "Ps": People, Planet, Prosperity, Peace, and Partnership. Of these, Peace and Partnership are not just goals—they are the indispensable preconditions for progress. Without peace, there can be no sustained effort toward sustainability. Without partnership, knowledge remains siloed, innovation fragmented, and progress uneven.

Today, both peace and partnership seem under siege. From geopolitical tensions and protracted conflicts to the erosion of multilateral cooperation, the conditions for meaningful global action are deteriorating. The specter of climate breakdown, compounded by social inequality and distrust in democratic institutions, casts doubt on whether the sustainability agenda will survive the coming decade in its current form. In many places, skepticism toward the very idea of sustainability is being fomented, and in some cases, academic inquiry itself is targeted, restricted, or dismissed.

And yet, in this climate of uncertainty and regression, the need for informed, courageous, and collaborative research is greater than ever. It is precisely now, as the world stumbles on its path to sustainability, that the scientific and academic communities must respond—not just by continuing their work, but by amplifying it, defending its value, and ensuring it reaches policymakers, practitioners, and citizens alike. Knowledge must not retreat into the margins. On the contrary, it must serve as a bulwark against disinformation and a beacon of hope for future generations.

The contributions gathered in this volume—drawn from the 9th International Conference on Urban Planning and Architectural Design for Sustainable Development (UPADSD), held in Florence from October 22 to 24, 2024—reflect this spirit of engaged scholarship.

Structured into four thematic sections—Green Infrastructure and Urban Climate Resilience, Urban Design and Mobility Strategies for the Regeneration of Cities and Communities, Urban Heritage and Cultural Identity: Methods and Applications for Conservation, and Design Approaches for Community Engagement and Social Sustainability—this volume illustrates the breadth of innovation and inquiry currently underway in urban studies.

From strategies for nature-based solutions and participatory design, to methods for cultural preservation and mobility planning, these papers show how sustainable development can be locally rooted and globally relevant. They also reaffirm the importance of context-sensitive approaches, adapted to the diverse challenges and potentials of different territories.

In this uncertain moment, sharing knowledge is itself a political act. It is a statement of belief in cooperation, in evidence-based decision-making, and in the capacity of cities to lead by example. We hope this book will contribute to a renewed momentum, and that it serves as a modest but meaningful step toward a future that still remains within our collective reach—if we choose to build it together.

Francesco Alberti

Acknowledgments

The editors wish to express their sincere gratitude to the chairs of the UPADSD 2023 Conference, held by the Department of Architecture at the University of Florence, as well as to the academic partners working alongside IEREK, for their valuable support and contribution throughout the editorial process.

We are also grateful to the reviewers from a range of disciplinary backgrounds who provided thoughtful and constructive evaluations of the papers selected for inclusion. Their contributions have helped ensure the academic quality and coherence of both volumes.

It is our hope that this publication will encourage sustained scholarly dialogue and foster new collaborations, while also serving as a platform to share critical advancements in urban and architectural research aimed at sustainable development.

Scientific Committee

Ayse Buckley	Michigan State Giresun Univ, USA
Mona Helmy	British University, Egypt
Valeria Valeriano	Polytechnic University of Bari, Italy
Sheila Holz	Centre for Social Studies, Coimbra, Portugal
Chiara Belingardi	PHD in Urban, Territorial and Landscape Design
Aida Kesuma Azmin	International Islamic University, Malaysia
Caterina Frettoloso	University of Campania, Italy
Silvia Vitale	University of Florence, Italy
Cecilia Brunetti	University of Florence, Italy
Giovanni Marino	CNR—National Research Center, Florence, Italy
Alessandra Donato	University of Florence, Italy
Marta Buoro	PHD in Landscape Architecture
Fabiana Landi	University of Perugia, Italy
Antonia Sore	University of Florence, Italy
Matteo Scamporrino	TPS Pro, Perugia, Italy
Jacopo Ammendola	University of Bolzano, Italy
Sergio Serra	University of Cagliari, Italy
Luciano de Bonis	University of Molise, Italy
Caterina Giannattasio	University of Cagliari, Italy
Andrea Ugolini	University of Bologna, Italy
Ilaria Agostini	University of Bologna, Italy
Francesca Conti	Università La Sapienza, Rome, Italy
Monica Cannaviello	University of Campania, Italy
Elisa Mazzoni	University of Florence, Italy
Roxana Georgiana Aenoai	University of Campania, Italy
Giuseppe Mangano	University of Reggio Calabria, Italy
Giampiero Lombardini	University of Genoa, Italy
Annalisa Giampino	University of Palermo, Italy
Claudia Pipos	Ion Mincu University of Architecture and Urban Planning, Bucharest, Romania
Erica Bruno	University of Trento, Italy
Eleonora Giannini	University of Sheffield, United Kingdom
Margherita Vacca	University of Florence, Italy
Marica Merola	University of Campania, Italy
Simone Porfiri	University of Camerino, Italy
Elena Tarsi	University of Florence, Italy
Maria Rita Gisotti	University of Florence, Italy

The Editors warmly thank all the Reviewers who have contributed their authority to the double-blind review process, to ensure the quality of this publication.

Contents

Green Infrastructure and Urban Climate Resilience

Assessment of Green Infrastructure in the Context of Urban Sustainability: A Study in Northern Nicosia 3
Ipek Yaralioglu and Buket Asilsoy

The "Mugnone River Contract": A Landscape Infrastructure for the Florence Metropolitan Area 17
Claudia Mezzapesa, Elena Moretti, Valeria Lingua, and Elisa Caruso

Pocket Parks in Florence: Small-Scale Landscape Projects for Big Urban Challenges 35
Claudia Mezzapesa and Elena Moretti

Ruralizing the Urban Envelope: Experimental Tools to Cultivate the City 51
Matteo Clementi, Marco Migliore, Silvia Falasco, and Paola Caputo

Impact of Street Trees on Thermal Comfort and Air Quality 61
Carolina Girotti, Luiz Fernando Kowalski, Alessandra R. Prata Shimomura, Fernando Akira Kurokawa, Érico Masiero, and António Lopes

From Margin to Urban Interface: Designing a Network of Open Spaces as a New Public Park Around the Heritage City Walls of Siena 71
Francesco Alberti, Eleonora Giannini, Sara Naldoni, and Lorenzo Nofroni

The Development of Urban Parks as Green Public Space in Colonial Java 81
Mutiah Amini

New Design Approaches for Urban Voids: Developing a Framework for Analysing Innovative Projects in Terrain Vague Spaces 93
Lorenzo Stefano Iannizzotto, Alexandra Paio, and Camilla Perrone

Evaluating the Ecosystemic Function of the "Model Cities" to Lead Contemporary Urban Sustainability 101
Xanthippi St. Dereka

Ecosystem Services to the Test of the Territory: Perspectives for the Urban Bioregion Approach 111
Daniela Poli and Giulia Luciani

Enhancing Nature-Based Solutions Through the Bioregional Approach: The Case Study of Agri Park of Florentine Piana, Italy 121
Maria Rita Gisotti and Elena Tarsi

Urban Design and Mobility Strategies for the Regeneration of Cities and Communities

Enhancing Urban Connectivity: A Comprehensive Study of Transit-Oriented Development 131
Salma Safi and Hadeel El Naggar

Beyond Transit-Oriented Development: Linear Transit Cities, an Urban Structure for the Twenty-First Century 145
Halvard Dalheim

Transformation of Highways to Reduce GHG Emissions and Improve Social Equity: Syracuse and Brooklyn, NY 165
Marc Wouters

Research on Sustainable Urban Design Based on Small-Scale and Progressive Concept 179
Hu Ruotong, Liu Yuxuan, Song Yunzi, and Hou Zhulin

Revitalization Strategies for Inner Areas in Italy: Constraints and Opportunities from an Analysis of Best Practices 187
Maria Rita Gisotti, David Fanfani, Claudia Mezzapesa, Eletta Naldi, Giulia Biagi, Matteo Mengoni, and Paola Scarpellini

Urban Form and Housing Evolution: The Sha'abiat Model in Al-Ain, UAE 203
Boshra Hassan and Sahera Bleibleh

The Central Area of Ho Chi Minh City from the Perspective of Sustainable Urban Design 211
Le Ngoc Ha Mai, Nguyen Huu Khang Bui, and Ngoc Nhi Nguyen

The Built Environment Policy Making: The Egyptian New Administrative Capital Model 221
Mahmoud Khalifa and Amany A. Khodair

Urban Heritage and Cultural Identity: Methods and Applications for Conservation

Revival of Urban Heritage as an Engine of Social and Economic Recovery in Post-War Homs, Syria 233
Abeer Abdelal and Nadia Kandakji

Revitalizing Urban Spaces: The Role of Minga and Community Engagement 247
Daniela Hidalgo Molina

The Medieval Aqueduct of Perugia: From Abandonment to Reuse as a Pedestrian Path 253
Pietro Matracchi, Mariano Sartore, and Maddalena Branchi

From Cultural Preservation to Cultural Tourism: Examining the Heritagization of Rijal Almaa's Vernacular Architecture Within the Al Soudah Peaks Development 265
Saad Hanif

Knowledge and Valorisation of Sepulchral Cavities in the Neapolitan Underground 281
Federica Itri and Simona Scandurra

Participatory Heritage: Reflection on the Elaboration of the Atlas of Territorial Heritage of Playa, La Habana, Cuba 289
Valentina D'Ippolito, Elena Tarsi, and Raffaele Paloscia

Cosmopolitan Paradox: Dynamic Egalitarianism in the Redevelopment of the Cultural Center of the Philippines Complex 297
Chris Jehram M. Morallos and Leonido M. Gines Jr.

The Paradoxical Heterogeneity of Al-Abdali's Imageability and the Dilemmatic Identity of Amman 315
Omar M. Amireh, Dania H. Al-Harasis, Zaineh S. Abu Omar, and Saja A. Zeadeh

Spatial Identities of Sarajevo's Residential Neighborhoods 323
Alma Hudović Kljuno and Leila Krivošić Dizdarević

Exploring Urban Elements in a Postindustrial Destination: A Case Study of Bilbao 333
Jaramillo Estefanía, Alvarez Irantzu, Gandini Alessandra, Quesada Laura, and Garmendia Leire

Bridging Affordability and Resilience in African Housing: A Case Study of Housing Design Guidelines in Kenya 345
Garret Gantner

Design Approaches for Community Engagement and Social Sustainability

Investigating Universal Design Within the Framework of Social Sustainability in Northern Nicosia with Analyses and User Views 357
Ümran Duman and Buket Asilsoy

Community Engagement as an Innovative Urban Design Strategy: Conservation and Revitalization of Pantang Historic District, Guangzhou 373
Dingran Chen

Conflictual and Insurgent Planning: The Case of *Mondeggi Bene Comune*: Tuscany (Italy) 383
Naggila Taissa Silva Frota, Elena Tarsi, and Clarissa Figueiredo Sampaio Freitas

Shared Living with Local Connections: Creating a New Neighborhood Through the Participation of Young People 389
Yuno Tanaka

Normalizing Neglected Public Spaces: The Case of Portoviejo's Ground Transportation Node 397
Paúl Intriago and Jaire Cagigal

Enhancing Quality of Life for Older Adults in Rural Thailand: The Role of Community Residential Environments in Ageing-in-Place Strategies 411
Alisa Nutley

Green Infrastructure and Urban Climate Resilience

Assessment of Green Infrastructure in the Context of Urban Sustainability: A Study in Northern Nicosia

Ipek Yaralioglu and Buket Asilsoy

Abstract

Urban green infrastructure is a significant feature in the sustainable development of cities. Green infrastructure refers to the network of natural, semi-natural, public, and privately owned green, gray, and blue spaces to provide ecosystem services for the biodiversity of the cities. Northern Nicosia in Cyprus, as a rapidly urbanizing city, is suffering from the consequences of global warming and climate change, and also needs to implement and support the system of green infrastructure. Within this framework, the Dereboyu region, including Mehmet Akif Street and Pedieos River along the street, as one of the most crowded commercial and public hubs of the city, was chosen as the study area. The street and its surroundings as a region were analyzed, and urban corridors were determined in order to make a qualitative assessment based on several features for developing a network of green infrastructure. Regarding the findings, the traffic load needs to be decreased in the region. Further, urban landscaping, including the hardscaping such as urban furniture and permeable surfaces, and softscape materials such as street trees, must be deployed to improve both ecological function and aesthetic appeal. In addition, Pedieos River needs to be integrated with the region after a rehabilitation and design process. As concluding remarks, more scientific, theoretical, and evidence-based approaches are required in order to integrate and evaluate the concept of urban green infrastructure within the framework of sustainable development.

Keywords

Urban sustainability · Green infrastructure · Urban corridors · Qualitative assessment · Dereboyu region · Northern Nicosia · Pedieos river

1 Introduction

The increase in individuals relocating from rural regions to cities is referred to as urbanization, which has spiked significantly over the last 100 years due to industrialization and globalization. Expanding cities often emerge as centers for innovation and cultural exchange, attracting a mixture of people and promoting economic advancement (Glaeser 2011). However, this rapid urban growth brings challenges like overcrowding, heightened demand for housing and services, and environmental deterioration (Angel et al. 2011). As reported by the United Nations in the 2014 World Urbanization Prospects, the worldwide urban population is projected to rise from 54% to 66% by 2050. This will result in further urban sprawl and increase pressure on both natural and cultural landscapes. In other words, it can be stated that as number expected to rise, more than half of the global population currently resides in urban areas, underscoring the critical need for sustainable approaches.

Urbanization is a global phenomenon with far-reaching implications for economic development, social interactions, and environmental sustainability (Seto et al. 2012). The trend of urbanization has a significant influence on land use, so it creates urban sprawl to extend existing cities with high pressure on the environment and agricultural lands (McDonald

I. Yaralioglu (✉)
Faculty of Architecture, Near East University, Nicosia, Cyprus
e-mail: ipek.yaralioglu@neu.edu.tr

B. Asilsoy
Faculty of Agriculture, Near East University, Nicosia, Cyprus
e-mail: bukey.asilsoy@neu.edu.tr

F. Alberti et al. (eds.), *Towards Sustainable Urban Regeneration: Utilizing Environmental, Cultural, and Social Assets*,
Advances in Science, Technology & Innovation, https://doi.org/10.1007/978-3-032-02091-8_1

et al. 2018). Good urban planning and policies, therefore, remain key to achieving sustainable development; i.e., development that balances economic growth with environmental protection and social equity (United Nations 2018). Hence a growing interest in sustainable urban development, that is a broad concept that not only concerns about ecologic issues but also social, economic, and cultural problems in cities.

Sustainable urbanism is a multidimensional word, where integration of environmental, cultural, social, and economic sustainability in the field of urban planning and design is a different reality of urbanism. Recognized to a significant extent as sustainable urbanism addresses towards solutions for the challenges faced due to climate change, depletion of resources, and overpopulated urban migration. Moreover, as the World Bank (2020) shows, sustainable urban planning raises the resilience of cities to climate-related disasters such as flooding or heat waves, which leads to an enhanced quality of life for citizens. In addition, mixed-use developments, better public transport, and green infrastructure can significantly decrease greenhouse gas emissions (Newman et al. 2017).

Green infrastructure (GI) is the planning and development method of integrating natural systems to provide ecosystem services (ES). Which is exactly why it matters in terms of enhancing the sustainability of urban spaces. As a growing portion of the world population is turning into urban, the idea of GI has received increased attention. The concept of green infrastructure is considered an essential component of sustainable urban planning because of its wide-ranging ecosystem services, including temperature regulation, flood mitigation, and pollutant filtration. Providing these services helps increase city resilience, decreasing the vulnerability of urban people to climate-related risks, including heat waves and floods. As such, GI increases the urban quality of life by enhancing ecological sustainability.

GI involves both natural and semi-natural systems and therefore entails a wide array of techniques in improving air and water quality, mitigating urban heat effects, and increasing biodiversity, which include parks, urban forests, green roofs, and permeable pavements. These solution techniques adopted by GI can be many in planning and design methods, as well as engineering methods. Effective management of stormwater through green infrastructure decreases the vulnerability of cities to flooding and enhances their resilience against climate change issues. For example, Gill et al. (2007) discovered that green infrastructure can reduce surface runoff by 65% or more, which significantly lessens urban flood risks (Fletcher et al., 2015).

Ahern J. (1995) also described GI as a multifunctional resource to enhance an urban ecology by managing stormwater, ensuring biodiversity, and building areas for recreation. These areas of recreation promote the health and social cohesion of its residents. Apart from the environmental aspects, green infrastructure enriches the economic point of view by increasing property value and reducing health budgets related to pollution in cities. It follows that, in addition to ecological benefits, green infrastructure can have economic ones, too, as land around areas of greenery generally appreciates in value and draws visitors. Wheeler (2013) corroborates this. Given that green infrastructure plays a vital role in the creation of sustainable urban environments for future challenges, investing in it is something every city should pursue. Moreover, the development of these solutions within urban planning will allow for healthier and more sustainable development of the residents of cities.

In order to improve urban green spaces, an ecological approach to urban planning and design must also be established in Northern Nicosia (Bolkaner and Asilsoy 2023). As a city facing rapid urbanization along with the adverse effects of global warming and climate change, it seems that implementing green infrastructure has become imperative to enhance the city's environmental quality and adaptive capacity. Within this framework, the study discusses green infrastructure in the context of urban sustainability and evaluates the term with its definitions and functions in the next section. Later, in the methodology section of the study, the Dereboyu region, including Mehmet Akif Street and the adjacent Pedieos River corridor, was selected as the study area. Afterwards, the paper involves the findings and conclusions.

2 Literature Review

2.1 Green Infrastructure in the Context of Urban Sustainability

As highlighted in numerous studies, the integration of both gray and green infrastructure is essential, with green infrastructure offering a resilient approach to creating sustainable urban environments. It contributes to the production of high-quality, livable spaces, recognizing that the quality and management of neighborhoods, streets, and parks are intrinsically tied to community identity and shared values. Furthermore, green infrastructure supports the preservation of historical, archaeological, and architectural heritage, while simultaneously fostering ecological sustainability and reinforcing local landscape character (Tzoulas et al., 2007; Davies & Lafortezza, 2017). In other words, GI is part of the sustainability discourse used by a wide range of agencies, organizations, companies, community groups, and planners (Wang and Banzhaf 2018).

Green infrastructure planning plays a crucial role in enhancing regional image, urban identity, local distinctiveness, and the aesthetics of urban spaces, contributing to the creation of a stronger "sense of place" and fostering social interaction. The contribution of open and green spaces to cities and their inhabitants, from an ecological, socio-cultural, and economic perspective, is undeniable. The loss of these areas not only diminishes the quality of life but also results in the depletion of vital ecosystem services they provide at no cost. Traditional infrastructure planning alone is insufficient to address the multifaceted challenges faced by urban environments, necessitating the incorporation of more environmentally sensitive approaches. Green infrastructure, as described by Lucius et al. (2011), functions as a network of interconnected green spaces that protect ecological principles and provide community benefits, in contrast to gray infrastructure, which includes technical systems such as roads, railways, and utilities.

GI evaluation in relation to urban sustainability has something to do with various SDGs defined within the 2030 Agenda. Among those, one of the most relevant may be considered SDG 11: Sustainable Cities and Communities, since GI works toward improving air and water quality, enhancing biodiversity, and reducing climate-related risks-developing urban resilience. It is also linked with SDG 13, Climate Action, whereby GI helps capture carbon and manage climates —a very important contribution in the fight against climate change. Further, under SDG 3, Good Health and Well-being, green areas ensure betterment in both physical and mental health, while SDG 15, Life on Land, points to the contribution that GI makes to biodiversity conservation and restoration of ecosystems. See Fig. 1.

GI is defined as: "a strategically planned network of natural and semi-natural spaces that provide ecosystem services enhancing urban resilience, biodiversity, and human well-being" (Benedict and McMahon 2006a). Unlike traditional gray infrastructure, GI contributes to sustainability through the promotion of natural system processes for stormwater management, climate regulation, and carbon sequestration. It improves the quality of air and water, reduces the effects of a heat island in cities, and, in turn, promotes public health. The challenges to implementing GI in cities are conflicts over land use, limited policies, and financial constraints, according to Hansen and Pauleit (2014).

2.2 Green Infrastructure: Definitions and Functions

GI was first introduced in the mid-1990s (Pauleit et al. 2011). The concept of green infrastructure has evolved to encompass a variety of functions and scales, from localized urban parks to regional ecosystems. The definition of green infrastructure, which is more than just a green space, is vary considerably (Nieuwenhuijsen 2021). This definition can include not only landscape patches and corridors but also other representations of nature (green roofs, community gardens, forests, wetlands, street trees), forming an interconnected network of green, gray, and blue space in the urban environment.

Fig. 1 Sustainable development goals (United Nations SDGs 2015)

Table 1 Urban green infrastructure (UGI) encompasses several key components (Cvejić et al. 2015)

Green spaces (parks and gardens)	Public parks, gardens, and urban forests provide recreational areas, improve air quality, and support biodiversity
Green corridors	Linear green spaces like greenways, riverside paths, and vegetated roadsides, which connect parks, natural reserves, and other green spaces to promote ecological connectivity
Green roofs and green walls	Vegetation layers installed on building rooftops and walls contribute to energy efficiency, reduce the urban heat island effect, and support biodiversity
Urban forests	Trees and woodlands in and around cities mitigate air pollution, reduce heat, and offer ecological, aesthetic, and social benefits
Community gardens and allotments	Small-scale urban agriculture areas where residents grow their own food, improving food security and encouraging community engagement
Permeable surfaces	Materials used in pavements, walkways, and parking areas that allow water to percolate into the ground, reducing flood risk and managing stormwater runoff
Wetlands	Urban wetlands help filter pollutants, control floods, and support rich biodiversity while providing recreational and educational spaces
Sustainable urban drainage systems (SUDS)	Systems designed to reduce runoff and improve water quality by mimicking natural drainage processes. These include features like rain gardens, bioswales, and detention ponds
Urban rivers and waterways	Managed and restored rivers, streams, and canals that enhance water management, improve biodiversity, and provide leisure opportunities
Biodiversity hubs	Areas that serve as biodiversity hotspots within the urban context promote species conservation and enhancing the ecological network

In other words, as a wide concept, GI involves a considerable spectrum of components. It can be argued that there are three main types as gray, blue, and green, forming the key features of GI. Wetlands and urban drainage systems such as bioswales, rain gardens, etc., are among the blue surfaces listed under the network of GI. Permeable surfaces, open spaces are among the gray components of the concept. All kinds of green spaces, such as neighborhood parks, vertical gardens, and green corridors are the green items of GI. See Table 1, see Fig. 2.

2.3 Urban Green Corridors

Urban green corridors are a specific application of green infrastructure, designed to connect fragmented habitats and provide continuous ecological and recreational spaces within cities. Forman and Godron (1986) were among the first to articulate the concept of ecological corridors, emphasizing

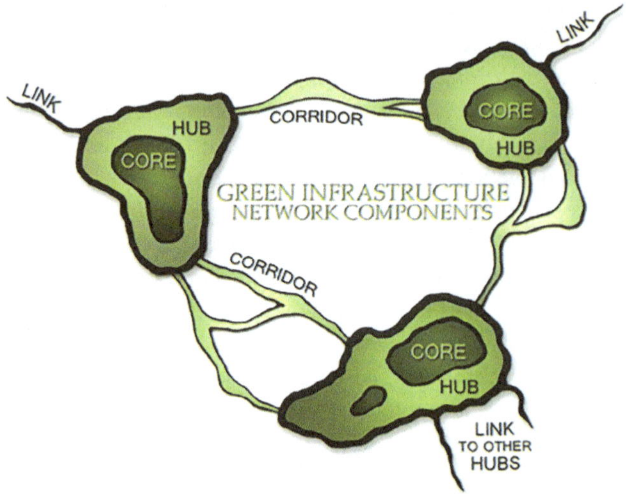

Fig. 2 Green infrastructure components (Benedict and McMahon 2006b)

their role in biodiversity conservation by facilitating species movement and reducing habitat fragmentation (Forman & Godron, 1986; Li et al., 2005). In urban contexts, green corridors serve as vital links between parks, rivers, and other natural areas, creating a network of connected spaces that support wildlife and offer recreational opportunities for residents. Jim and Chen (2003) argue that green corridors are particularly valuable in dense urban areas, where they serve as "green lungs" that improve air quality, reduce noise pollution, and provide shaded pathways for walking and cycling. Moreover, urban green corridors contribute to social sustainability by offering equitable access to green spaces for diverse urban populations, fostering social cohesion, and promoting public health. As part of a broader green infrastructure network, these corridors play a key role in enhancing urban resilience and promoting sustainable land use practices.

2.4 Challenges and Opportunities in Green Infrastructure Implementation

While the benefits of green infrastructure are widely recognized, the literature points to several challenges in its implementation. Davies and Lafortezza (2017) identifies issues such as limited funding, competing land uses, and a lack of political will as key barriers to widespread GI adoption. Additionally, the complexity of managing multifunctional spaces often requires interdisciplinary collaboration between urban planners, ecologists, and policymakers, which can be difficult to achieve (Hansen & Pauleit, 2014; Wang & Banzhaf, 2018).

Nonetheless, opportunities for green infrastructure implementation continue to grow. Kabisch et al. (2016) suggest that the increasing awareness of climate change and environmental degradation is driving demand for more sustainable

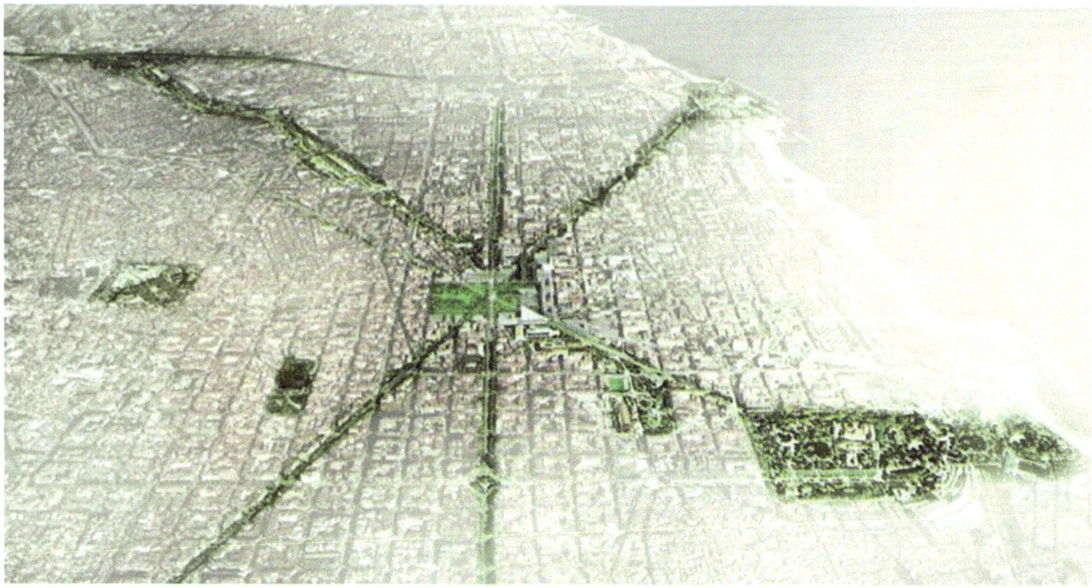

Fig. 3 Plaça de les Glòries Catalanes in Barcelona: Green Corridors and Urban Fabric (Baró Planella 2019)

Fig. 4 Barcelona, Catalonia plan and green infrastructure images (Ajuntament de Barcelona. 2020)

urban development practices, with green infrastructure at the forefront. Moreover, innovative funding mechanisms, such as public–private partnerships and ecosystem service valuation, are emerging as potential solutions to overcome financial barriers. Some green infrastructure projects are given later in the text. See Figs. 3, 4, 5, 6, 7, and 8.

3 Methodology

3.1 Materials

3.1.1 Study Area

The study area encompasses the boundaries of Mehmet Akif Street and the adjacent Pedieos River, forming one of the most vibrant and functionally dynamic urban zones in Northern Cyprus. See Fig. 9. Mehmet Akif Street, extending in a northwest direction, serves as a crucial commercial and social hub, featuring a high concentration of retail spaces, dining establishments, and various service-oriented businesses. This area stands out for its significant pedestrian and vehicular density, making it the most frequently traversed street within the city. The heavy pedestrian traffic, coupled with an intense flow of vehicles, underscores the street's role as a primary axis of movement and interaction within the urban landscape. Additionally, the proximity of Pedieos River enhances the area's potential for developing green infrastructure by providing an opportunity for ecological connectivity and recreational spaces. This unique combination of commercial, social, and natural elements positions the study area as a key focal point for examining the integration of green infrastructure within a bustling urban environment. See Fig. 10.

The material of the study is the Dereboyu region in Northern Nicosia. The main physical features of the region can be grouped in four headlines—A: Pedieos River, B: MehmetAkif Street, C: Osman Orek Street, and D: Kumsal Park. See Fig. 11.

The selected study area includes a combination of natural, infrastructural, and recreational zones within the *Dereboyu*

Fig. 5 The *Chicago City Hall* rooftop garden, one of the earliest large-scale green roof projects in the United States

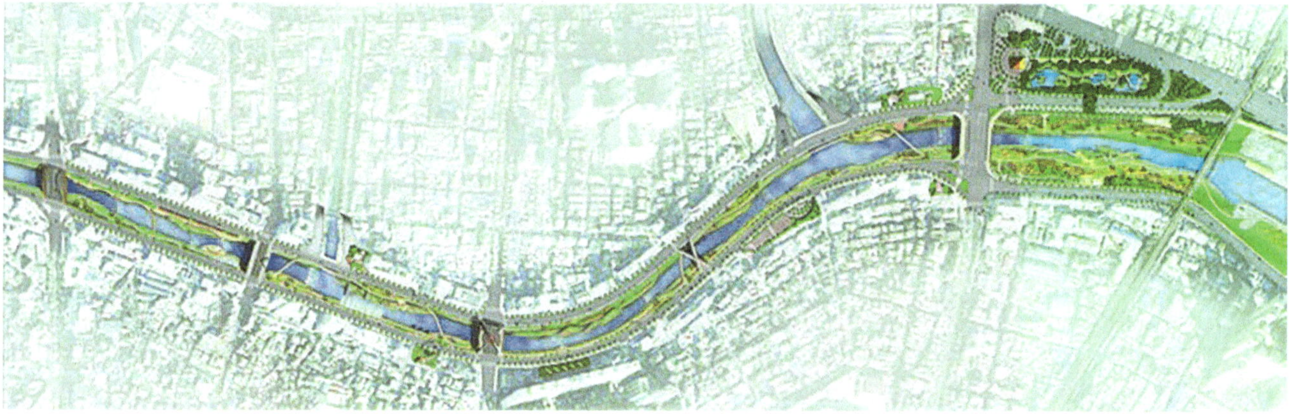

Fig. 6 Cheonggyecheon Stream Restoration general site plan (Di Carlo et al. 2020)

region, making it a dynamic and diverse urban space. The views of the selected areas are given in Fig. 12. *The Pedieos River (Corridor A)* runs through the area and is classified as "*streams and their surroundings*," providing an important natural element that enhances the ecological value of the region. Adjacent to the river are significant transportation corridors, such as Fig. 13. *Mehmet Akif Street (Corridor B)* and Fig. 14. *Osman Örek Street (Corridor C)*, both categorized as "*roads and surroundings*." These roads act as primary arteries for movement and connectivity within the urban fabric. Additionally, Fig. 15. *Kumsal Park (Corridor D)*, classified as a "*green area*," serves as a key recreational space for the community, offering opportunities for leisure and relaxation. Together, these components reflect a well-integrated mix of natural, infrastructural, and public green spaces, highlighting the area's role as a central hub for residential, social, and environmental interactions.

3.2 Research Design

At this stage of this study, first on-site observations and analyses were made. After determining the corridor areas in terms of their width, connectivity, integrity, and impact on the city, an evaluation was made in terms of aesthetic and functional qualities. Corridor areas are divided into two main groups: "natural" and "cultural" corridors. Based on these two main types, four urban corridors were determined under three headings: "roads and their surroundings," "green areas," and "streams and their surroundings."

Fig. 7 Cheonggyecheon Stream Restoration Project – pre-restoration condition and implementation phase, Seoul, South Korea (Kim and Jung 2009)

Fig. 8 The Cheonggyecheon Stream Restoration Project in Seoul, South Korea, revitalized a concrete-covered stream into a natural waterway, enhancing biodiversity, flood control, and recreational space (Kim and Jung 2009)

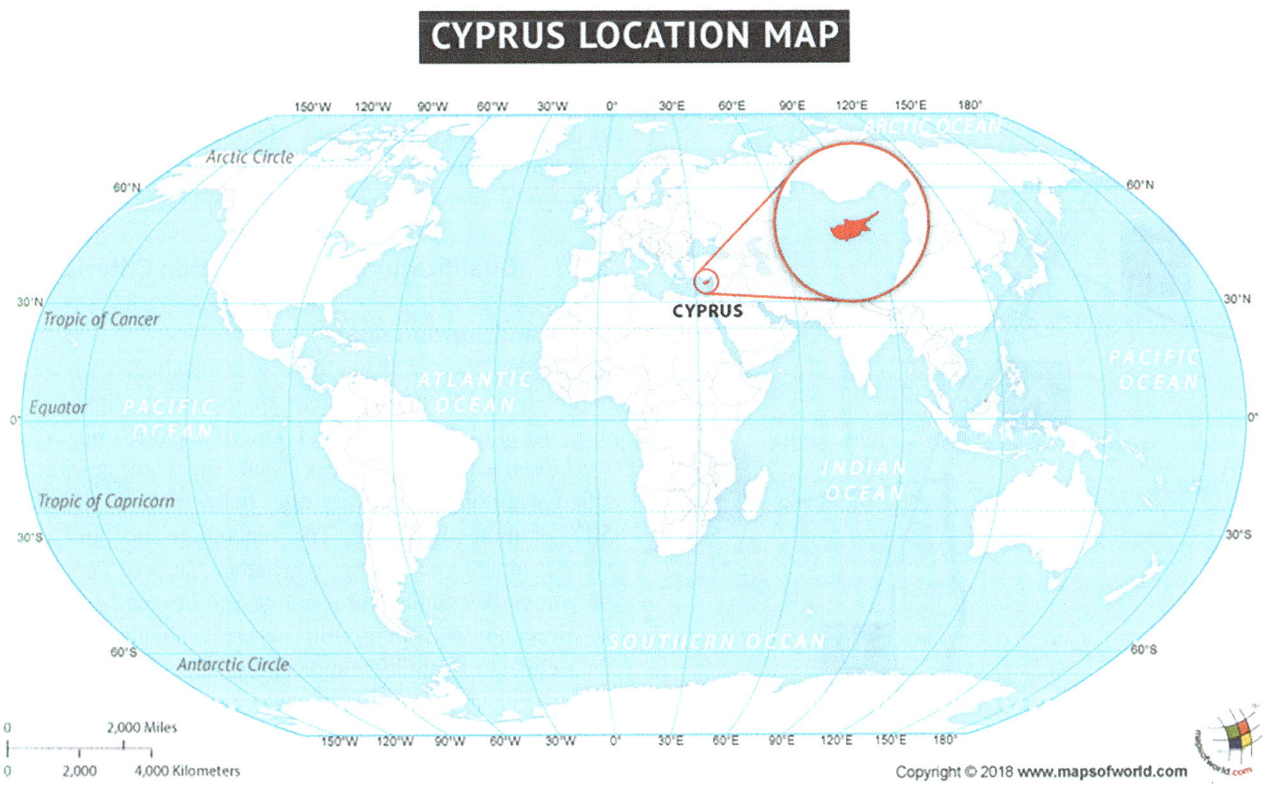

Fig. 9 Cyprus location on the map Source: Maps of World (2018)

Fig. 10 Dereboyu region as the study area. Source: Google Earth (2024)

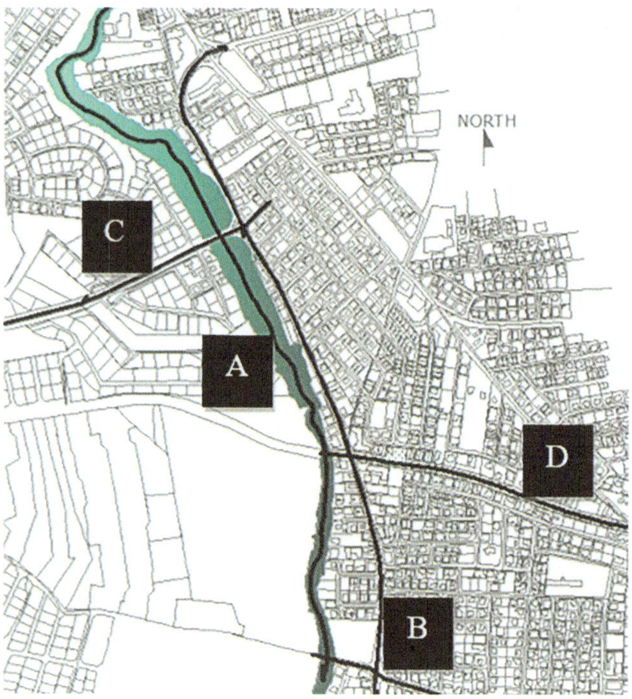

Fig. 11 The Dereboyu region as study area with the determined urban corridors (Source: Author's own drawing)

The criteria table for the evaluation of urban corridors of the Dereboyu region was created by a literature review and evaluated within this scope. Measurable parameters were determined and analyses were made within the scope of these parameters, and an evaluation table was created.

3.2.1 Qualifications and Evaluation Criteria

Rivers and Surroundings

Rivers and their surroundings were evaluated on various parameters. The improvement status was evaluated within the scope of landscape and environmental improvement works carried out in the areas around the rivers and was classified according to recovery rates below 50%, at the level of 50% or more than 50%. The vegetation and flora criteria were evaluated according to whether the area consisted of only trees, trees and shrubs, or trees, shrubs, and ground covers. Recreational opportunities were evaluated in terms of the richness of activity and recreation areas available around the rivers, and this situation was classified as offering an inadequate (poor), medium, or high level of opportunity. Intensity of use refers to how often and to what extent the space is used by users and is rated at similarly poor, medium, and high levels. Connectivity to other corridors represents the integration and accessibility of the area with other corridor systems within the city, and the adequacy of connectivity was determined at weak, medium, or high levels.

Fig. 12 Corridor A; Pedieos River view

Fig. 13 Corridor B; Mehmet Akif Street view

Fig. 14 Corridor C; Osman Örek Street view

Fig. 15 Corridor D; Kumsal Park view

Roads and Surroundings

Roads and their surroundings are considered as an important category within urban corridors. Road afforestation was evaluated within the scope of planting works carried out along the road and was classified according to afforestation rates below 50%, at 50% or above 50%. Usage intensity refers to the frequency of use of roads by both pedestrians and vehicles and has been examined at weak, medium, or high levels. Pedestrian mobility was evaluated by the quality and suitability of roads that facilitate the movement of pedestrians. Pavement quality refers to the surface conditions that roads provide for pedestrian access and is classified as poor, medium, or high-quality levels. While the presence and width of the central median determine whether the areas in the middle of the roads are available and wide enough, the central median planting and its quality were evaluated in terms of planting works and plant diversity in these areas. Lighting refers to the safe and effective illumination of roads and has been analyzed at poor, medium, or high levels. Connectivity with other corridors was evaluated in terms of integration of roads with the urban corridor network.

Green Areas

Green areas have been considered as important elements that improve the quality of urban life and support environmental sustainability. The presence of plant material was analyzed according to whether the area consisted only of trees, trees and shrubs, or trees, shrubs, and groundcovers. Lighting refers to the adequacy of lighting systems to provide a safe environment for the users of the area. Other equipment was evaluated for the quality and presence of physical equipment such as seating areas, play equipment, or sports facilities. While the presence and amount of plant material represent the flora density and diversity of the area, drainage refers to the adequacy of green areas in water management. The connection with other corridors shows the level of integration of these areas into the urban corridor system.

These evaluation criteria were developed to ensure a detailed analysis of urban corridors from a functional and aesthetic perspective and were measured at weak, medium, and high levels under all headings. See Table 2.

Table 2 Evaluation criteria of urban corridors. Source: Author's own elaboration inspired by previous corridor assessment approaches

	Qualifications	Evaluation criteria		
Streams and its surroundings	Improvement status	1: Less than %50 2: %50 3: More than 50%		
	Vegetation/flora	1: Only from trees 2: Trees and bushes 3: Trees, shrubs, and ground covers		
	Excessive recreation opportunities	1: Poor	2: Medium	3: High
	Intensity of use	1: Poor	2: Medium	3: High
	Connectivity with other corridors	1: Poor	2: Medium	3: High
Paths and surroundings	Road afforestation	1: Less than %50 2: %50 3: More than 50%		
	Intensity of use	1: Poor	2: Medium	3: High
	Pedestrian movement	1: Poor	2: Medium	3: High
	Pavement quality	1: Poor 2: Medium 3: High		
	Refuge presence and width	1: Nope refuge 2: There is a median but insufficient width 3: Refuge exists and has enough width		
	Refuge planting and quality	1: Only from trees 2: Trees and bushes 3: Trees, shrubs, and ground covers		
	Lighting	1: Poor	2: Medium	3: High
	Connected with other corridors	1: Poor	2: Medium	3: High
Green space	Presence of plant material	1: Only from trees 2: Trees and bushes 3: Trees, shrubs, and ground covers		
	Lighting	1: Poor	2: Medium	3: High
	Other fittings	1: Poor	2: Medium	3: High
	Presence and quantity of plant material	1: Poor	2: Medium	3: High
	Drainage	1: Poor	2: Medium	3: High
	Connected with other corridors	1: Poor	2: Medium	3: High

4 Results of the Study

4.1 Findings

Analyses were made within the scope of these parameters, and an evaluation table was created. Areas analyzed;

A: The Pedicos River is classified as "streams and their surroundings."

B: Mehmet Akif Street's roads and surroundings.

A: Osman Örek Street is classified as "roads and surroundings."

D: Kumsal Park is classified as a "green area."

Each of the four corridors titled under three headlines was assessed according to the several determined features involved in the evaluation criteria table. See Table 3

Table 3 Evaluation of Dereboyu region corridors. Source: Author's own elaboration

		Natural	Cultural Natural		
		Streams and surroundings	Paths and surroundings		Green spaces
	City corridors	A	B	C	D
Stream and surroundings	Improvement status	1			
	Vegetation/flora	1			
	Excessive recreation opportunities	1			
	Intensity of use	1			
	Connectivity with other corridors	1			
Paths and curroundings	Road afforestation		1	1	
	Intensity of use		3	3	
	Pavement quality		2	2	
	Pedestrian movement		2	1	
	Refuge presence and width		1	2	
	Planting		1	1	
	Lighting		1	2	
	Connectivity with other corridors		2	2	
Green space	Presence of plant material				2
	Lighting				2
	Other urban furnishings				1
	Presence and quantity of plant material				2
	Drainage				2
	Connectivity with other corridors				1

4.2 Discussion

In general, all corridors have remarkable insufficiencies in terms of determined qualities such as the presence of plant material, pedestrian movement, and urban furniture. Regarding the A: Pedieos River, all determined features are qualitatively assessed as "poor." Recreation opportunities such as landscaping functions (walking/cycling paths, etc.), vegetation, and connectivity with other corridors can all be suggested to be low scored.

Regarding B: Mehmet Akif Avenue, it can be argued that most of the qualities for a qualified GI corridor are relatively inadequate. The quality of the corridor can be increased by limiting the load of the vehicular movement in order to improve pedestrian access and urban landscaping features such as urban furniture, street trees, and plantations. Integrating the Pedieos River corridor with the area will further strengthen the connectivity between the green infrastructure network and the city. The river's rehabilitation and the establishment of pedestrian and cycling paths along its banks will provide an ecological transportation alternative, promoting social interaction and encouraging sustainable mobility among residents. In doing so, the street will establish vital linkages with both natural and urban areas, enhancing the integration of green corridors within the city. Regarding the Osman Örek Avenue, as the most vehicular corridor, pedestrian movement can be increased, such as overpass access and plantations, street trees, and urban furniture can also be improved. In addition, regarding D: Kumsal Park as the green space, its connectivity with the surrounding needs to be improved, and other qualities need to be supported for creating an adequate green corridor.

5 Conclusion

The study first discusses the importance of green infrastructure in the context of sustainable urbanism. The study then focuses on the Dereboyu region as a study area for a qualitative evaluation according to various green infrastructure criteria. The findings highlight the need for sustainable urban development through the integration of green infrastructure and urban corridors. This area, one of the city's most congested urban environments, currently lacks the environmental and social functionality that green infrastructure can provide. Therefore, a strategic framework focusing on ecological, social, and aesthetic functions is essential to increase the region's contribution to the urban fabric. Green corridors play an important role in strengthening ecological connectivity and supporting biodiversity in urban areas, while improving the quality of urban life.

Recommendations based on the findings include the integration of urban corridors identified along Mehmet Akif Street. To achieve such integration, plantations of trees, shrubs, and ground covers need to be improved in all corridors to create a microclimate that can reduce temperature and improve air quality through carbon sequestration. Urban furnishings and pedestrian access in the area also need to be improved. Successful implementation of this strategy will ensure that the Dereboyu district and Mehmet Akif Street become an integral part of the city's green infrastructure system, promoting urban resilience and ecological sustainability. Ultimately, redesigning the area around green infrastructure and green corridors will not only improve the quality of urban life but also serve as a model for creating a more resilient and sustainable urban environment in Nicosia.

In this regard, the plantation strategy to be implemented in the Dereboyu region can be adapted to other parts of the city in the future and serve as an example for the revitalization of the city ecosystem.

The proposed strategy should be aimed at creating an economically sustainable model for the cities of the future by contributing to the economic development of the region in the long term. As a result, this green infrastructure and urban corridor strategy proposed for the Dereboyu region, as a transformation model that integrates ecological, social, and economic dimensions, will create an urban development that is resilient to climate change, environmentally sustainable, and increases the quality of social life in Nicosia as well as other cities. Will encourage the approach. Such projects are of great importance in terms of adapting cities to the future and creating a more livable and sustainable urban environment.

In order to support the strategic framework proposed in this study, the following recommendations for urban green infrastructure should be implemented:

Creation of Green Corridors: Vegetation consisting of trees, shrubs, and ground covers along Mehmet Akif Street should be improved, and ecological connections should be strengthened by adding walking and cycling paths.

Urban Parks and Pocket Gardens: Existing green areas, such as Kumsal Park, should be improved within the region, and accessible green areas should be increased at the neighborhood scale by creating new pocket gardens.

Rainwater Management: Sustainable water management solutions such as rainwater gardens and permeable pavements should be planned to ensure that water meets the soil and flood risks are reduced.

Vertical and Horizontal Greening: The urban heat island effect should be reduced by planning vertical gardens on building facades and green roofs on rooftops.

Smart Green Infrastructure Systems: Air quality, temperature, and plant health should be monitored regularly with sensor-supported systems, and data-based maintenance and improvement should be carried out.

Social Participation and Education: Training programs should be organized to ensure the participation of local people in green infrastructure projects, and environmental awareness should be increased in the region.

Urban Furniture and Pedestrian Access: Social functionality should be increased by planning urban furniture such as pedestrian-friendly designs, seating areas, shades, and lighting.

Planning and implementation of these recommendations will increase the ecological, social, and economic sustainability of the Dereboyu region and enable it to become an effective part of the urban green infrastructure system.

References

Ahern J (1995) Greenways as a planning strategy. Landsc Urban Plan 33:131–155

Ajuntament de Barcelona (2020) Barcelona green infrastructure and biodiversity plan 2020. Ajuntament de Barcelona. https://ajuntament.barcelona.cat/ecologiaurbana/sites/default/files/Barcelona%20green%20infrastructure%20and%20biodiversity%20plan%202020.pdf

Angel S et al (2011) Making room for a planet of cities. Lincoln Institute of Land Policy

Balmon E (n.d.) The new urban fabric. https://issuu.com/ebalmon/docs/the_new_urban_fabrik

Baró Planella A (2019) Barcelona gives way to green infrastructure: Les Glories urban transformation as a case study of citizen participation (Master's thesis). Universität Polytechnical de Catalunya. https://upcommons.upc.edu/handle/2117/171990

Benedict MA, & McMahon ET (2006) Green infrastructure: Linking landscapes and communities. Island Press.

Benedict MA, McMahon ET (2006a) Green infrastructure. Island Press

Benedict MA, McMahon ET (2006b) Green infrastructure: linking landscapes and communities. Island Press

Bolkaner MK, Asilsoy B (2023) Reinventing the urban neighborhood green index in the context of urban ecology as a conceptual framework in Northern Nicosia, Cyprus. Sustainability 15(18):13880

Boverket S (1992) Storstadsuppdraget: En förstudie om storstädernas miljö (A preliminary study of the environment in big cities). The National Board of Housing, Building and Planning

City of Seattle (2015) Green stormwater infrastructure strategy. https://www.seattle.gov/Documents/Departments/OSE/GSI_Strategy_Nov_2015.pdf

Clackmannanshire Council (2013) Supplementary guidance 6: green infrastructure. http://www.clacksweb.org.uk

Cvejić R, Eler K, Pintar M, Železnikar Š, Haase D, Kabisch N, Strohbach M (2015) A typology of urban green spaces, ecosystem services provisioning services and demands. GREEN SURGE project, European Commission FP7. http://greensurge.eu/working-packages/wp3/

Davies C, Lafortezza R (2017) Urban green infrastructure in Europe: is greenspace planning and policy compliant. Land Use Policy 69:93–101

Di Carlo F, Giancotti A, Reale L (2020) Re-inventing water–ground relations in landscape architecture projects. Sustainability 12(24):10358. https://doi.org/10.3390/su122410358

Environmental Protection Agency (EPA) (2021) Green infrastructure. [EPA website]

European Commission (2020) Green infrastructure (GI)—enhancing Europe's Natural Capital. [European Commission website link]

Fabos JG (2004) Greenway planning in the United States: its origins and recent case studies. Landsc Urban Plan 68:321–342

Fletcher TD et al (2015) Sustainable urban water management: a review of the role of green infrastructure. Water 7(6):4656–4684

Flink C, Searns R (1993) Greenways: a guide to planning, design, and development. The Conservation Fund
Forman RTT, & Godron M (1986) Landscape ecology. Wiley.
Garner GW et al (2020) Green infrastructure: an emerging approach to urban sustainability. Environ Sci Pol 108:26–34
Gill SE, Handley JF, Ennos AR, Pauleit S (2007) Adapting cities for climate change: the role of the green infrastructure. Built Environ 33(1):115–133
Glaeser EL (2011) Triumph of the city: how our greatest invention makes us richer, smarter, greener, healthier, and happier. Penguin Press
Google Earth (2024) Dereboyu, Cyprus [Satellite image]. Google Earth
Hansen R, Pauleit S (2014) From multifunctionality to multiple ecosystem services? A conceptual framework for multifunctionality in green infrastructure planning for urban areas. Ambio 43:516–529
Jim CY, & Chen WY (2003) Comprehensive green space planning based on landscape ecology principles in compact Nanjing city, China. Landscape and Urban Planning, 65(3), 95–116.
Kabisch N, Frantzeskaki N, Pauleit S, et al. (2016) Nature-based solutions to climate change mitigation and adaptation in urban areas. Springer.
Kim KH, Jung JY (2009) Rehabilitation of urban ecosystems: the case of Cheonggyecheon, Seoul. Restor Ecol 17(6):963–968
Li F, Wang R, Paulussen J, Liu X (2005) Comprehensive concept planning of urban greening based on ecological principles: a case study in Beijing, China. Landsc Urban Plan 72:325–336
Lucius I, Dan R, Caratas D (2011) Green infrastructure. European Regional Development Fund
Maps of World (2018). World map showing Cyprus [Harita]. Maps of World. https://www.mapsofworld.com
Maps of World (n.d.) Cyprus location map. https://www.mapsofworld.com/cyprus/cyprus-location-map.html
McDonald RI et al (2018) Global urban expansion and the declining value of future urban land. Glob Environ Chang 48:88–98
Newman P, Beatley T, Boyer H (2017) Resilient cities: overcoming fossil fuel dependence. Island Press
Nieuwenhuijsen MJ (2021) Green infrastructure and health. Annu Rev Public Health 42(1):317–328
Pauleit S, Liu L, Ahern J, Kazmierczak A (2011) Multifunctional green infrastructure planning to promote ecological services in the city
Seto KC et al (2012) Global redefinition of urbanization. Environ Res Lett 7(4):1–10
Tan KW (2004) A greenway network for Singapore. Landsc Urban Plan
Texas A & M University and Sea Grant (2012) Coastal watershed program. http://tcwp.tamu.edu/files/2012/06/urban_growth_primer_5.pdf (Accessed on June 5, 2019)
The Countryside Agency (2006) The countryside in and around towns: the green infrastructure of Yorkshire and the Humber. https://www.leedsbeckett.ac.uk/as/cudem/CIAT_final.pdf
Tzoulas K et al (2007) Promoting ecosystem and human health in urban areas using green infrastructure: a literature review. Landsc Urban Plan 81(3):167–178
U.S. Environmental Protection Agency (2012) Green infrastructure. http://www.epa.gov/green-infrastructure
United Nations (2015) Sustainable development goals. United Nations. https://sdgs.un.org/goals
United Nations (2018) The world's cities in 2018—data booklet. United Nations Department of Economic and Social Affairs
United Nations (2021) World urbanization prospects: 2021 revision. Department of Economic and Social Affairs, Population Division. [UN website link]
Wang J, Banzhaf E (2018) Towards a better understanding of green infrastructure: a critical review. Ecol Indic 85:758–772
Wheeler SM (2013) Planning for sustainability: creating livable, equitable, and ecological communities. Routledge
World Bank (2020) Building urban resilience: principles, tools, and practice. [World Bank website link]

The "Mugnone River Contract": A Landscape Infrastructure for the Florence Metropolitan Area

Claudia Mezzapesa, Elena Moretti, Valeria Lingua, and Elisa Caruso

Abstract

Lungo il Mugnone is a public engagement process aimed at rethinking the relationship between the city and the river, proposing a collective vision for the future of this crucial landscape infrastructure between city and countryside. Activated thanks to the initiative of *Le Curandaie* association, with the support of the Architecture Department of the University of Florence and the landscape architects of *Memoscape*, the bottom-up participatory process envisions connecting parks and green areas through pedestrian and bicycle paths, trails along the basin, and water access points, creating a continuous public realm 17.5 km long.

The project transcends a mere "green and blue infrastructure" framework by embracing the broader and more transformative approach of **Landscape Infrastructure Design**. As previously laid out by Hung in *Landscape Architectural Design, Landscape Infrastructure* represents an interdisciplinary and transdisciplinary design methodology, where various disciplines collaborate to create a common language, both theoretically and practically. This approach rethinks infrastructure beyond its utilitarian function, integrating ecological processes and fostering community interaction.

The initiative, aimed at supporting environmental protection, planning, and urban land development policies, has established the *Mugnone River Contract*, a pact that brings together local authorities, residents, and various stakeholders to regenerate the river basin as a vibrant, living environment and a shared resource (in line with the European Landscape Convention, 2000). Landscape architects play a pivotal role in leading these strategies, ensuring that ecological restoration and urban regeneration are conducted through blended knowledge and perceptions.

Within this systemic vision, the project aligns with the dual notion of Landscape Infrastructure: conceiving *infrastructure as landscape*, which enriches infrastructure by incorporating social, ecological, and imaginative dimensions, while considering *landscape as infrastructure*, which regards living systems and ecological flows as fundamental infrastructural components capable of shaping urban environments and enhancing biodiversity (Bélanger, Landscape infrastructure. Urbanism beyond engineering. Wageningen University, 2013; Buoro, Tracce segni e rappresentazioni del paesaggio che cambia. Il Poligrafo Editore, 2023; Cortesi, Il paesaggio al centro. Lettera Ventidue Edizioni, 2017; Girot and Imhof, Thinking the contemporary landscape. Princeton Architectural Press, 2017; Hung, Landscape infrastructure: case studies by SWA. The Infrastructure Research Initiative at SWA, Birkhäuser, 2013; Mossop, The landscape urbanism reader. Princeton Architectural Press, pp 164–167, 2006; Weilacher, Tasting the landscape. 53rd IFLA world congress, April 20–22, 2016. Edifir, 2016). This holistic approach promotes the creation of active, functional landscapes that regenerate degraded areas and foster greater community interaction and resilience.

The future steps envisioned amplify the project's impact. The *Mugnone River Contract* could become part of the broader Pact for the Arno, a "contract of contracts" for the entire Tuscan river system. This integration aims to position the *Mugnone River Contract* as a pilot project within regional, metropolitan, and local policy frameworks, setting a precedent for similar initiatives.

C. Mezzapesa (✉)
Department of Architecture, University of Florence, Florence, Italy

Landscape Architect, Memoscape, Florence, Italy
e-mail: claudia.mezzapesa@unifi.it

E. Moretti
Landscape Architect, Memoscape, Florence, Italy

V. Lingua · E. Caruso
Department of Architecture, University of Florence, Florence, Italy
e-mail: valeria.lingua@unifi.it; elisa.caruso@unifi.it

F. Alberti et al. (eds.), *Towards Sustainable Urban Regeneration: Utilizing Environmental, Cultural, and Social Assets*,
Advances in Science, Technology & Innovation, https://doi.org/10.1007/978-3-032-02091-8_2

Keywords
Landscape infrastructure · Community engagement · The "Mugnone River Contract"

1 Context

1.1 The Mugnone Creek

The Mugnone Creek is a common good of historical, economic, cultural, and ecological significance, and a resource of extraordinary value for the metropolitan area of Florence. Along its course, the creek connects the northern hills of Fiesole with the city of Florence. Traversing a rich countryside, it offers numerous opportunities as a linear ecosystem; a hybrid space between the natural and artificial; and as a recreational area, essential for fostering a new sense of community between the inhabitants and the river itself (Fig. 1).

The initial concept connected the parks and green spaces along the Mugnone through a pedestrian pathway and, where possible, a bicycle path, with water access points, creating a river park—a continuous public space. Designing a river park today involves addressing multiple requirements, including reconciling the river's natural dynamics with flood risk mitigation, enhancing water quality and natural habitats, and fostering a new sense of belonging.

The project that led to the vision of a continuous river park along the Mugnone, stretching from Fiesole to Florence, is the result of a process started in 2020 in collaboration with *Le Curandaie*, an association active in the Cure neighborhood. The association is committed to providing a network of connections and services to families, women, and promoting best practices for the enhancement of the local area. The foremost goal was to promote, through a petition, a participatory process as the first step toward the establishment of a "Mugnone River Contract," aimed at building a riverine community around the creek to support environmental protection, urban planning, and land development policies.

The participatory process laid the groundwork for the establishment of the "Mugnone River Contract." On April 13, 2024, the Mugnone River Contract was officially signed (Mezzapesa and Moretti 2024). This agreement will contrib-

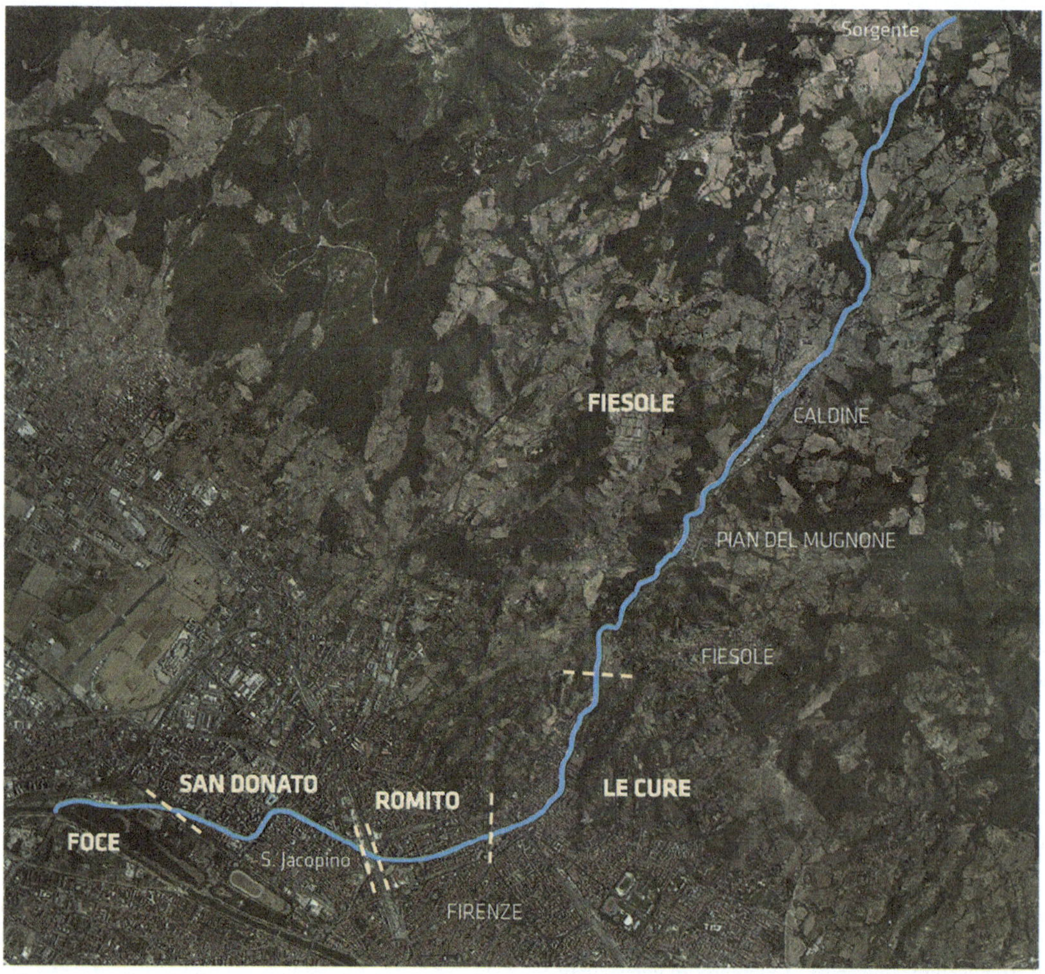

Fig. 1 The Mugnone Creek and the river Arno (in Mezzapesa and Moretti 2024)

ute to the revitalization and regeneration of the hydrographic basin, ensuring the proper use of the creek while enhancing the area and supporting the local development of the surrounding regions. The contract aims to promote slow and green tourism, create a safe and connective corridor and active public space, improve well-being and sustainable mobility, and enhance biodiversity and ecosystem services.

1.2 River Contracts

River Contracts are voluntary instruments of participatory democracy aimed at strategic and negotiated planning. Their goal is to build a *bottom-up* process that integrates sectoral policies for hydraulic risk mitigation with those for ecological regeneration, as well as social and landscape utilization (Bastiani 2011; Brun 2010; Brusarosco and Visentin 2023; D'Alessandro and Pistocchi 2014; Leona 2024; Massarutto 2008; Mazzoni 2015; Vismara and La Rocca 2013).

The River Contracts, or Creek Contracts, as defined in Italy by the National Charter of River Contracts (V National River Contracts Forum, Milan 2010), seek to bring together the various stakeholders in a given area: the entities and authorities directly involved in river and land management, local residents, and all other relevant stakeholders. This pact for the revitalization of the watershed calls upon both public institutions and private actors to adopt an integrated approach, rather than a sectoral one, recognizing the river as a living environment (European Landscape Convention, 2000) and thus as a common good to be managed collectively.

Professor Massimo Bastiani, coordinator of the National River Contracts Forum, describes the River Contract as 'a participatory democratic process focused on enhancing the environmental quality of rivers and their surrounding areas, while addressing the challenges posed by climate change and biodiversity loss' (Bastiani 2011).

River Contracts are structured through the following phases:

- Development of a "Statement of Intent": This document outlines the motivations and general objectives of the initiative.
- Preparation of an integrated "Preliminary Overview": This involves a comprehensive assessment of environmental, social, and economic aspects.
- Formulation of a "Strategic Document": This document defines the long-term scenario and vision for the project.
- Definition of an "Action Plan": This plan typically spans a 3-year period, after which the contract may be updated or a new "Action Plan" may be approved based on monitoring outcomes.
- Implementation of open and inclusive participatory processes: These processes allow for the sharing of intentions, commitments, and responsibilities among the participants in the River Contract.
- Signing of a "Formal Commitment Act": The River Contract formalizes the shared decisions made during the participatory process and specifies the specific commitments of the parties involved.
- Activation of a "Monitoring and Control System": This periodic system evaluates the implementation status of the various phases and actions.
- Publication of data and information on the River Contract: Ensuring transparency and accessibility of information related to the contract.

The process of establishing a River Contract has already been initiated in several river basins across both Italy and Europe (Massarutto 2008).

Studies on River Contracts have primarily focused on three key areas: participatory governance, environmental sustainability, and territorial integration. In terms of governance, research examines how these contracts can foster inclusive and multi-level decision-making processes, ensuring equitable involvement of various stakeholders (Vismara and La Rocca 2013). Regarding sustainability, River Contracts are studied as tools for promoting sustainable water management practices, aimed at reducing pollution and protecting river ecosystems. Finally, in the realm of territorial integration, research has concentrated on the interaction between water management policies and territorial planning, highlighting the importance of an integrated approach to natural resource management (Magnaghi 2000; D'Alessandro and Pistocchi 2014).

For River Contracts, the local context becomes the starting point for a dialogue that links different scales of action and operational methods. This approach aims to move beyond the hierarchical and often short-term perspective of urban and architectural planning by integrating structural and strategic decisions with project choices for individual nodes from the very beginning. It seeks to integrate scales and thematic sectors that are traditionally addressed in isolation during the planning process. The goal is to raise awareness of the heritage value of the riverine environment and to establish a social and cultural identity path that empowers local communities and administrations. In this way, initiating a River Contract becomes a truly grassroots-driven necessity.

1.3 Mugnone River Contract

Considering these factors, the Mugnone River Contract has endeavored, from its inception, to experiment with a *bottom-up* approach by engaging the community and associational networks in a process of reclaiming a historically, culturally, and naturally significant asset that has been undervalued.

The Mugnone Creek is a right-bank tributary of the Arno River, originating in the municipality of Fiesole, at Vetta alle Croci, approximately 518 m above sea level, not far from the Medici Park of Pratolino. The watercourse flows into the Arno near the Ponte all'Indiano and the Parco delle Cascine, after crossing 17.5 km through rural areas, suburbs, and skirting the historic center of Florence. For centuries, the Mugnone has been important to the history of Florence (Fig. 3); yet, despite its presence, most people feel disconnected and unable to appreciate this remarkable natural resource. Anthropogenic activities have led to substantial alterations in the characteristics of watercourses, which, especially in urban areas, have lost their recreational and restorative functions for the population (Benucci 2022; Cecconi 1980; Coli and Pranzini 2023; Detti 1977; Fanelli 1980; Ferretti 2016; Gasperini et al. 2019; Poggi 1992; Pranzini 2008). Nevertheless, its location could be strategic for the development of a slow mobility network to facilitate alternative territorial use (e.g., the Via degli Dei and Ciclovia dell'Arno).

Thanks to the Tuscan Regional Authority for Participation funding, the partnership of the municipalities of Fiesole and Florence, the Consorzio di Bonifica 3 Medio Valdarno, and the support of the University of Florence's Department of Architecture (DIDA) and the *Memoscape* Landscape Architecture Studio, it has been possible to launch a public participation process in which citizens have been engaged in a journey of awareness and understanding to promote and envision the Mugnone River Contract (Fig. 2).

The project is also supported by: European University Institute, Legambiente Toscana, Q1-Q2-Q5 Comune di Firenze, UISP Comitato Territoriale Firenze, PIA Palazzina Indiano Arte, Centro di rilevante interesse per la danza Virgilio Sieni, Teatro Solare, Nordic Walking Toscana, Laboratorio del Cammino, G.I.R.O.S. Gruppo Italiano per la ricerca sulle Orchidee Spontanee APS Sezione Fiorentina, Associazione Il Crinale, Ranch Lory Quarter Horse, Fratellanza Popolare Valle del Mugnone, Sebastian Schweizer, Toscana SUP, CNGEI, Centro Storico Lebowski, Associazione Onouka, Tommaso Lisa, L'Erbacanta, Fondazione Angeli del Bello, Cantiere Obraz.

The project's approach has been structured around four general objectives, which have contributed to defining a framework that equally prioritizes connectivity, inclusivity, ecology, and the identity of the Mugnone Creek. Each objective has been translated into three main strategies, which in turn have led to specific and targeted actions along the river's course.

- **A safe and connective corridor:** The primary objective is to create a safe and connective corridor along the Mugnone, a "slow line" as defined by Paolo Pileri, "because investing in excellent slow lines is a prerequisite for providing substance and citizenship to slowness," and "slowness can become the backbone of a genuine territorial development model of potential interest" (Careri 2006). The related strategies, therefore, focus on integrating existing pathways, employing nature-inspired design solutions (NbS in the EEA Nature-based Solutions in Europe Report No 1/2021) to conserve and enhance water and soil resources, and designing an effective and integrated communication system for the project (Careri 2006; Pileri 2021).
- **A common good for all:** The second objective focuses on the river as a shared resource for all, emphasizing design that fosters inclusive cities and communities by addressing the diverse needs of different groups. Viewing public space as a fundamental right involves valuing heritage and connecting open spaces to strengthen green and blue infrastructure.
- **An ecological refuge**: The protection of nature and the river ecosystem is realized through three strategies, which are underpinned by a quest for balance between safeguarding flora, fauna, and fish, and the progressive appropriation and enjoyment of the river by the public.

Fig. 2 The Mugnone River Contract (in Mezzapesa and Moretti 2024)

Fig. 3 The Mugnone River Contract, Historical timeline (in Mezzapesa and Moretti 2024)

Enhancing the ecological corridor, enriching the biodiversity mosaic, and implementing differentiated management as a landscape practice are the three strategies aimed at ensuring the project's sustainability.

– **A living legacy for future generations:** The project considers nurturing and unfolding the complex relationship between community and landscape, as well as between memory and both material and immaterial heritage along the Mugnone throughout history and the contemporary context, as both a starting point and an end goal. Strategies implemented include transmitting the river's history, promoting art as a catalyst for new perspectives and insights, and enhancing sports and well-being.

Following these four general objectives, the project aim is to design a Landscape Infrastructure: conceiving infrastructure as landscape, which enriches infrastructure by incorporating social, ecological, and imaginative dimensions, while considering landscape as infrastructure, which regards living systems and ecological flows as fundamental infrastructural components capable of shaping urban environments and enhancing biodiversity (Bélanger 2013; Buoro 2023; Cortesi 2017; Girot and Imhof 2017; Hung 2013; Mossop 2006; Weilacher 2016).

This holistic approach promotes the creation of active, functional landscapes that regenerate degraded areas and foster greater community interaction and resilience (Fig. 3).

1.4 The Article Structure

In terms of content, this paper specifically examines the River Park project along the Mugnone Creek. To this end, Sect. 2 outlines the methodology of the participatory process that led to the successful signing of the Mugnone River Contract. Section 3 provides a synthesis of the key outcomes, focusing on the community participation events conducted and the design strategies that emerged during the public consultation. Finally, Sect. 4 discusses the lessons learned, with the aim of informing the future development and direction of the project.

2 The Methodology

The participatory project primarily aimed to initiate a process of awareness and understanding, beginning with a group of citizens and an established network of local associations.

This approach enabled the mobilization of interest, actions, and resources across the entire river basin, moving beyond a view of the creek only focused on hydraulic and hydrogeological risk, and fostering, through renewed awareness, the revival of the creek as a component of cultural identity and a tool for territorial development.

The main objective of the participatory process was to promote a River Contract through the synergy and integration of various activities undertaken by different entities and associations. This included the general works of the Basin Authority, the ongoing maintenance by the Consorzio di Bonifica 3 Medio Valdarno, the integration of municipal visions of Firenze and Fiesole (Piano Strutturale, Piano Operativo, and Piano del Verde), and the recreational and sports initiatives organized by numerous local sports and cultural associations, in collaboration with local administrations.

The project was primarily aimed at citizens, river users, and associations. Criteria for representation were employed both in relation to the diverse segments of the population (age, gender, origin, profession, etc.) and with regard to the various interest groups present in the area (neighborhood committees and sports, religious, environmental, and socio-health volunteer associations, etc.). To promote gender participation, the main participatory events were organized at suitable times and included support for families through activities for children within the workshops. Recognizing that workshops held in specific territorial areas often result in the self-selection of citizens who have the time and resources to participate, the association *Le Curandaie*, which promoted the participatory process, played a key role in mobilizing and engaging its members as well as the members of the local associative network to involve a broader range of stakeholders.

The neutrality and impartiality of the participatory process were also ensured by the central role of facilitators from the DIDA (Department of Architecture, University of Florence). These facilitators provided the professional expertise necessary to ensure that participants had access to information, appropriate methods and timing to express themselves, specific communication skills for internal and public engagement, and a comprehensive sociological understanding of the area to effectively select participants.

The goal of experimenting with a collaborative, proximity-based approach among stakeholders was to activate, through the promotion of collective learning, a process capable of generating an organized riparian community composed of citizens, public and private entities, and interest holders (Pisano and Lingua 2024). This approach materialized into a relational proximity process aimed at creating shared visions between the community and institutions. It was guided by the principles of "Place of Proximity" (PoP), a registered trademark filed by the University of Florence, that introduces an innovative approach to participatory urban planning, focusing on transforming citizens into active participants in decision-making processes (Caruso 2024). By enabling territorial governance through reciprocity and mutual learning, PoP represents a new paradigm in urban planning, evolving grassroots communities into organized ones that generate intangible relational spaces for co-creating shared visions with institutions. The participatory process was so organized following the five stages of the Place of Proximity model: (1) initiation; (2) engagement; (3) organization; (4) collective learning; (5) visioning.

The process was structured in multiple phases to educate and organize a conscious community capable of collaborating with institutions in the development of future visions.

2.1 Stage 0: Signature Collection and Application for the APP (Regional Authority for the Guarantee and Promotion of Participation)

The idea emerged in 2020 from a dynamic and socially conscious association. *Le Curandaie* has been active since 2015 as a local association in the Florence area, initiating cultural, economic, and social initiatives. The association focuses on enhancing the richness of the social fabric, particularly emphasizing the role of women and fostering virtuous practices of social engagement.

The project originated as a generative effect of the "Tracce sul Mugnone" which aimed to enhance a stretch of the Mugnone Creek as the initial step toward creating an urban public river space and as an experiment in the revitalization of a significant environmental and landscape resource. Additionally, the Mugnone Creek is part of the Arno River basin, which has recently embarked on the process of establishing the Arno River Contract.

The innovative aspect of this creek project, namely the *bottom-up* approach led by local associations and riparian citizens, could serve as a useful model for ensuring the replicability of the process and the sustainability of its outcomes over time and across different contexts.

The network established during this initial phase, dedicated to signature collection, engaged various entities and social actors:

- Local stakeholders: individuals and experts recognized by the community as references, including shopkeepers and economic operators.
- Informal groups: active citizens, makers, sports enthusiasts who utilize the river, urban regenerators, and urban innovators.
- Elderly citizens: individuals who are knowledgeable about the history and traditions of the area and who preserve the memory of the place and its transformations.

- Younger generations: children and teenagers who represent the future of the area.
- New residents: individuals bringing diverse cultures, traditions, and needs.
- People with disabilities and vulnerable groups: valuable subjects who can help identify aspects and issues that are often overlooked.

At the end of May 2023 approximately 1800 signatures were collected, which enabled the proposal for a participatory listening and idea-generation process to be submitted to the call promoted by the APP (Regional Authority for the Guarantee and Promotion of Participation) of the Tuscany Region thanks to the Regional Law 46/2013 (Participation Law), which recognizes citizens' right to actively participate in the development of regional and local public policies. The proposal received the requested funding, and the participatory process started in September 2023.

2.2 Stage 1: Promotion, Engagement, Community Activation

The initial stage was divided into three sub-phases concerning the promotion of the project, the investigation, and community activation:

(a) *Promotion of the Project*

The project began with preparatory outreach activities, including the presentation of the *Lungo il Mugnone* initiative within the targeted area and the engagement of citizens, associations, and institutions through:

- Traditional communication methods (web, print, notices).
- Creation of a dedicated project page on Open Toscana and the establishment of dedicated social media pages (Facebook, Twitter, etc.)
- Division of the territory into working areas and organization of interactive meetings with the public for each area.

(b) *Listening*

This sub-phase involved investigative moments (gathering information from all stakeholders related to the creek) and direct listening to the public and administrators through:

- Local meetings involving civil society and existing organized groups (Fig. 4).
- Semi-structured interviews with stakeholders.
- Establishment of the Coordinator Table, composed of members from entities at all levels and associations participating in the project.

(c) *Community Activation*

This sub-phase engaged local residents and existing associations or organized groups. Following the proposed relational proximity process, the community was involved in field activities, creating a cohesive group of individuals focused on the creek-related issues through:

- Territorial animation events.
- Walks along the creek.
- Training sessions on specific topics identified by the group.

Fig. 4 Key moments of Stage 1 (in Mezzapesa and Moretti 2024)

The entire project was accompanied by continuous communication efforts aimed at informing and raising awareness among the public and institutions.

2.3 Stage 2: Definition of Territorial Identity and Community Mapping

The second phase addresses the need to delineate the general state of the Mugnone by identifying the environmental and anthropogenic resources that characterize the creek in terms of local identity. The objective was to co-create a comprehensive community map of the Mugnone's entire course (Fig. 5), highlighting historical, architectural, and landscape heritage, as well as existing or potential public green spaces. Defining the identity elements not only structured a knowledge framework but also led to a shared representation of the region's issues and opportunities. This included a particular focus on evolutionary dynamics, thereby identifying the region's strengths, weaknesses related to the hydrological themes, and broader opportunities and threats.

To ensure this phase represented an expression of the living identity of the riparian communities, a map was created in collaboration with residents, associations, and stakeholders through workshops for both adults and children along the Mugnone. The outcome of this phase is a map of the natural, architectural, and landscape heritage, which serves both as a foundation for the subsequent phase of developing the river

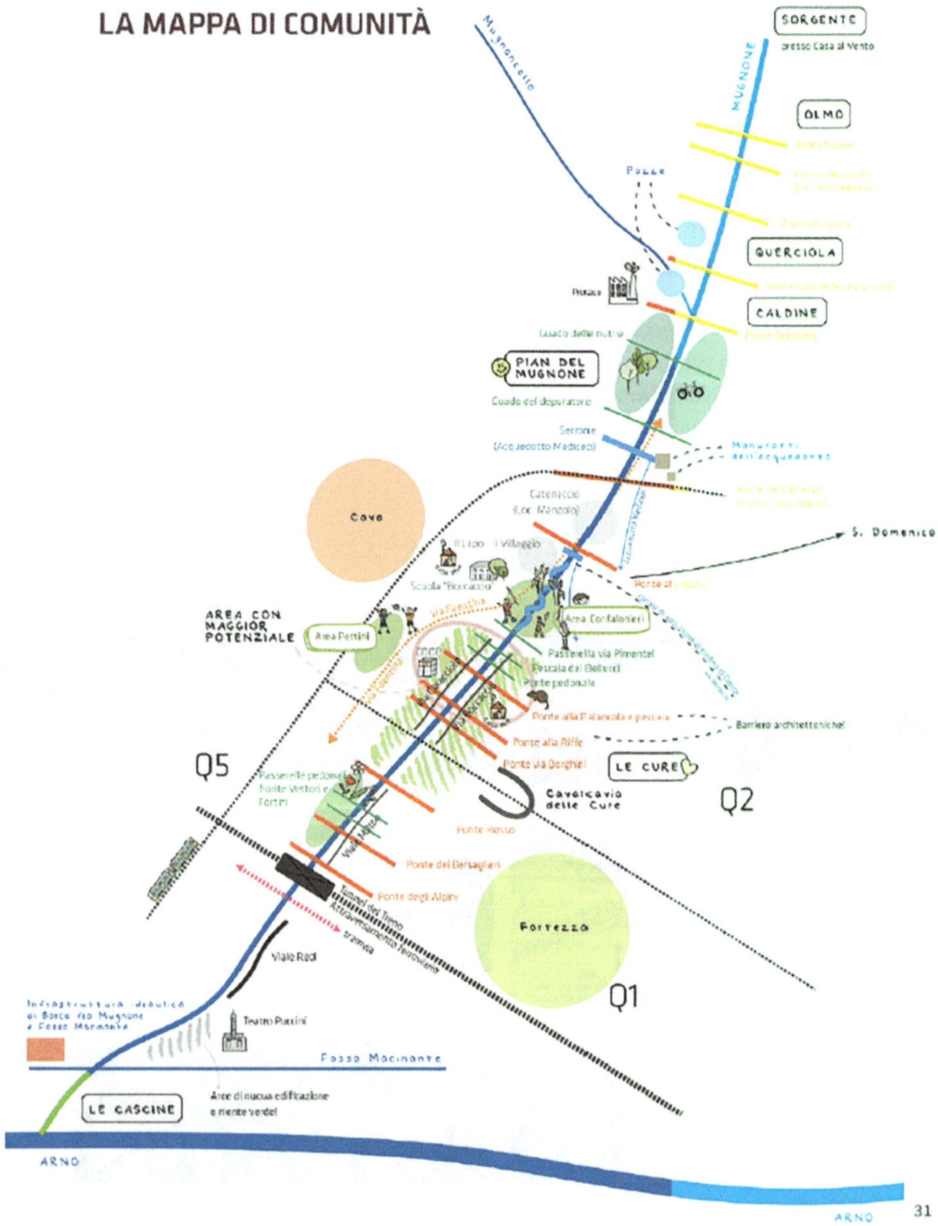

Fig. 5 Community Map (in Mezzapesa and Moretti 2024)

area's use and as a tool to activate the community through a collective learning process.

2.4 Stage 3: Co-design for the Activation of the Project Community

Following the analysis and diagnosis, this phase aimed to develop a project vision for the Mugnone Creek, through actions designed to enhance its usability while also improving its environmental, landscape, social, and economic quality.

(a) *Project Systematization*

Given that these aspirations cannot be separated from the ongoing institutional projects, this phase involved preliminary work to systematize the various projects already underway along the creek. This was done not only by the municipalities of Florence and Fiesole but also through the coordination of projects planned by the authorities responsible for managing the creek (the Region, the Metropolitan City, the Municipal Office, and the Consorzio di Bonifica).

(b) *Co-design Workshops*

Following this preliminary phase, three co-design workshops were organized, led by community facilitators identified in Stage 1 and based on the analysis and community map developed in Stage 2. These workshops were intended to integrate the various elements related to the creek's quality (from usability to ecosystem quality, from water quality to biodiversity) in order to design a comprehensive usability plan. The goal of these activities was not only to arrive at a shared project but also to foster a sense of ownership among the community, which would also be encouraged to take responsibility for the maintenance of the spaces through active citizenship initiatives (e.g., cleaning the banks, organizing periodic events along the creek, etc.).

To this end, the co-design workshops were supported by collateral activities organized in collaboration with local associations and institutions, such as walks along the banks, on-foot or SUP (stand-up paddleboard) excursions, and so forth. These events were strategic for identifying existing connections to be restored or new paths to be designed, to enhance the existing mobility network and envision a route connecting the mouth of the Mugnone to its source. From the development of the overall vision for the route, it was possible to focus on the most critical points where access to the creek is obstructed by complex orographic features (for example, between Pian del Mugnone and Ponte alla Badia) or by densely urbanized areas (such as in the Novoli area). The targeted co-design actions fit within and detail the overall vision and represent the pilot projects of the action plan (Fig. 6) aimed at implementing the Mugnone River Contract and detailed in the "Preliminary Overview," the "Strategic Document," and the "Action Program."

2.5 Stage 4: River Contract Protocol Signing

The final phase involved the signing of the "Memorandum of Understanding" and the formalization of the Mugnone River Contract. In this context, the essential documents for the implementation of the River Contract were presented: the "Preliminary Overview," the "Strategic Plan," the "Action Program," and the "Memorandum of Understanding." This phase also led to the completion of content on a larger watershed scale and initiated actions scheduled for 2024/2025. The phase concluded with a final event ("*Lungo il Mugnone in Festa*," April 13, 2024) during which the official documents were presented to the public, and the "Memorandum of Understanding" was signed.

This final phase played a pivotal role in establishing the foundation for the future of the Mugnone River Contract. While it signifies the culmination of the participatory process, it also serves as the starting point for a shared journey, one that citizens, associations, and political institutions are collectively committed to pursuing moving forward.

Throughout every phase of the participatory process, approaches and methods were adopted to ensure the maximum communication of the information necessary to achieve a comprehensive understanding of the participatory process. Full equality of expression for all viewpoints was promoted through techniques and tools used in the collaborative proximity approach.

3 Research Outcomes

3.1 The Impact of the Participatory Process

The direct impact of the participatory process primarily concerned the key thematic areas of the project:

- Development of an action and intervention framework: This involved designing a set of actions and interventions aimed at enhancing the creek as a safe and connective corridor, which was shared among various participants and stakeholders, highlighting a prioritized order of interventions.
- Creation of a new sense of awareness and community: This is related to increasing awareness and a sense of community connected to water accessibility.
- Identification of a vision-strategy for ecological and environmental restoration: This involved formulating a vision-strategy for the ecological and environmental restoration of the creek as a sensitive ecosystem.

Fig. 6 Results of the Stage 3 (in Mezzapesa and Moretti 2024)

- Definition of an adaptive model for cultural landscape interpretation: This included developing a model for interpreting the cultural landscape associated with the Mugnone, which can be integrated into various planning and maintenance tools.
- Formulation of a strategic vision for the programmatic document and action plan: This aimed at guiding the programmatic document and action plan for the Mugnone River Contract.

Additionally, the participatory process had a cultural impact on the involved communities and the administrations, which may amplify its long-term effects. For the riverside populations, anticipated future effects include:

- Reclaiming a sense of community and territorial identity: enhancing local identity and community cohesion.
- Increased awareness in civil protection: improved understanding and involvement in civil protection.
- Wider participation: greater public engagement in community activities.
- Renewed interest in public affairs and common goods: increased concern for public and communal resources.

For the administrations and entities involved in water management and land governance within the watershed, expected effects include:

- Redefinition of interaction methods with citizens: new approaches to engaging with the public.
- Rediscovery of the need for horizontal and vertical interaction methods: enhanced methods for territorial and water management.
- Integration of guidelines for the Arno River contract: incorporation of guidelines into the management framework.

The participatory process (Fig. 7) facilitated the organization of:

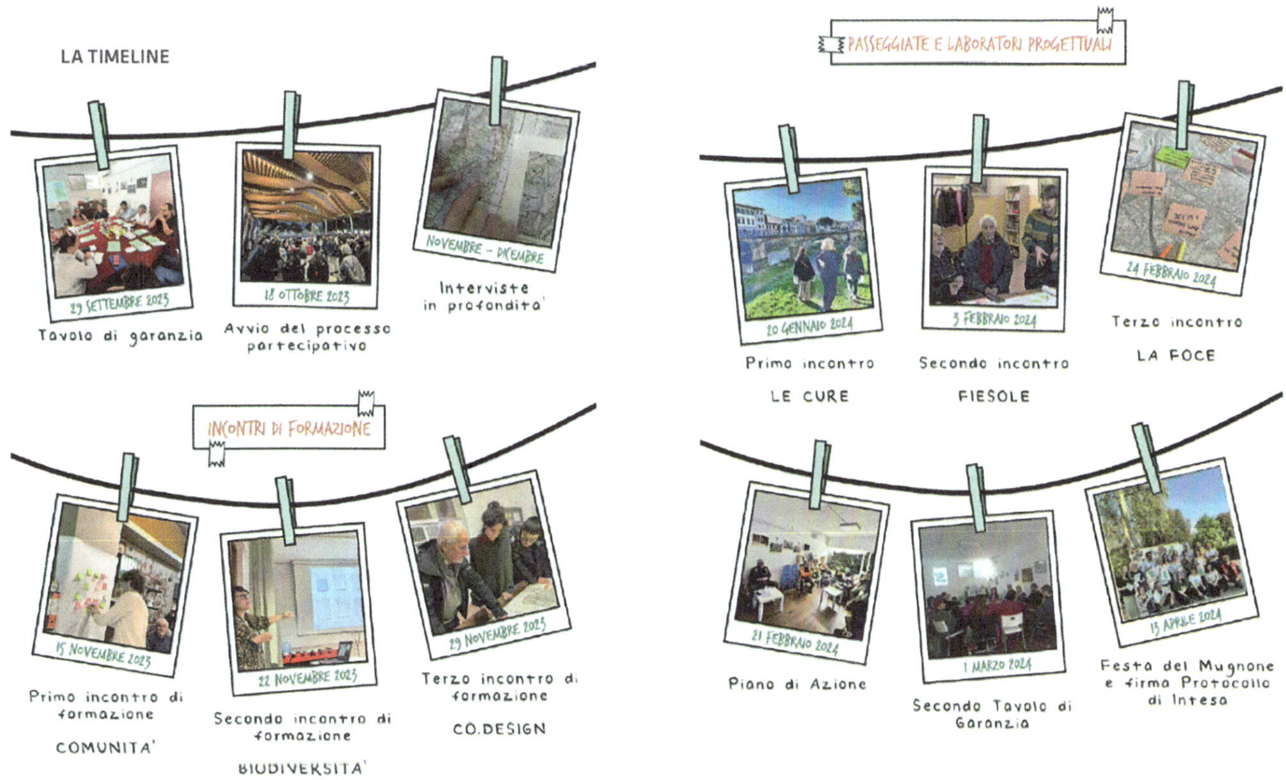

Fig. 7 Participatory process timeline (in Mezzapesa and Moretti 2024)

- 3 thematic seminars and meetings: held in different sections of the creek (Fiesole, Florence Q2-Q5)
- 3 design walks: held in various sections of the creek (Fiesole, Florence Q2-Q5), resulting in the creation of community maps on paper and fabric
- 3 coordination tables: held at the beginning, middle, and end of the process

3.2 "Lungo il Mugnone in Festa"

At the conclusion of the participatory process, a closing event was organized on April 13, 2024, to celebrate the Mugnone River Contract with walks and activities along the creek from Fiesole to Florence (Fig. 8).

The celebration featured a satellite event of the New European Bauhaus Festival, centered on the theme "Walking with Water: Exploratory Walks across Evolving Waterscapes." This involved a collective walk from the source of the Mugnone to its confluence with the Arno. The walk was divided into stages, each offering various activities ranging from sports like SUP, river trekking, and Nordic walking to citizen science, landscape studies, historical, artistic, and botanical tours, as well as artistic events and workshops. The day concluded with the signing of the "Memorandum of Understanding" for the Mugnone River Contract at the Palazzina Indiano Arte, located at the creek's mouth into the Arno.

This event marked the first opportunity for the partners to collaborate in organizing and executing an event intended to become an annual celebration, highlighting the values of the creek.

3.3 The Mugnone River Park

With the valuable contribution of landscape architects from *Memoscape* studio, the following documents for the Mugnone River Contract have been prepared:

- "Preliminary Overview" for the Mugnone Creek: an atlas of planning and programming at the watershed scale and across various specialized sectors, including emerging visions at the watershed scale, contributions from awareness-raising activities, and thematic discussion tables
- "Strategic Document": a proposal for a river park that incorporates ideas emerging from the participatory process, envisioning a new configuration for slow travel along the Mugnone Valley while respecting the ecological balances of the river basin
- "Action Plan": a collection of descriptive sheets of shared ideas/projects, organized into various categories, outlin-

Fig. 8 Key moments of "*Lungo il Mugnone in Festa*" (in Mezzapesa and Moretti 2024)

ing a comprehensive framework that involves multiple entities and local realities at different scales (regional, metropolitan, municipal, etc.) to be realized in the medium term during the 2023/2024 period

– "Memorandum of Understanding": a proposal to continue the ongoing process, develop the Mugnone River Contract with an expanded focus on a broader watershed scale, and begin implementing actions for the 2024/2027 period.

These materials are available to all parties involved in the River Contract. This agreement aims to contribute to the revitalization and regeneration of the hydrographic basin, ensuring proper enjoyment of the creek, promoting slow and green tourism, activating a safe and connective corridor, and providing an active public space, improving well-being, soft mobility, biodiversity, and ecosystem services.

The Mugnone Creek is a typical Mediterranean river with a highly torrential regime, characterized by alternating periods of flooding and drought. This hydrological reality, along with current climate change projections, represents one of the primary challenges of the project. Consequently, strategies and actions have been implemented not only to enhance water resources but also to improve drainage capacity and create space for the river itself.

The project proposes a multimodal green and blue infrastructure around the Mugnone Creek to make the river accessible to all, improving connections both north-south between Fiesole and Florence, and transversal connections within the city itself (Fig. 9). The Mugnone river park will thus form the spine of a broad system for reconnecting built environments and open spaces and will serve as an ecological corridor between the city and the countryside, where biodiversity is

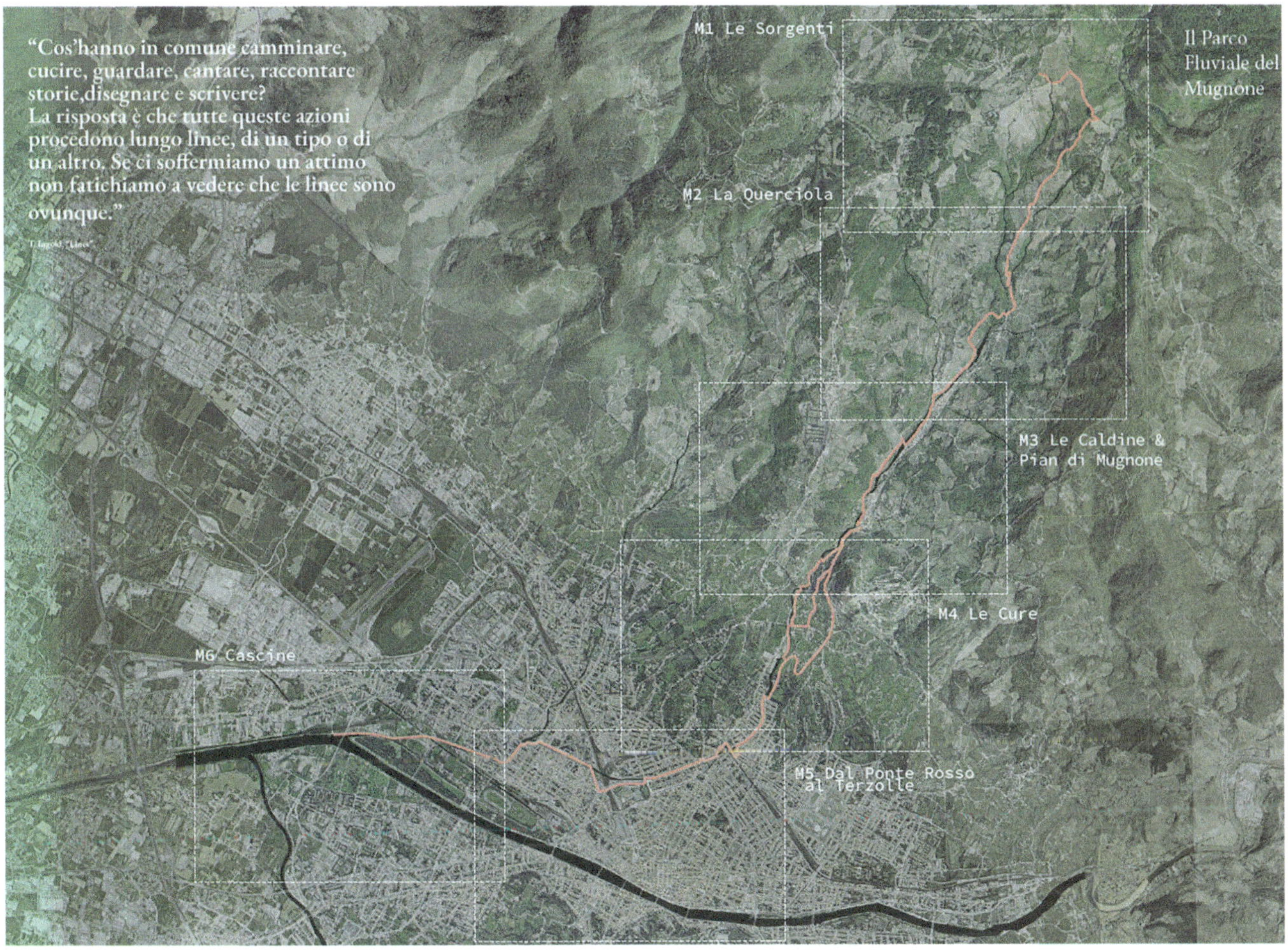

Fig. 9 Mugnone Creek map (in Mezzapesa and Moretti 2024)

enhanced through differentiated maintenance practices and actions aimed at strengthening ecosystem services.

The plan proposes starting the route from the northernmost source, the Fosso Cafaggio in the Pratone di Monte Senario, creating a connection with the Via degli Dei along the Poggio Capanne trail (Figs. 10 and 11). This section includes creating a new access area to the river park around the source, with an information point, and the restoration of historic access routes. A crucial aspect of the project will be an effective communication system, continuous and readable signage, and co-branding to guide people through the metropolitan area while developing a shared identity for the Mugnone Creek.

The route continues to the Querciola, following existing CAI trails, with the aim of enhancing the connection with the Medici Park of Pratolino at the confluence of the Mugnone with the Mugnoncello. This section of the Mugnone passes through an exceptionally beautiful agricultural landscape and the route could also serve as a catalyst for promoting high-quality local products and activating the network of producers and farmers. To preserve the creek's ecological functions, a natural management of the riverbed and banks would be desirable, allowing the river to reclaim its space. This can be achieved through experimental re-naturalization practices that allow the small river to naturally define its course by adopting dynamic flow as a starting point and providing it with a space of freedom for continuous processes of erosion and deposition.

Between Caldine and Pian di Mugnone, there is already a widely used pedestrian and cycle path along the river. The project focuses on enhancing a green area, previously created through cleaning actions and new plantings by local associations. This garden can be transformed into a natural playground area for active learning, telling the story of the Mugnone, which from Caldine onwards winds into the Monte Rinaldi and is accompanied by the Medici aqueduct of Montereggi. This area will be dedicated to play and memory, serving as a natural classroom for meeting and interpreting history and nature.

In the final stretch where the Mugnone Creek retains its natural form, from Ponte alla Badia to the Ponte alla Palancola at the Cure, the project foresees the creation of a

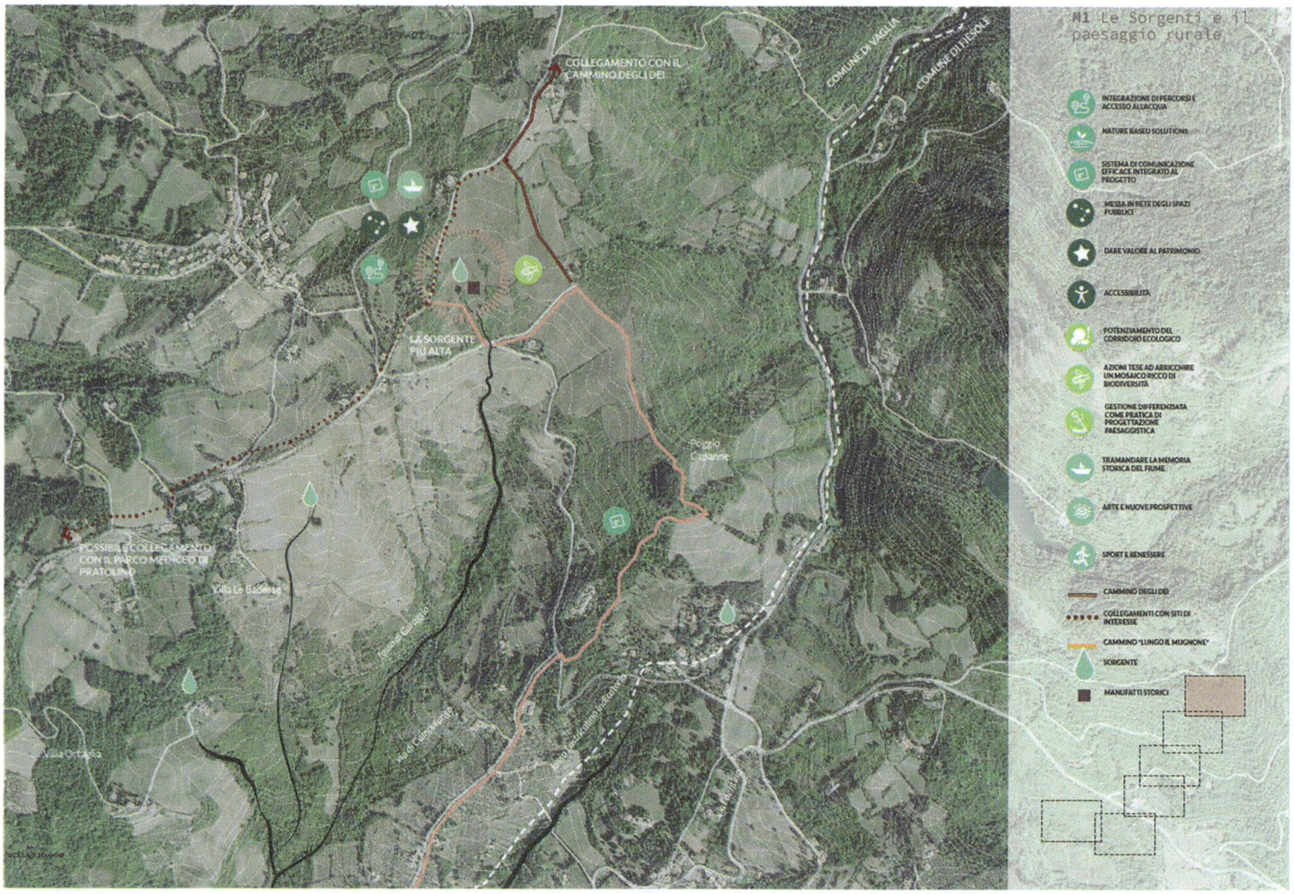

Fig. 10 Mugnone river park, starting point at the creek source, design map (in Mezzapesa and Moretti 2024)

Fig. 11 Mugnone river park, new access area at the creek source, design view (in Mezzapesa and Moretti 2024)

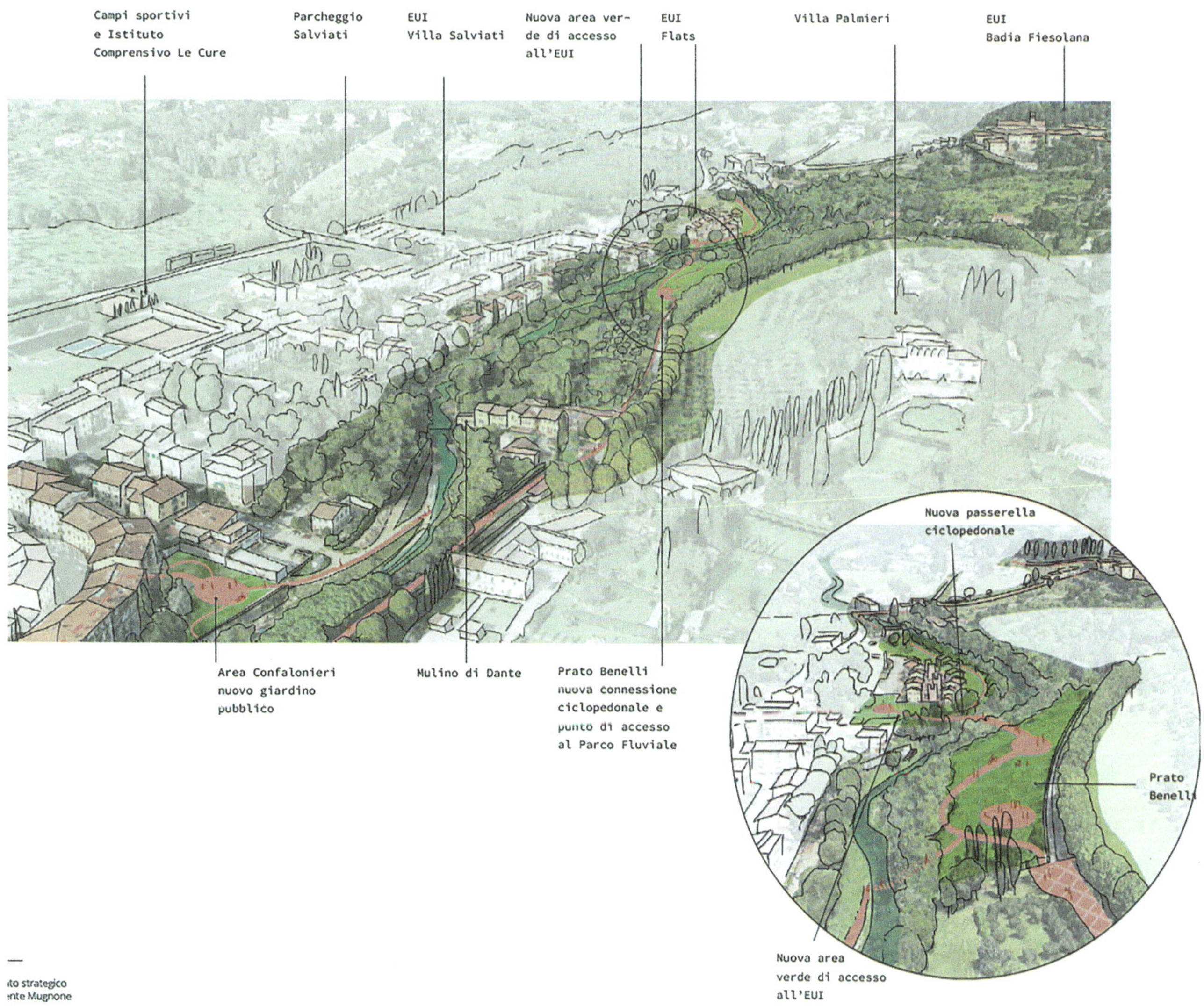

Fig. 12 Mugnone river park, Le Cure, design views (in Mezzapesa and Moretti 2024)

pedestrian and cycling path to facilitate comfortable and safe walking and sports activities, connecting the Cure neighborhood with the Ponte alla Badia area and simultaneously stitching together fragments of green areas along the river (Fig. 12).

Enhancing these connections between open meadows, areas managed with differentiated practices for public enjoyment, and peripheral riverine public spaces is the focus of this proposal, which includes assessing the feasibility of a pedestrian and cycle bridge connection between the European University (EUI) residences on Via Faentina, including the partially wooded small area, and the so-called Prato Benelli (currently private property) on Via Boccaccio. New routes will allow for integrated reconsideration of accessibility in the neighborhood and incorporation of new green areas, such as the one planned in the former Confalonieri area. Additionally, the overall strategy aims to improve existing connections through specific projects to make the bridges in the Cure neighborhood fully accessible. Currently, three of the four creek crossings are characterized by numerous architectural barriers with significant implications for safety and inclusivity in the urban environment.

From Ponte Rosso, the Mugnone increasingly takes on the artificial form of a narrow channel between high embankment walls, eventually flowing into the Arno near Ponte all'Indiano at the Cascine. The route is thus structurally already present. However, the river's role in generating new spaces needs to be enhanced. In continuity with the interventions already carried out by the Consorzio di Bonifica 3 Medio Valdarno in this section of the creek, ecosystem services should be improved with the protection of flora and fauna; the creation of new refuges for animals and insects; and, where possible, by enhancing the presence of wetlands or areas sufficiently quiet and protected for riverine flora and

Fig. 13 Mugnone river park, the confluence with the Arno, design views (in Mezzapesa and Moretti 2024)

fauna. Visual accessibility to the river is crucial in this stretch, creating "windows" from which the river can be seen while walking or cycling along the path.

The final significant proposal concerns the stretch from Piazza Puccini to the confluence with the Arno (Fig. 13). Here, it is essential to expand the riverbank section near major public or publicly accessible spaces to create access ramps to the service path, facilitating connection with the River Park, such as with the Castelnuovo Tedesco Park and the La Trave sports center.

In this scenario, the Mugnone river park becomes a strategic connection and development axis, linking slow routes between the Sentiero degli Dei (which connects Bologna with Florence) and the Arno cycle path (Ciclovia dell'Arno, which links Casentino with Pisa), as well as serving as a linear ecosystem between Fiesole and Florence.

4 Lessons Learned

The process of developing the *Lungo il Mugnone* project has provided valuable insights into effective landscape management and community engagement. This initiative, which aimed to enhance the Mugnone Creek's ecological and social value, has generated a series of key takeaways that highlight best practices and strategic approaches. The lessons learned from this project offer a comprehensive understanding of how multi-scale planning, participatory processes, and strategic documentation could contribute to successful river park development.

The principal **lessons learned** are:

- Integration of multi-scale planning: Combining various planning scales and specialized sectors creates a comprehensive framework for managing and enhancing river systems. Effective integration of local, regional, and municipal perspectives ensures that both immediate and long-term goals are addressed (Pisano and Lingua 2024);
- Participatory engagement: Engaging diverse stakeholders, including local communities, authorities, and specialized entities, through workshops, public walks, and collaboration tables, fosters a sense of ownership and enriches the planning process. This approach also helps in prioritizing actions that are broadly supported.
- Strategic vision and flexibility: Developing a strategic vision that incorporates feedback from participatory processes is crucial for creating a responsive and adaptable river park plan. This vision should balance ecological integrity with human use and accessibility.

- Actionable documentation: Producing clear and actionable documents, such as the "Preliminary Overview", the "Strategic Document", the "Action Plan", and the "Memorandum of Understanding", helps in translating broad goals into specific, implementable actions. These documents should guide both immediate interventions and long-term strategies.
- Ecological and community benefits: Enhancing both ecological functions and community access through a multimodal green and blue infrastructure can create significant environmental and social benefits. Such improvements contribute to biodiversity, local development, and sustainable tourism.
- Implementing a robust communication strategy helps in guiding users and creating a unified identity for the project. This aids in understanding the area and reinforces the project's branding.
- Preservation and enhancement of natural spaces: Natural management of riverbeds and banks helps preserve ecological functions while providing enjoyable experiences for users.
- Addressing accessibility challenges: Focusing on removing architectural barriers and improving accessibility in urban environments is crucial for ensuring inclusivity. This includes making bridges and pathways more accessible to all users.
- Promoting sustainable use and conservation: Integrating conservation efforts with sustainable recreational use can enhance public engagement and support for river management. Actions such as creating natural play areas and supporting local agriculture contribute to both conservation and community well-being.
- Collaboration and shared goals: Collaboration among various stakeholders and adherence to shared goals is essential for the success of river park projects. Continuous engagement and joint efforts in organizing events and activities can strengthen partnerships and community involvement.

These lessons highlight the importance of comprehensive planning, stakeholder involvement, ecological sensitivity, and effective communication in creating successful and sustainable river park projects.

The main challenges encountered by the project, as well as those it will continue to face, are deeply interconnected, reflecting the complex nature of large-scale participatory initiatives. One significant hurdle has been the resistance from certain segments of the local population. Many citizens expressed skepticism about the feasibility and realism of a project of such scale, fearing that it might be too ambitious or unattainable. This reluctance has, in part, stemmed from previous experiences with large projects that failed to materialize or were not executed as promised. Overcoming this skepticism will require continued dialogue, transparent communication, and tangible results that demonstrate the project's potential.

On the other hand, the greatest challenge the project faces at this stage is securing adequate funding. Without sufficient financial support, the project cannot proceed to the detailed design phases necessary for the successful development of the various areas. Funding is not only crucial for the technical and infrastructural aspects of the project but also for maintaining stakeholder engagement, as it provides the resources needed for ongoing participation, workshops, and other collaborative activities. As such, the ability to attract and secure investments will be pivotal in determining the future trajectory of the project, ensuring its sustainability and long-term success.

To conclude, we wish to emphasize that *Lungo il Mugnone* is a landscape project rooted in community engagement, envisioned with and for the community. The concept of landscape cannot be separated from the community; landscape is a right, and caring for it means allowing the landscape to shape our ideas of community as well as our dreams and projects. This is particularly true for a river landscape, with its continuous and unpredictable dynamics.

Thus, the project's realization has been conceived through specific actions, or "confetti" as landscape architect Martí Franch terms them—intensive nodes of experience, space, and memory intended to invite a fresh perspective on the landscape.

A popular saying goes that "water has memory," suggesting that a river never forgets its course. The Mugnone Creek has been altered numerous times; with each new city wall, the river was shifted further west, and in recent years, changes have been driven not by walls but by new infrastructures, from railways to high-speed lines, increasingly separating the river from the city, a mere barrier to cross.

Lungo il Mugnone aims to be a collaborative and empirically validated project, designed to overcome historical transformations and to collectively establish a new legacy for the river and its surrounding community.

Acknowledgments We acknowledge all the entities and authorities directly involved, local residents, and all other relevant stakeholders and participants in the working group of *Lungo il Mugnone* for their contribution to the results presented in this paper.

Authors' Contribution This paper is the result of a joint reflection by the authors. Section 1 was written by C. Mezzapesa and E. Moretti. Section 2 by C. Mezzapesa, E. Moretti, V. Lingua, E. Caruso. Sections 3 and 4 by C. Mezzapesa and E. Moretti.

References

Bastiani M (ed) (2011) Contratti di fiume. Pianificazione strategica e partecipata dei bacini idrografici. Dario Flaccovio Editore

Bélanger P (2013) Landscape infrastructure. Urbanism beyond engineering. Wageningen University

Benucci P (2022) Il Mugnone e i torrenti tombati di Firenze. Sarnus

Brun A (2010) Les contrats de rivière en France: un outil de gestion concertée de la ressource en local. In: Lasserre L, Molle F (eds) L'eau mondialisée: La gouvernance en question. Éditions La Découverte, pp 305–321

Brusarosco A, Visentin F (2023) Costruire Contratti di Fiume: riflessioni, percorsi, pratiche. Forum

Buoro M (2023) Landscape infrastructure. Tautologia per le città del XXI secolo. In Cianci MG, Rabazo Martin M (eds) Tracce segni e rappresentazioni del paesaggio che cambia. Il Poligrafo Editore

Careri F (2006) Walkscapes. Einaudi

Caruso E (2024) Verso nuove forme di comunità. Un approccio sperimentale per la pianificazione di area vasta. Labsus Edizioni

Cecconi A (1980) Il Mugnone attraverso i secoli. Cappelli

Coli M, Pranzini G (2023) Firenze e le sue acque. Angelo Pontecorboli

Cortesi I (2017) Il paesaggio al centro. In: Cortesi I, Cappiello V (eds) Il paesaggio al centro. Lettera Ventidue Edizioni

D'Alessandro E, Pistocchi A (2014) La gestione integrata dei bacini idrografici: I Contratti di Fiume come strumento di pianificazione partecipata. Alinea Editrice

Detti E (1977) Firenze scomparsa. Vallecchi

Fanelli G (1980) Firenze, Laterza

Ferretti E (2016) Acquedotti e fontane del Rinascimento in Toscana. Olschki

Gasperini F, Pucci B, Lasso A, Cencetti T (2019) Studio per la riqualificazione ambientale dei torrenti Mugnone e Terzolle in collaborazione con il Consorzio di Bonifica 3 Medio Valdarno

Girot C, Imhof D (2017) Thinking the contemporary landscape. Princeton Architectural Press

Hung Y (2013) Landscape infrastructure: systems of contingency, flexibility, and adaptation. In: Landscape infrastructure: case studies by SWA. The Infrastructure Research Initiative at SWA, Birkhäuser

Leona C (2024) I Contratti di Fiume. Riflessioni interdisciplinari. Mimesis Edizioni

Magnaghi A (2000) Il progetto locale. Verso la coscienza di luogo, Bollati Boringhieri

Massarutto A (2008) Gestione delle risorse idriche e contratti di fiume: Un'analisi comparata tra esperienze europee. Franco Angeli

Mazzoni R (2015) I Contratti di Fiume: uno strumento per la gestione integrata dei bacini idrografici. Edizioni Ambiente

Mezzapesa C, Moretti E (eds) (2024) Mugnone river contract, 'preliminary overview' and 'strategic document'

Mossop E (2006) Landscape of infrastructure. In: Waldheim C (ed) The landscape urbanism reader. Princeton Architectural Press, pp 164–167

Pileri P (2021) Progettare la lentezza. People

Pisano C, Lingua V (2024) The impact of regional design on river agreements: the case of the Ombrone river in Tuscany. Plan Pract Res 39(1):32–53

Poggi G (1992) Relazione sui lavori per l'ingrandimento di Firenze (1864–1877). Giunti

Pranzini G (2008) Idrogeologia della Piana Fiorentina. In: Atti convegno "Un piano per la piana: idee e progetti per un parco", 9–10 Maggio 2008

Vismara R, La Rocca M (2013) I contratti di fiume in Italia: Strumenti di governance e gestione delle risorse idriche. Quaderni di Ecologia

Weilacher U (2016) Revisiting structuralism—revitalizing living networks. In: Tasting the landscape. 53rd IFLA world congress, April 20–22, 2016. Edifir

Pocket Parks in Florence: Small-Scale Landscape Projects for Big Urban Challenges

Claudia Mezzapesa and Elena Moretti

Abstract

Several studies highlight that in large urban centers, the absence of vegetation and widespread soil de-sealing intensify the effects of climate change, negatively impacting human health. Local governments play a crucial role in promoting public space regeneration models centered on resilience to achieve climate adaptation goals.

This article examines the theoretical and practical development of guidelines for implementing Pocket Parks, positioning them within the broader framework of Landscape Infrastructure as effective tools for addressing urban challenges and mitigating climate change. By analyzing significant European initiatives, it underlines the relevance of these interventions in fostering systemic approaches to urban sustainability. It also presents a case study led by the Municipality of Florence, which has implemented a structured methodology for designing and creating nine Pocket Parks within the city.

Developed by the landscape architecture firm *Memoscape*, the project demonstrates how small-scale interventions can function as operational components of Landscape Infrastructure by enhancing biodiversity, improving stormwater management, and reducing the urban heat island effect. Aligned with Law 10/2013, Pocket Parks focus on transforming paved and neglected areas into vibrant public spaces that encourage social interaction and community well-being.

The project's guidelines, articulated into ten key points, adhere to Directive 42/2001/EC and address themes such as sustainability, biodiversity, and urban livability. This paper highlights how the theoretical framework underpinning these guidelines can be strengthened through the practical implementation of the nine examples described.

Within this framework, dynamic simulations conducted in collaboration with the Department of Architecture at the University of Florence demonstrate the effectiveness of these interventions in enhancing microclimatic conditions and community comfort.

By connecting theory with practice, this research enhances the understanding of Landscape Infrastructure as a versatile strategy operating across scales and levels to transform urban environments, effectively integrating ecological, social, and cultural dimensions.

Keywords

Pocket Park · Landscape infrastructure · Urban resilience · Climate adaptation · Nature-based solutions

1 Context

1.1 The City of Florence and the Climate-Adapted Measures for Landscape Infrastructure

In numerous scientific publications (European Environment Agency 2020; Heat—Topos 127 2024; Macintyre et al. 2018), the lack of green spaces, permeable surfaces, and natural watercourses, combined with heat-retaining materials, amplifies climate change effects and negatively impacts human health. Local governments must adopt resilient models for regenerating public spaces that address climate adaptation, urban quality, sustainable mobility, public health, and social inclusion (United Nations 2015). Cities like Florence, facing extreme weather events, demonstrate how historic

C. Mezzapesa (✉)
Department of Architecture, University of Florence, Florence, Italy

Landscape Architect, Memoscape, Florence, Italy
e-mail: claudia.mezzapesa@unifi.it

E. Moretti
Landscape Architect, Memoscape, Florence, Italy

F. Alberti et al. (eds.), *Towards Sustainable Urban Regeneration: Utilizing Environmental, Cultural, and Social Assets*, Advances in Science, Technology & Innovation, https://doi.org/10.1007/978-3-032-02091-8_3

centers can integrate green infrastructure and sustainable mobility to enhance livability and fight climate change.

This paper highlights the role of landscape design in urban regeneration, focusing on biodiversity, desealing, and new green spaces (MUR 2023). The term "Landscape Infrastructure" is introduced to emphasize the holistic integration of landscape and infrastructure, avoiding the reductionist view of landscape as merely vegetative. Unlike the ambiguous concepts of "Green Infrastructure" or "Green & Blue Grids", Landscape Infrastructure encompasses material and symbolic dimensions, fostering systemic approaches to adaptability and resilience (Buoro 2023; Cortesi 2017; Weilacher 2016, 2017).

Structural networks, such as water, traffic, green spaces, and ecological systems, are essential for livable landscapes. These networks are interdependent and mutually reinforcing (Weilacher 2016), positioning infrastructure as an intrinsic element of the landscape. Landscape Infrastructure design interweaves processes, flows, and spaces, creating functional landscapes that restore habitats, enhance community interaction, and build resilience (Bélanger 2013; Hung 2013).

Pocket Parks represent a practical application of Landscape Infrastructure principles. By transforming small urban spaces into multifunctional green areas, they address biodiversity loss, provide permeable surfaces, and foster social interaction. These micro-scale interventions contribute to urban resilience by reconstituting habitats, mitigating heat islands, and offering accessible public spaces, ultimately exemplifying how Landscape Infrastructure can improve ecological and social systems in densely built environments.

1.2 Pocket Parks

Pocket Parks are small public green spaces typically created by regenerating residual or marginal areas within or near the most densely populated parts of the city. The importance of these spaces is the redevelopment of the smaller parks, reserves, and street closures that makes a difference to the local community. Thus, Pocket Parks are small-scale public spaces situated within large urban environments. These spaces are created with limited budgets and often include elements such as vegetation, children's play areas, relaxation zones, and more (Dong et al. 2023). This spatial concept became common in Western countries towards the end of the twentieth century as a response to the need for ecological and cultural oases in metropolitan areas.

The concept of the "vest-Pocket Park" emerged in Europe during the post-war era as a way to reclaim and revitalize urban spaces that had been devastated by bombings. These spaces, which were not particularly attractive for residential or commercial uses due to their location, size, or shape, were transformed into green areas. In the 1950s, the idea spread to the United States, where Robert Zion played a key role in promoting this innovative vision: creating small urban spaces among skyscrapers that could serve as temporary refuges for city dwellers. This vision materialized in 1967 with the creation of the first Pocket Park, Paley Park, in the heart of Midtown Manhattan. Featuring a water wall, vine-covered walls, and trees, the park provides visitors with a moment of respite from the hustle and bustle of city life. Tables and chairs allowed people to interact and socialize in a relaxing environment. Since then, several Pocket Parks have been created in various American and European cities, following the example set by Paley Park.

Philadelphia was one of the first cities to begin developing Pocket Parks within its neighborhoods. These parks were built on vacant or abandoned lots, with the community actively involved in both the design and construction processes. The primary uses of these spaces include play areas and resting spots, particularly for children and the elderly. Pocket Parks in Philadelphia are characterized by features such as climbing structures, exploration areas, vibrant colors, community involvement, basketball courts, flower gardens, or vegetable plots.

The "London's Great Outdoors—Manifesto for Public Space," established in 2009 by Mayor Boris Johnson, promotes the creation of 100 Pocket Parks throughout London's urban fabric, recognizing them as essential spaces for respite and relief from the fast-paced city environment. This initiative is part of a broader program aimed at enhancing London's public spaces as a strategy to improve the quality of life and the services offered by the city.

The Eco-Metropole project for Copenhagen was conceived with the goal of ensuring that every citizen could reach a park, square, pier, or dock within a 15-min walk from any point in the city. Pocket Parks play a crucial role in this vision by operating on a small scale to meet the everyday needs of residents, providing accessible and effective green spaces, and bringing positive significance to otherwise anonymous locations. In Copenhagen, five key elements are considered essential in the creation of a Pocket Park: size (less than 5000 m^2), a prominent green feature, a welcoming and positive atmosphere, a sense of security through clear boundary lines, and a reflection of the local community's identity.

The Pocket Parks project in Athens was launched in 2020 by the municipal administration in collaboration with local and international organizations. This innovative initiative aims to create urban green spaces in a densely populated city with limited green areas. The need for open spaces has been heightened by the economic crisis and the COVID-19 pandemic. The project has been highly successful, receiving acknowledgement for its positive impact on residents and its contribution to the city's climate resilience.

In recent years, Pocket Parks have frequently been the focus of studies in landscape architecture, urban planning and environmental design, examining various aspects of these urban green spaces, which are recognized as effective tools for improving city life quality:

- **Environmental impact and sustainability**: several studies highlight how Pocket Parks contribute to reducing air pollution and the urban heat island effect. Research indicates that even small green spaces can have a significant impact on the urban microclimate.
- **Health and well-being**: research underlines the benefits of Pocket Parks for both the mental and physical health of urban residents. Accessible green spaces are associated with lower levels of stress and anxiety and increased physical activity among inhabitants (Zhang and Gobster 2018).
- **Social cohesion**: studies demonstrate that Pocket Parks promote social interaction and a sense of community among residents. These spaces provide meeting places that foster inclusion and dialogue (Francis et al. 1984; Littke 2016).
- **Urban regeneration**: the transformation of underutilized or degraded urban areas into Pocket Parks is a key theme, with a particular focus on participatory processes and collaborative design with local communities (Carmona and Magalhães 2020).

Today, one of the principal and perhaps most urgent objectives of these projects is to mitigate urban heat islands, thereby addressing climate emergencies and improving health and well-being in urban areas. This goal is achieved through the incorporation of various elements such as water, vegetation, permeability, and materiality. The addition of trees helps regulate environmental hygrometric conditions, relative humidity and the sensation of air freshness. These small open spaces, as critical structural elements of the city, have the potential to become bioclimatic areas capable of regulating temperature and humidity. Connecting these spaces into networks and Landscape Infrastructure is a key aspect that can significantly contribute to reducing temperatures and improving urban liveability.

Moreover, optimizing the capture and retention of rainwater in green areas is crucial. Hard, paved surfaces should be minimized, and synergies with adjacent natural spaces should be created through green corridors to promote urban biodiversity. Maintaining natural soil with native species that prevent erosion and create habitats for biodiversity, combined with the use of permeable materials, helps absorb heat and reduce albedo.

Functions accommodated in these spaces may include small event areas, children's play zones, relaxation spaces, and spots for socializing or taking lunch breaks. They can provide a refuge from the hustle and bustle of the surrounding urban environment, offering opportunities for rest and relaxation. However, due to limited space and the varying needs of users throughout the day, conflicts often arise among different user groups. The design must implement effective processes to ensure harmonious coexistence among the various uses and functions.

In conclusion, Pocket Parks represent a model of sustainable urban regeneration and urban acupuncture, capable of enhancing citizens' well-being and the city's appearance while simultaneously encouraging community engagement in the management of public spaces.

1.3 Pocket Parks in Florence

Among European cities investing in Pocket Parks as a model for urban regeneration, Florence distinguishes itself with an ambitious initiative funded through the REACT-EU PON METRO program under the "Public Gardens" category. This project aligns with Florence's broader strategy to achieve decarbonization by 2030 and reflects its commitment to innovation as a member of the EERA JP Smart Cities program. Leveraging the city's rich cultural heritage and modern urban challenges, the project establishes Florence as a permanent laboratory for experimentation, in line with the "Smart City Plan" and the structural plan "Firenze Prossima," launched in 2023.

What sets Florence apart from other urban contexts is its integration of advanced nature-based solutions (NbS) into the creation of nine pilot Pocket Parks (Spazi Verdi Urbani Diffusi). These interventions are conceived as critical nodes within the city's *Landscape Infrastructure*, a systemic network that interweaves ecological, social, and spatial dimensions to enhance resilience and adaptability. Funded through European structural and investment resources managed by the Agency for Territorial Cohesion, the initiative emphasizes the restoration, redevelopment and reuse of underutilized or degraded urban spaces. Guided by Law 10/2013, the project promotes the enhancement of urban green areas through diverse ecological components such as lawns, trees, and shrubs, contributing to the creation of interconnected and functional landscapes.

By embedding *Landscape Infrastructure* principles, Florence's Pocket Parks operate as multifunctional spaces that connect ecological systems, foster social cohesion, and improve urban liveability. These spaces serve as bioclimatic areas, mitigating urban heat islands and enhancing the city's capacity for rainwater management through permeable surfaces and natural soil preservation. The deliberate design of these green nodes aligns with broader efforts to integrate urban biodiversity and reinforce ecological corridors, creating synergies with adjacent natural spaces.

Drawing inspiration from successful projects in cities like Copenhagen and Athens, Florence incorporates key elements such as community-driven design, permeability, and urban biodiversity. However, it also introduces unique approaches, such as integrating NbS with the city's historical and architectural context. The Preliminary Design Document highlights the strategic use of small urban gardens not only to address climate adaptation but also to strengthen the systemic coherence of the city's *Landscape Infrastructure*.

In comparison to similar initiatives, Florence's Pocket Parks not only serve as green refuges but also function as hubs for experimental urban planning, connecting ecological systems with cultural narratives. This holistic vision exemplifies how small-scale interventions can generate far-reaching impacts on urban sustainability and resilience, while simultaneously reinforcing the city's identity through the principles of *Landscape Infrastructure*.

1.4 The Article Structure

This article explores the strategies implemented by the city of Florence through the analysis of nine Pocket Parks designed to reduce the city's vulnerability and improve the quality of life. To this end, Sect. 2 presents the methodology used for selecting the intervention areas, while Sect. 3 summarizes the main design outcomes. Finally, Sect. 4 discusses the lessons learned, with the aim of guiding the initiative into the future (Sect. 5).

2 The Methodology

The selection of areas for the experimental implementation of the Pocket Parks began in July 2022, when the landscape architecture firm *Memoscape*[1] was commissioned by the City of Florence for the project. At that stage, 25 areas suggested by the municipality were identified, evenly distributed across the city's five districts (Figs. 1 and 2).

Among these 25 areas, many consisted of unused and paved urban spaces, while others were small squares with no identity within the dense historic center. A third category included traffic islands, as well as parking areas for cars, motorcycles and bicycles.

Following the analysis of current urban planning strategies and the study of "Pocket Park networks" in similar urban contexts, it became clear that car parking functions are not included in the design of these green oases. On the contrary, the current trend is to repurpose several parking spaces into green areas for public use (Parklets). Additionally, some areas presented challenges due to regulatory constraints in the historic center and due to the difficulty of completing the design stages within the specified timelines. These excluded areas were further supplemented by additional sites likely to undergo transformations due to imminent urban development projects, particularly related to the construction of new tram lines (Lauria et al. 2020).

As a result of this selection process, nine areas were chosen sharing similar characteristics for the redevelopment and creation of Pocket Parks (*SVUD*), distributed across various districts of the city of Florence (Fig. 3).

Most of these areas were selected for their proximity to schools (PP1 via Monteverdi and PP11-PP12 via Isonzo) and cafes, bookstores, or shops, which can provide daily oversight and potential maintenance and care in the future. These areas are predominantly impermeable, with asphalt surfaces, located in high-traffic zones near major roadways and lacking trees or green spaces. The only exception is the two areas PP11–PP12 in Sorgane, where the interventions are within spaces already designated as public green areas, involving portions of these spaces. Consequently, they have a different, more temporary character compared to the other Pocket Parks.

Based on the observation of physical, spatial, and perceptual characteristics, the nine areas were classified into three main categories (Fig. 4):

- **Paved pocket parks:** areas where installation of furniture, artistic installations, water features, and planters was feasible, and where small trees, shrubs, and perennials could be planted.
- **Mixed pocket parks:** partially permeable areas where drainage could be improved through depaving and planting of trees, shrubs, and perennials. These spaces could be transformed into social areas with the addition of furniture, fountains, and artistic installations.
- **Permeable Pocket Parks:** asphalted areas where removing impervious surfaces would enable the integration of bioretention systems.

3 Research Outcomes

3.1 Nine Pocket Parks for the City of Florence

The project was developed according to the principles of:

- **Design for connections:** the design approach focuses on connecting small urban gardens, or Pocket Parks, with the larger scale of Landscape Infrastructure to create a new ecosystem for living. The Pocket Parks can be organically distributed throughout the city, forming networks of connected spaces that offer residents opportunities to spend time outdoors, supporting the city's ecosystem services.

[1] *Memoscape*, born in 2016, is a project that landscape architects Claudia Mezzapesa and Elena Moretti carry out in parallel to their own professional practice, teaching, and research. It is a project that combines landscape design with research and learning activities.

Fig. 1 The 25 areas suggested by the municipality of Florence (in Mezzapesa and Moretti 2022)

Fig. 2 Aerial views of the 25 areas suggested by the municipality of Florence (in Mezzapesa and Moretti 2022)

Fig. 3 The nine selected areas (Mezzapesa and Moretti 2022)

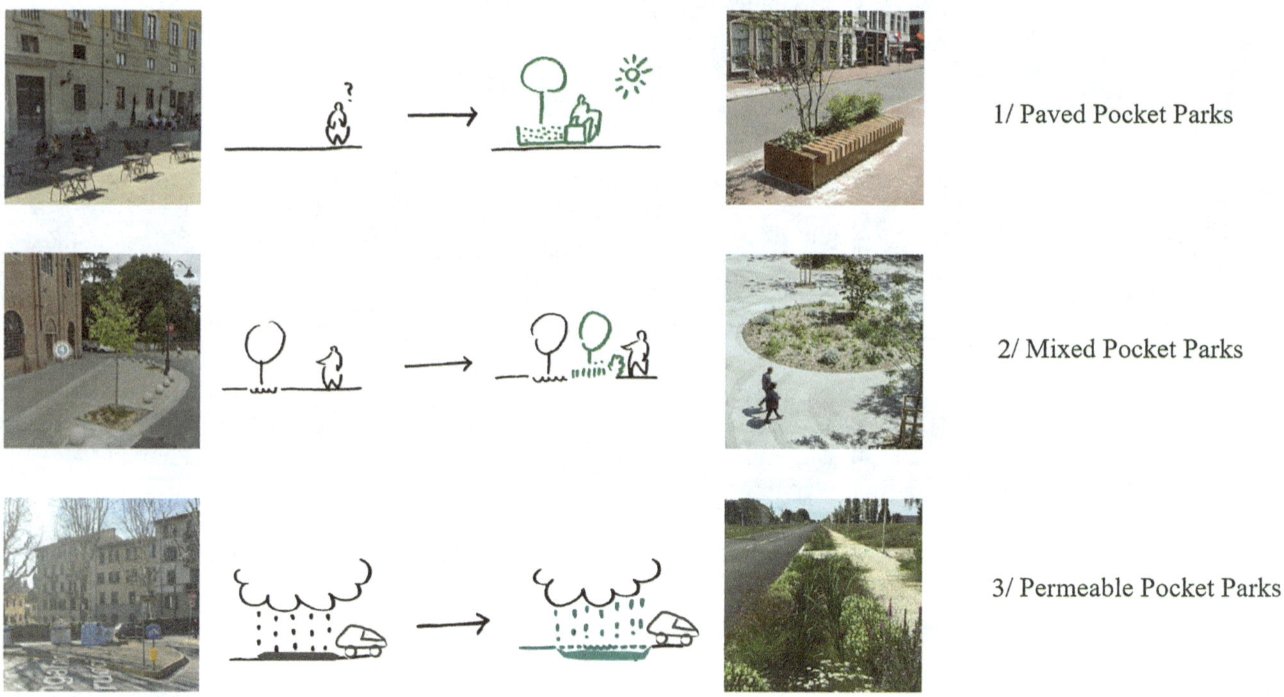

Fig. 4 Pocket Parks for Florence and the three main categories (Mezzapesa and Moretti 2022)

– **Design for wellness:** the principles of biophilia guide the design in selecting shapes, textures, and vegetation to maximize the connection with nature, promote the community's well-being, and ensure conditions of thermal, visual, and acoustic comfort. The goal is to foster social interaction and create a more welcoming urban landscape.

- **Design for sustainability:** circular supplies, renewable energy, and products made entirely of recyclable materials are emphasized. The selection of plants is focused on those tolerant to climatic stress and resistant to drought, with a focus on maintenance to minimize resource use and ensure long-term eco-efficiency. Greater emphasis is placed on the efficient use and management of water resources, as well as on the preservation and protection of soil quality.

Moreover, since this project aims to enhance urban biodiversity, particular attention has been given to the selection of vegetation, utilizing:

- For large trees: deciduous species that display a variety of autumn colors and small multi-stem trees with spring blossoms. These species were chosen for their medium to high tolerance to abiotic stress, good resistance to major pathogens, diseases, and pests, and their moderate tendency to produce litter.
- For borders of shrubs, herbaceous plants, and grasses: species with high hardiness were selected, offering medium to high tolerance to both drought and excessive water conditions. These plants require low maintenance and have textures and blooms that are effective in attracting pollinators.

To coordinate all interventions, a design guideline was developed to support the project's main objectives, addressing sustainability, increased biodiversity, and stormwater bioretention, as well as improving livability, social interaction, and reducing the urban heat island effect. This guideline directs the design process without aiming to impose stylistic uniformity; rather, its primary goal was to enhance the unique characteristics and peculiarities of the local context (Fig. 5).

The design covered the period from July 2022 to February 2023, site works began in June 2023 and were completed in December 2023.

The interventions, categorized into three distinct types, can be distinguished by their technological solutions and by the way they shape shared spaces. However, each area is carefully adapted to its urban context, ensuring that it maximizes benefits in terms of environmental, social, and economic impact.

3.1.1 1/Paved Pocket Parks

The PP5 area on Viale dei Mille (Fig. 6) was originally an open space that had unfortunately become a makeshift parking spaces for bicycles and scooters. The goal of the project was to enhance the greenery and amenities by adding a bench that also serves as a planter, along with a bicycle rack.

Similarly, the PP2 area on Via del Ponte alle Mosse was primarily used for transit and was completely covered in asphalt, making it fully exposed to the sun.

To improve these areas, the project introduced several key changes:

- Replacing the impermeable asphalt with thinner, permeable, and environmentally-friendly paving.
- Removing sections of pavement to create new green spaces with trees, shrubs, and herbaceous plants that are both climate-resistant and drought-tolerant. These spaces require minimal upkeep, thanks to a modular system that incorporates Corten steel edging and benches made from durable hardwood slats.

Fig. 5 Pocket Parks for Florence design guidelines (Mezzapesa and Moretti 2022)

Fig. 6 PP5/Pocket Park in Viale dei Mille before and after (Mezzapesa and Moretti 2022)

3.1.2 2/Mixed Pocket Parks

The PP8 area on Viale Redi was almost entirely paved with asphalt (Fig. 7a, b), while the PP7 area on Via Benedetto Marcello had become an informal resting spot for bar-goers who used an old bench (Fig. 8).

The project aimed to improve these spaces by enhancing soil permeability, removing asphalt where possible, and creating areas that allow rainwater to filter through, making room for new plants and trees.

In both PP7 and PP8, several improvements were made:

- Bioretention systems were introduced to help reduce the amount of stormwater flowing into the sewer system and to boost the area's urban ecosystem services.
- Some traffic islands were depaved to create new planting beds filled with trees, shrubs and herbaceous plants that can withstand climate stress and drought, requiring only minimal maintenance.
- New seating was added, crafted from sustainable, recycled and recyclable materials.

In the first area, between Viale Redi and Via Benedetto Marcello, a new resting spot was created, featuring a flower bed with deciduous trees. The adjacent traffic island was transformed into a bioretention zone. In the second area, between Via Cassia and Viale Redi, two new planting beds were added, with hedges and large deciduous trees providing shade for the new seating.

The placement of the planting beds was determined by the presence of numerous underground utilities near the buildings, as well as the need to maintain open space for circulation, given that these are heavily trafficked areas. Additionally, the choice of vegetation was made with consideration of the species already present in the green area surrounding the Belfiore tram stop.

Fig. 7 (**a**, **b**) PP8/Pocket Park in Viale Redi, the project (Mezzapesa and Moretti 2022)

Fig. 8 PP7/Pocket Park in Via Benedetto Marcello, before and after (Mezzapesa and Moretti 2022)

The PP1 area on Via Monteverdi (Fig. 9) was previously an unremarkable entrance to a school building, often used for parking. The project's objective was to expand the permeable surface and plant additional trees. This intervention serves as an example of the application of tactical urbanism principles. The asphalted area was painted to create a playful transition space between the school and the city.

3.1.3 3/Permeable Pocket Parks

In the PP3 area on Via Marconi (Fig. 10), it was possible to depave the traffic island, thereby reactivating the soil and enhancing the space by strengthening the rainwater filtration and bioretention systems (SuDS) to create an integrated ecological system. This allowed for the planting of new trees, shrubs, and herbaceous plants to increase biodiversity.

The following interventions were implemented in this area:

- Creation of bioretention systems to reduce the burden of stormwater on the sewer system and enhance urban ecosystem services.
- Depaving of traffic islands to establish planting beds with trees, as well as shrubs and herbaceous plants that are climate-resilient, drought-tolerant, and require minimal maintenance.

The PP11 and PP12 areas are located in the southern district of Sorgane and were originally asphalted spaces featuring raised flower beds with concrete walls and porphyry coping, creating seating areas (Fig. 11a, b). These areas were dominated by evergreen trees such as pines, cedars and magnolias.

Due to the large size of PP11 and PP12 and the need for extensive restoration, the decision was made to focus on specific areas with semi-permanent solutions. These interventions are designed to be easily removed or adjusted when the

Fig. 9 PP1/Pocket Park in Via Monteverdi, before and after (Mezzapesa and Moretti 2022)

Fig. 10 PP3/Pocket Park in Via Marconi, before and after (Mezzapesa and Moretti 2022)

Fig. 11 (**a**, **b**) PP12/Pocket Park in Sorgane, before and after (Mezzapesa and Moretti 2022)

entire area is renovated, while still providing the space with a unique character and opportunities for play and exploration.

The following actions were implemented in the PP11 and PP12 areas:

- Depaving and the creation of a rustic lawn area (PP11).
- Depaving to establish planting beds with trees, as well as shrubs and herbaceous plants that are resilient to climate stress and drought-resistant, requiring minimal maintenance. The cuts in the asphalt were made by following and utilizing the existing cracks where spontaneous vegetation had already taken root (PP12).
- Planting of new trees and the enhancement of the long rectangular flower bed with ground-cover species that are climate-resilient and drought-tolerant, requiring minimal maintenance.
- Restoration of the lower sections of the walls through new plastering.

– Installation of new semi-spherical concrete furniture-sculptures, designed to create a continuity of language with the architectural context.

4 Lessons Learned

The experience of designing nine Pocket Parks in Florence offers several valuable lessons, and these projects hold significant potential for future evolution, especially in light of new urban planning tools and national and European legislation.

4.1 Context-Sensitive Design

One of the key takeaways is the importance of adapting interventions to the specific urban context. Each Pocket Park was designed with careful consideration of its surroundings, which maximized the environmental, social, and economic benefits. This approach underscores the need for context-sensitive design in urban projects, ensuring that solutions are not only technically sound but also socially and ecologically appropriate.

4.2 Flexibility

The application of semi-permanent solutions and tactical urbanism principles, particularly in the PP1, PP11, and PP12 areas, highlights the value of temporary, flexible interventions. These can serve as experimental solutions that can be easily modified or expanded based on the evolving needs of the community and the urban environment. This flexibility is crucial for accommodating future changes and ensuring long-term sustainability.

4.3 Integration of Landscape Infrastructure

The projects emphasized the integration of Landscape Infrastructure, such as bioretention systems and permeable surfaces, which are vital for managing stormwater and enhancing urban ecosystems. These interventions not only address environmental concerns but also contribute to the overall resilience of urban areas, making them better equipped to handle climate change impacts.

4.4 Urban Climate Resilience

A study conducted by the University of Florence, as part of the National Biodiversity Future Center's Spoke Five initiative, examined the effectiveness of Pocket Parks in climate change adaptation. The research validated the designs of 4 out of the 9 Pocket Parks implemented by the Municipality of Florence using ENVI-met software (Romano et al. 2024).

The findings demonstrate that small public spaces incorporating well-designed features, such as green areas, trees, shrubs, and materials optimized for thermal comfort and permeability, can effectively mitigate the impacts of climate change, particularly the urban heat island effect. Furthermore, the study underscores the critical role of developing regulatory frameworks to support and standardize these innovative design strategies, ensuring measurable benefits in urban environments.

The monitoring and evaluation of comfort-related outcomes were explored through microclimatic simulations conducted with ENVI-met software (Fig. 12). These simulations assessed parameters such as air temperature, relative humidity, and perceived thermal comfort. The analyses revealed significant improvements in the microclimatic conditions of the targeted areas, confirming the effectiveness of Pocket Parks as Landscape Infrastructure interventions to enhance urban resilience to climate change.

4.5 Community Engagement and Multifunctionality

The creation of multifunctional spaces, as seen in several of the Pocket Parks, demonstrates the importance of designing areas that serve multiple purposes, providing green spaces, recreational opportunities, and enhancing urban aesthetics. Engaging the community in the design and use of these spaces is essential for their success, fostering a sense of ownership and ensuring that the spaces meet local needs.

5 Prospective Developments and Strategic Integration for Landscape Infrastructure

Looking ahead, several pathways exist for expanding and refining the lessons learned from the Pocket Park projects in Florence.

Firstly, while the initial successes suggest the potential for scaling up and replicating this model in other neighborhoods or cities, challenges related to implementation in historically significant areas must be addressed. These areas, characterized by exceptional architectural and artistic value, often impose constraints on design interventions due to strict preservation requirements. Balancing the integration of green spaces with the principles of Landscape Infrastructure, while protecting Florence's cultural heritage, will require

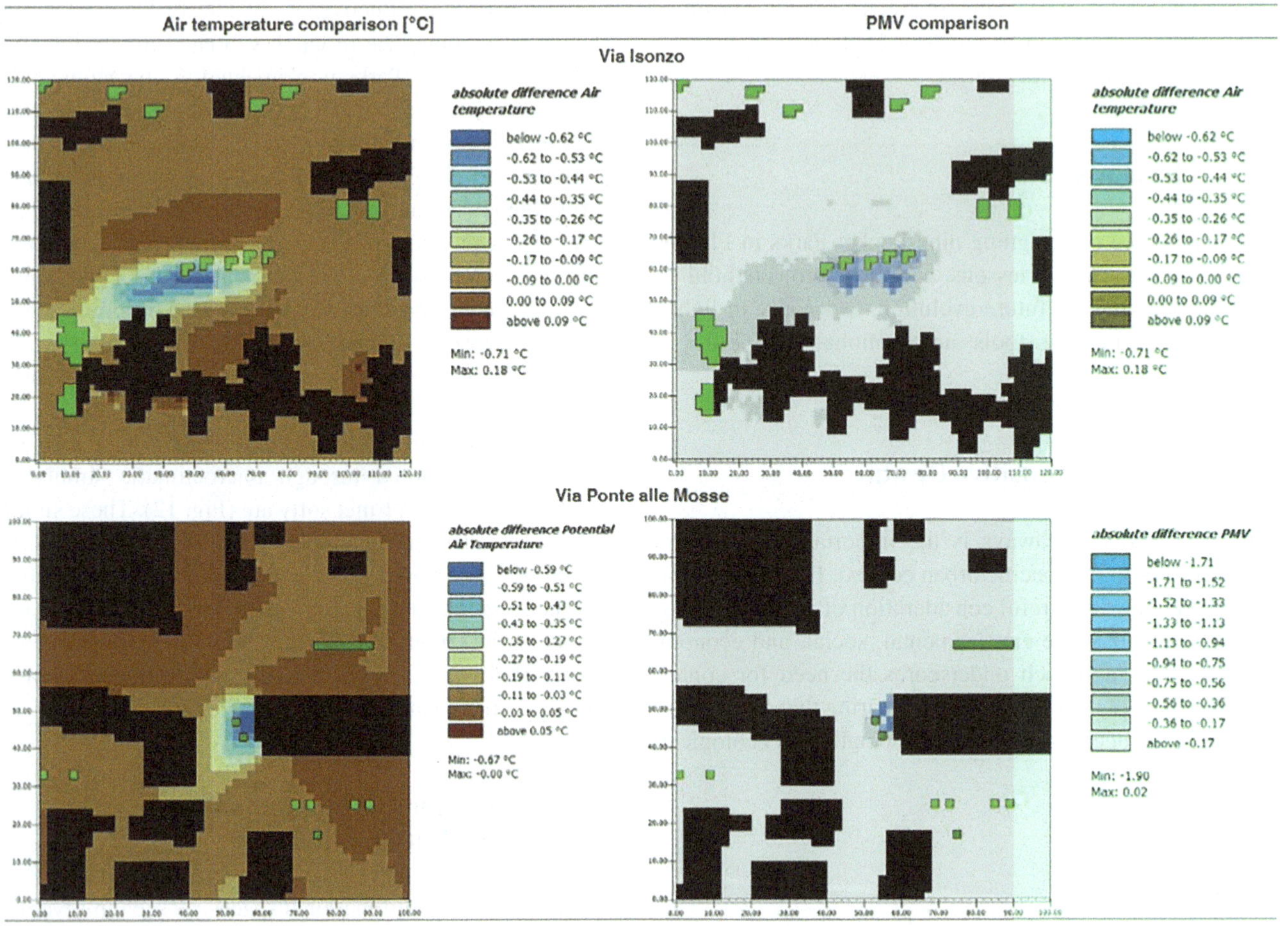

Fig. 12 Pocket Parks in Sorgane and Ponte alle Mosse, microclimatic simulations with ENVI-met software (in Romano et al. 2024)

innovative and context-sensitive approaches, as briefly highlighted in the methodology section.

Additionally, Pocket Parks offer an opportunity to align with broader urban sustainability initiatives through the concept of Landscape Infrastructure. By incorporating these green interventions into a comprehensive network of interconnected urban elements, addressing ecological, social, and functional needs, they can serve as vital components of a resilient urban system. Specifically, aligning these spaces with goals such as reducing urban heat islands, enhancing local biodiversity and promoting active transportation ensures their multifunctionality, and strategic integration into the urban fabric.

National and European legislations[2] related to climate action and urban sustainability presents another key opportunity. These frameworks often provide funding, policy support, and guidance that can strengthen the expansion of Landscape Infrastructure projects. By aligning future initiatives with legislative priorities, such as those outlined in the European Green Deal or national climate adaptation plans, Florence can secure additional resources and ensure these projects contribute to wider environmental and societal goals.

Ongoing monitoring and adaptability will be critical to the long-term success of the Pocket Parks within the context of Landscape Infrastructure. Regular assessments of their environmental performance (mitigating urban heat islands, improving air quality), social usage (fostering community interaction), and maintenance needs will provide valuable data. This information can guide adjustments to optimize their functionality and ensure they remain responsive to emerging urban challenges.

In summary, the Pocket Park initiative has laid a solid foundation for future urban regeneration efforts through the application of Landscape Infrastructure principles. By

[2]European Legislation: European Green Deal (2019), EU Biodiversity Strategy for 2030, EU Climate Law (2021), EU Urban Agenda (2016), Nature-based Solutions (NBS) for Climate Adaptation (Horizon 2020 and Horizon Europe).

[Italian] National Legislation: Piano Nazionale di Adattamento ai Cambiamenti Climatici (PNACC), Legge 14/2022 (Legge Clima), Decreto Legislativo 152/2006 (Codice dell'Ambiente), Legge 10/2013: Also known as the "Law on Green Spaces.

addressing implementation challenges in sensitive areas, integrating these spaces into a broader, multifunctional urban system, leveraging legislative support, and adopting a dynamic and adaptive approach, Florence can further enhance its resilience. Such efforts will contribute significantly to improving the city's environmental quality, urban connectivity, and the well-being of its citizens.

References

Bélanger P (2013) Landscape infrastructure: urbanism beyond engineering. Wageningen University

Buoro M (2023) Landscape infrastructure. Tautologia per le città del XXI secolo. In: Cianci MG, Rabazo Martin M (eds) Tracce segni e rappresentazioni del paesaggio che cambia. Il Poligrafo Editore

Carmona M, Magalhães C (2020) Designing urban pocket parks: the case of Lisbon. J Urban Des 25(2):146–164

Cortesi I (2017) Il paesaggio al centro. In: Cortesi I, Cappiello V (eds) Il paesaggio al centro. Lettera Ventidue

Dong J et al (2023) Pocket Park—a systematic literature review. Environ Res Lett 18(8)

European Environment Agency (2020) Urban adaptation in Europe: how cities and towns respond to climate change. Publications Office of the European Union

European Environment Agency (EEA) (2020) Urban adaptation to climate change in Europe 2020: transforming cities in a changing climate. Publications Office of the European Union

Francis M, Cashdan L, Paxson L (1984) Community open spaces: greening neighborhoods through community action and land conservation. Island Press

Heat—Topos 127 (2024). https://toposmagazine.com/heat-127/

Hung Y (2013) Landscape infrastructure: systems of contingency, flexibility, and adaptation. In: Landscape infrastructure: Case studies by SWA. The Infrastructure Research Initiative at SWA, Birkhäuser

Lauria A, Romagnoli M, Vessella L, Bravi E (2020) The Pocket Park system as a regeneration strategy for the historic city. Green Lines Institute for sustainable development—PRT—Barcelos, Portugal

Littke H (2016) Revisiting the San Francisco Parklets program: interpreting citizens' motivation and satisfaction. Urban For Urban Green 15:165–174

Macintyre H, Heaviside C, Taylor J (2018) Potential benefits of cool roofs in reducing heat-related mortality during heatwaves in a European city. Environ Int 121:121–130

Mezzapesa C, Moretti E (2022) Relazione tecnica: Realizzazione e riqualificazione Spazi Verdi Urbani Diffusi

MUR (2023) National Biodiversity Future Center—NBFC

Romano R, Di Monte E, Sore A (2024) New perspectives and moving targets to improve biodiversity and urban resilience. TECHNE J Technol Architect Environ 27:151–166

United Nations (UN) (2015) Transforming our world: the 2030 agenda for sustainable development. United Nations General Assembly

Weilacher U (2016) Revisiting structuralism—revitalizing living networks. In: Tasting the landscape: 53rd IFLA world congress, April 20th–22nd, 2016. Edifir

Weilacher U (2017) Green infrastructure and landscape architecture. www.landezine.com

Zhang Y, Gobster PH (2018) Pocket parks as urban green spaces for health and well-being in densely populated cities: lessons from Beijing and Shanghai. Urban For Urban Green 29:122–133

Ruralizing the Urban Envelope: Experimental Tools to Cultivate the City

Matteo Clementi, Marco Migliore, Silvia Falasco, and Paola Caputo

Abstract

Promoting the integration of agriculture into urban areas could represent an ethical and social achievement in the short term; in this way, consistently with the provisions of the UN SDG (Sustainable Development Goals), the new generations will not be disconnected from rural reality and the resulting micro-productions could represent a contribution to overcoming the population social gap (UN SDG 2 "Zero hunger," 11 "Sustainable cities and communities" and 12 "Responsible consumption and production"). Starting from this assumption, the paper presents some results of an ongoing research project, called BioLoop, aimed at laying the foundations for the activation of local micro-economies in the southern periphery of Milan. The paper presents a method and tools to support hyperlocal design, i.e., based on the energy and material resources available within a radius of 1 km. The results presented show the application of such tools to an urban portion of the southern outskirts of the city, taking as a case study a square that hosts student housing and other public buildings. In particular, the paper focuses on the first steps of the application and implementation of experimental tools: the local Atlas of Energy and Materials (AEM) and some of the proposals inside the Open-Source Projects Hub (OSPH). AEM and OSPH are products of the Bioloop project, and the applied design solutions of the OSPH are based on the use of locally available resources, both in terms of secondary material and renewable energies. These solutions refer, in particular, to devices for cultivation on both vertical and horizontal impervious surfaces and to a mobile collective kitchen for the transformation of food potentially produced by the same devices. In addition to the representation of the design solutions, some quantitative results on the impacts of these devices on the urban metabolism of the area are presented, both in terms of material cycles (water and nutrients) and renewable energies.

Keywords

Circular economy · Peri-urban agriculture · Green walls · Building integrated agriculture · Recycle · Local renewable energies · Waste reduction

1 Introduction

The growth of the world population, the tendency towards human settlement concentration in large cities (UN 2018), and the need to have greater resources available for human sustenance are leading states to update sustainable development policies. It is believed that promoting alternative forms of cultivation, especially in urban areas, will help achieve the sustainability goals (Pueyo-Ros et al. 2024) of Agenda 2030 (UN 2015). For example, urban agriculture is a form of cultivation that does not imply furthermore land use, rather it favors the reconversion of unused surfaces and in many cases, it can also take place on impervious surfaces. Furthermore, urban agriculture offers a wide range of ecological (e.g., increase of biodiversity, micro-climate control, etc.) and social (e.g., recreational, self-sustaining, and micro-entrepreneurial activities, etc.) functions, which benefit the proximal community and the entire society (De Oliveira Alves et al. 2024; Pradhan et al. 2023). An important aspect of urban agriculture is that by producing locally, it is possible

M. Clementi (✉)
Department of Architecture and Urban Studies, Politecnico di Milano, Milan, Italy
e-mail: matteo.clementi@polimi.it

M. Migliore
DASTU—Department of Architecture and Urban Studies, Politecnico di Milano, Milan, Italy
e-mail: marco.migliore@polimi.it

S. Falasco · P. Caputo
DABC—Department of Architecture, Built Environment and Construction Engineering, Politecnico di Milano, Milan, Italy
e-mail: silvia.falasco@polimi.it; paola.caputo@polimi.it

F. Alberti et al. (eds.), *Towards Sustainable Urban Regeneration: Utilizing Environmental, Cultural, and Social Assets*, Advances in Science, Technology & Innovation, https://doi.org/10.1007/978-3-032-02091-8_4

to balance production with consumption; in this way, the energy embodied in the food used to support cities is reduced by a shorter transport distance, lower packaging, and greater efficiency in production (Bohn and Viljoen 2005). The reduced energy requirement contributes to decreasing greenhouse gas emissions (Deelstra and Girardet 2000), and the recovery and local reuse of biodegradable waste to produce compost and rainwater contributes to energy conservation (Goddard 2006; Holmer and Drescher 2005). Furthermore, the reuse and valorization of scraps and waste offers a measurable advantage both in reducing transport (Midmore and Jansen 2003) and in closing the waste cycle.

The term "ruralization" is here intended as "making rural," and the term "rural" refers specifically to the productive activities typical of rural areas. In the Bioloop research, "ruralization" refers to the surfaces of the city and specific reference is made to the urban envelope (the external surfaces of buildings and the impermeable urban open spaces), focusing on the agricultural productive potential of these surfaces. Urban agriculture is, in fact, seen in this study as an important opportunity to activate circular micro-economies to support the fragile categories that inhabit the outskirts of the city. In particular, the term circular refers to the possibilities offered by the local-scale closure of the emission flows of urban activities. When we talk about circular economies in this text, we refer to local circular flows, referring to those studies conducted on the topic that have led to the recognition of the relationship between territory and issues of sustainable development, resilience, and climate change, discussing the concept of circular city and territory (Christensen 2021). This paradigm can be effectively summarized in the concept of "the circular city." formulated by the Ellen McArthur Foundation and underlying interesting urban policy initiatives (ICLEI 2022). In short, it can be said that a circular city assumes the principles of the circular economy as its foundations and establishes an urban system that will be regenerative by definition. Furthermore, as stated in the scientific literature (Lopes De Morais et al. 2021; Sajjad et al. 2024; Souza Piao et al. 2023), in a circular city, a real social engagement is established with its stakeholders (citizens, businesses, administrators, associations, research and training centers, communities, etc.).

The paper presents a case study on the application of experimental tools for the activation of circular economies based on agricultural production on vertical and horizontal public urban surfaces and located in a square, close to a stable covered market and a weekly outdoor market.

These tools have been developed with the aim of building up scenarios oriented toward a local and participatory circular economy to support the fragilities of the territory: the local Atlas of Energy and Materials (AEM) and the Open-Source Project Hub (OSPH). Starting from the zero-cost materials available and renewable energy sources mapped in the atlas (AEM), specific solutions have been developed to support urban agricultural supply chains capable of optimizing the use of these resources in the form of open-source projects archived in the OSPH. Among the information mapped in the atlas, those that most influenced the choice of the site and aroused interest in this space are:

- The possibility of having a large amount of urban surfaces available that are not used to their full potential
- The proximity to regular producers of edible and non-edible scraps/waste
- The proximity to an urban area with a high population density

In this specific case, AEM and OSPH are intended to support the development of scenarios that maximize the production capacity of available surfaces (roofs, facades, impervious surfaces, or contaminated soils). To activate this "ruralization" process, the project includes the installation of a mobile laboratory, demonstrating the possibilities of transforming surpluses from local markets and capable of hosting and supporting the creation of the first cultivation and production devices on an impervious area. It is a reused ship container, which houses a kitchen for transformation activities and, in the external part, open-source solutions to support cultivation. This mobile laboratory is designed as an activator of local production and transformation processes, starting from what has been mapped in the AEM and some of the design solutions proposed in the OSPH. It aims to be an activator of the application of these devices in an urban context, presenting the proposed solutions as prototypes and hosting training workshops on the construction of production devices and on the possibilities of transforming edible surpluses through a small mobile kitchen. The current state of research is related to the design and first prototyping of the mobile laboratory and cultivation devices. The following paragraphs illustrate the methodology adopted and some of the solutions mentioned above.

2 Methodology Notes

The work presented in the paper is a part of a research project called Bioloop. The project was supported by Polisocial Award 2023, the social responsibility program of Politecnico di Milano. The research is structured according to the following preparatory phases:

- Phase 1: Knowledge of the territory and identification of possible contexts of experimentation. The research work is hosted in one of the four "Off-Campus" of the Politecnico di Milano, Cascina Nosedo, located on the border between the southeastern outskirts of the city and the Parco Agricolo Sud (wide rural area in south Milan).

The Off-Campuses are special laboratories that have been created with the aim of establishing a connection between research and the needs of the communities that live in the most disadvantaged areas of the city. In line with this aim, the Bioloop research aims to equip the Cascina Nosedo with open-source tools to support local circular design. Once these tools have been developed, a phase of involvement of the local community is expected through associations already active in the area, through involvement activities, mapping of the area, co-design, and sharing and implementation of the tools. The goal, in the long term, is to involve the most fragile sectors of local communities in local circular microeconomies that use the resources available in the area starting from local waste materials and renewable energy resources. This activity starts from the study of the urban area near the Off-Campus, which coincides with the Corvetto-Chiaravalle neighborhood (Fig. 1). The context assessment is carried out by detecting the possible flows of people (high or low presence of resource flows), therefore, the population distribution is mapped by identifying the most densely populated areas. This is possible using the data on the resident population of ISTAT, the National Institute of Statistics, available as open data, and through their mapping with a GIS on the territory of the city.

- Phase 2: Prototyping of the setup. The prototyping of the mobile laboratory for cultivation and transformation is realized starting from resources recovered in the urban area. For the definition of its set-up, the resources are identified through the AEM. The tracking of resources is carried out for the resident population, the agricultural sector, and the productive sector, starting from ISTAT data. They are codified, quantified, and localized, to facilitate their faster redistribution, in a circular economy logic, and the results become some of the projects that enrich the OSPH.
- Phase 3: Functional implementation of the prototype. In this phase, the possible functional arrangements to support the laboratory activities are evaluated, the experimentation starts from ideas that consider the ease of operation, the integration in the urban context and the utility.

3 Urban Agricultural Structures: Design and Relationships with the Urban Context

The results presented show the application of the experimental tools mentioned above to an urban portion of the southern outskirts of the city, taking as a case study a square that hosts student housing and other public buildings, Ferrara Square (Fig. 1). The choice of the site is motivated by the presence of both horizontal and vertical surfaces of public property with high productive value (Fig. 2) and by the presence of edible and non-edible waste from the local covered market and weekly outdoor local market.

The area identified for the installation of the experimental prototype coincides with the free space in front of the covered market, the surface has an extension of approximately 300 m^2, and currently has no specific use. It is suitable for the purpose because it was characterized by a series of aspects functional to the success of the project: the first one is the presence of the covered market and the weekly open-air market (Fig. 3—map 2); the second one is the high population density that characterizes this neighborhood (Fig. 3—map 3), condition that extends the opportunities for combining food production with cultural functions on urban space (Dubbeling et al. 2009; Johansson et al. 2024; Mattijssen

Fig. 1 On the left, satellite image of the territorial area analyzed in the research work; on the right, a glimpse of Ferrara square. (Source: Google Maps)

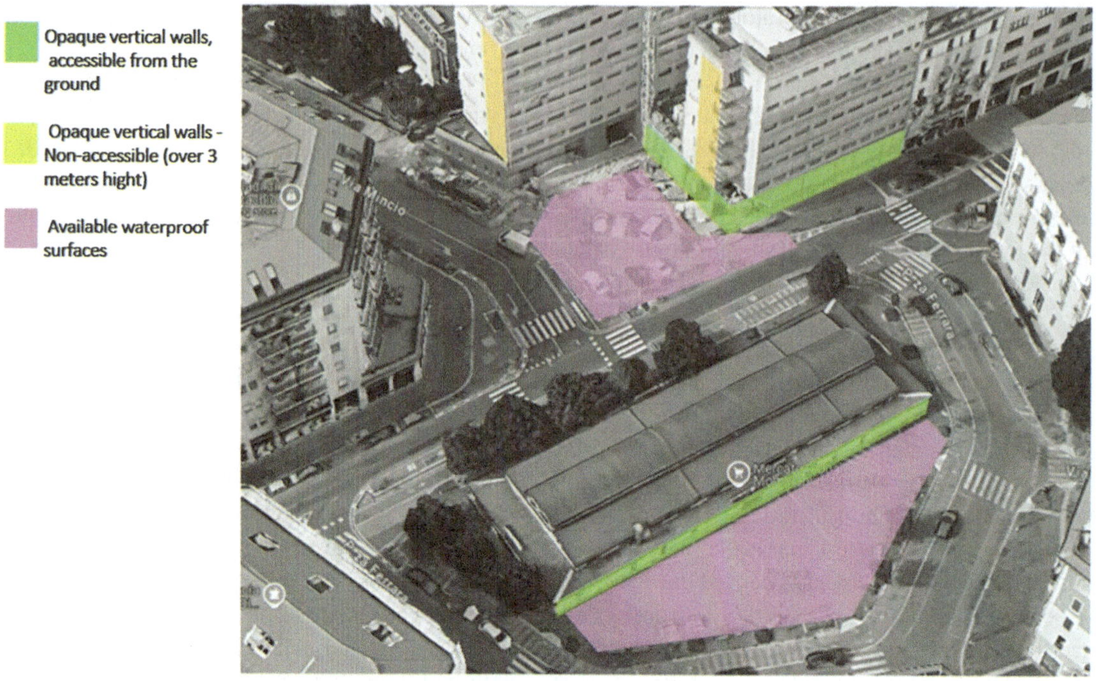

Fig. 2 Glimpse of Ferrara square with the areas potentially usable for plant production highlighted in different colors

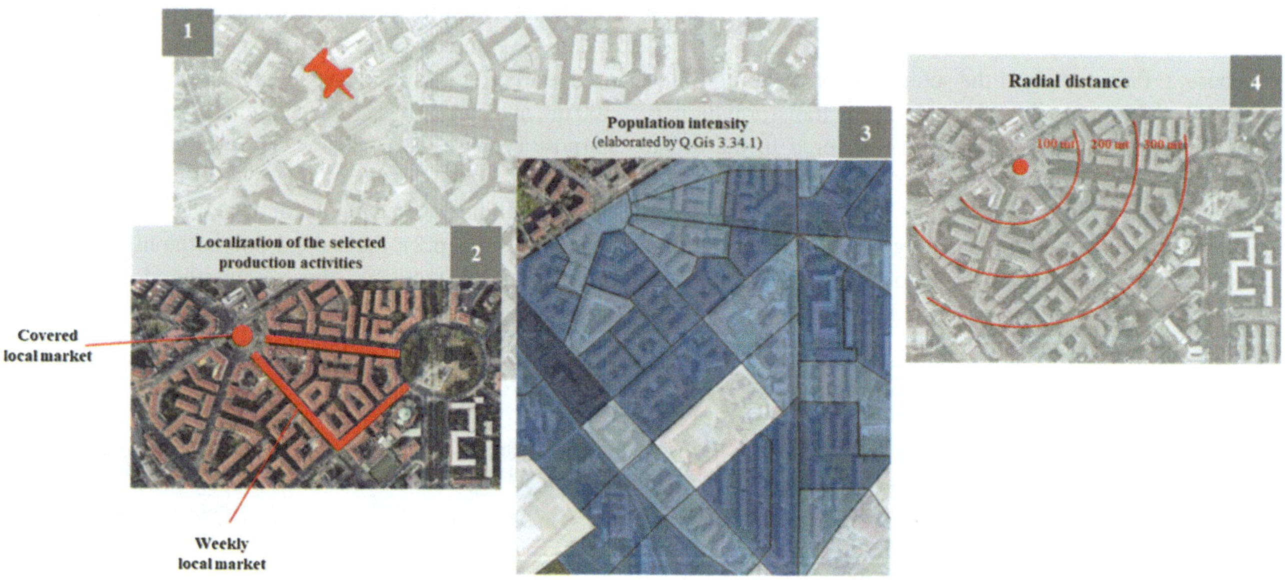

Fig. 3 Map 1 relates to the identification of the area. Map 2 relates to the localization of the productive activities considered in the experimentation (covered market and weekly outdoor market). Map 3 relates to the intensity of the resident population. Map 4 relates to the radial distances from the point of interest

et al. 2024) and create the opportunities for the entire communities to socialize and cooperate with other people; the third is availability of resources from other activities such as bars, restaurants, etc., within a radius of less than 300 m from the operations center (Fig. 3—map 4). The maps created to support the choice of the area can be implemented in order to be used to support urban planning activities by the municipal authorities (Alkhaja et al. 2024; Du and Tanaka 2024; Jansma and Wertheim-Heck 2022). From this point of view, the map is configured as a shared planning tool, since it identifies the areas that can be the object of social initiatives related to the theme of city "ruralization" and urban cultivation. The usefulness of this tool is also demonstrated by the growing interest in the topic of urban agriculture that is spreading in various European municipalities (Kirby et al. 2021), an interest that can also be found in the research projects co-financed by Europe through the Horizon program (INFARM 2016; JUSTGREEN 2022; PROTEAN 2019; SiEUGreen 2017; URBAG 2019; WaysTUP! 2019).

4 Prototype of Agricultural Modules

The prototype currently under development starts from the reuse of a ship container as a social kitchen and combines this activity with self-production of energy, through a system of photovoltaic panels on the roof and on the facade, horticultural cultivation on vertical surfaces or in tanks, and the production of quality compost.

The prototype presented in Fig. 4 shows the design of the exterior of the container to have both agricultural production and self-production of energy (to be used inside the container for the transformation of products). In this setup, the design attention was paid to both the vertical surfaces and the roof.

The vertical surfaces were set up respecting a regular module that allows the assembly and disassembly of the elements used. The module is the result of the type of wooden pallet made available by neighborhood activities, anchored and reinforced by a structure made of the pruning logs available from the urban green maintenance (Fig. 5), photovoltaic panels have been positioned next to the pallets (used for vertical cultivation), and both elements are fixed to a wooden frame created close to the container.

Inside the pallets, a system of jute containers has been created, useful for inserting soil and therefore being able to cultivate. On the rear side, to cover the opening area of the container access doors, a trellis (in metal or wood) was created, once again useful for vertical cultivation. The roof of the container is instead used as a receiving surface for the photovoltaic panels, except for an area dedicated to the positioning of the plastic tanks for storing the water, which will be used for irrigating the crops. The total surface area of the container is equal to approximately 40 m^2 of vertical wall (partially intended for cultivation and partly for energy production) and approximately 15 m^2 of roofing (partially intended for solar energy production and partly for water storage in tanks).

Figure 6 shows some scenarios relating to horticultural and energy production, compared to the usable surface of the container. With regard to horticultural production, at least three consecutive annual productions are considered (one production between February and May, one production between May and September, and one production from September to December) of both leafy vegetables and vegetables such as tomatoes, peppers, or eggplants. Agricultural production is facilitated through seasonal crop rotation and by using plants produced in a nursery to accelerate growth and production times.

The prototype configuration presented in Fig. 4 instead has the possibility of being able to expand both the cultivation and catchment surface through the creation of modular greenhouses that can be placed against the container. The advantage deriving from the expansion of the container module can also be found in the possibility of having sheltered and covered spaces in which to grow horticultural plants out of season.

The container, functional to agricultural production, generates output in terms of food production, but the real advantage of positioning the container in this public space comes from the possibility of using edible waste (to be transformed into the social kitchen) and non-edible waste (to be used as a component of compost) produced by the weekly street market and local activities. Furthermore, excess agricultural production and processed foods could be sold (De Oliveira Alves et al. 2024) during the weekly market (with micro-entrepreneurial initiatives), as well as feeding the kitchen of the nearby student residence. It should be noted that in this urban context, there is already an active network of associations that deals with the recovery of food surpluses and their

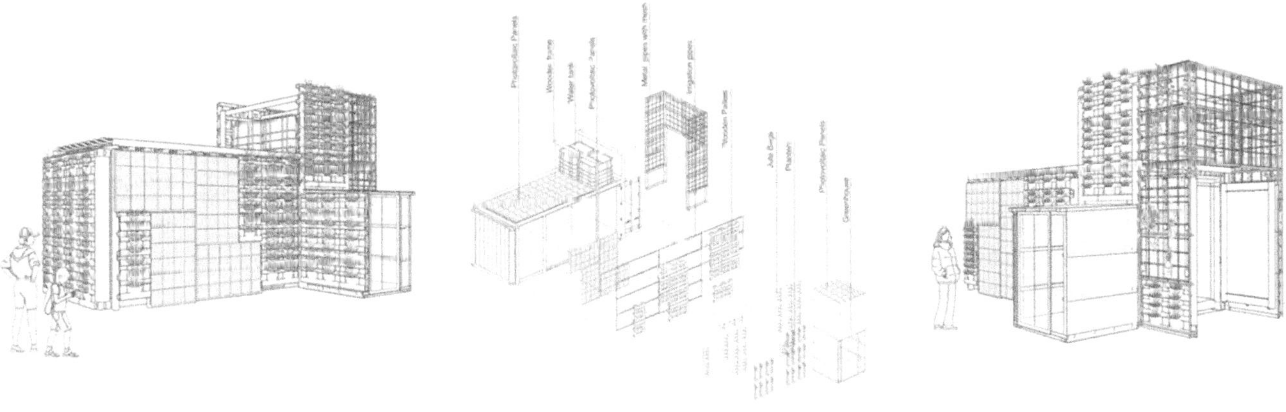

Fig. 4 The figure shows the container with integration of agricultural production and solar energy production

Atlas AEM	RESOURCES	TIPOLOGY	EWC CODE	
Residential	Wood pallet	Pallet 200*100 cm	15.01.03	01
Agricultural and green	Wood pallet	Europallet 200*100 cm	15.01.03	02
Local activities	Pruning logs	Robinia pseudoacacia	20.02.01	01

Atlas AEM	RESOURCES	RESOURCES	EWC CODE	
Residential	Wood pallet	Pallet 200*100 cm	15.01.03	01
Agricultural and green	Wood pallet	Europallet 80*120 cm	15.01.03	02
Local activities	Pruning logs	Robinia pseudoacacia	20.02.01	01

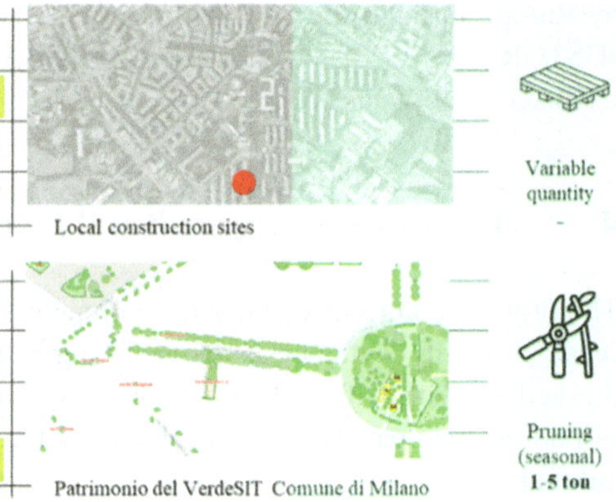

Fig. 5 The figure presents a schematic extract of the AEM tool. In the resource tracking and coding system, it is possible to observe the origin (resident population, agricultural and green sector, and local economic activities, from the ISTAT database), the resource and its characteristics, its coding through the EWC code (EC 2014) and the Atlas code, its mapping, and the estimable or detected quantity. For example, the recovered wooden pallets have different characteristics and dimensions, as they come from local construction sites (with variable quantities) and economic activities in the area (with regular amounts). The pallets are coded with the same EWC code but with a different Atlas code (01 and 02)

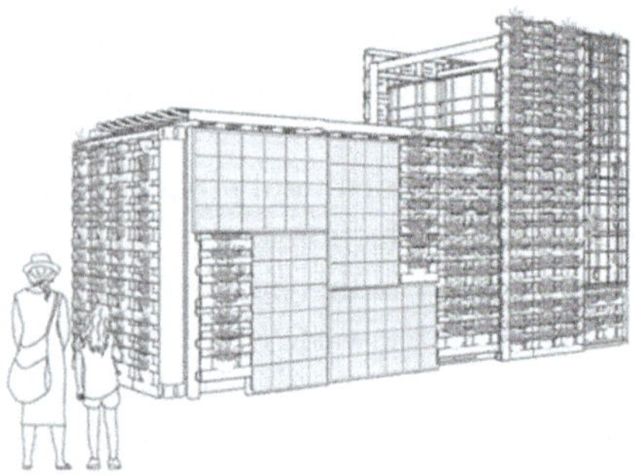

Scenario 1	surfaces	production	
2/3 cultivation production	32 m²	600/1.000 kg	
1/3 energy production	16 m²	1.185 KWh/m²	1.120 KWh/m²
		Vertical surface	Orizontal surface
Scenario 2	**surfaces**	**production**	
1/2 cultivation production	24 m²	450/750 kg	
1/2 energy production	24 m²	2.240 KWh/m²	1.120 KWh/m²
		Vertical surface	Orizontal surface

Fig. 6 Scenarios relating to horticultural and energy production

redistribution (Karki et al. 2021; Yu et al. 2024). Bioloop aims to optimize the activities already underway, starting from the recognition and recovery of those edible wastes that can no longer be used by the practices in place and that need to be transformed, through the experimentation of transformative practices and the distribution of what is produced or processed.

5 Prototype of Functional Implementations to Support Activities in the Container

Design activities are in progress for the reuse of the old container (Fig. 7). This regards the study of the best and cost-effective solutions for integrating PV systems on the roof and facades, adding also mobile shading devices and other components to be installed outdoors. These new installations will be equipped with PV systems to make all the implemented activities mostly self-sufficient from an energy point of view.

The first configuration (Fig. 7) is related to compounds to be placed near the container, to serve as a useful place to reclaim currently unused open space; it is intended to be a space for staying, with the added utility of allowing users to charge their electronic devices. The roof is ideally sized to accommodate a photovoltaic panel and, to optimize solar gain, a system for fixing and adjusting its inclination has been designed. The second configuration in Fig. 7 focuses on the transformation of the container into a kitchen equipped with professional appliances. A "double skin" exterior provides better insulation and supports photovoltaic panels on the roof and long walls, which also create shading when open. The third configuration concerns additional modules to

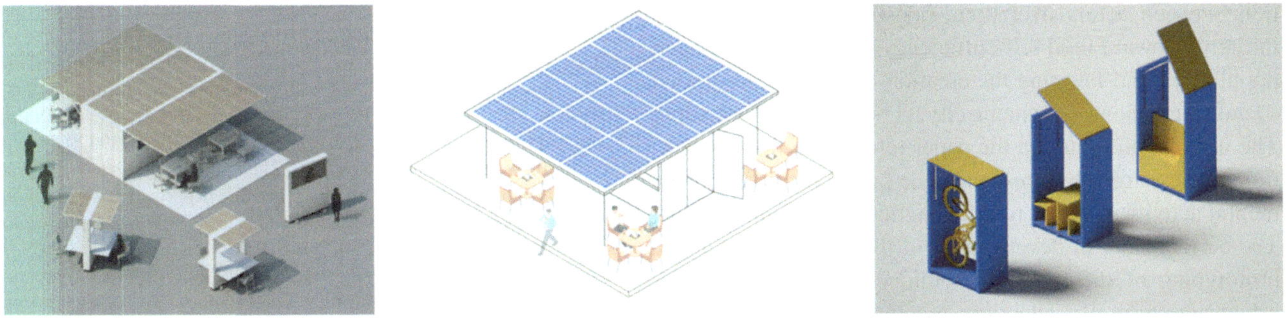

Fig. 7 Potential configurations for the reuse of the old container and predesign of other components. (Images provided by students involved in the project)

Table 1 Potential PV power on the container and the other PV element

Container	Other elements in the square
5–6 kW	5–10 kW

The ranges indicated depend on the surfaces dedicated to building-integrated PV systems and on the types of PV technologies adopted. Further, the actual PV options have to be evaluated based on the existing constraints

support the activities carried out in the container, e.g., areas in which it is possible to recharge cargo bikes, which can be used for short trips in the neighborhood aimed at recovering and transporting the resources to be exploited (waste from commercial activities), both as a useful system for local distribution of the products made in the social kitchen.

6 Energy Self-Sustaining Systems of Agricultural Modules

The solar radiation and the PV production in several configurations and surfaces selected have been evaluated in the area using the tools BimSolar (https://www.bim-solar.com/) and PVGIS (https://pvgis.com/).

Even if it is not possible to perfectly balance the PV production with the electricity needs of the different activities, a relevant percentage in terms of energy self-sufficiency has been reached. The PV powers that can be installed are reported in the following Table 1, considering that these are only options.

These systems are assumed to be connected to the electric grid, but other configurations, e.g., with accumulators, can also be forecasted. A first hypothesis of electricity needs has been elaborated considering the possible uses, e.g., a mobile kitchen, a laboratory for creating products from agricultural wastes, stops for charging bikes and other appliances, and so on. The total power can be estimated at around 6–8 kW.

Referring to the possible case of the mobile kitchen, a total power of 6 kW can be justified by the following appliances: 2 kW for the oven, 2 kW for the cooker, 0.7 kW for a big fridge (more in the case of a blast chiller), and 1.3 kW for the other appliances. A part of the fridge that will work all the time, for the other equipment, use of about 2–3 h/day could be forecasted, i.e., an electricity demand of about 4000–5500 kWh/year, according to the working days per year. The demand for other uses, such as the laboratory, has to be better investigated according to the needed appliances and schedules of use. This may allow for covering about 70% of the possible electricity consumption by the PV installations, reducing the impact of the socio-economic new activities.

7 Conclusions

The Bioloop research activities represent a pilot experience of the circular economy in peri-urban areas. Addressing inequalities in marginalized communities offers an opportunity for economic development, environmental enhancement, and social redemption through the creation of an innovative hub that could attract other stakeholders and investments post-project. The first results of the work briefly described in this paper represent the current state of development of tools such as the local Atlas of Energy and Materials and the Open-Source Project Hub. In particular, the project scenarios outlined in the paper demonstrate that such tools could be effective to support project activities conducted in collaboration with local actors. The current level of development of the project solutions is related to phase 3 of the methodology, i.e., oriented to the evaluation of the effectiveness of the project solutions proposed in the OSPH. At the moment, the verification of the effectiveness of such solutions is oriented to measure the productivity of the devices in terms of the quantity of food and renewable energy. The prototyping of the devices, currently underway, will allow economic indicators to be combined with the first quantitative estimates in order to support possible activation processes of local circular economy scenarios. The activities already completed and still in progress provide a sustainable connection between actors over time, boosting urban cultivation initiatives toward food self-sufficiency among disadvantaged populations and favoring the creation of new neighborhood social connections, enhancing city recovery and valorization. The insights presented in the paper refer to the first application of these tools by presenting the progress of the experimental activities related to urban agricultural supply chains. The current state of development of the work related to the prototyping of the mobile laboratory and the devices for urban agricultural production, together with the data processed so far, highlight a dual value: if on the one hand, the use of locally available resources favors the activation of local economies starting from investments at very low costs on the other hand the same solutions can be included among the possible strategies for climate change mitigation and adaptation. Future developments of the work intend to deepen these aspects by processing quantitative data to verify the effects.

References

Alkhaja N, Alawadi K, Almemari K, Alshehhi G (2024) How is urban agriculture practiced, institutionalized, implemented, and sustained? A literature review. Prog Plann:100917. https://doi.org/10.1016/j.progress.2024.100917

Bohn K, Viljoen A (2005) More space with less space: an urban design strategy. In: Viljoen A (ed) CPULS: continuous productive urban landscapes. Designing urban agriculture for sustainable cities. Elsevier, Amsterdam, pp 11–16

Christensen TB, (2021) Towards a circular economy in cities: Exploring local modes of governance in the transition towards a circular economy in construction and textile recycling. Journal of Cleaner Production 305, 127058. https://doi.org/10.1016/j.jclepro.2021.127058

De Oliveira Alves D, de Oliveira L, Durante Mühl D (2024) Commercial urban agriculture for sustainable cities. Cities 150:105017. https://doi.org/10.1016/j.cities.2024.105017

Deelstra T, Girardet H (2000) Urban agriculture and sustainable cities. In: Bakker N, Dubbeling M, Gundel S, Sabel-Koschela U, de Zeeuw H (eds) Growing cities, growing food: urban agriculture on the policy agenda. Deutsche Stiftung fur Internationale Entwicklung (DSE), Feldafing, pp 43–65

Du S, Tanaka K (2024) Residents' preferences for urban agriculture in Shanghai. Heliyon 10(10):e30974. https://doi.org/10.1016/j.heliyon.2024.e30974

Dubbeling M, Bracalenti L, Lagorio L (2009) Participatory design of public spaces for urban agriculture, Rosario, Argentina. Open House Int 34:36–49. https://doi.org/10.1108/OHI-02-2009-B0005

EC—European Commission (2014) 2014/955/EU: Commission Decision of 18 December 2014 amending Decision 2000/532/EC on the list of waste pursuant to Directive 2008/98/EC of the European Parliament and of the Council Text with EEA relevance. http://data.europa.eu/eli/dec/2014/955/oj

Goddard M (2006) Urban greywater reuse at the D'LUX development. Desalination 188:135–140. https://doi.org/10.1016/j.desal.2005.04.110

Holmer RJ, Drescher AW (2005) Allotment gardens of Cagayan de Oro: their contribution to food security and urban environmental management. In: Knie C (ed) Urban and peri-urban developments structures, processes and solutions; Southeast Asian-German Summer School Program, Cologne, Germany, pp 149–155

ICLEI—Local Governments for Sustainability (2022) Circular Cities Declaration Report 2022. www.circularcitiesdeclaration.eu

INFARm (2016) Horizon Project—the vertical farming revolution, urban farming as a service. https://doi.org/10.3030/739479

Jansma JE, Wertheim-Heck SCO (2022) Feeding the city: a social practice perspective on planning for agriculture in peri-urban Oosterwold, Almere, The Netherlands. Land Use Policy 117:106104. https://doi.org/10.1016/j.landusepol.2022.106104

Johansson J, Roitto M, Steiner B, Alakukku L (2024) Co-creation of urban agriculture through participatory processes in residential building environment: insights from Finland. Cleaner Responsible Consumption 13:100197. https://doi.org/10.1016/j.clrc.2024.100197

JUSTGREEN (2022) Horizon Project—just and green cities: an intersectional perspective on urban agriculture and Roma inclusion in Europe. https://doi.org/10.3030/101025100

Karki ST, Bennett ACT, Mishra JL (2021) Reducing food waste and food insecurity in the UK: the architecture of surplus food distribution supply chain in addressing the sustainable development goals (Goal 2 and Goal 12.3) at a city level. Ind Mark Manag 93:563–577. https://doi.org/10.1016/j.indmarman.2020.09.019

Kirby CK, Specht K, Fox-Kämper R, Hawes JK, Cohen N, Caputo S, Ilieva RT, Lelièvre A, Poniży L, Schoen V, Blythe C (2021) Differences in motivations and social impacts across urban agriculture types: case studies in Europe and the US. Landsc Urban Plan 212:104110. https://doi.org/10.1016/j.landurbplan.2021.104110

Lopes de Morais LH, Costa Pinto D, Cruz-Jesus F (2021) Circular economy engagement: altruism, status, and cultural orientation as drivers for sustainable consumption. Sustainable Prod Consumption 27:523–533

Mattijssen TJM, Hennen W, Buijs AE, De Dooij P, Van Lammeren R, Walet L (2024) Urban greening co-creation: participatory spatial modelling to bridge data-driven and citizen-centred approaches. Urban For Urban Green 94:128257. https://doi.org/10.1016/j.ufug.2024.128257

Midmore DJ, Jansen HGP (2003) Supplying vegetables to Asian cities: is there a case for peri-urban production? Food Policy 28:13–27. https://doi.org/10.1016/S0306-9192(02)00067-2

Pradhan P, Callaghan M, Hu Y, Dahal K, Hunecke C, Reusswig F, Lotze-Campen H, Kropp JP (2023) A systematic review highlights that there are multiple benefits of urban agriculture besides food. Glob Food Secur 38:100700. https://doi.org/10.1016/j.gfs.2023.100700

PROTEAN (2019) Horizon Project—prospective environmental assessment of urban agriculture emerging-systems. https://doi.org/10.3030/842460

Pueyo-Ros J, Škerjanec M, Castellar JAC, Atanasova N, Comas J, Corominas L (2024) Beyond food: a stochastic model to estimate the contributions of urban agriculture to sustainability. Landsc Urban Plan 241:104930. https://doi.org/10.1016/j.landurbplan.2023.104930

Sajjad A, Zhang Q, Asmi F, Azfar Anwar M, Bhatia M (2024) Identifying the motivating factors to promote socially responsible consumption under circular economy: a perspective from norm activation theory. J Retail Consum Serv 76:103544

SiEUGreen (2017) Horizon Project—Sino-European innovative green and smart cities. https://doi.org/10.3030/774233

Souza Piao R, Braga de Vincenzi T, Fernandes da Silva AL, Chinen de Oliveira MC, Vazquez-Brust D, Montei ro Carvalho M (2023) How is the circular economy embracing social inclusion? J Clean Prod 411:137340

UN, General Assembly (2015) Transforming our world: the 2030 agenda for sustainable development. https://www.refworld.org/legal/resolution/unga/2015/en/111816

UN—Department of Economic and Social Affairs, Population Division (2018) The world's cities in 2018. Data Booklet (ST/ESA/SER.A/417). www.unpopulation.org

URBAG (2019) Horizon Project—integrated system analysis of urban vegetation and agriculture. https://doi.org/10.3030/818002

WaysTUP! (2019) Horizon Project—value chains for disruptive transformation of urban biowaste into biobased products in the city context. https://doi.org/10.3030/818308

Yu M, Principato L, Formentini M, Mattia G, Cicatiello C, Capoccia L, Secondi L (2024) Unlocking the potential of surplus food: a blockchain approach to enhance equitable distribution and address food insecurity in Italy. Socio-Econ Plan Sci 93:101868. https://doi.org/10.1016/j.seps.2024.101868

Impact of Street Trees on Thermal Comfort and Air Quality

Carolina Girotti, Luiz Fernando Kowalski,
Alessandra R. Prata Shimomura,
Fernando Akira Kurokawa, Érico Masiero,
and António Lopes

Abstract

In urban environments, street trees influence air quality by altering wind flow and the dispersion of atmospheric pollutants. They also provide shade, impacting thermal comfort and the Universal Thermal Climate Index. This article investigates the influence of street trees on thermal comfort indices, focusing on the Universal Thermal Climate Index (UTCI), and the dispersion of atmospheric pollutants, specifically nitrogen oxides (NOx) and Particulate Matter (PM10), within Brazil's cold and humid Z1 climate zone during December's hot summer conditions. Simulations using ENVI-met software were performed under three scenarios: no trees, 8-m trees, and 16-m trees of average foliage density, in three wind directions. The results showed that street trees had a significant positive impact on thermal comfort, reducing UTCI values. However, they also influenced wind flow and tended to concentrate atmospheric pollutants, with a greater impact on gases like NOx than on particles like PM10. These findings highlight the complex impact of street trees on urban microclimates and pollutant dispersion, emphasising the need for strategic urban planning to optimise their benefits.

Keywords

Atmospheric pollution · UTCI · Street trees · ENVI-met simulation · Urban microclimate · Pollutant dispersion

C. Girotti (✉)
Faculty of Architecture and Urbanism, University of São Paulo, São Paulo, Brazil

Center for Geographical Studies, IGOT—Institute of Geography and Spatial Planning, University of Lisbon, Lisbon, Portugal
e-mail: cgirotti@usp.br

L. F. Kowalski
Center for Geographical Studies, IGOT—Institute of Geography and Spatial Planning, University of Lisbon, Lisbon, Portugal

Center for Acoustic and Thermal Research in Buildings and Road Networks, Federal University of São Carlos, São Carlos, Brazil

A. R. Prata Shimomura
Faculty of Architecture and Urbanism, University of São Paulo, São Paulo, Brazil
e-mail: arprata@usp.br

F. A. Kurokawa
Polytechnic School of the University of São Paulo, São Paulo, SP, Brazil
e-mail: fernando.kurokawa@usp.br

É. Masiero
Center for Acoustic and Thermal Research in Buildings and Road Networks, Federal University of São Carlos, São Carlos, Brazil
e-mail: erico@ufscar.br

A. Lopes
Center for Geographical Studies, IGOT—Institute of Geography and Spatial Planning, University of Lisbon, Lisbon, Portugal
e-mail: antonio.lopes@edu.ulisboa.pt

1 Introduction

In an urban environment, street trees impact air quality by altering wind flow and consequently affecting the dispersion of atmospheric pollutants (Janhäll 2015). Additionally, street trees provide shading that influences thermal comfort and the Universal Thermal Climate Index (UTCI) (Coutts et al. 2016).

Regarding street trees and their relationship with atmospheric pollutant dispersion, authors Gromke and Ruck (2007) conducted experiments in a wind tunnel. They inserted small trees with spherical canopies on thin stems into the urban canyon to simplify the study of different vegetation densities. The trees were replaced by metal cages filled with varying amounts of fibrous material and indicates that when empty cages are used and every two cages are filled, there is little effect on wind flow (Gromke 2011). However, the presence of larger trees alters wind flow and increases atmospheric pollutant concentration (Gromke and Ruck 2007).

F. Alberti et al. (eds.), *Towards Sustainable Urban Regeneration: Utilizing Environmental, Cultural, and Social Assets*, Advances in Science, Technology & Innovation, https://doi.org/10.1007/978-3-032-02091-8_5

Studies on microscale atmospheric models using the ENVI-met software (Bruse et al. 1998) are increasingly being employed to project the thermal benefits of urban heat mitigation strategies, such as tree planting programmes or the use of high-albedo materials (Ouyang et al. 2022), and are gaining attention for developing adaptation strategies (Tsoka et al. 2018). Tsoka et al. (2018) evaluate ENVI-met's ability to model diurnal microclimatic variables and assess the cooling potential of urban greenery and cool materials. The study highlights that urban greenery is the most popular mitigation strategy, with combined strategies enhancing cooling potential (Tsoka et al. 2018). Crank et al. (2018) address the grid independence of ENVI-met and its ability to represent air temperature changes from heat mitigation strategies. Results show that grid dependence does not invalidate the model for heat mitigation, but concerns remain regarding its application in more complex simulations and its ability to capture vertical mixing (Crank et al. 2018).

Studies on atmospheric pollutant dispersion and street trees using simulation with the ENVI-met software (Bruse et al. 1998) have shown higher pollutant concentrations due to vegetation in urban canyons. In the studies by Wania et al. (2012), larger and denser trees considerably reduced atmospheric pollutant dispersion. Authors Vos et al. (2013) recommend green barriers between vehicular traffic and pedestrians and the placement of vegetation near the source of atmospheric pollutant emissions, as this increases pollutant deposition (Vos et al. 2013; Wania et al. 2012). The study by Acero and Arrizabalaga (2018) compares diurnal meteorological data from four urban spaces with ENVI-met, using measurements taken over 3 days in Bilbao. Urban areas modify meteorological variables, affecting thermal comfort and pollutant dispersion, which in turn influences pedestrian comfort (Acero and Arrizabalaga 2018).

In terms of thermal comfort, it has been demonstrated that street trees significantly decrease the maximum mean radiant temperature. Additionally, street trees can reduce daytime UTCI by 0.2–0.6 °C, with even greater reductions of up to 1.5 °C observed in shallow canyon environments (Coutts et al. 2016; Jia and Wang 2021). A study conducted in Curitiba, Paraná, Brazil, analysed the thermal comfort of passers-by in pedestrian streets. The study defined ranges of thermal comfort and discomfort for Curitiba, using the UTCI for local climatic conditions and acceptable percentages of dissatisfied people in open spaces. The analysis suggested that the UTCI index was an effective predictor of thermal sensation, with differences regarding gender, age, and body mass being negligible (Rossi et al. 2012).

This study examines the impact of street trees on thermal comfort indices, focusing on UTCI, and the dispersion of atmospheric pollutants, especially nitrogen oxides (NO_x) and particulate matter (PM_{10}), in Brazil, considering a cool and humid climate zone (Z1) during December's hot summer conditions.

2 Methodology

ENVI-met is a simulation tool based on the principles of computational fluid dynamics (CFD), designed to model the complex interactions between surfaces, vegetation, and air in urban microclimates (Bruse & Fleer 1998). It uses an Eulerian approach to simulate the dispersion of pollutants, applying the finite difference method to solve the non-hydrostatic Navier–Stokes equations. ENVI-met incorporates the k–ε model (Yakhot and Orszag 1986; Yakhot and Smith 1992) a member of the Reynolds-Averaged Navier-Stokes (RANS) family, to simulate turbulent kinetic energy (k) and its dissipation rate (ε), enabling the prediction of averaged turbulent flow behavior and its interaction with urban features.

The governing equations for Navier–Stokes and pollutant transport are solved on a structured, orthogonal, and non-uniform grid mesh, where topographical elements such as soil are represented by discretised cells (Alsaad et al. 2022). The time-stepping approach varies by subsystem, utilising either explicit or partially implicit schemes. The advection and atmospheric diffusion equations are solved using the alternating direction implicit (ADI) method to ensure numerical stability, particularly for long-term simulations (Huttner 2012).

ENVI-met is widely recognised for its application in assessing atmospheric pollutant dispersion in urban microclimates (Rodrigues et al. 2024; Wania et al. 2012) he reliability of numerical models is critical for accurate simulations of pollutant behaviour in urban environments. Validation studies have reported results ranging from acceptable to good agreement with observed data (Bruse et al. 1998; Crank et al. 2018; Detommaso et al. 2021). However, it is important to note that ENVI-met is sensitive to grid configuration, which may affect accuracy in complex urban scenarios (Crank et al. 2018).

For particle deposition analysis, ENVI-met integrates gravitational flux with local wind-particle interactions. Studies have shown that the model effectively predicts local particulate matter concentrations by incorporating pollution sources and sedimentation fluxes, making it a reliable tool for urban air quality assessments (Bruse et al. 1998). In another study, simulations using meteorological and traffic flow data from a specific day demonstrated good agreement between ENVI-met's modeled pollutant concentrations and measurements from monitoring stations (Ferrari et al. 2024).

One of ENVI-met's main features is its ability to simultaneously model the release, dispersion, and deposition of up to six pollutants, including both passive and reactive gases and particles. This study focuses on the dispersion of NO_x

and PM10, as these pollutants are key indicators of urban air quality. The selection of ENVI-met was based on its comprehensive capability to simulate the urban microclimate, including detailed 3D vegetation structures and pollutant behaviour. For this research, ENVI-met version 5.7 was used.

2.1 Case Study and Domain Configuration

This study investigates the impact of street trees on thermal comfort indices and the dispersion of air pollutants, focusing particularly on Brazil's Z1 climate zone during hot summer conditions in December. The aim is to evaluate the effectiveness of urban trees in mitigating environmental stressors by analysing UTCI and the dispersion of NO*x* and PM10. The simulation was conducted using ENVI-met, which generated three scenarios for a generic neighbourhood: the first scenario without street trees, the second with 8-m-high trees of average foliage density, and the third with 16-m-high trees of average foliage density.

Regarding the buildings, the scenario is generic, with structures reaching a height of 9 m and spaced 12 m apart along the main street and 6 m apart on the surrounding streets. Three wind directions were simulated, as shown in Fig. 1.

The simulation domain covers an area of 600 m × 532.80 m horizontally, with a uniform Cartesian grid, and a vertical height of 56.75 m with a variable grid configuration. The horizontal resolution of 3 m × 3 m was maintained constant throughout the domain. For the vertical resolution, due to the importance of pedestrian height, the first cell was divided into five cells of 0.4 m each; above this, a resolution of 3 m was set, and above 60 m, a telescopic factor of 25% was applied. To avoid numerical issues caused by the interference of model boundaries with internal dynamics and to ensure adequate spacing between obstacles and model boundaries, extra grids were added on each side of the computational domain, totalling 200 × 160 × 15 grid cells.

For the soil and surface parameters, asphalt pavement (Asphalt—ENVI-met database—ID 0200ST) with an albedo of 0.12 and emissivity of 0.9 was used in street areas. In other areas, a standard concrete pavement (Pavement—ENVI-met database—ID 0200PG) with an albedo of 0.3 and emissivity of 0.9 was used.

A total of 155 street trees were considered in Scenarios 2 and 3. Their characteristics of the tree are listed in Table 1.

Table 1 presents the characteristics of the trees used in the simulation. Leaf coverage influences the amount of sunlight filtered through the foliage, affecting the cooling effect provided by evapotranspiration and shading. Foliage albedo determines how much solar radiation is reflected, with higher albedo values reducing heat absorption. Emissivity impacts the tree's ability to release heat through longwave radiation, influencing nighttime cooling.

The analysis focuses on data collected in December, at 5 pm, coinciding with peak vehicle traffic and high temperatures. Pollutant emissions were calculated using ENVI-met, assuming 8000 daily vehicles on an inner urban road with two lanes. The composition of vehicles included 5% light commercial vehicles, 2.5% heavy commercial vehicles, 0.5% motorcycles, 3% buses, 1% coaches, and 88% light vehicles. Emission factors for PM10 and NO*x* were deter-

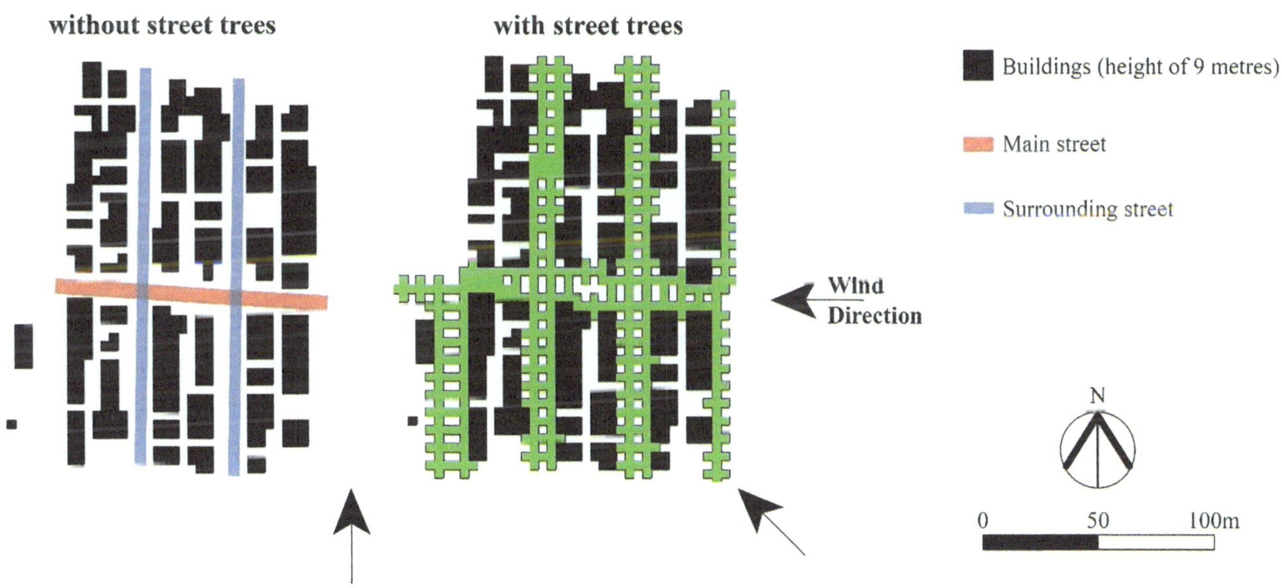

Fig. 1 Case study scenarios

mined using averages from INMETRO (2024). The distribution of NO*x* and PM10 emissions is depicted in Fig. 2.

For the analysis, climate data from Curitiba, Brazil, was obtained from the Curitiba 838420 INMET.epw file (energyplus 2024), was used. This file format provides weather data in the EnergyPlus Weather format, containing comprehensive information such as temperature, humidity, wind speed, solar radiation, and precipitation. Curitiba's climate data is classified under climate zone Z1, characterised by cold and humid conditions typical of Brazil. Figure 3 shows the graphs with climate data extracted from the .epw file, considering December 7.

The calculation of UTCI was performed using the Biomet module within ENVI-met software, which estimates UTCI from parameters including wind speed, air temperature, mean radiant temperature, and air vapor pressure.

3 Results

The findings underscore the role of street trees in improving thermal comfort and mitigating atmospheric pollution in urban environments during hot summer conditions. The results for PM10 and NO*x* concentrations are illustrated in Figs. 4 (PM10) and 5.5 (NOx).

Table 1 Characteristics of trees

Scenario	2	3
Specie	Heart-shaped, medium trunk, sparse, medium	
Stem height (m)	2.0	4.0
Total height (m)	8.0	16.0
Leaf coverage	100.0%	100.0%
Foliage shortwave albedo	0.18	0.18
Foliage shortwave transmittance	0.30	0.30
Emissivity of leaves	0.96	0.96
Isoprene capacity	12.0	12.0

Figures 4 and 5 demonstrate that, in the scenario without trees, PM10 and NO*x* concentrations are relatively homogeneous across the three wind directions analysed (0°, 135°, and 180°). PM10 levels remain low, below 14.10 µg/m³, indicating effective pollutant dispersion due to the absence of barriers obstructing ventilation. Conversely, NO*x* exhibits peaks along the roads aligned with the wind direction, up to 24 µg/m³. This suggests that these wind directions promote pollutant accumulation in areas with higher vehicular traffic.

In the scenarios with 8-m and 16-m trees, a moderate increase in PM10 concentrations and a more pronounced increase in NO*x* are observed. The presence of 8-m trees causes a slight rise in pollutant levels, whereas 16-m trees significantly intensify concentrations, especially in areas beneath the canopies, where airflow is more restricted. For wind directions of 0° and 135°, PM10 peaks exceed 14.20 µg/m³, while NO*x* levels surpass 24 µg/m³, reflecting the reduced ventilation and the entrapment of pollutants in these scenarios.

With winds from the 180° direction, the longitudinal configuration of the urban canyon promotes a tunnel effect, increasing wind speed within the corridor. This effect enhances pollutant dispersion compared to the other wind directions. However, the presence of trees, particularly those 16 m tall, partially reduces this efficiency, creating zones of lower ventilation beneath the canopies, where pollutants tend to accumulate.

The results for UTCI index are illustrated in Fig. 6.

The UTCI results (Fig. 6) across the three scenarios—without trees, with 8-m trees, and with 16-m trees—reveal significant differences in thermal comfort within the urban canyon. In the scenario without trees, the UTCI values are predominantly high, exceeding 28.80 °C, indicating thermal stress, particularly in areas exposed to direct sunlight and with limited ventilation. This pattern is consistent across all wind directions (0°, 135°, and 180°).

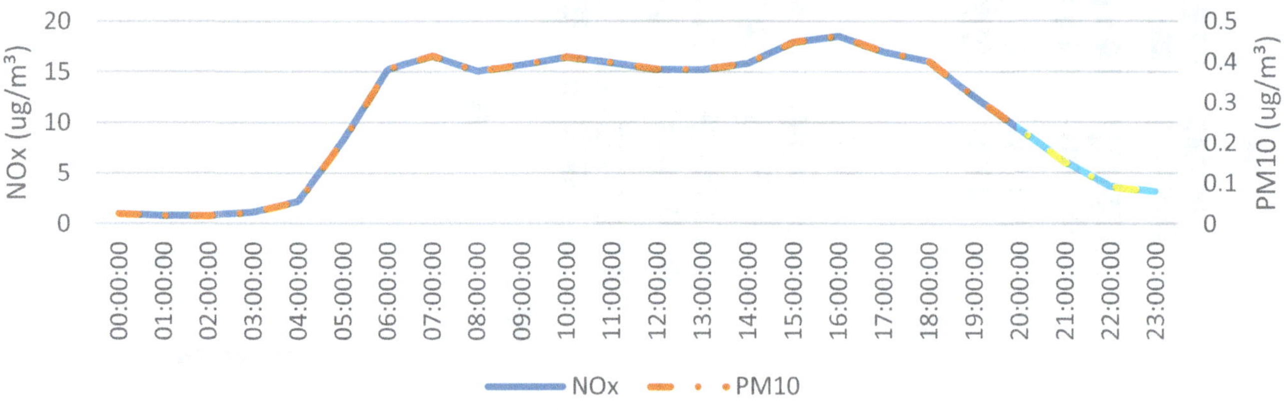

Fig. 2 Distribution of NO*x* and PM10 emissions

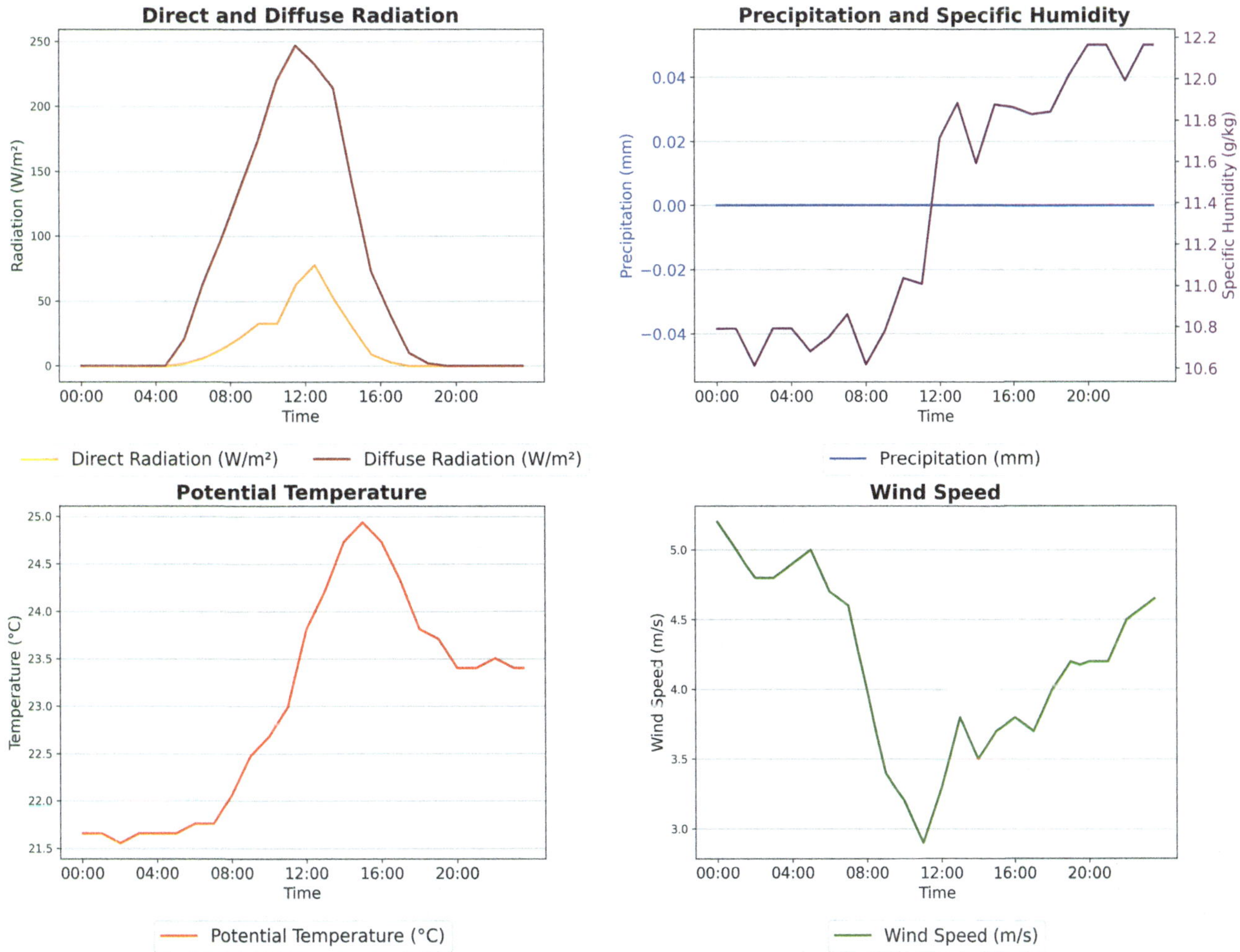

Fig. 3 Climate data from December 7, Curitiba (838420 INMET.epw) used for simulation

In contrast, the scenarios with street trees, particularly those with a height of 16 m, show a substantial reduction in UTCI values, below 27.20 °C. The cooling effect is most pronounced beneath the tree canopies, where shading and reduced solar radiation contribute to lower thermal stress. The presence of 8-m trees provides moderate cooling, but the impact is less significant compared to the 16-m trees. The wind direction of 180° further enhances the cooling effect, as the longitudinal canyon configuration facilitates air circulation.

4 Discussion

The influence of street trees on pollutant dispersion varies between NO*x* and PM10 due to their differing properties. While PM10 concentrations remained relatively homogeneous across different wind directions in the scenario without trees, the presence of trees, especially 16-m tall trees, led to slight localised increases in PM10 concentrations. In contrast, NO*x* concentrations showed more significant changes. In the scenarios with 16-m trees, NO*x* levels increased by more than 1.5 μg/m^3.

Vranckx et al. (2015) highlight that PM10 concentrations can range from 0.2% to 2.6% depending on tree cover, while elemental carbon levels can vary between 1% and 13% (Vranckx et al. 2015). Grylls and Van Reeuwijk (2022) argue that the impact of trees on air quality is pollutant-specific: for background particles such as PM2.5, deposition is the dominant mechanism, whereas for locally emitted pollutants like NO*x*, dispersion plays a more critical role (Grylls and van Reeuwijk 2022).

NO*x*, being primarily a locally emitted pollutant, is highly dependent on effective ventilation to disperse. This explains the variability observed in NO*x* concentrations across different scenarios. On the other hand, PM10, which is largely composed of background particles, tends to accumulate in zones with low air circulation, such as the areas beneath tree canopies. This accumulation is consistent with the slight increases in PM10 concentration observed near tree canopies

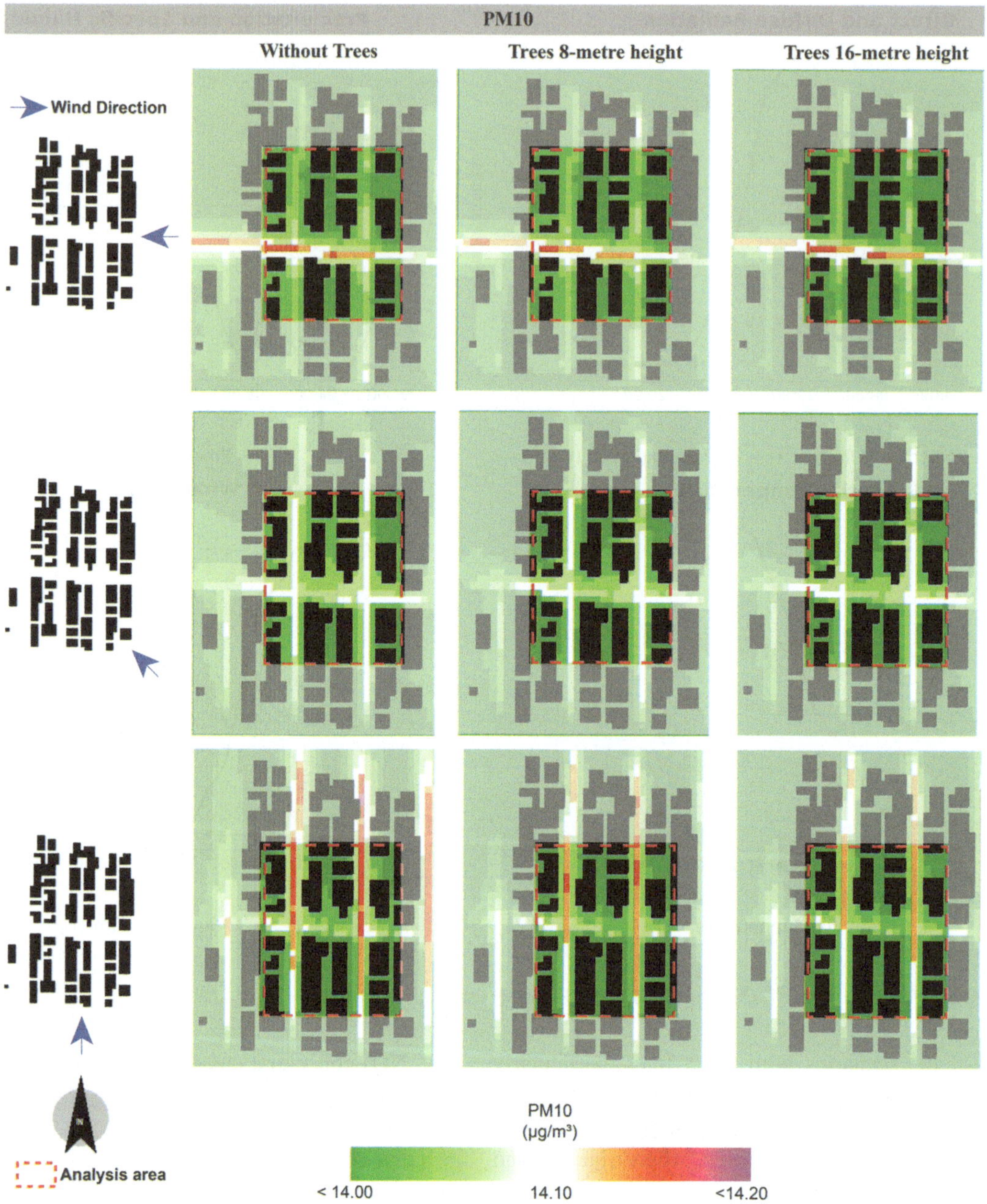

Fig. 4 PM10 concentration results across the three scenarios analysed

in our study (Gromke 2011; Gromke and Ruck 2007; Vos et al. 2013; Wania et al. 2012).

Regarding thermal comfort, the inclusion of street trees led to a UTCI reduction of over 1.0 °C at pedestrian height in all scenarios. This reduction underscores the importance of shade and localised cooling provided by urban greenery (Coutts et al. 2016; Jia and Wang 2021). The findings align with those of Tsoka et al. (2018), who reported a median peak air temperature reduction of 1.0 °C due to urban trees, further emphasising the effectiveness of combined strategies for urban heat mitigation (Tsoka et al. 2018).

The results align with findings by Rossi et al. (2012), who identified the significant influence of open space planning and urban requalification on outdoor thermal comfort in pedestrian streets in Curitiba. Their study suggests that the UTCI index is an effective predictor of thermal sensation, with temperature ranges between 15 °C and 27 °C considered comfortable for Curitiba's climate (Rossi et al. 2012).

It is important to acknowledge the limitations associated with using the ENVI-met model in this research. While ENVI-met is widely employed to assess the impact of vegetation on pollutant dispersion within urban microclimates (Rodrigues et al. 2024; Wania et al. 2012), its performance

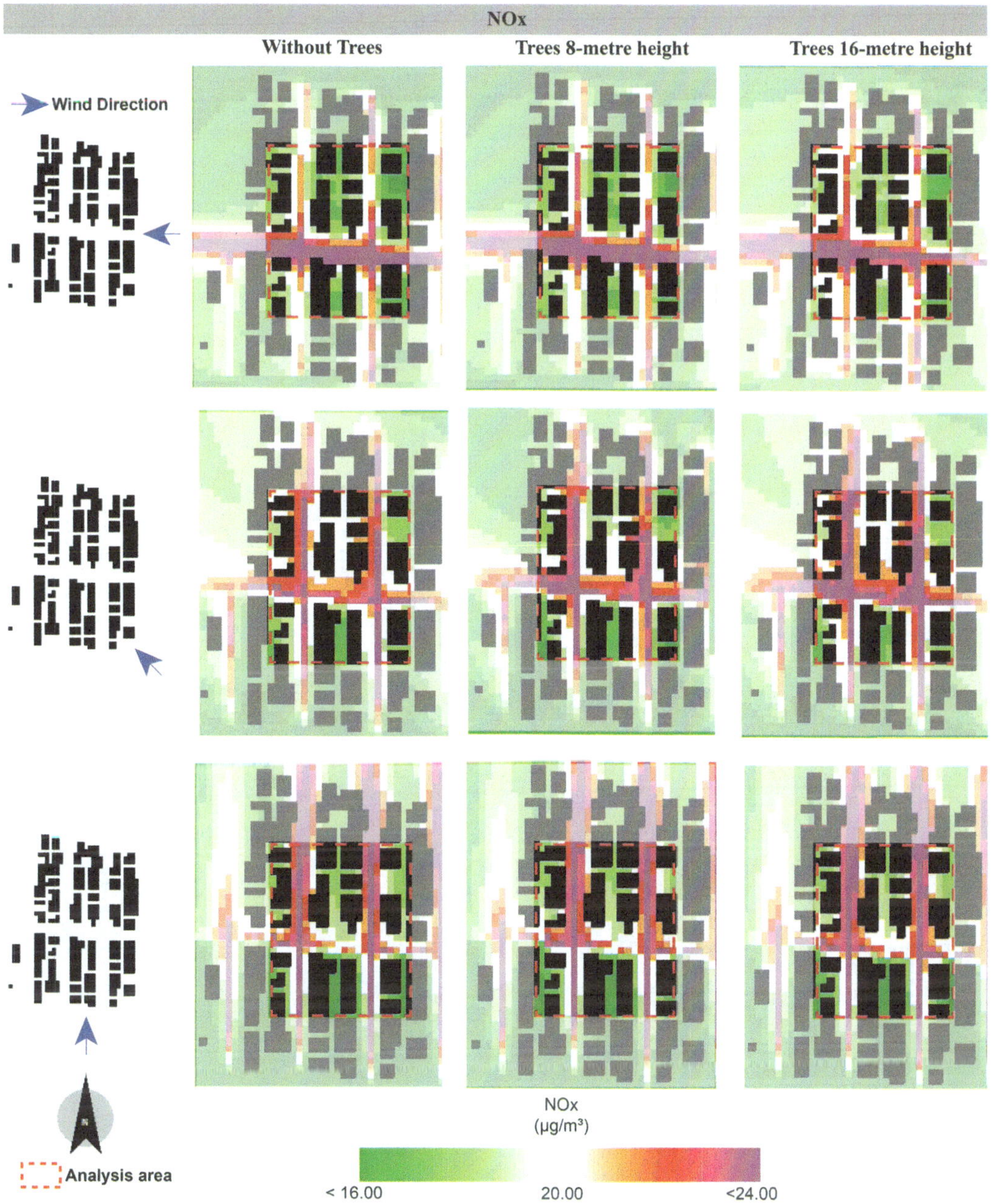

Fig. 5 NO*x* concentration results across the three scenarios analysed

varies from acceptable to good, as indicated by validation studies (Crank et al. 2018; Detommaso et al. 2021). These studies highlight the model's sensitivity to simulation grid configuration, which can affect the accuracy of results, particularly in complex urban scenarios.

Despite these limitations, ENVI-met remains a valuable tool for urban climate studies, especially when evaluating the cooling and air quality benefits of vegetation (Crank et al. 2018).

5 Conclusion

This study examined the effect of urban street trees on reducing high summer temperatures and pollutant concentrations in Brazil's climate zone Z1. Using ENVI-met simulations, different scenarios with and without trees were compared, along with varying wind directions. The results showed that street trees had a significant positive impact on thermal comfort, particularly reducing UTCI values. However, they also

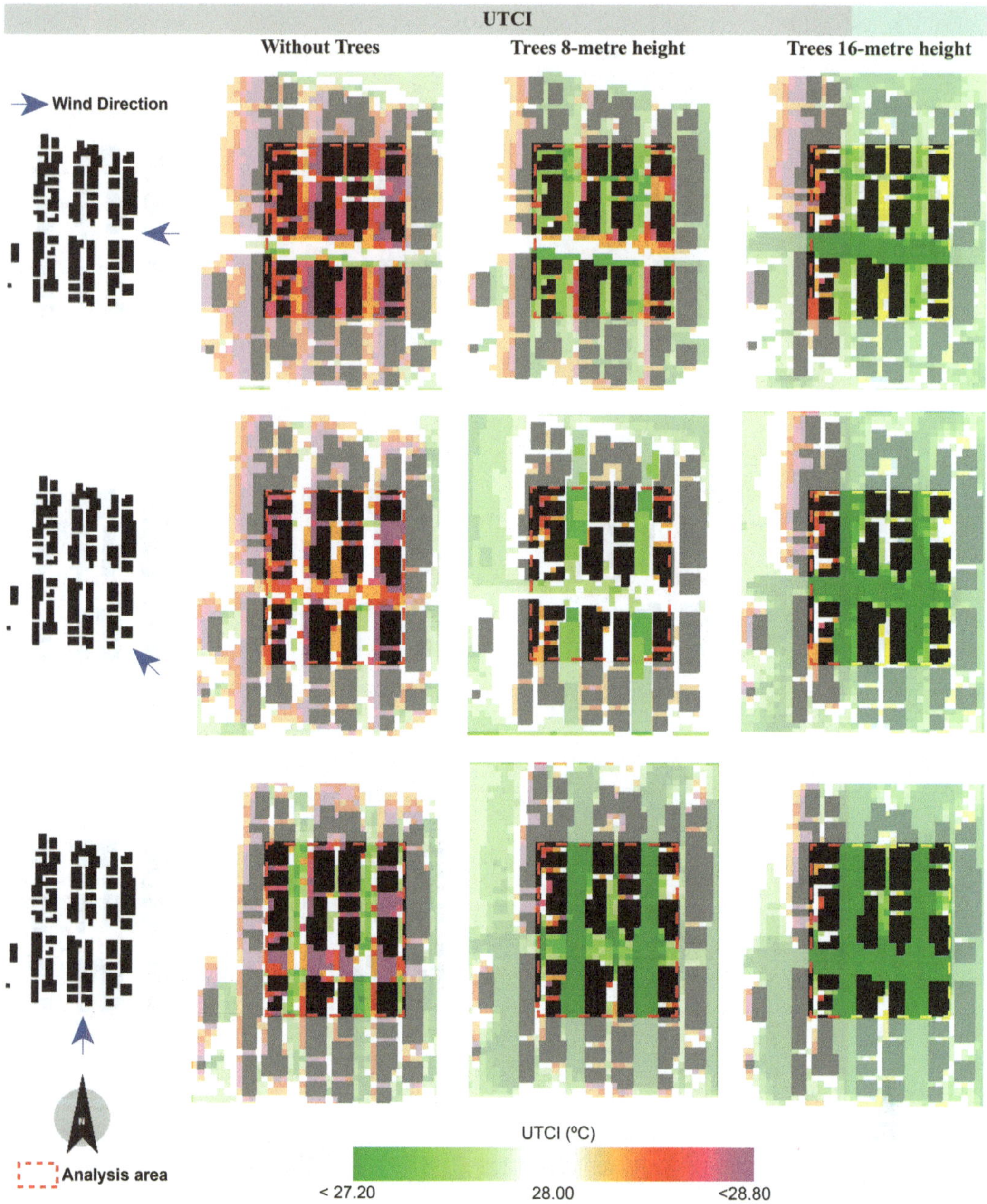

Fig. 6 UTCI results across the three scenarios analysed

influenced wind flow and tended to concentrate atmospheric pollutants, with a greater impact on gases like NOx than on particles like PM10.

Further validation of the model through direct observations and real-world data is essential to generalise these findings to other urban contexts. While ENVI-met has proven to be a reliable tool for simulating urban microclimates, direct chemical and physical analyses in real environments are necessary to validate the model's predictions and enhance its applicability for urban planning. This validation will ensure that the model can serve as a robust guideline for urban greening initiatives aimed at improving both thermal comfort and air quality across diverse climatic zones and urban forms.

By optimising tree placement and understanding the complex interactions between vegetation, airflow, thermal comfort and pollutants, urban planners can effectively leverage street trees as a tool for sustainable urban development, improving both the livability and environmental quality of cities.

Acknowledgements Grant #2021/14533-7, São Paulo Research Foundation (FAPESP).

Grant #2023/01130-7, São Paulo Research Foundation (FAPESP).

References

Acero JA, Arrizabalaga J (2018) Evaluating the performance of ENVI-met model in diurnal cycles for different meteorological conditions. Theor Appl Climatol 131(1–2):455–469. https://doi.org/10.1007/s00704-016-1971-y

ALSAAD, H et al. (2022) ENVI-met validation data accompanied with simulation data of the impact of facade greening on the urban microclimate. Data in Brief, v. 42, p. 108200, fev.

Bruse M, Fleer H (1998) Simulating surface–plant–air interactions inside urban environments with a three dimensional numerical model. Environ Model Softw 13(3–4):373–384. https://doi.org/10.1016/S1364-8152(98)00042-5

Bruse D, Bruse M, Simon H (1998) ENVI-met. https://www.envi-met.com/pt-pt/

Coutts AM, White EC, Tapper NJ, Beringer J, Livesley SJ (2016) Temperature and human thermal comfort effects of street trees across three contrasting street canyon environments. Theor Appl Climatol 124(1–2):55–68. https://doi.org/10.1007/s00704-015-1409-y

Crank PJ, Sailor DJ, Ban-Weiss G, Taleghani M (2018) Evaluating the ENVI-met microscale model for suitability in analysis of targeted urban heat mitigation strategies. Urban Climate 26:188–197. https://doi.org/10.1016/j.uclim.2018.09.002

Detommaso M, Costanzo V, Nocera F (2021) Application of weather data morphing for calibration of urban ENVI-met microclimate models. Results and critical issues. Urban Climate 38:100895. https://doi.org/10.1016/j.uclim.2021.100895

energyplus (2024) Curitiba 838420 INMET. https://energyplus-weather.s3.amazonaws.com/south_america_wmo_region_3/BRA/BRA_PR_Curitiba.838420_INMET/BRA_PR_Curitiba.838420_INMET.zip

Ferrari S, Santus A, Tendas L (2024) Validation of a numerical software for the simulation of the pollutant dispersion from traffic in a real case: some preliminary results. EPJ Web Conf 299:01010. https://doi.org/10.1051/epjconf/202429901010

Gromke C (2011) A vegetation modeling concept for building and environmental aerodynamics wind tunnel tests and its application in pollutant dispersion studies. Environ Pollut 159(8–9):2094–2099. https://doi.org/10.1016/j.envpol.2010.11.012

Gromke C, Ruck B (2007) Influence of trees on the dispersion of pollutants in an urban street canyon—experimental investigation of the flow and concentration field. Atmos Environ 41(16):3287–3302. https://doi.org/10.1016/j.atmosenv.2006.12.043

Grylls T, van Reeuwijk M (2022) How trees affect urban air quality: it depends on the source. Atmos Environ 290:119275. https://doi.org/10.1016/j.atmosenv.2022.119275

Huttner S (2012) Further development and application of the 3D microclimate simulation ENVI-met [Dissertation]. Johannes Gutenberg-Universität Mainz.

INMETRO (2024) Programa Brasileiro de Etiquetagem (PBE) 2024. https://www.gov.br/inmetro/pt-br/assuntos/avaliacao-da-conformidade/programa-brasileiro-de-etiquetagem/tabelas-de-eficiencia-energetica/veiculos-automotivos-pbe-veicular

Janhäll S (2015) Review on urban vegetation and particle air pollution—deposition and dispersion. Atmos Environ 105:130–137. https://doi.org/10.1016/j.atmosenv.2015.01.052

Jia S, Wang Y (2021) Effect of heat mitigation strategies on thermal environment, thermal comfort, and walkability: a case study in Hong Kong. Build Environ 201:107988. https://doi.org/10.1016/j.buildenv.2021.107988

Ouyang W, Sinsel T, Simon H, Morakinyo TE, Liu H, Ng E (2022) Evaluating the thermal-radiative performance of ENVI-met model for green infrastructure typologies: experience from a subtropical climate. Build Environ 207:108427. https://doi.org/10.1016/j.buildenv.2021.108427

Rodrigues V, Augusto B, Oliveira K, Ascenso A, Rafael S, Nascimento D, Miranda AI (2024) Setting up a CFD model to evaluate the impact of green infrastructures on local air quality. Air Qual Atmos Health. https://doi.org/10.1007/s11869-024-01567-6

Rossi FA, Krüger EL, Bröde P (2012) Definition of thermal comfort and discomfort ranges for open spaces in Curitiba, PR, with the UCTI index. Ambiente Construído 12(1):41–59

Tsoka S, Tsikaloudaki A, Theodosiou T (2018) Analyzing the ENVI-met microclimate model's performance and assessing cool materials and urban vegetation applications—a review. Sustain Cities Soc 43:55–76. https://doi.org/10.1016/j.scs.2018.08.009

Vos PEJ, Maiheu B, Vankerkom J, Janssen S (2013) Improving local air quality in cities: to tree or not to tree? Environ Pollut 183:113–122. https://doi.org/10.1016/j.envpol.2012.10.021

Vranckx S, Vos P, Maiheu B, Janssen S (2015) Impact of trees on pollutant dispersion in street canyons: a numerical study of the annual average effects in Antwerp, Belgium. Sci Tot Environ 532:474–483. https://doi.org/10.1016/j.scitotenv.2015.06.032

Wania A, Bruse M, Blond N, Weber C (2012) Analysing the influence of different street vegetation on traffic-induced particle dispersion using microscale simulations. J Environ Manag 94(1):91–101. https://doi.org/10.1016/j.jenvman.2011.06.036

Yakhot V, Orszag SA (1986) Renormalization group analysis of turbulence. J Sci Comput 1:3–5

Yakhot V, Smith LM (1992) The renormalization group, the e-expansion and derivation of turbulence models. J Sci Comput 7:35–61

From Margin to Urban Interface: Designing a Network of Open Spaces as a New Public Park Around the Heritage City Walls of Siena

Francesco Alberti, Eleonora Giannini, Sara Naldoni, and Lorenzo Nofroni

Abstract

The city centre of Siena, a UNESCO World Heritage site since 1995, is surrounded by 7-km-long city walls which helped to preserve large open spaces wedging into the historic town, the so-called 'green valleys'. This potential green belt is currently poorly accessible: fragmented by different uses and properties on the inside of the walls, scarcely connected to the outside, especially in terms of soft mobility. The project Connessioni Verdi Siena (ConVerSi) developed by Urban LIFE, design practice and spin-off company of the University of Florence, on behalf of the Municipality of Siena, networked this mosaic of diverse spaces—which include two botanical gardens, areas belonging to the local communities of the *contrade* (historical quarters of Siena), agricultural spaces and various public or private gardens—to form a park around the city walls. The park is conceived as an interface between the heritage centre and the districts outside the walled city, and as a circular green infrastructure radially connecting the city with its territory. The project, developed also through community engagement activities, resulted in a masterplan for the park of the city walls which brings together existing and new elements of historical, cultural and environmental value, identifying a network of internal-external connections and finally making the green valleys accessible again.

Keywords

Green belt · Public space · Historical landscape · Walled cities · Accessibility · Public engagement · Ecosystemic urbanism

1 Introduction: Urban Development and Walled Cities

The heritage of walled cities in Tuscany is a strong identifying element of a globally-known landscape and it includes architectural structures from the Etruscan-Roman age to the nineteenth century. In many cases, this legacy has been preserved intact—like in Lucca, Grosseto or Siena—thanks to the re-functioning of the historical walls over the centuries and, more recently, thanks to laws for preservation which have started to give the necessary attention not only to conservation, but also to the strategic role that these ancient artefacts can play in contemporary sustainable urban development. A reflection on urban transformations in contexts of high historical and architectural value necessarily evokes the concept of valorisation, which has assumed a broader connotation thanks to the European Landscape Convention (2000) and *Codice dei Beni Culturali e del Paesaggio* (2004), extending the concept beyond the idea of static conservation of single artefacts to aspects related to the management of cultural heritage, to the actors taking part in the process and to the importance of cooperation between the different actors involved, both public and private. The *Codice* also emphasises how processes of knowledge, reuse and fruition contribute to valorisation and enhance the role of the community, which can be involved through forms of direct participation. In this light, developing the project of a park around the historical walls of Siena represented an opportunity to imagine forms of active protection of this historical heritage, rethinking it with new uses and making it accessible as a network of public open space. This planning approach

F. Alberti
Department of Architecture (DIDA), University of Florence, Florence, Italy
e-mail: francesco.alberti@unifi.it

E. Giannini (✉) · L. Nofroni
Department of Architecture, University of Florence, Florence, Italy

Urban LIFE Srl, Florence, Italy
e-mail: eleonora.giannini@unifi.it; lorenzo.nofroni@unifi.it

S. Naldoni
Urban LIFE Srl, Florence, Italy
e-mail: sara.naldoni@urbanlife.city

F. Alberti et al. (eds.), *Towards Sustainable Urban Regeneration: Utilizing Environmental, Cultural, and Social Assets*, Advances in Science, Technology & Innovation, https://doi.org/10.1007/978-3-032-02091-8_6

Fig. 1 M.C. Escher, Italian landscape in the surroundings of Siena (1923). (Credits: CC-BY-NC-ND 4.0)

was inspired by the work of Bernardo Secchi (1988a, b, 1993), the visionary urban planner behind Siena's forward-thinking *Piano Regolatore Generale* of 1996.

Siena is an emblematic example of a Tuscan walled city (Detti et al. 1968) (Fig. 1). The historical settlement stands in an elevated position overlooking the open territory and surrounded by the walls, which contributed to maintaining the image of this urban landscape basically unaltered over the centuries, as described by nineteenth-century Tuscan geographer Emanuele Repetti: '*the iconographic Figure of Siena, standing in the valleys that descend from the bifurcated hillock, along which rise the major temples, the prompt Torre del Mangia and the main streets of this city*' (Repetti 1843, p. 356).

The circle of fortifications was built in the early medieval period and was expanded to accommodate the growth of the city, delimiting the settlement while leaving open space for future urban expansion at a time of flourishing economy and political power. This growth was interrupted by the Black Death of 1348 and bequeathed large open spaces between the built fabric and the boundary of the city walls, namely the green valleys which are still clearly visible today, especially south of the historic centre, and represent 40% of the urban territory within the walls (Balestracci and Piccinni 1977; Hook 1979). Ambrogio Lorenzetti's infamous fresco *Allegory and Effects of Good and Bad Government* (1338–39) clearly describes the spatial and functional relationship between the walled city and the countryside, preserved to the present day (Fig. 2). Siena benefits from a unique situation in which extensive green wedges of nature have been maintained within the city walls, in a sinuous topography that alternates hills and deep valleys in physical continuity with the built environment (Redon 1999): this collection of open spaces is rich in gardens, small woods, former agricultural plots, abandoned areas and artefacts of historical and cultural value that can have a fundamental strategic value if networked within a vision of public utility. Today, the town is a lively city with a medieval urban structure facing contemporary needs related to resource management, climate change, social cohesion, sustainable tourism and cultural development. The presence of this vast resource of open space can play a key role in addressing many of the challenges for urban well-being and livability, e.g. in relation to the availability of short-range services (15-min city), walkability, circularity of resource production, energy management and the availability of an inclusive and varied cultural offer.

The numerous European funding lines are often a catalyst for research on innovative forms of urban regeneration. In fact, the idea of a project vision for a park of the city walls originated from the Municipality's participation in the Tuscany Region's *Manifestazione di Interesse* (Expression of Interest) for the identification of urban areas and territorial

Fig. 2 Ambrogio Lorenzetti, *Allegoria ed effetti del buono e del cattivo governo* (1338–39). (Credits: CC-BY-NC-ND 4.0)

strategies to support within the framework of the POR-FESR 2021–27 Funding Scheme. This programme is the tool with which the Region contributes to the pursuit of the EU objectives for a climate-neutral economy and a cohesive and inclusive society, consistent with the European Green Deal and the Sustainable Growth Strategy and the Millennium Development Goals promoted by the UN 2030 Agenda. Within this picture, the Municipality of Siena, with the support of the academic spin-off Urban LIFE, presented a territorial strategy that develops a number of project proposals, to transversally address the challenges of social inclusion, security, environmental sustainability and cooperation between territories, as well as to foster innovative ways of providing services for the community through the recovery and valorisation of urban spaces for social and cultural purposes, the reduction of emissions in the environment and the enhancement of ecosystem benefits in the urban context. These proposals were developed in an overall project for a new park of the city walls, a strategic document to coordinate the projects financed by the Region and, in a long-term vision, support for a systemic and coherent transformation of the Sienese territory along three development pillars: urban regeneration and quality of living, culture and environment. Currently, the Municipality is progressing in the development of some of the masterplan proposals, implementing and defining in detail the design solutions for six green valleys.

2 The Centre of Siena, a UNESCO World Heritage Site

The historic centre of Siena was included in the list of World Heritage Sites by UNESCO in 1995, due to its urban and architectural characteristics representing a testimony to the creativity of humans and a tangible expression of their artistic and aesthetic abilities. The UNESCO site Management Plan explicitly refers to the need to actively protect the system of green valleys with a return to their original use as orchards, the system of walls with their redevelopment and cleaning, the system of pedestrian and cycle paths with the creation of connections between the valleys and the historic buildings, and a better wayfinding infrastructure to signal routes to residents and tourists. Going beyond the idea of UNESCO sites as areas defined by a rigid boundary, the position of the walls, intermediate and barycentric with respect to the entire open spaces of the valleys, was understood by the project as a key factor to re-semantise the city walls as a new centrality, with the role of interface between inside and outside lands. Firstly, it is necessary to recognise the potential that open spaces adjacent to historical walls may have to enrich urban public space, taking into account the different roles that walls play in different contexts. The walled city of Siena has a green belt protected from building for reasons linked to geomorphology, historical evolution and far-sighted planning policies implemented in the last century. Today, there is the opportunity to activate this green belt as an area open to the public by identifying solutions for the physical accessibility of the spaces, shared management methods, maintenance plans and the recovery of existing artefacts. Secondly, such design occasions make it possible to intervene on walls and fortifications in historic cities by proposing a sustainable recovery and enhancement method centred on reuse, beneficial from a social and economic point of view, through the participation and involvement of local communities. Active participation means that citizens and users recognise the value of the heritage walls as a cultural-historical memory and as a resource for activating economic sustainability through the attribution of new uses. Finally, the symbolic values of walls have changed over time and may still change in the light of the increasing pressure of urban growth and the global tendency to erect new barriers; in this sense, the objective of developing new ideas to redefine the role of walls as an element not of separation and exclusion, but of inclusion and integration, as an active interface between inside and outside and not as a margin separating cities from their landscape, becomes the main design challenge for urban regeneration in walled cities (Fig. 3).

3 The Masterplan as a Flexible and Adaptive Planning Tool

The design methodology—combining fieldwork, desk research and community participation—informed a strategic masterplan, the planning tool considered most appropriate to address the rethinking of the green wedges as a new urban park. Although masterplans do not have the same prescriptive value as other planning tools in the Italian context, they allow to design a complex project vision which can then be implemented through more standardised planning instruments and procedures or through specific agreements between the actors interested by single actions. In this sense, the masterplan stands as a reference framework for the specific projects and helps to build a coherent overall vision over time. It is a collection of the strategic orientations and project scenarios for the future and defines the time sequence to achieve the planning objectives (Llewelyn-Davies et al. 2000).

The masterplan is divided in two parts, a knowledge and a strategic framework. The knowledge framework is built around five topics—accessibility; habitats; mobility; historical heritage, culture and innovation; ongoing projects on the open spaces of the city walls—and was developed with an applied research methodology, collecting information from fieldwork and deskwork. Fieldwork allowed to analyse the current state of the art and verify the conditions of mainte-

Fig. 3 Siena, the green valleys today (2023). (Credits: Eleonora Giannini)

Fig. 4 Pania Valley during an urban trekking organised by the association Le Mura (2023). (Credits: Sara Naldoni)

Fig. 5 One of the activities of the participatory process by Codesign Toscana (2023). (Credits: Codesign Toscana)

nance of the valleys; it was also an opportunity to activate contacts with the network of local actors with the aim of understanding their needs and defining the possible ways of their involvement in the project. Surveys, field and online interviews to selected stakeholders, explorations, meetings with local representatives and citizens were conducted throughout the whole design process, all aimed at a critical-interpretative reading of this context and the elaboration of an overall vision. The field research was also enriched by the participation to activities guided by local actors, e.g. an urban trekking inside the green valleys organised by Associazione Le Mura (Fig. 4). Regular meetings with the Municipality were organised on a monthly basis, to share the progress of the project and gradually assess the steps of the work. In parallel, deskwork consisted in the systematisation of the existing information database, from textual and cartographic sources, with a specific focus on the system of open spaces within the walls of Siena. The existing planning tools were analysed to understand urban transformation previsions for the valleys, in order to think of how the proposed regeneration strategy could make use of the numerous projects underway at all levels, from structured plans to bottom-up initiatives.

The elaboration of the masterplan was also accompanied by a participatory process, which enriched its contents and tested some of the design hypotheses (i.e. the conditions for the re-opening of some valleys in terms of management and possible use of informal and private existing paths), sharing them with citizens and various stakeholders through public engagement initiatives during the different phases of development. The collaboration with the project partners—CoDesign Toscana and Sinloc, working respectively on the participatory process (Fig. 5) and on the economic development strategies of selected relevant urban polarities—was also a fundamental part of the project.

Such activities contributed both to building the knowledge database and to identifying recurring topics in the different contributions. The topics were organised into seven thematic macro-categories, called Key Themes—culture and tourism, mobility and accessibility, biodiversity, territorial marketing, agricultural production, safeguarding, governance—which have oriented the second part of the masterplan, the strategic framework, containing planning guidelines divided for each valley according to their character and potential, operational indications based on the existing spatial features of the different areas, and a framework of actions, actors and tools for the project implementation.

4 Connessioni Verdi Siena: The ConVerSi Masterplan for the Historical City Walls

The Sienese landscape, in brief, can thus be considered the result of a dual identity: ancient and modern, urban and rural. Operating in such a complex landscape implies an articulated and holistic approach, able to look forward to orient the transformation processes, adapting them to the needs of the local community, but also able to look backwards to grasp all the heritage and cultural aspects of the long history of the city and pass them to future generations. For these reasons, the masterplan is articulated into specific and in-depth investigations that attempt to identify all critical and potential aspects of the historic centre context and of the green wedges themselves. The transfer of contemporary urban quality objectives onto the walled city of Siena has sought to actively valorise the unique characteristics of its landscape formed by the historical walls, the green valleys and the existing multifunctional urban nodes, developing a strategy to improve urban usability and liveability.

4.1 An Integrated Planning and Design Approach for a Heritage Urban Ecosystem

In order to work with the complexity of this historic landscape, the adopted design methodology used the critical lens of ecosystemic urbanism, focusing on the environmental perspective to develop urban development models suited to the challenges of the contemporary city (Agency for Urban Ecology 2022).

The planning guidelines for the regeneration of the green valleys firstly address the topic of accessibility. The concept was studied primarily as accessibility to the whole urban organism: the gates to the city haven't changed from the Middle Ages to the present day, but there are also other possible accesses at the level of the valley floor whose reopening could increase the internal-external permeability at the scale of the entire urban system, with a general benefit for the infrastructural and ecological connections. Another level of accessibility is that of the valleys in relation to the urban fabric since, to date, none of the areas is freely accessible as a public space. Most valleys have internal pathways shaped by current or historical uses, with conditional accessibility depending on their ownership or management (Fig. 6). In addition to providing for the opening of these routes, the project guidelines identify other potential connections, existing informal routes or possible crossings to create a full circuit of pedestrian paths around the green belt. Another accessibility topic is related to the fragmentation of land property around the walls, an alternation of private plots and areas belonging to public bodies or foundations, which leads to differing conditions of accessibility and for which the masterplan guidelines suggest collaborative management tools, like *patti di collaborazione* (collaboration pacts) or conventions to manage the diverse parts of this future urban park. On the other hand, the conditions of limited accessibility have contributed to maintaining the green valleys as an important biodiversity reserve for the city, so the mosaic of open spaces and vegetation structures were mapped identifying habitat categories and main ecological features for each valley: slopes with spontaneous tree and shrub cover; grassland glades on the valley floor with sporadic tree and shrub vegetation, mostly located on the margins; olive-dominated meadows with colonies of pioneer trees and shrubs; areas of water bodies and channels, riparian cover and wetland environments; orchards and gardens with hydraulic-agricultural systems, soil containment structures and vegetation for domestic and hobby management; cultivated areas with a

Fig. 6 Existing paths in the valleys, only occasionally accessible (2023). (Credits: Sara Naldoni)

prevalence of seasonal arable crops, with plots left fallow, frequent mixed cultivation with crops such as orchards and vineyards; areas with buildings, structures, paved surfaces and roads with sporadic presence of vegetation for ornamental purposes. In order to promote the use of the spaces as a public green infrastructure, the project aims to respect and put into value this biodiversity, defining differentiated management for the different habitats, with varying degrees of accessibility. The project also foresees the reuse of the ancient water supply system, a network of underground water channels and historical fountains which, in addition to being an element of historical-architectural attractiveness, can return to its hydraulic function and contribute to the sustainable management of the valleys.

The analysis of current projects and initiatives in Siena—implemented or underway, and financed with different types of funds—has provided an overview of present urban transformations and has highlighted how the context is already very active in sustainable urban regeneration strategies. The stakeholders involved in each of the initiatives were identified, as well as existing management tools, as examples of virtuous governance between different local stakeholders. Finally, the masterplan includes project proposals presented by the *contrade,* collected during the public engagement phase with an open call. The historical quarters of the town were involved by the Sienese municipality and participated in the project by sharing proposals for their areas of interest.

4.2 Planning Guidelines and Design Strategy for a New Park of the City Walls

The planning guidelines of the masterplan are based on four pillars: the maintenance or implementation of spatial, visual and perceptive continuity between urban open space and built space, between the city and the open territory; the networking of existing heritage (cultural, historical, artistic, environmental, etc); the improvement of the functionality of the public space network through new connections in the valleys, new uses, opening of some of the currently closed accesses; the promotion and support of cooperation among local actors, first involved in the co-design of the masterplan strategy and then possibly involved in the everyday co-management of the areas in the future. Such guidelines intersect with the design strategy, which relates to the main emerging elements of Sienese urban space, identifying four macro-systems: the system of the valleys, which includes interventions on the city walls, the ancient gates, the towers, the structures for water collection; the system of functional nodes, which includes interventions to make the valleys suitable for public uses (such as new public spaces around the ancient water sources, the historical gardens, building complexes or architectural artefacts) already characterised or to be characterised as important aggregation and functional nodes; the system of connections, which includes interventions for urban connectivity—mainly pedestrian and cycle paths—and more generally the connections between the valleys with the historic urban fabric and the open territory outside the walls; the system of the valleys, which includes interventions for the redevelopment and re-functionalisation of the open spaces of the valleys themselves (for example, for the agricultural production, production of vegetables, temporary artistic exhibitions) and more generally the system of initiatives aimed at the regeneration of the valleys and the promotion of their attractiveness.

The system of open spaces enclosed in the heritage walls is a mosaic of landscape units with different characteristics, for which the masterplan proposes interconnection as an indispensable and preliminary action on which all the others depend. All the spaces of the park of the city walls were connected with new routes, existing paths made accessible and walkways on the walls, the latter as connections at high altitude but also as a historical-artistic itinerary, alternative to more usual tourist routes. Widening the project scale, this network of paths and accesses along the walls relates to the broader system of territorial connections. In this respect, the masterplan locates new internal axes as possible connections developing radially from the centric urban layout, new paths between the valley floors and the walled city, and new connections transversal to the walls to connect the green valleys inside the historical centre with external valleys.

The masterplan systematised the existing elements of historical, cultural and social quality not only in terms of physical characteristics but also with respect to the activities, uses and initiatives which shape everyday urban life (Gehl 2003): new uses for the open space of the valleys, the reuse of the artefacts of the city walls (such as the towers or the ancient gates) or the regeneration of some open space nodes that acquire a strategic role in the system(such as new public spaces around the historical fountains).

The park is composed of eleven character areas, ten with homologous features due to the geomorphologic conformation, the urban layout or their spatial relation to the walls and a transversal connection. In some cases, the areas coincide with the historical valleys—i.e. green valleys bordered by the walls and included between two ancient gates and the urban fabric of the historic centre—in other cases new networks of valuable urban spaces external to the walls have been identified, which contribute to the overall development vision of the city through the strengthening or creation of new relations and a new axis permeating the characteristic Y-shaped structure of Siena (Fig. 7). Each of these areas is defined by specific characteristics that determined a site-specific design approach and the definition of possible

Fig. 7 10 valleys +1 transversal connection for the city walls park (2023). (Credits: Urban LIFE)

specific management and implementation tools, but at the same time, each of the areas contributes to the global vision of transformation for Siena.

A fundamental condition for a good implementation of these guidelines is a management and maintenance of both the safety and accessibility of the spaces and the vegetation, consistently with the uses (recreational, productive, etc.), with the geomorphologic characteristics (i.e. planting of new trees and shrubs to contrast erosion), with the specificity of the habitats recognising and respecting floral-faunal biodiversity and operating with diversified project actions (i.e. removal of weeds, care of spontaneously grown vegetation). For this reason, project actions and local actors who can support the transformation strategy with different roles have been defined for each area. These tools are useful for the implementation of the masterplan strategy (Fig. 8) and, more specifically, for the implementation of the individual design actions, taking into account the complexity and fragmentary nature of the various contexts investigated and the necessary identification of transformation and management methods alternative to the traditional tools, adaptable over time, based on a form of public-private collaborative governance in order to ensure the economic, social and environmental sustainability of the entire system in the long term.

For each area, the masterplan identified the main value elements and typological interventions (such as possible opening gates, possible paths, etc.), divided into intervention lots. Each lot is characterised by an evaluation of complexity level based on the expected complexity in terms of:

- Administrative-regulatory processes (related to property fragmentation).
- Implementation processes (related to technical complexity).
- Management processes (related to the presence of different stakeholders).

The design strategy also provides a flexible range of design solutions and defines, for each of the six selected implementation areas, suggested solutions in terms of street furniture, pavements, lighting and vegetation management. On the one hand, the intention is to provide a flexible design tool, on the other hand, to ensure a consistent and homogeneous development of the overall park project.

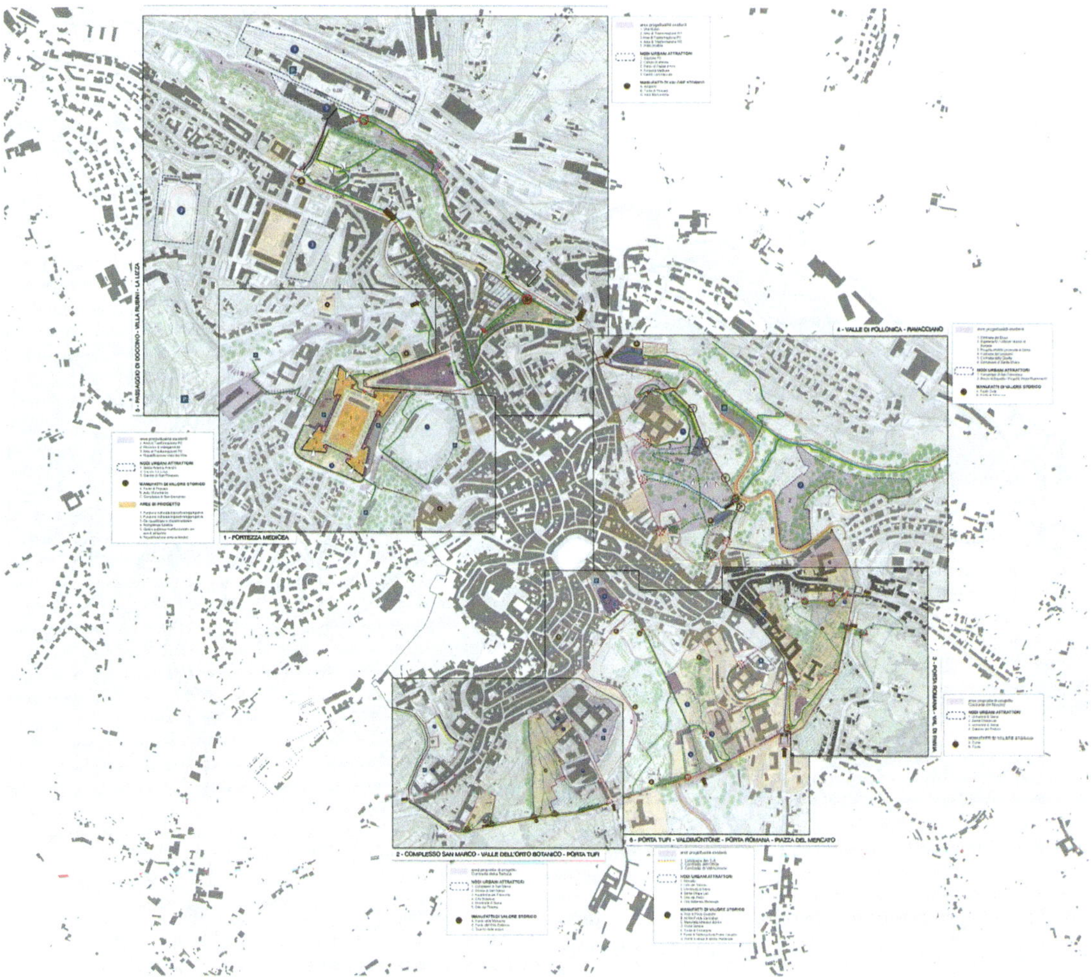

Fig. 8 Implementation plan with interventions for six specific valleys (2023). (Credits: Urban LIFE)

The identification of these innovative tools, in turn, needs close collaboration between the public decision-maker, planners and the actors involved, to be engaged in the design, definition, management and monitoring of results.

5 Results and Conclusions

The themes of sustainable urban regeneration have become structured in Siena's urban development agenda and strategy, with a focus on the cultural uniqueness of the city and its urban landscape. The project attempted to combine the philosophy of urban regeneration, which underlies the idea of renewal and transformation, with the necessary safeguard of the cultural elements of the territory's tradition, which by their nature need to be treated with attention, sharing and participation. The whole process, a creative tension between the impulse to renovation and the focus on tradition, lands in a territory with specific and peculiar settlement characteristics, where there is a close and historical correlation between urban and agricultural settlement systems, from which derives the particular urban configuration of the historical villages built on the ridges of the gentle Sienese hills and surrounded by the mixed territorial layout of the agricultural landscape of the Tuscan countryside.

Historic cities can be resilient and adaptable to contemporary needs, oriented towards sustainability and architectural heritage must be understood as a living part of the city. The results of this experience show that a sustainable project starts firstly from careful and attentive listening to the places, putting into value what is already there. The masterplan provides the framework for the transformations of the green valleys with the awareness that it is not always necessary to intervene drastically, but in most cases it is enough to make the existing and large open spaces accessible and ignite the process of involvement of local actors, irreplaceable carers, when the planners' work is completed.

Since their erection, the walls have been a defence structure but also a connection between the city and the countryside, regulating the balance between the two to guarantee a

Fig. 9 Siena, the Follonica Valley and its spatial continuity with the outside territory (2023). (Credits: Eleonora Giannini)

buon governo (good government). From the idea of heritage walls as an urban device to enclose the city or to regulate its development, to the idea of historical walls as a built element around which a system of diversified open spaces has been maintained to form a green belt, the aim of the project was to turn the walls from margin to urban interface using the availability of open space—as a provision of quality public open space in the dense walled city and as a connection with the open territory—in which the walls are an active device of exchange between inside and outside (Fig. 9).

Acknowledgements The design team Urban LIFE would like to thank the Municipality of Siena, the Associazione Parco delle Mura, Codesign Toscana and Sinloc for almost 2 years of joint work and shared enthusiasm for the project.

References

Agència de Ecologia Urbana de Barcelona (2022) BCN Ecología. 20 anys de l'Agència d'Ecologia Urbana de Barcelona. Arquines Editores, Barcelona

Balestracci D, Piccinni G (1977) Siena nel '300. Assetto urbano e strutture edilizie. Edizioni CLUSF, Firenze

Council of Europe (2000) European Landscape Convention, Florence 20. 10. 2000. European Treaty Series, No. 176

Decreto Legislativo n. 42 of 22nd January 2004. Codice dei Beni Culturali e del Paesaggio. Gazzetta Ufficiale (45). https://www.gazzettaufficiale.it/dettaglio/codici/beniCulturali

Detti E et al (1968) Città murate e sviluppo contemporaneo. Centro internazionale per lo studio della cerchia urbane, Roma

Gehl J (2003) Life between buildings. Danish Architectural Press

Hook J (1979) Siena. Una città e la sua storia. Nuova Immagine Editrice, Siena

Llewelyn-Davies, Baxter and Associates, English Partnerships, Housing Corporation (2000) Urban design compendium. English Partnerships, London

Redon O (1999) Lo spazio di una città. Siena e la Toscana Meridionale. Ed. Viella, Roma

Repetti E (1833–1843/2005) Dizionario Geografico Fisico Storico della Toscana, vol 5. Firenzelibri, Firenze

Secchi B (1988a) Siena. Casabella 545:35–45

Secchi B (1988b) Un progetto per l'urbanistica. Einaudi, Torino

Secchi B (1993) L'importanza della forma. In: Campos Venuti F, Oliva F (eds) Cinquant'anni di urbanistica in Italia. Laterza, Bari, pp 348–360

The Development of Urban Parks as Green Public Space in Colonial Java

Mutiah Amini

Abstract

This paper examines the development of urban parks built in Java from the 1910s to the 1940s. For Java, the construction of urban parks is not entirely novel. Traditionally, Javanese kingdoms have established urban parks since the seventeenth century. These parks were conceived as spaces connecting humans with the Creator, thereby fostering harmonious terrestrial existence. Examples include Taman Sari in the Yogyakarta palace complex and Taman Balekambang in the Surakarta palace complex. However, unlike their royal predecessors, urban parks during the colonial period represented a significant evolution in park development tradition. They not only served as spiritual connective spaces but also symbolized environmental consciousness in response to increasing ecological degradation caused by modernization. Consequently, various parks were designed with hydrological concepts, incorporating infiltration wells or artificial lakes as integral components. The development of hydrologically informed parks gained momentum particularly after the colonial government implemented the municipalities system (*swapraja*) in 1903, granting special authority to several Javanese cities such as Batavia and Semarang to independently plan and manage their regional budgets. To analyze the park planning process and the ensuing contestations, this paper utilizes diverse archival sources from Binnenlands Bestuur, Burgerlijke Openbare Werken (BOW), and Tweede Waterstaat, complemented by contemporaneous newspapers and magazines.

Keywords

Colonial Java · Public Spaces · Urban Planning · Green Spaces · Indonesia

M. Amini (✉)
Department of History, Universitas Gadjah Mada, Yogyakarta, Indonesia
e-mail: mutiah.a@ugm.ac.id

1 Introduction

Cities in Java underwent significant transformations from the nineteenth to the early twentieth century, marked by a strategic shift in urban centrality from traditional Javanese kingdoms such as Surakarta and Yogyakarta to colonial administrative centers like Batavia and Semarang. This transition coincided with intensified industrialization of plantation products and extensive infrastructural development (Nas 1986). The implementation of decentralization policies, beginning with the Decentralization Law (*Decentralitatie Wet*) of 1903 and the Decentralization Act (*Decentralisatie-besluit*) of 1905, proved pivotal in this transformation (Anonymous 1908; Kakebeeke 1932). These administrative reforms empowered colonial municipal governments (*gemeente*) to independently manage their regions and budgets, including the democratic election of council members (*locale raad*), thereby enabling more responsive local planning initiatives. The colonial central government in Batavia envisioned these administrative reforms as a mechanism to develop colonial cities as faithful reproductions of Dutch urban centers.

During the early twentieth century, newly empowered municipal governments undertook comprehensive development programs addressing multifaceted urban needs. These initiatives encompassed organized housing, transportation networks (highways and railways), and essential infrastructure, including bridges, drainage systems, water supplies, markets, hospitals, fire departments, street lighting, and cemeteries (Wertheim 1958). Significantly, the development of public spaces and green areas, particularly urban parks, emerged as an integral component of this holistic approach to urban planning, reflecting both European urban ideals and responses to local environmental conditions.

Scholarship on colonial Javanese urbanism has approached these cities from various perspectives. Jean Gelman Taylor examined Batavia through the lens of European social dynamics in the eighteenth century (Taylor

F. Alberti et al. (eds.), *Towards Sustainable Urban Regeneration: Utilizing Environmental, Cultural, and Social Assets*, Advances in Science, Technology & Innovation, https://doi.org/10.1007/978-3-032-02091-8_7

2009), while James L. Cobban analyzed Semarang's development from the nineteenth to the twentieth century through the prism of political contestation between governmental actors and non-European communities (Cobban 1988). Pauline K.M. van Roosmalen's comprehensive study of colonial urban planning included substantial analysis of Batavia's urban development, including its park systems (van Roosmalen 2008). Environmental dimensions of colonial urbanism in Java have been explored by Peter Boomgaard, who documented how resource exploitation led to deforestation and ecological degradation (Boomgaard et al. 1997; Boomgaard 1999), while Robert Cribb investigated the environmental consequences of industrial expansion in colonial urban centers (Cribb 1988). However, these studies have not specifically addressed how urban parks functioned as both environmental interventions and public spaces within the colonial urban framework.

This article examines the planning and development of urban parks as multifunctional spaces in colonial Javanese cities following decentralization reforms. The study focuses on Batavia and Semarang, selected from among 18 municipalities and 70 townships established under decentralization policies (van der Heiden 1990), due to their distinctive development trajectories. These cities represent contrasting models of colonial urban planning—Batavia evolved as an administrative center aspiring to replicate Amsterdam (Putri 2019), while Semarang developed as an industrial city that pioneered *kampungverbetering* (village improvement) initiatives emphasizing public health and environmental management. Semarang's development notably involved multidisciplinary collaboration among professionals, including H.F. Tillema (pharmacist), W.T. de Vogel (physician), and Thomas Karsten (urban designer) (Amini 2023; van Roosmalen 2008). By analyzing how urban parks were conceptualized and implemented in these distinct contexts, this study illuminates the intersection of colonial governance, environmental consciousness, and public space creation in early twentieth-century Java.

2 The Origin of Urban Parks in Java

At the beginning of the twentieth century, Batavia and Semarang emerged as prominent colonial urban centers, in contrast to earlier Javanese royal cities that were centered around the *keraton* or palace. The *keraton* functioned as the residence of the king, his family, and their retainers (*abdi dalem*). As the heart of the royal city, the *keraton* was complemented by structural elements, including gates leading to the royal center, fortresses, surrounding moats, road networks, markets, mosques, public squares, gardens, residential areas, granaries, and burial grounds, particularly during the Islamic kingdoms era (Adrisijanti 2000).

Currently, several royal cities with discernible historical remnants include Cirebon, Surakarta, and Yogyakarta. These three significant Islamic kingdoms experienced substantial growth between the sixteenth and eighteenth centuries. However, their development gradually diminished with the increasing dominance of colonial administration, which established new colonial cities in the nineteenth century. Within these royal centers, alongside the *keraton*, the park constituted a highly significant public space.

In Javanese kingdoms, particularly during the Islamic period, parks traditionally served as sites for rest and recreation, similar to their counterparts in other major kingdoms worldwide. However, park construction in Javanese kingdoms incorporated a profound spiritual dimension. A distinctive characteristic of Javanese royal parks was the presence of diverse ornamental and fruit-bearing plants alongside a water feature, which typically included a centrally or peripherally located mound or wall, aesthetically arranged to symbolize a mountain. Atop this symbolic elevation stood a small pavilion known as the *bale kambang* (floating pavilion), where the king would engage in meditation (*semedi* or *tapa*) to cultivate a spiritual connection with the Creator (Lombard 2008).

Several remnants of historical parks that symbolically illustrate the connection between water and mountains remain visible today in Tasik Ardi (Keraton Surosowan in Banten), the royal parks of Cirebon (Keraton Kasepuhan and Keraton Kanoman), Sunyaragi (Cirebon), Baleratna (Kasunanan Surakarta), and Taman Sari (Kasultanan Yogyakarta) (Behrend 1984). Examining the nomenclature and design patterns of these parks, which served as the king's spiritual retreats, it is evident that they functioned as private royal spaces specifically constructed for royal use.

The concept of parks as private royal spaces has persisted. However, with modernization's influence, Javanese kings gradually began establishing parks for public use. These public parks were typically constructed separate from the royal grounds. A notable example is Taman Sriwedari, whose name in Javanese conveys the concept of a celestial park (*taman kahyangan*), serving as a space to experience worldly tranquility while fostering a connection with the Creator. This park was officially inaugurated in 1899 during King Paku Buwono X's reign (Swastika et al. 2022).

As illustrated in Fig. 1, Taman Sriwedari features diverse flowering plants and thriving trees, with lotus plants adorning the clear-water pool (*segaran*) that contains a centrally positioned pavilion. This pavilion functions both as a visitor resting area and as a meditation site. Unlike the exclusive royal parks, public spaces such as Taman Sriwedari host various activities throughout the year, including religious celebrations, art festivals, and sporting events (Asal-usul Taman Sriwedari 1928).

Fig. 1 The pool of Taman Sriwedari, with a relaxation pavilion. (Source: Asal-usul Taman Sriwedari [Origins of Taman Sriwedari], Kajawèn, 487)

3 Urban Parks in Colonial Cities

The policy changes in land ownership and indigenous labor utilization in the mid-nineteenth century indirectly transformed colonial economic policies. The colonial government granted European private companies opportunities to manage indigenous lands for plantation purposes, coinciding with the Agrarian Law (*Agrarische Wet*) enactment in 1870. This period marked the beginning of land liberalization in the Dutch East Indies (Boomgaard 2004). The colonial government also permitted Europeans to establish processing factories for plantation produce. The plantation industry in the Dutch East Indies produced major export commodities, including sugar, coffee, tea, rubber, cinchona, and tobacco for European, American, and Asian markets (Kartodirdjo and Suryo 1991).

The increase in plantation commodities for export necessitated transportation infrastructure development, such as roads and ports. The colonial government collaborated with private companies to construct roadways and railway lines for transporting industrial goods to major ports (Geertz 1966). Consequently, the colonial government established the Bureau van Openbare Werken (Public Works Bureau), which later evolved into the Departement van Burgerlijke Openbare Werken (BOW) or the Department of Civil Public Works in 1866. As a government agency, the BOW planned and constructed government buildings, residential areas, roads, railway lines, bridges, ports, water supply systems, sanitation facilities, irrigation systems, and dams (Arsip Nasional 2015).

With the delegation of authority from the central government to local governments in 1905, the BOW established offices in colonial cities, including Batavia and Semarang. Through decentralization, regional BOW offices under local government jurisdiction gained independent authority to manage construction and maintenance budgets. By the end of 1909, the Surabaya City government had paved roads totaling 64,100 m with an approximate area of 299,900 m^2. Meanwhile, the Semarang City government allocated *f*11,796.35 for road paving and existing road repairs (Anonymous 1911).

Industrial growth attracted foreigners, particularly Europeans, to the Dutch East Indies for skilled labor positions or professions requiring specialized expertise. Decentralization led to significant population increases in the four major cities from 1900 to 1925: Batavia (including Meester-Cornelis area) by 130% and Semarang by 100% (Kerchman 1930). This gradually transformed traditional cities into colonial urban centers. The expanding foreign population required housing when their employing companies did not provide accommodation. Housing provision became a critical urban planning issue. City governments collaborated with architects to develop residential areas with supporting facilities, such as the Candi hillside neighborhoods (Fig. 2)

Fig. 2 New city in Candi, Semarang, with a row of native residents' houses at the bottom. (Photo: Charls & Co., in *Nieuw-Semarang De Heuvelstad* [New Semarang the Hill City], 1918)

in South Semarang (Anonymous 1938). Architect Thomas Karsten, for example, designed a new city in Candi, Semarang, by integrating European and traditional Javanese elements (Cobban 1992).

Decentralization granted autonomous regions the authority to govern and manage their cities and financial budgets, including urban development plans. However, construction implementation did not always align with paper plans. In Semarang, for instance, many new buildings could not be immediately occupied, or design concepts failed to fully meet prospective residents' expectations. According to Karsten, successful development required urban construction to reflect and express the social life of both groups and individuals (Karsten 1919; Kerchman 1930).

To evaluate decentralization implementation, the colonial government organized a congress in Semarang from May 26 to 28, 1938. During the congress, the government documented various infrastructures in the municipalities, as shown in Table 1.

In the early twentieth century, local governments began to change their approach to urban development, considering

Table 1 Infrastructures in Batavia and Semarang in 1938

No.	Infrastructure	Batavia	Semarang
1.	Area (in hectares)	18,260	9940
2.	Built-up area (in hectares)	8300	3637
3.	Average population density in built-up area (per hectare)	65.39	62.49
4.	Area under maintenance (in hectares)	64.66	1640
5.	Road area (in hectares)		
	a. Main roads	222.7	–
	b. Sidewalks	13	8.5
	c. Roadside verge	159.50	–
	d. Squares and public parks	–	5.9
6.	Length of drainage channels and flushing pipes (in kilometres)		
	b. Unlined flushing pipes	50	4.35
	c. Idem	50	27.49
	d. Unlined drainage pipes	400	76.67
	e. Idem	–	226.90
	f. Sewer	140	25.38

Source: H. Westbroek, Openbare Reining van Bevolkingscentra in Nederlandsch-Indie [Public Cleansing of Population Centers in the Dutch Indies], 1938 not only physical and health aspects but also factors of comfort. This means that creating habitable and prosperous residential areas became a consideration in city planning. As an ecosystem, urban development should consider a forward-looking vision: dense office and commercial spaces, alongside quiet, healthy, and well-maintained residential areas; lush vegetation; availability of sports facilities and gathering places; and wide roads with busy traffic (Kerchman 1930).

Critical facilities that must be included in the planning and development of a city are open spaces where its residents can socialize. In this regard, urban parks play a crucial role, even though residential areas may have yards or gardens. Public spaces like city parks are not merely recreational areas but also places for relaxation, music, sports, gardening, and water absorption areas (Budiman 2021). One essential requirement for improving the welfare of residents is the availability of extensive recreational spaces. The development of city parks and sports facilities is concentrated in several easily accessible points for the entire population (Anonymous 1938).

4 Parks as Urban Landscapes

The Dutch East Indies' natural beauty prompted Treub, an Amsterdam City Council member, to dub it "Mooi Indie" (literally "pleasant Indie") during his visit (Treub 1923). The region's flourishing plants and trees enhanced urban areas, exemplified by the Ciliwung River basin in Meester-Cornelis (in Batavia) and the hills of Candi in Semarang's expansion. Prioritizing environmental preservation, city governments meticulously planned serene residential areas, with careful consideration given to building placement, infrastructure development, land zoning, tree planting, and public space provision (Anonymous 1938).

In more developed colonial cities, municipal governments planted trees along streets extending to the periphery, creating landscapes where trees visually dominated buildings (Kerchman 1930). To enhance urban aesthetics, local governments of autonomous regions developed squares and parks as public spaces, including Wilhelminapark and Waterlooplein in Batavia, and Raadsplein and Vogelplein in Semarang.

4.1 Batavia

Wilhelmina Park (Wilhelminapark) was a prominent public space in Batavia, situated in Weltevreden (Fig. 3). Weltevreden emerged as an expansion of Batavia, where the colonial government center of the Dutch East Indies was established, initially planned during Governor General Willem Daendels' tenure. Eventually, Weltevreden was connected to the Meester-Cornelis area south of Batavia (De Stadsgemeente Batavia 1937). Before Weltevreden's development, a defense center named Fort Prins Frederik (Citadel Prins Frederik) existed there, commemorating Prince Willem Frederik Hendrik's visit to the Dutch East Indies from February to September 1837 (Prins Willem Frederik Hendrik Der Nederlanden in Indië 1937). Prince Frederik himself laid the foundation stone on August 30, 1837. The naming of this fort coincided with the naming of Prince Hendrik Fort in Surabaya, Prince of Orange Fort in Semarang, and Willem I Fort in Ambarawa (Oost-Indie 1837).

During its era, Wilhelmina Park stood as the largest park in the Dutch East Indies (Fig. 4). Located along a tributary of the Ciliwung River, it featured diverse plants and trees. Beyond aesthetic enhancement, the park functioned as a water absorption area. To mitigate flooding in Weltevreden, the government constructed a sluice gate (*sluis*) on the park's northern edge (Fig. 5).

> *Hier zijn wij weder aan het water, dat de gracht straks van Noord- en Rijswijk, tevens van Molenvliet, vult, niet echter dan voor, door een sluis te zijn geloopen, waarover een brug geworpen is die natuurlijk de Sluisbrug heet. Onstuimig bruischt hier opeens de stille stroom tusschen de steenen muren. Alweder een aardig, levendig gedeelte van Weltevreden, de Sluisbrug, waaraan een klein plein, het groene Wilhelminapark, grenst, eender nieuwste parken van Weltevreden, gevormd van de oude Citadel.*

Here we find a waterway that will soon fill the canals of Noordwijk and Rijswijk, as well as Molenvliet, passing through a water gate, over which a bridge called Sluisbrug has been built. Suddenly, the calm flow of the river rushes vigorously between the stone walls. Once again, it is a beautiful and bustling part of Weltevreden, the Sluis bridge, bordered by a small square, the green Wilhelmina Park, one of the newest parks of Weltevreden, formed from the old citadel (Verster and Zeggelen 1921).

Opposite Wilhelmina Park lay an open field named Waterlooplein (Waterloo Square). On Waterlooplein's eastern side, the colonial government constructed a palace (*groote huis*) to replace Batavia's existing town hall. This building later became the Department of Finance offices. The government also built a new palace for the governor-general north of Koningsplein, as part of a complex that included the former Hotel Rijswijk (later housing the General Secretariat office). Additional palaces were established in Bogor and country retreats in Cipanas, Priangan (present-day West Java) (Anonymous 1894). While Wilhelminapark served as a recreational area for relaxation and free music performances, Waterlooplein functioned as a venue for ceremonies and military parades.

In response to rapid urban growth, particularly the proliferation of city parks, the City Works Department established a public park supervisor following a proposal by the City

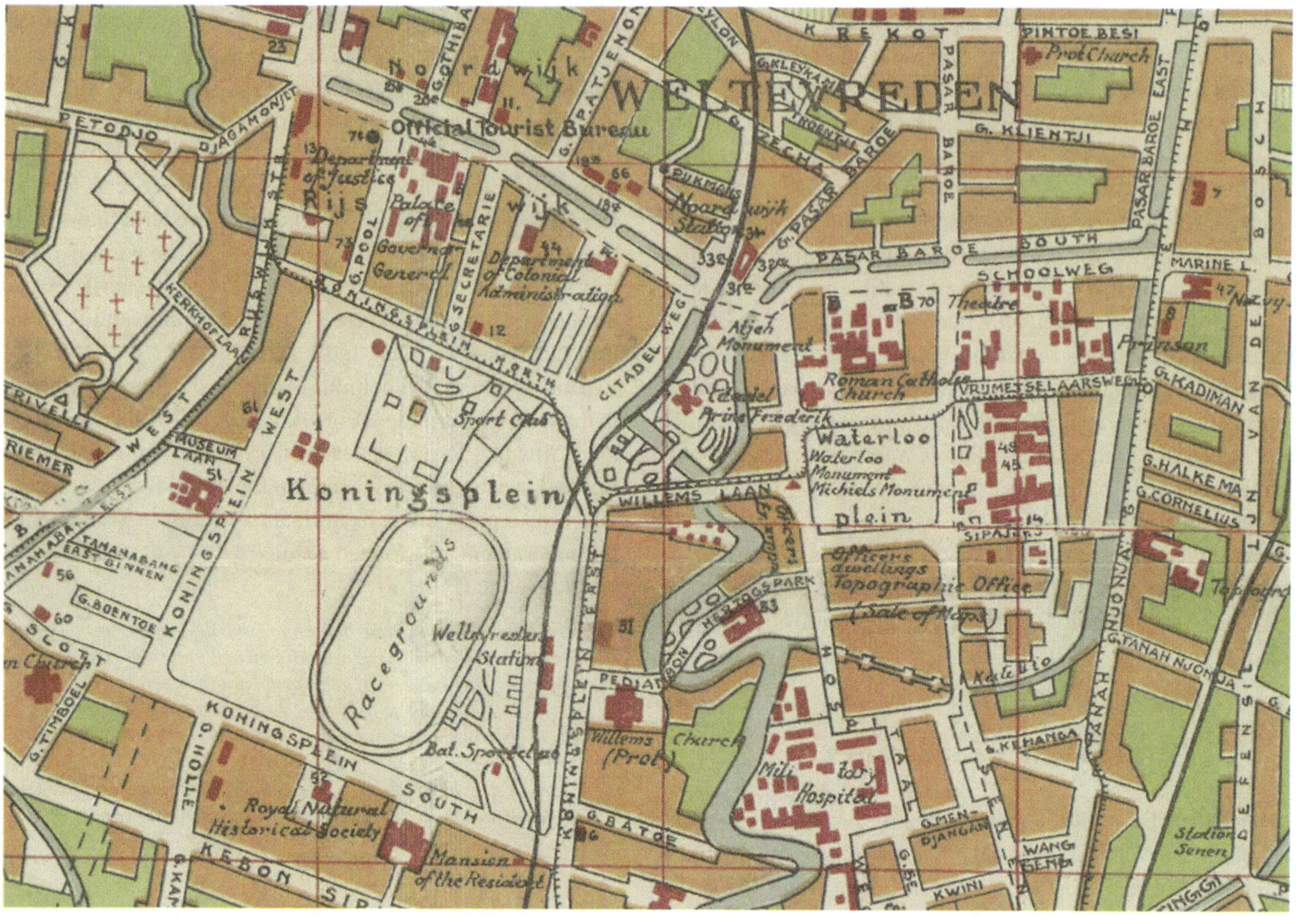

Fig. 3 Map of Weltevreden, Batavia, 1930. (Source: Leiden University Libraries Digital Collections, D E 20.4)

Council during the 1921 municipal budget discussions. Beyond Wilhelminapark's development, the government created new parks to enhance the city's aesthetic appeal, such as those at Burgemeester Bischopplein (Menteng) and Koningsplein. They further planted shade trees along residential streets and adorned lawns and green spaces with colorful flowers and ornamental shrubs (Eggink 1930).

4.2 Semarang

While the colonial government developed the new city of Weltevreden, the Semarang City government initiated the construction of a new settlement in the hilly region of Candi (Fig. 6). During this period, like many cities in the Dutch East Indies, Semarang faced challenges with slums characterized by poor sanitation, exacerbated by outbreaks of cholera, malaria, and dysentery. Well water used for drinking was contaminated with cholera bacteria. In response, Dr. W. de Vogel and H.F. Tillema, members of the City Council, proposed constructing a new, hygienic, and planned city called Nieuw-Semarang. This proposal generated debate among council members. Despite anticipated challenges, the City Council ultimately endorsed Dr. Vogel's proposal to build the new area (van Geuns, 1910a, b).

> *Een ieder, die Semarang van voor 1915 kent en er thans komt, wordt, zoodra hij het Zuidelijke gedeelte der stad bezoekt, getroffen door de reusachtige veranderingen, die daar in de laatste jaren hebben plaats gehad. Op de plaats van een, slechts voor een wandelaar to bereiken braak liggend terrein, is daar een nieuwe stad, het "Nieuw-Semarang" verrezen.*

Anyone familiar with Semarang before 1915 who visits today, particularly the southern part of the city, will be surprised by the significant changes that have taken place there in recent years. What was once barren land accessible only by footpaths has now been transformed into a new city known as "Semarang Baru" (Anonymous 1918).

To realize Nieuw-Semarang, the Semarang City government, with city council approval, engaged the architectural bureau of Henri MacLaine-Pont, where architect Ir. Thomas Karsten worked after completing his education in the Netherlands and returning to Semarang. Planning the new city in the hilly Candi area, with its varying slopes, required careful consideration, as developing a hygienic and economically viable residential area necessitated thorough technical studies for effective implementation. Urban development

Fig. 4 Wilhelmina Park, 1908. (Source: E.J. Eggink, *"Na 25 jaar" Beknopt Gedenkschrift ter gelegenheid van het 25-jarig bestaan der Gemeente Batavia 1905–1930* ["After 25 years," Concise Memoir on the Occasion of the 25th Anniversary of the Municipality of Batavia], 110)

Fig. 5 Water gate (*sluis*) to prevent flooding in the Weltevreden area 1908 and 1925. (Source: Leiden University Libraries Digital Collections, KITLV 1406241)

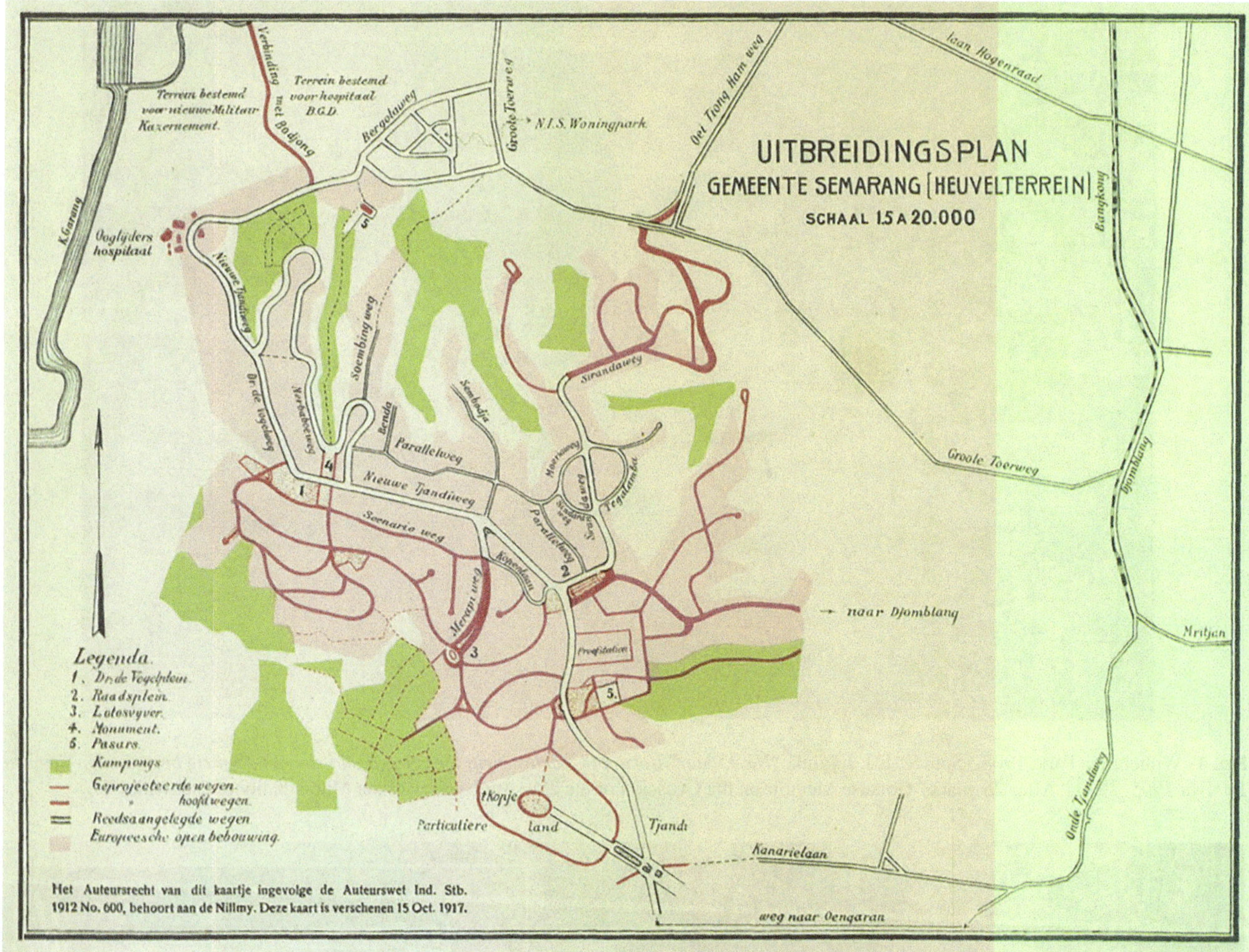

Fig. 6 Expansion plan for the Municipality of Semarang (hill site), 1917. (Source: *Nieuw-Semarang De Heuvelstad* [New Semarang the Hill City], 1918)

encompassed not only aesthetics but also technical factors and solutions. Planning and constructing housing, roads, drainage, and other supporting infrastructure required integration into a comprehensive plan (Anonymous 1917; Cobban 1992).

Ir. A. Plate initially drafted the technical plan for developing the new city across approximately 400 ha, which he subsequently discussed with Karsten. After several revisions, Karsten formulated the development plan for the Candi area and new settlements for indigenous people, Chinese ethnic groups, and less affluent Europeans in Lower Semarang, implemented by the city government. In his planning, Karsten anticipated future needs, particularly focusing on road access from Lower Semarang to Upper Semarang (Candi). Four main wide roads leading to Candi would converge at a central point known as Raadsplein, encircled by a five-way intersection. From the Candi area, a road would lead toward the main Semarang-Ungaran-Ambarawa road, with several connections to Lower Semarang.

Raadsplein was designed as a green open space and public area featuring a fountain pond, plants, and flowers. Shade trees were planted along the roads surrounding Raadsplein. Water from the Raadsplein pond flowed through pipes to the lotus pond (Lotosvijfer) located in the valley, where it emerged as a fountain. An open grassy field named after Dr. Vogelplein was constructed on the northwest side along the road towards Bojong (Fig. 7). Banyan trees, pine trees, and flowering plants adorned Vogelplein, alongside a monument inscribed with the name of Dr. De Vogel, one of the initiators of the new city (Anonymous 1917).

The residential area featured rainwater drainage and wastewater channels from houses. Rainwater was directed through specialized channels to fill ponds in parks and supply fountain sources. As development progressed, maintaining water purity in the Raadsplein and Lotosvijfer ponds required additional effort. During heavy rainfall, water entering the ponds often carried mud, complicating the separation of rainwater from household wastewater. Sediment and the

Fig. 7 (**a**) Raadsplein; (**b**) Vogelplein; (**c**) Lotosvijfer; (**d**) Trapaanleg. (Source: *Indisch Genootschap*: *Vergadering van 16 October 1917* [Indisch Society: Meeting of 16 October 1917])

mixture of rainwater and wastewater polluted the pond water. Consequently, channels leading to the ponds were sealed to prevent them from becoming mosquito breeding grounds (Stadsnieuws: Het Heuvelterrein 1921).

To ensure the health standards of the residential area, the water supply system did not rely on well water. The water company piped clean water to every house through the Ungaran-Semarang water pipeline network. Specifically for the New Candi area, clean water sourced from mountain springs in Ungaran was channeled through a distribution reservoir located in Kepoh. From the Kepoh Reservoir, clean water was distributed through the Candi area to the Siranda Reservoir, which supplied clean water to western Semarang (Schuil 1932).

5 Conclusion

The local governments in Batavia and Semarang demonstrated significant commitment to colonial urban planning. A vision of urban landscapes that harmonized human habitation with environmental considerations constituted a central focus in their planning endeavors. This approach substantially influenced the creation of green open spaces, utilizing effective air circulation where water bodies and vegetation played crucial roles in shaping urban environments. Such systematic planning was inextricably linked to municipal governance structures. This historical approach provides valuable insights for contemporary urban studies, emphasizing the importance of structured and responsive city planning by local authorities to create comfortable urban spaces for residents, yielding long-term benefits that remain evident today.

Furthermore, the arrival of colonial architects in various colonial cities, particularly Batavia and Semarang, significantly influenced detailed urban planning. Supported by robust regulatory frameworks, these architects were empowered to design urban spaces alongside the strengthening autonomy of local governments in city planning. These conditions ultimately facilitated the development of city parks in Batavia and Semarang, forming integrated urban ecosystems whose legacies persist in the present. This demonstrates that urban development, when supported by a comprehensive planning system and implemented holistically, holds profound implications for environmental sustainability. Regrettably, the planning initiatives established during the colonial period were not maintained by subsequent administrations, resulting in the diminishment of environmental health benefits that originated from colonial urban planning in Batavia and Semarang. This serves as a critical lesson for future urban studies, underscoring the importance of state involvement in comprehensive and sustainable urban planning. Without governmental engagement across all planning dimensions, meaningful benefits for natural ecosystems remain unattainable.

Acknowledgement This article is the result of research funded by the Ministry of Education, Culture, Research, and Technology (RISTEK DIKTI), 2023–2024.

References

Adrisijanti I (2000) Arkeologi Perkotaan Mataram Islam. [Urban Archeology of Mataram Islam]. Jendela

Amini M (2023) Clean water supply and urban hygiene control in colonial Semarang. Plan Perspect 39(4):835–851. https://doi.org/10.1080/02665433.2023.2283836

Anonymous (1894) Verslag over de Huisvesting van Zijne Excellentie den Gouveeneük-Generaal in Nederlandsch-Indie 1892–1893 [Report on the housing of his excellency the governor-general in the Dutch Indies 1892–1893]. Martinus Nijhoff

Anonymous (1908) Verzameling van Decentralisatie Bepalingen Tweede Druk [Collection of decentralization provisions second edition]. E. Fuhri & Co.

Anonymous (1911) Verslag over de Burgerlijke Openbare Werken in Nederlandsch-Indië over het Jaar 1909 [Report on civil public works in the Dutch Indies for the year 1909]. Landsdrukkerij

Anonymous (1917) Indisch Genootschap Vergadering van 16 October 1917 Bl. 1–29: Uitbreidingsplan de Indische Gemeente [Indisch Society: Meeting of 16 October 1917, Bl. 1–29, The expansion plan of the Indische Municipality]. Martinus Nijhoff

Anonymous (1918) Nieuw-Semarang De Heuvelstad [New Semarang the hill city]. Nillmij

Anonymous (1938) Toelichting op de Stadsvorming Ordonnantie Stadsgemeenten Java [Explanation of the urban development ordinance for Java City municipalities]. Landsdrukkerij

Arsip Nasional Republik Indonesia [National Archives of the Republic of Indonesia] (2015) Sejarah Organisasi Departement van Burgerlijke Openbare Werken [History of Department of Civil Public Works]

Asal-usul Taman Sriwedari [The Origin of Sriwedari Park] (1928) Kajawèn, 485–488

Behrend TE (1984) Kraton, Taman, Mesjid: a brief survey and bibliographic review of islamic antiquities in Java. Indonesia Circle. School Orient Afr Stud Newsl 12(35):29–55. https://doi.org/10.1080/03062848408729594

Boomgaard P (1999) Oriental nature, its friends and its enemies: Conservation of nature in late-colonial Indonesia, 1889–1949. Environ Hist 5(3):257–292. https://doi.org/10.3197/096734099779568245

Boomgaard P (2004) Anak Jajahan Belanda: Sejarah Sosial dan Ekonomi Jawa 1795–1880 [Children of the Dutch colony: history of Javanese social and economy 1795–1880]. Djambatan dan KITLV

Boomgaard P, Colombijn F, Henley D (1997) Paper landscapes: exploration in the environmental history of Indonesia. KITLV Press

Budiman, H. G. (2021). Jejak Manajemen Air di Masa Hindia-Belanda: Identifikasi Saluran Air di Kawasan Stasiun Bogor [Trace of water management system in the Netherland-Indie period: identification of water channels in the Bogor Station Area]. Jurnal Panalungtik, 4(2), 97–114. https://doi.org/10.24164/pnk.v4i2.67

Cobban JL (1988) Kampungs and conflict in colonial Semarang. J Southeast Asian Stud 19(2):266–291. https://doi.org/10.1017/S0022463400000564

Cobban JL (1992) Exporting planning: the work of Thomas Karsten in colonial Indonesia. Plan Perspect 7(3):329–344. https://doi.org/10.1080/02665439208725753

Cribb RB (1988) The politics of environmental protection in Indonesia. Monash University

De Stadsgemeente Batavia [The City Municipality of Batavia] (1937) Batavia als Handels-, Industrie- en Woonstad [Batavia as A Commercial, Industrial and Residential Center]. G. Kolff & Co.

Eggink EJ (1930) "Na 25 jaar": Beknopt gedenkschrift ter gelegenheid van het 25-jarig bestaan der Gemeente Batavia ["After 25 years," Concise memoir on the occasion of the 25th anniversary of the municipality of Batavia]. Gemeente Batavia

Geertz C (1966) Agricultural involution: the processes of ecological change in Indonesia. University of California Press

Kakebeeke HC (1932) Gedenkboek der Gemeente Semarang 1906–1931: Uitgegeven ter Gelegenheid van het VIJF en Twintig Jarig Bestaan der Gemeente [Memorial book of the municipality of Semarang 1906–1931: published on the occasion of the five and twenty years of existence of the municipality]. De Locomotief

Karsten T (1919) Opmerkingen over de ontwikkelingsmogelikheid der inheemsche bouwkunst [Remarks about the possibilities of developing an indigenous architecture]. Bandung Institute of Technology

Kartodirdjo S, Suryo D (1991) Sejarah Perkebunan di Indonesia: Kajian Sosial Ekonomi [History of plantations in Indonesia: socio-economic study]. Aditya Media

Kerchman FWM (1930) 25 Jaren Decentralisatie in Nederlandsch-Indië 1905–1930 [25 years of decentralization in the Dutch Indies 1905–1930]. G. Kolff & Co.

Lombard D (2008) Garden in Java. Pusat Penelitian dan Pengembangan Arkeologi Nasional

Nas PJM (1986) The Indonesian city: studies in urban development and planning. KITLV Press

Oost-Indie [East-Indie] (1837) Leydsche Courant, 2

Prins Willem Frederik Hendrik der Nederlanden in Indië [Prince Willem Frederik Hendrik of the Netherlands in the Indies], 10 Februari–30 September 1837 (1937) Historia: Maandschrift voor Geschiedenis en Kunstgeschiedenis, 34

Putri PW (2019) Sanitizing Jakarta: Decolonizing Planning and Kampung Imaginary. Plan Perspect 34(5):805–825. https://doi.org/10.1080/02665433.2018.1453861

Schuil BJ (1932) De Waterleiding [The Clean Water]. Locale Techniek 3:35

Stadsnieuws: Het Heuvelterrein [City news: The Heuvelterrein] (1921). De Locomotief, 5

Swastika ND, Aliyah I, Yudana G (2022) Kajian Perkembangan Ruang Publik Bersejarah di Pusat Kota, Studi kasus: Taman Sriwedari sebagai Kebun Raja di Kota Surakarta [Study on the development of historical public space in the city center, Case study: Taman Sriwedari as King's Garden in Surakarta]. Region: J Reg Dev Participatory Plan 17(1):43–54. https://doi.org/10.20961/region.v17i1.34239

Taylor JG (2009) Kehidupan Sosial di Batavia: Orang Eropa dan Eurasia di Hindi Timur [Social life in Batavia]. Komunitas Bambu

Treub MWF (1923) Nederland in de Oost [Nederland in the East]. Tjeenk Willink & Zoon

van der Heiden CN (1990) Town planning in the Dutch Indies. Plan Perspect 5(1):63–84. https://doi.org/10.1080/02665439008725695

van Geuns M (1910a) Koloniaal-staatkundige vlugschriften III: De nooden van Semarang en de werking der decentralisatie [Colonial-political pamphlets III: the needs of Semarang and the effect of decentralization]. Soerabajasch Handelsblad

van Geuns M (1910b) Semarangsche Kronieken [Semarang Chronicles]
van Roosmalen PKM (2008) Summary, Ontwerpen aan de stad: Stedenbouw in Nederlands-Indië en Indonesië 1905–1950 [Summary, planning the city, town planning in the Dutch East Indies and Indonesia 1905–1950]. Doctoral thesis, Delft University of Technology, p 226
Verster JFLB, Zeggelen MCK (1921) Ons Mooi Indië: Batavia Oud en Nieuw [Our beautiful Indies: Batavia old and new]. J.M. Meulenhoff
Wertheim WF (1958) The Indonesian town: studies in urban sociology. The Hague
Westbroek H (1938) Openbare Reiniging van Bevolkingscentra in Nederlandsch-Indie [Public Cleansing of Population Centers in the Dutch Indies]. In Technische Mededeeling Vereeniging voor Locale Belangen, No. 11 Prae-Advies, published for the 28th Decentralization Congress in Semarang, 26–28 May, annex 21. Technisch Tijdschrift

New Design Approaches for Urban Voids: Developing a Framework for Analysing Innovative Projects in Terrain Vague Spaces

Lorenzo Stefano Iannizzotto, Alexandra Paio, and Camilla Perrone

Abstract

In recent years, there has been a growing interest from researchers, academics, and architects in a new type of space generated by the invasiveness of urbanization processes, the lack of an adequate scale of regional planning, and the cycles of sprawl and shrinking. Many authors have attempted to define these spaces: Terrain Vague, Vacant Land, Urban Voids, Wasteland, or Brownfield, are just some of the terms used to define and describe abandoned, underdeveloped, and unutilized urban spaces outside the productive cycles of the city. In addition to the interest of the academic world, there is a growing emergence and multiplication worldwide of alternative projects, practices, and activities based precisely in these in-between or terrain vague spaces, which aims to both change and transform these spaces, and to preserve and enhance some of their intrinsic characteristics. In fact, by alternative projects, the authors refer to projects that both act by transforming the space and preserving the value and potential of terrain vague spaces, aiming to address the challenges of cities and achieve sustainable urban development. However, it is possible to state that there is a lack of systematic research with the aim of collecting, analyzing, and categorizing these emerging projects. The objective of this paper is the creation of a theoretical framework to identify, collect, and analyse alternative projects and practices emerging in the Terrain Vague context, and the development of the criteria to analyse for conducting a future comparative analysis, this would support the understanding of a list of differences, similarities, approaches, objectives, impacts, strategies, challenges, and difficulties. To achieve that, the paper proposes: (i) A brief theoretical introduction and definition of the concept of terrain vague inclusive of the ecological and social value; (ii) Introduction to the concepts and theoretical lenses chosen as criteria to select alternative and innovative projects, which can both enhance the values of Terrain Vague spaces and contribute to urban challenges (urban commons, everyday urbanism, temporary urbanism); (iii) Definition of the characteristics and aspects to be analysed for conducting a comparative analysis of these innovative projects (Scale/Duration, socio-ecological aims, level of participation, Existing/Flexibility).

Keywords

Terrain vague · Urban voids · Everyday urbanism · Urban common · Tactical urbanism · Participatory processes

1 Introduction

In contemporary cities, there are abandoned open spaces without a specific function that seem to lie in a perpetual state of waiting and in-between, ambiguous and difficult to define. This interest has led to a re-evaluation of these spaces, previously perceived negatively or merely as gaps to be filled. Due to the highly diverse characteristics and uncertain status of these spaces, numerous definitions and concepts have been developed to describe and define them: Terrain Vague (Mariani and Barron 2014; de Solà-Morales 1995), Vacant Land (Bowman and Pagano 2004), Urban Voids (Lopez-Pineiro 2020), Wasteland (Gandy 2013, 2022), or Third Landscape (Clément 2022). Additionally, a wide range of disciplines has shown interest in studying them (Lopez-

L. S. Iannizzotto (✉)
DINÂMIA'CET-Iscte, ISCTE-University Institute of Lisbon, Lisbon, Portugal
e-mail: Lorenzo_Stefano_Iannizzotto@iscte-iul.pt

A. Paio
DINÂMIA'CET-Iscte, Department of Architecture and Urbanism (ISTA), ISCTE – University Institute of Lisbon, Lisbon, Portugal
e-mail: alexandra.paio@iscte-iul.pt

C. Perrone
Department of Architecture, University of Florence, Florence, Italy
e-mail: camilla.perrone@unifi.it

F. Alberti et al. (eds.), *Towards Sustainable Urban Regeneration: Utilizing Environmental, Cultural, and Social Assets*, Advances in Science, Technology & Innovation, https://doi.org/10.1007/978-3-032-02091-8_8

Pineiro 2020). For the sake of brevity, the authors adopt the term Terrain Vague (de Solà-Morales 1995), intentionally encompassing a broad variety of spaces under this definition. This approach reflects the focus of the article, which lies not so much in the physical and spatial characteristics of individual spaces, but rather in the potential that is activated by innovative projects. It is possible to define these spaces as underdeveloped and underutilized open areas, often in a state of abandonment and waiting, lacking an official productive function, and frequently awaiting future development. More recently, increasing research has focused on demonstrating the significant social and ecological value and potential of these spaces, describing them as rare opportunities for addressing urban challenges (Anderson and Minor 2017; Gandy 2022; Kamvasinou and Roberts 2014; Kim 2016). Paradoxically, part of the great value of these spaces lies in their state of abandonment and the absence of human control, which allows for the emergence of spontaneous and informal activities, as well as the spread of weeds and species not permitted elsewhere, providing refuge for biodiversity. However, as Ignasi de Solà-Morales (1995) noted in his definition of *Terrain Vague*, a new approach is necessary for designing and planning in these spaces without destroying or devaluing their essence, an approach that manages to preserve and enhance the value of existing spontaneity and informality. Over the past 20 years, it has been possible to observe the emergence of formal projects carried out in Terrain Vague spaces (Fig. 1) with an innovative approach aimed at valuing and preserving the existing value of these spaces (Bertrando and Imma 2021). These projects, such as urban community gardens created in previously abandoned spaces (Fig. 1), operate a subtle transition between what already exists and the future project design, challenging traditional categories of formal and informal, spontaneous and planned, temporary and long-term, and bottom-up and top-down. Although these projects and practices are often well-known and widely publicized, perhaps not enough has been done or published in terms of analytical study and systematization of innovative projects in Terrain Vague spaces. We believe that such systematization is essential. It could make a significant contribution to advancing knowledge in the field of urban planning. This process would help to: identify the types of practices and activities most carried out; it would also allow for an understanding of how these vary based on the type and condition of the original space; additionally, it would provide insights into how these practices are implemented (that includes examining the legislative context, the relationship with the public sector, funding sources, strategies used, actors involved and intended objectives). Finally, it could help establish methods to evaluate the impact and benefits of these practices. Such an evaluation would also consider the original value of Terrain Vague spaces.

This paper aims to lay the groundwork for an initial attempt to formulate a theoretical framework that describes, analyses, and systematizes innovative projects and practices in Terrain Vague spaces, to be used for the future creation of a project taxonomy. This paper includes: a brief theoretical overview of terrain vague spaces, their value, and the theoretical concepts used to interpret practices (from which the selection criteria for projects and practices will be derived); a hypothesis on the information deemed relevant to analyse for each project (which, along with the theoretical framework, will provide the criteria for project analysis); a conclusion and suggestions for future research necessary to develop the framework and taxonomy.

2 Towards a New Socio-Ecological Approach

With the key characteristics of Terrain Vague spaces established, their social and ecological value can be highlighted. Drawing on academic theory and literature, we will propose indicative principles for a new design approach to these spaces that seeks to preserve some of their essential features while enhancing and activating their potential through formal interventions, activities, and projects. The idea is that these types of projects, which operate via a gentle transition from a state of informality and spontaneity to a planned one, can play an important role in addressing the urban challenges of the future (Lokman 2017; Németh and Langhorst 2014). As Terrain Vague spaces are intersections of social and ecological values and interests, authors from different disciplines have outlined the principles of a new approach to these spaces, favouring either a social or ecological approach. A new approach capable of truly activating the enormous potential of these spaces should consider and balance the two components: a new socio-ecological approach, having "different priorities: indeterminacy, less control, layers of memory of previous activities" (Kamvasinou 2006). By social approach, we broadly refer to a range of values and interests that are diverse but nonetheless connected to social and human dimensions. These include, for instance, cultural, creative, educational, historical, and heritage aspects, as well as aesthetic considerations and even economic interests (such as food production or the creation of green spaces in disadvantaged neighbourhoods). These aspects are understood in a social context, such as urban metabolism or spatial justice. Similarly, when we broadly refer to ecological aspects, we use the term to encompass all elements related to the environment, ecosystem services, biodiversity, and the prevention of environmental disasters. First, a new approach to these spaces should start from the knowledge and valorization of what already exists and takes place in these spaces

Fig. 1 Festival in a community urban garden in Lisbon, or *Regador*, created in a former abandoned space

daily, in line with the principles of everyday urbanism (Chase et al. 1999; Speaks et al. 2005), without considering them empty spaces, devoid of life, a tabula rasa. In this way, a relationship of continuity with the space, with the activities and actors present, would be favoured. This may include various aspects: the residents or communities that use it daily should be actively involved in the diagnosis and analysis phases, or in the choice of future functions; likewise, the plants and spontaneous species present on the site could be integrated into the new project, instead of being destroyed.

This would allow the future of the project to maintain a high rate of diversity, understood as diversity of functions and users, but also as biodiversity. Time should be regarded as a crucial factor in two key ways: both temporary intervention and a flexible approach. First, these spaces can be activated, or their potential explored, through temporary and ephemeral interventions. Examples include activities rooted in tactical or insurgent urbanism (Hou 2010). Such initiatives can stimulate locally based projects or grassroots associations, which may later be integrated into long-term planning (Kamvasinou 2017, 2020). Second, preserving the indeterminacy and vagueness of Terrain Vague spaces requires a flexible approach. The project should avoid predefining and predicting everything. Instead, it can focus on designing for the indefinite, leaving room for future appropriations that are unpredictable at the time of planning. Additionally, these spaces could continue not to be oriented for profit or market, but rather to serve the community and residents, for example by enhancing or incentivizing community uses and by granting the management of these spaces directly to local residents and associations, conceiving them as *Urban Commons* (Dellenbaugh-Losse et al. 2018; Urban Commons Research Collective 2022), as in the case of urban community gardens.

3 Methodology

The methodology consists of a selection of key concepts and texts, based on a literature review of Terrain Vague spaces, aimed at developing a theoretical framework and criteria for analysis and observation of projects in these spaces. The goal of the paper is to propose a theoretical framework that addresses the three research questions: (i) What characteristics of the terrain vague have been preserved? (ii) How have they been preserved? (iii) What characteristics have been enhanced? The framework should be regarded as a preliminary outcome of the research. It aims to provide an initial draft that addresses the research questions and to explore and describe if and how the projects have enhanced the potential of Terrain Vague spaces. For the review of academic literature, the authors initially conducted a systematic review by performing keyword searches in databases such as Scopus. However, due to the wide variety of terms and definitions used to describe these spaces—each valid and often referring to the same types of spaces—and the broad range of disciplinary fields and branches of knowledge involved, the authors opted for an alternative methodology. The authors selected key texts or collections recognized as essential for the study of these spaces. Based on these, they identified additional texts by the same authors, works cited in the bibliographies of these key texts, and related publications referencing them in databases such as Scopus, employing a methodology known as snowballing. Through this approach, the authors were able to curate a sample of texts representing diverse authors, perspectives, and disciplines relevant to the subject of study. The first part of the literature review aims to select the theory and identify the criteria for both project selection and analysis. The characteristics and essence of Terrain Vague spaces are briefly described, highlighting three fundamental characteristics of these spaces. Following this, the value and potential of these spaces are discussed, particularly emphasizing the relationship between social and ecological interests, which should be preserved or enhanced through formal project intervention. Based on the literature, some key design principles for these spaces are outlined. The resulting framework proposed includes: (i) Project selection criteria; (ii) Criteria and parameters for comparative analysis and comparison between projects; (iii) A theoretical framework for an initial systematization of these spaces, towards a future possible taxonomy, visualized in graphical form. The aim of the framework is to conduct a content analysis of projects and practices in Terrain Vague spaces. It seeks to understand and describe if and how the selected projects have enhanced and preserved the value and potential of these spaces by implementing an innovative approach that integrates both ecological and social aspects.

4 Review of the Literature

4.1 Terrain Vague

Over the past 30 years, many disciplines (social sciences, architecture and urbanism, geography, biology, economics, art) have shown increasing interest in a new type of abandoned urban open spaces. Numerous authors have described their characteristics, proposing various definitions, such as: Terrain Vague (Lévesque 2001; Mariani and Barron 2014; de Solà-Morales 1995), Vacant Land (Bowman and Pagano 2004), Urban Voids (Careri 2006; Lopez-Pineiro 2020), Brownfield or Wasteland (Gandy 2013), and Third Landscape (Clément 2022), in-between spaces or urban interstices (Brighenti 2013). Among these definitions, the authors opt for the term Terrain Vague because, for the first time, it considers these spaces positively, evoking the need for a new approach to design and outlining their characteristics. Terrain Vague spaces are undesignated and abandoned open spaces without any specific and productive function, without human control, which seem to lie in a perpetual temporary state of waiting for new plans, projects, or developments. In these spaces, which may have different origins (Kamvasinou 2011), the spontaneous, the ephemeral, and the unexpected prevail over the controlled, the planned, and the lasting. Due to their characteristics and status, Terrain Vague spaces are sites of spontaneous appropriation and informal use by very different residents and communities, and they are also rich in biodiversity. Terrain Vague spaces host very different communities, species, functions, and activities that coexist. Moreover, they are the site of temporary and ephemeral informal activities and uses, and always remain open and available for new future uses. Finally, these spaces are not recognized as productive from a market and profit point of view, but they can be productive, generating spontaneous appropriations that encourage informal contacts between residents and users of the space, and often lead to its common management. To conclude, we summarize the three characteristics of terrain vague spaces: (i) Diversity or Continuity; (ii) Openness or Temporary; (iii) Unproductiveness and Commoning.

4.2 Social and Ecological Value

One of the most interesting aspects of Terrain Vague spaces is that they are a rare convergence of both social and ecological values and interests in an urban context (Lopez-Pineiro 2020). In fact, the absence of human control and designated productive functions generates a condition of greater freedom than other traditional urban spaces, e.g., squares or gardens. There is an extensive academic literature on the value and ecological benefits generated by the presence of Terrain Vague spaces and their benefits on the urban ecosystem (Anderson and Minor 2017; Gandy 2013, 2022; Kim 2016; McPhearson et al. 2013; Soares et al. 2017; Twerd and Banaszak-Cibicka 2019). For instance: it is shown how these spaces are among the richest in biodiversity, due to the absence of human control; they are often green spaces with vegetation, which perform a number of ecological functions and ecosystem services (such as rainwater absorption, air quality improvement, cooling, and resistance against landslides); although small, they can also function as a green corridor system, which allows the passage and communication of animal and plant species. Finally, it is now known that the presence of green spaces increases the quality of urban life and generates social and psychological benefits, especially in dense neighbourhoods or those with fewer public green spaces (Nunes et al. 2021). Due to their immediate availability, these spaces are sites of spontaneous appropriation by different residents and communities, of informal, ephemeral and temporary activities and functions, sometimes not possible in other public open spaces (Anderson and Minor 2017; Kamvasinou 2011; Kim 2016; Mariani and Barron 2014; Németh and Langhorst 2014; Pagano and Bowman 2000).

The use of space by different users and communities, in the absence of control and pre-established rules, often encourages forms of contact and self-organization of local communities, which can lead to the strengthening of neighbourhood ties. Furthermore, the presence of certain functions and activities in these spaces can be an indicator of the lack of these functions in other dedicated spaces in the neighbourhood. It is as if these spaces absorb the immediate needs of the community, as a kind of catalyst (Kamvasinou and Roberts 2014). Finally, there is a growing interest and aesthetic appreciation for these spaces, which may also have an important didactic and educational value (Gandy 2013; Soares et al. 2017).

5 Results

As a result, a preliminary drafted theoretical framework that selects, analyses, and systematizes innovative projects in Terrain Vague spaces is developed and presented. Once the selection criteria for the projects are established and outlined, the framework includes criteria and aspects necessary to analyse and compare in the chosen projects, with the aim of deriving lessons, principles, and innovative strategies for projects in Terrain Vague spaces. As depicted in Fig. 2, the analytical framework derives from the theory of project selection criteria and then proceeds with the criteria to analyse and compare in the selected projects. The criteria for selecting projects considered innovative are: (i) Projects implemented in terrain vague spaces that aim to preserve and enhance at least one of the essential characteristics of these spaces (namely diversity, flexibility, openness, and *commoning* practices); (ii) Projects that explicitly combine social, ecological, economic, and cultural objectives with the aim of contributing positively to sustainable urban development and future urban challenges; (iii) Projects that realize and implement some of the principles identified as guiding principles for a new socio-ecological approach to these spaces. After defining the selection criteria, it will be necessary to review publications, articles, websites, and journals to find projects that meet these criteria. Once the projects are selected, they must be studied, described, and then analysed and compared to provide useful information on their replicability, introduced strategies, and to carry out a preliminary systematization. The theory presented in the theoretical background will provide guidance on the most important aspects to observe. The criteria and parameters for a comparative analysis of the selected projects are: the state of the Terrain Vague space (derelict land, abandoned space, previous uses); scale and duration; social, ecological, and economic aims and objectives (in line with sustainable development goals and main functions and activities); maintenance and enhancement of existing activities or elements; the level of openness and spatial flexibility planned for unforeseen future activities; the relationship between local authorities and public institutions and the active participation of residents or local associations (specifying the type of public support, for instance funding); the type and level of participation in the different phases (diagnostic, proposal, implementation, and management). The diagram in Fig. 2 represents the stated criteria and their relationships.

6 Conclusion and Discussion

Terrain Vague or Vacant urban spaces represent a rare opportunity for regeneration projects and future urban challenges, but they are also fragile areas at risk of disappearing (Lopez-Pineiro 2020). Urban projects and the transition to formal projects in these spaces must adopt a new approach that enhances existing assets and integrates social and ecological interests to avoid erasing or diminishing their potential (de Solà-Morales 1995). In recent years, a series of innovative projects have implemented this new approach, applying some of the principles outlined in theory. However, there is insufficient academic literature on the categorization and systematization of these projects, which would allow for the deduction of their potential, strategies, and benefits. Bridging this gap is both urgent and highly significant. The traditional approach to these spaces is often destructive, disregarding their qualities and undermining the potential of Terrain Vague. The emergence of innovative practices signals promising new possibilities and opportunities. However, without a systematic collection and study of these practices, it becomes more challenging for them to be adopted as standard practices by local political authorities and to be integrated into conventional planning tools. This article seeks to take a preliminary step in this direction by proposing an initial theoretical framework that selects, analyses, and categorizes innovative projects and practices in vacant urban spaces, along with criteria for project selection and comparative analysis. The framework should not be considered complete or definitive, but rather a first attempt at systematizing these projects. The authors believe that these spaces can play an important role in addressing future urban challenges and that the development of this framework can aid in a greater understanding of these emerging practices. The framework presented in this article is part of a broader research effort by the authors on innovative projects and practices in these spaces and will need to be tested, validated, and possibly revised and refined by more scholars in the future. The intention is that it will eventually lead to a comprehensive taxonomy and categorization of innovative practices in terrain vague spaces.

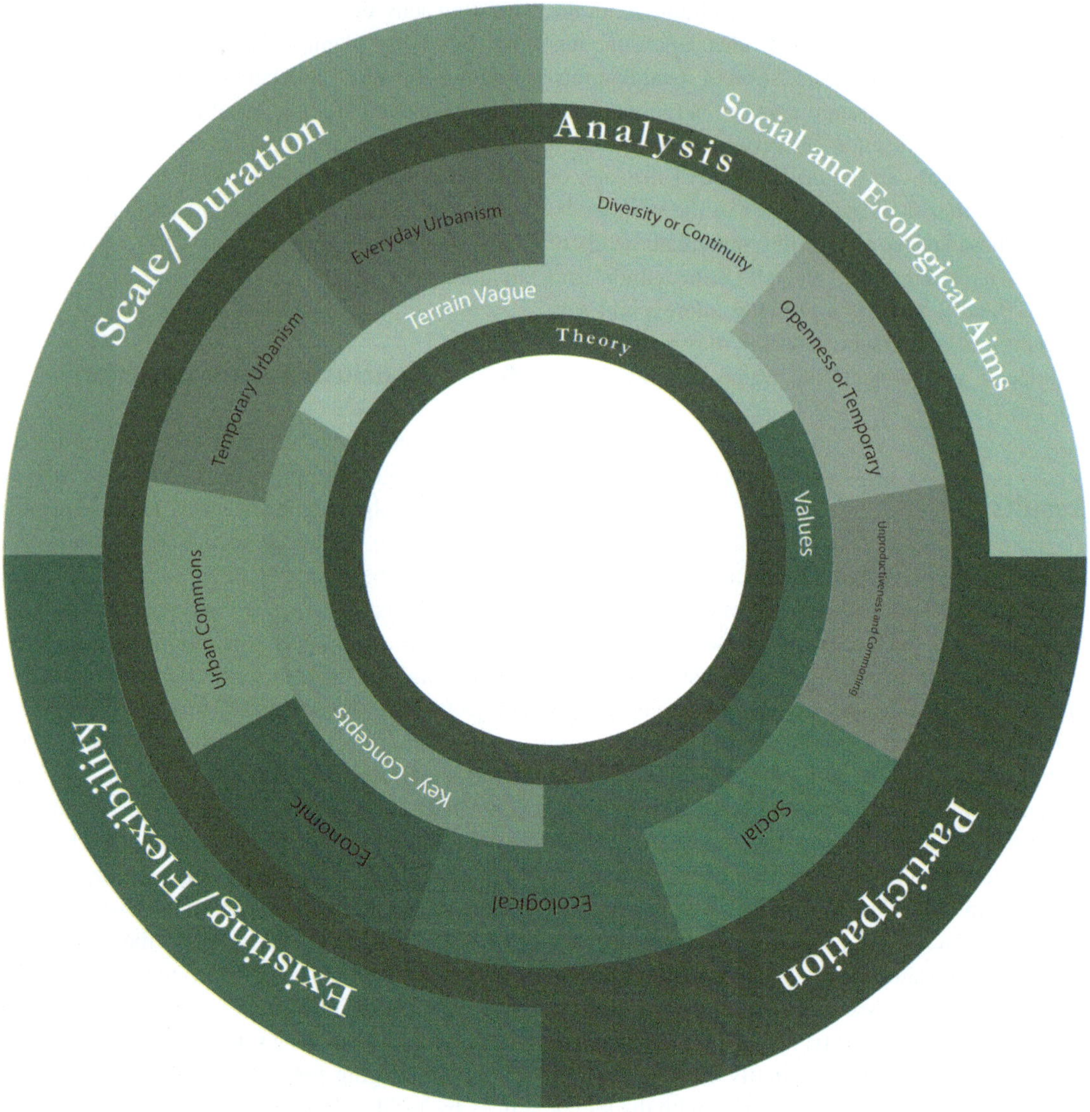

Fig. 2 Framework diagram: visualization of the concepts presented in the paper

Acknowledgements This work was supported by FCT – Fundação para a Ciência e Tecnologia, I.P. by project reference 2022.11783.BD and DOI identifier https://doi.org/10.54499/2022.11783.BD and the project UID/3127/2025 and supported by DINÂMIA'CET-Iscte – Centre for Socioeconomic and Territorial Studies. The author sincerely thanks The Foundation for Science and Technology (FCT) for funding the author's research grant with the reference 2022.11783.BD, and DINÂMIA'CET-Iscte—Centre for Socioeconomic and Territorial Studies for supporting the author's research activities. In addition, the authors would like to thank Krystallia Kamvasinou for the valuable insights on Terrain Vague spaces.

References

Anderson EC, Minor ES (2017) Vacant lots: an underexplored resource for ecological and social benefits in cities. Urban For Urban Green 21:146–152. https://doi.org/10.1016/j.ufug.2016.11.015

Bertrando B, Imma F (2021) Urban interstices in Italy. Design experiences. LetteraVentidue

Bowman AO, Pagano MA (2004) Terra incognita: vacant land and urban strategies. Georgetown University Press

Brighenti AM (2013) Urban interstices: the aesthetics and the politics of the in-between. Ashgate

Careri F (2006) Walkscapes. Camminare come pratica estetica. Einaudi

Chase J, Kaliski J, Crawford M (1999) Everyday urbanism. The Monacelli Press

Clément G (2022) Manifesto of the third landscape. Trans Europe Halles

de Solà-Morales I (1995) Terrain vague. In: Davidson C (ed) Anyplace. MIT Press, pp 118–123

Dellenbaugh-Losse M, Zimmermann N-E, de Vires N (2018) The urban commons cookbook

Gandy M (2013) Marginalia: aesthetics, ecology, and urban wastelands. Ann Assoc Am Geogr 103(6):1301–1316. https://doi.org/10.1080/00045608.2013.832105

Gandy M (2022) Natura Urbana. Ecological constellations in urban space. MIT Press

Hou J (ed) (2010) Insurgent public space: Guerrilla urbanism and the remaking of contemporary cities. Routledge

Kamvasinou K (2011) The public value of vacant urban land. ICE Municipal Eng 164(3):157–166. https://doi.org/10.1680/MUEN.9.00020

Kamvasinou K (2017) Temporary intervention and long-term legacy: lessons from London case studies. J Urban Des 22(2):187–207. https://doi.org/10.1080/13574809.2015.1071654

Kamvasinou K (2020) Experiencing interim landscapes: ephemerality and place-making. In: Revealing change in cultural landscapes: material, spatial and ecological considerations, pp 34–45

Kamvasinou K, Roberts M (2014) Interim spaces: vacant land, creativity and innovation in the context of uncertainty. In: Mariani M, Barron P (eds) Terrain vague: interstices at the edge of the pale. Routledge, pp 187–200

Kamvasinou K (2006) Vague parks: The politics of the late twentieth-century urban landscapes. *Architectural Research Quarterly*, 10 (3–4): 255–262.

Kim G (2016) The public value of urban vacant land: social responses and ecological value. Sustainability (Switzerland). https://doi.org/10.3390/su8050486

Lévesque L (2001) The "terrain vague" as material: some observations. Paysages, (Newsletter of the Association Des Architectes Paysagistes Du Québec), Montréal, pp 16–18

Lokman K (2017) Vacancy as a laboratory: design criteria for reimagining social-ecological systems on vacant urban lands. Landsc Res 42(7):728–746. https://doi.org/10.1080/01426397.2017.1355446

Lopez-Pineiro S (2020) A glossary of urban voids. Jovis

Mariani M, Barron P (2014) Terrain Vague. Interstices at the edge of the pale. Routledge

McPhearson T, Kremer P, Hamstead ZA (2013) Mapping ecosystem services in New York City: applying a social-ecological approach in urban vacant land. Ecosyst Serv 5:11–26. https://doi.org/10.1016/j.ecoser.2013.06.005

Németh J, Langhorst J (2014) Rethinking urban transformation: temporary uses for vacant land. Cities 40:143–150. https://doi.org/10.1016/j.cities.2013.04.007

Nunes N, Björner E, Hilding-Hamann KE (2021) Guidelines for citizen engagement and the co-creation of nature-based solutions: living knowledge in the URBiNAT project. Sustainability (Switzerland) 13(23). https://doi.org/10.3390/su132313378

Pagano MA, Bowman AO (2000) Vacant land in cities: an urban resource. Brookings Institution, Center on Urban and Metropolitan Policy, December, 1–9

Soares AL, Azambuja ST, Brito-Henriques E, Simões AR (2017) Vacant land in city: potential functional, ecological and aesthetic role in the urban landscape. In: ECLAS conference 2017/proceedings, creation/reaction/

Speaks M, Crawford M, Kelbaugh D (2005) Everyday urbanism: Michigan debates on urbanism I. Mehrotra R (ed). University of Michigan, Taubman College of Architecture and Urban Planning

Twerd L, Banaszak-Cibicka W (2019) Wastelands: their attractiveness and importance for preserving the diversity of wild bees in urban areas. J Insect Conserv 23(3):573–588. https://doi.org/10.1007/s10841-019-00148-8

Urban Commons Research Collective (U. C. R.) (2022) Urban commons handbook, 1st edn. Dpr-Barcelona

Evaluating the Ecosystemic Function of the "Model Cities" to Lead Contemporary Urban Sustainability

Xanthippi St. Dereka

Abstract

The scope of the research is the creation of comparable standards to face urban sustainability. The research adapts the guiding lines of the European Environmental Agency (EEA), concerning the proper way, to a holistic approach to urban challenges. An analysis should be conducted within a five-chain effect framework known as Drivers, Pressures, State, Impacts, and Responses (DPSIR). The suggested approach of EEA involves comparisons of indicators between cities to promote urban sustainability. The method of city comparisons is promoted by EEA and EC (European Commission), also applied by Urban ecology that uses greener areas as standards for comparisons to evaluate the environmental condition of an area. Identifying the non-existence of sustainable models for the proper comparisons, the research is exploring the model cities through time, which aimed to serve common causes, sustainability as a solution to various social, environmental, economic, and political issues. The method applied concerns the evaluation of the ecosystemic function of the known "city models" or Utopias. The function is analyzed in a frame in the following fields: aim, duration, role of municipality, morphology of the model city, flows, out-flows, products, social characteristics, intellectual and social development, living standard, population, number of houses, area size, and causes of implementation or failure. Selective utopian city models are applied in the DPSIR framework of EEA and evaluated. The five-face chain reaction framework will actually enlighten the reasons why urban sustainability, through time, failed or not, and the ways or standards to get there. The conclusion, among others, refers to the perennial need and effort of mostly non-experts in solving social, environmental, economic, and political issues, through urban proposals and creation of sustainable city models.

Keywords
Urban sustainability · EEA tools · Urban ecosystem · City model · Urban evaluation

1 Introduction

The research aims to find the common characteristics of urban ecosystems, implemented or just envisioned, to create standards for comparisons with contemporary cities, targeting urban sustainability. The evaluation of the function will be done with contemporary tools of the EEA (European Environmental Agency).

Sustainability is a widespread and misunderstood term, even though it concerns a large scale of interest, from social groups to political "round tables" at municipal and state level (EEA 1997).

Sustainability is the ability of a system to self-sustain. According to EEA (1997) and Urban ecology (Sukopp 2002), sustainability and the urban circular ecosystem are identical terms. An ecological approach is the consideration of a city as an ecosystem, which, however, is characterized by its own history and functions (Sukopp 2002; Dereka 2023a, b)

The EEA, responsible for the environmental reports of the European local authorities, proposes tools to analyze and respond to urban challenges (EEA, 2005, 2020), (Dereka 2023a, b). The basic tool of EEA is a framework of the 5 faces chain effect, which is based on the ecosystemic function. It started with 3 phases in 1979 as a "Stress-State-Response" (EEA 2003). The EEA framework to which the proper indicators should be applied by the local authorities may differ according to the challenges.

The method of indicator comparison is promoted by EC (European Commission), (2018), promoted and imple-

X. St. Dereka (✉)
Department of Architecture, University of Patras, Patras, Greece

F. Alberti et al. (eds.), *Towards Sustainable Urban Regeneration: Utilizing Environmental, Cultural, and Social Assets*, Advances in Science, Technology & Innovation, https://doi.org/10.1007/978-3-032-02091-8_9

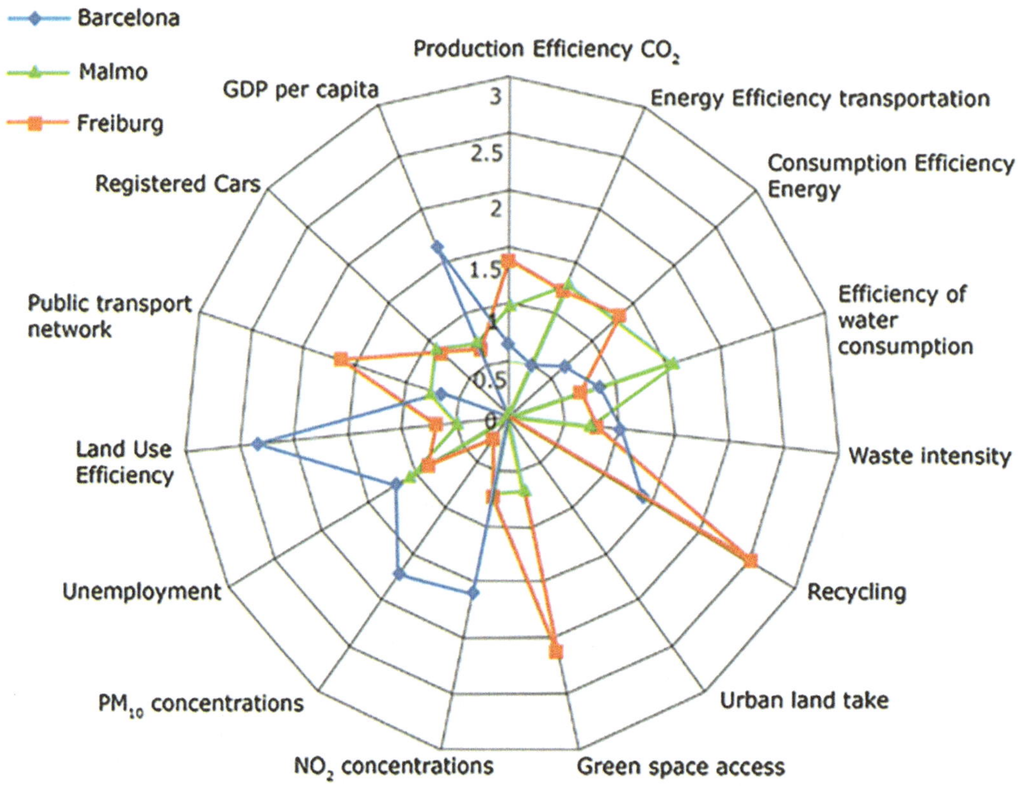

Fig. 1 The urban metabolism framework, EEA. (Source: EC 2018)

mented by the EEA (Fig. 1) (1999, 2003, 2014, 2020), promoted and implemented by Urban Ecology which uses areas with exceptional environmental conditions as standards for comparisons (Sukopp 2002).

The EEA states the difficulty in creating and analysing sustainable indicators (EEA 2014) that connect environment, society, and economic issues (Dereka 2023a).

Identifying the need for sustainable standards for comparisons, to lead the contemporary cities to sustainability, the research is investigating and evaluates 21 of the known "model-cities" or else known Utopias, from Plato (380 B.C.) to Frank Lloyd Wrigh (1932), as ecosystems using the EEA, DPSIR (Drivers, Pressures, State, Impact, Response) framework and typology.

2 Methodology

The selected 21 model cities are categorized and analyzed in 6 chronic periods and 11 fields. The need to create models initiated from ancient times. The research is investigating their common characteristics through time to reach the common goal, sustainability.

The 21 models are analyzed as ecosystems, applied to the 5-chain DPSIR framework of EEA.

The five phases of the EEA (Fig. 2) framework are (EEA 1999, 2003, 2014), (Dereka 2023a):

2.1 First Phase: Drivers (D)

The forces (D) push or create the occasion for the subsequent pressure (P). They can be social, demographic movement, economic development, change in lifestyles, consumption, or resource production.

The Utopists, model visioners, down the centuries were educated, highly politicized, wealthy, restless, well-travelled, some factory owners, and rarely architects or urban planners.

The ideal model city could concern a description without urban drawings.

The (D) *driving force* (Table 1) of the visioners is the "sustainable community." The forces are social and political. Aiming to improve the quality of life, to declare a philosophy that surely influenced the next generations of utopians and the contemporary urban fabric. There are cases, like Le Corbusier, where the function of the urban fabric has a competitive role.

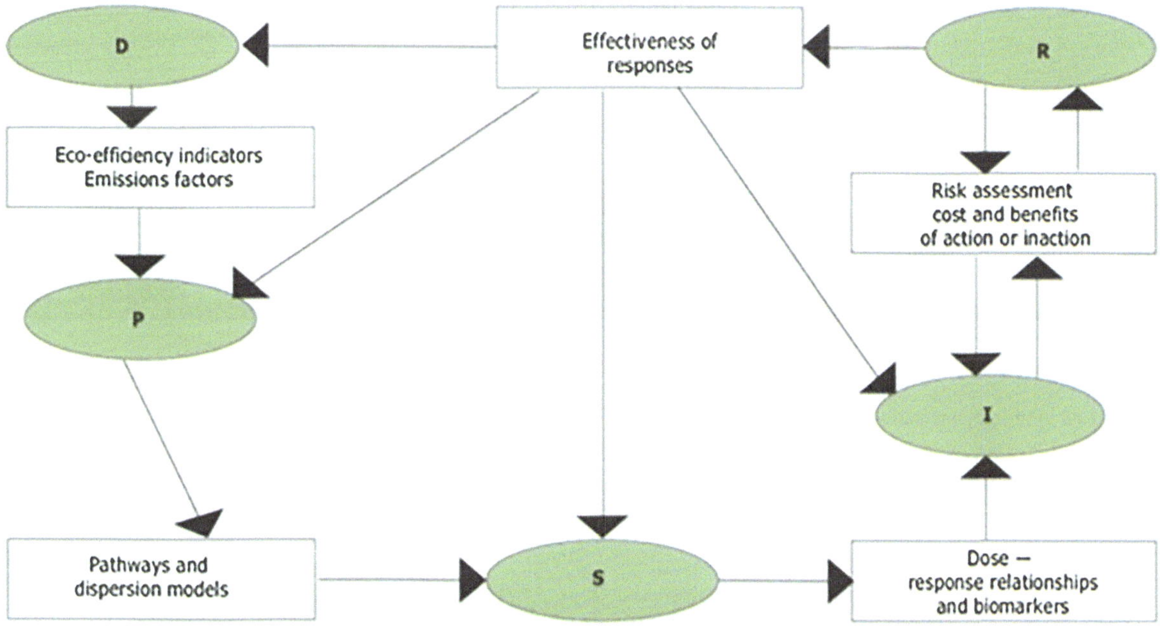

Fig. 2 The five phases of the EEA. (Source: EEA 1999, 2003, 2014)

Table 1 The aim of each visioner. From ancient to modernism. Own processing

	Ancient times	Middle Ages	14th and 15th "Late Middle Ages" the poverty increased greatly	Renaissance "humanism tradition" engagement with the "ideal city", strict geometry					18th century Pre-industrial era	19th century, the golden age of utopia- Utopian socialism								20th century - Modernism: the evolution of utopias.			
	1. Plato (380 B.C.) Greek philosopher	2. Marcus Vitruvius Pollio (80–70 B.C. – after 15 B.C.) Roman architect & military engineer	3. Leon Battista Alberti (1404–1472), 1452, Renaissance humanist author, artist, architect, poet, priest, linguist, philosopher, cryptographer	4. Leonardo da Vinci (1452 – 1519), painter, draughtsman, engineer, scientist, theorist, sculptor, architect	5. Thomas More (1477-1535), 1516, lawyer, judge, social philosopher, author, statesman, amateur theologian, Renaissance humanist	6. Tommaso Campanella (1568-1639), italian Dominican friar, philosopher, theologian, astrologer, poet	7. Johann Valentin Andreae (1586–1654), 1619, Theologian, social works, prominent member of the Protestant utopian movement	8. Francis Bacon (1561 –1626), 1626, English philosopher & statesman	9. Claude-Nicolas Ledoux (1736 - 1806), salt mine inspector, architect, utopian	10. Owen (1771 – 1858), 1800, owner of a textile factory, Welsh textile manufacturer, philanthropist & social reformer, a founder of utopian socialism & co-operative movement	11. Owen, 1824	12. Charles Fourier (1772-1837), 1822, French philosopher, influential early socialist thinker, one of the founders of utopian socialism	13. Jean Baptiste Andre Godin (1817 – 1888) factory owner, French industrialist, writer, political theorist, social innovator, manufacturer of cast-iron stoves	14. Etienne Cabet (1788-1856), 1840, French philosopher, utopian socialist founder of the Icarian movement.	15. James Silk Buckingham (1786-1855), 1849, journalist, publisher, politician	16. William Morris (1834-1896), 1890-factory owner, English textile designer, poet, artist, writer, socialist activist associated with the British Arts & Crafts movement.	17. George Cadbury (1839-1922), 1893, industrialist, English Quaker businessman and social reformer	18. Ebenezer Howard,(1850-1928), 1898 & 1902, English urban planner and founder of the garden city movement	19. Tony Garnier (1869- 1948), 1918, 1920. French architect and city planner	20. Le Corbusier (1887 -1965), 1924 pioneer of the Modern Movement in architecture	21. Frank Lloyd Wrigh, (1867-1959), 1932. American architect, designer, writer, educator
Aiming for	Well-being of the whole community	Principles Dreamed of rebuilding the city of Rome, realize a fragment of his visionary plans through his book	One of the most important humanist scholars. Stylistic principles of classical art aesthetic theory proportionality & harmony	Ideal city. Social causes address living conditions, humanist sensibilities	Political issues & social	Political-Social causes total social change and the creation of an Ecumenical State	Social reform. Educational system	Political teaching	Social- workers- to live in harmony with their fellow residents and nature	Social Early example of an industrial "model" village	Decent and healthy life was the aim	The perfect harmonious society through autonomous (socialist) communities	Social & economic utopia-housing an industrial community- (socialist values)	Social goal through a communist political system. For a society of peace & justice.	Social goal. Due to urbanization environmental & social problems- Solving the social issue of housing by a "model city"	Social goal. Due to governance policies. Environmental and social problems- vision of an ideal society. utopian socialism	A "model" village to house their workers- to improve the lives of those living and working in Birmingham	Social goals-quality of life- Environmental & social problems- urbanization	Ideal city of the 20th century- economic development	Modernist city model- holistic approach-Provide residents with a new lifestyle.	Social & political position as much as an urban plan- decentralization- - social goal a better quality of life
Evaluation	√																	100%			√
Social & Political is the (D) driving Force approximately 80%, the rest concern function and aesthetics																					

2.2 Second Phase: Pressures (P)

Pressure (P) is the result of the Drivers. It may be indicators that show growth in emissions, biological agents, consumption, resource use or the cause of the (S) State.

As (P) *pressure* to the next face (S) *state*, (Table 2) is the political authorities, giving the ability to the models to self-sustain. The political trends and systems of the time became an influence in favor of socialist idealism before its very existence. The city-state, the independent, autonomous, powerful local authority, is a necessity to achieve sustainability in a "green" area.

2.3 Third Phase: State (S)

The state (S) may concern a quantitative, qualitative description of natural phenomena, biological, chemical, and social characteristics of an area.

From the eighteenth century, along with the birth of political ideas and parties, the implementation of the "models" began (Table 3). The power of wealth, strongly politically colored, with social sensitivities, influencers, made it possible. Although they were non-implemented, ideas influenced other idealists. Notable is the fact that some visioners had no intention to implement, but just to make a proclamation, as Wright did (Gray 2018), among others.

The utopian models were presented with formulations, sketches, layout plans, and were usually not detailed (Gospodini 1995; Siolas 2015).

Common formation of the urban models (Table 4) that had a description of morphology (until the nineteenth century) is the island formation, as a proposal of urban- isolation from the city, through strong geometric forms. Afterwards, the formations concern zoning with architectural influences of their time.

Contrary to the contemporary way of building is a declaration of Jean Baptiste Andre Godin (1871): *"There is no*

Table 2 The role of local authorities. From ancient to modernism. Own processing

	Ancient times	*Middle Ages*	*14th and 15th "Late Middle Ages" the poverty increased greatly*	*Renaissance "humanism tradition" engagement with the "ideal city", strict geometry*					*18th century Pre-industrial era*	*19th century, the golden age of utopia-Utopian socialism*								*20th century - Modernism: the evolution of utopias.*			
	1. Plato (380 B.C.)	*2. Marcus Vitruvius Pollio (80–70 B.C. – after 15 B.C.)*	*3. Leon Battista Alberti (1404–1472), 1452*	*4. Leonardo da Vinci (1452 – 1519)*	*5. Thomas More (1477-1535), 1516*	*6. Tommaso Campanella (1568-1639)*	*7. Johann Valentin Andreae (1586–1654), 1619*	*8. Francis Bacon (1561 –1626), 1626*	*9. Claude-Nicolas Ledoux (1736 - 1806), salt mine inspector*	*10. Owen (1771 – 1858), 1800, owner of a textile factory*	*11. Owen, 1824*	*12. Charles Fourier (1772-1837), 1822.*	*13. Jean Baptiste Andre Godin (1817 – 1888) factory owner*	*14. Etienne Cabet (1788-1856), 1840*	*15. James Silk Buckingham (1786-1855), 1849*	*16. William Morris,(1834-1896), 1890 factory owner*	*17. George Cadbury (1839-1922), 1893, industrialist*	*18. Ebenezer Howard,(1850-1928), 1898 & 1902*	*19. Tony Garnier (1869-1948), 1918, 1920*	*20. Le Corbusier (1887 -1965), 1924*	*21. Frank Lloyd Wrigh, (1867 –1959), 1932*
Role of local authorities	City-state	Dedicated to his patron, the emperor Caesar Augustus	Pope Nicholas V, to whom Alberti dedicated the whole work	1487 - 1490 a new city planned for Ludovico il Moro, Duke of Milan	City-state	An absolute communist model Ecumenical State, the government is managed by a priestly class.	City-state	City-state, the State does not hold the monopoly	Self-sustaining organization. The state system of the utopia is described as "closed"	Self-governing & autonomous communities	Self-governing & autonomous communities with characteristics of agricultural cooperatives.	Self-sufficient community-He considered the State incapable of managing the affairs of the Community	Self-sufficient community industrial community	Self-sustaining State. A communist political system, decisions on all issues through special committees, city owns production and industry	The ownership of the model city belonged to a Company as well as the management with rules for smooth operation	Common ownership. The government is unnecessary, communities with autonomy. There is a denial of any type of power	Self contained communit homes were never pupic owned	Self-sustaining - the municipal authorities should own the land	Socialist theory & self -sustained	Prominence & prestige given to him by the "ruling circles". Rejected by the relevant state authorities. "Brasilia" owned entirely by the government	Local government-each county one minor government-weakening of political power
Evaluation:																					
The (P) pressure force to create the desirable State (S) is politically colored 100% & 62% concern a city- state & local governance																					

Table 3 The duration-implementation. From ancient to modernism. Own processing

	Ancient times	*14th and 15th "Late Middle Ages" the poverty increased greatly*	*Renaissance "humanism tradition" engagement with the "ideal city", strict geometry*	*18th century Pre-industrial era*	*19th century, the golden age of utopia-Utopian socialism*							*20th century - Modernism: the evolution of utopias.*			
	1. Plato (380 B.C.)	*3. Leon Battista Alberti (1404–1472) 1452*	*4. Leonardo da Vinci (1452 – 1519)*	*9. Claude-Nicolas Ledoux (1736 - 1806), salt mine inspector*	*10. Owen (1771 – 1858), 1800, owner of a textile factory*	*11. Owen, 1824*	*12. Charles Fourier (1772-1837), 1822.*	*13. Jean Baptiste Andre Godin (1817 – 1888) factory owner*	*14. Etienne Cabet (1788-1856), 1840*	*16. William Morris,(1834-1896), 1890 factory owner*	*17. George Cadbury (1839-1922), 1893, industrialist*	*18. Ebenezer Howard,(1850-1928), 1898 & 1902*	*19. Tony Garnier (1869- 1948), 1918, 1920*	*20. Le Corbusier (1887 -1965), 1924*	*21. Frank Lloyd Wrigh, (1867 –1959), 1932*
Duration	The end of the utopia came with divine intervention due the behavior of the citizens	A town modelled on this theory was Zamosc founded in the 16th century in Poland.	Never actually came to fruition. some ideas were applied during the 19th & 20th centuries, with the emergence of modernist and avant-garde movements, the renovation of the layout of some cities such as Paris.	Implemented. The royal saltworks closed in 1790 just after the French Revolution	Operated, the factory ceased to operate, 10 years	Implemented 4 years duration. Organizational problems administrative matters, where regulations and prohibitions were imposed on workers.	Implementation of a community of 600 people, by the utopian Victor Prosper Considérant in Conté sur Vesgre, 1832 & Brook Farm (1841–47) both failed due technical difficulties & management.	Small-scale labor society functioned successfully for a century the co-operative was dissolved in 1968. Successful application of F's utopia and perhaps the only one. Failed to adapt to changed economic conditions and social tensions after the 2nd World War.	Implemented - Organizational problems after a few years led to its dissolution.	In 1881 he gave to his workers (of his factory) part of the income but not to all of them. Realizing that a social company is impossible to be competitive in a capitalistic economy	Implemented- 1900, the Bournville model- was a big influence for the future garden cities-blueprint for many other model villages in Britain	Applied, 1903- today, degenerate or not. Hamstead, Letchworth Garden City, Welwyn ...were built with the garden city ideology	He influenced Le Corbusier & the industrial town planning of 20th	Unrealized urban masterplan. Realization of Le Corbusier's ideas can be witnessed in the conception Brasilia. Principels are applied.	Unrealized urban masterplan- to address pressing social, economic and environmental issues.

Table 4 Morphology. From ancient to modernism. Own processing

	Ancient times	*14th and 15th "Late Middle Ages" the poverty increased greatly*	*Renaissance "humanism tradition" engagement with the "ideal city", strict geometry*					*18th century Pre-industrial era*	*19th century, the golden age of utopia-Utopian socialism*								*20th century - Modernism: the evolution of utopias.*			
	1. Plato (380 B.C.)	*3. Leon Battista Alberti (1404–1472), 1452*	*4. Leonardo da Vinci (1452 – 1519)*	*5. Thomas More (1477-1535), 1516*	*6. Tommaso Campanella (1568-1639), 1602*	*7. Johann Valentin Andreae (1586–1654), 1619*	*8. Francis Bacon (1561 –1626), 1626*	*9. Claude-Nicolas Ledoux (1736 - 1806), salt mine inspector*	*10. Owen (1771 – 1858), 1800, owner of a textile factory*	*11. Owen, 1824*	*12. Charles Fourier (1772-1837), 1822.*	*13. Jean Baptiste Andre Godin (1817 – 1888) factory owner*	*14. Etienne Cabet (1788-1856), 1840*	*15. James Silk Buckingham (1786-1855), 1849*	*16. William Morris,(1834-1896), 1890, factory owner*	*17. George Cadbury (1839-1922), 1893, industrialist*	*18. Ebenezer Howard,(1850-1928), 1898 & 1902*	*19. Tony Garnier (1869- 1948), 1918, 1920*	*20. Le Corbusier (1887 -1965), 1924*	*21. Frank Lloyd Wrigh, (1867 –1959), 1932*
Morphology	*"Politia" & "Atlantis"* Island Rich natural environment with variety of animals & plants	*"De Re Aedificatoria"* Choosing the location of the town first, careful setting up the size & direction of streets, then location of bridges and gates, & finally a building pattern ruled by perfect symmetry.	*"the Ideal Renaissance City"* Rigorously geometric urban planning. well-organised and in-keeping with the aesthetics of the time, but always with functionality as the priority.	*"Utopia"* Island, semicircular figure, cut off from the rest of the mainland	*"The City of the Sun"* -His model is located on the slope of a hill with a perfect climate. The form follows the trend of the time the radiant city	Island *"Christianopolis"* The form of the city has a tetragonal geometry, with fortifications.	*"New Atlantis"* Island	Principles were applied to *"City of Chaux"* Influences on the morphology of the facilities from the geometric radial Renaissance model of the city.	*"New Lanark" in Scotland*	*"New Harmony" Indiana United States* zoning of the houses according to the number of children of each family. Merge city function into one building.	"Phalanstery" *"Traité de l' association domestique agricole"*. Distance from urban areas in a phthisic environment, zoning for children in the building. Merge city function into one building, a model of the Versailles type.	*"Le Familistère"*, or the "Palais social", small scale - concerning his foundry, typical of mid 19th-century French institutional architecture.	*"Voyage en Icarie"* is located somewhere isolated. Morphologically it concerns a classicist form.	*"Victoria"*, geometric, square shape, placed in the countryside -zoning in terms of public buildings -around the perimeter are located the residences of the economically and socially weaker. Influenced by the City of Chaux of Claude-Nicolas Ledoux.	*"News from nowhere"*. Communities based on rural life and production, located in the countryside	*"Bourneville Village"- out of the bustling city centre- typically built for factory workers* - relatively high-quality housing- physical environment- The "Bourneville model" was realized for plenty of villages in G. Britain- *a big influence for the future garden cities*	*"Garden Cities of Tomorrow"*- Depends on the site. The city occupy up to one-sixth of the total, farms and civil society facilities. In the center a vast park, a series of smaller cities, growing in concentric circles around a central one. zoning of uses, combine the advantages of urban centers & rural perimeter agricultural.	*"Une Cité Industrielle"* located in a flat area and at a short distance from the city-core of (the old), industrial zone, Hippodamian system, ventilation and the sun for the urban fabric	*"Standard city model"* or *"Ville Radieuse"*. A functionalist proposal based on operating Zoning, cartesian grid, geometry & repetition, pre-fabricated apartment buildings. Against urban disorder.- Rational urban planning	*"Broadacre City"*, suburban development- spacious highways. No traffic problems. No glaring cement roads or walks- mix of uses-re-thinking of the urban & rural, harmony humanity & environment, organic architecture- small-scale farming & manufacturing.
Evaluation:	√																			√
The morphology differs, the models stand on a white canvas. The main common characteristic is the "isolation" like the islands or distanced from urban centers, 76%																				

point in creating cheap housing, because cheap housing is the most expensive for people; what needs to be built is housing that allows real domestic economies, a place where human well-being and happiness can be nurtured" (Hidden Architecture 2021, Le Familistere Guise).

The most common characteristic of the 21 models researched and the comparisons applied is the ability to self-sustain (Table 5). Based on the agriculture sector, from ancient times, and the combination of light industry, as Charles Fourier proposed (Obaindou 2011) in the nineteenth century. For George Cadbury (Bournville Village Trust 2024) and Wright (Frank Lloyd Wright Foundation 2017), the ability to sustain begins from the house garden, where the individual sustainability accrues. Le Corbusier, as a prophet, lacking political support in his ideas (Sarigiannis 2011), proposed living in dependence on mineral resources.

Aiming the social perfection, each visionary gave detailed descriptions. For Plato (Table 6) education was a dominant element for the operation of an idealized State; only through it can harmony and cohesion be achieved. An important factor that will give protection to the region is the family (Plato 2008).

For Thomas Moore, the social characteristic should also concern the family and monogamy as a core for society. The rights of women are declared, since both women and men are in the same work, giving equality in employment (More

Table 5 Flows, out flows, products. From ancient to modernism. Own processing

	Ancient times	**Renaissance** "humanism tradition" engagement with the "ideal city", strict geometry				*18th century Pre-industrial era*	*19th century, the golden age of utopia- Utopian socialism*								*20th century - Modernism: the evolution of utopias.*			
	1. Plato (380 B.C.)	*5. Thomas More (1477-1535), 1516*	*6. Tommaso Campanella (1568-1639)*	*7. Johann Valentin Andreae (1586–1654), 1619*	*8. Francis Bacon (1561 –1626), 1626*	*9. Claude-Nicolas Ledoux (1736 - 1806), salt mine inspector*	*10. Owen (1771 – 1858), 1800, owner of a textile factory*	*11. Owen, 1824*	*12. Charles Fourier (1772-1837), 1822.*	*13. Jean Baptiste Andre Godin (1817 – 1888) factory owner*	*14. Etienne Cabet (1788-1856), 1840*	*15. James Silk Buckingham (1786-1855), 1849*	*16. William Morris,(1834-1896), 1890-factory owner*	*17. George Cadbury (1839-1922), 1893, industrialist*	*18. Ebenezer Howard,(1850-1928), 1898 & 1902*	*19. Tony Garnier (1869-1948), 1918, 1920*	*20. Le Corbusier (1887 -1965), 1924*	*21. Frank Lloyd Wrigh, (1867 –1959), 1932*
Flows	City-state, a self-sufficient state	City-state, a self-sufficient state	Autonomy, Ecumenical State	City-state, a self-sufficient state	City-state, a self-sufficient state	self-sufficiency, a self-sustaining organization	self-governing & autonomous communities with characteristics of agricultural cooperatives.	self-sufficient & self-governing settlements.	self-sufficiency & self-sustaining organization	self-sustaining industrial community	self-sustaining - communist community	self-sustaining and self-governing city.	Autonomy	self-sustaining community	self-sufficient urban formations,, the model would be autonomous...	self -sustained	influenced by the "linear city" function	self-sufficient
Out-Flows, Products	Local products in abundance & polythelia	Resources are based on land and cultivation	Agriculture also as employment, preoccupation with sciences	Agriculture, livestock & industry- Regardless of profession compulsory municipal duties	Science and research are given with great importance	agricultural production	agriculture & light industry	agriculture & light industry	agriculture & light industry	mainly industrial	production & industry	agriculture & industry	rural life and production	aiming independent living, a land to grow vegetables	agricultural, industrial and commercial operation,	industrial	planning a society based on energy consumption	agricultural, industrial and commercial operation, grow their own food
Evaluation:	√															100%		√
The flows concern the ability to self-sustain, with autonomy by 81% for those who envisioned a holistic approach																		
The production to gain the ability to self-sustain is agriculture and combined by industry after the 19th century																		

Table 6 The Social characteristics of the "model cities". From ancient to modernism. Own processing

	Ancient times	*Renaissance "humanism tradition" engagement with the "ideal city", strict geometry*					*18th century Pre-industrial era*	*19th century, the golden age of utopia- Utopian socialism*								*20th century - Modernism: the evolution of utopias.*			
	1. Plato (380 B.C.)	*4. Leonardo da Vinci (1452 – 1519)*	*5. Thomas Moore (1477-1535), 1516*	*6. Tommaso Campanella (1568-1639)*	*7. Johann Valentin Andreae (1586–1654), 1619*	*8. Francis Bacon (1561 –1626),1626*	*9. Claude-Nicolas Ledoux (1736 - 1806) mine inspector*	*10. Owen (1771 – 1858), 1800 manager of a textile factory*	*11. Owen, 1824*	*12. Charles Fourier (1772-1837), 1822.*	*13. Jean Baptiste Andre Godin (1817 – 1888) factory owner*	*14. Etienne Cabet (1788-1856), 1840*	*15. James Silk Buckingham (1786-1855), 1849*	*16. William Morris,(1834-1896), 1890-factory owner*	*17. George Cadbury (1839-1922), 1893, industrialist*	*18. Ebenezer Howard,(1850-1928), 1898 & 1902*	*19. Tony Garnier (1869- 1948), 1918, 1920*	*20. Le Corbusier (1887 -1965), 1924*	*21. Frank Lloyd Wrigh, (1867 –1959) 1932*
Language & Religion	God-fearing		Emphasis on Christian spirit & education.	Religion plays a vital role, leading to harmony and collective welfare	Christian ideology, church has an important role in the functioning of the community.	*Salomon's House* superior to the State					The project contained no churches				√		No churches		
Social characteristics intellectual & social development	Law-abiding & prudence. Class separation, freedoms & rights- education through it harmony & cohesion- Important role, protection to the region is the family.	Social division between the upper and working classes.	Basis of society the family & monogamy. Women & men the same work. Equality in employment. Same types of plain clothes.	Preoccupation in sciences- Equality in political affairs men and women, no private property and goods, women and children are in common.	The family as social unit. Its constitution was based on godliness, instructive plays were put on in the churches, luxury and lavish clothing were immoral.	Generosity and enlightenment, dignity and splendour, piety and public spirit. Science and research are given with great importance	Social equality. He offered to the workers free food and housing. Sustainability would shape the social shapes & behaviors of the inhabitants	New living, working, sanitary, educational and other standards	Decent and healthy life. Equality between residents even in the resources produced. Emphasis was placed on education.	Free moral system, with its own institutions and principles.co-owner, they share the income-profits & benefits also concern those who do not work.	Thanks to the Co-operative Association of Capital & Labour where foundry workers became shareholders. The family as the core of the social organization.	The political system with the state in organization and decisions sought to discipline citizens by good functioning, equality, happiness	Influenced from the social organization of Fourier's Phalanstere. Social class segregation in housing. the weakest are placed in the periphery, ten social classes	Harmony in a lifestyle with the natural world, no private property, no big cities, no authority, no monetary system, no marriage or divorce, no courts, no prisons, no class systems.	Annuity, a deposit account & education for everyone, pioneered pension schemes, work committees, full staff medical service	Low rents- meeting the needs of the residents	A socialist and socially disciplined society	Improve the quality of life on a socio-economic level, housing was categorized. "Brasilia", been harshly criticized for ignoring residents' habits or desires	Public own-ership of utilities: power, transportation, mediums of ex-change. Ideas inspired by *spiri-tual integrity to the modern world*
Evaluation:																			√
The social characteristics interest the 76% of the visioners, all who applied the holistic approach.																			

1984). The description for simple clothes, of Thomas Moore, concerning the same type, is also presented in pre-industrial settlements with self-sustaining capacity (Dereka 2023b), where the local identity of an area was also created. Johann Valentin Andreae has common social characteristics like Moore, concerning the family as a unit and the clothing idealism (Mumford 1998).

In the era of the Renaissance, humanism became a "tradition" of engagement with the "ideal city." Concerning Tommaso Campanella, science was the core of his model city. Equality in political affairs for men and women. Though the ethics is questionable since women and children are common in his model (Obaindou 2011).

For Francis Bacon, generosity and enlightenment, dignity and splendor, piety and public spirit are the social characteristics of the ideal city. Bacon's political teaching is indirect; actually it is the core of his utopia (Weinberger 1976)

The characteristics of the Claude-Nicolas Ledoux model are about social equality with an emphasis on farmers; some of his ideas he was able to implement in Chaux, France, due to his position as salt mines inspector. He offered the workers free food and housing, an effort to achieve autonomy and self-sufficiency. The implemented project concerns workers' residences. The salt mines of Chaux, a self-sustaining organization, could shape the social characteristics of the inhabitants (Soumas and Papadaki 2013).

The utopias of the nineteenth century are associated with the birth of the socialist political movement, Karl Marx (1818–1883) and Friedrich Engels (1820–1895). This period concerns the French Revolution, socialism, and the Industrial Revolution. The utopias of this period, apart from the overthrow and improvement of everyday life, as is their history, achieve application. Marx and Engels consider the functioning of the city to be a political, economic, and social matter. On this basis, the solution of the ideal city as a design model is underestimated (Berneri 1999).

For Owen, the purpose was to raise the living standards of 3.000 workers in New Lanark from the moment he became the manager of a textile factory. Equality between residents, even in the resources produced. Special emphasis was also placed on education. A decent and healthy life was the aim (Berneri 1999; Obaindou 2011).

Charles Fourier is considered one of the important representatives of utopian socialism. His ideal community was about living and working in a model of Versailles type. Impressive for the time is the free moral system proposed by him (Hidden Architecture 2021). He believed that the ideal community concerns cooperation and social cohesion, also that the perfect communion concerns self-reliant ability, but with its own institutions and principles. The location refers to a distance from urban areas, in a phthisic environment.

Godin also had the family as the core of social organization on the small scale that he was involved in concerning his factory (Hidden Architecture 2021, *Le Familistere Guise*). The living conditions of the workers were excellent and of a high standard, having good lighting and ventilation in the residences (Benevolo 1978).

Etienne Cabet's utopia is based essentially on a communist political system. Once again, utopia is located somewhere isolated. Through analysis of the healthy functioning of the city and political organization, the aim is to ensure the comfortable life of the inhabitants. The political system with the state in organization and decisions sought to discipline citizens through good functioning, equality, and happiness for the inhabitants (Mumford 1998).

James Silk Buckingham's effort was about addressing social problems of the time through architecture (Charles Holland 2017). For William Morris, people find harmony in a lifestyle with the natural world. He was associated with the political ideal of "*Liberalism*" and later joined the "*National Liberal League*," founder of "*Socialist League*" (Mackail 1899). He is also the founder of "*Society for the Protection of Ancient Buildings*" (Mackail 1901). In 1881 he gave to the workers of his factory a part of the income. Realizing that a social company is impossible to be competitive in a capitalistic economy (Mackail 1899).

George Cadbury was an Englishman, an industrialist who was also involved in politics. The need to provide houses to his employees and their families created the "Bourneville Village." The goal of Cadbury was to improve the quality of life for the residents of Bourneville, with better, bigger houses and the advantage of a green environment. The residence of the workers concerned good quality, at an affordable price. Under consideration was the natural lighting and the large open space for socializing (Bourneville Village Trust).

In the period of the Industrial Revolution, a new social class, the working class, was created. Utopians, as educated, multi-talented, several factory owners or managers, envisioned or mostly implemented socialist utopias on a small scale, for the working class.

Ebenezer Howard's aim was the quality of life and communication between work and accommodation, at the same time, slowing down the spread of industrial centers. Hamstead, Letchworth, and Welwyn were built with the garden city ideology (Siolas 2015). The low rents, among the design principles, were supposed to meet the needs of the residents, having a quality of living in an affordable way (*Utopias through time*). Tony Garnie's socialist theory is also applied in his utopia, for the social characteristics achievements, he believed in no private ownership (Sennott 2004).

Although the model of Le Corbusier had as a social goal to give a better quality of life to the lower working industrial class (Gili Merin n.d., AD Classics, Ville Radieuse), often his ideas are characterized as antisocial, since his main target is the form and the function. According to Gili Merin the *Brasilia*, the realized model city of Le Corbusier, ignored social needs and wants. In the twentieth century, the city models focused more on the core of the person, the individuality, instead of a strong political coordination (Historiou Theodora 2015).

"*Broadacre City*" for Wright is a city model that can be applied universally. Although it is thought that the modernist generation was politically colorless, for F. L. Wright and the "Broadacre City," there is a statement that "*it was both a planning statement and a socio-political scene*" influenced by the political beliefs of Henry George. The theory of George, also known as "*Georgism*" is that the natural resources should belong equally to all (Pfeiffer 1993). Private ownership is an important factor for Wright, a principal contrary to other utopias, the inhabitant should own a house, a farm, and a place to work. The land of each house would give economic independence and personal sustainability. His utopia was an urban plan with a social and political base and idealism (Gray 2018).

The term is "*Limit to growth*" for the models researched, 14 out of 20 visioners (Table 7), ether proposed or applied a limit for their model to be able to self-sustain, as Howard's model involved a limit on the size (Laskow 2016), which would be permanently surrounded by open agricultural land. Le Corbusier's models are basically the maximum capacity to function (Gili Merin, Ville Radieuse).

A green fabric, often isolated, with a self-maintenance capacity is sought. Self-preservation is achieved in the green tissues based mainly on agricultural production. Education is a dominant element for the operation of an idealized city; only through it will there be harmony and cohesion. An important role in the structure of society is the family, as mentioned. Sustainability starts from small-scale, family farming with gardeners, something that was proposed and applied also in pre-industrial societies (Dereka 2023b).

2.4 Fourth Phase: Impact (I)

The Impact (I) concerns the effects that came from the State (S), which may concern natural ecosystems, economy, health, or well-being. It can be the result of chain effects, local to global.

> *A map of the world that does not include utopia is not worth even glancing at, for it leaves out the one country at which humanity is always landing. And when humanity lands there, it looks out, and seeing a better country, sets sail. Progress is the realization of utopias* Oscar Wilde, ("The Soul of Man under Socialism"), (Hodgkinson 2017).

The impact surely concerns the influence on other visionaries. Francis Bacon was influenced by Plato (Weinberger

Table 7 The living standard and the limits of the "model cities." From ancient to modernism. Own processing

	Ancient times	*Renaissance "humanism tradition" engagement with the "ideal city", strict geometry*					*18th century Pre-industrial era*	*19th century, the golden age of utopia- Utopian socialism*								*20th century - Modernism: the evolution of utopias.*			
	1. Plato (380 B.C.)	*4. Leonardo da Vinci (1452 – 1519)*	*5. Thomas Moore (1477-1535), 1516*	*6. Tommaso Campanella (1568-1639)*	*7. Johann Valentin Andreae (1586–1654), 1619*	*8. Francis Bacon (1561 -1626),1626*	*9. Claude-Nicolas Ledoux (1736 - 1806) mine inspector*	*10. Owen (1771 – 1858), 1800 manager of a textile factory*	*11. Owen, 1824*	*12. Charles Fourier (1772-1837), 1822*	*13. Jean Baptiste Andre Godin (1817 – 1888) factory owner*	*14. Etienne Cabet (1788-1856), 1840*	*15. James Silk Buckingham (1786-1855), 1849*	*16. William Morris,(1834-1896), 1890-factory owner*	*17. George Cadbury (1839-1922), 1893, industrialist*	*18. Ebenezer Howard (1850-1928), 1898 & 1902*	*19. Tony Garnier (1869- 1948), 1918, 1920*	*20. Le Corbusier (1887 -1965), 1924*	*21. Frank Lloyd Wrigh, (1867 –1959), 1932*
Living standard	harmony and cohesion in the state	*well-being*	no private property	no private property	The houses belong to the municipality	√	Live in harmony with their fellow residents and nature	Enhancing his workers' environment resulted in increased productivity and profit	Decent and healthy life	√	The living conditions of the workers were very good & high standard.	Comfortable life of the inhabitants	The right living conditions, would protect the inhabitants from what he called the 'national evil. Safety, harmony and co-operation	No private ownership, transactions are made with products	Houses were never privately owned	Municipal authorities should own the land	No private ownership	√	Individual own-ership of home, farm & place of employment
Population, n. of houses, area size	limit 5.000 inhabitants	limit of 30.000 inhabitants or 5.000 houses	44 cities -6.000 homes each, identical cities in social characteristics and in terms of architecture		Densely populated, 400 inhabitants			3.000 workers	Each village would consist of about 1.200 persons living on 1.000 to 1.500 acres	Each convoy involves 1.600 people.	2,000 people, 475 apartments	limits to the growth of industry. There is the center - city & is surrounded by 100 provinces and these in 10 communities.	Maximum number of inhabitants 10.000	Common ownership & democratic control of the production	300 homes on 330 acres (1.3 km2) of land	limited size, belt agricultural, area of 24.000 acres 32.000 inhabitants, divided into 6 equal and autonomous sections, max 250.000 inhab.	35.000 inhabitants	A single Unité could accommodate 2.700 inhabitants and function as a vertical village-Density of 240 inhab/ acre, for 3 million inhab	1,400 families, 2.5 people per acre "larger villages" 10,000 individuals, home individual 4,000 sq.m. land. Community of 10 km2.
Evaluation:																			
Raising the living standards 100%, A percent of 35% of the visioners and employers believed in no private ownership																			
72% gave a limit to growth for the models																			

1976). Fourier's social views and proposals inspired an entire movement of "conscious communities." Godin's application of F.'s theories was completed with the creation of the *Co-operative Association* of *Capital and Labour,* where foundry workers became shareholders (Hidden Architecture 2021, Le Familistere Guise).

For James Silk Buckingham, the shape of his model city is influenced by the *City of Chaux* of Claude-Nicolas Ledoux. As well as influences from the social organization of Fourier's "*Phalanstere.*" George Cadbury's model, the "*Bourneville model,*" was realized for plenty of villages in Great Britain. It is believed that it was a big influence on the future garden cities (Bourneville Village Trust).

Howard's utopian model still inspires urban planners, who were not town planners or architects, was influenced by Edward Bellamy's (1888) utopian vision of transforming America into a socialist society (Laskow 2016). The first garden city that was implemented concerns an area outside of London, the suburb still operates to this day with specific rules (Letchworth Garden City 2023).

The goals of architecture in the twentieth century were to create utopias that would embrace new technology. For Tony Garnier, the "*industrial city"* is a heritage, which is, however indiscernible today. It is a utopia in an industrial environment with the goal of economic development (Martek 2018).

Tony Garnier influenced Le Corbusier, who published parts of the "industrial city" (Sennott 2004). Le Corbusier's proposals greatly influenced urban planning; he featured minimum dimensioning and the industrial materials. Sarigiannis (2011) claims that the prominence and prestige given to him by the "ruling circles" was not by accident, since he maintained relations with Moscow and the OCA, while his ideology follows a parallel course with "fascistic" concepts. The political activities at the time had a connection with Le Corbusier's proposals, in the Soviet Union, the political need was to respond to the housing demands, to the new large number of industrial inhabitants, who were living in improper living conditions. On the other hand, in Europe, the politicians, had the revolutionary fear of the working class with recent cases from Vienna, Budapest, or Germany. For that reason, as a social offer, large quantities of housing for the working class were given. So, the production of "as many as possible," of the modern construction, was a political task and wish simultaneously. But also, convincing the world that "minimum housing" is also a trend (Sarigiannis 2011). F. L. Wright's model has a lot in common with Howard's. The mix of use, the agricultural, industrial, and commercial operations, but also the link between the cities. Among important thinkers, Wright was also influenced by the "*Looking Backward*" (1888) utopia of Edward Bellamy (Gray 2018), where the effects of the capitalistic system are solved through social organization (Pfeiffer 1993).

2.5 Fifth Phase: Response (R)

The Response concerns the efforts for prevention, treatment, and improvement. The political response to the issues applied in the chain framework.

At this phase, generally the DPSIR tool—framework gives the local authorities the evaluation of the urban system. In that way, a holistic approach should be applied to deal with the city challenges (Dereka 2023a, b).

For the 21 models, the Response (R) in the models examined concerns the effort by the visioners to sustain the models.

For Plato (Table 8), the end of his utopia came with divine intervention due to the changed behavior of the citizens (Plato 2008). For Owen, the duration of operation was 4 years due to organizational problems, administrative matters, where regulations and prohibitions were imposed on workers (Berneri), (Ohaindou 2011).

Settlements based on Fourier's ideas were "*Brook Farm,*" in Massachusetts, a short-lived utopian experiment in communal living (1841- 47) and the "*North American Phalanx*" at Red Bank, New Jersey (Charles Fourier, Britannica 2021). The community "*Brook Farm"* involved intellectuals of the time, but also ordinary workers. A newspaper published "*The Harbinger*" which dealt with political and social issues. The end of the community came after a fire, in a newly built embroidery building, "*The Phalanstery,*" gradually became financially exhausted (*Brook Farm*, Britannica).

Table 8 The causes of the function or failure of the model cities. From ancient to modernism. Own processing

	Ancient times	***14th and 15th "Late Middle Ages"** the poverty increased greatly*	***Renaissance** "humanism tradition" engagement with the "ideal city", strict geometry*	***18th century Pre-industrial era***	***19th century**, the golden age of utopia- Utopian socialism*							***20th century -** Modernism: the evolution of utopias.*			
	1. Plato (380 B.C.)	*3. Leon Battista Alberti (1404–1472) 1452*	*4. Leonardo da Vinci (1452 – 1519)*	*9. Claude-Nicolas Ledoux (1736 - 1806), salt mine inspector*	*10. Owen (1771 – 1858), 1800, owner of a textile factory*	*11. Owen, 1824*	*12. Charles Fourier (1772-1837), 1822.*	*13. Jean Baptiste Andre Godin (1817 – 1888) factory owner*	*14. Etienne Cabet (1788-1856), 1840*	*16. William Morris,(1834-1896), 1890 factory owner*	*17. George Cadbury (1839-1922), 1893, industrialist*	*18. Ebenezer Howard,(1850-1928), 1898 & 1902*	*19. Tony Garnier (1869-1948), 1918, 1920*	*20. Le Corbusier (1887 -1965), 1924*	*21. Frank Lloyd Wrigh, (1867 –1959), 1932*
Duration & Cause	The end of the utopia came with divine intervention due the behavior of the citizens	A town modelled on this theory was Zamosc founded in the 16th century in Poland.	Never actually came to fruition. some ideas were applied during the 19th & 20th centuries, with the emergence of modernist and avant-garde movements, the renovation of the layout of some cities such as Paris.	Implemented. The royal saltworks closed in 1790 just after the French Revolution	Operated, the factory ceased to operate, 10 years	Implemented 4 years duration. Organizational problems administrative matters, where regulations and prohibitions were imposed on workers.	Implementation of a community of 600 people, by the utopian Victor Prosper Considérant in Conté sur Vesgre, 1832 & Brook Farm (1841–47) both failed due technical difficulties & management.	Small-scale labor society functioned successfully for a century the co-operative was dissolved in 1968. Successful application of F's utopia and perhaps the only one. Failed to adapt to changed economic conditions and social tensions after the 2nd World War.	Implemented - Organizational problems after a few years led to its dissolution.	In 1881 he gave to his workers (of his factory) part of the income but not to all of them. Realizing that a social company is impossible to be competitive in a capitalistic economy	Implemented- 1900, the Bournville model- was a big influence for the future garden cities- blueprint for many other model villages in Britain	Applied, 1903- today, degenerate or not. Hamstead, Letchworth Garden City, Welwyn ...were built with the garden city ideology	He influenced Le Corbusier & the industrial town planning of 20th	Unrealized urban masterplan. Realization of Le Corbusier's ideas can be witnessed in the conception Brasilia. Principels are applied.	Unrealized urban masterplan- to address pressing social, economic and environmental issues.
Evaluation:								√			√	√			
The implementation of the ecosystems concern 38%															

For Jean Baptiste Andre Godin, the "*Cooperation Organization*" was dissolved in 1968, due lack of adaptation to different economic conditions after the world wars and social tensions (Hidden Architecture 2021, Le Familistere Guise). Godin's small-scale labor society functioned successfully for a century (Hidden Architecture 2021). It is considered a successful application of Fourier's utopia (Benevolo 1978).

The "*Bourneville model*" of George Cadbury was realized for plenty of villages in Great Britain. Ebenezer Howard's model is applied, from 1903 to today, but surely the aim of the responsible authorities is completely different now. According to J. Gray, Wright's intention was just to give high-pressure concerning socio-economic, and environmental matters (Gray 2018).

3 Conclusions

Summing up the procedure of the ecosystemic function in the DPSIR framework of EEA:

The *(D) "Driving force" or "Aiming for"* is social and political, approximately 80%, the rest concern function and aesthetics.

The *(P) "Pressure force"* is *"The Role of local authorities,"* it is politically colored 100% & 62% concern a city-state & local governance with autonomy.

In the *(S) "State"* of the ecosystems, the "*morphology*" differs, but all models stand on a white canvas. The main common characteristic is the isolation, like the islands or being distant from urban centers, by 76%. The *"flows"* concern the ability to self-sustain, with autonomy by 81% for those who envision a holistic approach. The "*products*" to gain the ability to self-sustain are based in agriculture, combined with industry after the nineteenth century. The "*social characteristics*" interest 76% of the visioners, all of whom applied the holistic approach.

Raising the "*living standard*" concerned all of them 100%. A percentage of 35% of the visioners and employers believed in no private ownership. The common goal was the quality of living, especially for the implemented models by the visioners–employers who were influenced by socialism and donated rights and benefits to the working class. A high percentage of 72% gave a "*limit to growth*" for the models.

The *(I) "Impact"* certainly concerns universal effects of the visioned or implemented models to our day.

The *(R) "Response"* is the effort to sustain the models, concerning the *implementation* of 38%. The end of the realized models concerns economic and organizational problems. The models that were not realized in their time had the meaning of a declaration, objection, and demonstration of social issues. The models of George Cadbury and Ebenezer Howard are the most elaborate ones.

Overall, the common approach to solving social and political issues through time is a model city.

The highly common characteristic and condition is the ability to (self) sustain through a powerful autonomous local political authority in a green environment.

A strong conclusion is that urban sustainability is a complex term, that requires holistic approaches, thought strong, talented, educated and powerful people as the visioners of the model cities.

References

Antonia Obaindou (2011) Utopia & ideal city in a macro-historical perspective: Theoretical investigation of their socio-spatial components. Department of Spatial Planning. Aristotle University, 2011.

Benevolo L (1978) Histoire de l'architecture moderne. Dunod, 1 Jan 1978. ISBN-13: 978-2040103040

Berneri ML (1999) A Journey through Utopia. Publisher Nisides. March 1999.

Bournville Vilage Trust (2024) Bournville's story. Foundation and legacy

Britannica, T. Editors of Encyclopedia (2021) Brook Farm. Encyclopedia Britannica. 15 Aug 2021

Charles Holland (2017) Nobody's perfect. The RIBA Journal. Culture Utopic. 31 January 2017

Dereka X (2023a) Targeting Sustainability to face Urban challenges. Using EEA & EC tools. IN FOLIO, Issue no. 41—"Urban challenges in Mediterranean cities: sustainable development and architectural heritage" University of Palermo, (June 2023), Palermo, Italy

Dereka X (2023b) Modeling pre-industrial urban ecosystems, to aim sustainability. In: 4th International Conference on Environmental Design, (ICED2020), (October 2023) Athens, Greece

EEA Mehra M (1997) Towards Sustainable Development for Local Authorities. European Environmental Agency. Part 1. IMSA Amsterdam. The Netherlands.

EEA Bosch P, Buchele M & Gee D (1999) Environmental Indicators: Typology and Overview. EEA.

EEA (2001) Reporting on environmental measures: are we being effective? Environmental issue report No 25. European Environment Agency, Copenhagen

EEA (2005) EEA core set of indicators guide. EEA. Technical report No 1/2005

EEA (2020) The European environment—state and outlook 2020: knowledge for transition to a sustainable Europe

Frank Lloyd Wright Foundation (2017) Revisiting frank Lloyd Wright's vision for "Broadacre City". The Whirling Arrow News and updates from the Frank Lloyd Wright Foundation September 8, 2017

Gabrielsen P, Bosch P (2003) Environmental indicators: typology and use in reporting. EEA

Gospodini A (1995) Stories of the city and urban planning. University of Thessaly, Volos/Department of Spatial Planning and Development Engineers

Gray J (2018) Reading Broadacre. Frank Lloyd Wright Foundation

Hidden Architecture (2021) Le Familistere Guise

Historiou T (2015) In search of the optimal organization of cities. From utopias to nowtopias. Research paper—September 2015. Aristotelio University of Thessaloniki Department of Engineering, Planning & Development

Hodgkinson T (2017) The importance of being utopian. IDLER

International Garden Cities Institute (2023) Letchworth Garden City. Heritage Foundation Letchworth Garden City

Jean Baptiste Andre Godin (1871) Solutions Sociales. Kessinger Publishing Co. ISBN-13: 978-1437157147.

Laskow, S. (2016). These Utopian city maps have influenced urban planners for over a century. Ebenezer Howard's Garden Cities are now embedded into London and New York. Atlas Obscura

Mackail JW (1901) The life of William Morris, vol 1. Longmans, Green & Co, London, New York, and Bombay

Maguire C (2014) Digest of EEA indicators 2014. EEA. Technical report, No 8/2014

Martek I (2018) The sustainability narrative in contemporary architecture: falling short of building a sustainable future. ResearchGate. https://doi.org/10.3390/su10040981

Merin G (n.d.) AD classics. Ville Radieuse/Le Corbusier

More T (1984) De optimo statu reipublicae deque nova in sula utopia. Publisher Kalvos

Mumford L (1998) The story of Utopias. B. Tomanas, Publisher NISIDES. January 1922. ISBN13

Pfeiffer BB (1993) The disappearing city. In: Wright FL (ed) Frank Lloyd Wright Collected Writings, vol 3: 1931–39. Rizzoli International Publications, Inc, New York

Plato (2008) Timaeus & Critias. Penguin Classics. ISBN139780140455045

Sarigiannis G (2011) Harta of Athens & her historic framework (B μέρος). Greek architects

Science for Environment Policy (2018) Indicators for sustainable cities. In-depth report 12. Produced for the European Commission DG Environment by the Science Communication Unit. UWE, Bristol

Sennott RS (2004) Encyclopedia of 20th century architecture. Fitzroy Dearborn, New York

Siolas A (2015) Urban planning methods, applications and tools. Publications Kallipos, Open Academic Publications. ISBN: 978-960-603-075-8

Soumas M, Papadaki E (2013) C.N. Ledoux, the realized salt pipes & his ideal society. National Technical University, Athens

Sukopp H (2002) On the early history of urban ecology in Europe. Preslia, Praha

Weinberger J (1976) Science and rule in Bacon's Utopia: an introduction to the reading of the New Atlantis. Am Polit Sci Rev 70(3):865–885 (875)

Ecosystem Services to the Test of the Territory: Perspectives for the Urban Bioregion Approach

Daniela Poli and Giulia Luciani

Abstract

A growing body of research is focusing on ecosystem services as a useful framework to include the biophysical processes and their reproduction in urban and territorial decision-making. Originally conceived as a heuristic metaphor, its operationalisation has raised critiques, especially regarding the universalising abstractions, standardised procedures and service monetisation involved. Within the urban bioregion approach, the concept of ecosystem services can prove effective as a tool to draw attention to the socio-ecological dynamics related to the valorisation through community design of the territorial heritage, which includes the natural resources, as long as it is brought back to the physical and social dimensions of the territory. Starting from a consideration of the relationship to nature that the ecosystem services imply, the critiques the approach has raised, and some possible alternatives, the paper aims to reinterpret the concept and redevelop the approach from a bioregional perspective. A case study from the Florence metropolitan area, concerning integrated strategies for water management, provides some insights to shed light on both the opportunities and the critical aspects of a revisited approach to the ecosystem services, within the urban bioregion perspective.

Keywords

Ecosystem services · Nature · Territorial heritage · Urban bioregion

D. Poli · G. Luciani (✉)
Department of Architecture, University of Florence, Florence, Italy
e-mail: daniela.poli@unifi.it; giulia.luciani@unifi.it

1 Introduction

Modernisation kills itself. Today, when the dependence of human survival on the interconnectedness of the entire ecosystem has become manifest, it prompts reflection that 80% of the earth's biodiversity and most of the 200 places with the highest biodiversity lie in territories that are still inhabited by indigenous peoples (Staid 2022). When realising that indigenous peoples know how to take care of their environment and their survival better than anyone else, we cannot help but reflect on the fact that European thinking and modernising developmentalist acceleration entail the gene of destruction. Where have we gone wrong? Where to start again?

It is certainly with modernity that the process of domestication of nature has gained significance. Science, first with Galileo then with Descartes, defined nature as something different and external to the human, inert matter, an object that thought can circumscribe, understand, measure and use to its own ends. Such a process was supported by the definition of cartography, refined with the birth of sovereign states, which introduced a language and at the same time a perspective capable of controlling from above the other-than-oneself, the external reality no longer described in an imaginative and narrative manner, as in the splendid T-shaped globes, but rather in increasingly certain and universal forms and measures (Dematteis 1985; Poli 2019).

The inertia of the landscape and the world of life in adapting to the new socio-economic drives, as Emilio Sereni (1961) saw so well, has for a long time resulted in the permanence of complex cities in which nature coexisted with the urban. Nature, although managed, domesticated and controlled, represented an alterity that was recognised and respected, and often feared, with a clear feeling of dependence for human survival. Even after World War II, we could still read the polycentric structure of the territory that, thanks to the containment of centres and the balance between urban population and resource availability, allowed for intercon-

F. Alberti et al. (eds.), *Towards Sustainable Urban Regeneration: Utilizing Environmental, Cultural, and Social Assets*,
Advances in Science, Technology & Innovation, https://doi.org/10.1007/978-3-032-02091-8_10

nection, ecological connectivity and careful use of resources. Modern urban planning progressively excludes nature from the conceptual horizon of urban design. However, it is only in the second half of the Twentieth century that, due to a combination of several phenomena, the dialogue between the urban and nature becomes increasingly difficult. The polarisation of major centres, the abandonment of the countryside, population growth and increased extractive capacity are leading to a collapse of settlement structures that urban planning struggles to contain and manage, lacking cultural horizons to refer to.

2 Chasing New Forms of Relation with Nature

After structural and increasingly devastating ecological and economic crises, a new idea of nature has been taking shape and gaining space in the urban planning disciplines, which attempts to depart from the Eurocentric conception, rediscovering other cosmovisions. It seems now widely accepted that there is continuity between the biological and human dimensions, one constantly influencing the other. For instance, the mutations studied by epigenetics, which occur with age and exposure to environmental factors, dependent in turn on cultural factors, can modify gene expression and thus be transmissible. In many non-European cultures, the distinction between human and non-human simply does not exist; on the contrary, there is a perception of the interconnectedness of the different elements of the earth, rocks, plants and animals. In the Hawaiian culture, for example, a term that can be translated as 'person' is used indifferently for animal, human and non-human, plant and mineral elements (Borgnino 2022). Both the way 'nature' is interpreted and the way 'culture' is understood are, in fact, cultural expressions, influenced by the values, choices and policies of populations.

It is therefore not possible to objectivise and abstract the conception of nature. We act within semiospheres that have constructed the nature-culture dialectic. The concept of nature to which we refer is therefore part of a purely Eurocentric conception (Staid 2022, p. 18), to such an extent that it is untranslatable in many languages. It is from this separation between humans and nature, from this conception of domination over nature, that the Anthropocene originates: a condition where the intrinsic value of nature, reduced to an object, is denied and resources are 'extracted' to generate human wellbeing, a wellbeing now criticised by many precisely because of the anthropocentric and colonial vision it implies (Mies 2014). Because of this extractivist vision, many areas in the global south are still being depleted of resources that generate well-being elsewhere.

It is therefore necessary to embark on a path away from modernisation, towards a re-sematisation of the concept of nature that recentres the dimension of interconnectedness and raises awareness about how much the political process of 'naturalisation' has contributed to making certain visions—such as the rational, colonial and patriarchal ones—experienced by European culture as 'normal'. In some contexts, such as Ecuador, touched by the oil boom of the 1970s, urbanised and modernised in a short time, but left poorer and ecologically devastated, constitutional rights have been conferred on nature: 'the right to the integral respect of its existence and to the maintenance and regeneration of its life cycles, structures, functions and evolutionary processes' (Art. 7). It was a sort of compromise between different visions: on one side the idea of a unitary nature of native peoples, on the other western thought, which uses rights as a means of defence (Biemann and Tavares 2020), born in interconnection with the historical struggles for human rights and collective land rights.

The theme of the rights of nature and their representation in decision-making arenas, as is well known, has long been explored (Latour 2002; Merchant 2012). Beyond its actual effectiveness, this positioning highlights the current need to recalibrate the relationship with nature in spatial governance instruments. Although it is now commonly accepted that 'we are all nature' and that there is only one health, that of the planet we are a part of, an unknown and worrying phenomenon is taking place that reverses the stated objective. If, with the overcoming of dualism, the desired objective is to reintegrate human activity within ecosystem logics, to erase extractivism, colonialism and patriarchy (Staid 2022), the neoliberal politics designed to achieve such goals does not dampen the power of human subjectivity, but rather amplifies it and 'blends the idea of a complete plasticity of the real as bendability to the will of the actor and as fluidity, contingency and unpredictability, drawing from this an entirely unprecedented logic of government' (Pellizzoni 2023, p. 18, authors' translation). In a phase of fusion-confusion between nature and humanity, in a regime of unequal power, nature is once again in danger of being dispossessed, only in other forms, that is, by resorting to the metaphors of hybridisation, cross-fertilisation and interconnection. 'All that in a game where who is an actor and what is acted upon become relative, but the endpoint (at least so it is assumed) is marked in favour of instrumental reason. Operating in a condition of ontological fluidity brings into full light the latter's engine: a will to power in its pure, unbridled state, devoid of efficient or final cause' (ibid., p. 51, authors' translation).

3 Study Aim

Some of the recent operational concepts linked to the search for new forms of relationship with nature, such as biomimesis (Pedersen Zari 2018) or ecosystem services (Costanza et al. 2017), may offer relevant advantages for spatial planning, but at present, they overly rely on instrumental logics.

To read nature as a territorial heritage, a living subject, outcome of the dialogue between nature and culture in the long time of history, in which values and knowledge are enclosed and constantly reinterpreted, requires acceptance and mutual recognition, in a balance between relationality and otherness. Without the recognition of otherness, only the strongest will benefit from the relationship, just as in a loving relationship. We believe that the lens of the territory as a living subject and the device of the urban bioregion (Magnaghi 2020a, b) can help point towards the redefinition of non-instrumental modes of dialogue between settled communities and nature.

Our aim is therefore to explore the potential for an integration of the ecosystem services (ESS) concept into the urban bioregion approach to territorial planning. First, a critique of the ESS is briefly outlined through a literature review. Then, an experiment of bioregional ESS application is described with the help of a model case study from the Florence metropolitan area, concerning integrated strategies for the management of water.

4 Ecosystem Services, or the Invisibility of Nature

The operational concept of ecosystem services (ESS), defined as the direct and indirect contributions of ecosystems to human wellbeing (TEEB 2010), gained recognition as a corrective mechanism to address the distortions of the global economy, that appeared unable to economically account for the human dependence from nature (Costanza et al. 1997). By the end of the Millennium, publications by Daily and Costanza sparked the interest of researchers and policymakers and popularised the concept to a wide audience (Craig et al. 2019); since then, the ESS has rapidly become a widely accepted and valued analytical and operational framework. However, the new concept of ESS, which aimed at 'correcting, expanding' the economic dimension, has resulted in a reinforcement of the modern 'economism'—that is, the hegemonic worldview, or cosmovision, of neoclassical economics (Scott Cato 2013). If the Nineties, the decade before the ecosystem services concept upsurge, were years of disputes and contestations over the paradigm of economic growth, the diffusion of market-based language and tools in the following decades testifies for a relegitimisation of growth as a frame and basis for the management of natural resources (Levidow 2020). The notion, together with coeval and synergistic concepts such as that of 'natural capital' (Fig. 1), often employed to refer to the ecosystems that provide the services—capital is meant as a stock which, over time, yields a flow of services (Costanza et al. 2017)—appears to have even intensified the centuries-long process of domestication and objectification of nature. Natural capital assets include species and genes as well as the ecosystems, which are described as its least recognised and most undervalued component (UNEP 2011). Paradoxically, the lens of natural capital and economic accounting is trumpeted as a strategy to recognise and protect nature. Since 'economic invisibility has been a major reason for the neglect of natural capital' (NCC 2014), accounting for the economic value of nature through procedures of Natural Capital Accounting and ESS assessment is proposed as a valid framework to manage and protect natural resources (TEEB 2010).

Despite the global institutional support that the ESS concept has encountered, such an operational strategy has, in fact undergone consistent critique. Among the issues raised is the excessive simplification of the complex relations within natural systems, with a reduction of the many ecological ways of thinking to a single stock-flow model, which produces the paradoxical result of turning the ESS 'from eye-opening metaphor to ecological blinders' (Norgaard 2010). Many, especially because of the practice of monetisation of the ESS, foresee a perspective of degradation of nature through an expansion of capitalism to ecosystems: a neoliberalisation (Castree 2008), a commodification or 'assetization' (Birch and Muniesa 2020) of nature. Other strains of critique concern the biases intrinsic to the concept of service, which does not acknowledge disservices and trade-offs, the abstraction from social issues, the difficulties in measuring and comparing the services, and the fact that variations of value over time are not accounted for. On a more general level, Ernstson and Sörlin (2012) interpret the ESS as a technology of globalisation characterised by being abstract, transferable and spread homogeneously all over the world thanks to the international frameworks and the work of multinational consultancy companies. Not so different, one could observe, from the technology of normalised cartography in its endeavour to objectively investigate, describe and measure the world with a single universal language (Poli 2019). To clarify their claim, Ernstson and Sörlin (2012) place the ESS approach against the background of a global-wide managerial turn in public administration, underlining that

> These are technologies precisely because they offer certain prescribed routines, techniques, and practices whereby the standardization and the ensuing comparability is achieved, which in turn opens up the possibility to move issues and dilemmas, for example of controversial urban planning, from the mess of local

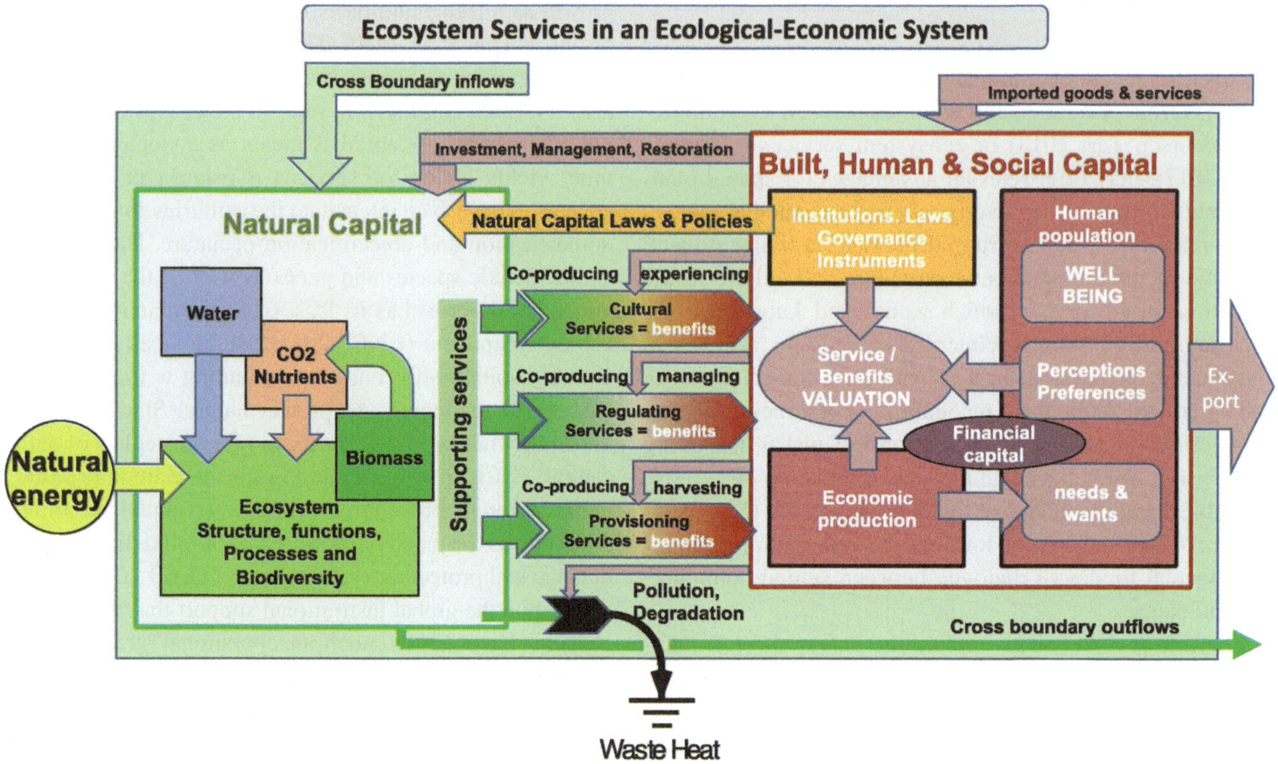

Fig. 1 The interactions and feedbacks between built, human, social, and natural 'capital' implied in the production of ecosystem services, as depicted by the currently employed models. (Source: Costanza et al. 2017)

> claims and into the clarity of numbers, that is from the sphere of politics to that of science (or quasi science) (ibid., p. 281).

The neoliberal process in which we are immersed tends to compress and deny value to the political dimension of human action to privilege and expand the procedural domain. The move from the terrain of politics to that of management is achieved through de-historicisation, de-contextualisation, de-ecologisation and silencing of alternative voices. The violence of the attempts of abstraction-extraction—i.e. an abstraction that makes it impossible to encounter nature in other forms than that of a commodity for the extraction of surplus value from the world of life (Robertson 2012)—resonates with the violence imposed on nature with the original formulation of the scientific method (Merchant 1980) which established a knowledge system based on mastery and domination, on 'controlling reproduction for the sake of production' (Shiva 1992, p. 232) thus turning living nature into a bare 'resource'.

The ontological shift from living nature to resource and now to capital asset or commodity should therefore be read as an evolving domination technique. If modernity achieved domination by separating, organising into fixed categories (such as nature and culture), predicting and controlling, the new paradigm draws its strength from indeterminacy, emergency, resilience and fluidity (Table 1). In the present turn of neoliberal capitalism, such a technique takes advantage of an epistemic-ontological confusion among things, information and concepts (Pellizzoni 2023). Through the abstractions of the markets and payments for ecosystem services, natural 'things' appear not to acquire but to simply reveal the commodity nature they already originally had. This way, everything that modern capitalism excluded from the economic rationality is now absorbed within its boundaries through a blurring of the categories of commodity/non-commodity, production/reproduction and so on, which also explains the diffusion of 'the lexicon of capital, to which nothing is alien anymore' (ibid., p. 67, authors' translation).

5 A Bioregional ESS Approach for the Mugello Area, Tuscany

The concept of urban bioregion describes a local territorial system whose main elements are (a) a reticular settlement system characterised by synergistic rather than centre-periphery relations, (b) hydrogeomorphological and environmental systems that coevolve with the settlement and the agro-forestry systems, and (c) forms of self-government of the living environment by the local community, aimed at

Table 1 The table summarises and confronts the main characteristics of the opposed anthropocentric and ecocentric discourses, adapted from Ventura Pujolar et al. (2002), those of the emerging neoliberal discourse, blending and blurring the first two and those of an alternative discourse, named relational and based on a balance between relationship and otherness, derived from the territorialist approach that Magnaghi (2001) indicates as a neither dissipative nor conservative attitude towards the territory

Discourse	Anthropocentric	Ecocentric	Fluid	Relational
Society–nature relationship	Demand/consumption	Preservation and restoration	Nondualist relationship	Democratic care and enhancement
Aim	Increase in material wellbeing	Natural balance	Ambivalent	Natural and human coevolutionary balance
Reference	Fluxes	Pristine nature	Indetermined, fluid nature	Human environment
Scientific discipline(s)	Water regulation (quality and quantity forecast and control)	River ecology (natural values preservation and enhancement)	Complexity sciences	Territory//landscape sciences (human and natural values enhancement)
Time	Linear progress	Retrospective	Non-linear progress	*Longue durée*
Space	Anonymous	A (new) previous place	Fluid	Identitarian, historical, relational place
Social aspects of the river	Economic-rational	Identitarian-emotional	Economic, instrumental	Relational/dialogical
Management model	Managerial (expert-led)	Impact minimisation// contemplation (expert-led)	Continuous experimentation (expert-led)	Integrated, cooperative (expert and experiential knowledge)
Perception of conflicts	Caused by ecocentric discourse	Caused by anthropocentric discourse	Conflict and emergency are an opportunity	Rooted in the disconnection of human and natural realms

assuring the self-sustainability of the territorial ecosystem and the wellbeing of the inhabitants. Such well-being does not derive from the extraction of resources and profit but rather from a wealth-producing enhancement of territorial heritage and a tendency to the closure of the territorial metabolic cycles (Magnaghi 2020a).

In the last years, territorialist scholars have focused on a quest for an operational framework into which the ESS can be pragmatically applied, to achieve wider bioregional goals within a local, territorialised and socially shared strategy (Poli 2020). Their research has highlighted the main issues to address to achieve such integration.

First, even though many of the ecosystem benefits derive from portions of territory not subjected to human action, such as areas of natural forest, humans draw most of their services from a 'second nature', the 'territorial ecosystem' (Saragosa 2001), which is the result of a coevolutionary process. Moreover, community actions to preserve 'territorial figures', which express and transmit the rules of settlement rationality, have a primary role in the (re)production of the ESS. Second, the abstraction from the site's landscape and layout, brought by functional design and standardised accounting techniques, can lead to unintended consequences where any solution can be accepted as long as it is functional to the delivery of the services. Thus, 'service' is itself an ambiguous word, reflecting a predominant performance-based view towards the territory. Third, sectoralisation must be avoided in favour of a multifunctional perspective and the integration of systems and services. Fourth, and more generally, introducing the ESS without challenging the current development model can increase the reliance on exogenous fluxes: territorialising the fluxes according to the limits of local endowments is instead key to the territorialist approach to the ESS (Marino et al. 2023). The parameters on which the population's well-being is defined are also not neutral to the project and to the evaluation of the ESS. Fifth, the pricing of services as a form of evaluation mediated by monetary values is to be rejected in the way it is predominantly used, especially as a form of an all-encompassing assessment of a 'capitalised nature'. Instead, a form of payment can be envisaged as a recognition and an incentive for those who take care of the territorial endowments and their reproducibility over time through community design. Payments for ecosystem services (PES), understood in this light, have positive implications if they activate forms of civil and community economy that strengthen the local market, redistribute benefits to the community and increase local awareness (Poli et al. 2020).

With this in mind, a qualitative methodology was outlined that, by resorting to the territorial heritage, its forms and rules, has overcome the tendency towards pure functional and quantitative accounting inherent in the concept of 'natural capital' and 'ecosystem service' (Figs. 2 and 3). The framework is based on identifying the endowment of heritage elements, describing their existence value and the functions they perform, and subsequently translating these elements into resources and services capable of producing benefits for the population, which can be assessed both in quantitative and qualitative terms of increased well-being and quality of life. The territorialist approach to the ESS,

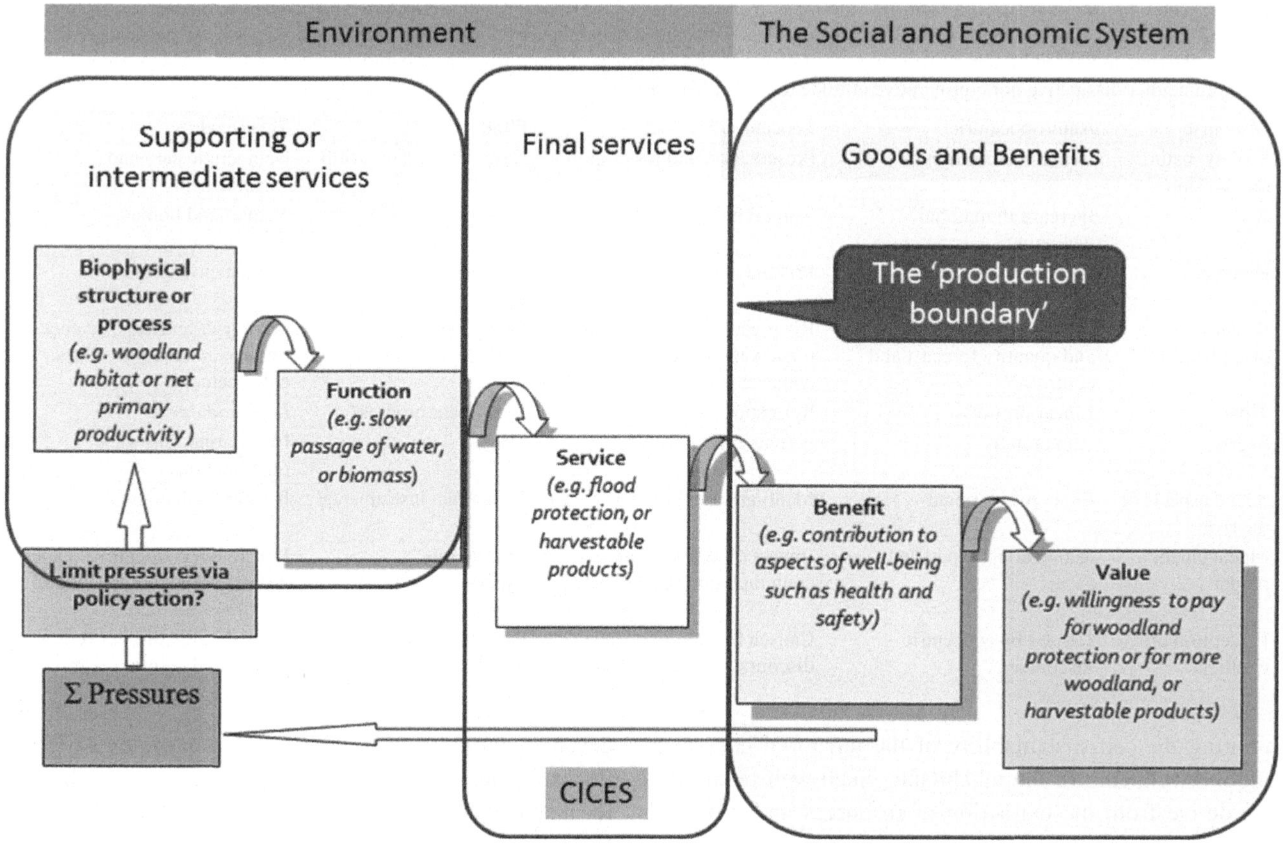

Fig. 2 The cascade diagram, describing the process leading from structure to functions, to services, to benefits, and finally to value (Source: Potschin-Young et al. 2018). The cascade model was adopted by the CICES classification of ESS (EEA 2024). Here, the 'supporting services' defined in the MA are not included but rather treated as part of the underlying ecosystem structures, process and functions

therefore, on one side introduces the territorial heritage as a source of the services and goods—as opposed to the financial definition of natural capital. Since it stems from the synergetic and co-evolutionary relationship between local population, settlements and environment, the heritage value of the territory's elements is dynamic, incremental and measurable only subjectively and relationally (Magnaghi 2020b). On the other side, by binding the ESS assessment to an analysis of the consistency of the territorial heritage and its rules of reproducibility, the territorialist perspective uses the ESS as a way to shed new light on the new centrality of those parts of the urban bioregion—as, for instance, mountain areas—that are usually considered marginal in comparison to the urban areas (Dematteis 2020). The 'vibrancy of the margins', the 'new centrality of the mountains' (Bonomi 2013; Bolognesi and Corrado 2021) indicate this new way forward. To sustain this 'return to the mountain' it is also necessary to establish conscious alliances between the rural inner or mountain areas and the city, based on a vision of mutual interdependence, within which resources and services can be exchanged.

This framework has been tested in the Mugello area, in the North of the Metropolitan City of Florence (Fig. 4), an intermountain basin with large wooded and rural areas and abundant water reservoirs that feed the metropolitan city (Marino et al. 2023). The ESS were part of a strategy aimed at preserving the qualitative and quantitative qualities of the water resources through the sustainable management of the agricultural and forestry areas. To achieve this goal, a comprehensive strategy of local design and patrimonial management of the ESS related to water was elaborated (Poli and Butelli 2023). The strategy is integrated, territorialised and shared with the locally settled community (Fig. 5). A PES system was outlined with the aim of increasing groundwater storage to avoid summer droughts through the realisation of drainage and infiltration systems and specific tree planting, also used for wood and fruit production with a multifunctional approach. The landowners bordering the Sieve River and the land reclamation authority ('Consorzio di Bonifica'), as well as the Tuscany Region, were among the designated actors involved in the draft of a public-private agreement tool. Lump sum payments for the creation of water infiltration forestry areas and annual payments for their maintenance were proposed, on the model of the Brenta River water infiltration project, which has already started to show positive results (Villamagna et al. 2013). Appropriate funding

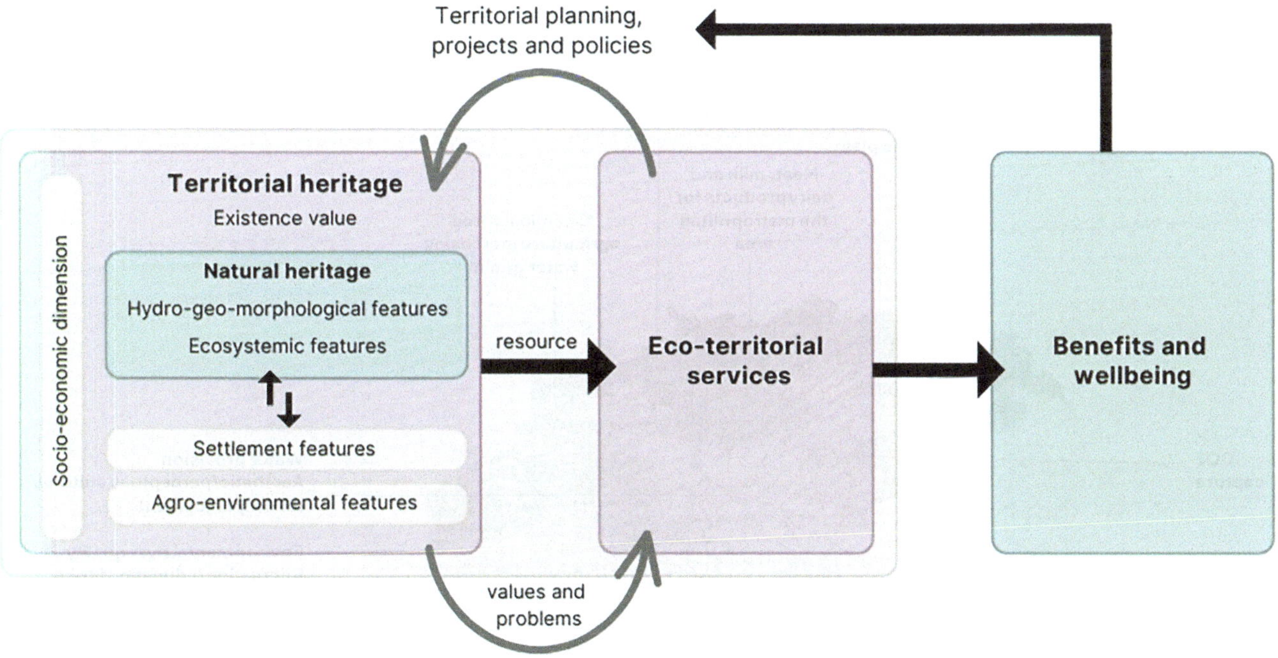

Fig. 3 The eco-territorial model overcomes the tendency to abstraction and quantitative-functional accounting implicit in the previous models, as well as the definition of 'natural capital' and 'ecosystem service'. The territorial heritage is highly dependent on territorial forms, landscape and social practices and has existence value regardless of its use as a resource or service producer. Moreover, the services are distinguished but never separated from the specific territorial heritage and its 'rules of reproducibility'. (Source: research 'The metropolitan city of Florence: a system of polycentric, self-sustainable and resilient urban bioregions', coordinator D. Poli, image by M. Chiti and G. Granatiero)

sources were identified, including the already existing water service rate component related to the environmental costs for the regeneration of the resource.

6 Bioregional Communities and the Management of Nature

The case here described features an attempt to use the now mainstream instruments provided by the ESS to bioregional ends. Adapting the ESS concept to the territorialist frame of the urban bioregion has meant reframing their meaning, revisiting the terminology, reorienting goals and methods, reconsidering the role of the actors and more.

Pellizzoni and Centemeri (2022), in framing various agricultural styles as related to different accounts of human-nature material dependency, describe the bioregional approach as an experimentation that seeks to combine the ESS approach with peasant agroecology, building on accounts of weak comparability between different ES values. 'Key conceptual move here is replacing 'nature' with 'territory' as a backdrop for conceiving ES, which in this way results in being entangled with human settlements and the host of interventions whereby communities take care over time of their homeland; hence typically multifunctional, integrated and localized' (ibid., p. 364). This entails taking distance from the abstract, standardised, universalising and ultimately depoliticising application of the ESS, mainly witnessed at a global level.

Actually, within the urban bioregion, the natural environment plays the role of material precondition for the very existence of the human settlement, and is deeply tied to the participation institutions which guarantee the local community's active involvement in the reproduction of the living systems. Such an active involvement is not limited to common forms of participation: rather, it takes the form of social co-design and co-production, and points to the reactivation of a territorialising relationality between the settled community and the natural matrices. The Mugello project provides a model case of a tailor-made management solution oriented towards a renovated territorial balance, whereby the city-mountain exchanges are marked by mutuality and reciprocity. Gaining new centrality for the marginal parts of metropolitan areas, however, still requires governance innovations to strengthen the self-management capacity of the local communities and achieve a balance between the 'strong' urban subject and the 'others'. A necessary step in this direction will be to effectively introduce a strategic instrument to federate a network of public and private local and territorialised initiatives able to represent a more powerful collective actor in the co-design and dialogue with the institutions on one side, and on the other side with the many actors and groups that aspire to grab, extract and capitalise on the territorial and natural resources.

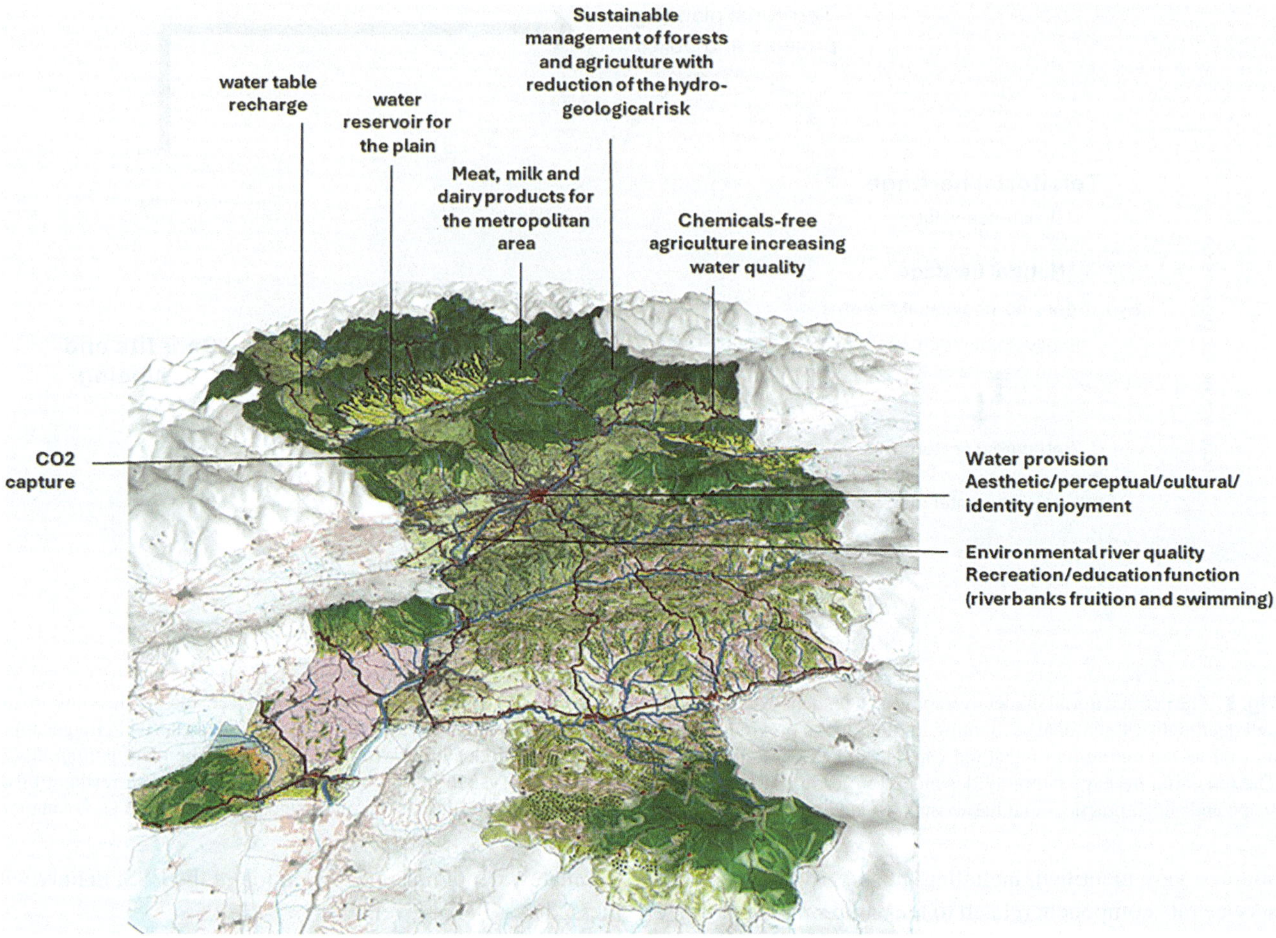

Fig. 4 Experimental qualitative representation of ecosystem/ecoterritorial services between mountain and urban areas in the metropolitan area of Florence. (Source: research 'The metropolitan city of Florence: a system of polycentric, self-sustainable and resilient urban bioregions', coordinator D. Poli, image by M. Chiti and G. Granatiero)

In the debate that followed the presentation of the Mugello strategy, a need was stressed to further develop the governance model through agreement tools between providers and beneficiaries of ESS. In the proposed scheme, the agreements are guaranteed and supported by a regulatory framework established by public administrations, while the facilitation of the contract construction process is entrusted to structured local associations, such as community cooperatives, which can guarantee a role of mediation and cooperation between the parties. Moreover, a guarantee committee representing the various stakeholders would supervise and monitor the correct implementation of the entire strategy.

Overall, the redefinition of the urban-rural exchanges and metabolisms within the urban bioregion is not simply a matter of economic compensation: rather, it is deeply tied to a recentring of the settled communities' role as protagonists of the territorial resources management, to ultimately recover that nature-community nexus that modernity attempted to cut and post-modernity is still attempting to blur and dissolve.

Acknowledgements This paper presents some of the preliminary results of the research project BioCode—Bioregional planning tools to co-design life places. Empowering local communities to manage and protect natural resources (prot. P2022NSAEJ), funded by the Italian Ministry of University and Research and the European Union—NestGenerationEU under the PRIN 2022 PNRR programme.

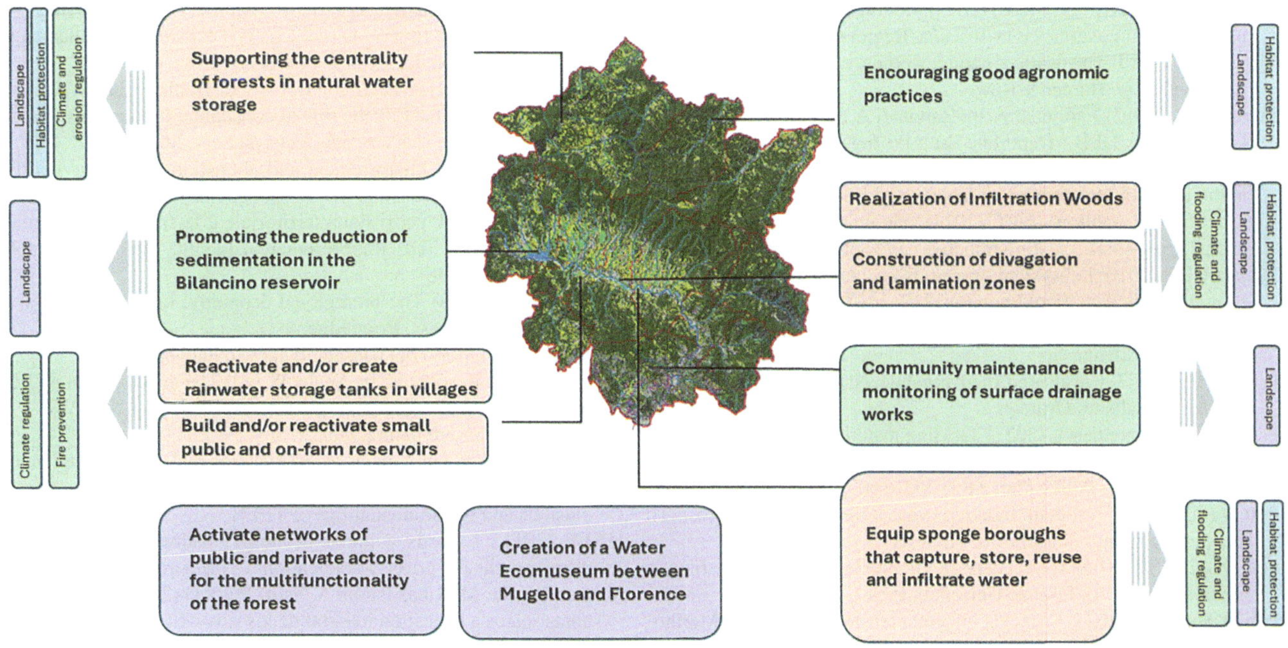

Fig. 5 Integrated and territorialised strategies for the Mugello territory. (Source: research "A support for the definition of a socio-territorial production model of ecosystem services for the integrated water service' (2021), coordinated by D. Poli; image by D. Poli and E. Butelli)

Finanziato dall'Unione europea NextGenerationEU
Ministero dell'Università e della Ricerca
Italiadomani PIANO NAZIONALE DI RIPRESA E RESILIENZA

Attributions The text is the result of the joint work of the two authors. However, parts 1, 2 and 3 should be attributed to Daniela Poli, while parts 4, 5 and 6 to Giulia Luciani.

References

Biemann U, Tavares P (2020) Forest law—Foresta giuridica. Nottetempo

Birch K, Muniesa F (eds) (2020) Assetization: turning things into assets in technoscientific capitalism. The MIT Press

Bolognesi, M., & Corrado, F. (Eds.). (2021). La centralità della montagna. Scienze del territorio 9

Bonomi A (2013) Il capitalismo in-finito. Indagini sui territori della crisi. Einaudi

Borgnino E (2022) Ecologie native. Elèutera

Castree N (2008) Neoliberalising nature: processes, effects, and evaluations. Environ Plan A 40:153–173

Costanza R, dArge R, de Groot R, Farber S, Grasso M, Hannon B, Limburg K, Naeem S, Oneill RV, Paruelo J, Raskin RG, Sutton P, van den Belt M (1997) The value of the world's ecosystem services and natural capital. Nature 387(6630):253–260

Costanza R, de Groot R, Braat L, Kubiszewski I, Fioramonti L, Sutton P, Farber S, Grasso M (2017) Twenty years of ecosystem services: how far have we come and how far do we still need to go? Ecosyst Serv 28:1–16

Craig MPA, Stevenson H, Meadowcroft J (2019) Debating nature's value: epistemic strategy and struggle in the story of 'ecosystem services'. J Environ Policy Plan 21(6):811–825

Dematteis G (1985) Le metafore della Terra. Feltrinelli

Dematteis G (2020) I servizi ecosistemici nella riproduzione dei sistemi territoriali. Osservazioni da una ricerca sugli scambi montagna-città. In: Poli D (ed) I servizi ecosistemici nella pianificazione bioregionale. Firenze University Press, pp 47–57

EEA—European Environment Agency (2024) CICES: towards a common classification of ecosystem services. https://cices.eu/

Ernstson H, Sörlin S (2012) Ecosystem services as technology of globalization: on articulating values in urban nature. Ecol Econ 86:274–284

Latour B (2002) Una sociologia senza oggetto? Note sull'interoggettività. In: Landowski E, Marrone G (eds) La società degli oggetti. Problemi di interoggettività. Meltemi, pp 203–229

Levidow L (2020) Turning nature into an asset: corporate strategies for rent-seeking. In: Birch K, Muniesa F (eds) Assetization: turning things into assets in technoscientific capitalism. The MIT Press, pp 225–257

Magnaghi A (ed) (2001) Rappresentare i luoghi. Metodi e tecniche. Alinea

Magnaghi A (2020a) The territorialist approach to urban bioregions. In: Fanfani D, Matarán Ruiz A (eds) Bioregional planning and design, vol I. Perspectives on a transitional century. Springer, pp 33–62

Magnaghi A (2020b) Un'introduzione ai servizi eco-territoriali. In: Poli D (ed) I servizi ecosistemici nella pianificazione bioregionale. Firenze University Press, pp 37–45

Marino D, Poli D, Rovai M (eds) (2023) Montagna, servizi ecosistemici e strumenti di governance in Toscana. Regione Toscana

Merchant C (1980) The death of nature: women, ecology, and the scientific revolution. Harper & Row

Merchant C (2012) Partnership. In: Zabonati A (ed) Ecofemminismo/Ecofeminism, DEP—Deportate Esuli Profughe, vol 20, pp 34–52

Mies M (2014) Patriarchy and accumulation on a world scale: women in the international division of labour. Zed Books

Natural Capital Coalition—NCC (2014) Valuing natural capital in business. Taking stock: existing initiatives and applications. ICAEW

Norgaard RB (2010) Ecosystem services from eye-opening metaphor to ecological blinders. Ecol Econ 69:1219–1227

Pedersen Zari M (2018) Regenerative urban design and ecosystem biomimicry. Routledge

Pellizzoni L (2023) Cavalcare l'ingovernabile. Natura, neoliberalismo e nuovi materialismi. Orthotes

Pellizzoni L, Centemeri L (2022) Tackling material dependency in sustainability transition: rationales and insights from the agriculture sector. J Environ Policy Plan 24(3):355–366

Poli D (2019) Rappresentare mondi di vita. Radici storiche e prospettive per il progetto di territorio. Mimesis

Poli D. (Ed.). (2020). I servizi ecosistemici nella pianificazione bioregionale (pp. 1–34). Firenze University Press

Poli D, Butelli E (2023) Strategia integrata territorializzata. In: Marino D, Poli D, Rovai M (eds) Montagna, servizi ecosistemici e strumenti di governance in Toscana. Regione Toscana, pp 243–247

Poli D, Chiti M, Granatiero G (2020) L'approccio patrimoniale ai servizi ecosistemici. In: Poli D (ed) I servizi ecosistemici nella pianificazione bioregionale. Firenze University Press, pp 1–34

Potschin-Young M, Haines-Young R, Görg C, Heink U, Jax K, Schleyer C (2018) Understanding the role of conceptual frameworks: reading the ecosystem service cascade. Ecosyst Serv 29(Pt C):428–440

Robertson M (2012) Measurement and alienation: making a world of ecosystem services. Trans Inst Br Geogr NS 37:386–401

Saragosa C (2001) L'ecosistema territoriale e la sua base ambientale. In: Magnaghi A (ed) Rappresentare i luoghi. Metodi e tecniche. Alinea, pp 55–138

Scott Cato M (2013) The bioregional economy: land, liberty and the pursuit of happiness. Routledge

Sereni E (1961) Storia del paesaggio agrario italiano. Laterza

Shiva V (1992) Resources. In: Sachs W (ed) The development dictionary. A guide to knowledge as power. Zed Books, pp 206–218

Staid A (2022) Essere natura. Utet

TEEB (2010) The economics of ecosystems and biodiversity: mainstreaming the economics of nature: a synthesis of the approach, conclusions and recommendations of TEEB

UNEP (2011) Towards a green economy: pathways to sustainable development and poverty eradication. UN Environment Programme

Ventura Pujolar M, Ribas Palom A, Saurí Pujol D (2002) Dos discursos antagónicos a la gestión integral de los ríos—El río antropocéntrico 'versus' el río ecocéntrico. Estud Geogr 63(246):119–141

Villamagna AM, Angermeier PL, Bennet EM (2013) Capacity, pressure, demand and flow: a conceptual framework for analyzing ecosystem service provision and delivery. Ecol Complex 15:114–121

Enhancing Nature-Based Solutions Through the Bioregional Approach: The Case Study of Agri Park of Florentine Piana, Italy

Maria Rita Gisotti and Elena Tarsi

Abstract

Nature is increasingly recognized as a key ally in addressing a wide range of societal challenges, including microclimate regulation, climate change mitigation, water management, the creation of green jobs, and the promotion of health and well-being. Nature-based solutions (NbS) are defined as approaches that are inspired by and work with nature, providing cost effective strategies that deliver simultaneous environmental, social, and economic benefits. However, as highlighted in the literature, significant challenges remain in fully realizing the transformative potential of NbS. This paper explores how current NbS methodologies and practices can be enhanced by critically addressing the limitations in their conception, design, and implementation. Presenting the case study of the Agri Park of the Florentine Piana, in Italy, we propose the adoption of the Urban Bioregion approach—a design framework aimed at rebalancing the relationship between built and open spaces—as a means to enrich NbS strategies and foster a renewed co-evolution between human communities and the natural environment.

Keywords

Ecosystems · Landscapes · Urban bioregion · Societal challenges

1 Introduction

Nature-based solutions (NbS) are defined as approaches inspired and endorsed by nature, offering cost-effective solutions with simultaneous environmental (EC 2015; Keesstra et al. 2018), social, and economic benefits (Albert et al. 2017). NbS has been defined by IUCN—International Union for Conservation of Nature—as "actions to protect, sustainably manage, and restore natural or modified ecosystems, that address societal challenges effectively and adaptively, simultaneously providing human well-being and biodiversity benefits" (Cohen-Shacham et al. 2016). They encompass a broad range of actions, based on the ecosystem approach, which involve working with nature for societal benefits (Seddon et al. 2021). They emerged as a comprehensive concept enclosing various ecosystem-based approaches to address the impacts of the multifaceted environmental crisis on both humans and non-human entities (Liu et al. 2021). However, as emerged in literature, there are still important challenges to overcome to reach its approach's powerful capability.

The paper examines ways to enhance the current methodology and practice of NbS in Europe by addressing the persistent challenges in its conception, design, and implementation. The authors propose adopting the Urban Bioregion approach (Fanfani and Mataran 2020; Magnaghi 2020; Colavitti and Serra 2022), a design methodology aimed at rebalancing the relationship between built and open spaces, fostering a new co-evolution between humans and the environment. Developed by the Italian Territorialist School, this approach can help overcome the limitations of the sectoral and fragmented methods often associated with NbS interventions by embedding them within an ecosystemic vision of the territory. Furthermore, by emphasizing the role of local communities, it offers new opportunities for incorporating citizens' perspectives and values into the design process of such interventions.

The paper is organized as follows: the next section reviews the current state of research and projects, particularly within the European context, that have focused on NbS, highlighting the remaining critical issues and challenges that need to be addressed. The subsequent section presents the Urban Bioregion approach, developed by the Territorialist School of Florence (Italy), and outlines its key structural aspects. Following this, the authors present the case study of the Agri

M. R. Gisotti · E. Tarsi (✉)
Department of Architecture, University of Florence, Florence, Italy
e-mail: mariarita.gisotti@unifi.it; elena.tarsi@unifi.it

F. Alberti et al. (eds.), *Towards Sustainable Urban Regeneration: Utilizing Environmental, Cultural, and Social Assets*, Advances in Science, Technology & Innovation, https://doi.org/10.1007/978-3-032-02091-8_11

Park of the Florentine Piana in the Tuscany Region (Italy) and in the discussion examine how the Urban Bioregion framework can help address the existing challenges in the implementation of NbS from a multidimensional and cross-scale perspective. Finally, the conclusion offers potential avenues for future research and practical applications.

2 State of the Art

The scientific literature on NbS has expanded significantly in recent years, paralleling the development of practical experiences and subsequent evaluations that highlight both positive outcomes and ongoing challenges. The European Commission has undeniably played a central role in these efforts by funding numerous projects and research initiatives on the topic. One notable outcome of this focus is the OPPLA repository, also funded by the Commission and organized with the participation of 60 entities from across Europe, aimed at consolidating knowledge developed within the European context and making it widely accessible.

Despite this extensive experimentation, the literature continues to highlight several challenges in ensuring that NbS lead to meaningful transformations, not only in physical space but also in social dynamics. These challenges are diverse, ranging from governance issues and the integration of NbS into planning processes to the absence of an integrated approach due to the sectoral nature of these initiatives. Other issues include difficulties in effectively engaging stakeholders and citizens, as well as the lack of studies examining the distribution of the positive impacts of NbS interventions.

Below, we will attempt to briefly outline this framework of issues, organizing them into three main challenges:

2.1 Governance

NbS are not yet widely integrated into urban development practices (Fastenrath et al. 2020; Dorst et al. 2022). They tend to be on the margins of the daily operations of many stakeholders and, as a result, are not fully embedded in planning and governance systems. One of the perhaps most critical aspects concerns limited information regarding NbS integration across levels and scales of urban governance, particularly the lack of studies that assess such integration in urban policy and planning (Kauark-Fontes et al. 2023). As the aforementioned authors point out, "NbS integration in urban policy and planning is novel in all cities but advancing considerably with initial dialogues among the public sector, academia, and local actors of various provenances" (p. 1).

Knowledge about NbS is often fragmented and isolated within different departments, disciplines, sectors, and jurisdictions, which frequently have conflicting priorities (Clar et al. 2013; Pasquini and Cowling 2015; Timboe and Pharr 2021). The lack of an integrated approach can undermine not only the proper implementation of NbS but also the interactions and interdependencies of ecosystems (Cowling et al. 2008; Hauck et al. 2013). Moreover, various authors have highlighted a series of Temporal, Spatial, Functional, Social equity, and Species trade-offs (Chelleri et al. 2015; Mexia et al. 2018; Turkelboom et al. 2018; Parris et al. 2018). While these trade-offs can be seen on the one hand as the result of a sectoral vision, on the other hand, they are also the outcome of sporadic actions with a primarily technical nature; in other words, they lack a deep understanding of the place in which they are implemented, its environmental and social dynamics, and a long-term vision.

2.2 Inclusion

The strongly technical approach that characterized the first generation of NbS underestimated the need to include citizens' perspectives in the design and management of interventions (Haase et al. 2014). It has become increasingly evident that, in order for interventions to bring about not only physical but also social transformations, it is essential to first and foremost consider community needs and expectations, as well as to pay particular attention to local vulnerabilities and minorities. While this aspect has been progressively addressed by many international practices, a gap still remains regarding a poor incorporation of local cultural values, both socially and environmentally (Combes and Viles 2021). This aspect, in fact, requires specific attention during the pre-design phase, necessitating the reconstruction and understanding of the local context in which the project is to be implemented, where history and innovation, cultural representations, and environmental characteristics can be harmonized. Finally, for the sake of completeness, it is necessary to mention the criticisms directed at NbS experiences, which highlight a poor consideration of the intrinsic values of nature beyond an anthropocentric view (Folke et al. 2011; Schröter et al. 2014; Eggermont et al. 2015). The recognition of nature's intrinsic value is at the core of the debate on Ecosystem Services and NCP (IPBES 2019; Díaz et al. 2018) and suggests the need to fundamentally rethink the relationship between humans and nature. The participatory processes that can be developed for the design and implementation of NbS offer this collective opportunity for addressing such radical rethinking.

2.3 Management, Monitoring, and Evaluation

The aspect of long-term management and monitoring of any infrastructural intervention is crucial, especially when it integrates plant species. The absence of both has predominantly characterized an early wave of interventions through NbS (Van Herzele et al. 2005), in many cases leading to counterproductive effects. Another aspect highlighted in the literature is the existence of a still weak assessment framework (Raymond et al. 2017; EC 2021) and a specific difficulty in evaluating the improvement of biodiversity and ecosystem resilience (Key et al. 2022; Calliari et al. 2019). It has been emphasized how poorly designed NbS can cause adverse impacts on climate mitigation, local communities, biodiversity, and ecosystem services (Seddon et al. 2021).

Although presented as three categories of issues, it is clear that the challenges still present for NbS to reach their full transformative potential are, in fact, strongly interconnected. Addressing them requires joint efforts from researchers across various disciplines (agronomists, botanists, geographers, urban planners, architects, engineers, sociologists, and social scientists), as well as from stakeholders working in the field (professionals, practitioners, local communities, and citizens), and local government technicians.

3 The Urban Bioregion Approach

To contribute to overcoming the above-mentioned critical issues and to achieve a more mature and innovative use of NbS, we believe it is useful to include them in the theoretical and application framework of the Urban Bioregion (Magnaghi 2020; Fanfani and Mataran 2020; Colavitti and Serra 2022). It is a planning approach developed since the 1990s in the Italian context, particularly within the Territorialist School, and oriented to establish a new co-evolutionary relationship between man and environment. This approach is based on the assumption that places are the result of a continuous interaction process, developed in history, between human and natural components. This interaction, after a long series of adaptive tests, has given birth to some territorial figures, landscapes, and settlement forms that have historically guaranteed a balanced and sustainable use of natural resources by mankind and, at the same time, their reproduction. Based on this premise, the Urban Bioregion approach is proposed as a radical alternative to the model of planetary urbanization that consumes irreversibly natural resources and produces serious environmental and social imbalances.

From a spatial point of view, the urban bioregion is a polycentric system, "a local territorial system characterized by: (a) the existence of a plurality of urban and rural centres, organized into reticular and non-hierarchical systems of small and medium-size cities; each city has a unique, synergic and multifunctional connection with its own rural territory for the production of ecosystem services; (b) the existence of complex and differentiated hydrogeomorphological and environmental systems, related in coevolutionary and synergic forms with the urban settlement and the agro-forestry system" (Magnaghi 2020, p. 38). The urban bioregion is centred around unartificial open spaces (agricultural areas, forests, wetlands, parks, urban greenery, wasteland etc.) for their role in territorial and urban regeneration; they are capable of closing the water cycle, production of food and renewable energy, fighting the effects of global warming, contributing to maintaining hydrogeological balance and water risk containment, supporting the creation of self-sustainable economies on a local basis, etc. In short, the urban bioregion approach is a design method to rebalance the relationship between built and open spaces and to achieve a new co-evolution between mankind and the environment.

The main design experiments concerned with the application of this approach are some regional and landscape plans in Italy (PIT Toscany, PPTR Apulia) and numerous regional projects focused on patent forms, as river contracts and agricultural parks developed mainly in Italy and France (Albrechts et al. 2020; Carta et al. 2022; Barbanente and Grassini 2022a). Landscape plans in Italy play the role of regional plans that are superior to all others, and for this reason, they define a framework of coherence for policies and transformations related to the entire regional territory (Gisotti and Tarsi 2023). In particular, the plans of Puglia and Tuscany (approved in 2015) present significant innovations because they adopt a structural approach: they identify themes or "layers" of analysis and design of the regional territory, and subsequently synthesize them within the so-called "territorial heritage," which highlights the elements of greatest value and design role. Multifunctional ecological networks play a fundamental role in the construction of the urban bioregion in both plans.

Including NbS within the urban bioregion framework could lead to a more mature use of them, contributing to addressing the challenges that still exist. Following, we tried to summarize such contributions, namely (1) trans-scalar and systemic, (2) place-based, and (3) co-designed.

1. Urban bioregional plans and projects foresee the creation of multifunctional green infrastructures (with ecological, aesthetical, and functional values) that connect open spaces from the regional to local scale. NbS, infiltrated within the built fabrics, can be connected with such green and blue infrastructures of the urban bioregion, contributing to the incremental implementation of regional strategies and to the morphological, ecological, and functional redevelopment of the relationship between settlements and open spaces (Marot and Catsaros 2020; Choi et al.

2021). In this way, each urban regeneration intervention using NbS can be combined into a larger overall design (conceived at an urban and regional scale) which gives them consistency and produces additional benefits, shifting from a discrete and episodic logic to a systemic and diffuse one (Barbanente and Grassini 2022b).

2. Urban bioregional plans and projects are based on a structural concept of the territory, a stratified product of thematic "layers" (hydrogeomorphology, ecological network, settlements, agricultural spaces) that have defined a specific identity. This latter is not only morphological but functional: the position of settlements with respect to the ground, for example, is regulated by principles of intrinsic rationality, which ensure their stability and safety. The NbS used in this type of project will therefore have a strongly place-based approach, which will identify site-specific solutions and be adapted to the characteristics of the place not only in terms of performance but also in terms of aesthetics and landscape. Furthermore, framing NbS in a structural approach (which breaks down the bioregional territory into key components) can support the development of their multifunctionality, cooperating to identify solutions that benefit each of the above layers or components at the same time (think, for example, of the renaturalization of river banks which, simultaneously, contributes to the reduction of hydraulic risk, acts as a buffer zone with respect to pollutants in the river, improves the livability also microclimatic of the settlement, increases the biodiversity of the agricultural landscape) (Alves et al. 2024).
3. In the urban bioregion approach plans and projects, the component of involving local communities is fundamental. Participation is understood as a form of empowerment of the populations in expressing their vision and as a way of mutual learning between designers and inhabitants on values and criticality of the bioregional territory. Through such processes, NbS can be developed inspired by "retroinnovation" (Stuiver 2006), namely, all those principles historically well known to local communities and often forgotten in favor of homologous and standardized solutions. Some examples of the application of this approach are the *rivers' contracts*, voluntary and contractual instruments that bring together various stakeholders around shared goals of safeguarding and managing watercourses and their associated territories, reducing hydraulic risk, promoting landscape values, and fostering forms of local development (Gisotti et al. 2025). The involvement of local communities can provide significant contributions to the maintenance of riverbanks through agricultural activities and riparian vegetation management, which are considered Nature-based Solutions. Another example of the use of NbS in participatory processes with a bioregionalist approach is found in agricultural parks, voluntary instruments for the shared management of rural territories. The involvement of the inhabitants, also consulted and involved in the possible uses of public spaces within the redevelopment interventions using NbS, can contribute to the definition of a regeneration network not only ecological but also socio-ecological, indeed more complex, and multifunctional (Viganò 2024).

4 The Case of the Agri Park of the Florentine Piana

The case study analyzed is a project developed by the Region of Tuscany (Italy) within a bioregional planning framework, starting in 2007. This is the Agricultural Park of the Florentine Plain, which extends across the river basin of the middle and lower Arno Valley in Tuscany. The park covers the municipalities of Florence, Sesto Fiorentino, Campi Bisenzio, Signa, Calenzano, Prato, Poggio a Caiano, and Carmignano. It is characterized by the coexistence of urban and rural spaces and a dense network of enterprises.

The concept of Agricultural Park emerged in Europe in the 1990s, aiming to safeguard peri-urban agricultural areas from urban development pressure while serving as an active tool for environmental protection. Numerous initiatives have since been implemented across various European countries.

An Agricultural Park is a strategic and voluntary agreement-based tool initiated by a coalition of local stakeholders—including farmers' unions, environmental, cultural, and sports associations, civic groups, public administrations, universities, and schools—brought together around the recognition of agriculture's multifunctionality and the shared goal of fostering a self-sustaining model of local development. Peri-urban areas perform strategic functions for urban systems: water supply, hydraulic safety, waste treatment, recreation, environmental education, and more.

The evolution of the park concept—from a protected area to an agricultural park—is a crucial shift, as it moves beyond the restrictive zoning that often meets resistance from farmers. Instead, it promotes a living "laboratory area" for innovative models of urban–rural interaction and development. In Tuscany's case, the Agricultural Park represents a "territorial project" of regional significance, involving multiple actors and governance levels. Between 2007 and 2010, the Tuscany Region established a series of memoranda of understanding and developed the Park's Master Plan (Fig. 1). In 2013, the initiative was incorporated into a formal amendment to the Regional Territorial Planning Guidelines, which later became the Regional Landscape Plan (PIT/PPTR).

Between 2014 and 2020, through the ERDF Operational Programme (POR FESR), the Region funded projects for sustainable mobility and renewable energy upgrades for

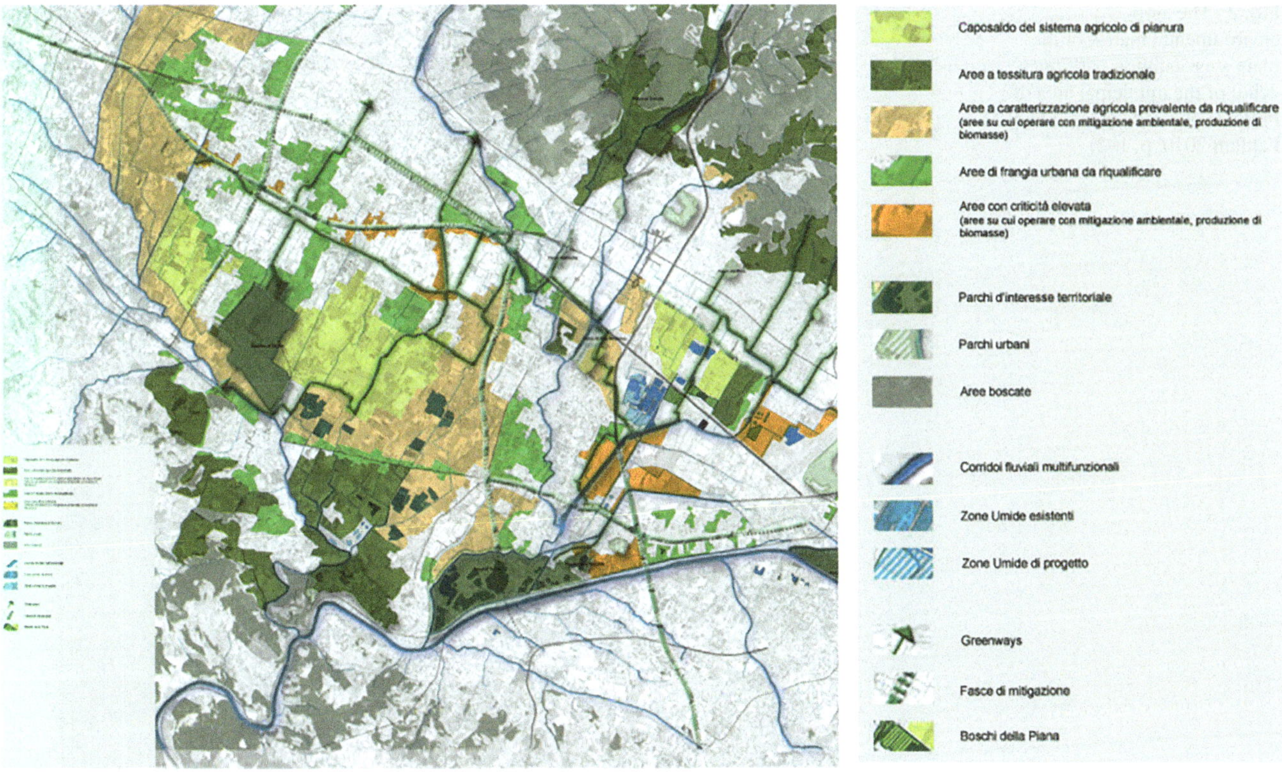

Fig. 1 Master Plan of the Agri Park of the Piana: Project Summary and legend. (Source: Regione Toscana 2013, https://www.regione.toscana.it/-/proposta-deliberazione-al-c-r-n-10-del-14-2-2011-e-allegati)

industrial and public buildings. Additionally, €3 million from the Rural Development Programme (RDP—EAFRD) was allocated to the Park project to support measures primarily aimed at mitigating climate change effects and promoting biodiversity.

In the design of the Agricultural Park, open agricultural and natural spaces are not merely areas to be protected from urban expansion. Instead, they play a central role in the functioning of the broader ecological system, acting as structural elements of the territorial framework. The open space strategy aims to reorganize the local territorial metabolism in a self-sustaining way, closing resource cycles (water, waste, food, and energy), reducing the ecological footprint, and enabling active and collective land stewardship. The park thus aspires to shape a territorial organization capable of sustaining its own life cycle, enhancing urban quality, and harmonizing productive, social, environmental, cultural, and aesthetic dimensions.

To achieve these objectives, a territorial vision was developed based on the role of the so-called "agro-environmental matrix"—a diversified system of open spaces interwoven with the settlement fabric. Within this matrix, various types of spaces have been identified, which function as NbS due to their spatial characteristics and ecosystem services. These spaces can be grouped into the following main categories:

- Natural or linear non-cultivated areas, which include lowland forests—key ecological nodes in the heavily urbanized plain; grasslands and pastures; riparian vegetation, which serves as a buffer against water pollution and provides ecological connectivity; hedgerows, tree lines, isolated trees; and woody biomass plantations.
- Agricultural areas, including both arable crops and woody perennials (such as orchards, olive groves, vineyards). Urban gardens and horticultural zones, as well as areas dedicated to beekeeping—vital for supporting pollinator populations—are also included.
- The water system, which is highly articulated and includes the main and secondary hydrographic networks (with the Bisenzio River and the Iolo stream playing key structural roles), naturally occurring wetlands, constructed wetlands for phytoremediation, and rainwater collection basins. In the Prato plain, the historical system of gore (man-made irrigation canals) plays a particularly significant role in water management and irrigation.

These components of the agro-environmental matrix (Fig. 2) are spatially organized into a complex territorial structure comprising: an external green belt beyond the densest urban settlements; an inner green belt closer to the urban core; a multifunctional urban-rural fringe; and fluvial and hydrographic corridors.

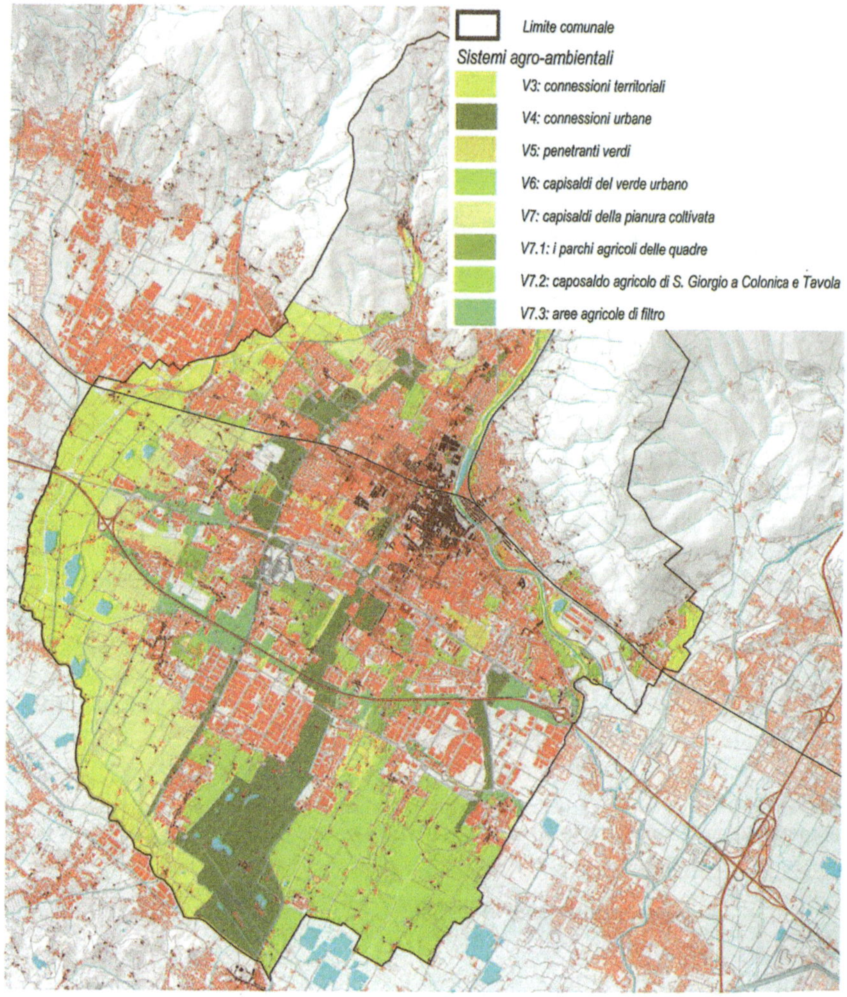

Fig. 2 The agro-environmental matrix of the plain's agricultural park, map detail of the municipal area of Prato. (Source: Magnaghi and Fanfani 2010, p. 192)

Clearly, these are not only nature-based solutions but also culturally grounded strategies that reinterpret historically rooted territorial characteristics.

For the purposes of this study, it is important to highlight one of the most innovative aspects of this project from an NBS perspective: the maintenance of continuity within the agro-environmental matrix for ecological, recreational, and landscape functions.

5 Discussion: How to Enrich NbS?

From the case study presented it clearly emerges that using a bioregional approach to design and implement NbS could make a significant contribution to addressing the challenges outlined in paragraph 2, particularly in terms of governance and inclusion. In the case presented, the agricultural park constitutes a territorial project integrated into the regional planning framework. It overcomes the issues of sectoral fragmentation and sporadic implementation that often characterize NbS initiatives, as highlighted in the literature. Another key aspect is the active participation of local stakeholders, which also ensures the long-term care and maintenance of the interventions.

However, to effectively adopt the proposed approach, it is essential to work simultaneously on several fronts.

A fundamental prerequisite for facilitating the implementation of NbS within the urban bioregional framework is the strengthening of public action. This would help correct and prevent the distortive effects generated by market-driven urban regeneration, which, as is well-known, exacerbates inequalities even while improving the ecological quality of the areas it targets. Public action can be bolstered by fostering dialogue between regulatory urban planning (which impacts land legal status), bottom-up participatory practices, and strategic plans. These three dimensions could be more coherently interconnected, with regulatory urban planning serving as a bridge between strategies and grassroots practices. While these dimensions are central to the current disciplinary debate, they often fall short in producing systematic and lasting changes, as they are frequently linked to the "human" context that generates them, and therefore subject to its inherent variability.

A second key aspect concerns the integration of NbS into governance. To support this integration, several strategic actions could be considered, including:

- Developing regional and large-scale planning frameworks that establish a coherent vision for the transformations needed for the transition. The regional scale also provides access to community funding sources that facilitate implementation (Gisotti and Tarsi 2023).
- Encouraging the systematic incorporation of NbS into urban plans by creating catalogues, guidelines, and pilot projects designed for "sensitive" areas due to their fragility and/or exposure to climate change risks, starting from regional and large-scale knowledge frameworks.
- Focusing on the existing stock of buildings and public spaces as part of strengthening public action.
- Educating technicians and professionals on NbS through a trans-scalar, systemic, place-based, and co-designed approach.
- Promoting cross-sectoral dialogue in the conception and implementation of NbS.

Finally, considering issues related to the distribution of wealth, power, equity, and access to urban ecosystems when designing and implementing NbS will not only strengthen the role of urban planning and governance (Anguelovski et al. 2018; Kull et al. 2015) but will also contribute to achieving environmental justice (Moha et al. 2009).

6 Conclusion

This paper has argued that while NbS hold significant promise for addressing a range of urban and regional challenges, their full transformative potential remains underrealized due to persistent issues in their conceptualization, design, and implementation. Through the case study of the Agri Park of the Florentine Piana, we have demonstrated that integrating the Urban Bioregion approach into NbS planning processes can offer a more systemic and place-based framework, capable of addressing these limitations.

By embedding ecological, social, and productive functions within a broader territorial vision, the Urban Bioregion perspective enables a shift from fragmented, sector-specific interventions toward coherent strategies of territorial regeneration. Moreover, the involvement of local actors in the co-design and long-term stewardship of the park highlights the importance of participatory governance in ensuring the durability and legitimacy of NbS.

Ultimately, this study suggests that aligning NbS with territorial planning logics grounded in bioregional principles not only enhances their effectiveness but also promotes a deeper, more reciprocal relationship between human settlements and ecological systems. Future research and policy should further explore the potential of bioregional design as a critical lens for scaling up and contextualizing Nature-based Solutions in diverse territorial settings.

Attributions This paper is the result of a joint reflection by the authors. Sections 1 and 2 were written by Elena Tarsi, Sects. 3 and 4 by Maria Rita Gisotti, Sects. 5 and 6 by the two authors.

References

Albert C, Spangenberg JH, Schröter B (2017) Nature-based solutions: criteria. Nature 543:315. https://www.nature.com/articles/543315b

Albrechts L, Barbanente A, Monno V (2020) Practising transformative planning: the territory landscape plan as a catalyst for change. City Territory Architect 7(1). https://doi.org/10.1186/s40410-019-0111-2

Alves A, van Opstal C, Keijzer N, Sutton N, Chen WS (2024) Planning the multifunctionality of nature-based solutions in urban spaces. Cities 146. https://doi.org/10.1016/j.cities.2023.104751

Anguelovski I, Connolly J, Brand AL (2018) From landscapes of utopia to the margins of the green urban life: for whom is the new green city? City 22(3):417–436

Barbanente A, Grassini L (2022a) Fostering transitions in landscape policies: a multi-level perspective. Land Use Policy 112:105869. https://doi.org/10.1016/j.landusepol.2021.105869

Barbanente A, Grassini L (2022b) Integrating nature-based solutions into urban planning and policies: learning from the Apulia case study. In: La Rosa D, Privitera R (eds) Innovation in urban and regional planning. INPUT 2021. Lecture notes in civil engineering, vol 242. Springer. https://doi.org/10.1007/978-3-030-96985-1_2

Calliari E, Staccione A, Mysiak J (2019) An assessment framework for climate-proof nature-based solutions. Sci Total Environ 656:691–700. https://doi.org/10.1016/j.scitotenv.2018.11.269

Carta M, Gisotti MR, Lucchesi F (2022) Settlements and urban morphological quality in landscape planning: analytical models and regulating tools in the landscape plan of Regione Toscana. Sustainability 14:1–31. https://doi.org/10.3390/su14031456

Chelleri L, Waters JJ, Olazabal M, Minucci G (2015) Resilience trade-offs: addressing multiple scales and temporal aspects of urban resilience. Environ Urban 27(1):181–198. https://doi.org/10.1177/0956247814550780

Choi C, Berry P, Smith A (2021) The climate benefits, co-benefits, and trade-offs of green infrastructure: a systematic literature review. J Environ Manag 291. https://doi.org/10.1016/j.jenvman.2021.112583

Clar C, Prutsch A, Steurer R (2013) Barriers and guidelines for public policies on climate change adaptation: a missed opportunity of scientific knowledge-brokerage. Nat Res Forum 37(1):1–18. https://doi.org/10.1111/1477-8947.12013

Cohen-Shacham E, Walters G, Janzen C, Maginnis S (eds) (2016) Nature-based solutions to address global societal challenges. IUCN

Colavitti, A. M., & Serra, S. (Eds.). (2022). Building the urban bioregion governance scenarios for urban and territorial planning. SdT Edizioni. https://iris.unica.it/retrieve/b9d73080-a6fc-470f-9604-c419991627ba/Floris_Serra_Building%20the%20Urban%20Bioregion%202022.pdf

Combes M, Viles H (2021) Integrating nature-based solutions and the conservation of urban built heritage: challenges, opportunities, and prospects. Urban For Urban Green 63. https://doi.org/10.1016/j.ufug.2021.127192

Cowling RM, Egoh B, Knight AT, O'Farrell PJ, Reyers B, Rouget M, Roux D, Welz A, Wilhelm-Rechman A (2008) An operational model for mainstreaming ecosystem services for implementation. Proc Natl Acad Sci U S A 105(28):9483–9488. https://doi.org/10.1073/pnas.0706559105

Díaz S, Pascual U, Stenseke M, Martín-López B, Watson RT, Molnár Z, Hill R, Chan KMA, Baste IA, Brauman KA, Polasky S, Church A, Lonsdale M, Larigauderie A, Leadley PW, Van Oudenhoven APE, Van Der Plaat F, Schröter M, Lavorel S et al (2018) Assessing nature's contributions to people. Science 359(6373):270–272. https://doi.org/10.1126/science.aap8826

Dorst H, van der Jagt A, Toxopeus H, Tozer L, Raven R, Runhaar H (2022) What's behind the barriers? Uncovering structural conditions working against urban nature-based solutions. Landsc Urban Plan 220:104335. https://doi.org/10.1016/j.landurbplan.2021.104335

Eggermont H, Balian E, Azevedo JMN, Beumer V, Brodin T, Claudet J, Fady B, Grube M et al (2015) Nature-based solutions: new influence for environmental management and research in Europe. GAIA Ecol Perspect Sci Soc 24(4):243–248. https://doi.org/10.14512/gaia.24.4.9

European Commission (2015) Nature-based solutions and re-naturing cities. Final report of the Horizon 2020 Expert Group. Directorate-General for Research and Innovation https://ec.europa.eu/newsroom/horizon2020/document.cfm?doc_id=10195

European Commission (2021) Evaluating the impact of nature-based solutions: a handbook for practitioners. Directorate-General for Research and Innovation https://op.europa.eu/en/publication-detail/-/publication/d7d496b5-ad4e-11eb-9767-01aa75ed71a1

Fanfani D, Mataran A (eds) (2020) Bioregional planning and design: perspectives on a transitional century. Springer, pp 33–62. https://doi.org/10.1007/978-3-030-46083-9

Fastenrath S, Bush J, Coenen L (2020) Scaling-up nature-based solutions: lessons from the Living Melbourne strategy. Geoforum 116:63–72. https://doi.org/10.1016/j.geoforum.2020.07.011

Folke C, Jansson Å, Rockström J, Carpenter SR, Chapin FS III, Crépin AS, Daily G et al (2011) Reconnecting to the biosphere. AMBIO 40:719–738. https://doi.org/10.1007/s13280-011-0184-y

Gisotti MR, Tarsi E (2023) Regional spatial planning for implementing the European Green Deal: a new method of assessment applied to the metropolitan area of Florence. Plan Pract Res 38(5):581–611. https://doi.org/10.1080/02697459.2022.2142837

Gisotti MR, Lucchesi F, Morelli E (eds) (2025) La forma dell'acqua. CONTESTI, Special Issue

Haase D, Larondelle N, Andersson E, Artmann M (2014) A quantitative review of urban ecosystem services assessment: concepts, models and implementation. AMBIO 43:413–433. https://doi.org/10.1007/s13280-014-0504-0

Hauck J, Hansjürgens B, Volk M (2013) The promise of the ecosystem services concept for planning and decision-making. GAIA Ecol Perspect Sci Soc 22(4):232–236. http://www.ingentaconnect.com/content/oekom/gaia/2013/00000022/00000004/art00006

Intergovernmental Science-Policy Platform on Biodiversity and Ecosystem Services (IPBES) (2019) Global assessment report on biodiversity and ecosystem services. IPBES Secretariat

Kauark-Fontes B, Marchetti L, Salbitano F (2023) Integration of nature-based solutions (NBS) in local policy and planning toward transformative change: evidence from Barcelona, Lisbon, and Turin. Ecol Soc 28(2):25. https://doi.org/10.5751/ES-14182-280225

Keesstra S, Mol G, de Leeuw J, Okx J, Molenaar C, de Cleen M, Visse S (2018) Soil-related sustainable development goals: four concepts to make land degradation neutrality and restoration work. Land 7(4):133. https://doi.org/10.3390/land7040133

Key IB, Smith AC, Turner B, Chausson A, Girardin CAJ, Macgillivray M, Seddon N (2022) Biodiversity outcomes of nature-based solutions for climate change adaptation: characterising the evidence base. Front Environ Sci 10:905767. https://doi.org/10.3389/fenvs.2022.905767

Kull CA, Arnauld de Sartre X, Castro-Larrañaga M (2015) The political ecology of ecosystem services. Geoforum 61:122–134. https://doi.org/10.1016/j.geoforum.2015.03.004

Liu H-Y, Jay M, Chen X (2021) The role of nature-based solutions for improving environmental quality, health and well-being. Sustainability 13(19):10950. https://doi.org/10.3390/su131910950

Magnaghi A (2020) The territorialist approach to urban bioregions. In: Fanfani D, Mataran A (eds) Bioregional planning and design: perspectives on a transitional century. Springer, pp 33–62. https://doi.org/10.1007/978-3-030-45870-6

Magnaghi A, Fanfani D (eds) (2010) Patto città-campagna: Un progetto di bio-regione urbana per la Toscana. Alinea

Marot S, Catsaros C (2020). Agriculture et architecture: Trajectoires communes 3. Architecture Aujourd'hui. https://hal.archives-ouvertes.fr/hal-03521348/

Mexia T, Vieira J, Príncipe A, Anjos A, Silva P, Lopes N, Pinho P (2018) Ecosystem services: urban parks under a magnifying glass. Environ Res 160:469–478. https://doi.org/10.1016/j.envres.2017.10.037

Moha P, Pellow D, Timmons RJ (2009) Environmental justice. Annu Rev Environ Resour 34:405–430. https://doi.org/10.1146/annurev.environ.34.013007.103154

Parris KM, Amati M, Bekessy SA, Dagenais D, Fryd O, Hahs AK, Williams NSG (2018) The seven lamps of planning for biodiversity in the city. Cities 83:44–53. https://doi.org/10.1016/j.cities.2018.06.006

Pasquini L, Cowling RM (2015) Opportunities and challenges for mainstreaming ecosystem-based adaptation in local government: evidence from the Western Cape, South Africa. Environ Dev Sustain 17(5):1121–1140. https://doi.org/10.1007/s10668-014-9594-x

Raymond CM, Frantzeskaki N, Kabisch N, Berry P, Breil M, Nita MR, Calfapietra C (2017) A framework for assessing and implementing the co-benefits of nature-based solutions in urban areas. Environ Sci Pol 77:15–24. https://doi.org/10.1016/j.envsci.2017.07.008

Schröter M, van der Zanden EH, van Oudenhoven APE, Remme RP, Serna-Chavez HM, de Groot RS, Opdam P (2014) Ecosystem services as a contested concept: a synthesis of critique and counter-arguments. Conserv Lett 7(6):514–523. https://doi.org/10.1111/conl.12091

Seddon N, Smith A, Smith P, Key I, Chausson A, Girardin C, House J, Srivastava S, Turner B (2021) Getting the message right on nature-based solutions to climate change. Glob Chang Biol 27(7):1518–1546. https://doi.org/10.1111/gcb.15513

Stuiver M (2006) Highlighting the retro side of innovation and its potential for regime change in agriculture. In: Marsden T, Murdoch J (eds) Between the local and the global (Research in rural sociology and development), vol 12. Emerald Group Publishing Limited

Timboe I, Pharr K (2021) Nature-based solutions in international policy instruments. In: Cassin J, Matthews JH, Gunn E (eds) Nature-based solutions and water security: an action agenda for the 21st century. Elsevier, pp 125–147. https://doi.org/10.1016/B978-0-12-819871-1.00015-4

Turkelboom F, Leone M, Jacobs S, Kelemen E, García-Llorente M, Baró F, Rusch V (2018) When we cannot have it all: ecosystem services trade-offs in the context of spatial planning. Ecosyst Serv 29:566–578. https://doi.org/10.1016/j.ecoser.2017.09.016

Van Herzele A, De Clercq EM, Wiedemann T (2005) Strategic planning for urban green infrastructure: a conceptual framework for challenges and opportunities. Urban For Urban Green 3(3–4):195–206. https://doi.org/10.1016/j.ufug.2004.11.003

Viganò P (2024) Biopolitical garden: space, life, transition. Actar

Urban Design and Mobility Strategies for the Regeneration of Cities and Communities

Enhancing Urban Connectivity: A Comprehensive Study of Transit-Oriented Development

Salma Safi and Hadeel El Naggar

Abstract

Urban sprawl and population growth are pressing global challenges that contribute to a fragmented urban fabric and create inefficiencies in transportation, land use, and environmental sustainability. In response to these issues, transit-oriented development (TOD) has developed as an effective urban planning strategy aimed at enhancing connectivity, accessibility, and mobility. This approach focuses on creating compact, mixed-use communities that are walkable and centered around public transit systems. This paper examines the concept of TOD, outlining its fundamental principles and components, as well as the factors that are driving its growing adoption in cities across the globe. The paper examines two case studies from China that showcase successful implementations of TOD in the face of rapid urbanization. These cases demonstrate how TOD can effectively manage urban growth, encourage efficient transit use, and foster more livable communities

The paper concludes by emphasizing the importance of adopting TOD principles in urban planning, particularly in rapidly growing cities. Through well-designed urban layouts and land use policies, TOD can lead to more sustainable, efficient, and connected urban environments, serving as a model for future urban growth.

S. Safi (✉)
Urban Design and Digital Environments, American University in Dubai, Dubai, UAE
e-mail: salma.safi@mymail.aud.edu

H. El Naggar
Architecture, American University in Dubai, Dubai, UAE
e-mail: helnaggar@aud.edu

Keywords
Transit-oriented development · Mobility · Connectivity · Urban fabric · Land use · Urban growth · Adaptability

1 Introduction

Urban sprawl and population growth have given rise to a number of urban planning issues, and one important solution is transit-oriented development, or TOD. Building walkable, compact, mixed-use districts around efficient public transportation systems is known as transit-oriented development (TOD), and it offers an alternative to unsustainable urban growth and development that is predicated on automobiles. The trend toward the adoption of Transit-oriented development (TOD) is being driven by a number of factors, including the need for more sustainable land use, enhanced urban connectivity, and the increasing demand for accessible public transit options in growing cities.

Urban sprawl and rapid population growth have placed a strain on city infrastructure, leading to fragmented city layouts, longer commutes, and deteriorating environmental conditions. TOD promotes the creation of denser, transit-oriented communities that maximize land utilization, reduce the need for private automobile use, and support environmental sustainability in order to address these problems.

One of the many goals and objectives of TOD is improved mobility. Transit-oriented development (TOD) seeks to create more livable, resilient, and equitable communities by combining basic ideas like accessibility, connectedness, and efficient land use. Its primary components are mixed land use, high-density development near transit hubs, pedestrian-friendly zones, and improved transit access for all residents.

An overview of the fundamental concepts and components of transit-oriented development (TOD), along with an analysis of the optimal urban form and layout for TOD and how the built environment impacts transit behavior, has been focused on. It also examines connectivity in TOD, emphasizing the integration of urban spaces to ensure easy access and

F. Alberti et al. (eds.), *Towards Sustainable Urban Regeneration: Utilizing Environmental, Cultural, and Social Assets*, Advances in Science, Technology & Innovation, https://doi.org/10.1007/978-3-032-02091-8_12

mobility. Through successful case studies, the paper demonstrates how TOD has been effectively applied to create more sustainable, connected, and vibrant urban landscapes.

2 Research Methods

2.1 Research Problem

Urban areas today face a significant challenge: a fragmented urban fabric caused by urban sprawl and rapid population growth. Creating central anchors that connect and support neighboring areas lacking transportation options and access to essential amenities is vital to maintaining a cohesive urban fabric. Many communities experience issues with poor connectivity, insufficient transportation infrastructure, and limited access to services, all of which contribute to fragmentation. Addressing these challenges requires the development of central "anchors" that link and support surrounding areas, promoting better connectivity, accessibility, and mobility. This research will examine how transit-oriented development (TOD) principles can be applied to transform key sites into focal points that improve overall connectivity and accessibility within the urban fabric.

2.2 Research Aim

This paper aims to explore how urban design can address the shortcomings of fragmented urban developments that disrupt connectivity. By focusing on transforming urban areas into effective transit-oriented development (TOD) focal points, the study seeks to strengthen connectivity across the broader urban fabric. Additionally, it aims to mitigate the effects of urban sprawl and population growth, which place increasing pressure on the urban fabric.

2.3 Research Methodology

This paper employed a quantitative research methodology, grounded in a thorough review of literature and the analysis of case studies, with a specific focus on urban contexts in China. A wide array of secondary sources—such as academic publications, journal articles, conference papers, reports, and research studies—were systematically reviewed to build a solid foundation of knowledge on the subject.

The quantitative approach centered on the use of literature reviews and case studies as primary methods of data collection and analysis. Case studies from urban areas in China were carefully selected, and key urban elements such as layout, land use, surroundings, and transportation systems were examined. Data analysis was conducted using quantitative methods, including urban mapping, comparative analysis, and statistical evaluation, ensuring a robust investigation of the study area.

The insights gathered from the quantitative analysis informed the development of a framework aimed at addressing fragmented urban fabrics. The proposed framework emphasizes solutions such as improved connectivity and the application of transit-oriented development (TOD) principles, offering recommendations for future urban planning and design initiatives.

3 Literature Review

3.1 Introduction to Transit-Oriented Development (TOD)

Transit-oriented development (TOD) is defined as a planning strategy aimed at aligning land use with transportation planning. While numerous authors provide varying definitions of this term, the central idea behind TOD is the integration of diverse transportation modes. It promotes walking, cycling, and public transit usage, reducing reliance on private vehicles. Similarly, Hannawi et al. (2019), mentioned that transit-oriented development (TOD) is fundamentally a pattern of land use development centered around transit stations or networks, designed to promote the use of public transportation. These urban developments are distinguished by several planning principles aimed at fostering environmentally, socially, and economically sustainable urban growth.

Papa et al. (2017), also agreed and stated that the integration of transportation, land use, and environmental policies is an effective strategy for achieving sustainable and balanced development in urban areas. This approach can serve as a means to mitigate the negative externalities associated with car dependence, urban sprawl, population growth, and the spatial and temporal dispersion of mobility trips.According to Harmadi et al. (2022), transportation serves two primary roles: it acts as a tool to guide development in urban areas and facilitates the movement of people and goods generated by activities within those areas. Urban regions are characterized by substantial flows of goods and services. A frequent issue associated with transportation in these areas is congestion, which arises from the heavy reliance on private vehicles as the primary mode of transport. Therefore, the integration of a high-quality public transport is essential.

Additionally, Harmadi et al. (2022), mentioned the crucial need to address the challenges posed by high levels of mobility among individuals using private vehicles. Their findings indicated that it is essential to implement an integrated public transportation system that meets the community's needs. Transit-oriented development (TOD) is an urban planning concept that emphasizes the role of public transportation. Several major cities in developed countries have adopted the TOD principle in their regional development initiatives, but several cities have yet to integrate and develop.

According to Hannawi et al. (2019), three primary factors have driven this process of the fragmented urban fabric: economic prosperity, the adoption of modern technology in transportation and public utilities, and government land use policies. Additionally, this process is associated with six specific phases of spatial urban development that cities have undergone. These phases can be characterized by examining the unique economic and demographic changes over time alongside the evolution of urban planning policies related to land use and transportation.

3.1.1 Definition of TOD

The term "transit-oriented development" comes with various definitions. According to Stojanovski (2019), transit-oriented development (TOD) is typically defined as a development characterized by comfortable walking distances to designated transit stops. Transit professionals have established guidelines indicating that walking distances to transit stops should be between 5 and 10 min. This translates to a radius of approximately 300–400 m or 600–800 m around bus stops and rail stations. Urban designers and planners use these guidelines to create catchment zones around transit stops, organizing transit-supportive land uses—such as a mix of dense residential, retail, commercial, and public spaces—within these walkable zones.

Moreover, Singh (2015) defines transit-oriented development (TOD) as a relatively high-density form of urban development that incorporates a balanced mix of land uses within a walkable or cyclable environment. Consequently, the TOD concept serves as an approach to sustainable development that emphasizes the integration of land use and transportation. Similarly, Salat and Ollivier (2017) defined TOD as a planning and design approach focused on achieving compact and mixed-use urban spaces that prioritize pedestrian and bicycle accessibility, while being closely linked to transit stations.

Likewise, Yang et al. (2022) define Transit-oriented development (TOD) as a concept that emphasizes "transit," focusing on public transportation; "orientation," which centers on an area with a radius of 400–800 m (approximately 5–10 min walking distance); and "development," which promotes mixed-use zones for work, business, residence, education, and culture. Furthermore, Kamruzzaman et al. (2015) defined TOD as a development having the ability to promote the use of more sustainable transportation modes (such as buses, trains, walking, and cycling) while reducing car dependency, thereby decreasing greenhouse gas emissions and alleviating congestion levels.

Zhou et al. (n.d.) highlighted that studies on transit-oriented development (TOD) generally focus on two variables: attributes and outcomes. TOD attributes include measurable qualities like density, diversity, and design—known as the "3Ds"—which indicate alignment with transit use. In contrast, TOD outcomes refer to effects such as increased transit ridership and a greater share of transit usage.

The various definitions of transit-oriented development (TOD) arise from different perspectives and end goals of a vision. This diversity reflects the complexity of urban environments and the multifaceted challenges they face. However, all definitions converge on a central focus: creating walkable, transit-oriented, and mixed land uses that enhance urban livability and sustainability. Figure 1 shows the process of implementation of the TOD framework.

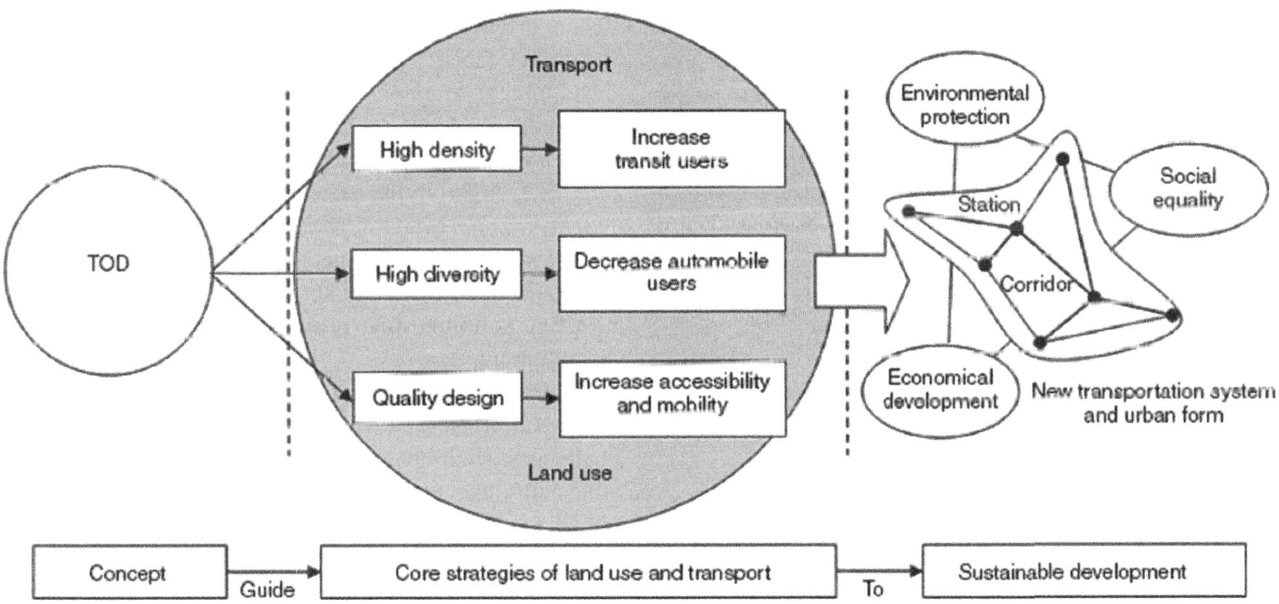

Fig. 1 TOD and sustainable development—the transit-oriented development framework progresses from the conceptual stage to strategic planning and ultimately to implementation

3.1.2 Factors Driving the Trend Toward Implementing TOD

Wang et al. (2018) studied how transportation infrastructure serves as a complex network that links cities and supports human activities, intertwining social, economic, and environmental systems with urbanization and population growth. Furthermore, this transportation network plays a crucial role in socioeconomic development and enhances the quality of life by facilitating inter- and intra-city connections throughout the urbanization process. This is one reason TOD has become a crucial part of an urban area. Moreover, TOD plays a crucial role in the movement and the flow of populations during urban agglomeration and diffusion. As noted in the definitions provided previously, TOD is an essential component of environmental, social, and economic development across all income levels, fostering personal well-being and driving growth.

According to Salat and Ollivier (2017), the trend toward implementing TOD has become more focused on as it brings many benefits to the overall urban fabric. By establishing transport nodes in areas where they are needed, connectivity and accessibility can be enhanced both within an urban zone and beyond its borders. By concentrating on the "3 V's," transit-oriented development (TOD) leverages node values and place values to boost and enhance market values. In this manner, it addresses the overall connectivity and accessibility challenges within the urban fabric to improve and enhance these aspects.

Several factors are driving the trend toward transit-oriented development (TOD). Worsening nationwide traffic congestion, with 2-h commutes becoming routine, has made expanding highways ineffective, often leading to more sprawl and congestion. Simultaneously, there's growing dissatisfaction with suburban sprawl and strip developments, as people increasingly seek walkable urban lifestyles away from traffic. Changes in family structures, like more singles and empty nesters, are also influencing demand for compact, accessible urban living. Rising national support for Smart Growth and shifting federal policies further promote TOD as a sustainable solution for growing urban populations (Transit Oriented Development 2024).

3.2 Overview of Urban Sprawl and Population Growth

3.2.1 TOD Approaches for Urban Sprawl and Population Growth

According to Brueckner (2000), the term "urban sprawl" carries various meanings depending on perspective, often complicating discussions. Urban sprawl refers specifically to the excessive spatial expansion of cities. The emphasis here is on the word "excessive." While cities need to expand to accommodate growing populations, the concern is that this spatial growth can become disproportionately large, leading to inefficiencies and other urban issues. Additionally, Brueckner stated that urban spatial expansion is primarily driven by three key factors: population growth, increasing incomes, and decreasing commuting costs. Population growth refers to the increase in the number of individuals in a specific population over a defined period of time.

Moreover, Dieleman and Wegener (2004) stated that there is an increasing awareness and concern regarding urban development, particularly the issue of urban sprawl. The expansion of low-density urban areas, extensive outward growth, spatial separation of land uses, leapfrogging development, and the prevalence of commercial strip developments are generally seen as detrimental to the quality of life in cities. These patterns often contribute to inefficiencies and fragmented urban landscapes. Urban growth and expansion lead to several unintended consequences, including disinvestment in urban core areas and the decline of central cities, increased reliance on private cars, causing more vehicle miles traveled, road congestion, and the loss of open spaces and a fragmented urban fabric.

The urbanization in several cities extends far beyond mere urban sprawl and population increases, encompassing significant shifts in the region's economic and social frameworks. The city's rapid expansion has triggered profound environmental and societal changes, resulting in challenges such as overcrowding, transportation inefficiencies, social unrest, and a fragmented urban fabric (Fazli and Faridi n.d.).

According to Shandas et al. (2017), rapidly expanding urban zones place significant pressure on both nearby and distant urban areas, impacting the overall sustainability of the urban fabric. The debate around sustainable urban development among scholars highlights the economic, environmental, and social challenges posed by growing cities. While sprawling suburban development offers benefits like larger living spaces, lower housing costs, and reduced crime rates, it also results in significant drawbacks, such as a disconnected urban fabric. Identifying the issues and impacts of sprawl is crucial for developing effective solutions that promote more sustainable urban development.

Dieleman and Wegener (2004) emphasized that multifunctional urban land use is essential for addressing urban challenges. High-density areas, especially those near accessible hubs like train and metro stations, often adopt this model, known as transit-oriented developments (TOD). TOD integrates various land uses around public transportation, contrasting with the rapid, spatially fragmented growth typical of urban sprawl. TOD strategies, characterized by high-density, mixed-use development, aim to manage population growth, mitigate urban sprawl, and unify disconnected urban landscapes.

3.2.2 Benefits of Implementing TOD

Transit-oriented development (TOD) offers a range of benefits that contribute to improved urban living. By creating higher-quality environments for living, working, and playing, TOD enhances overall quality of life. This approach facilitates greater mobility, making it easier for residents to navigate their communities and resulting in increased transit ridership. As a consequence, traffic congestion, car accidents, and injuries are significantly reduced. Moreover, households can spend less on transportation, leading to more affordable housing options and fostering a healthier lifestyle that encourages walking and minimizes stress (Transit Oriented Development 2024).

Additionally, TOD often leads to higher and more stable property values, benefiting local homeowners. Increased foot traffic brings more customers to area businesses, boosting the local economy. By reducing dependence on gas emissions, TOD helps lower pollution levels and mitigates environmental damage. Furthermore, it discourages urban sprawl and promotes compact development, making it a more cost-effective solution compared to building new roads and sprawling neighborhoods. Ultimately, TOD enhances a region's ability to maintain economic competitiveness, making it a valuable approach to sustainable urban growth (Transit Oriented Development 2024).

3.2.3 Goals and Objectives of TOD

Petrucci (2021) states that TOD goals rely on diverse transportation modes and essential infrastructure like lanes, parking, transit stops, and sidewalks to ensure safe, convenient, and comfortable travel for all users.According to the TOD Standard—ITDP (2024), the objectives of transit-oriented development (TOD) focus on:

1. To create a pedestrian realm that is safe, accessible, and complete for all users. This realm should also be active and vibrant, fostering community engagement and enhancing the overall urban experience. Furthermore, it aims to provide a temperate and comfortable environment, encouraging walking and outdoor activities.
2. The cycling network is designed to be safe and comprehensive, ensuring secure and ample cycle parking and storage facilities for users.
3. Walking and cycling routes are designed to be short, direct, and varied, offering pathways that are often shorter than those for motor vehicles.
4. High-quality transit is conveniently accessible on foot.
5. Opportunities and services are located within a short walking distance from where people live and work, and public spaces are activated for extended hours. The community reflects diverse demographics and income levels among its local residents.
6. High residential and job densities promote high-quality transit, local services, and vibrant public space activity.
7. The development is situated within or adjacent to an existing urban area, ensuring convenient travel throughout the city.
8. The land used by motor vehicles is minimized.

3.3 Principles and Components of TOD

3.3.1 TOD Principles

According to Hermansyah et al. (2024), TOD 3.0 encompasses several key principles: walking, cycling, connecting, mixing, condensing, compressing, and shifting. Pedestrian facilities in TOD areas are designed to ensure safety and completeness, providing protection from the elements and direct access to buildings. Cycling infrastructure includes a safe and comprehensive network with facilities shielded from motorized vehicles. Mixed land use integrates residential and non-residential functions within the same block or nearby areas. High density supports efficient transportation, local services, and active public spaces, ensuring all residents have access to public transit. The compact design emphasizes built-up areas and urban transitions, preparing global cities, particularly in developing countries, for future growth. Residents are encouraged to live near schools, offices, and service centers to reduce traffic congestion.

Similarly, Petrucci (2021) stated that to enable urban development in line with cities' regulations, it is essential to ensure they are clearly communicated. The eight TOD principles that should be incorporated into an urban development are walking, cycling, connecting, transit, mixing, condensing, compressing, and shifting.

A key element of transit oriented development (TOD) is human-centered design, prioritizing walkability, safety, and comfort through complete streets. Secure cycling and micro-mobility options connect first and last miles, activating streets and boosting transit use. The "connect" principle integrates pedestrian and cycling networks with short, direct routes linking zones. Reliable, high capacity transport connects neighborhoods to the city, acting as civic hubs. The "mix" principle creates inclusive, vibrant spaces where diverse activities and residents thrive, supporting transport services and access to local destinations. Compact design works at both city and neighborhood levels, supported by transit, density, and cycling networks. The "shift" principle reclaims space from private vehicles, reducing harmful emissions and noise (Petrucci 2021). The TOD principles are highlighted in Figure 2, and it show the prominent symbols used.

3.3.2 Key Components and Elements of TOD

Components of transit-oriented development (TOD) include essential elements that create a cohesive urban environment. Walkable design prioritizes pedestrians, with the train station as a focal point in the town center, complemented by a public square. The development acts as a regional node, featuring a mix of office, residential, retail, and civic uses in close proximity. A high-density district lies within a 10-min walk of the train station, supported by collector transit systems like streetcars, light rail, and buses. Easy access for bicycles and scooters is also emphasized, with large ride-in bicycle parking areas, an integrated bikeshare rental system, and a comprehensive bikeway network. Parking is reduced and managed within the 10-min walk radius to promote pedestrian activity. Additionally, specialized retail at the stations serves commuters and locals, providing essential services such as cafes and grocery stores (Transit Oriented Development 2024).

Petrucci (2021) mentioned that the elements of TOD come from its name, transit-oriented development. Transit components of an effective transportation system prioritize high-quality station architecture and public spaces that harmonize with the surrounding environment. Amenities like retail options enhance the commuter experience, making it more comfortable and seamless. Thoughtful design in transit stations fosters a welcoming atmosphere, encouraging public transport use and community engagement. Multi-modal integration focuses on seamlessly connecting various transit modes, ensuring efficient links for all users and abilities to facilitate smooth access to and from stations. This integrated approach enhances mobility, making navigation easier and promoting public transit use. Moreover, complete streets are designed to ensure safe access for all users—pedestrians, cyclists, motorists, and transit riders—by equitably distributing road space, thus prioritizing safety and accessibility for everyone. Lastly, traffic management includes safe speed strategies around transit stops and measures to manage traffic demand and reduce parking needs, promoting sustainable mobility choices and minimizing reliance on private vehicles.

Additionally, oriented open space components play a vital role in enhancing the urban environment around transit areas. Transit plazas promote congregational activities by incorporating inclusive and context-sensitive architectural and landscaping features that encourage community interaction. Building on this, walkability focuses on creating attractive pedestrian environments that are continuous and interconnected, offering a variety of experiences and amenities. Transitioning from walkable spaces, the public realm emphasizes visual interest at the pedestrian scale through thoughtful landscaping and building design, fostering an active street life that invites people to engage with their surroundings. Furthermore, the creation of urban parks and open spaces adds value by providing amenity spaces, green areas, playgrounds, and civic squares within a 5-min walking radius of

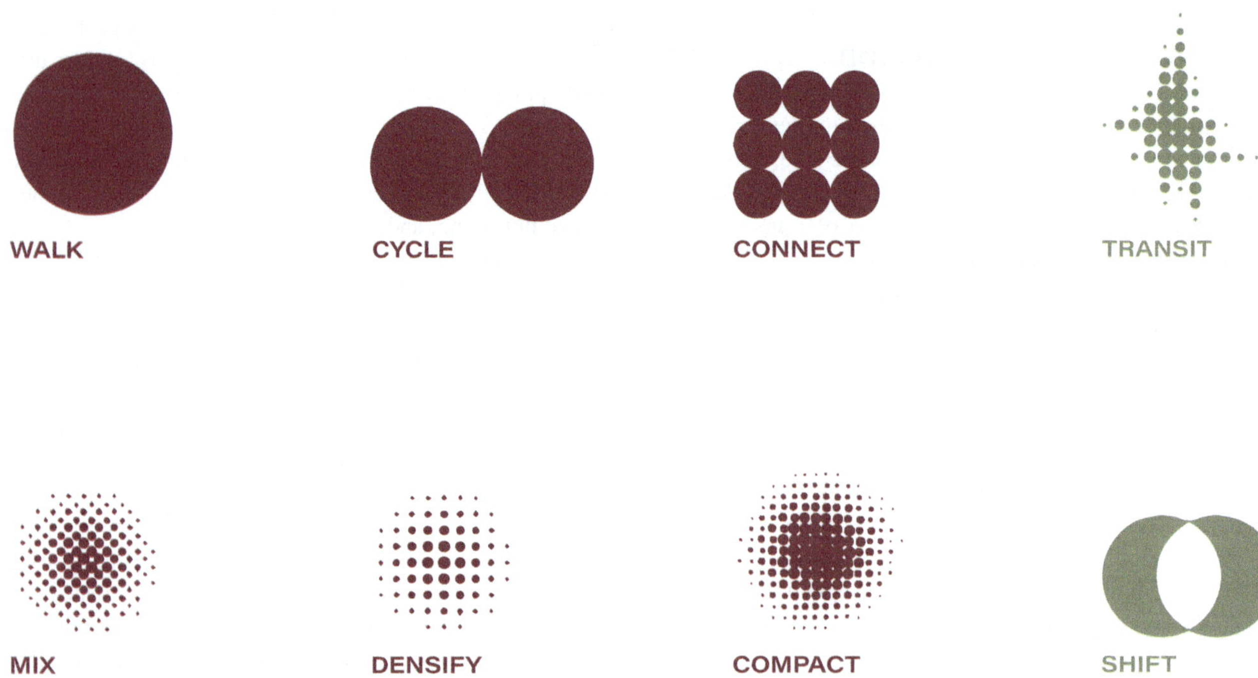

Fig. 2 TOD principles—the symbols and imagery of TOD principles

residents. This comprehensive approach enhances community connectivity and promotes a vibrant, accessible urban landscape (Petrucci 2021).

Lastly, the development-built environment components aim to create vibrant, accessible communities around transit areas. compact development optimizes employment and residential densities along transit corridors, enhancing the capacity of transit and non-motorized transport (NMT) to promote walking and transit use. This leads to a mix of uses that ensures residents can access retail, commercial, and civic services, as well as employment and recreational facilities, without relying on cars. Additionally, fostering housing diversity within a 10-min walk of transit stations supports equitable transit-oriented developments (TODs) by offering various housing types and price ranges. Finally, integrating the informal sector is vital for inclusive development, addressing the needs of street vendors and settlements in planning and design. Together, these components create a cohesive built environment that enhances residents' quality of life (Petrucci 2021). To integrate the transit-oriented development efficiently, the TOD key components need to be focused on as shown in Fig. 3.

3.3.3 TOD Typologies

The TOD typology's codes can serve as either a replacement for or an overlay on the existing base zoning framework. All development parcels within a specific station area typology must comply with the form-based regulations relevant to that typology. When applying the model template to a city, it is essential to clearly define the TOD station area typologies and their boundaries to eliminate any ambiguity in property selection (Petrucci 2021).

According to Petrucci (2021), each station area, whether existing or proposed, faces unique challenges that require tailored strategies for effective TOD development. This form-based code addresses seven key station area typologies based on common planning considerations, including existing land use, transportation functions (right-of-way, multimodal availability, and citywide connectivity), land availability, susceptibility to change (e.g., building age), mobility networks (block size and barriers), and infrastructure capacity.In the context of city planning for TOD, common station area typologies include but are not limited to:

- Intermodal gateways
- Employment centers
- Destination nodes
- Transit neighborhoods
- Urban cores (CBD)
- Infill neighborhoods
- Residential areas

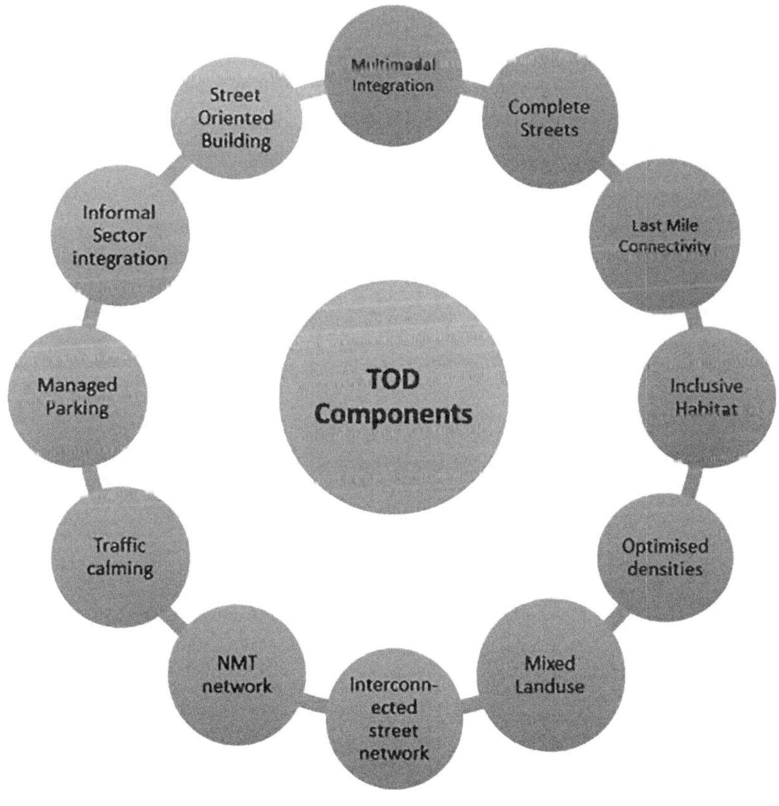

Fig. 3 TOD key components (*Drishti IAS*, 2021)

3.3.4 Scale of TOD

The scales of transit-oriented development (TOD) plans range from city-region level to corridor level, station area, and site level, as shown in Table 1.

3.4 Influence of Built Environment on Transit Behavior

3.4.1 TOD's Ideal Urban Layout and Land Use

Floor area ratio (FAR) and density standards are crucial for ensuring the strategic distribution of density across urban layouts, particularly to promote compact development near transit hubs. FAR reflects the intensity of land use by calculating the total floor area of a building as a ratio of the plot's total area. This tool helps control urban density by enforcing maximum permissible FAR limits. In transit-oriented development (TOD) areas, these limits often need to be adjusted to allow for higher-density structures (Petrucci 2021).

Additionally, according to Petrucci (2021), the ideal land use is mixed land use that encourages more efficient land development by providing residents with convenient access to retail, commercial, civic services, employment, and recreational facilities within walking distance. This concept is incorporated into zoning regulations through permissible and non-permissible land uses or building functions. Permissible uses should support the integration of complementary activities, optimizing shared resources like streets and parking. Non-permissible uses, on the other hand, should limit automobile-dependent developments, such as large industrial sites, car showrooms, and cemeteries.

Urban layouts and land use should prioritize the 3D framework—density, diversity, and design—developed to assess and enhance TOD characteristics. According to Nyunt and Wongchavalidkul (2020), the framework focuses on:

- Density: Concentration of population and employment in relation to the size of the area, ensuring a compact urban form around transit hubs.
- Diversity: A balanced mix of land uses and development types, fostering various activities and interactions near transit stations.
- Design: Creation of pedestrian- and bicycle-friendly environments, improving transit accessibility and encouraging walking, cycling, and public transit use through better connectivity and mobility.

3.4.2 TOD's Ideal Urban Forms

Urban forms should incorporate both vertical and horizontal layers of development to create dynamic, multifunctional environments that promote land efficiency and connectivity. These layers play a crucial role in shaping how space is utilized, both within individual buildings and across larger urban areas, enabling cities to support a wide range of activities and services while maintaining a compact and sustainable form.

According to Nyunt and Wongchavalidkul (2020), vertical layering refers to the integration of different land uses within a single building or development. For example, retail spaces may occupy the ground floor, offices can be located on the middle floors, while residential units are placed on the upper levels. This stacking of functions optimizes land use, particularly in dense urban areas, by consolidating activities within a smaller footprint. It supports transit-oriented development (TOD) by concentrating population and employment near public transportation, reducing the need for long commutes and encouraging walkability. Vertical layers also help maintain a constant flow of people and activity throughout different times of the day, creating vibrant urban hubs.

Horizontal layering, on the other hand, involves the distribution of various land uses across an area. A well-planned urban form might include residential, commercial, office, and recreational spaces spread throughout a neighborhood or district. This horizontal arrangement ensures that key services and amenities are within reach, improving access and enhancing the quality of life for residents. By organizing land uses in proximity to one another, horizontal layering

Table 1 Transit-oriented development (TOD) plans range from city-region level to corridor level

Scale	City region	Corridor level	Station area	Site level
Purpose	This approach involves integrating land uses with transit system planning to support analysis and decision-making related to citywide growth management. It serves as a point of intervention for transit-oriented development (TOD) within policy frameworks, such as master plans or development plans	This approach ensures that development at each station complements that of others, creating a network of transit-oriented areas. At this scale, specific transit ridership goals can be assessed in relation to the development potential surrounding transit stations	This scale concentrates on the areas surrounding transit stations within a 5–10-min walking	This scale focuses on individual developments within a station area, including targets for net intensity and density, internal circulation, building design, and parking
Boundary definition	Administrative boundaries/transit systems	A 10-min (800 m to 2 km) walking or cycling distance on either side of an existing or planned transit corridor	A 5 to 10-min (400 m to 1 km) walking distance from station facilities	An individual parcel located within a 5 to 10-min (800 m to 1 km) walking distance from the station facility

promotes seamless movement between different parts of the urban fabric, supporting walking, cycling, and public transit as primary modes of transportation (Nyunt and Wongchavalidkul 2020).

Vertical and horizontal layers work together to produce a balanced urban shape that optimizes the use of available land, promotes environmentally friendly transportation, and cultivates vibrant, connected communities. In transit-oriented developments, where the objective is to lessen reliance on automobiles and build areas that are easily navigable by foot, bicycle, or public transportation, this integrated approach is extremely beneficial.

3.5 Connectivity in TOD

3.5.1 Definition of Connectivity

According to Asriana and Koerniawan (2020), connectivity refers to the relative accessibility or spatial interconnection of a network.

Cities are often characterized as intricate networks from which various locations emerge. These locations are formed through the interactions within these networks. This perspective allows us to utilize network science to study urban areas, helping us uncover the underlying order within a city's structure and effectively plan for its inherent complexity.

As studied by Darren et al. (2019), the flow of goods and information within a network is determined by its connectivity, which is influenced by how its components are linked, the number of connections, and the strength of these connections. Connectivity can be defined as the minimum number of network components (such as nodes or edges) that need to be removed to disconnect a connected network. It serves as a measure of the network's resilience, as networks with high connectivity offer more routing options and are more robust against failures.

Connectivity plays a vital role in building resilience in urban environments. A lack of connectivity can lead to the failure of specific functions following disruptions, leaving areas vulnerable with multiple potential points of failure. This is often observed in urban design, where permeability and choke points create vulnerabilities—if movement is restricted to single points of failure, traffic congestion can occur, compromising circulation networks (Darren et al. 2019).

3.5.2 Integration of TOD with Urban Connectivity

Connectivity is essential as federal and state agencies work to expand the public transit system to accommodate the needs of a multimodal transportation network.

In the context of transit-oriented development (TOD), connectivity encompasses the coordination of transit routes, coverage, scheduling, speed, operational capacity, and urban form characteristics. It is a critical factor in shaping the overall perception of any transit network (Mishra et al. 2012).

Hadas and Ceder (2010) highlight that the stochastic nature of various public transit attributes—such as travel time, dwell time, demand, and connectivity—affects the reliability and usage of transit systems. A key focus is the transit network's connectivity, particularly regarding the reliability of transfers and its integration with urban areas. Transfers are classified into three types based on the spatial relationship between transit routes and adjacent land: nonadjacent, adjacent, and shared transfer points, each contributing to overall system connectivity. This classification informs the analysis and improvement of transit systems to enhance accessibility and integration within transit-oriented development (TOD), ensuring stronger connections with surrounding urban areas

4 Case Studies

4.1 Case Study One: Tianhui TODTOWN

Ronald Lu & Partners has launched Tianhui TODTOWN, China's first transit-oriented development, situated in Shanghai with a built-up area of 740,000 m^2. This innovative project redefines the TOD concept by emphasizing sustainability, mass transit, and community cohesion, marking a significant step forward in urban development.

A key challenge for this site was the presence of two railways, which reduced land use efficiency and created a disconnect between adjacent urban zones. To overcome this, TODTOWN in Xinzhuang was developed on a newly constructed second-floor platform, effectively linking two previously separated plots of land divided by railway lines. This integration has transformed the area into a city sub-center, combining a transit interchange hall and shopping mall into a multifunctional transportation hub.

The project embodies the latest TOD 4.0 concept, emphasizing high-density, mixed-use development that integrates harmoniously with a transportation hub and its surrounding communities. This approach fosters a comprehensive living environment that meets the diverse needs of residents and visitors alike.

By offering a fully integrated living environment, the development features a subway station, an intercity high speed railway station, various residential options, a Grade-A office building, commercial spaces, a hotel, and a community center. Once completed, it is poised to stimulate further growth in the greater Shanghai area.

The mixed-use design includes a retail mall as its centerpiece, along with amenities such as offices, residences, service apartments, and open spaces. The project capitalizes on station traffic to create seamless connections throughout the area. By prioritizing pedestrian desire lines, the traffic network, the station concourse, and a public transport interchange, it establishes a new type of complex development that spans both existing and new railway lines. This "air rights" strategy effectively utilizes the space above transit corridors and facilitates connectivity between the west and east.

In conclusion, the project overcomes previous challenges by decking over the railway, adding 20 ha of land, and improving connectivity. These changes create a more integrated urban environment, enhancing mobility and quality of life for both residents and visitors. Figure 4 shows the site before and after implementation of the TOD components and principles.

4.2 Case Study Two: Guangming Hub

Foster + Partners has recently revealed its winning design scheme for a new urban destination in China, known as the Guangming Hub, situated along the high-speed rail connecting Hong Kong, Shenzhen, and Guangzhou.

The project's primary challenges stemmed from the three major metro lines and stations, which operated as isolated zones lacking cohesion and significance. Each line functioned independently, creating a disjointed experience for users and hindering the overall effectiveness of the transportation network.

This Transit-oriented development proposal envisions a city that facilitates the movement of people and goods through a strong infrastructure, efficient transport networks, reliable public services, and abundant greenery (Urbanitarian n.d.)

The transit-oriented development features a contemporary urban center centered around an integrated and sustainable transport hub. The high-speed rail station, oriented in a north-south direction, is strategically located to serve as the focal point of the hub. A sunken green spine runs east-west, traversing the site and passing through the high-speed rail station at ground level, creating a seamless connection with nature. To the west of the high-speed rail station, a new metro station straddles the green spine, enhancing accessibility. Further west, an intercity link station is positioned near the site's edge, connecting the Guangming Hub to the broader city. The design prioritizes easy wayfinding, minimal walking distances, and shaded walkways, facilitating smooth movement for pedestrians, cyclists, and autonomous vehicles (Harrouk 2020).

In terms of land use and architecture, a large retail podium is situated atop the underground metro and intercity link stations, providing a vibrant commercial space. Stepped green terraces from the retail podium blend harmoniously with the surrounding forest while offering views of the green spine. Office towers are strategically positioned at cardinal points across the site, drawing inspiration from traditional Chinese gateways. Ground-level openings in the towers provide direct access to the central high-speed rail station, further enhancing the connectivity and functionality of the urban environment. Figure 5 shows the implementation of TOD components and principles at the broader scale of Guangming Hub.

4.3 Comparative Analysis

The comparative analysis examines two distinct case studies, focusing on urban development challenges and solutions in different contexts. The analysis is structured around key comparative points, including location, area, challenges, solutions, vision, classification, and land use. By exploring the unique characteristics of each case, this study highlights how diverse urban environments address their respective

Fig. 4 Site before and after images

Fig. 5 Site before and after images

Table 2 Case study comparison (authors)

Case study	Tianhui TODTOWN	Guangming hub
Location	Shanghai, China	Shenzhen, China
Area	740,000 m^2	1,405,024 m^2
Challenges	Reduced land use efficiency and disconnected urban zones	The three major metro lines and stations were disconnected zones that had no meaning, all working on their own
Solutions	Urban solutions—Deck over railway added 20 ha of land Improved connectivity within and beyond boundaries	Urban solutions Connected three metro lines and a rail station Unity of scattered urban zones
Vision	The intermodal hub produces a complete living environment. Highlighting the principles of the TOD 4.0 concept, the project puts in place – High-density – Mixed-use – Transit hub – Connectivity	This project envisioned a smart city that supports – Flow of people – Infrastructure – Transportation – Public areas
Classification	Mixed-use development	Mixed-use development
Land use	– Mixed-use development: The project features a combination of residential, commercial, and office spaces, fostering a vibrant urban environment – Public transit integration: It includes a subway station and an intercity high-speed railway station, ensuring seamless connectivity – Amenities: The development offers various amenities, such as a community center, retail spaces, and recreational areas – Green spaces: Incorporates open areas and green terraces to enhance the urban landscape and provide recreational opportunities for residents – Civic spaces: Public plazas and gathering areas that promote community interaction and social cohesion	– Transport-oriented development: This project emphasizes a transport hub with a high-speed rail station, metro station, and intercity link station at its core – Commercial areas: Includes a large retail podium that serves as an anchor for shopping and dining, supporting economic activity – Office spaces: Strategically positioned office towers to cater to businesses and professionals – Residential development: Incorporates residential units to provide housing within a walkable distance from the transit hub – Green infrastructure: Features a sunken green spine that runs through the site, enhancing sustainability and providing recreational space

(continued)

Table 2 (continued)

challenges through tailored approaches. The analysis aims to uncover patterns in how urban visions shape the development of land use strategies and classifications, while identifying common and divergent solutions to similar urban issues. This comparative framework provides valuable insights into how these cities adapt to spatial and infrastructural constraints to enhance their urban fabric and meet future development goals. Table 2 shows a comparison of the two case studies, focused on the challenges and solutions implemented to create a overall TOD framework.

5 Conclusions and Findings

Urban sprawl and population growth are critical global challenges that contribute to fragmented urban fabrics and inefficiencies in transportation, land use, and environmental sustainability. Transit-oriented development (TOD) has emerged as a viable solution, offering an urban planning strategy focused on enhancing connectivity, accessibility, and mobility. By creating compact, mixed-use, walkable communities centered around public transit, TOD addresses these challenges while promoting more sustainable urban development. Through the analysis of two case studies from China, this paper demonstrates the effectiveness of TOD in managing rapid urbanization, optimizing transit use, and fostering livable communities.

The findings underscore the importance of incorporating TOD principles into urban planning, particularly in fast-growing cities, to create more connected, efficient, and sustainable urban environments that can serve as a blueprint for future growth. This study has highlighted the transformative potential of TOD principles in creating central "anchors" that strengthen connectivity within the urban fabric. By integrating diverse land uses and fostering accessibility through well-placed transit hubs, TOD can counteract the fragmentation and help reshape urban sites into cohesive focal points. These central anchors promote urban cohesiveness, reduce reliance on private vehicles, and establish pedestrian-friendly environments, supporting the city's goals for sustainable urban growth.

Additionally, addressing the main research challenge, this study has provided a framework or criteria for assessing urban deficiencies and applying TOD principles to urban design by creating a case study guideline. By focusing on vertical and horizontal connections, mixed land uses, and adaptable urban spaces, TOD can enhance connectivity and create urban environments that support connectivity, accessibility, and continuity of the urban fabric as a whole.

To guide these developments, establishing TOD guidelines tailored to a city's specific context would provide developers with a clear path toward building an urban environment that balances growth with public accessibility and connectivity. By embedding TOD principles into urban design and planning, a city could transform into a more connected, accessible, and vibrant city, where the built environment actively promotes sustainable growth and enhances residents' quality of life.

References

Asriana N, Koerniawan MD (2020) TOD model through low carbon city concept in urban design (case study: Palembang, Indonesia). IOP Conf Ser Earth Environ Sci 532(1):012017. https://doi.org/10.1088/1755-1315/532/1/012017

Brueckner JK (2000) Urban sprawl: diagnosis and remedies. Int Reg Sci Rev 23(2):160–171. https://doi.org/10.1177/016001700761012710

Darren N, Gerhard B, Christopher H (eds) (2019) Urban design, connectivity and its role in building urban spatial resilience. XXV ISUF International Conference

Dieleman F, Wegener M (2004) Compact city and urban sprawl. Built Environ 30(4):308–323. https://doi.org/10.2148/benv.30.4.308.57151

Drishti IAS (2021). (Transit Oriented Development, Website, https://www.drishtiias.com/images/uploads/1611737136_image1.png

Fazli R, Faridi R (n.d.) Urbanization in Dubai: process, problems and challenges. J West Asian Stud. https://www.publishingindia.com

Hadas, Y., & Ceder, A. (2010). Public Transit Network Connectivity. Transportation Research Record Journal of the Transportation Research Board, 2143(1), 1–8. https://doi.org/10.3141/2143-01

Hannawi N, Jones P, Titheridge H (2019) Development of transit-oriented development in Dubai City and the Gulf states. Edward Elgar Publishing eBooks. https://doi.org/10.4337/9781788971713.00010

Harmadi CK, Sulistyono S, Listyawati RN, Hayati NN, Alfiah R (2022) Development strategy of Malang Kota Baru station area using transit-oriented development (TOD) principal approach. IOP Conf Ser Earth Environ Sci 1000(1):012026. https://doi.org/10.1088/1755-1315/1000/1/012026

Harrouk C (2020) Foster + Partners wins competition to design Guangming Hub, a new transport- oriented development in China. ArchDaily. https://www.archdaily.com/951836/foster-plus-partners-wins-competition-to-design-guangming-hub-a-new-transport-oriented-development-in-china

Hermansyah A, Samudra A, Satispi E (2024) Policy dialogue: key factors for the success of transit-oriented development infrastructure 3.0 in big cities in Indonesia. J Infrastruct Policy Dev. https://doi.org/10.24294/jipd.v8i5.3400

Kamruzzaman M, Baker D, Turrell G (2015) Do dissonants in transit-oriented development adjust commuting travel behaviour? Deleted J. https://doi.org/10.18757/ejtir.2015.15.1.3059

Mishra S, Welch TF, Jha MK (2012) Performance indicators for public transit connectivity in multi-modal transportation networks. Transp Res Part Policy Pract 46(7):1066–1085. https://doi.org/10.1016/j.tra.2012.04.006

Nyunt KTK, Wongchavalidkul N (2020) Evaluation of relationships between ridership demand and transit-oriented development (TOD) indicators focused on land use density, diversity, and accessibility: a case study of existing metro stations in Bangkok. Urban Rail Transit 6(1):56–70. https://doi.org/10.1007/s40864-019-00122-2

Papa E, Coppola P, Angiello G, & Carpentieri G (2017). The learning process of accessibility instrument developers: Testing the tools in planning practice. Transportation Research Part a Policy and Practice, 104, 108–120. https://doi.org/10.1016/j.tra.2017.03.010

Petrucci A (2021) Transit oriented development as a smart city's interface, a GCC perspective. Sustainability 11(3):786. https://openknowledge.worldbank.org/handle/10986/34870

Salat S, Ollivier G (2017) Transforming the urban space through transit-oriented development: the 3V approach. International Bank for Reconstruction and Development/The World Bank

Shandas V, Makido Y, Ferwati S (2017) Rapid urban growth and land use patterns in Doha, Qatar: opportunities for sustainability? Eur J Sustain Dev Res 1(2). https://doi.org/10.20897/ejosdr.201711

Singh YJ (2015) Measuring transit-oriented development (TOD) at regional and local scales: a planning support tool. https://doi.org/10.3990/1.9789036539982

Stojanovski T (2019) Urban design and public transportation—public spaces, visual proximity and transit-oriented development (TOD). J Urban Des 25(1):134–154. https://doi.org/10.1080/13574809.2019.1592665

TOD Standard—ITDP (2024). https://tod.itdp.org/tod-standard/tod-standard-framework.html

Transit Oriented Development (2024) Transit Oriented Development Institute. http://www.tod.org/

Urbanitarian (n.d.) https://www.urbanitarian.com/masterplans_post?id=430

Wang L, Xiu X, Zhaou Z, Wang Z (2018) The impacts of transportation infrastructure on sustainable development: emerging trends and challenges. Environ Res Public Health 15(1172). https://doi.org/10.3390/ijerph15061172

Yang B, Zhaou D, Wang X (2022) Transit oriented development (ToD) integration of a sustainable residential building model. Politecnico Milano

Zhou J, Webster C, Yang Y (n.d.) Using big and open data to analyze transit-oriented development: new outcomes and improved attributes. The University of Hong Kong

Beyond Transit-Oriented Development: Linear Transit Cities, an Urban Structure for the Twenty-First Century

Halvard Dalheim

Abstract

Since the first emergence of cities, there have been changing philosophies for how to plan cities. Linear cities, put forward by Arturo Soria y Mata, have had limited attention in the planning of cities. Since the 1980s, urban consolidation has become a focus for accommodating growth in Melbourne and Sydney. Recently, the NSW Government announced the *Transport Oriented Development (TOD) Program* for Sydney, which is directed at delivering the well-understood benefits of TODs.

However, what if there is a way to understand the benefits of TODs in a broader context? That is, by recognising a single TOD as part of a wider urban area, specifically the rail (transit) corridor on which it is located. With the aim that a household could access most of their economic and social needs, such as health care, shops, schools and open space along a single corridor——a linear transit city.

This paper provides case studies of transit corridors in Melbourne and Sydney, examining accessibility to economic and social assets. The case studies reveal that location matters for investments in TODs and supporting economic and social infrastructure. Suggesting TODs can be re-imagined as linear transit cities, a new approach for how cities plan for urban renewal.

Keywords

Transit-oriented development · Transit cities · Urban renewal · Linear cities · Urban structure · Rail corridors · Economic and social assets · Planning policy · Urban renewal model · Urban benefits and efficiency

1 Introduction

In Australia, high population projections and increasing recognition of the substantial costs of developing new communities on the fringe of metropolitan areas have resulted in governments focusing on urban renewal initiatives directed at centres on train lines in line with the principles of transit-oriented development (TOD). However, the rationale for why one location is chosen over another is unclear to the community, which creates local angst due to the perceived impacts of change in a local area. There is, however, a potential alternate approach to identifying where change should occur that would provide both clarity to the community and highlight wider cost-of-living benefits, so a household can change its travel behaviour and shift to a one-car household. This paper seeks to provide a 'proof of concept' for re-imagining how governments plan for TODs, including outlining an urban renewal model based on the findings.

Transit-oriented development (TOD) or similar centre-based initiatives have been the mainstay for urban renewal planning for the last three decades. In Australia, the narrative has focused on centres, with multiple objectives for providing jobs closer to home, transport network efficiency, including congestion management and providing multi-purpose trips. Direct mention of TODs in policy is a very recent development, with the release of the *Transport Oriented Development Program* (New South Wales Government 2023a), noting slightly different nomenclature.

However, the emphasis on centres (TODs) has not been part of an integrated and overarching approach to the urban structure of established areas. There is guidance on elements, such as health and education (knowledge) precincts, industrial areas, a diversity of housing opportunities and transport networks, but how best to structure their inter-relationships

H. Dalheim (✉)
School of Architecture, Design and Planning, University of Sydney, Darlington, NSW, Australia
e-mail: halvard.dalheim@sydney.edu.au

F. Alberti et al. (eds.), *Towards Sustainable Urban Regeneration: Utilizing Environmental, Cultural, and Social Assets*, Advances in Science, Technology & Innovation, https://doi.org/10.1007/978-3-032-02091-8_13

is missing. The importance of this missing approach is amplified when the issue of the relative accessibility of a household to jobs, goods and services is considered, noting that there are a variety of tools that can show relative levels of accessibility.

This paper outlines the issue of accessibility to goods, services and jobs from the standpoint of accessibility along a rail (transit) corridor and from a specific TOD location. The aim is to determine whether an overarching approach can be developed to guide decisions, such as where is the best place to locate new TODs and where to locate new social and economic infrastructure to best improve accessibility for those on a linear transit corridor. If so, could the approach then be adapted for rapid transport networks to maximise accessibility across a wider area of a city? To explore this approach, data from the Australian cities of Melbourne and Sydney have been utilised, noting that both have populations of over five million people. This approach, therefore differs from concepts such as the TOD Index, which seeks to understand the potential or 'TOD-ness' of the walkable catchment of a station (Singh et al. 2017). Instead this approach seeks to utilise several indicators that reveal the accessibility of the walkable catchment of a station to the range of goods, services and jobs that a person would generally require access to over a lifetime, and where accessibility explicitly includes those nearby stations that may provide access to some of the desired locations, such as a university or major centre (a concentration of jobs, goods and services).

Therefore, this paper builds on and informs Principle 14 of the Charter of the New Urbanism—*Transit corridors, when properly planned and coordinated, can help organize metropolitan structure and revitalize urban centres. In contrast, highway corridors should not displace investment from existing centres* (Congress of the New Urbanism 2024).

The TOD Index approach is central to understanding the potential of any given location; its use does form part of the urban renewal model discussed in Sect. 5 of this paper.

In exploring opportunities of transit corridors, the concept of linear cities is also examined, from its emergence as a concept, developed by Arturo Soria y Mata in 1882, resulting in the Ciudad Lineal of Madrid (Collins 1959). Transport efficiency is central to the concept and its philosophy can be seen in the post-war 'Finger Plan' for Copenhagen (Sørensen and Jacob 2019), the 1971 plan for metropolitan Melbourne (Melbourne Metropolitan Board of Works 1971) and the 1968 plan for metropolitan Sydney (State Planning Authority for New South Wales 1968).

1.1 Problem Statement

Australia is a highly urbanised country with 68% of its population of 25.4 million (2021) living in the eight capital cities (Australian Bureau of Statistics 2021), with the two largest cities, Melbourne and Sydney, both having populations over 5 million. This settlement structure is vastly different from that of Europe (excluding Russia), where there are only four cities with a population of over 5 million, although the combined population of all countries is around 450 million (World Population Review 2024).

Over the last 50 years, the population of Australia's eight capital cities doubled, and they continue to grow rapidly (Australian Bureau of Statistics 2022), including the population growing by 0.5 million in the 12 months to June 30, 2023, the highest on record (Australian Bureau of Statistics 2024a). In managing this growth, there is increasing recognition of the relatively higher cost of delivering new greenfield development as opposed to urban renewal. The recent *Victorian Housing Statement* indicates a four-fold difference (Department of Premier and Cabinet 2023, p. 15). Multiple studies have also been undertaken on this issue since the work of Newman and Kenworthy (1989) on the impact of urban structure on vehicle travel, such as the assessment of settlement choices for Sydney (Centre for International Economics 2010), which reaffirmed the need to facilitate urban renewal.

Current metropolitan strategies for both Melbourne and Sydney support urban renewal activities, particularly connected to centres, and the delivery of additional rail (transit) corridors. At the same time, there is a recognition across Australia of a significant mismatch between housing supply and demand. For example, for Sydney in the last decade, housing completions peaked in the 2 years 2017–18 and 2018–19, with approximately 41,000 and 42,000 net dwelling completions, though in the year 2022–23, the total delivered was only 21,100 dwellings (New South Wales Government 2024a, b), yet the population growth in the same year was 146,700, suggesting a latent housing demand of around 54,300 dwellings (based on an average dwelling occupancy rate, Australian Bureau of Statistics 2024b). Consequently, all states have signed up to a *National Housing Accord* that aims to deliver 1.2 million dwellings over the next 5 years, including affordable housing targets (Australian Government 2022).

Considering the scale of the housing task and the limited resources of a government at any one time, city planning would benefit from the delivery of an ongoing supply of urban renewal development based on a practical land use model that identifies the optimal locations for additional urban renewal activity as well as optimal locations for new economic and social assets, such as schools, health facilities and business precincts.

1.2 Study Aim

This study seeks to build on the lessons learnt from transit-oriented development, a place-based approach to urban management, by exploring the potential to develop a citywide urban renewal framework that informs decisions about where, across a city, are the optimal locations to facilitate growth. Underpinning this approach is an assumption that one of the central tenets of planning a city is to improve the accessibility of residents to a wide range of economic, social and lifestyle needs.

In this context, the purpose of this study is to determine whether there are differences in the level of economic and social assets along different rail lines (transit corridors) across a city (Melbourne and Sydney) and to establish whether there are differing levels of access to economic assets from different TOD locations in the context of a 30-min city. The overarching objective is to establish a practical land use model to better inform government decision-making for the location of new TODs and the location of the economic, social and community assets to support population growth—a linear transit cities urban renewal model.

2 Literature Review

This literature review explores several areas with the aim of understanding the proposed hypothesis that *an urban renewal model can be developed to better inform the decision-making of government in managing growth and change.* The areas covered include transit-oriented development, linear cities, accessibility and walkability and urban management models.

2.1 Transit-Oriented Development

The concept of transit-oriented development (TOD) emerged in the late 1980s through the work of Peter Calthorpe, with a central thrust of removing the dependency on private vehicles for travel (Calthorpe 1993). An understanding of TODs in terms of benefits and planning considerations is well documented, reflecting its extensive use as a tool to manage urban development across the globe. Suggested benefits of TODs include: a substantial shift towards increased transit use (Crowley et al. 2009); increased residential densities (Ratner and Goetz 2013); positive perceptions of increased housing choices, enhanced sense of community and greater pedestrian and child safety (Brown and Werner 2011); increased development activity when land use planning is integrated with transport planning ((Ratner and Goetz 2013), and increased property prices, which in turn attracts more development (Ibraevaa et al. 2020).

Identified considerations for the planning of TODs include: the implications of planning controls, the roles of local government, urban design qualities and affordable housing incentives (Mathur and Gatdula 2023; Tsivanidis 2023); the balance between amenity richness and accessibility with density and land use diversity (Zandiatashbar and Laurito 2023); and the potential for TODs to facilitate commercial activities depends on the scale of the centre to its catchment and the complex interactions between a train station and neighbourhood characteristics (Searle et al. 2014).

Not long after Calthorpe (1993) articulated the concept of transit-oriented development, Bertolini (1996) conceptualised the dynamics of redevelopment around a station in terms of—'The station is at the same time a node of networks and a place in the city'.—which he formalised into the 'node-place model' that was seen as a tool to understand the potential for a TOD (Bertolini 1999). The model is still utilised to understand change in cities, such as the work of Zhao et al. (2024), which explored vulnerability and imbalanced development in transit-oriented areas with the model variant node-place people. From the node-place model emerged the TOD indices framework with a focus on identifying an extensive list of success factors (Cucuzzella et al. 2022). The most widely referenced work is that of Singh et al. (2017), where their TOD Index is based on eight 'rules' that seek to reflect the principles of transit-oriented development. Their approach focuses on measurement, not evaluation, to, in part, inform policymakers of the actions required to enhance the potential of a place as a TOD.

In Australia, TODs can be seen as an evolution of the centres' policy, which became part of metropolitan plans for Melbourne and Sydney in the 1980s. In Melbourne, the 'district centre' policy was introduced to reduce the need for travel and facilitate the development of centres as a focal point for administrative, commercial, retail, cultural and entertainment activities (MMBW 1981). In Sydney, the centres' policy emerged with the designation of two regional centres (CBDs) and 17 sub-regional centres, all on rail lines, two on the Central Coast. The aims were to have retail and office jobs closer to home and provide access to a full range of retail, recreation and community services (New South Wales Department of Environment and Planning 1988). The recent designation of TODs in Sydney was, in part, informed by feasible capacity (Department of Planning, Housing and Infrastructure 2024).

2.2 Linear Cities

The concept of linear cities, as a tool for shaping urban development, has been part of urban planning approaches, particularly urban growth corridors, over the last 70 years.

Lessons learnt from its use as an urban management tool include: linear cities can be used as a tool for connecting more complex structures (Antyufeev and Antyufeeva 2019); linear cities can emerge due to natural and topographic circumstances, as well as reflecting transport options (Antyufeev and Antyufeeva 2019); linear cities should not be seen as an all-encompassing urban form, but more as an element of a wider complex urban form (Doxiadis 1967; Batty 2022); travel times of a transit corridor/linear city need to be faster than that of a private vehicle (Curtis 2008); the mix of activities available at a TOD will influence the extent to which other car based trips are required to satisfy the full range of needs of a household including employment (Wolański 2023); attraction of commercial activities in linear corridors can contribute to economic growth (Knowles 2012); and lack of control for managing activity outside a transit corridor can impact on its success (Ryckewaert 2010; Batty 2022).

The need for transport corridors to include a wide range of activities was recognised in the 1971 metropolitan plan for Melbourne (MMBW 1971), where it stated: '*The growth corridors must not simply be new dormitory areas where residential settlement will occur, …*' (p. 68). The opportunity to deliver on this policy intention has been eroded over time as the green wedges between the urban corridors were progressively reassigned to urban development, resulting in the tensions as described by Doxiadis (1967) and Batty (2022).

2.3 Accessibility and Walkability

Across all seven plans for Australia's capital cities are several common policy positions, including the need to support walkability, that urban renewal is linked to public transport, that infrastructure is optimised, and that infrastructure meets the needs of the community (Dalheim 2023). Considerations needed when planning for accessibility and walkability in TODs and linear cities include: an understanding that there is limited data on the propensity for people to walk to a station and then to a destination (Park et al. 2015); models that test accessibility to public transport are available (Saghapour et al. 2016); transit networks with headways of less than 10 min can support interchange transfers (Currie and Loader 2010); changes in accessibility, such as new motorways or railway lines, have significant impacts on urban structure overtime, compared to land use change (SGS Economics and Planning 2012); research undertaken in Melbourne suggests that people will walk 400 m to a bus service and 800 m to a train service, in addition, travel distances are influenced by neighbourhoods with gross residential densities exceeding 20 dwellings per hectare, a well-connected street network, and access to nine or more local living destinations (Boulange et al. 2017); low-cost 'feeder' bus systems that complement mass rapid transit by providing 'last-mile' service for passengers' and easy access to a system's terminals, yield high returns (Tsivanidis 2023), and improved micro-walkability (quality of the pedestrian links) may increases the size of the catchment of a railway station, particularly along major pedestrian corridors (Park et al. 2015).

2.4 Urban Management Models

Planning approaches (models) evolve (Hamnett and Freestone 2000), and urban models typically build on the lessons learnt from past planning activities (Yudi et al. 2023). Their application can be seen in the varied urban structures of cities across the globe. Kostof (1991) argues that at a broad scale, these structures have resulted in displaying one of five urban patterns: organic, the grid, the city as a diagram, the grand manner and the urban skyline. Therefore, a necessary task in planning for urban renewal is an understanding of the underlying urban structure of a city and how it influences city dynamics (Dalheim 2023).

Modern planning, recognised as starting in the late 1800s, brought new urban models to manage the growth of cities, principally in response to the 'urban ills' of the cities of the Industrial Revolution (Hall 1996, pp. 14–23). Most recognisable of these models was the Garden City of Ebenezer Howard (1898), with Letchworth Garden City and Hampstead Garden Suburb as renowned exemplars.

Post World War II (WWII) came the slum clearance programs and the emergence of suburbia, the latter becoming the panacea for the outward growth of cities in North America and Australia and still is today. Jane Jacobs, in her seminal work *The Death and Life of Great American Cities* (1961), attacked the slum clearance philosophies and provided a foundation to understand how different elements of the urban fabric create life in a city. In *Suburban Nation: The Rise of Sprawl and the Decline of the American Dream* (2000), the authors Andrés Duany, Elizabeth Plater-Zyberk and Jeff Speck provide a similar negative critique of suburbia, specifically its lifelessness.

This rejection of the post-WWII urban planning approaches gave rise to the new urbanism planning model and transit-oriented development (TOD). The concept of the TOD was originally outlined by Peter Calthorpe in his work *The Next American Metropolis, Ecology, Community and the American Dream* (1993). Importantly, the concept of the TOD was intrinsically part of the new urbanism model and much of Calthorpe's book is devoted to guidelines for the new urbanism approach. Duany et al. (2000) also guide the planning of the traditional neighbourhood through a development checklist. Emerging from these works is the highly recognised publication *The Smart Growth Manual* (2010), which provides detailed guidance in the areas of the region, the neighbourhood, the street and the building.

In addition, one of the leading organisations advocating for new urbanism, the *Congress of New Urbanism*, has a Charter, that provides 27 overarching principles to guide public policy, development practice, urban planning and design within three scales: the region: metropolis, city and town; the neighbourhood, the district and the corridor; and the block, the street and the building.

More recently (2016), the concept of a 15-min city has been introduced as an urban model. Conceptually, it emphasises the importance of proximity; it also suggests that city planning should, at the same time, address issues of the smart and sustainable city (Allam et al. 2022).

2.5 Infrastructure Planning

Development usually requires supporting infrastructure, whether that be utility services for a new greenfield development or expanded school facilities to support urban renewal. Consequently, across Australia, planning jurisdictions have legislated requirements that development part-fund the need for required infrastructure. Such as in *Part 7, Infrastructure contributions and finance* of the New South Wales *Environmental Planning and Assessment Act 1979 No 203*. In jurisdictions of Victoria (Melbourne) and New South Wales (Sydney), infrastructure agencies have been established with a core role to advise the government on infrastructure decisions.

Metropolitan planning highlights that the need for infrastructure investments, such as schools, utilities and transport, is generally linked to population or household thresholds. The demand for major retail anchors such as supermarkets, department stores and large bulky goods stores is also linked to population thresholds. Consequently, assessments of residential development opportunities for a new TOD can be converted to an understanding of the demand for a range of infrastructure and commercial investments when existing provisions (excess capacity or not) are considered.

3 Methodology

The methodology aims to determine the level of economic, social and foundational community assets within an 800-m catchment area of train stations across the metropolitan areas of the cities of Melbourne and Sydney, as well as whether the full range of assets is accessible by public transport and walking from any given station in the context of a 30-min city. The approach taken is a series of inter-related steps, including identification of the rail lines to be assessed, identification of the stations (TODs) to be assessed in terms of a 30-min city, identification of what economic, social and foundational community assets will be the focus for the analysis and how the 800-m catchment areas and 30-min concepts have been defined.

3.1 Rail Lines

Rail lines (corridors) were assessed based on the published networks for each city, including the published timetables (PTV 2024; Transport for NSW 2024). In total, nine rail lines were assessed in Melbourne, including the Melbourne Metro line, currently under construction and due to open in 2025, refer to Fig. 1. The new line will connect the Sunbury and Pakenham/Cranbourne lines with a new tunnel and stations under the Melbourne CBD. For Sydney, 11 lines were assessed, including the Sydney Metro line, anticipated to be open August–September 2024, refer to Fig. 2. This new line will connect a new line from Tallawong in Sydney's northwest to the Bankstown line with a tunnel under Sydney Harbour. In both cities, several rail lines have spur lines; in these cases, only the principal corridor was assessed.

3.2 Transit-Oriented Development (TOD) Assessment

The assessment of TODs was limited to centres in Sydney where there is a specific Transport Oriented Development program, and the eight-state government-led initiatives assessed were Bankstown, Bays West, Bella Vista, Crows Nest, Homebush, Hornsby, Kellyville and Macquarie Park (Department of Planning, Housing and Infrastructure 2024). The attributes assessed only included the economic and social assets accessible within 30 min of a home within a TOD. See walkability for how this was measured. Scoring is based on the total count of assets accessible. The Bays West proposed TOD is on the committed but yet-to-be-built Sydney Metro West, for which the station locations have been committed, though completion is a decade away.

3.3 Economic and Social Assets

The economic and social assets (attributes) utilised for the analysis of this study are based on a review of common policy areas of the metropolitan plans for the major state capital cities—*A Metropolis of Three Cities*, (Sydney) (Greater Sydney Commission 2018), *Perth and Peel @3.5 million* (Department of Planning, Lands and Heritage 2018), *Plan Melbourne 2017—2050* (Department of Environment, Land, Water and Planning 2017), *ShapingSEQ* (Brisbane) (Department of Infrastructure, Local Government and Planning 2017) and *The 30-Year Plan for Greater Adelaide* (Department of Planning, Transport and Infrastructure 2017;

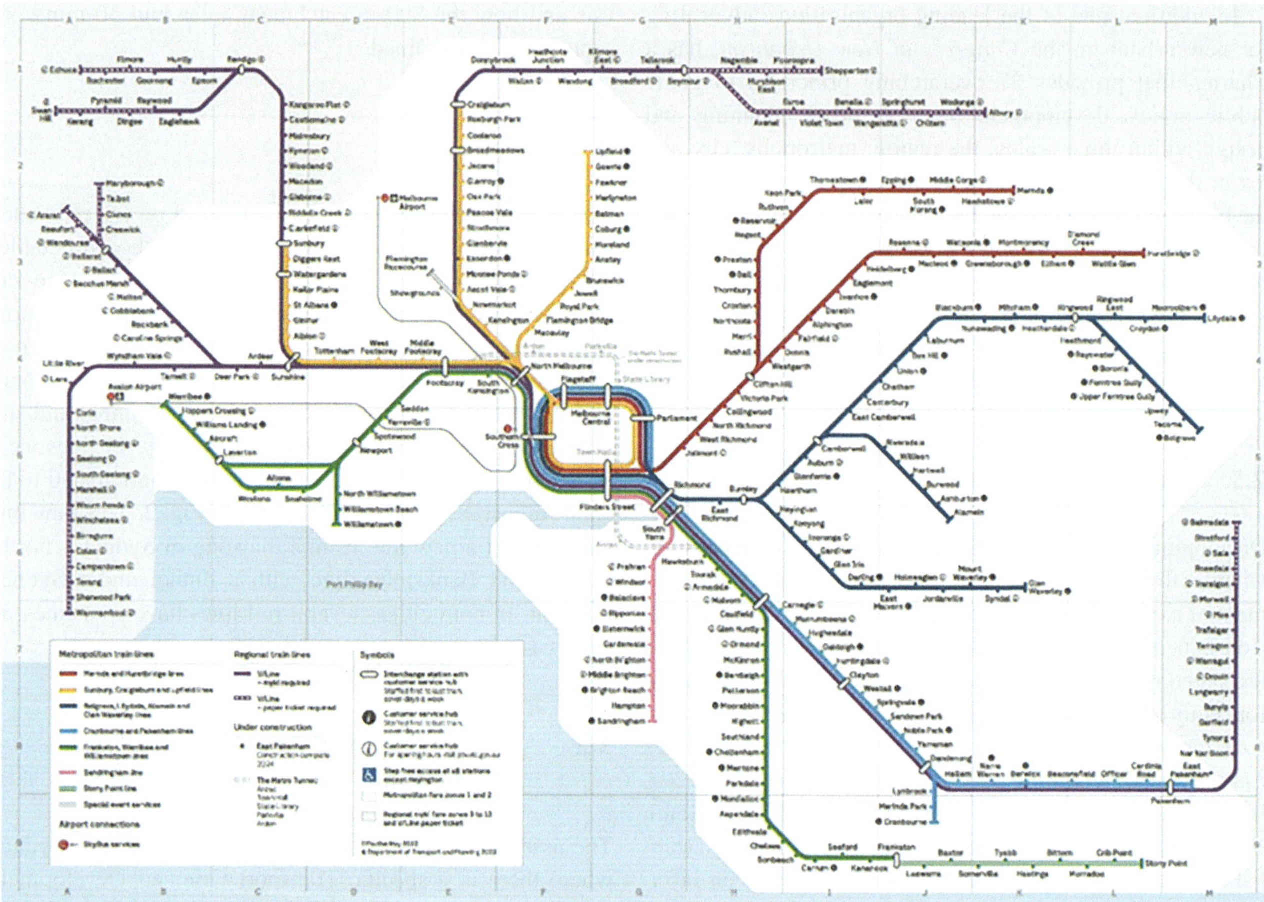

Fig. 1 Melbourne rail network. (Source: PTV 2024)

Dalheim 2023). The review reveals policy consistency in three areas: health and education precincts; major centres, both the CBD and major suburban centres, with states having differing nomenclature, such as metropolitan activity centres in Melbourne and strategic centres in Sydney; and access to community facilities and services—the issue of health and wellbeing. These three areas form the basis for the economic and social assets to be identified within the 800-metre catchments of the station precincts. The latter area is covered under the foundational community assets section.

For health and education facilities, the metropolitan plans for both Melbourne and Sydney identify specific health and education precincts (Department of Environment, Land, Water and Planning 2017; Greater Sydney Commission 2018). For this research, which has a local area focus, the health and education facilities identified included all hospitals (public and private), all university campuses, all single university facilities together with all Technical and Further Education facilities (TAFEs), the latter being state government managed (TAFE NSW—https://www.tafensw.edu.au/ and TAFE Victoria—https://www.vic.gov.au/tafe).

The metropolitan plans for Melbourne and Sydney, as referenced above, identify a top tier of centres in addition to CBDs. The criteria utilised for each city differ, such that for Melbourne, only centres in the outer areas are designated. In addition, over the last 70 years, the number of centres identified in the metropolitan plans has evolved significantly; for example, at one point, there were over 100 centres identified as significant for Melbourne. Consequently, this assessment of major centres is complemented by the presence or absence of top retail centres in each state, which are located on a rail line. The top centres were classified in terms of the size of a centre, recognising that size relates to the breadth of the retail offer and the scale of the jobs within the centres, with the source for Melbourne proprietary data of the consulting firm Urbis and Sydney, Wikipedia (2024) and cross-checked with individual centre websites, noting that for a fee, the same data can be purchased from several peak bodies and consultancies.

In addition, data was obtained on population and employment levels within 800 m of every station on the 11 Sydney rail corridors, with the principal aim of identifying the level of jobs accessible in the context of a 30-min city. This data

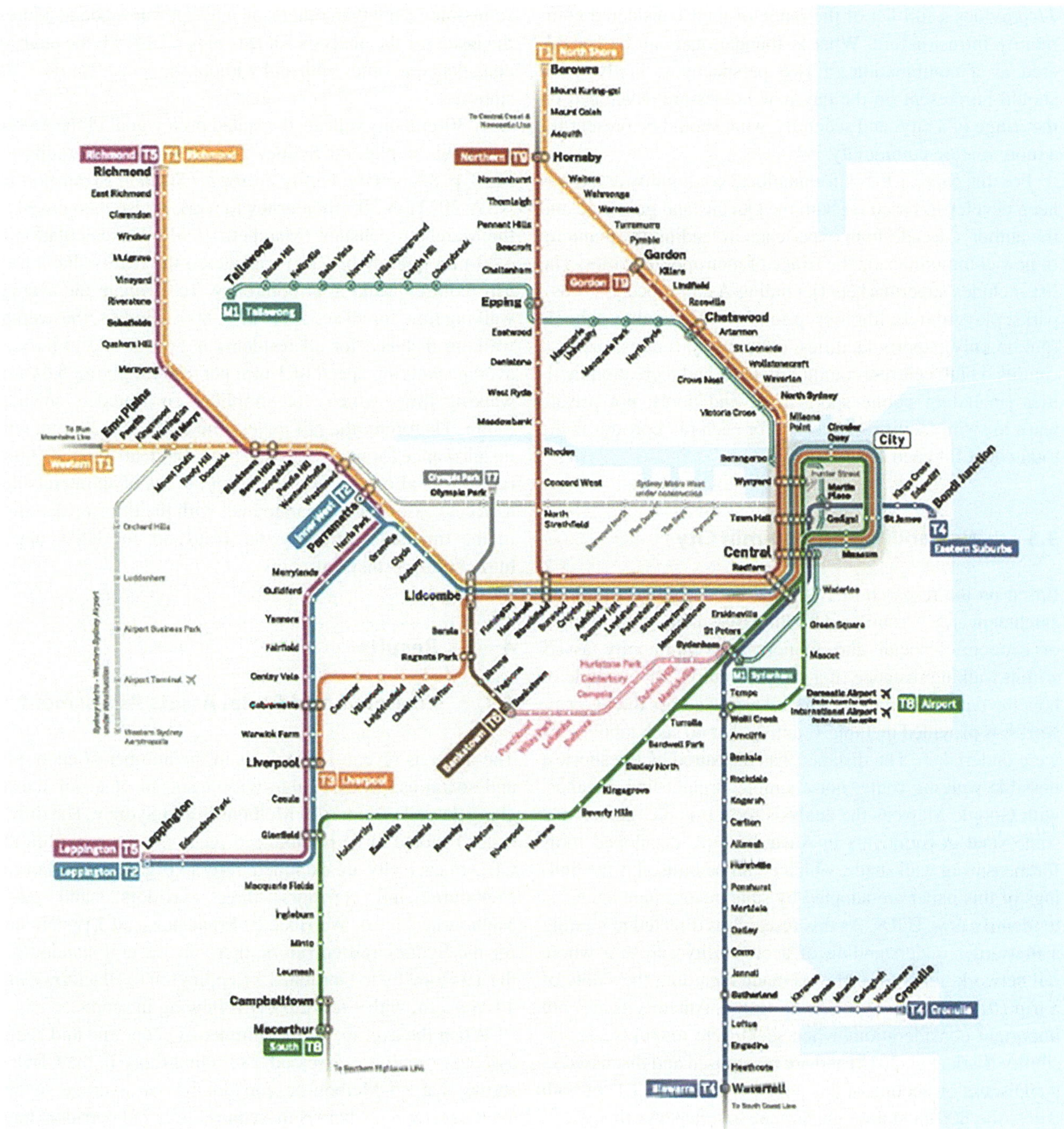

Fig. 2 Sydney rail network. (Source: Transport for NSW 2024)

was provided by the consultancy Mecone, utilising a bespoke assessment of their data tool—Mosaic. Data on the size of retail centres as well as the level of retail at the 20 largest centres in Melbourne was provided by the consultancy Urbis from their internal retail database.

3.4 Foundational Community Assets

There is no consistency in the identification of what is the required community infrastructure when planning a city. The Queensland Department of Infrastructure, Local Government and Planning (2015) *Community facilities PDA guideline no.*

11 provides a full list of the range of their considered community infrastructure. What is foundational can be considered as a combination of two perspectives: firstly, what should be present on the day new homes are developed on the fringe of a city; and secondly, what should be present for a more mature community.

For this paper, a list of foundational community assets has been developed based on both the Queensland guideline and the author's decade-long experience in leading the planning of new communities on the fringe of metropolitan cities. The list includes supermarkets (including Asian grocery stores), parks, playgrounds, libraries, primary and secondary schools (public only), sports facilities, community centres, medical centres, youth centres, community halls and recreation facilities (including public sports halls and pools, not private learn to swim facilities). Scoring for each rail corridor is the total count for each asset.

3.5 Walkability and a 30-min City

Based on the research of Boulange et al. (2017), an 800-m catchment area was utilised for the assessment of the number of economic, social and foundational community assets within walking distance of a railway station. An example of how the data was tabulated, for all assets across the two networks, is provided in Table 1, noting that 40 such tabulations were undertaken. The distance was measured by the shortest possible walking route: not a simple straight-line distance, with Google Maps as the analysis tool. It is recognised that some state governments in Australia have developed tools for measuring walksheds, which could be utilised if the findings of this paper are adopted by state government agencies to identify new TODs. As this research is directed at gaining a first-order understanding of accessibility across a whole rail network, no allowance was made regarding the safety of a trip (road crossings) or terrain (hills). Amenity issues are important considerations when seeking to maximise accessibility (Park et al. 2015) and are recognised and discussed in the Discussion section of this paper. Within the CBD of both cities, the 800-m station catchments overlap. As this was a consistent anomaly for all rail lines, though it did favour some lines more than others, no attempt was made to address the issue for the analysis for this paper. Instead, the analysis considers outcomes with and without the scores for the CBD stations.

A 30-min city reflects the stated policy goal in the current metropolitan plan for Sydney (Greater Sydney Commission 2018, p. 84) and the *Future Transport Strategy* (Transport for NSW 2022). A 30-min journey to work is also recognised by the Marchetti constant (Marchetti 1994). The calculation of a 30-min journey has also considered the reality that a train trip requires walking at both ends. To estimate the average walking time for all activities in an 800-m radius, the average walking distance for all residents is 564 m and utilises an average walking speed of 1 min per 100 m, giving 5.64 min walking time, which, for simplicity, is rounded down to 5 min. Therefore, the rail travel time utilised is 20 min, with no allowance for wait time for a train. In identifying a 20-min rail trip, an allowance was made for the use of an interchange to access assets on a separate line, with the time as identified in the timetable, utilising the Transport for NSW app—https://transportnsw.info/apps .

4 Results

4.1 Economic and Social Assets Assessment

The analysis reveals differences in the number of economic and social assets accessible within 800 m of a rail station along the rail corridors in Melbourne and Sydney. The differences become more pronounced when the assets within the CBD of each city are excluded, refer to Figs. 3 and 4. For the Melbourne rail corridors, three corridors stand out—Melbourne Metro, Werribee to Frankston and Lilydale and for the Sydney rail corridors, there are several standouts—the T9 Hornsby to Gordon, T5 Leppington to Blacktown and T1 Western, with a few closely following in scores.

When the average of the number of economic and social assets per station is assessed, excluding those in the CBD, it shows that no Melbourne corridor has an average of one asset per station, whereas in Sydney, three rail corridors have

Table 1 Tabulation of foundational community assets T5 Leppington to Blacktown (Part only)

Station	Supermarket-based centre	Secondary school	Primary school	Library/community centre	Recreation centre	Playground	Park ≥ 0.5 ha	Medical centre
Liverpool	1	1	2	2		1	2	12
Warwick farm	1		1			1	4	
Cabramatta	1		2	2	1	1	2	2
Canley Vale		1	1	1	1		4	1
Fairfield	1	1	2	5	1	1	3	16

Source: By the author

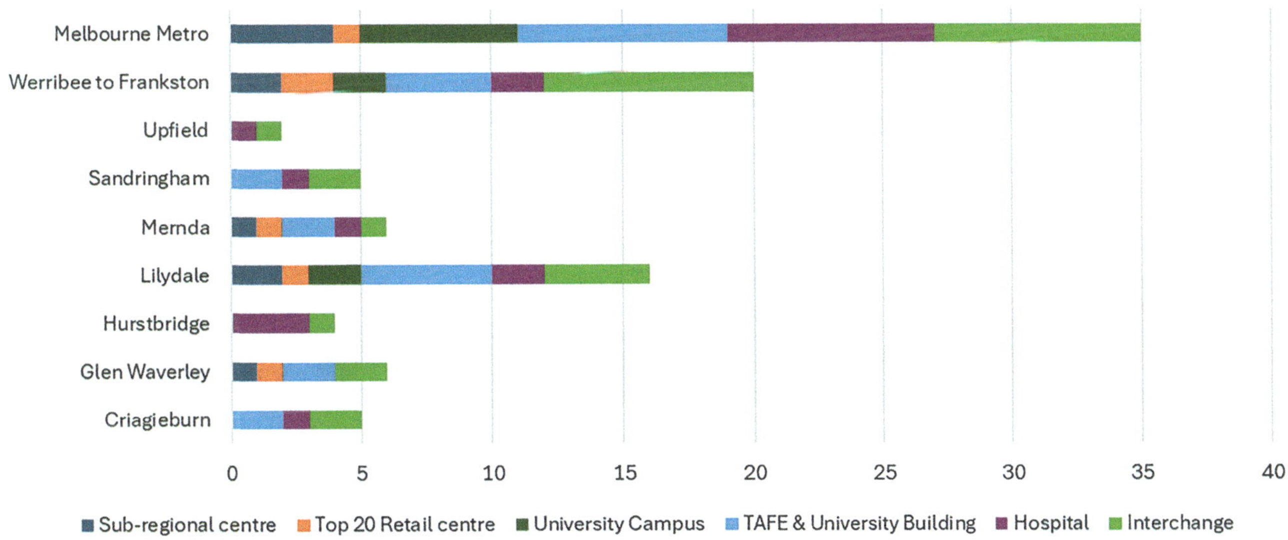

Fig. 3 Melbourne rail corridors' relative levels of economic and social assets, excluding the CBD assets. (Source: By the author)

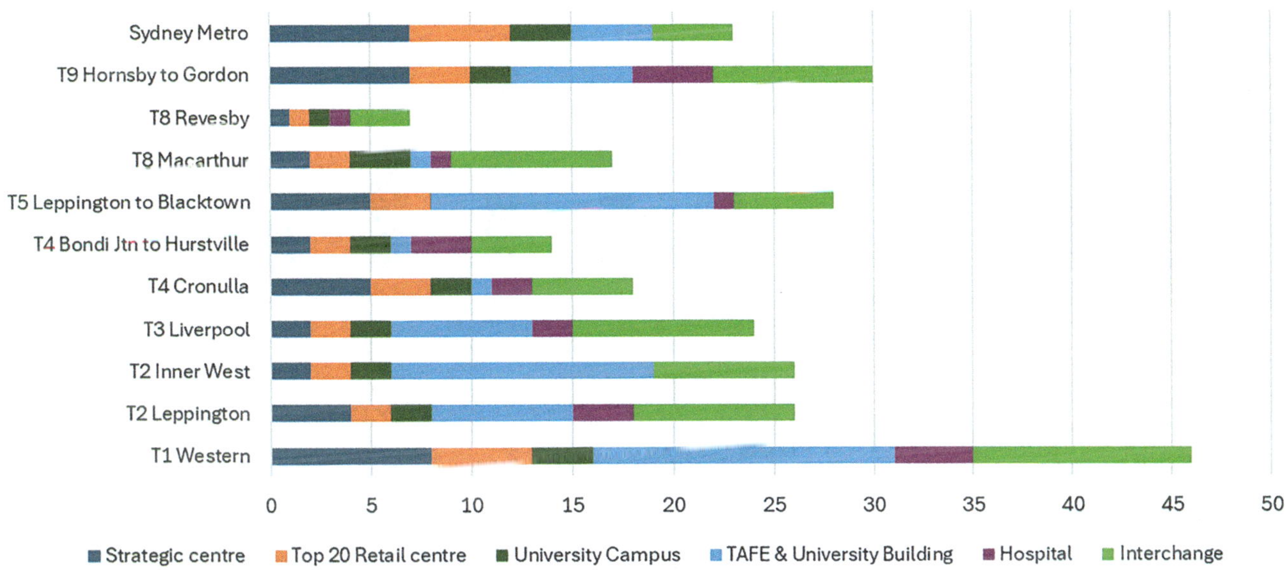

Fig. 4 Sydney rail corridors' relative levels of economic and social assets, excluding the CBD assets. (Source: By the author)

averages above one and seven above 0.8, compared to Melbourne's one above 0.8, refer to Figs. 5 and 6.

The assessment of economic and social assets accessible within 30 min from each of the eight state-led TODs, in Sydney, reveals differences between each TOD, where Crows Nest is the 'standout' with high scores also for Macquarie Park, Bella Vista, Homebush and Bay West. The tabulated data for each TOD is shown graphically in Fig. 7.

An assessment was made of the level of population and employment along each rail corridor. For employment, the assessment highlights the influence of CBD as a focus; when excluded, different corridors are revealed as providing access to the most employment, refer Table 2. One example of a concentration of suburban employment is on the Sydney Metro Line between Macquarie University and Victoria Cross stations, refer Table 3. For population, the CBD makes little difference to the total population accessible along a rail corridor, also refer Table 2.

4.2 Foundational Community Asset Assessment

The assessment of the level of foundational community assets present along individual rail corridors, for both cities, reveals a common pattern for each rail corridor, but differences in variation for individual assets between the Melbourne and Sydney networks, as shown in Figs. 8 and 9,

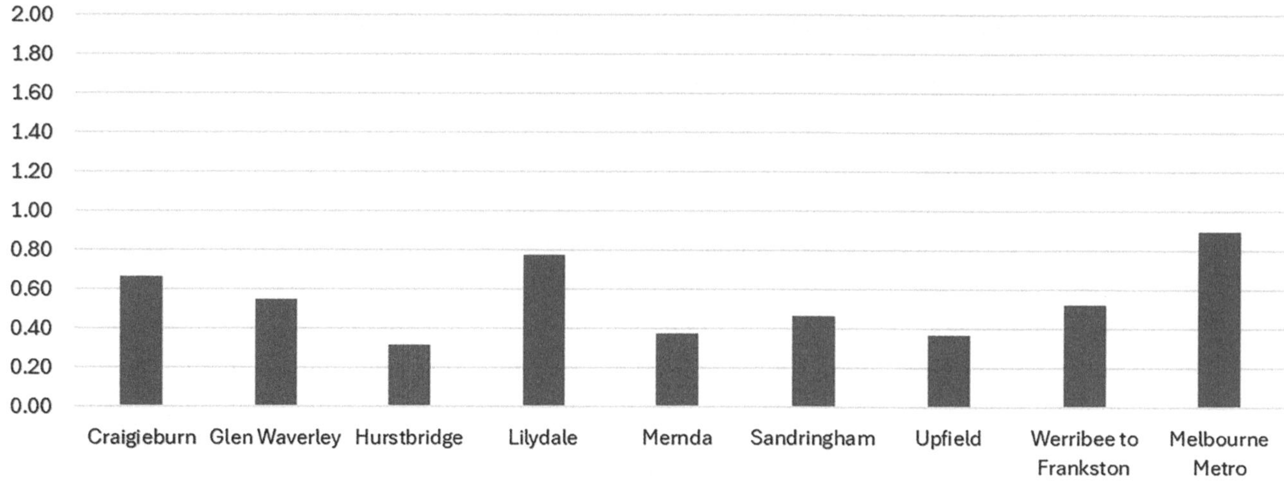

Fig. 5 Melbourne rail lines, average economic and social assets per station, excluding CBD. (Source: By the author)

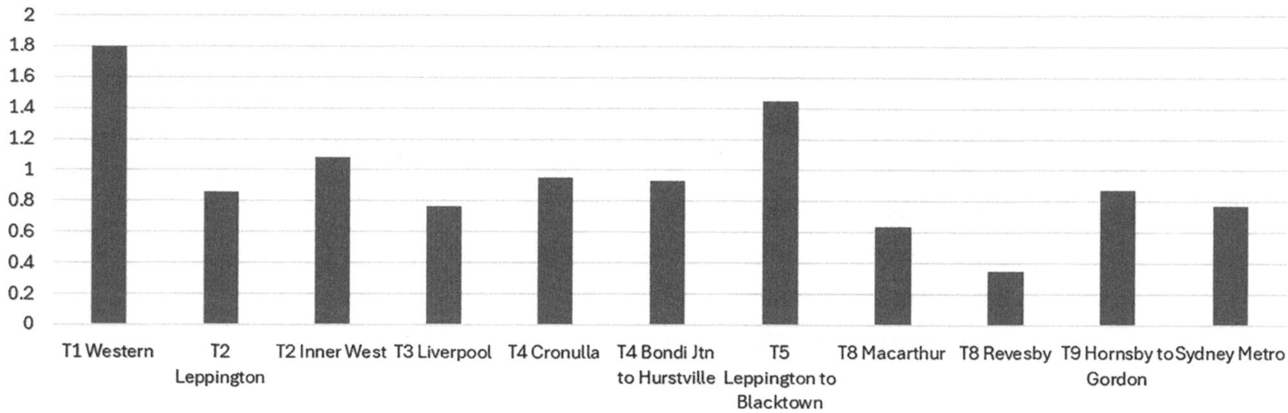

Fig. 6 Sydney rail lines, average economic and social assets per station, excluding CBD. (Source: By the author)

respectively. Along the Melbourne rail corridors, the biggest variation is for primary schools, at approximately 30%. The Sydney rail corridors show a higher variation at around 40% for primary schools, recreation centres and playgrounds. The scores for secondary schools, primary schools and recreation centres are the lowest for both cities, with Melbourne generally below 40% and Sydney generally below 50%, with primary schools the exception.

Walkable communities will be encouraged when foundational community assets can be accessed locally. Regarding the Sydney network (183 rail stations), this has generally been found to be the case for local parks and playgrounds. However, access to public schools and supermarkets, within 800 m of a station, is more varied, with the following results:

- Public primary school—88 stations precincts without
- Public secondary school—147 station precincts without
- Supermarkets—39 station precincts without

An assessment of the average number of foundational community assets per station, excluding those in the CBD, shows that Sydney rail corridors generally have a higher number of assets than the Melbourne corridors refer to Figs. 10 and 11.

A more detailed examination of the data for both cities reveals a lack of public primary and secondary schools within their respective CBDs, except for a primary school in an isolated location in the Sydney CBD. Consequently, access to a public school from within a CBD requires a journey by train. Figure 12 shows one example of the distribution of primary and secondary schools along a single rail corridor and their lack of presence in the CBD of Melbourne. In addition, for some stations close to a CBD, access to a primary or secondary school requires changing trains, such as, in Sydney, to get from Kings Cross to Waterloo (the closest station with a secondary school), requires a change at Martin Place, refer Fig. 11.

In the context of a 30-min city (20-min rail trip), there are also stretches along a few rail corridors where there is no access to a secondary school, within the walkable catchment of a station (800 m). For Sydney, this includes the five stations passed Kirrawee towards Cronulla on the T4 Cronulla

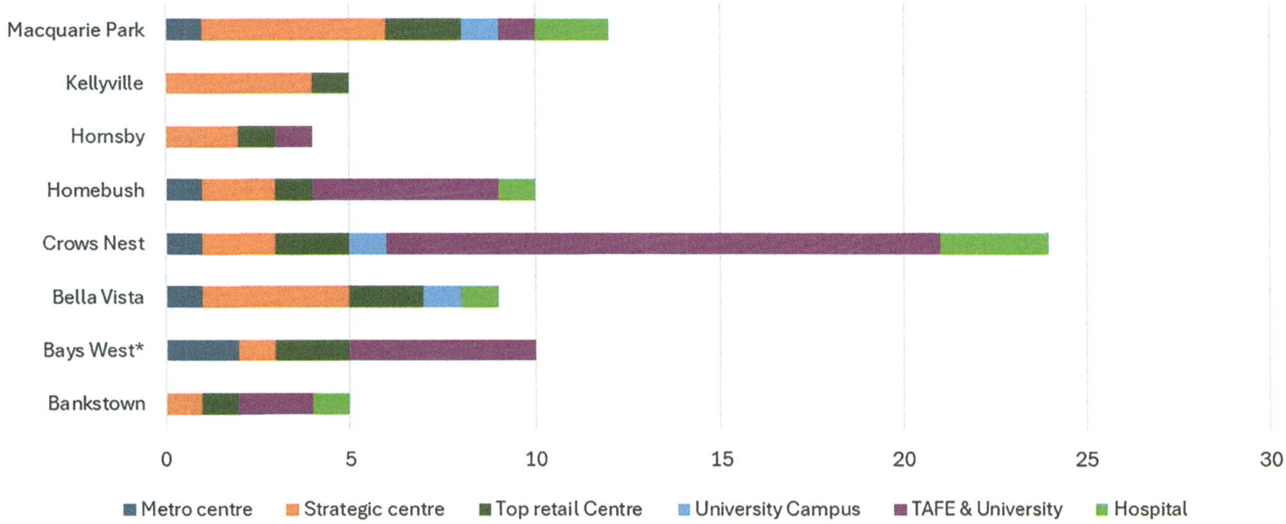

Fig. 7 Sydney State led TODs' relative levels of accessibility to economic and social assets. (Source: By the author)

Table 2 Sydney: total population and employment by rail line

Rail line	Population	Employment	Population ex CBD	Employment ex CBD
T1 Western	262,716	880,251	193,866	309,789
T2 Leppington	317,758	1,198,126	205,540	144,735
T2 Inner West	361,551	1,283,984	249,333	230,593
T3 Liverpool	342,683	1,196,077	269,743	239,786
T4 Cronulla	267,149	706,767	197,952	136,878
T4 Bondi Jtn to Hurstville	248,192	702,936	178,995	133,047
T5 Leppington to Blacktown	152,659	161,154	152,659	161,154
T8 Macarthur	231,937	1,196,901	91,364	105,828
T8 Revesby	216,311	1,131,833	104,093	78,442
T9 Hornsby to Gordon	339,509	846,458	270,659	275,996
Sydney Metro	306,087	829,310	236,397	298,045

Source: By the author

line, refer to Fig. 13, and in Melbourne, two stations on separate lines—Frankston and Sandringham.

Very few stations have all the foundational community assets with an 800-m catchment area. For Melbourne, only two stations, Watsonia and Glenferrie, have all assets. The presence of recreation facilities is low for both cities, so when this asset is excluded, several Melbourne stations have all other assets in their 800-m catchment: Footscray, Glen Huntly, Mordialloc, Newmarket and Northcote. For Sydney, one station has all its assets within its catchment—Fairfield. When a recreation centre is excluded, several stations have all other assets within their 800-m catchment—Parramatta, Liverpool, Auburn, Flemington, Punchbowl, Bankstown, Hurstville, Kogarah, Glenfield, Meadowbank and Victoria Cross.

The high concentration of economic and social assets in the CBDs of both cities and the regular distribution of secondary and primary schools along each rail corridor imply that, in general terms, rail stations within 20 min of the CBD have good access to the broad range of economic, social and foundational community assets that a household would benefit from access by public transport. The stations within a 20-min trip from the CBD for each city are:

– Melbourne stations, with either Flinders Street or Melbourne Central stations as the starting point: Alphington, Canterbury, Caulfield, Coburg, Croxton,

Table 3 Example of a concentration of employment at suburban stations—Sydney Metro Line

Station	Employment
Macquarie University	22,994
Macquarie Park	34,224
North Ryde	11,603
Chatswood	21,446
Crows Nest	34,029
Victoria Cross	58,058

Source: By the author

where each station has 30-min accessibility to all the analysed assets, and many also have 30-min accessibility to the Sydney CBD, noting some overlap in the Sydney CBD. In summary, the sections of individual rail corridors that include the 51 stations are as follows (refer Fig. 14).

- Sydney metro line: 18 stations from Bella Vista to Marrickville, the first six without access to the Sydney CBD in terms of a 30-min city.
- T1 North Shore line: 8 stations from St Leonards to Central.

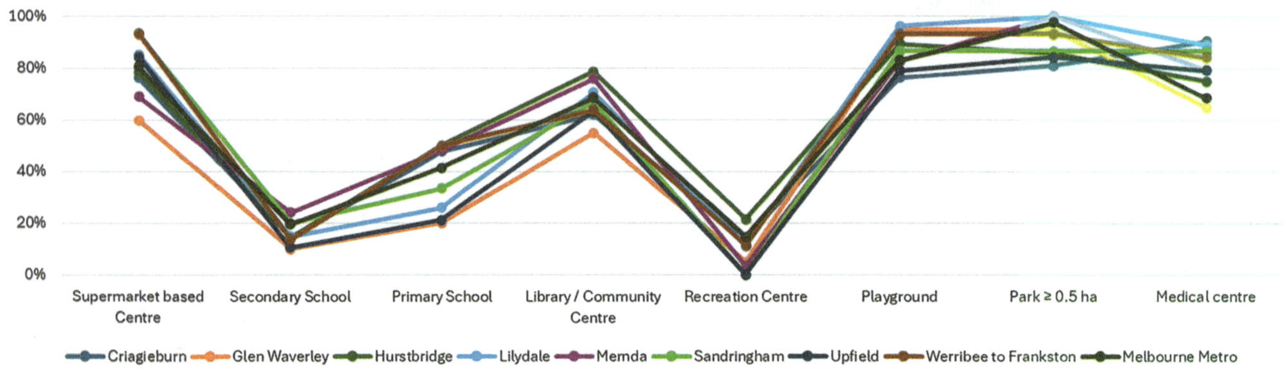

Fig. 8 Melbourne: Percentage of foundational community assets present by rail corridor. (Source: By the author)

Gardenvale, Glen Iris, Middle Footscray, Moone Ponds and Spotswood.

- Sydney stations, with Town Hall station as the starting point: Ashfield, Bondi Junction, Dulwich Hill, St Leonards and Wolli Creek.

4.3 A 30-min City

Collectively, these results confirm differences in the level of accessibility to economic, social and foundational community assets between and within rail corridors. However, this research is directed at determining whether stations along a rail corridor, or a section of it, can provide access to the full range of economic and social assets that a community would anticipate being available, within a 30-min travel time. The final assessment examines this for all 183 train stations across the Sydney network and Table 4 shows an example of the analysis for a few stations.

The assessment reveals that 38 train stations have access to all the economic, social and foundational community assets analysed, as well as access to the Sydney CBD in terms of 30-min accessibility. Thirteen train stations similarly have access to all economic, social and foundational community assets analysed in terms of 30-min accessibility, but not to the Sydney CBD, within 20 min by rail.

When all 51 stations are plotted on the rail network map for Sydney, it reveals that six of the rail lines, linear sections

- T2 Leppington line: 2 stations, Newtown and Macdonaldtown.
- T4 Cronulla line: 11 stations from Hurstville to Martin Place (excluding Allawah), the first four without access to the Sydney CBD in terms of a 30-min city.
- T8 Macarthur line: 6 stations from Sydenham to Town Hall.
- T8 Macarthur line: 5 stations from Mascot to St James.
- T8 Macarthur line: 3 stations, Ingleburn, Minto and Glenfield, without access to the Sydney CBD in terms of a 30-min city.
- T9 Northern line: 2 stations, Strathfield and Burwood.

5 Discussion and Conclusion

The results show that the location of a TOD or related supporting infrastructure matters if accessibility to goods, services and jobs via public transport or walking is a principle for guiding urban renewal activities across a city. Furthermore, in the context of a 30-min city, the results for Sydney show that only a limited number of places (rail line segments), can be seen as linear clusters, which provide access to the full range of goods, services and jobs, that a person/household would anticipate having access to over their lifetime.

Concerning the distribution of foundational community assets across the rail networks, the results show close to

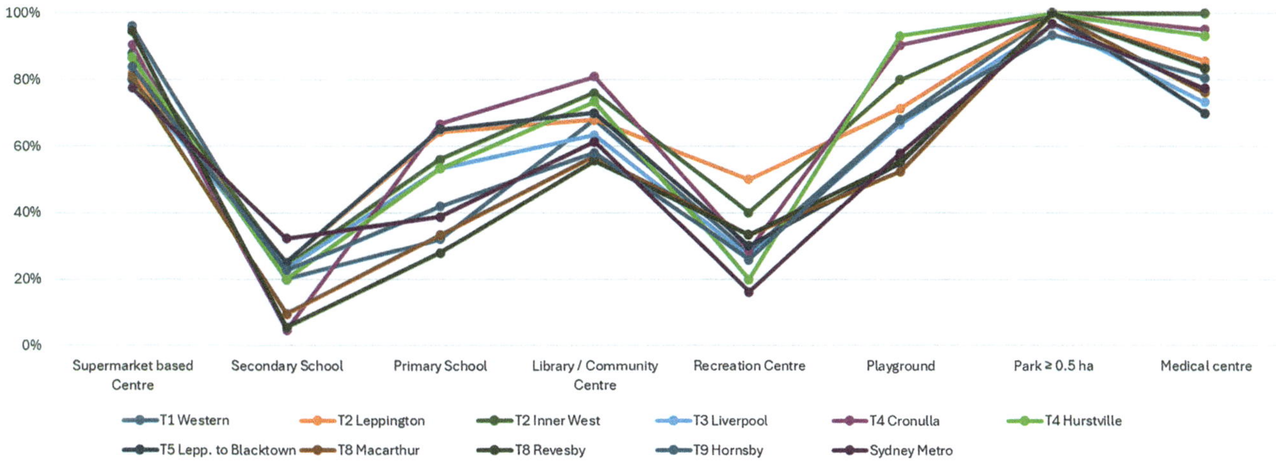

Fig. 9 Sydney: Percentage of foundational community assets present by rail corridor. (Source: By the author)

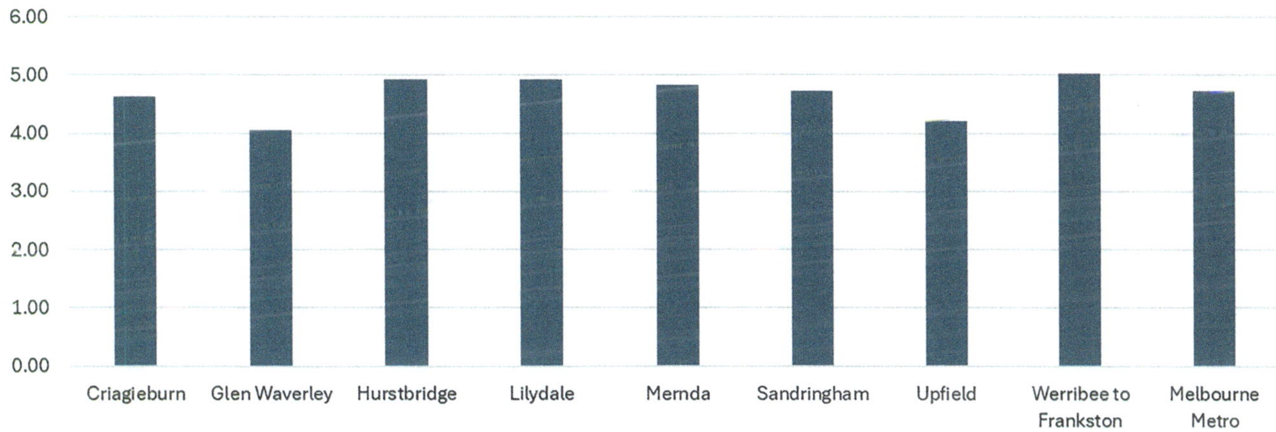

Fig. 10 Melbourne rail lines, average foundational community assets per station, excluding CBD. (Source: By the author)

100% presence of parks, playgrounds and medical facilities at all train stations, suggesting that where assets are absent, the gap should be addressed if the station precinct is to be supported as a TOD.

An interesting finding is that the CBDs of both Melbourne and Sydney have high levels of jobs, goods and services, though they have limited to no access to public schools within the catchments of the stations that serve them. This may or may not influence the attractiveness of the CBDs as family-friendly locations.

The assessment of the spatial location of individual assets was simply whether the asset was accessible within a strict 800-m walking distance, noting that numerous times an asset was found 50 m outside the catchment area limit and thus excluded from the list. Also, the location of an asset within the catchment could have been near the edge of the 800-m boundary, this would mean a person on the other side of the catchment would need to walk around 1.5 km to access it. These issues suggest further research is required into how far people will walk to a variety of destinations (assets) within their local area, as well as how far someone will walk to the same destinations if they have had to travel by train.

The high variations in the distribution of public schools suggest that if they form part of a 30-min city policy approach, further research is required on their catchments and whether a train trip can be considered as the mode of travel for students. Intuitively, it is unlikely to be the case for primary school students unless accompanied by an adult or older sibling. In addition, understanding dwelling thresholds for primary and secondary schools in the context of local capacity will be an important prerequisite for understanding the infrastructure implications of projected growth resulting from a TOD initiative.

The research for this paper on access to goods, services and jobs is based on policies of numerous governments and the author's experience in metropolitan and growth area planning. With the findings showing that location matters, and that there are parts of the rail network, in Sydney, where a 30-min city concept could be delivered, further research is required to better understand what economic, social and

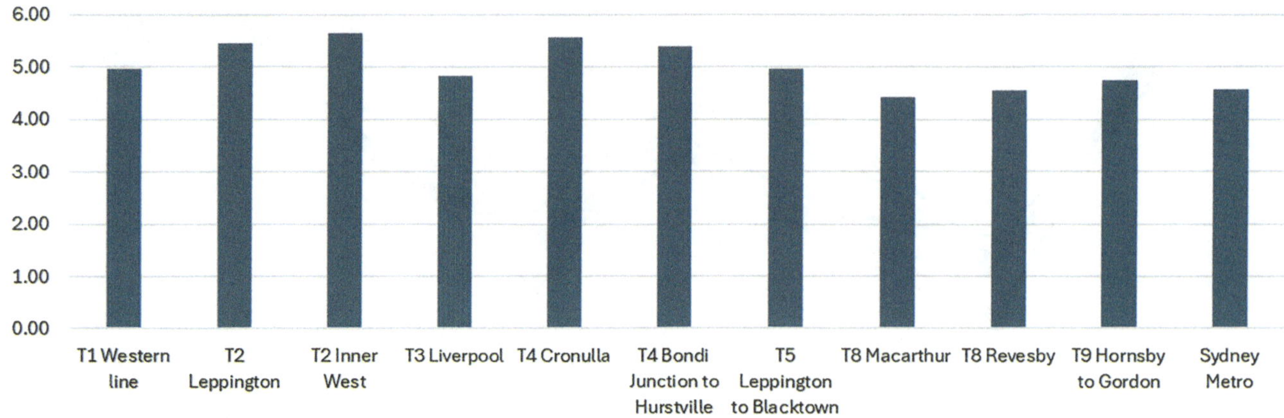

Fig. 11 Sydney rail lines, average foundational community assets per station, excluding CBD. (Source: By the author)

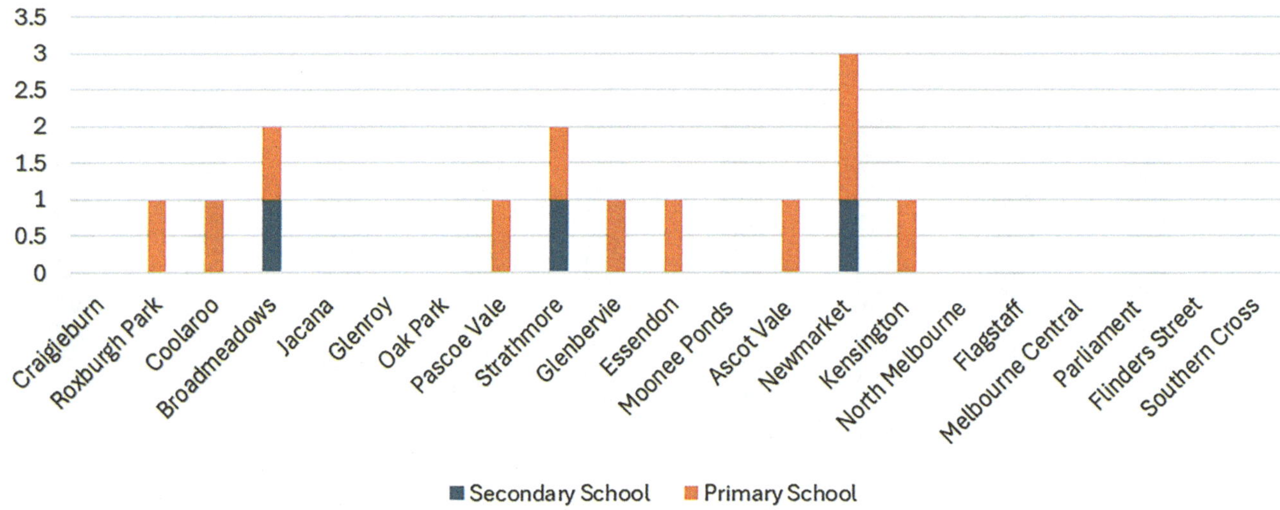

Fig. 12 Distribution of primary and secondary schools on the Craigieburn line—Melbourne. (Source: By the author)

foundational community assets should form the basis of developing a 30-min linear transit city policy, particularly what sub-categories are to be included. There is also a need to further understand how to deal with overlapping catchments, such as in CBDs. Learnings in this area may also provide insights as to whether a weighting system is used in any multi-criteria analysis based on differences in the importance of different assets to the community.

Collectively, the analysis suggests that TODs can be re-conceptualised in terms of the community's accessibility to a range of economic, social and foundational community assets along a transit corridor. The level of accessibility will determine whether a TOD can be re-imagined to be part of a 'linear transit city'. That is, where most of the needs of a person, over their lifetime, can be accessed by walking or public transport. Such a conceptualisation requires recognition that location matters, and that the spatial distribution of new economic, social and foundational community infrastructure investments can support a linear transit city concept and the data highlights where there are gaps that could be addressed over time to expand options for the delivery of linear transit cities.

The findings suggest that government decisions regarding the location of new TODs are equally decisions about the location of new economic, social and foundational community infrastructure investments—collectively an urban renewal model.

5.1 An Urban Renewal Model

In conceptualising a model several assumptions have been made including: the underlying importance of accessibility to goods, services and jobs as a policy outcome; that the locations of a TOD and its supporting infrastructure matters; the need for governments to support urban renewal in the context of limited finances to fund outward growth; investments in infrastructure can influence accessibility patterns;

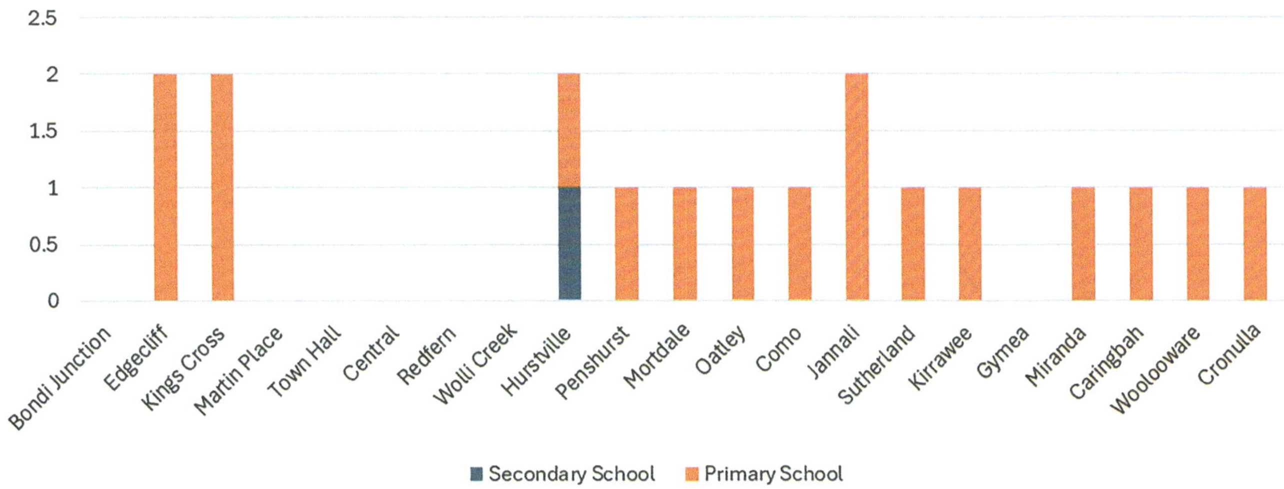

Fig. 13 Distribution of primary and secondary schools on the T4 Cronulla line—Sydney. (Source: By the author)

Table 4 Sydney rail network—assets gap analysis by station

Station	Assets not within 30 minutes		No Town Hall (CBD) station connection in 20 minutes	Assets not within the station catchment		
	University Campus	Hospital		Secondary school	Primary school	Supermarket
Central						
Chatswood***						
Cheltenham						
Cherrybrook						
Chester Hill						
Circular Quay						
Clarendon						

Legend

No access to a strategic centre or a TAFE / university facility
Full access to all assets and within 20 minutes of the Sydney CBD
Full access to all assets but not within 20 minutes of the Sydney CBD
No access to a hospital and within 20 minutes of the Sydney CBD
No access to a hospital but not within 20 minutes of the Sydney CBD
No access to a university campus and hospital and not within 20 minutes of the Sydney CBD

Source: By the author

the scale of growth projected will require urban renewal activity into the long term; the TOD index approach, including development feasibility and capacity, provides a systematic approach to understand and compare the potential of station precincts as TODs (Cucuzzella et al. 2022); that government investments in infrastructure are a signal to the private sector and their investment decisions; the potential role of rapid bus networks as a location for second tier linear transit cities (Glazebrook and Newman 2018; Tsivanidis 2023); access to models to test development feasibility for a range of dwelling types, such as the NSW Government's development assessment model (Department of Planning, Housing and Infrastructure 2022); the demand for additional economic, social and foundational community assets and enabling infrastructure can be measured in terms of known thresholds (dwelling ratios); the time scale from identifying an opportunity for additional housing to when a pipeline starts to be delivered is 10 years; and that a model must be practical and repeatable in application and build on the data bases generally held by governments.

The proposed model aims to provide a framework for managing growth in the established areas of cities by identifying the optimal locations for targeting urban renewal development and the infrastructure investments to support it, based on the premise of maximising a community's access to goods, services and jobs by public transport and walking. The model involves seven interrelated steps, refer Fig. 15:

- Step 1: Overarching principles and benefits: The purpose is to explicitly document the wide range of principles and benefits of TODs, linear cities, access and walkability to

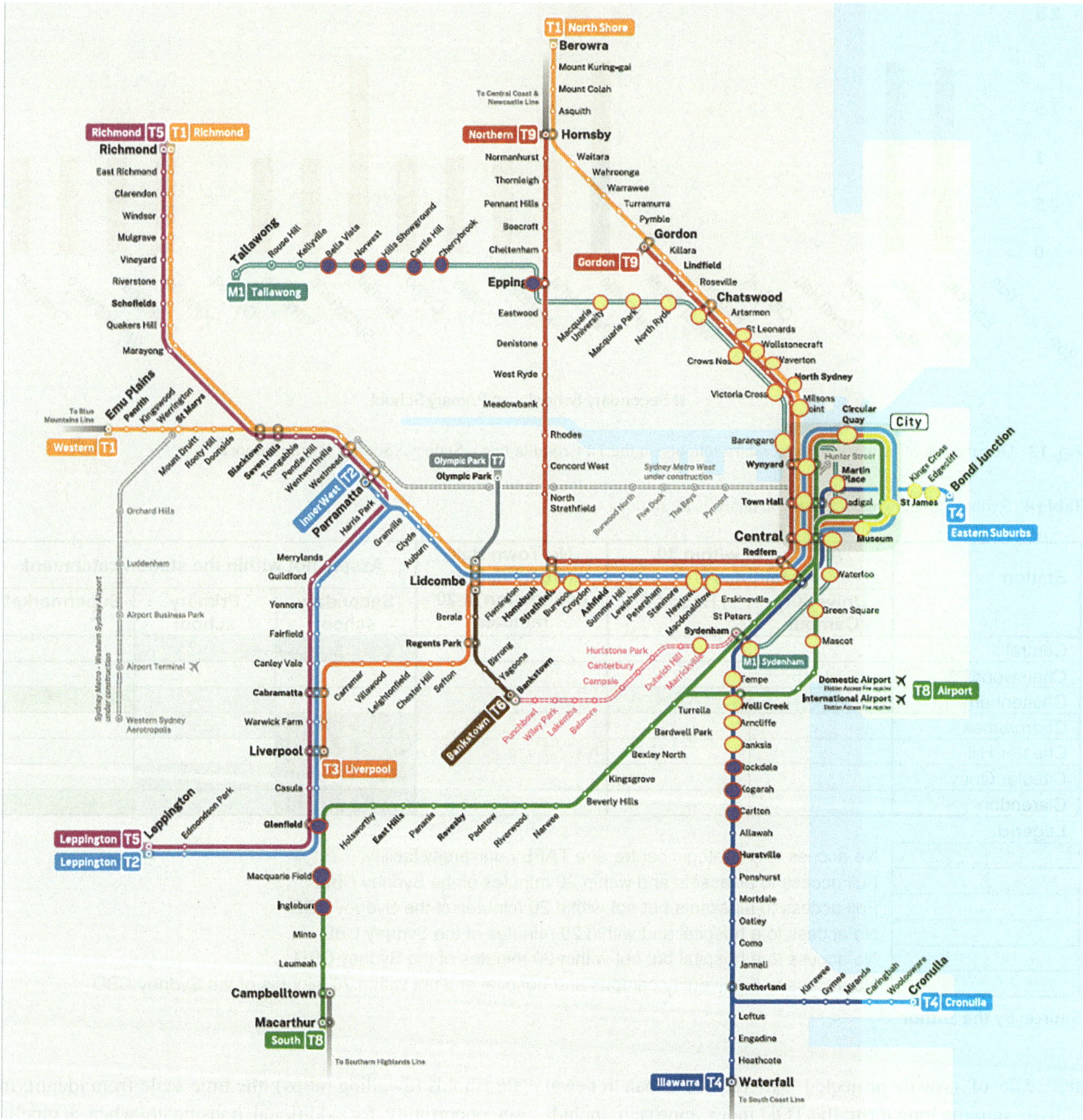

Fig. 14 Train stations with full access to all economic, social and foundational community assets—potential TOD locations. Yellow dot—full access to all economic, social and foundation assets within 30 and 20 min by rail from the CBD. Purple dot—full access to all economic, social and foundation assets within 30 min. (Source: Base map Transport for NSW (2024) and the author's analysis)

provide a context for all the remaining steps, particularly steps five and seven.

- Step 2: Initial rail network assessments: The aim is to prepare a comprehensive GIS database of the distribution of economic, social and foundational community infrastructure within the 800-m catchment of each station of the rail network of a city and what is accessible from each station in the context of a 30-min city. Outputs should include the identification of clusters of all assets as well as gaps in the distribution of assets that can inform the coordination of decisions across government agencies. The assessment could also be expanded to include a rapid bus network (existing or proposed), informing its alignment with potential TODs.
- Step 3: Annual audit: An annual process primarily aimed at identifying new infrastructure investments and any

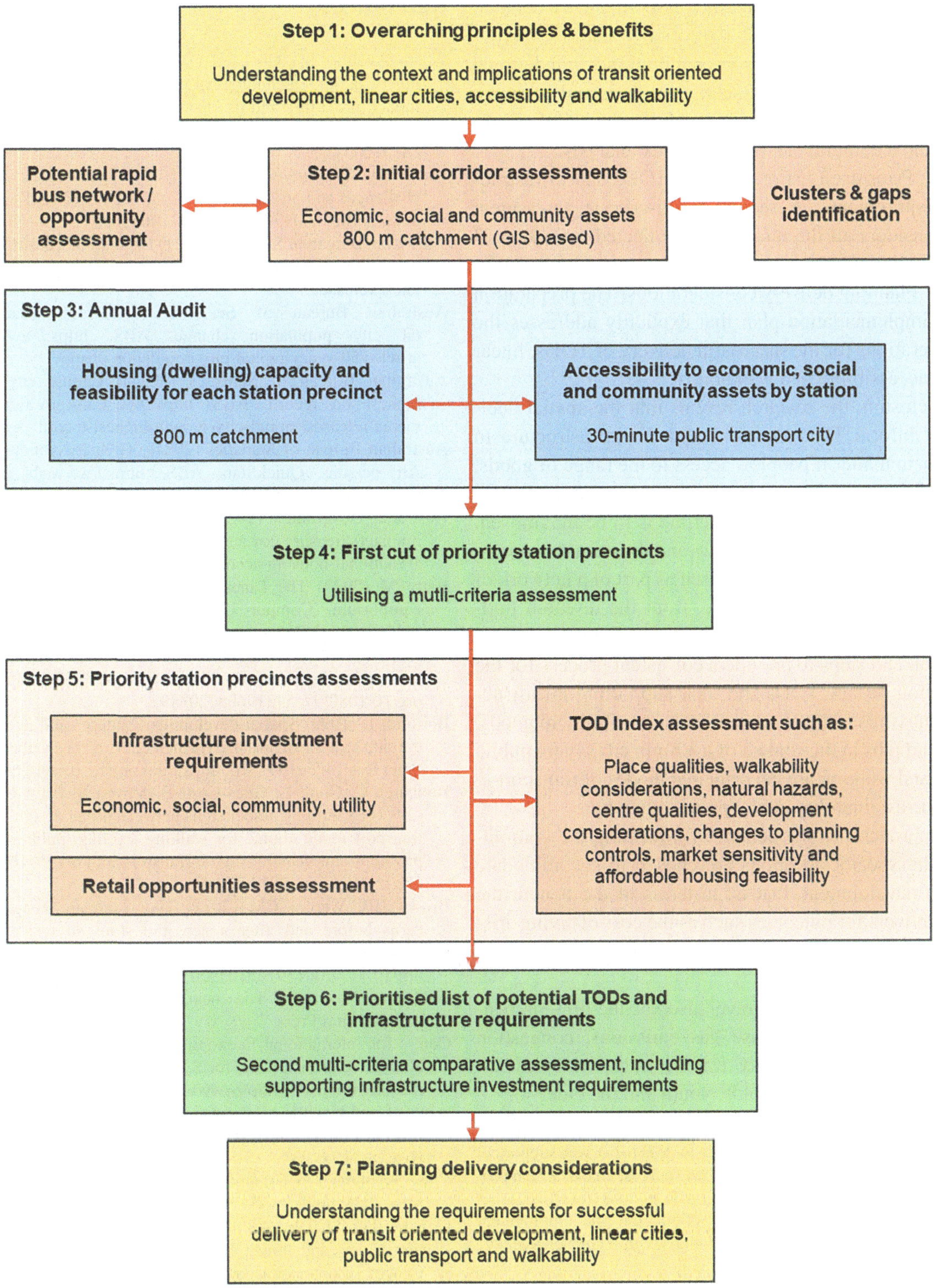

Fig. 15 A proposed linear transit cities urban renewal model.(Source: By the author)

changes in development potential (capacity and feasibility).

– Step 4: First cut of priority station precincts: An initial multi-criteria assessment based only on accessibility and development potential considerations, to identify a list of potential new TOD locations to address the latest understanding of medium to long-term dwelling demand.

- Step 5: Priority station precinct assessment: A detailed assessment of the first cut list of priority stations, including identification of required enabling and foundational community infrastructure (such as utilities and schools), assessment of the potential for additional retail development and a comprehensive TOD index analysis.
- Step 6: Prioritised list of potential TODs and infrastructure requirements: A second multi-criteria assessment that considers all the information collected from steps 3 and 5, with the output a prioritised list of potential TODs.
- Step 7: Planning delivery considerations: The preparation of an implementation plan that explicitly addresses the considerations for the successful delivery of TODs, linear cities, accessibility and walkability.

In conclusion, the research reveals that the spatial decisions that inform TOD locations and the infrastructure to support them matter if people's access to the range of goods, services and jobs (economic, social and foundational community assets) expected over a lifetime is to be maximised. In addition, the distribution of TODs and the infrastructure to support them can be reconceptualised as part of a network of 'linear transit cities'. Furthermore, from the research findings, an urban renewal model can be developed to codify the information and steps to provide a consistent process for the identification of new TODs and their supporting infrastructure that maximises the communities' accessibility to goods, services and jobs in the context of a 30-min city where public transport and walking are the principal modes of transport—equity and sustainability public policy outcomes.

This approach has the benefit of informing the community of why specific locations are chosen to accommodate additional development, that is, in terms of the benefit the location delivers for outcomes such as the cost-of-living arising from a shift away from the need for multiple private vehicles.

Reducing private vehicle travel also has the flow-on benefits of mitigating greenhouse gas emissions, congestion management and accessibility for the transport disadvantaged (single-vehicle households, youth and the elderly).

Funding The undertaking of the analysis for this paper was supported, pro bono, by the provision of data from two sources: **Urbis**, which provided data on the scale of retail in centres and at stations (for Melbourne) and **Mecone**, which provided data on population and employment within 800 m of rail stations across all stations within Greater Sydney.

Informed Consent Statement Not applicable.

Conflicts of Interest The author declares no conflict of interest.

References

Allam Z, Bibri SE, Chabaud D, Moreno C (2022) The theoretical, practical, and technological foundations of the 15-minute city model: proximity and its environmental, social and economic benefits for sustainability. Energies 15:6042. https://doi.org/10.3390/en15166042

Antyufeev AV, Antyufeeva OA (2019) Linear cities: controversies, challenges and prospects. IOP Conf Ser Mater Sci Eng 687. https://doi.org/10.1088/1757-899X/687/5/055025

Australian Bureau of Statistics (2021) Population: census. ABS. https://www.abs.gov.au/statistics/people/population/population-census/latest-release

Australian Bureau of Statistics (2022) 50 years of capital city population change. ABS. https://www.abs.gov.au/articles/50-years-capital-city-population-change

Australian Bureau of Statistics (2024a) Capital city growth the highest on record. ABS. https://www.abs.gov.au/media-centre/media-releases/capital-city-growth-highest-record

Australian Bureau of Statistics (2024b) Greater Sydney 2021 Census All persons QuickStats ABS. https://www.abs.gov.au/census/find-census-data/quickstats/2021/1GSYD

Australian Government (2022) National Housing Accord 2022. https://ministers.treasury.gov.au/sites/ministers.treasury.gov.au/files/2022-10/national-housing-accord-2022.pdf

Batty M (2022) The Linear City: illustrating the logic of spatial equilibrium. Comput Urban Sci 2(8). https://doi.org/10.1007/s43762-022-00036-z

Bertolini L (1996) Nodes and places: complexities of railway station redevelopment. Eur Plan Stud 4(3):331–345. https://doi.org/10.1080/09654319608720349

Bertolini L (1999) Spatial development patterns and public transport: the application of an analytical model in the Netherlands. Plann Pract Res 14(2):199–210. https://doi.org/10.1080/02697459915724

Boulange C, Gunn L, Giles-Corti B, Mavoa S, Pettit C, Badland H (2017) Examining associations between urban design attributes and transport mode choice for walking, cycling, public transport and private motor vehicle trips. J Transp Health 6:155–166. https://doi.org/10.1016/j.jth.2017.07.007

Brown BB, Werner CM (2011) The residents' benefits and concerns before and after a new rail stop: do residents get what they expect? Environ Behav 43(6):789–806. https://doi.org/10.1177/0013916510392030

Calthorpe P (1993) The next American metropolis. Princeton Architectural Press

Centre for International Economics (2010) The benefits and costs of alternative growth paths for Sydney economic, social and environmental impacts. https://www.thecie.com.au/publications-archive/costs-and-benefits-of-alternative-growth-paths-for-sydney

Collins GR (1959) Linear planning throughout the world. J Soc Archit Hist 18(3):74–93

Congress of the New Urbanism, Charter of the New Urbanism, citied 19 May 2024. https://www.cnu.org/who-we-are/charter-new-urbanism

Crowley DF, Shalaby AS, Zarei H (2009) Access walking distance, transit use, and transit-oriented development in North York City Center, Toronto, Canada. J Transp Res Board 2110:96–105. https://doi.org/10.3141/2110-12

Cucuzzella C, Owen J, Goubran S, Walker T (2022) A TOD index integrating development potential, economic vibrancy, and socio-economic factors for encouraging polycentric cities. Cities 131:103980. https://doi.org/10.1016/j.cities.2022.103980

Currie G, Loader C (2010) Bus network planning for transfers and the network effect in Melbourne, Australia. J Transp Res Board 2145. https://doi.org/10.3141/2145-02

Curtis C (2008) Evolution of the transit-oriented development model for low-density cities: a case study of Perth's new railway corridor. Plann Pract Res 23(3):285–302. https://doi.org/10.1080/02697450802423559

Dalheim H (2023) Planning better cities: a practical guide. Palgrave Macmillan, Cham

Department of Environment, Land, Water and Planning (2017) Plan Melbourne 2017–2050. The State of Victoria. https://www.planmelbourne.vic.gov.au/

Department of Infrastructure, Local Government and Planning (2017) ShapingSEQ, South East Queensland Regional Plan 2017. The State of Queensland. https://planning.dsdmip.qld.gov.au/planning/better-planning/state-planning/regional-plans/seqrp

Department of Planning, Housing and Infrastructure (2022) 2022 Sydney housing supply forecast—methodology: assumptions and inputs. https://www.planning.nsw.gov.au/sites/default/files/2023-06/2022-sydney-housing-supply-forecast-methodology-assumptions-and-inputs.pdf

Department of Planning, Housing and Infrastructure (2024) Transport Oriented Development (TOD) Program—assessment criteria. https://www.planning.nsw.gov.au/sites/default/files/2024-03/transport-oriented-development-program-assessment-criteria.pdf

Department of Planning, Lands and Heritage (2018) Perth and Peel @ 3.5 million. Government of Western Australia. https://www.dplh.wa.gov.au/perth-and-peel-@-3-5-million

Department of Planning, Transport and Infrastructure (2017) The 30-year plan for greater Adelaide, 2017 update. Government of South Australia. https://livingadelaide.sa.gov.au/the_plan

Department of Premier and Cabinet (2023) Victorias Housing Statement The decade ahead | 2024–2034 Victoria Government. https://www.vic.gov.au/housing-statement

Doxiadis CA (1967) On linear cities. Town Plan Rev 38(1):35–42

Duany AM, Plater-Zyberk E, Speck J (2000) Suburban nation: the rise of sprawl and the decline of the American dream. North Point Press

Duany A, Speck J, Lydon M (2010) The smart growth manual. McGraw Hill

Glazebrook G, Newman P (2018) The city of the future. Urban Plan 3(2):1–20. https://doi.org/10.17645/up.v3i2.1247

Greater Sydney Commission (2018) A metropolis of three cities (2018) State of New South Wales. https://www.greater.sydney/metropolis-of-three-cities

Hall P (1996) Cities of tomorrow: an intellectual history of urban planning and design in the twentieth century. Blackwell Publishers Ltd, UK

Hamnett S, Freestone R (eds) (2000) The Australian metropolis: a planning history. Allen & Unwin

Howard E (1898) Tomorrow: a peaceful path to real reform. Swan Sonnenschein & Co

Ibraevaa A, Correiab GHdA, Silvac C, Antunesa AP (2020) Transit-oriented development: a review of research achievements and challenges. Transp Res A Policy Pract 132:110–130

Jacobs J (1961) The death and life of great American cities. Random House, New York City

Knowles RD (2012) Transit oriented development in Copenhagen, Denmark: from the finger plan to Ørestad. J Transp Geogr 22(5):251–261

Kostof S (1991) The city shaped, urban patterns and meanings through history. Thames and Hudson Ltd, London

Marchetti C (1994) Anthropological invariants in travel behavior. Technol Forecast Soc Chang 47(1):75–88. https://doi.org/10.1016/0040-1625(94)90041-8

Mathur S, Gatdula A (2023) Review of planning, land use, and zoning barriers to the construction of transit-oriented developments in the United States. Case Studies on Transport Policy. https://doi.org/10.1016/j.cstp.2023.100988

Melbourne Metropolitan Board of Works (1971) Planning policies for the Melbourne metropolitan region. Melbourne Metropolitan Board of Works. https://www.planning.vic.gov.au/guides-and-resources/strategies-and-initiatives/melbournes-strategic-planning-history/planning-policies-for-metropolitan-melbourne-1971

Melbourne Metropolitan Board of Works (1981) Metropolitan strategy implementation. Melbourne Metropolitan Board of Works. https://www.planning.vic.gov.au/guides-and-resources/strategies-and-initiatives/melbournes-strategic-planning-history/metropolitan-strategy-implementation-1981

New South Wales Government (2023a) Transport oriented development program. https://www.planning.nsw.gov.au/policy-and-legislation/housing/transport-oriented-development-program

New South Wales Government (2024a) Sydney housing supply forecast. https://www.planning.nsw.gov.au/research-and-demography/sydney-housing-supply-forecast

New South Wales Government (2024b) Trends in housing supply. https://www.planning.nsw.gov.au/policy-and-legislation/housing/housing-supply-insights/quarterly-insights-monitor-q1/trends-in-housing-supply

New South Wales. Department of Environment and Planning (1988) Sydney into its third century: metropolitan strategy for the Sydney region. Dept. of Environment and Planning, Sydney

Newman PWG, Kenworthy JR (1989) Cities and automobile dependence: a sourcebook. Ashgate Publishing, Limited

Park S, Deakin E, Jang K (2015) Can good walkability expand the size of transit-oriented developments? Transp Res Rec 2519:157–164. https://doi.org/10.3141/2519-17

PTV (2024) Victorian Train Network. https://www.ptv.vic.gov.au/assets/PTV-default-site/Maps-and-Timetables-PDFs/Maps/Network-maps/Victorian-Train-Network-Map-May-2023-v3.pdf.pdf

Ratner KA, Goetz AR (2013) The reshaping of land use and urban form in Denver through transit-oriented development. Cities 30:31–46. https://doi.org/10.1016/j.cities.2012.08.007

Ryckewaert M (2010) The ten-year plan for the port of Antwerp (1956–1965): a linear city along the river. Plan Perspect 25(3):303–322. https://doi.org/10.1080/02665433.2010.481179

Saghapour T, Moridpour S, Thompson RG (2016) Public transport accessibility in metropolitan areas: a new approach incorporating population density. J Transp Geogr 54:273–285. https://doi.org/10.1016/j.jtrangeo.2016.06.019

Searle G, Darchen S, Huston S (2014) Positive and negative factors for transit oriented development: case studies from Brisbane, Melbourne and Sydney. Urban Policy Res 32(4):437–457. https://doi.org/10.1080/08111146.2014.931280

SGS Economics and Planning (2012) Long run economic and land use impacts of major infrastructure projects, Unpublished

Singh YJ, Lukman A, Flacke J, Zuidgeest M, Van Maarseveen MFAM (2017) Measuring TOD around transit nodes—towards TOD policy. Transp Policy 56:96–111. https://doi.org/10.1016/j.tranpol.2017.03.013

Sørensen E, Jacob T (2019) The Copenhagen Metropolitan 'Finger Plan': a robust urban planning success based on collaborative governance. In: Hart P't, Compton M (eds) Great policy successes, pp 218–243. https://doi.org/10.1093/oso/9780198843719.003.0012

State Planning Authority for New South Wales (1968) Sydney region outline plan 1970–2000 AD. State Planning Authority for New South Wales

The State of Queensland, Department of Infrastructure, Local Government and Planning (2015) Community facilities PDA guideline no. 11. https://www.livingstone.qld.gov.au/downloads/file/2192/pda-guideline-no-11-community-facilities-may-2015

Transport for NSW (2022) Future transport strategy. State of New South Wales (Transport for NSW)
Transport for NSW (2024) Routes and timetables, citied 27 April 2024. https://transportnsw.info/routes/train
Tsivanidis N (2023) Evaluating the impact of urban transit infrastructure: evidence from Bogotá's TransMilenio. University of California, Berkeley. https://static1.squarespace.com/static/55bb98e2e4b0ba843f39599e/t/62882b12d140ce5b51ff6e24/1653091092837/TsivanidisTransMilenio_5.2022.pdf
Wikipedia (2024) The largest shopping centres in Australia. https://en.wikipedia.org/wiki/List_of_largest_shopping_centres_in_Australia
Wolański M (2023) The potential role of railway stations and public transport nodes in the development of "15-minute cities". Infrastructures 8:141. https://doi.org/10.3390/infrastructures8100141
World Population Review (2024). https://worldpopulationreview.com/world-cities
Yudi L, Murayama A, Manabe R, Nitanai R (2023) Transit-oriented development in planning history: a west-east dialogue in regaining social goals. Urban Reg Plan Rev 10:94–120. https://doi.org/10.14398/urpr.10.94
Zandiatashbar A, Laurito A (2023) An empirical analysis of the link between built environment and safety in Chicago's transit station areas. J Am Plan Assoc 89(2):225–239. https://doi.org/10.1080/01944363.2022.2069590
Zhao Y, Hu S, Zhang M (2024) Evaluating equitable transit-oriented development (TOD) via the node-place-people model. Transp Res A Policy Pract 185. https://doi.org/10.1016/j.tra.2024.104116

Transformation of Highways to Reduce GHG Emissions and Improve Social Equity: Syracuse and Brooklyn, NY

Marc Wouters

Abstract

The City of Syracuse, in New York State, is tearing down an elevated highway in its city center and replacing it with a new tree-lined boulevard. As an urban planner working on this $2 billion project and other highway transformation projects, the goal is to reduce GHG emissions, foster economic growth, address social equity, and repair a divided city. Several highways around the world have already changed. Examples include the Embarcadero in San Francisco, the Cheonggyecheon in Seoul, and the Catharijnesingel in Utrecht. The highway transformation project in Syracuse will become one of these great international examples. The project has support from the US Department of Transportation.

Project Background: When the City of Syracuse constructed Highway I-81 in the 1950s, it demolished 27 blocks of the city center. Most of the housing demolished belonged to the African American community. The elevated highway had many negative impacts on the city. Air pollution and GHG emissions increased, and many city blocks became surface parking. In the USA, vehicles contribute 28% of total GHG emissions. Studies show residents who live within 500 ft (152 M) of a highway are more likely to develop health problems such as asthma. The highway also divided the city's neighborhoods and contributed to racial inequality.

New Design: The new at-grade boulevard by our design team will have enhanced sidewalks, rows of shade trees, bike lanes, transit stops, and shift the city to clean transportation. The new design was created through inclusive public workshops. The boulevard will have new buildings along it with retail spaces on the lower floors and apartments or offices above. The demolition of the highway's ramps creates nearly 20 acres (8 ha) of vacant land that can be used for new housing and businesses. A portion of this new land will help restore what the African American community lost. This new plan will result in economic growth for the City of Syracuse, reduced GHG, and improved social equity.

Keywords

Sustainable cities · Highway transformation · GHG reduction · Social equity · Transportation and mobility

1 Introduction

The City of Syracuse, New York, is tearing down a large elevated highway in the heart of its city center and replacing it with a new tree-lined urban boulevard. The goal of the project is to reduce GHG emissions, foster economic growth, address issues of social equity, improve air quality, and repair a divided city. In recent years, a public movement to tear down elevated highways that contribute to GHG emissions and divide neighborhoods has become a major initiative. Notable examples of successful highway transformations include the Embarcadero in San Francisco, the Cheonggyecheon Restoration Project in Seoul, the Catharijnesingel in Utrecht, and the Inner Loop East in Rochester, NY. The highway transformation project in Syracuse will become one of these great international examples. Marc Wouters Studios is one of the design partners for the project. A second highway transformation project by the studio for a segment of New York City's Brooklyn Queens Expressway is also under consideration. The Syracuse highway transformation project is supported by the USA's National Department of Transportation and has received acclaim throughout the USA by climate change leaders and social equity groups.

M. Wouters (✉)
Marc Wouters I Studios, New York, NY, USA
e-mail: marc@mwouters.com

F. Alberti et al. (eds.), *Towards Sustainable Urban Regeneration: Utilizing Environmental, Cultural, and Social Assets*, Advances in Science, Technology & Innovation, https://doi.org/10.1007/978-3-032-02091-8_14

2 Background

2.1 History

In the nineteenth and early twentieth centuries, Syracuse was a prosperous city with an economy focused on manufacturing that served the surrounding region (New York State Department of Transportation 2022a). It was a principal stop on the Erie Canal, a major regional canal for shipping freight between the Hudson River in New York City and Lake Erie, Cleveland, and Detroit (Rust 2023). During this era of prosperity, the city built a magnificent walkable downtown with elegant Romanesque and Art Deco buildings (Figs. 1 and 2). The city's urban plan has a grid of streets lined with many historic buildings and punctuated with several elegant public spaces (Fig. 3). Unfortunately, in the mid-twentieth century, industries began to close, and the economy significantly reduced in size.

2.2 Construction of Highway I-81

In the late 1950s, the New York State Department of Transportation (NYSDOT) began construction of elevated Highway I-81 through the center of the City of Syracuse (Mulcahy 2021) (Figs. 4 and 5). During this period in the USA, the new Interstate Highway Program was viewed by many as a way to improve the economy and bring in new businesses. The highway was designed to run through the city's 15th Ward, a large area in the city's downtown and largely occupied by the area's African American Community (New York Civil Liberties Union NYCLU 2021a). Residents settled in the 15th Ward in the early and mid-twentieth century to escape the discrimination found in other portions of the nation. Residents owned their homes, operated their own businesses, and created their own Main Street (New York Civil Liberties Union NYCLU 2021a). Therefore, they were able to have a life with less discrimination.

In the 1950s the construction of the elevated highway was approved by the city's government. Some members of the city viewed the 15th Ward as a "slum" (Onondaga Historical Association 2018). Approximately 27 blocks in the area were targeted for demolition to make room for the new highway and a new Civic Center (Figs. 4 and 5). Property was taken by eminent domain, a process by which the government can take private property for public use and compensate the owners. Protests against the demolition resulted in the arrests and imprisonment of several residents. Approximately 1300 families were displaced by the construction and several community landmarks were demolished (New York Civil Liberties Union NYCLU 2021b). Some residents relocated to the area immediately south, within an area known as Southside (New York Civil Liberties Union NYCLU 2021c), and some eventually moved to public housing campuses such as Pioneer Homes directly adjacent to the new highway. To this day, many residents in the area retain the deed document of their family homes that were taken and demolished. Residents point out where their former homes they grew up in were located.

Similar accounts are associated with the construction of the Brooklyn Queens Expressway I-278 (BQE) in New York City. The highway was constructed in the 1950s and 1960s to connect large portions of New York City's eastern communi-

Fig. 1 Syracuse's Niagara Mohawk Power Plant in the Art Deco style. (Credit: Marc Wouters Studios)

Fig. 2 Typical Romanesque style building in Syracuse from the late nineteenth century. (Credit: Marc Wouters Studios)

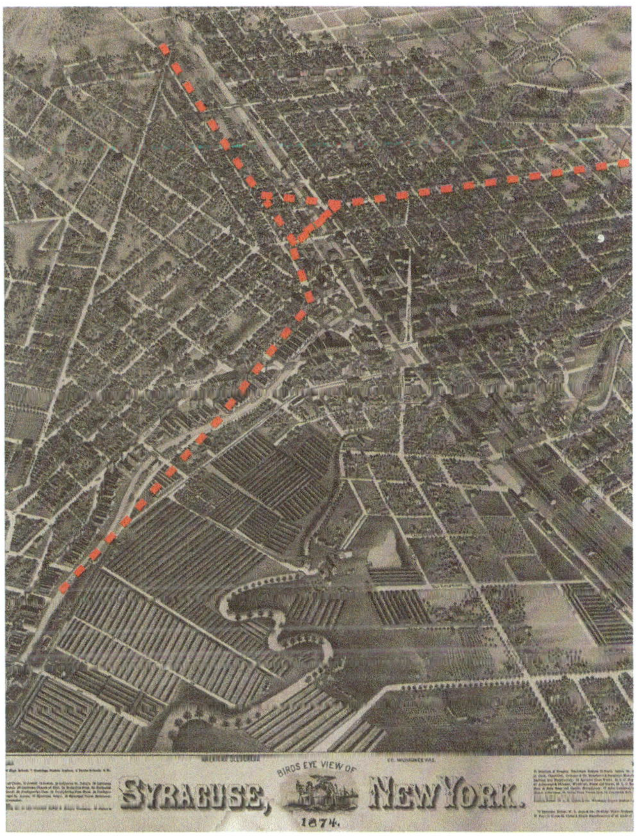

Fig. 3 Syracuse historic map with location of future highway shown in dashed red line. (Credit: City of Syracuse annotation by Marc Wouters)

ties. The highway was cut through portions of the city that had been "redlined." Redlining is a practice of that era in which the insurance industry rated properties often according to demographics (Jackson 2021). Disadvantaged communities, including Black, Hispanic, and low-income white working-class groups, were placed in these redline zones. Just as in Syracuse, homes and businesses in Brooklyn and Queens, NY, were demolished to build the BQE highway.

3 Impact of the Existing Elevated Highway

3.1 The Highway's Land Use and Economic Impact

The new highway in Syracuse not only demolished portions of the city but it also divided it. While the new elevated concrete highway allowed streets to pass underneath, the wide expanses of dark space under the large structure hindered pedestrian connections. Sightlines across the downtown were blocked. The newly created vacant land around the highway was challenging to use because of noise and fumes from vehicles on the new highway. While some elements of a new Civic Center were constructed, large portions of the demolished area were converted into surface parking (Fig. 6).

During the years following the construction of the highway, the City of Syracuse's economic growth was not strong. The population of the city declined from 220,000 residents in 1950 to 143,000 in 2018 (U.S. Bureau of the Census 2018). Many manufacturers left or closed by the 1970s often due to competition from companies overseas (New York State Department of Transportation 2022a). Retail shops in the downtown shopping district fell into decline as people shifted their purchases to auto-oriented suburban shopping centers.

Fig. 4 Photo of elevated highway Syracuse I-81. (Credit: Marc Wouters Studios)

Fig. 5 Existing elevated Highway I-81with surrounding parking lots. (Credit: NYSDOT 2022)

3.2 GHG Emissions

Carbon and other GHG emissions are associated with vehicles on the highway. According to the US Environmental Protection Agency (EPA), vehicles contribute 28% of total US GHG emissions (The US Environment Protection Agency 2022). Approximately 57% of these emissions come from passenger vehicles and 23% from trucks (The US

Fig. 6 Photo of 15th Ward with extensive surface parking lots adjacent to elevated highway. (Credit: Marc Wouters)

Environment Protection Agency 2022). The elevated highway carries about 52,000 vehicles per day, according to the New York State Department of Transportation (New York State Department of Transportation 2024). According to the EPA, "A typical passenger vehicle (in the US) emits about 4.6 metric tons of carbon dioxide per year. Every gallon of gasoline burned creates about 8,887 grams of CO_2" (The US Environment Protection Agency (US EPA) 2023).

Regional truck traffic in Syracuse is essential to the area's economy. The region's manufacturing and agricultural commerce are dependent on truck use. Many trucks use Highway I-81 as a through route to transport goods from one portion of New York State to the other, but they comprise a low percentage of overall vehicle traffic volume.

3.3 Health Impacts

The long-term impact of the highway also includes possible health issues. Numerous studies in recent years have provided data on respiratory-related illnesses found in communities adjacent to highways. The US Environmental Protection Agency (EPA) publication EPA-420-F-14-044 indicates that exhaust from vehicles on highways can travel approximately 500 ft (152 m) into adjacent neighborhoods (The US Environment Protection Agency (US EPA) 2014). "Examples of directly emitted pollutants include particulate matter (PM), carbon monoxide (CO), oxides of nitrogen (NOx), and benzene, though hundreds of chemicals are emitted by motor vehicles. Motor vehicles also emit compounds that lead to the formation of other pollutants in the atmosphere, such as nitrogen dioxide (NO_2), which is found in elevated concentrations near major roads, and ozone (O_3), which forms further downwind." The EPA report also states that "people who live, work or attend school near major roads appear to have an increased incidence and severity of health problems associated with air pollution exposures related to roadway traffic including higher rates of asthma onset and aggravation, cardiovascular disease, impaired lung development in children, pre-term and low-birthweight infants, childhood leukemia, and premature death." (The US Environment Protection Agency (US EPA) 2014). A similar study, Near-highway pollutants in motor vehicle exhaust: A review of epidemiologic evidence of cardiac and pulmonary health risks, published in 2007 (Brugge et al. 2007), suggests "elevated risk for development of asthma and reduced lung function." Similar reports have been assembled by the American Lung Association, The Centers for Disease Control (CDC), and more.

4 The Plan for a New Boulevard

4.1 Grass Roots Movement

Over the decades, concerns expressed by area residents over the highway grew. As the highway became older and required repairs, residents organized groups to advocate for its removal. A grassroots movement to demolish the highway and replace it with an at-grade urban boulevard became increasingly popular during the past two decades. Residents organized groups to assemble data to convince the New York State Department of Transportation, who control the highway, to demolish the highway instead of repairing it (Rethink I-81, Reconnect Syracuse 2014). Groups representing the Black Community also expressed a desire to remove the highway (NYCLU 2021). Eventually, Mayor Walsh, who endorsed the highway's removal, was elected to office.

With growing pressure from residents and political leaders, the New York State Department of Transportation eventually commissioned studies on possible transformation strategies. Their report of 2022 examined several options, including creating a new tunnel for the highway (New York State Department of Transportation 2022b). Community meetings were held so that the general public could advise on which option to pursue. Eventually, the at-grade urban boulevard alternative known as the "Community Grid Option" was selected. With a large amount of popular support for the proposal, the New York State Department of Transportation raised approximately $2 billion in funding from federal and state sources for the transformational project.

4.2 The New Boulevard Design

"With a Record of Decision announced on May 31st, 2022, the New York State Department of Transportation (NYSDOT) officially began the process of removing Interstate-81 through the heart of the City of Syracuse and replacing it with the Community Grid alternative" (Dover Kohl and Partners and Marc Wouters Studios 2024a). In 2024, the final Vision Plan for the highway transformation, which Marc Wouters Studios helped create, was approved by the City of Syracuse (Dover Kohl and Partners and Marc Wouters Studios 2024b). The elevated highway will be replaced with a boulevard located on the ground directly below the existing location of the elevated structure. The boulevard will have new buildings with shops, apartments, and offices along its length. The plan will have two moving traffic lanes in each direction with curbside parking along the sides. The city's network of streets will intersect the boulevard at several points, and the city's two halves, which had been separated for decades, will be reunited. Dedicated bicycle lanes and wide sidewalks will be located along the sides. A wide central median, big enough to hold a walking path, trees, public art, and benches, will create a linear park that connects the northern and southern portions of the downtown (Figs. 7, 8, 9, 10, 11, 12, and 13). The linear park includes green infrastructure to help sustainably manage stormwater. The park also improves the ease and safety of pedestrians crossing the boulevard. Four rows of street trees create a sense of enclosure and shade, cool the city streets in the summer, and absorb stormwater (Dover Kohl and Partners and Marc Wouters Studios 2024c). The result is a public space that reunites several sections of the city.

4.3 Transportation Considerations

This dramatic change is made possible by several transportation considerations. The existing daily vehicle trips on the highway, approximately 52,000, do not necessarily require the number of lanes that currently exist. To accommodate the amount of traffic, the roadway can be reduced from three lanes in each direction to two lanes (New York State Department of Transportation 2022c) (Fig. 7). Furthermore, the grid network of streets throughout the city has relatively few cars on it and has the capacity to absorb more traffic and disperse it across multiple streets. An existing ring road around the city can accommodate trucks passing through the region. This ring road will receive some improvements so trucks can navigate more easily.

The city did not want the highway replaced with another auto-oriented wide roadway that was on the ground. They did not want a new road that was challenging for pedestrians and new mixed-use development. The width of traffic lanes on the new boulevard and pedestrian patterns were studied extensively. The city is recommending narrower lane widths (11 ft or 3.35 m) because narrower lanes help slow down traffic and ease pedestrian crossings.

4.4 Alternative Transportation

The city's goal is to improve its pedestrian, transit, and bicycle networks so that fewer cars are required to commute to and around the city, and GHG emissions can be further reduced. The transformation of the highway is accompanied by an expansion of transit lanes, trails, and bike lanes throughout the downtown area (Dover Kohl and Partners and Marc Wouters Studios 2024d). Regional transit lines are planned to intersect a network of improved multimodal transportation systems.

4.5 Inclusive Public Process

The new plan was developed through a series of inclusive public workshops. One workshop included a 1-week-long charrette at a local school (Fig. 10). Residents could stop by every day and discuss community needs and goals. The pub-

Fig. 7 Proposed cross-section of new urban boulevard, named Almond Street, that will replace the elevated highway. It includes bike lanes, a wide median linear park, and places for tables and chairs. (Source: Marc Wouters Studios)

Fig. 8 Aerial of existing highway Syracuse I-8

Fig. 9 Blue areas on the map are new vacant land made available from the removal of the existing highway. New mixed use development within walking distance of jobs will reduce GHG emissions. (Source: Marc Wouters Studios, Aerial from Google Earth 2022)

lic charrette had large maps of the area on which people could write comments (Dover Kohl and Partners and Marc Wouters Studios 2024e). Consultants also met with a series of local community groups and organizations to hear their views on the transformation. The in-person workshops were supplemented by a series of online meetings with area stakeholders. The final plan was developed through discussions with the area stakeholders.

4.6 New Development

Removal of the elevated highway allows several acres (hectares) of land to be redeveloped. The former highway had several on-and-off ramps that connected to other major roadways. These ramps are no longer needed in the new design, and their removal creates several blocks of available land. The new boulevard is also narrower than the former

Fig. 10 Public Workshop with Marc Wouters standing at center. (Credit: Robert Haley Architect)

Fig. 11 Existing highway I-81 near Almond Street. (Source: Marc Wouters Studios)

elevated highway armature. As a result, an additional strip of land may become available to each of the property owners alongside the new boulevard. In all, about 20 acres (8 ha) of land are made available by the removal of the elevated highway (Wouters 2022) (Figs. 8 and 9).

The City's goal is to use this transformation of the highway as a source of new economic growth. This new land creates several opportunities for the city. The boulevard will have new buildings along it with shops, apartments, and offices (Figs. 11, 12, and 13). Preliminary tests on the new parcels of land conducted by Marc Wouters Studios indicate approximately 1000 apartments could be accommodated using simple, low-rise, and low-cost construction methods. With taller structures, even more development could be included. This new development is located in a walkable downtown and therefore allows more residents to live within walking distance of jobs and services. Existing surface parking lots may also be converted into new mixed-use developments over time and grow a pedestrian-oriented downtown. All of these development initiatives reduce auto-dependence and reduce GHG emissions for the City of Syracuse.

Fig. 12 Proposed urban boulevard, called Almond Street, in same location as Fig. 11 after demolition of highway. (Source: Marc Wouters Studios)

Fig. 13 View of new boulevard with linear park median at center and two lanes of traffic on either side. (Credit: Marc Wouters Studios)

4.7 Social Equity

This newly available land may help restore what was demolished when the highway began construction in the 1950s. New housing and places for new locally owned businesses may play a role in bringing back opportunities to the Black Community. Strategies to support this important community include job opportunities, housing, and cultural opportunities. These efforts to repair damage to the Black Community are essential to the success of the overall plan. The plan provides the opportunity to have a range of housing choices so that people from all income levels will be included (Dover Kohl and Partners and Marc Wouters Studios 2024f). Preliminary goals for new housing include supportive housing, affordable housing, workforce housing, and market-rate housing.

4.8 Community Land Trust

One new site of about 4 acres (1.6 ha) will likely be organized as a Community Land Trust, in which residents own the land and apartment buildings instead of a developer. This format allows residents to control the cost of rent and have control of housing that is permanently affordable. The format even accommodates low-rent commercial spaces for start-up Black-owned businesses and affordable homeownership units.

4.9 Building Design Guidelines

The City will likely develop guidelines for how the properties along the entire boulevard can be redeveloped. The Vision Plan calls for a boulevard with buildings that support the overall design. New buildings should not be set back, but meet the sidewalk edge. They should have storefronts that engage the pedestrians and have building facades that frame a grand new linear public space. A Form-Based Code has been discussed as a basis for these guidelines. This type of design code includes diagrams for building heights, building massing, storefront designs, the extent of glass surfaces, guidance on bay windows and balconies, materials, and more.

5 Highway Transformations in Brooklyn, NY

Several other highways are pursuing similar strategies. The Brooklyn Queens Expressway (BQE I-278) is one of New York City's largest arterials. It connects portions of the city east of Manhattan, including the boroughs of Brooklyn, Queens, and Staten Island, and conveys about 130,000 vehicles per day (New York City Department of Transportation 2022). It has a combination of elevated highway segments, on-the-ground highways, and trenches. New York is a densely built city, and many apartments and homes are located directly adjacent to this busy highway. Like many highways built in the 1950s and 1960s, it is deteriorating, and several areas are in need of regular maintenance. A growing group of local stakeholders and elected leaders are calling for a major transformation of this highway. Like Syracuse, residents experience poor air quality and increased rates of respiratory illnesses. Furthermore, New York City published its plan: Aligning NYC with the Paris Climate Agreement, in 2017. The plan cites that "Roughly 90% of NYC's current transportation GHG emissions come from private vehicle travel." The city has committed to an 80% sustainable mode share by 2050 (City of New York 2017).

In 2018, New York City's Department of Transportation proposed a full reconstruction and enlargement of an older segment of the BQE highway known as the Triple Cantilever at a cost of over $3 billion (New York City Department of Transportation 2018) (Fig. 14). The new plan would have altered significant landmarks like the Brooklyn Bridge and Brooklyn Heights Promenade. At that time, many residents gathered to oppose alterations to these landmarks and to the plan. They asked for more investments in sustainable transportation systems, commitment to the city's GHG reduction targets, and reduced investments in an auto-dependent culture. As in Syracuse, the call for progress came from the public, "grassroots." The city's first plan was halted due to this public opposition. The public's efforts succeeded in reducing this portion of the highway by one lane in each direction. However, the city's Department of Transportation has continued to develop proposals to reconstruct the highway with an enlarged roadway.

In 2018, an alternative plan was prepared by Marc Wouters Studios that reduced the width of the highway and permanently set a course for a slimmed-down, less auto-dependent transportation system (Figs. 15, 16 and 17) (Wouters 2018). It also reconnected the neighborhoods on either side. This alternative plan was presented to the Mayor of New York's Expert Panel on the BQE. The plan reduces GHG, improves air quality, requires significantly less construction and concrete, and requires less time to construct. It preserves the major landmarks valued by the residents. At the time of the writing of this paper, several local organizations have offered preliminary approval of this proposal.

6 Conclusion

The transformation of a highway is more than an infrastructure project. The removal of the elevated highway is a healing process for a city. The transformation allows the city to have a larger walkable downtown with a wide variety of new economic opportunities for all of its residents. In Syracuse, it creates ways to correct social inequities endured by the Black Community during the highway's construction in the 1950s. By increasing the number of housing and commercial uses in the downtown, the Syracuse plan creates a more pedestrian-oriented culture. By reducing auto-dependence, the plan reduces GHG emissions significantly. The plan reduces the negative health impacts of vehicle exhaust. This transformation is supplemented with new transit and bike

Fig. 14 Aerial view of existing BQE I-278 Triple Cantilever with three levels of traffic that has fallen into disrepair. (Source: Google Earth)

Fig. 15 Proposed adaptive reuse of existing structure with reduced number of traffic lanes under a landscape deck. (Source: Marc Wouters Studios)

lanes so that commuters can have more sustainable methods to get to the walkable downtown. Transforming these highways is much more than an infrastructure project; it's an entire city-wide plan for sustainable and equitable growth.

Acknowledgments The author wishes to acknowledge the City of Syracuse for their many years of work on Highway I-81. Acknowledgements include design partners Dover Kohl & Partners, Nelson Nygaard, and Ideas and Action.

Fig. 16 Proposed reuse of existing structure with new public space over reduced highway. It will offer some of the best views of New York City. (Source: Marc Wouters Studios)

Fig. 17 Existing and proposed adaptive reuse of the BQE Highway Triple Cantilever by Marc Wouters Studios. The proposal changes highway from six lanes to four lanes, covers the highway with a landscape deck, and adds bike lanes and pedestrian connections

References

Brugge D, Durant J, Rioux C (2007) Near-highway pollutants in motor vehicle exhaust: a review of epidemiological evidence of cardiac and pulmonary health risks. Environ Health

City of New York (2017) Aligning New York City with The Paris Climate Agreement. p 11

Dover Kohl & Partners, Marc Wouters Studios et al (2024a) City of Syracuse Community Grid Vision Plan, City of Syracuse. p 6

Dover Kohl & Partners, Marc Wouters Studios et al (2024b) City of Syracuse Community Grid Vision Plan, City of Syracuse. p 4

Dover Kohl & Partners, Marc Wouters Studios et al (2024c) City of Syracuse Community Grid Vision Plan, City of Syracuse. p 52

Dover Kohl & Partners, Marc Wouters Studios et al (2024d) City of Syracuse Community Grid Vision Plan, City of Syracuse. p 42

Dover Kohl & Partners, Marc Wouters Studios et al (2024e) City of Syracuse Community Grid Vision Plan, City of Syracuse. p 10

Dover Kohl & Partners, Marc Wouters Studios et al (2024f) City of Syracuse Community Grid Vision Plan, City of Syracuse. p 18

Jackson C (2021) What is redlining? New York Times

Mulcahy M (2021) The map: segregated Syracuse, CNY Central. pp 3–7

New York City Department of Transportation (2018) BQE I-278, Presentation Public Information Meeting
New York City Department of Transportation (2022) BQE Corridor Vision, BQE Central Workshop.
New York Civil Liberties Union NYCLU (2021a) The I-81 Story. p 4
New York Civil Liberties Union NYCLU (2021b) The I-81 Story. p 5
New York Civil Liberties Union NYCLU (2021c) The I-81 Story. p 6
New York State Department of Transportation (2022a) I-81 Viaduct Project final design report, final environmental impact statement, and final section 4(f) evaluation: Chapter 2, pp 2–3
New York State Department of Transportation (2022b) I-81 Viaduct project final design report, final environmental impact statement, and final section 4(f) evaluation
New York State Department of Transportation (2022c) I-81 Viaduct project final design report, final environmental impact statement, and final section 4(f) evaluation: Chapter 3
New York State Department of Transportation (2024) Traffic Data Viewer, web application
Onondaga Historical Association (2018) The destruction of the 15th ward. p 4
Rethink I-81, Reconnect Syracuse (2014) Impacts of DOT Viaduct and Boulevard
Rust R (2023) The Erie Canal-connecting New York to the great lakes. American History Central, R. Squared Communications, LLC
The US Environment Protection Agency (2022) Fast facts on transportation greenhouse gas emissions. p 1
The US Environment Protection Agency (US EPA) EPA-420-F-14-044 (2014) Near roadway air pollution and health: frequently asked questions. p 1
The US Environment Protection Agency (US EPA) (2023) Greenhouse gas emissions from a typical passenger vehicle. p 1
U.S. Bureau of the Census (2018) Annual estimates of the resident population for incorporated places. Accessed 22 Nov 2019
Wouters M (2018) I-278 BQE parallel method concept proposal, DOT submission
Wouters M (2022) Syracuse Parcel Study

Research on Sustainable Urban Design Based on Small-Scale and Progressive Concept

Hu Ruotong, Liu Yuxuan, Song Yunzi, and Hou Zhulin

Abstract

In China, the trend of urbanization is significant. Cities have become the concentration of population and economy, but they also face a series of "urban diseases" such as environmental degradation, traffic problems, and soaring housing prices. Large-scale urban expansion has made problems such as damaged ecosystems, chaotic spatial structures, and traffic congestion more serious. In order to promote high-quality and sustainable urban development and accelerate the revival of modern cities, the Chinese government has proposed a "double urban repair" policy, including "ecological restoration" and "urban repair."

This paper selects the core area of Nanjing as the research object, takes "urban repair and ecological restoration" as the guideline, starts from a small-scale, progressive design perspective, and combines a variety of urban design methods and concepts to propose a series of sustainable urban design strategies. These strategies aim to optimize the urban structure, improve the public transportation system, protect historical buildings, effectively integrate green spaces with public spaces, and achieve harmonious coexistence between nature and urban life. The ultimate goal is to create a more prosperous, livable, and sustainable Nanjing and provide experience and models for other cities to learn from. This study focuses on current problems and challenges and strives to provide feasible solutions and innovative design ideas for future development.

Keywords

Small-scale · Progressive design · Sustainable urban design · Urban repair · Nanjing

1 Introduction

According to current global development, more than half of the world's population already lives in cities, and future population growth will mainly be concentrated in urban areas. In the late Industrial Revolution, urban diseases began to emerge. Britain was one of the first European countries to initiate the industrial revolution and the urbanization process. With the advancement of industrialization, Britain's large cities became the concentration of population and economy, creating enormous wealth, but also accompanied by problems such as environmental degradation, traffic problems, and soaring housing prices. The history of urbanization in the United States has undergone three stages (Chudacoff et al. 2016). Japan has become a leader in urbanization with its unique development model. The urbanization development of Latin America is influenced by its colonial history. It mainly relies on foreign investment but is also accompanied by severe challenges, such as the gap between the rich and the poor, employment problems, and insufficient public facilities. Cities are the concentration of resources and the inevitable product of industrialization. "Urban diseases" mainly involve population, transportation, environment, poverty, housing, and health.

After decades of large-scale expansion and construction, the urban form of Chinese cities has undergone tremendous changes. However, with the development of cities, many "urban diseases" have also emerged, such as damaged urban ecosystems, chaotic urban spatial structure, severe traffic congestion, damaged historical heritage, and insufficient urban public space. In order to promote the higher quality development of Chinese cities, the Chinese government has launched the "urban double repair" policy, namely "ecologi-

H. Ruotong (✉) · L. Yuxuan · S. Yunzi
Department of Architecture and Design, Politecnico di Torino, Turin, Italy
e-mail: ruotong.hu@polito.it; yuxuan_liu@polito.it; s298874@studenti.polito.it

H. Zhulin
School of Architecture, Southeast University, Nanjing, China

F. Alberti et al. (eds.), *Towards Sustainable Urban Regeneration: Utilizing Environmental, Cultural, and Social Assets*, Advances in Science, Technology & Innovation, https://doi.org/10.1007/978-3-032-02091-8_15

cal restoration" and "urban repair," and implemented the policy across the country in response to the call of the central government. Against this background, this paper chooses Nanjing Xinjiekou as the design theme. This article aims to promote the high-quality development of Chinese cities and the revitalization of modern cities. Through the policy concept of "urban double repair," especially the guiding ideology of "urban repair," a series of specific urban design strategies are proposed for the problems existing in the core area of Nanjing. These strategies aim to improve the living environment of Nanjing, optimize the urban spatial layout, improve the quality of life, and further promote Nanjing's urbanization and economic development, making it more prosperous and sustainable.

2 Theoretical Basis of Urban Repair

Urban repair emerged as a response to shifting urban development priorities, focusing on internal regeneration rather than external expansion. It emphasizes small-scale, targeted interventions to revitalize urban spaces while preserving cultural heritage and ecological balance. This section reviews four foundational theories—Urban Acupuncture, Urban Catalyst Theory, Collage City Theory, and Ecological City Theory—highlighting their contributions to urban repair.

2.1 Urban Acupuncture: Targeted Urban Regeneration

Urban Acupuncture, introduced by Lerner (2014), applies targeted interventions inspired by Chinese acupuncture to revitalize small urban areas like abandoned factories or deteriorated neighborhoods. Reconnecting these spaces with streets and green belts stimulates rapid improvements in urban functionality. Compared to "tactical urbanism," it emphasizes long-term, high-impact solutions based on a holistic understanding of the urban environment (Lydon and Garcia 2015). However, its scalability to larger urban systems remains a challenge.

2.2 Urban Catalyst Theory: Triggers for Urban Transformation

Urban Catalyst Theory, proposed by Attoe and Logan (1989), draws from catalysis principles where small interventions trigger widespread systemic effects. Urban catalysts, characterized by hierarchy and linkage, initiate ripple effects that transform larger areas. For instance, developing a cultural hub can revitalize surrounding neighborhoods. This systemic approach complements urban acupuncture's precision, though its success depends on effective integration into broader urban strategies.

2.3 Collage City Theory: Balancing History and Modernity

In the 1970s, cities faced post-war reconstruction challenges requiring the restoration of urban structures and integration of historical and modern elements. Colin Rowe introduced the "collage" concept in urban design, using historical architectural and spatial elements to maintain continuity and original features while meeting current and future needs, thus preserving cultural and social memory and piecing together historical fragments into a cohesive whole that balances modernization with historical heritage (Rossi 1984).

2.4 Ecological City Theory: Sustainability and Harmony

Foreign scholars have defined eco-cities as ideal urban environments that satisfy human pursuits of natural health and vitality, and believe that this is what eco-cities are all about. Chinese scholars have proposed a more detailed definition of eco-cities, where "eco-cities" are defined as sustainable subsystems that coexist harmoniously within a broader regional or global ecosystem. It represents a complex approach that interweaves natural balance, social justice, and economic benefits, all guided by ecological rules. In addition, as an excellent habitat with inherent humanistic qualities, the unity of nature and humans dominates, allowing people to thrive in harmony (Huang and Yang 2001).

2.5 Synthesis of Theoretical Perspectives

The four urban repair theories—Urban Acupuncture, Urban Catalyst Theory, Collage City Theory, and Ecological City Theory—offer distinct approaches. Urban Acupuncture and Catalyst Theory focus on targeted and systemic interventions, while Collage City Theory emphasizes blending historical and modern elements. Ecological City Theory underscores sustainability and harmony with nature.

Together, these theories provide a multidimensional framework for addressing urban challenges. However, practical integration into cohesive strategies remains underexplored. Future research should develop actionable models combining their strengths to guide urban regeneration.

3 Current Situation Analysis Based on SWOT Method

The SWOT analysis concept is mainly used in management. Later, it was gradually applied to architecture and urban planning. SWOT analysis is a systematic method that helps decision-makers develop more effective strategies by identifying and evaluating the environment's strengths, weaknesses, opportunities, and threats. In architecture and urban planning, SWOT analysis can help architects and planners understand the environment they are designing more accurately to develop more effective design and planning plans based on the objective conditions on-site (Gürel and Tat 2017).

3.1 Strengths

Nanjing Xinjiekou has significant advantages in terms of function and spatial structure. Since the 1980s, the Nanjing Planning Bureau has been studying the development of the Nanjing Central System and Xinjiekou Central District, involving spatial form, industrial structure, railway transportation, and urban regeneration. With the efforts of many parties, Nanjing Xinjiekou has evolved from a retail and commercial center to a robust business agglomeration area, becoming Nanjing's economic, commercial, and cultural center. Nanjing Xinjiekou has developed public service facilities and convenient transportation. It is an important transportation hub in the city. The underground space is a subway transfer station and an area with frequent commercial activities. In addition, Nanjing Xinjiekou has a long history and rich cultural heritage, witnessing the city's rise and fall and changes. After years of urban renewal, many outstanding modern buildings have been left behind, a valuable legacy left by a generation of architects to future generations.

3.2 Weakness

Nanjing Xinjiekou has long had a substantial total traffic volume, and the limited road network cannot meet the needs of smooth travel, and traffic congestion becoming the norm. As the heart of urban transportation, Xinjiekou still needs to solve more traffic conflicts. The main problems include inefficient traffic structure, overlapping land use natural layout and main urban trunk road layout, serious mixing of motor vehicles and non-motor vehicles in traffic, unreasonable ratio structure of primary and secondary trunk roads, and severe shortage of parking facilities. At the same time, as the central business district of Nanjing, Xinjiekou has attracted many developers and enterprises, with the maximization of commercial interests as the primary development goal. However, this has led to separating public and building spaces and a need for sufficient public and green space for leisure use. The buildings have mixed functions, different appearances, and a lack of unity. There are problems of space waste and mixed functional zoning in the central space, and the building layout is chaotic and lacks order. In addition, the increasingly rich functions of the central area of Xinjiekou have brought prosperity to the old city while changing its historical style and destroying the characteristics of the ancient capital of Nanjing. High-rise buildings have had a vast and almost irreversible impact on historical and cultural resources, and multiple historical buildings are scattered and preserved in them, facing the problem of balancing protection and modern development (Fig. 1).

3.3 Opportunity

In 2015, China proposed the "urban repair" concept and issued relevant guidelines in 2017, requiring key cities to implement the "urban double repair" project. As one of the pilot cities, Nanjing received policy support and special funds. At the same time, the city center of Nanjing will highlight different functions such as commerce, finance, and culture through the construction of the urban framework of "Xinjiekou, Hexi Center, and Chengnan Center." This framework will contribute to the overall development of Nanjing, and it is expected that Xinjiekou will become a world-class commercial center. In addition, the Nanjing Municipal Planning Bureau has also formulated a general plan for the development and utilization of urban underground space, with subway stations and underground streets as the core, to create a multi-center, networked underground space structure. In the central area of Xinjiekou, an underground public pedestrian system has been established, which has promoted the development of commerce and service industries, enhanced the overall vitality of the pedestrian system, and formed a complete underground city network.

3.4 Threat

Nanjing Xinjiekou, as the commercial center of the old city, faces challenges due to limited land development opportunities and geographical constraints. The area has undergone high-density development, which has progressively diminished the potential for further site development and necessitates regional regeneration. Additionally, the rapid construction of high-rise buildings has altered the spatial pattern and negatively impacted environmental quality, leading to issues such as crowded streets, insufficient green space, and the presence of outdated buildings. To address these

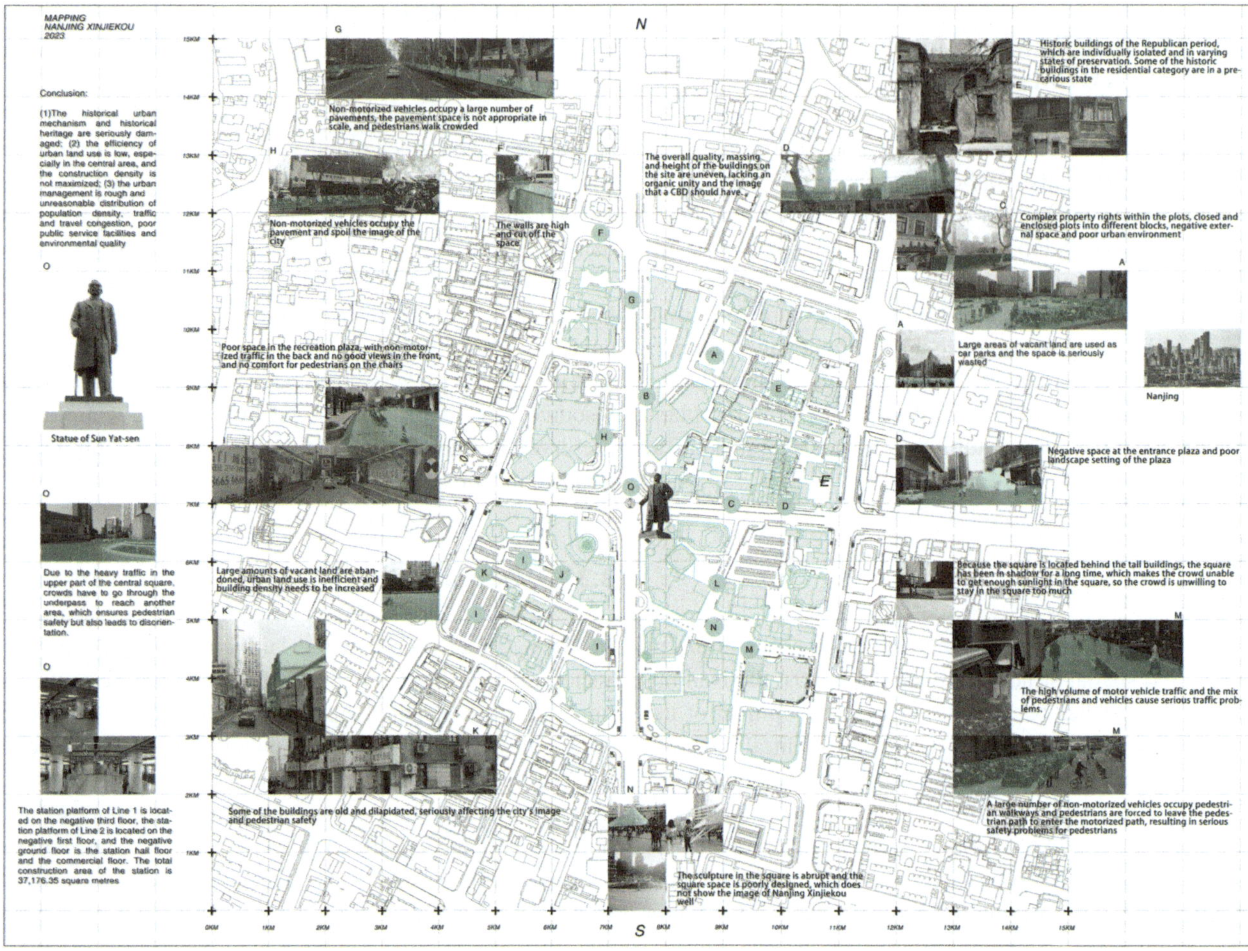

Fig. 1 Analysis of the current situation of Nanjing Xinjiekou. (Source: Hu Ruotong)

problems, it is essential to respect the traditional architectural style, enhance greening efforts, and focus on creating high-quality urban spaces.

4 Specific Urban Regeneration Strategy of Nanjing Xinjiekou

The urban regeneration strategy for Nanjing Xinjiekou revolves around four parts: urban space repair, transportation system repair, historical texture repair, and urban green space repair.

4.1 Overall Area

The strategies adopted for the whole area are installing skywalks to separate pedestrian paths from vehicle traffic and strengthening the internal connection of commercial spaces. Setting up skywalks to make the bronze statue of Sun Yat-sen the visual focus of the main street of Xinjiekou and provide an important transportation hub function (Fig. 2). Redesigning the entrance of the underground passage to make the dim entrance form a unified whole and promote the transformation of urban development to a three-dimensional model and redesigning pedestrian paths and facility strips to solve pedestrian and non-motor vehicle problems and beautify the city image (Fig. 3).

4.2 Northeast Region

Protect and repair the former Central News Agency and the buildings of the Republic of China on Qingshi Street, including the former sites of the Industrial and Commercial Bank of China and Zhejiang Industrial Bank, restore their original appearance, and continue the historical heritage of Xinjiekou. Repair and protect the main building and gate of the former Central News Agency, demolish some buildings on Qingshi Street, form an open public square of the former Central News Agency, and increase the atmosphere of the area; renovate the surrounding buildings of the former sites of the

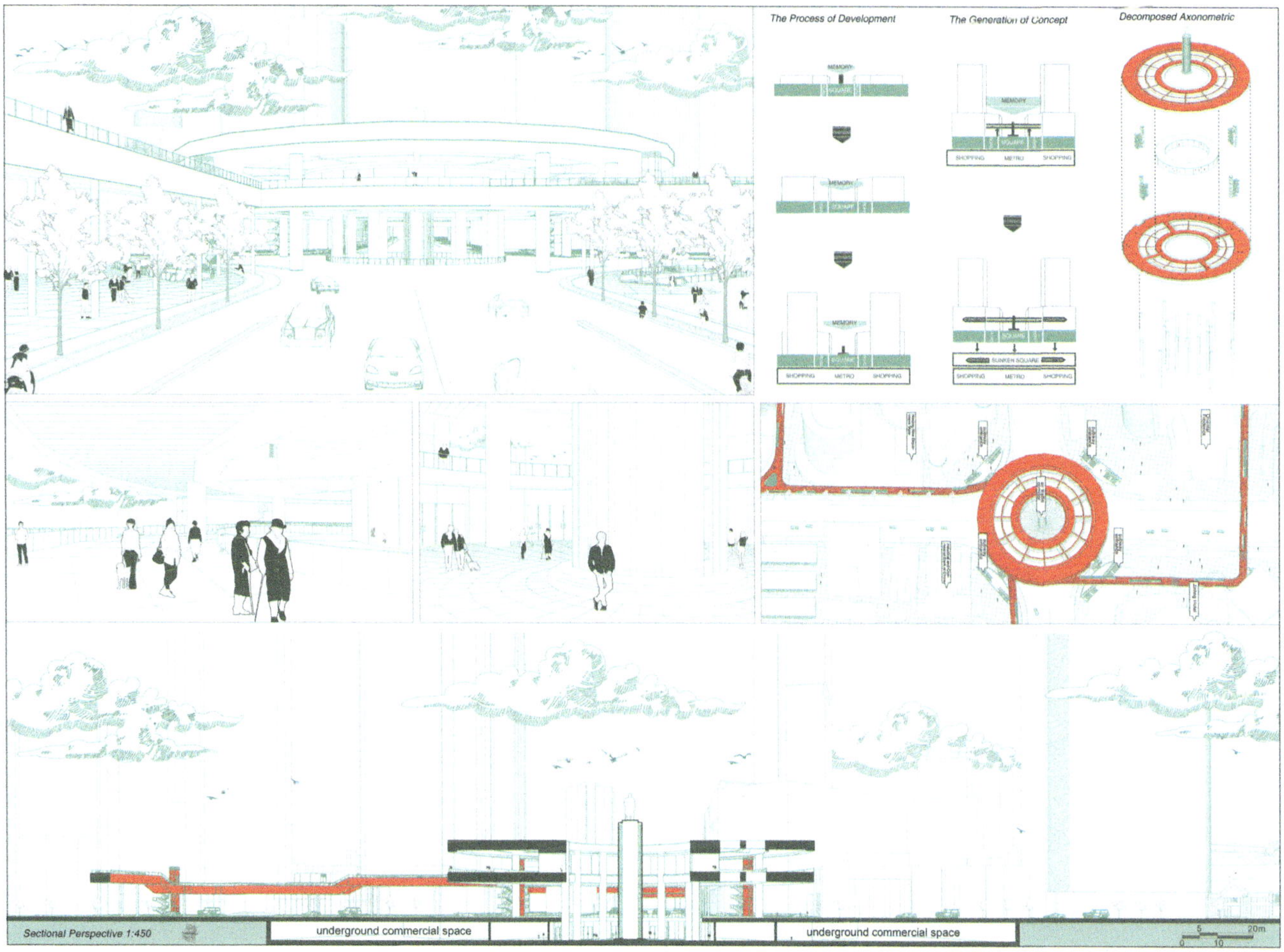

Fig. 2 Air traffic flow corridors. (Source: Hu Ruotong)

Industrial and Commercial Bank of China and Zhejiang Industrial Bank to form a complete square interface, connect the historical buildings in the area as a whole, and form a unique interface of historical buildings of the Republic of China. Based on protecting historical buildings as much as possible, give full play to the advantages of historical buildings and transform them into places for modern people's leisure, entertainment, culture, and other living needs. Repair and restore the buildings of the Republic of China on Qingshi Street, demolish and renovate some buildings in the base, and create a commercial block and cultural exhibition complex with the characteristics of the Republic of China. There are a large number of vacant urban spaces in the base, which are composed of closed interfaces and hard-paved areas. These urban spaces will be re-landscaped, and some small buildings will be built to provide urban living spaces for leisure and entertainment for the surrounding residents.

4.3 Northwest Region

Jinling Hotel was renovated; Jinling Hotel occupies the northwest corner of the core area of Xinjiekou and is an essential urban interface. Without affecting the regular operation of Jinling Hotel, the space above Jinling Hotel was fully utilized for landscape design. The roof of Jinling Hotel was used as an alternative entrance to provide leisure space for the public and improve the traffic convenience of the hotel (Fig. 4); at the same time, some facades of Jinling Hotel were renovated to enhance the image of the urban interface. The landscape design and refinement of the area alleviated the problem of insufficient urban greening quality and provided more high-quality green and natural urban space for citizens. By setting up a pedestrian bridge, the traffic and pedestrian flow are separated to ensure the safety of pedestrians. An aerial traffic corridor was set up. There is a bronze statue of Sun Yat-sen at the intersection of Zhongshan

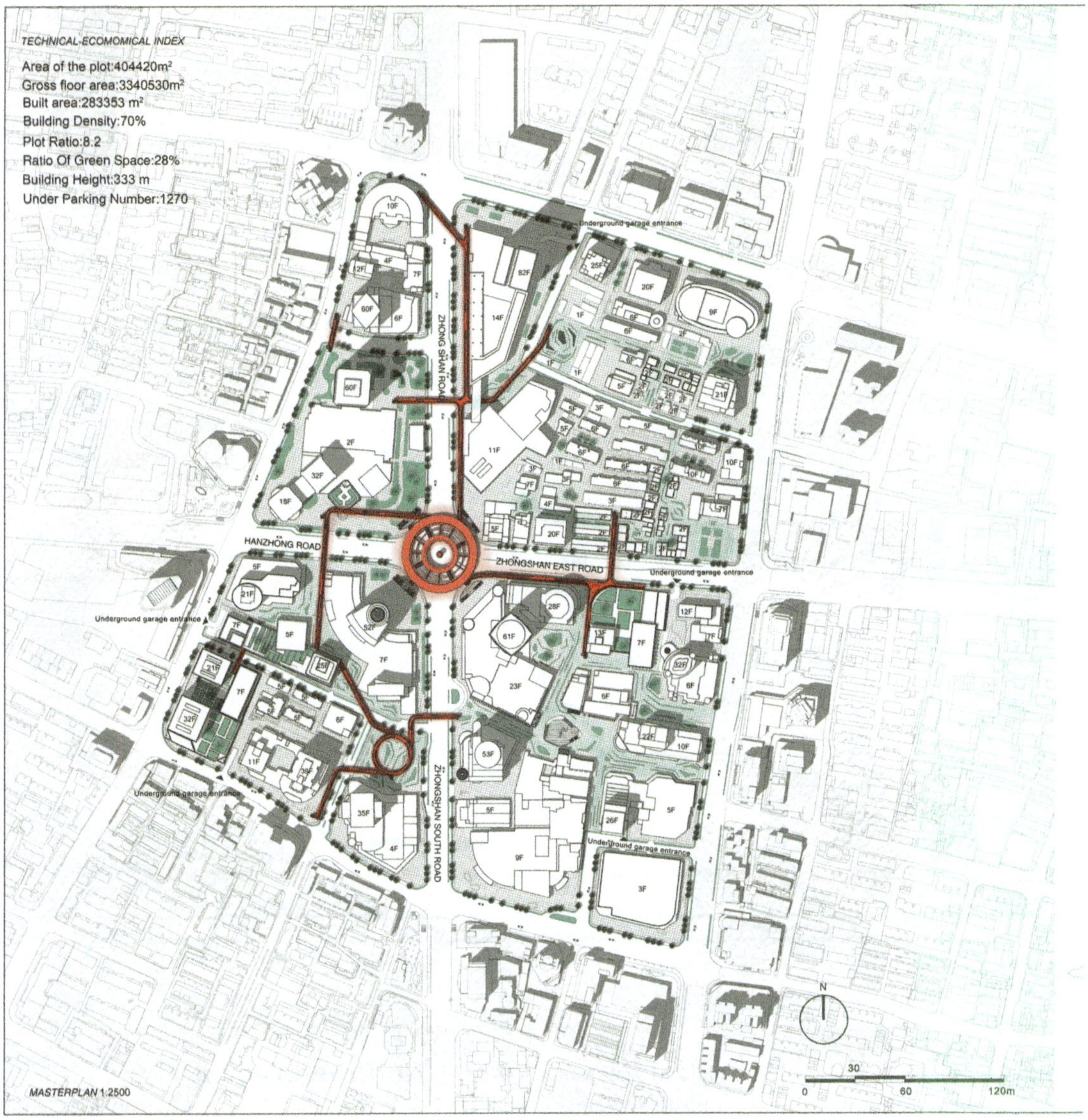

Fig. 3 Overall update strategy. (Source: Hu Ruotong)

Road and Zhongshan Road in Xinjiekou. However, the sculpture is surrounded by many high-rise buildings and traffic, and its role as a city landmark is minimal. Combined with the traffic corridor, the statue becomes the visual focus of the main street of Xinjiekou, evoking people's memories of the past. In addition, the corridor also serves as an essential transportation hub, connecting the four quadrants and allowing pedestrians to enter any area through the "roundabout." Redesign the underground passage entrance. Since the underground entrance around the intersection of Zhongshan East Road and Zhongshan Road is also the entrance to the underground passage, many people will enter the underground commercial space through the underground entrance. However, the current underground entrance is very dim. The five entrances are interconnected to form a unified whole. Organizing the city in a different way can promote the transformation of urban development from the traditional two-dimensional orientation to a three-dimensional mode. In addition, redesigning the sidewalk and facility strips solves the problems of pedestrian safety and non-motor vehicle parking and beautifies the city's image.

4.4 Southwest Region

Reuse the land initially used as a parking lot to improve land use efficiency. Build two high-rise commercial complexes in the original ground parking area, transfer the ground parking to the underground space of the commercial complex, and form a complex of multifunctional places and spaces. The volume design of the commercial complex takes into account the image of the urban interface, strictly controls the facade

Fig. 4 Jinling Hotel Renovation Project. (Source: Hu Ruotong)

and height of the commercial complex, and ensures that the interface on different roads in the city is orderly and beautiful. The design of the new building volume is fully integrated with the old building form to create a civic square with high-quality space. In addition, a former parking lot has been transformed into a new leisure node. In order to further increase the urban greening rate, make full use of the roof space of the commercial complex and connect it with the sky corridor to create a high-quality urban leisure space that is entirely different from the ground dimension.

4.5 Southeast Region

The land initially used as a parking lot was replanned to accommodate the new commercial complex, and the ground parking lot was moved to the lower part of the commercial complex. In addition, the roof of the new commercial complex was landscaped and connected to the pedestrian bridge to form an overall landscape (Fig. 5). The pedestrian street and square in Xinjiekou attract many people during holidays. However, they are often deserted during non-holiday periods, and the public is reluctant to stay in the square because of its poor spatial feel. In order to enhance the characteristics of the pedestrian street and square, some strategies were adopted: first, some "huts" were embedded in the buildings for people to relax and entertain, thereby livening up the monotonous interface of the square; second, the facades of the buildings around the square were unified to improve the integrity and quality of the square; finally, the upper part of the hollow square was demolished to connect it with the space of the ground square. Then, the ground square was landscaped, and movable seats were placed for people to rest and entertain on weekdays, while the seats could be removed during holidays to leave space for commodity stalls.

5 Conclusion

As urban development evolves to address increasingly complex needs, the focus has shifted to comprehensive, socially humanistic, and sustainable renewal. The Chinese "urban repair" concept integrates global insights with local characteristics, prioritizing ecological, cultural, and human-centered values over purely economic goals. This approach promotes gradual, collaborative processes involving residents, governments, and developers to address urban decay, historic preservation, transportation inefficiencies, and lack of green space.

Through strategies that optimize urban structures, enhance public transportation, protect historical heritage, and integrate green spaces, "urban repair" offers a multidimensional framework for creating harmonious and sustainable urban environments. Combining precise, localized interventions with systemic transformations provides scalable solutions that balance tradition with modernization.

Future efforts should operationalize these insights into adaptable models, ensuring tailored solutions for diverse urban contexts. Ultimately, "urban repair" represents a paradigm shift toward renewal strategies that align ecological sustainability with cultural and social priorities, paving the way for more resilient and inclusive cities.

Fig. 5 Green bridge and green roof. (Source: Hu Ruotong)

References

Book

Attoe W, Logan D (1989) American urban architecture: catalysts in the design of cities. Univ of California Press

Chudacoff HP, Smith J, Baldwin P (2016) The evolution of American urban society. Routledge

Lerner J (2014) Instructions for performing urban acupuncture. In: Urban acupuncture. Island Press, pp 62–64

Lydon M, Garcia A (2015) A tactical urbanism how-to. In: Tactical urbanism: how-to guide. Island Press/Center for Resource Economics, pp 171–208

Rossi A (1984) The architecture of the city. MIT Press

Journal

Gürel E, Tat M (2017) SWOT analysis: a theoretical review. J Int Soc Res 10(51):994–1006. https://doi.org/10.17719/jisr.2017.1832

Huang Z, Yang D (2001) A review of domestic and international research on ecological city theory. Urban Plan 25(1):59–66

Revitalization Strategies for Inner Areas in Italy: Constraints and Opportunities from an Analysis of Best Practices

Maria Rita Gisotti, David Fanfani, Claudia Mezzapesa, Eletta Naldi, Giulia Biagi, Matteo Mengoni, and Paola Scarpellini

Abstract

The revitalization of inner areas is a significant issue in the field of spatial and economic planning for sustainability and ecological transition. The settlement model based on the predominance of metropolitan areas has long shown its vulnerability in environmental, functional, economic, and socio-cultural terms. Strategies of revitalization and repopulation of inner areas can contribute to the construction of a better territorial arrangement because they structure a new reticular and polycentric settlement system in dynamic equilibrium with the natural resources present in the area of reference; they can be an opportunity to reorganize living according to "proximity" criteria; they can offer opportunities for maintenance of the rural territory, preserving its biodiversity and preventing hydrogeological and hydraulic risk; they contribute to the protection of essential heritage values. For some years now, a wide range of initiatives aimed at pursuing, among others, the above objectives have been activated in inner areas by institutional or bottom-up initiatives.

This paper presents the outcomes of a study conducted as part of the research project "Regenerating Cultural Landscapes of Inner Areas in a people-centered perspective" carried out by a multidisciplinary team of the University of Florence, involving researchers from the departments of architecture, economics, and educational sciences. In particular, the paper focuses on one of the first tasks of the project, namely the critical analysis of best practices related to regeneration projects in inner areas in Italy. The article presents the original analysis methodology developed for this research that produced a matrix of meta-criteria and criteria for selecting and evaluating best practices, which can also be used in other research contexts. Next, the article focuses on a selection of four best practices that were analyzed in more depth and summarizes the main findings regarding design strategies, highlighting limitations and perspectives for better regeneration of inner areas.

Keywords

Cultural landscape · Heritage · Sustainability · Marginal area · Integrated territorial development

1 Introduction

For some years, the question of "inner areas"—a term that literally refers to internal areas but broadly denotes marginal and fragile territories distant from essential services and experiencing depopulation—along with the criticalities and opportunities that characterize them, and the related development disparities, has been key themes in the European (ESPON 2017) and Italian context. This is a question not only of study and research, whose emergence is also reflected by the crisis of the forms of representation and consensus due to the polarized modes, urban-centered, and reductive construction of policies (Rodríguez-Pose 2018).

Relating to that, in Italy, in a way that was far-sighted at the time within the framework of national territorial cohesion policies department, the National Strategy for Inner Areas (SNAI) (Barca et al. 2014), was devised as a specific policy at national level to address the services availability

M. R. Gisotti · D. Fanfani · C. Mezzapesa · E. Naldi (✉)
Department of Architecture, University of Florence, Florence, Italy
e-mail: mariarita.gisotti@unifi.it; david.fanfani@unifi.it; claudia.mezzapesa@unifi.it; eletta.naldi@unifi.it

G. Biagi
Department of Education, University of Florence, Florence, Italy
e-mail: giulia.biagi1@unifi.it

M. Mengoni · P. Scarpellini
Department of Economics, University of Florence, Florence, Italy
e-mail: matteo.mengoni@unifi.it; paola.scarpellini@unifi.it

F. Alberti et al. (eds.), *Towards Sustainable Urban Regeneration: Utilizing Environmental, Cultural, and Social Assets*, Advances in Science, Technology & Innovation, https://doi.org/10.1007/978-3-032-02091-8_16

criticalities of inner areas as limit to enable the citizenship exercise, all the more relevant given the characteristics of plurality and widespread "marginality" of the Italian territory (Presidenza del Consiglio dei Ministri 2024).

This strategy allows shedding light on the broader theme of territorial disparities and differences that—"at large"—cross the country and that offers the tools for a new "outside-in" paradigm of observation of local settlement contexts "as a whole" (Carrosio 2019). A decisive and necessary reversal of the look in a phase of profound energy and socio-economic transition that challenges, even in a radical form, traditional "urban-centered" or center-periphery interpretations and images referring to spatial partitions and settlements polarizations that do not really reflect the complexity of the articulations that constitute the Italian territory and its dynamics (Cersosimo et al. 2018).

In this context, the lately elaborated Italian National Plan for Recovery and Resilience (PNRR), to be effective, is called necessary to cope with the general plurality and internal diversity of these territories to devise themes, interpretative frameworks, project methodologies, and development "place-based" scenarios and actions (Barca et al. 2012) relevant and fitting with the underlined manifold diversities.

1.1 The REACT Research Project: Goals, Cross-Disciplinary Adopted Methodology and Study Area

In this framework, the REACT research project addresses a fundamental issue related to the study of conditions for defining strategic guidelines and actions aimed at regenerating inner areas by leveraging the value of cultural landscapes. REACT is a competitive research project funded by the University of Florence within the framework of the Italian National Research Program (NRP), titled "Regenerating cultural landscapes of marginal areas with a people-centered perspective: historic villages and rural territories of Casentino area as a laboratory of creativity and innovation" (Università di Firenze 2024). The research explores the multifaceted nature of cultural landscapes, encompassing architectural heritage, cultivated landscapes, natural ecosystems, and communities' lived experiences and practices. By adopting an interdisciplinary approach, the project seeks to integrate these dimensions into a cohesive vision that can serve as a foundation for sustainable and inclusive development strategies that enhance the available cultural and human resources. A key focus of the REACT research is the application of its findings to a specific context: the Casentino, an area located within the sub-watershed of the upper valley of the river Arno in Tuscany (Fig. 1). This culturally and environmentally significant area serves as a case study for developing strategies to preserve its unique characteristics while positioning it as a model for sustainable development in similar contexts.

In particular, the context was also selected for its exemplary plurality, reflecting the characteristics of many Italian peripheral areas, where processes of abandonment, "return," and fragility coexist with latent potential and innovative energies (Lanzani and Curci 2018).

But also, a territory results from a "multiple genealogy" (De Rossi and Mascino 2020) and a long-lasting process of co-evolution between environmental peculiarities and anthropogenic action. A relationship that, particularly in the Italian context—in the close relationship between forms of living, care for the territory and places, and economic dimension production—has allowed to decline and "condense" through practices of sharing and community cooperation, the "morphological infrastructure" of territories (De Rossi and Mascino 2020, p. 500) in the forms of the "cultural landscape".

However, it is a process that has expressed itself in the form of precarious living in terms of socio-economic conditions such as jeopardizing its original environmental resilience, a living that has not withstood the impact of modernity and, more in particular, of urbanization throughout the twentieth century (Bevilacqua 2018), witnessed in Italy by the progressive decline and abandonment of hill and mountain colonization and its landscapes.

But it is precisely on these landscapes, on their plurality and fragility (Tarpino 2016), that in recent years, attention has again been focused, even before on behalf of the politics, on the one end on behalf of the "remaining" subjects (Teti 2011) who had never abandoned the "grip" on those territories and, on the other end, on behalf of some newcomers or "returning" inhabitants. The latter, who have put into practice alternative projects and actions of settlement and production, are certainly heterodox and multidimensional compared to the competitive homologation of forms of production and consumption of the "global" market. Projects and actions reveal themselves to be strategic in a more general framework set by the unsustainability and inadequacy of the prevailing strategies of settlement and economic development compared to the challenges posed by the effects of global warming. This is not only in environmental terms but also from the social and economic point of view and, in a broader sense, relating to the requirements to achieve human settlement regeneration (Magnaghi 2020).

This has been reflected, from the point of view of the research activity synthetically presented in this contribution, into the identification of an interpretative and evaluation methodology suitable for grasping the integrated and multidimensional, as well as multipurpose, value of re-inhabiting initiatives and innovative practices in the study area. Initiatives capable of expressing the close relationship between regenerative use of resources, equity of social relations, and

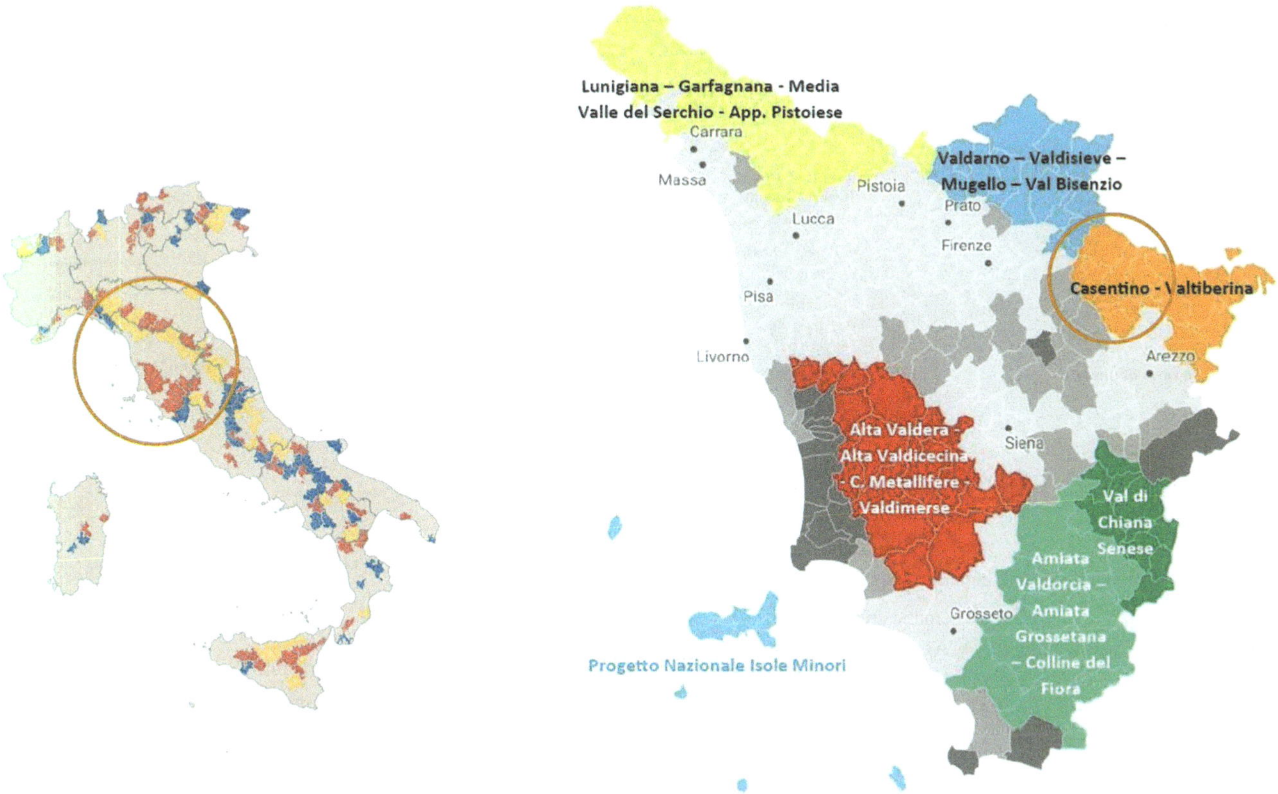

Fig. 1 The Casentino Valley in Tuscany, Italy (Fotina 2023; Regione Toscana 2021)

exchange and social cooperation with the "enhancement" and innovation of the cultural landscapes of Casentino. The methodology has also been devised with the aim of its transferability, that is, suitable for adapting to the plurality of territories of inner areas and to grasp, in any case, the specificities of the different territories in the Italian context.

This paper's specific objective is to illustrate the development and application of this interdisciplinary methodology. This includes analyzing best practices for evaluating the sustainability of cultural heritage regeneration interventions and critically assessing the success factors and limitations of regeneration processes and projects. By focusing on these elements, the paper aims to provide actionable insights that inform the research and contribute to designing strategies adaptable to other inner areas, ultimately supporting sustainable and inclusive development initiatives.

1.2 The Article Structure

In more specific terms of content, the contribution focuses on one of the first tasks of the REACT research project and, in particular, on defining a methodology relevant to guide the identification and critical analysis of best practices relating to projects for the regeneration of inner areas in Italy. To this aim, the article presents in the following section the original analysis methodology that was developed for this research, primarily characterized by a matrix of meta-criteria and criteria for the progressive selection and evaluation of best practices identified as being of relevance for the thematic areas on which the research is aimed to set up specific development guidelines. Section 3 then summarizes the main results regarding design strategies that emerged from this analysis, and finally, Sect. 4 presents and comments on the main findings earned and lessons learnt from the selected best practices. That mainly refers to limits and perspectives to outline—in the next REACT research steps—guidelines to devise an endogenous and multipurpose approach to regenerating inner areas' cultural landscapes.

2 The Methodology to Identify and Study the Best Practices

The methodological pathway that led to the definition of the best practices was a collaborative effort that ped in several stages, from the identification of 46 cases of interest to the study of 20 best practices, and finally to the selection of 4 best practices that were further investigated through interviews and on-site visits (Fig. 2).

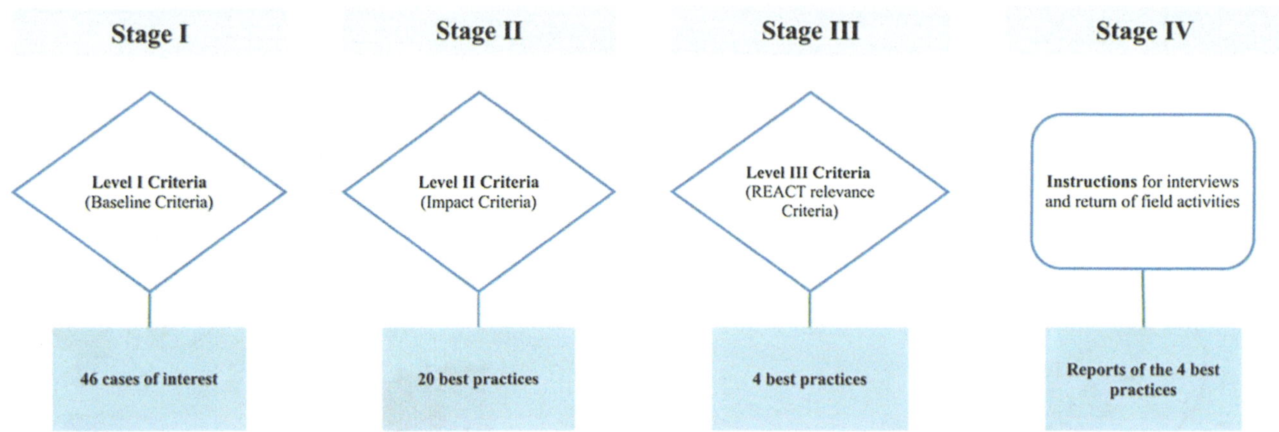

Fig. 2 Stages of the research (by authors)

Table 1 Operational units and thematic areas (by authors)

Operational units	Thematic areas
DISEI for the Department of Economics	A. Agri-food and forestry heritage and local crafts
FORLILPSI for the Department of Education	B. Social traditions and practices
DIDA for the Department of Architecture	C. Landscape and territory
	D. Settlements and buildings

The process involved three different operational units in parallel, which followed the methodological flow and identified the best practices for the thematic areas (TA) in which they had decided to work (Table 1).

2.1 Stage I. Delimitation of the Field of Investigation: Identification of the Pool of Cases of Interest

The work of delimiting the scope of the investigation and identifying the pool of cases of interest began by asking REACT Research Group members to report cases of interest, i.e., initiatives consistent with the scope and purpose of REACT research. Level I Criteria (Baseline Criteria) were applied to identify these cases of interest, which met the following priority parameters:

- Location: Italian inner areas.
- Territorial context: hilly and/or mountainous valley system and plural nature of the territory studied regarding economic vocation and socio-geographical diversity.
- Development strategy: endogenous forms of enhancement of socio-economic heritage and cultural landscape.

Specifically, REACT Research Group members reported 46 cases of interest (22 from the Department of Architecture (DIDA), 13 from the Department of Economics (DISEI), and 13 from the Department of Education (FORLILPSI) consistent with the above priority parameters.

These initiatives were the subject of the working group's in-depth critical work and information integration.

2.2 Stage II. Identification of Twenty Best Practices Based on Their Compliance with Selection Metacriteria and Criteria

At this stage, 20 best practices were selected for desk analysis from the 46 cases of interest indicated by the REACT Research Group members that met the Level I Criteria.

To identify the 20 best practices, an original procedure was prepared based on Selection Metacriteria and Criteria deemed appropriate to highlight the qualifying elements of the regeneration initiatives implemented in Italian inner areas. To this end, Level II Criteria (Impact Criteria) were identified. Specifically, 6 Metacriteria were identified, and each metacriterion was broken down into criteria (Table 2). The matrix is inspired to holistically support the multidisciplinary approach of the research holistically, contributing to a reflection on the cultural landscape, considering the urban, economic, organizational, and educational-cultural dimensions.

To evaluate, compare, and select the 20 best practices, each of the 46 cases of interest was studied and analyzed using the Metacriteria and Criteria and represented through a radar chart. The graph illustrates, for each case of interest, the simple average of the criteria calculated within each Metacriterion (Fig. 3).

The comparison of radar charts enabled the selection of the 20 best practices based on the following indicators:

- Initiatives that can respond to multiple metacriteria and, therefore, radar charts that cover a larger area.
- Initiatives that stand out because they respond to specific metacriteria and, therefore, radar charts that have higher peaks and values in particular metacriteria.

Through comparison, the territories where the 20 best practices are included were considered complex systems suitable for analysis from a multi-sectoral perspective. This

Table 2 Criteria and parameters for selecting best practices—Stage II (by authors)

Best practices react _ desk analysis		
Level II criteria (impact criteria)		
Stage II	**Metacriterion**	**Criteria**
Identification of 20 best practices	II.1 Forms of community organization	II.1.1 Community-based management of resources and services II.1.2 Associations for the promotion of the territory II.1.3 Bottom-up initiatives II.1.4 Self-help networks
	II.2 Forms of governance and policy processes	II.2.1 Multisectorality/integration II.2.2 Multiscalarity II.2.3 Contractual and/or partnership nature (PPP) II.2.4 Propensity towards interterritorial cooperation
	II.3 Mobilization of local resources, social capital, and cognitive assets	II.3.1 Productive retro-innovation in agriculture II.3.2 Recovery of contextual knowledge II.3.3 Creation of personal services II.3.4 Digital transition II.3.5 Multiculturalism II.3.6 Education for sustainable development II.3.7 Education/training/vocational education II.3.8 Creation of services for businesses II.3.9 Initiatives for active ageing II.3.10 Actions for self-reliant endogenous development
	II.4 Generation and distribution of economic value	II.4.1 Access to funding and financial resources: Financial sustainability II.4.2 Economic categories involved and governance modes: Inclusivity and participation II.4.3 Types of goods and services generated and their connection to specific local resources II.4.4 Value creation through the market: Qualification mechanisms II.4.5 Value creation through the market: Forms of sale II.4.6 Value creation through other mechanisms II.4.7 Distribution of value among categories of subjects and territories II.4.8 Ability to remunerate/regenerate the resources used (especially specific local resources) II.4.9 Territorial spillover effects
	II.5 Durable protection/activation of environmental resources	II.5.1 Initiatives for sustainable maintenance and care of the territory II.5.2 Protection and regeneration of biodiversity II.5.3 Development of renewable energy, emission neutrality II.5.4 Processes for circular use of resources
	II.6 Recovery, regeneration, and Innovation of the cultural landscape	II.6.1 Recovery of patterns from historic settlement structures (networks of slow mobility and rural settlements) II.6.2 Recovery and enhancement of buildings and works historically functional to forestry, agricultural, and/or artisanal and proto-industrial activities II.6.3 Recovery and enhancement of specific traditional agrarian and forest landscapes II.6.4 Recovery and enhancement of buildings and public spaces of significant cultural and community value II.6.5 Redevelopment of industrial/productive areas II.6.6 Contemporary interventions for the aesthetic and functional enhancement of the landscape II.6.7 Regeneration initiatives of heritage to reverse depopulation processes

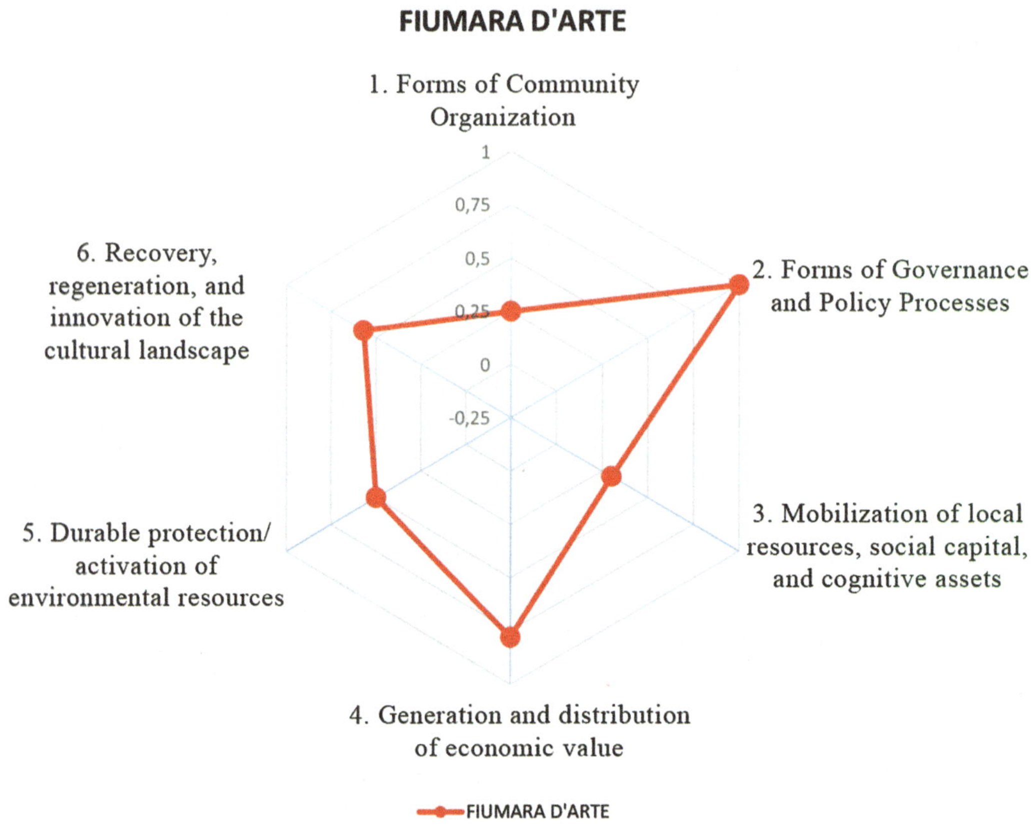

Fig. 3 Example of radar chart (by authors)

brings out endogenous processes, human capital development, enhancement of territorial resources, and examples of community activation that, better than other cases, provide valuable elements for their possible replicability (Table 3).

The 20 best practices identified in Stage 2 were finely analyzed by filling out a specially prepared Analysis Sheet. The Analysis Sheet comprises six parts divided into sections, as shown in Table 4.

2.3 Stage III. Identification of Four Best Practices Based on Their Relevance to the REACT Research

This stage assessed the relevance of the 20 analyzed best practices to the objectives of the REACT research. For this purpose, the best practices were compared with the 4 thematic areas and 7 intervention strategies of the REACT research, applying the Level III criteria (REACT relevance Criteria):

- 1.1 Strategies for education and training
- 1.2 Strategies for planning and project
- 1.3 Strategies for social cohesion
- 1.4 Strategies for improving public services
- 1.5 Strategies for knowledge and preservation of cultural heritage
- 1.6 Strategies for the enhancement of cultural heritage
- 1.7 Strategies for management

First, the prevailing thematic area and, if present, the other relevant ones were reported for each best practice. Next, the relevance of each best practice to the REACT Research Intervention Strategies was identified.

Based on the degree of coherence with the thematic areas and the ability to respond to multiple intervention strategies

Table 3 Complete list of cases of interest by the three operating unit (OU) REACT and, in bold, best practices selected through the metacriteria and criteria matrix (by authors)

OU	Case	Name of the initiative	Location
DIDA	**01**	**FIUMARA D'ARTE**	**Provincia di Messina, Sicilia**
	02	**RI-AMBIENTIAMOCI**	**Provincia di Belluno, Veneto**
	03	**COLLETTA DI CASTELBIANCO**	**Castelbianco (SV), Liguria**
	04	**BORGATA PARALOUP**	**Rittana (CN), Piemonte**
	05	**OSTANA**	**Ostana (CN), Piemonte**
	06	TEATRO DI ANDROMEDA	Santo Stefano Quisquina (AG), Sicilia
	07	UN PARCO MILLE SOGNI	Provincia de L'Aquila, Abruzzo
	08	SENTIERI APPENNINO MARCHIGIANI	Provincia di Ascoli Piceno, Marche
	09	SCUOLA DI ASTRA ZARINA	Civita di Bagnoregio (VT), Lazio
	10	MATer. ORME DI PIETRA NEL PAESAGGIO	Mamoiada (NU), Sardegna
	11	VAL DI SETTA A BASSA VELOCITÁ	Provincia di Bologna, Emilia-Romagna
	12	PROGETTO BORCA	Borca di Cadore (BL), Veneto
	13	**I LIVE IN VACCARIZZO**	**Montalto Uffugo (CS), Reggio Calabria**
	14	BOVA	Bova (RC), Reggio Calabria
	15	OLLOLAI CAPITALE	Ollolai (NU), Sardegna
	16	APPENNINO REGGIANO	Provincia di Reggio Emilia, Emilia-Romagna
	17	ALTA VALLE GRANA	Provincia di Cuneo, Piemonte
	18	FARANETO	Piacenza (RE), Emilia-Romagna
	19	CAIRANO	Cairano (AV), Campania
	20	**GANGI**	**Gangi (PA), Sicilia**
	21	**CASTEL DEL GIUDICE**	**Castel del Giudice (IS), Molise**
	22	VAL D'ULTIMO	Provincia di Bolzano, Trentino-Alto Adige
DISEI	**23**	**COMUNITÀ DEL CIBO DELLA GARFAGNANA**	**Provincia di Lucca, Toscana**
	24	**PULMINO CONTADINO**	**Provincia di Grosseto, Toscana**
	25	**CASEIFICIO SOCIALE ALPE DEL GARDA**	**Tremosine (BS), Lombardia**
	26	**TURSI DIGITAL NOMADS**	**Tursi (MT), Basilicata**
	27	**TUTELA E VALORIZZAZIONE DELLA CILIEGIA DI LARI**	**Lari (PI), Toscana**
	28	**VILLAGGIO GLOBALE DI RIACE**	**Riace (RC), Calabria**
	29	**COOPERATIVA DI COMUNITÀ "VALLE DEI CAVALIERI"**	**Ventasso (RE), Emilia-Romagna**
	30	TREBALLU RURAL COMMUNITY	Laconi (OR), Sardegna
	31	APT PAGANELLA	Provincia di Trento, Trentino-Alto Adige
	32	COOPERATIVA DI CONSUMO VESIO	Tremosine sul Garda (BS), Lombardia
	33	COOPERATIVA CERROTORRE	Flero (BS), Lombardia
	34	BISCOTTO SALATO DI ROCCALBENGA	Roccalbegna (GS), Toscana
	35	MARCHI COLLETTIVI GEOGRAFICI DELLA VALTELLINA	Provincia di Sondrio, Lombardia

(continued)

Table 3 (continued)

OU	Case	Name of the initiative	Location
FORLILPSI	**36**	**PRESIDIO DEL PATTO DEL FIUME SIMETO**	**Provincia di Catania ed Enna, Sicilia**
	37	**PROGETTO STUDY CIRCLES-UE**	**Provincia di Udine, Friuli-Venezia Giulia**
	38	**MIM. MONTAGNE IN MOVIMENTO**	**Gagliano Aterno (AQ), Abruzzo**
	39	**SISTEMA MUSEALE MONTAGNA PISTOIESE**	**Provincia di Pistoia, Toscana**
	40	**CONSORZIO MO.TO.R.E.**	**San Marcello Piteglio (PT), Toscana**
	41	ECOMUSEO DELLE ALPI APUANE	Provincia di Lucca, Toscana
	42	FALIA	Provincia di Brescia, Lombardia
	43	MUSEO DELLE MINIERE DEL MONTE AMIATA	Santa Fiora (GR), Toscana
	44	CASERMARCHEOLOGICA	Sansepolcro (AR), Toscana
	45	FONDAZIONE CDSE	Provincia di Prato, Toscana
	46	FONDAZIONE ARCHIVIO DIARISTICO NAZIONALE	Pieve Santo Stefano (AR), Toscana

Table 4 Structure of the best practices analysis sheet (by authors)

Part I	Anagraphy	
	I.1	Name of the initiative
	I.2	Location
	I.3	Inner area
	I.4	Author of the sheet
Part II	Description of the initiative	
	II.1	Brief descriptions
	II.2	Objectives
	II.3	Sponsoring parties
	II.4	Genesis
Part III	Selection metacriteria and criteria	
	III.1	Metacriteria-criteria matrix
	III.2	Radar chart
	III.3	Description of effective criteria
Part IV	Initiative analysis	
	IV.1	Strengths and weaknesses
	IV.2	Lessons learned
Part V	Relevance to REACT research	
	V.1	Relevance to REACT thematic areas
	V.2	Relevance with REACT intervention strategies
Part VI	Sources	
	VI.1	Bibliography
	VI.2	Sitography

among those identified by the REACT research, four best practices, one for each thematic area, were selected for further investigation, including through field analysis.

The identification of the four best practices was made by following two successive steps (Fig. 4):

- Breakdown of the 20 initiatives analyzed by prevailing thematic area.
- One best practice was selected for each Thematic Area. The best practice with the most relevant links to the seven REACT Intervention Strategies was selected.

Table 5 Presents a synoptic framework of the four best practices in relation to the operational units and thematic areas.

For each of the four selected best practices, themes that emerged during the desk analysis were explored through field analysis (field visits and in-depth interviews). In addition, cognitive integration was carried out through scientific literature and governance tools.

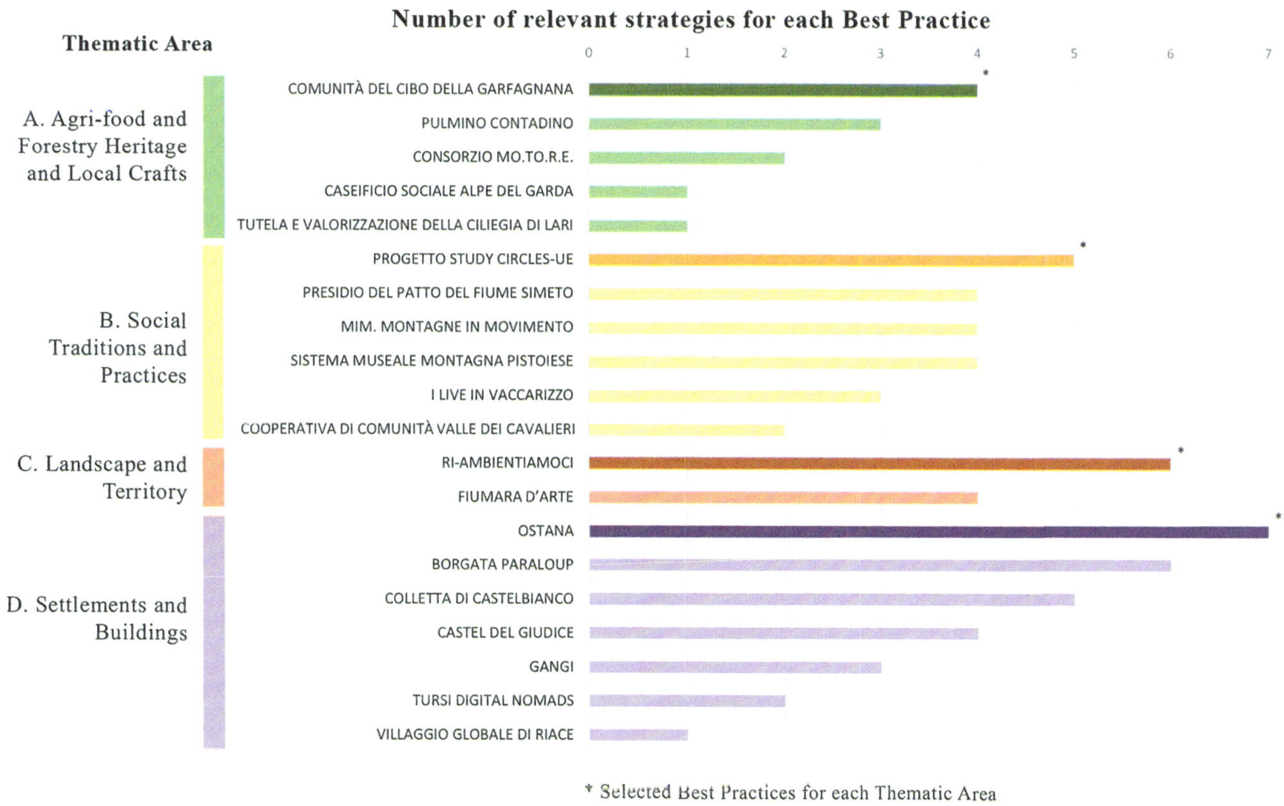

Fig. 4 Relevant strategies for each best practice (by authors)

Table 5 Operational units, thematic areas and the four selected best practices (by authors)

Operational units	Thematic areas	Four best practices
DISEI for the Department of Economics	Agri-food and forestry heritage and local crafts	Comunità del Cibo della Garfagnana
FORLILPSI for the Department of Education	Social traditions and practices	Progetto study circles-UE
DIDA for the Department of Architecture	Landscape and territory	Ri-Ambientiamoci
	Settlements and buildings	Ostana

3 Research Outcomes

3.1 The 20 Selected Best Practices

In examining revitalization strategies for the 20 best practices summarized in Fig. 5, several common characteristics and recurring project actions emerge from successful case studies:

- Community engagement: active involvement of local communities in planning and implementation processes.
- Cultural and natural heritage: emphasis on preserving and promoting local cultural, historical, and natural assets.
- Sustainability: focus on environmentally sustainable practices and renewable energy sources.
- Economic diversification: efforts to diversify the local economy, often through tourism, agriculture, and small-scale artisanal industries.
- Education and training: investment in education and vocational training to enhance local skills and employability.
- Public–private partnerships: collaboration between public authorities, private enterprises, and non-profit organizations.
- Innovation and technology: incorporating modern technologies and innovative approaches to traditional practices.

The thematic area "Settlements and Buildings" is where the best practices are distributed (seven best practices), whereas the "Landscape and Territory" area reports the fewest cases (two best practices). Regarding territorial distribution, the 20 selected best practices are spread across the entire peninsula, with a concentration in the northern and southern regions of the country (Fig. 5).

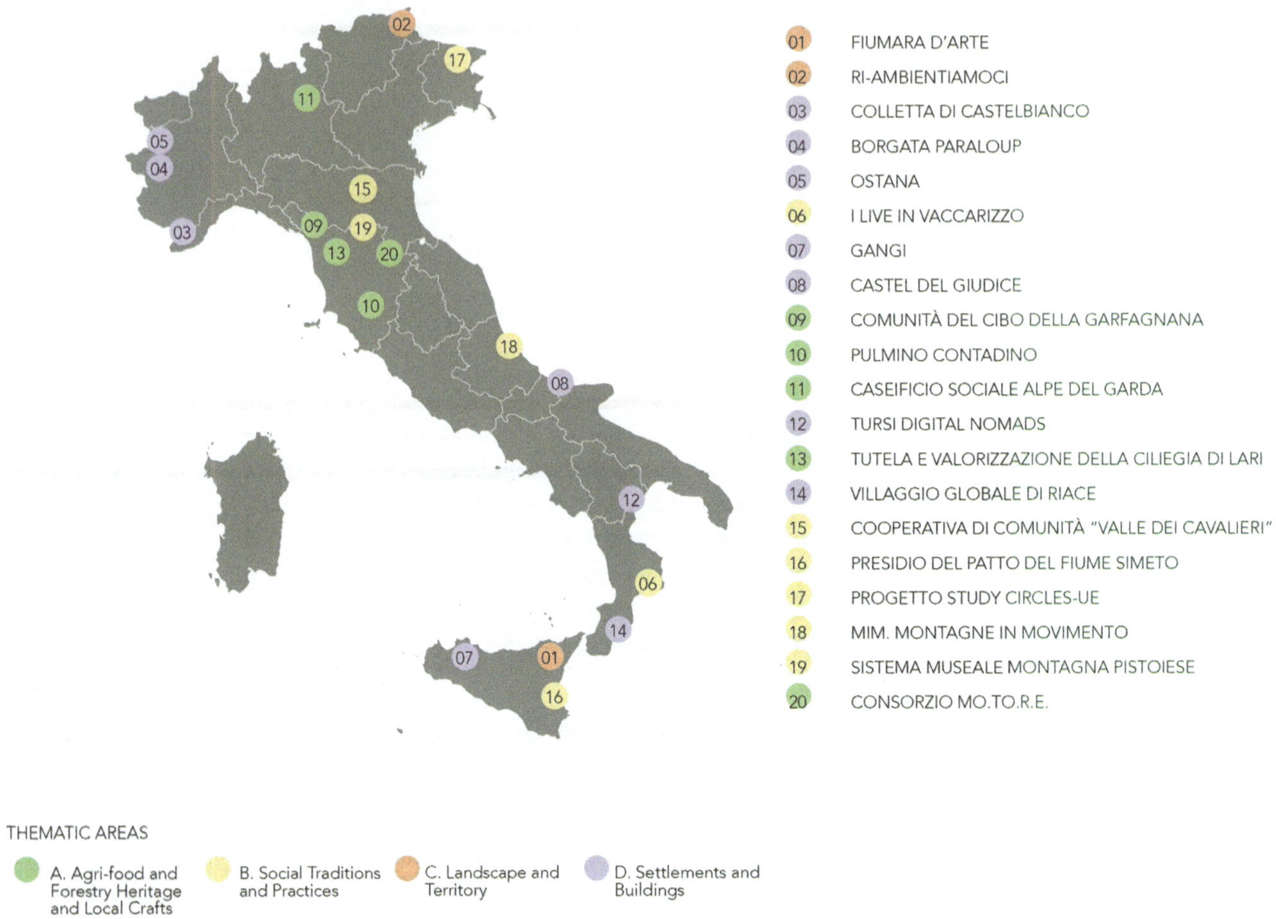

Fig. 5 Synoptic map of the territorial distribution of the 20 selected best practices (by authors)

Specifically, the following best practices were analyzed for the different thematic areas.

The following five best practices were analyzed for the thematic area "A. Agri-food and Forestry Heritage and Local Handicrafts." The "Garfagnana Food and Agrobiodiversity Community" (Tuscany) is a territorial governance system that brings together a diversity of local actors to enhance the Garfagnana's agri-food heritage and agrobiodiversity sustainably. "Pulmino Contadino" (Tuscany) is an association that brings together producers and consumers in the area between the "Colline Metallifere, Piombino and Grosseto", creating a circuit of small-scale organized distribution of agri-food products. "Caseificio Sociale Alpe del Gard" (Lombardy), a cooperative of dairy producers, which processes the milk of its members and handles marketing. "Tutela e Valorizzazione della Ciliegia di Lari" (Toscana), a shared valorization path of local cherry varieties, led to obtaining the PGI denomination. "I live in Vaccarizzo" encompasses projects aimed at repopulating the village of Vaccarizzo in the province of Cosenza. This initiative originated from a "Social Transformation Laboratory" conducted in 2019 as part of an international experiment by the Massachusetts Institute of Technology (MIT) in Boston.

At the same time, five best practices were selected for the Thematic Area "B. Social Traditions and Practices," which, better than others, highlight virtuous practices due to their emphasis on community engagement, intangible cultural heritage, innovation, and sustainable development, revitalizing rural communities. The "Presidio of the Fiume Simeto Pact" is a community initiative to safeguard, protect, and enhance the Simeto river valley in Sicily. It involves a network of associations and municipalities in the provinces of Catania and Enna. The "Study Circles-UE Project" is a cross-border initiative to develop human competencies and cooperation networks to promote local resources focusing on knowledge exchange. "Montagne in Movimento (MIM)" in Gagliano Aterno, a remote mountainous community in Abruzzo, involves the community in promoting endogenous energy transition processes. The "SIMOP (Sistema Museale Montagna Pistoiese)" in the mountainous region of Pistoia,

Tuscany, is a network of museums and cultural sites that preserve and promote the area's rich cultural and environmental heritage. Finally, the "Consorzio Mo.To.R.E. (Montagna Toscana Ricerca Energie)" is a business consortium that develops technologies and promotes knowledge related to renewable energies and the forest economy.

Two exemplary practices were selected for thematic area "C. Landscape and Territory," showcasing diverse strategies and measures to enhance landscape resources in Italy's inner areas. The first, "Fiumara d'Arte", involves the establishment of an open-air land art museum in the province of Messina, Sicily. This initiative, spearheaded by a private individual, commenced in 1982. The second practice, "Ri-ambientiamoci", is situated in Veneto near the Austrian border. Initiated in 2019, this project aims to transform the damage caused by the Vaia storm into an opportunity for community rejuvenation using art as a narrative tool.

Seven best practices have been selected for the thematic area "D. Settlements and Buildings". These practices address the theme of regenerating existing heritage and explore various methods to create social, economic, and environmental value around them. "Colletta di Castelbianco" (Piemonte) is the transformation of a medieval village into a telematic village, which was achieved while preserving the historical and architectural identity of the area. It serves as an example of how urban regeneration projects can address new needs and activate a new community of residents. Narration of memory can be a crucial strategy for enhancing marginal territories. An example of this is the "Borgata Paraloup" (Piemonte), which has utilized various fields to demonstrate how the history of a community can transform landscapes, convey ideas and values, and share modes of communication that transcend territorial boundaries. "Ostana" (Piemonte) exemplifies how a small, formerly abandoned village in an inner area has managed to attract new residents and counteract the negative depopulation trend, mainly due to the collaboration with the University of Turin. "Gangi" is in the interior of the Madonie mountains (Sicily). The municipality has undertaken several successful projects to counteract depopulation and develop a new economic pathway integrating agriculture, livestock farming, and tourism. "Castel del Giudice," a small municipality in the Molise Apennines, represents a virtuous and sustainable example of community-based entrepreneurship for inner areas. The implemented projects demonstrate a productive synergy that has led to the establishment of a public–private participatory management model capable of revitalizing the economic and social life of the community in the village. "Tursi Digital Nomads" (Basilicata) is a project for the valorization and repopulation of the small village of Tursi through the attraction of "digital nomads", i.e., people working remotely in smart working. "Villaggio Globale di Riace" (Calabria) is an innovative project for integrating refugees based on their involvement in restoring abandoned housing and work activities to repopulate and revitalize the village.

3.2 The Four Selected Best Practices

The four selected best practices are:

- **Garfagnana Food and Agro-biodiversity Community (A. Agri-food and Forestry Heritage and Local Crafts)**
  The Garfagnana Food and Agro-biodiversity Community (below FAC), founded in 2017 as a pilot project of the Union of Municipalities of Garfagnana and promoted by the Region of Tuscany within the 2014/2020 Rural Development Plan, was created within the framework of national law no. 194/2015 for the protection and enhancement of agricultural and food biodiversity. It brings together 54 member subjects, including agricultural farms, agri-tourism, operators of the agri-food system, associations, citizens, and local institutions. Food communities respond to the need for a territorial coordination and governance system related to food and agriculture. This generates an integrated and participatory approach to sustainable territorial development of inner rural areas. The FAC aim is the sustainable valorization of the Garfagnana's agri-food heritage and agrobiodiversity, and its objectives include the study, recovery and transmission of knowledge on local genetic agri-food resources; the creation of short supply chains and local marketing circuits for the exchange and purchase of agri-food products; the dissemination of organic and low-impact farming practices; the transmission of traditional knowledge on crops; the creation of educational and social farms as tools for valorizing local varieties, promote environmental education and social aggregation, and the redevelopment of abandoned and degraded farmland. The FAC is an interesting example of how it is possible, through collaboration between different actors, to hold together a territory and its various needs and to be able to coordinate and work together to intercept economic resources to reinvest in the territory itself, building processes of territorial protection and valorization from the bottom up, considering communities and their needs first.
- **Study Circles-EU Project: Study Circles-Cross Border Laboratory for the Development of Human Resources and Cooperation Networks Promoting Local Resources (B. Social Traditions and Practices)**
  The "Study Circles-EU" project, aimed at developing human resources and cooperation networks to promote local resources, was selected as a best practice among case studies related to "Social Traditions and Practices". It aligns with the REACT research intervention strategies. Funded under the Italy-Slovenia Cross-Border Cooperation Program 2007–2013, the project revitalized

the "study circles" model for cross-border education and local development, creating a network of educational and training agencies, local development agencies, and institutions from Slovenia, Veneto, and Friuli-Venezia Giulia (FVG). Themes included sustainable agriculture, local product promotion, heritage conservation, revitalization of traditional crafts, and enhancing local knowledge and identity values for territorial tourism and marketing. Led by the National Entity for Professional Training and Education (ENFAP) of FVG, in collaboration with the Department of Educational Sciences and Psychology of the University of Florence, the University of Venice, and the Slovenian Adult Education Centre, the project focused on the inner area located in the valleys of Torre and Natisone, provinces of Gorizia and Udine. The study circle, a pedagogical device used in non-formal adult education, is widely recognized as highly inclusive, supportive of the learner's active participation, and community-oriented (Bogataj and Del Gobbo 2015). A mentor leads each Study Circle; thanks to the project, the professional figure of the mentor has been defined and classified among the experts in training and curricular planning (Del Gobbo and Federighi 2021). An in-depth interview revealed the importance of ENFAP and the Local Action Group (GAL) synergy in activating the community and the significance of mentor training. The project's success lies in developing human capital, job creation, and cultural heritage enhancement, with academic involvement elevating the work quality.

- **Ri-Ambientiamoci (C. Landscape and Territory)**
  The "Ri-ambientiamoci" initiative has undergone comprehensive analysis within the Thematic Area "Landscape and Territory". Launched in 2019 in Veneto, near the Austrian border, "Ri-ambientiamoci" aims to transform the scars left by the Vaia storm into an opportunity for reflection on climate change and for enhancing the landscape's historical, cultural, and naturalistic understanding. The project focuses on improving forest resources through art, striving to balance economic value creation and environmental sustainability. Thanks to the collaborative efforts of the entire community, this initiative, which emerged from a catastrophic event, has successfully turned a disadvantage into an opportunity. The comprehensive analysis included direct participation of project leaders during site inspections in the affected areas. The initiative demonstrates how the complexity of the landscape dimension encompasses a wealth of aspects capable of uniting the community. Driven by pressing needs, the community is motivated to engage and develop new landscape management and enhancement methods. Field activities have highlighted how the expected outcomes during the planning phase conflicted with the realities of communities still entrenched in outdated approaches to forest resource utilization. The "*Regole*", traditional community-based organizations in Cadore, collectively own and manage agro-silvo-pastoral assets such as forests, meadows, and pastures. These properties, passed down through generations, are governed by democratic rules outlined in ancient rural codes called "*Laudi*". This model could offer a valuable framework for managing the silvo-forestry areas of inner regions like the Comelico valleys. However, the current approach remains too tied to tradition and insufficiently flexible to adapt to ongoing changes, preventing innovative initiatives like "Ri-ambientiamoci" from taking root. Another significant factor affecting the outcomes of similar initiatives in this area is regional policies, which have historically invested little in these peripheral territories. As a result, these areas struggle to be recognized as strategic for the country's development and consequently suffer from widespread depopulation.
- **Ostana (D. Settlements and Buildings)**
  Ostana is a small village located at the foot of the mountain Monviso, a virtuous example of rebirth and repopulation. The protagonist of the success was the Municipal Administration, which, starting in the 80 s, prevented the proliferation of new buildings by renovating existing ones and promoting local culture. Thanks to various grants and the involvement of the Politecnico di Torino, the quality of living in Ostana has increased through a series of projects that have improved the modernity of buildings and generated new cultures and businesses while respecting the landscape identity of the place. The numerous interventions of restoration of buildings, spaces, and public equipment for tourism, culture, or local entrepreneurship (e.g., cultural center, school, guest house, wellness center, museum, bakery, dairy, etc.) involved the community and reused local materials. "Viso a Viso", a community cooperative born in 2020 from a public participation process, manages most of the regenerated structures and guarantees goods and services to the community. It is an innovative model to produce livability, culture, and employment, building a contemporary model for the mountain. The case shows that revitalizing a place needs time, partnership, participation, and targeted investments in facilities, services, and communities. External subjects, carriers of different competencies or experiences, successfully mixed with the original inhabitants. The success of architectural design in the area's regeneration was due to complementary revitalization paths and a network of public interventions. The reactivation and reuse of heritage were combined with processes to revitalize local communities, establishing a welfare infrastructure (services, culture, tourism, production, and wellness facilities). Nearly

Table 6 Common strengths and weaknesses of the four selected best practices (by authors)

Four best practice: common strengths and weaknesses	
Strengths	Weaknesses
Community involvement – Participation of a diverse range of stakeholders with varying roles and expertise (BP1) – Inclusion and accessibility for different types of users (BP2) – Active engagement of the community (BP3) – Community involvement in decision-making processes (BP4) **Public–private collaboration** – Network of relationships with local entities (BP1) – Involvement of local public and private actors for integrated development (BP2) – Collaboration between public and private sectors (BP3) – Cooperation and co-design between public and private sectors (BP4) **Enhancement of local resources** – Richness of the territory in terms of agro-biodiversity and Agri-food heritage (BP1) – Promotion of the cultural heritage of the Valli del Torre and Natisone (BP2) – Creation of an artistic route and revitalization of the timber supply chain (BP3) – Enhancement of architectural and landscape heritage, along with the recovery of traditional knowledge (BP4) **Institutional and regional support** – Support from the Tuscany region and GAL Montagna Appennino (BP1) – Collaboration with the polytechnic University of Turin and promotion of community enterprises (BP4)	**Economic-financial sustainability:** – Difficult economic-financial sustainability (BP1) – Lack of stable funding (BP2) – Managerial issues due to external factors such as the pine beetle (BP3) – Dependence on public support and tourism (BP4) **Community participation and cohesion** – Low collaboration and participation among members, lack of a strong common identity (BP1) – Initial resistance to new learning methods (BP2) – Conflicts between residents and newcomers, difficulties in managing community dynamics (BP4) **Training and professionalism** – Lack of training for farmers (BP1) – Need for adequately trained professionals (BP2) – Economic difficulties for local businesses in a context with unsuitable regulations (BP4)

40 years of successful policies and strategies, a collective product of internal and external players, have resulted in an interweaving of bottom-up planning, local cultures, networks of expertise, and cooperative systems.

3.3 Strengths and Weaknesses of the Best Practices

The four best practices examined exhibit a range of strengths and weaknesses. These practices, implemented in various local contexts, provide valuable insights for replicating positive experiences and addressing everyday challenges. Table 6 outlines the main strengths and weaknesses of each practice, followed by a summary highlighting common elements.

The analysis of the four case studies highlights a complex landscape of strengths and weaknesses that shape the dynamics of community-based initiatives and territorial development. These case studies serve as valuable examples of how community involvement, public–private collaboration, enhancement of local resources, and institutional support can support inclusive growth and sustainable development. They reveal the transformative potential of diverse stakeholder participation, including various user groups and active community engagement in decision-making processes. Furthermore, the integration and collaboration of public and private actors, the valorization of local cultural and natural assets, and the strategic role of institutional support underline the multifaceted approach required for success.

However, these practices are not without challenges. Economic and financial sustainability remains a critical issue, with difficulties in securing stable funding and managing external factors that affect viability. Additionally, low collaboration, identity struggles, and conflicts among stakeholders hinder community participation and cohesion. Training and professionalism also emerge as pivotal factors, with gaps in expertise and regulatory barriers posing significant obstacles.

These aspects underscore the importance of integrated and sustainable strategies to ensure the continuity and success of community initiatives and territorial development, promoting inclusive growth and valuing local resources.

4 Lessons Learned and Future Perspectives

The research presented in this paper has shown that, despite many initiatives aimed at revitalizing inner areas and bridging the gap caused by their physical and socio-economic marginality, we are still far from implementing a systematic project on the whole national territory. Best practices often remain isolated, whereas moving from episodic individual interventions to real public policy would be necessary. As mentioned above, the National Strategy for Inner Areas was heading in this direction but has not yet produced the desired results. A place-based and integrated approach is therefore strongly encouraged.

The analysis of the 20 best practices (and of four studied in more detail) has revealed certain limitations attributable to the following points, from which it is worth starting to identify possible lines of action:

- A still significant demographic gap that characterizes the inner areas makes the social and economic fabric very fragile and less responsive to the systematic engraftment of policies and innovative practices on a large scale.
- The difficulty of developing effective and lasting strategies at the territorial and landscape scale (especially in agroforestry management), compared to settlements and buildings (we observe only two best practices against seven of those studied). This is probably because this scale of intervention covers larger territories and a broader and more complex set of potential actors (public and private) than the building scale.
- The difficulty of developing and managing integrated strategies over time, covering several sectors of economic and public administration (tourism, slow mobility, agriculture, crafts, environment, water management, cultural heritage, etc.).
- The lack of stable funding channels dedicated to developing such integrated strategies across all scales.
- The dependence of these virtuous projects on the initiative of individuals or small self-organized groups, often "outsiders" in relation to the context of intervention, allows them to see new opportunities and perspectives, especially in the field of common goods management. This may entail the risk of experience fading once the propulsive thrust of its promoters is exhausted.

To help overcome these limits, we believe it can be helpful to adopt an approach that puts the landscape and its quality at the center of public policies (Albrechts et al. 2020). In this perspective, the landscape is intended as a self-sustainable, integrated, and participatory local development engine. Moreover, it allows for the increase of consciousness in a new place (Magnaghi 2020). For this vision to be realized, it is essential that landscape policies are no longer seen as a sectoral field of public administration but as the place of integration of different spheres of intervention. Operationally, an essential first step towards making this route a reality could be to give impetus to the drafting of regional landscape plans, which could have a positive effect on the above limits concerning the following points:

- Draw polycentric scenarios in which the inner areas become less marginal in the region, starting from the redevelopment of existing settlements: improving accessibility to these inhabited centers and a better supply of basic services (starting from school and health) can be the first point of departure to draw a new geography no longer unbalanced on metropolitan areas but with a more balanced distribution of population. The regional scale is the most suitable for designing scenarios that can contribute to the repopulation of inner areas, as in a bioregional key (Fanfani and Matarán Ruiz 2020).
- Develop consistent analytical frameworks and intervention scenarios for territorial systems with specific reference to themes such as agro-forestry management and river system management. All these issues are highly complex both in terms of space and governance. Regional landscape plans could also stimulate the identification of forms of payment for ecosystem services for the fundamental functions that inner areas perform for the entire regional system (first, concerning hydrogeological balance and biodiversity preservation). Some examples can be traced in the experiences based on the co-management of common goods, which economically compensate farmers and breeders as "custodians of the territory" (Carrosio and De Renzis 2021).
- Federate sectoral policies and public administration fields of intervention, strengthening the intersectoral dialogue within the same body (e.g., the region) and between entities at different levels (Region, Municipalities, etc.). In Italy, the regional landscape plans have this role superordinate by law (D. Lsg. 42/2004) compared to other plans and programs that concern the regional territory. For this reason, they can be the place of synthesis of different policies (Gisotti and Tarsi 2023).
- Identify sources of funding for inner areas revitalization strategies independent of the administration's current spending, like EU funds that are distributed through the Regions. Strategic landscape projects, which are often included in regional landscape plans and deal with topics relevant to inner areas (such as slow mobility networks, green and blue infrastructures, and integrated cultural heritage systems), can be helpful for this purpose (Barbanente and Grassini 2021).
- Promote the development of a broad cultural and educational project on the landscape of the inner areas, conceived as an integrated common good, carrier of multifunctionality and values at the same time historical, ecological, economic, and aesthetic. This project could be developed within the participatory processes during regional landscape plan formation. It could promote dialogue between stakeholders involved in landscape transformations (public administrations, practitioners, entrepreneurs, farmers) and civil society (citizens, schools), and support transferring knowledge and innovations into current practice to move from episodic best practices to a real policy.

The REACT project, which identifies the cultural landscape as a key issue, will be able to develop some of these topics in its successive phases, dedicated to defining guidelines for the regeneration of inner areas.

Acknowledgments We acknowledge all the participants in the working group of the competitive research "REACT. Regenerating the cultural landscapes of inland areas in a people-centered perspective," CUP B55F21007810001, funded by the Next Generation EU and the Italian National Research Programme (NRP) 2021–2027, for their contribution to the results presented in this paper.

Authors' Contribution This paper is the result of a joint reflection by the authors. Section 1 was written by David Fanfani. Sections 2 and 3 by E. Naldi, C. Mezzapesa, M. Mengoni, G. Biagi, and P. Scarpellini. Section 4 by M.R. Gisotti.

The task T1.3 of the REACT research project "A survey of Italian inner areas based on the analysis of good practices" has been coordinated by David Fanfani and Maria Rita Gisotti and held by Giulia Biagi, Matteo Mengoni, Claudia Mezzapesa, Eletta Naldi, Paola Scarpellini, Leonardo Rosini, Elisa Betti, and Francesco De Maria.

References

Albrechts L, Barbanente A, Monno V (2020) Practicing transformative planning: the territory-landscape plan as a catalyst for change. City Territ Archit 7. https://doi.org/10.1186/s40410-019-0111-2

Barbanente A, Grassini L (2021) Fostering transitions in landscape policies: a multi-level perspective. Land Use Policy 112. https://doi.org/10.1016/j.landusepol.2021.105869

Barca F, McCann P, Rodríguez-Pose A (2012) The case for regional development intervention: place-based versus place-neutral approaches. J Reg Sci 52(1):134–152. https://doi.org/10.1111/j.1467-9787.2011.00756.x

Barca F, Casavola P, Lucatelli S (2014) Strategia Nazionale Per le Aree Interne. Definizioni, Obiettivi e Strumenti di Governance. Materiali UVAL 31:10

Bevilacqua P (2018) L'Italia dell'"osso": Uno sguardo di lungo periodo. In: De Rossi A (ed) Riabitare l'Italia. Le aree interne tra abbandoni e riconquiste. Donzelli, Roma, pp 111–122

Bogataj N, Del Gobbo G (eds) (2015) Lifelong learning devices for sustainable local development: the study circles experience in the cross-border area Italy-Slovenia. ETS Edizioni

Carrosio G (2019) I margini al centro: L'Italia delle aree interne tra fragilità e innovazione. Donzelli, Roma

Carrosio G, De Renzis A (2021) Nelle aree interne: Una corretta gestione e valorizzazione del capitale naturale. In: Coppola A, Del Fabbro M, Lanzani A, Pessina G, Zanfi F (eds) Ricomporre i divari: Politiche e progetti territoriali contro le disuguaglianze e per la transizione ecologica. Il Mulino, Bologna, pp 47–56

Cersosimo D, Ferrara AR, Nisticò R (2018) L'Italia dei pieni e dei vuoti. In: De Rossi A (ed) Riabitare l'Italia. Le aree interne tra abbandoni e riconquiste. Donzelli, Roma, pp 21–50

De Rossi A, Mascino L (2020) Valades Ousitanes, architettura e rigenerazione. ARCHALP 4

Del Gobbo G, Federighi P (2021) Professioni dell'educazione e della formazione: Orientamenti, criteri e approfondimenti per una tassonomia. FUP, Firenze

ESPON (2017) Inner peripheries: national territories facing challenges of access to basic services of general interest. Luxembourg

Fanfani D, Matarán Ruiz A (eds) (2020) Bioregional planning and design, vol I. Perspectives on a transitional century. Springer, Cham

Fotina C (2023) The new map of inner areas: 2 billion euros allocated for 1,904 municipalities [Map]. Il Sole 24 Ore. https://www.ilsole24ore.com/art/la-nuova-mappa-aree-interne-dote-2-miliardi-1904-comuni-AEg9cBDD

Gisotti MR, Tarsi E (2023) Regional spatial planning for implementing the European Green Deal: a new method of assessment applied to the metropolitan area of Florence. Plann Pract Res 38(4):581–611. https://doi.org/10.1080/02697459.2023.2230014

Lanzani A, Curci F (2018) Le Italie in contrazione, tra crisi e opportunità. In: De Rossi A (ed) Riabitare l'Italia: Le aree interne tra abbandoni e riconquiste. Donzelli, Roma, pp 79–107

Magnaghi A (2020) Il principio territoriale. Einaudi, Torino

Presidenza del Consiglio dei Ministri, Dipartimento per le politiche di coesione e per il sud (2024) Le aree interne 2021–2027. Politiche di Coesione. https://politichecoesione.governo.it/it/politica-di-coesione/strategie-tematiche-e-territoriali/strategie-territoriali/strategia-nazionale-aree-interne-snai/le-aree-interne-2021-2027/

Regione Toscana (2021) Strategia aree interne—PR FESR 2021–2027. https://www.regione.toscana.it/pr-fesr-2021-2027/strategia-aree-interne

Rodríguez-Pose A (2018) The revenge of places that don't matter (and what to do about it). Camb J Reg Econ Soc 11:189–209. https://doi.org/10.1093/cjres/rsx024

Tarpino A (2016) Il paesaggio fragile: L'Italia vista dai margini. Einaudi, Torino

Teti V (2011) Pietre di pane: Un'antropologia del restare. Quodlibet, Macerata

Università di Firenze (2024) REACT—Regenerating cultural landscapes of inner areas in a people-centered perspective. https://www.react-casentino.unifi.it/index.html?newlang=eng

Urban Form and Housing Evolution: The Sha'abiat Model in Al-Ain, UAE

Boshra Hassan and Sahera Bleibleh

Abstract

Urban form reflects governance, housing policies, and cultural heritage, influencing the social, economic, and environmental dynamics of cities. Key factors such as density and interconnectedness shape neighborhood layouts and development models, which evolve alongside urban growth. This study examines the transformation of housing policies in the UAE, focusing on the shift from the first national Emirati housing project, *Sha'abiat* (established in the 1960s), to the predominance of private villas. By analyzing *Sha'abiat* neighborhoods in Al-Ain City, the research conducts a physical assessment of block sizes, development patterns, and spatial organization, identifying both opportunities and challenges. Using a qualitative approach, the study starts with the physical mapping of neighborhoods to evaluate their morphology and spatial characteristics. It explores the relationship between these physical attributes and residents' lived experiences, providing insights into sustainable urban housing models. Additionally, the research adopts an intersectional perspective, considering how governance, housing policies, and cultural heritage intersect within urban planning. It also assesses governmental efforts to promote sustainability, offering insights to guide future housing developments. Ultimately, this study aims to support stakeholders in revitalizing and sustainably developing *Sha'abiat* neighborhoods within the broader urban context.

Keywords

Emirati housing models · Urban form · Sustainable development · *Sha'abiat* · Al-Ain · United Arab Emirates (UAE)

B. Hassan · S. Bleibleh (✉)
Department of Architectural Engineering, College of Engineering, United Arab Emirates University, Al Ain, United Arab Emirates
e-mail: 202190191@uaeu.ac.ae; sahera.bleibleh@uaeu.ac.ae

1 Introduction

Cities are shaped by a combination of housing regulations, cultural heritage preservation efforts, and governance decisions, all of which are physically represented in urban form. It includes how urban places are organized, perceived, and operated. It affects how neighborhoods are physically planned and their social, economic, and environmental characteristics (Haughton and Hunter 2003). Understanding how housing regulations and urban design interact is essential to comprehending how cities grow and change over time.

As cities expand and evolve, housing models face transformation through city development and contribute to urban form shifts. Two forms of development result from residential growth processes: densification, or the construction of new housing units within existing residential areas, and expansion, or the creation of new residential areas on formerly open land. These forms reflect broader trends in urban development, such as increasing density, compactness, and housing integration within urban areas (Broitman and Koomen 2015). These changes have profound implications for sustainable urban development, as they affect not only the physical attributes of neighborhoods but also their social cohesion, economic viability, and environmental sustainability. The urban shift can lead to changes in density, compactness, and the interconnection of housing within urban areas, affecting neighborhoods' physical layout and morphology. Hence, it can disrupt social cohesion, economic viability, and environmental sustainability, leading to fragmented communities and inefficient use of urban space. There is a pressing need to understand how these shifts impact the overall urban form and the daily lives of residents in neighborhoods.

The first national housing structures, the Sha'abiat, were built in the 1960s with the intention of giving Emirati families access to affordable, culturally suitable homes. The UAE has seen enormous transformations in its urban landscape in recent decades as a result of its fast economic expansion and

F. Alberti et al. (eds.), *Towards Sustainable Urban Regeneration: Utilizing Environmental, Cultural, and Social Assets*, Advances in Science, Technology & Innovation, https://doi.org/10.1007/978-3-032-02091-8_17

urbanization (Elsheshtawy 2004). The development of housing types, particularly shifting from the old Sha'abiat districts to modern private villas, raises different challenges that need to be addressed for sustainable urban development. However, existing literature discusses, and this research explores, the urban planning practices regarding the physical and social attributes of Sha'abiat neighborhoods. Comprehensive studies that combine physical mapping with qualitative assessments of lived experiences are lacking. Such studies are essential for developing informed urban planning and design interventions that can revitalize and sustainably develop Sha'abiat neighborhoods while preserving their cultural significance.

This study aims to provide a thorough physical examination of Al-Ain City's Sha'abiat communities to comprehend their growth patterns, morphology, and spatial organization in connection to changing housing regulations. It addresses the urban morphology by conducting a detailed physical analysis of Sha'abiat neighborhoods in Al-Ain City and exploring the lived experiences of their residents. The findings aim to inform sustainable urban planning practices that balance modernization with cultural heritage preservation. The provided physical analysis of Sha'abiat neighborhoods assesses several attributes, including block sizes, plot layouts, and development patterns in Sha'abiat communities, along with housing regulations, to understand the opportunities and risks that exist in the transformation. Utilizing Al-Ain City Sha'abiat as a case study, the study explores how physical attributes and lived experiences motivate comprehensive urban planning and design efforts.

2 Literature Review

The literature on housing forms and urban resilience in the context of rapidly urbanizing cities emphasizes a number of interconnected themes that are essential to comprehending the maintenance and adaptation of traditional communities within a modern urban framework. The purpose of this analysis is to examine these topics, with a particular emphasis on the field of urban planning, evolutionary and heritage-driven forms of resilience, and their consequences for the preservation of cultural heritage and urban growth. The topic of discussion centers on the necessity of incorporating old housing models—like the "Sha'abiat" in the United Arab Emirates—into contemporary urban plans. The review aims to provide a coherent understanding of the ways in which these ideas interact, especially with regard to the opportunities and problems that come with repurposing existing communities to meet the changing urban development of the United Arab Emirates.

2.1 Urban Form and Sustainability

Urban form, encompassing the arrangement, appearance, and functionality of urban spaces, is integral to understanding the development of cities. Theories of urban form emphasize the significance of spatial organization, density, and compactness in shaping sustainable urban environments (Lynch 1960; Alexander 1977). Kevin Lynch's seminal work, *The Image of the City*, underscores the importance of spatial legibility and coherence in urban design, advocating for a balance between connectivity and distinctiveness in urban layouts (Lynch 1960). Similarly, Christopher Alexander's *A Pattern Language* highlights the need for patterns in urban design that foster social interactions and community cohesion (Alexander 1977).

The relationship between urban form and sustainable development is a focal point in urban planning discourse. Scholars argue that compact, well-connected urban forms can reduce environmental impacts by promoting public transport, reducing reliance on private vehicles, and preserving open spaces (Samuelsson 2021; Jenks et al. 1996). Additionally, high-density developments are associated with economic efficiency and vibrant social interactions, contributing to the overall livability of cities (Burton 2000).

2.2 Housing Policies and Urban Developments

In the context of housing, cities and their neighborhoods are dynamic entities that must evolve to remain functional and relevant in the face of social, economic, and environmental changes. This perspective is reflected in efforts to emphasize the capacity for adaptation and transformation in response to changing conditions (Grant 2002).

Housing policies significantly influence urban form, particularly in rapidly urbanizing regions like the UAE. Since the 1960s, the UAE has implemented various housing policies to address the needs of its growing population and support national development goals. The establishment of Sha'abiat neighborhoods represented one of the first national efforts to provide affordable, culturally appropriate housing for Emirati families (Elsheshtawy 2019). These neighborhoods were designed to reflect traditional Emirati architectural styles while accommodating modern amenities, fostering a sense of community and cultural continuity. At the same time, the transformation of Sha'abiat demonstrates that housing policies are not merely mechanisms for shelter provision, but active drivers of urban development that reshape community dynamics, spatial practices, and the balance between modernization and the preservation of human-scale urban form (Bleibleh & Hassan, 2025a).

However, the UAE's housing model has evolved significantly over the decades. The shift towards private villas, driven by economic growth and changing lifestyles, reflects broader trends in urban development (Elsheshtawy 2013). This transformation has led to challenges such as increased urban sprawl, reduced social cohesion, and environmental concerns.

Governance plays a crucial role in shaping urban form and housing policies. In the context of the UAE, centralized governance structures have facilitated rapid urban development and large-scale infrastructure projects (MacLean 2017). The government's vision for modernity and economic diversification has driven the transformation of urban landscapes, often prioritizing high-end developments and iconic architectural projects.

This top-down approach to urban planning has also led to challenges. Critics argue that rapid urbanization, driven by economic imperatives, has sometimes overlooked the social and environmental dimensions of urban development (Elsheshtawy 2004). The emphasis on privatized, high-income housing models has contributed to socio-spatial segregation and the marginalization of traditional housing forms like Sha'abiat. Therefore, there is a growing recognition of the need for more inclusive and participatory approaches to urban governance that address the diverse needs of urban residents (Elsheshtawy 2004).

2.3 Sustainable Urban Development Practices

Sustainable urban development seeks to create cities that are environmentally responsible, socially inclusive, and economically viable (Bibri et al. 2020). The principles of sustainability emphasize the need for urban forms that minimize environmental impacts, promote social equity, and support economic resilience. Compact, mixed-use developments are often highlighted as exemplars of sustainable urban form, as they reduce the need for extensive transportation networks, preserve natural landscapes, and foster vibrant communities (Turner and Singer 2014).

In the UAE, the shift toward sustainable urban development is evident in various policy initiatives and planning frameworks. The UAE Vision 2021 and the Abu Dhabi Urban Planning Vision 2030 emphasize sustainability as a core objective, promoting green building standards, renewable energy, and efficient land use plans (Abu Dhabi: Urban Planning Council 2007). However, translating these policy visions into practice requires addressing the complexities of urban form and the specific needs of different communities, including those in traditional neighborhoods like Sha'abiat. The future vitality of Sha'abiat depends on integrating conservation with adaptive urban strategies that recognize its capacity to accommodate mixed-use transformation while maintaining its cultural and spatial identity. Such an approach positions Sha'abiat not only as a local heritage concern, but also as a relevant model for sustainable urban development in rapidly evolving cities facing similar tensions between heritage preservation and change (Bleibleh & Hassan, 2025b).

3 Methodology

In order to analyze the physical mapping and morphological features of the Sha'abiat neighborhoods inside the municipal limits of Al-Ain City, United Arab Emirates, this study uses a qualitative methodology. The goal is to figure out how urbanization forces have shaped these traditional communities and identify the current scenarios and key factors driving their physical growth and transformation. The analysis of physical mapping data to examine the physical growth and development of Sha'abiat neighborhoods is the primary qualitative component of this study. The process of spatial mapping entailed recording these communities' urban fabric through satellite images, existing city maps, and on-site surveys. A comprehensive understanding of the building typologies, public space utilization, and spatial organization was obtained through this procedure. After that, the maps were interpreted using the qualitative method, which provided insights into the physical traits and developmental patterns of the Sha'abiat neighborhoods. This approach is consistent with the ideas put out by (Kropf 2017) and (Oliveira 2016), who emphasize the value of qualitative interpretation in urban morphology research to reveal spatial and sociocultural changes.

As an essential part of this research, field observations were carried out to verify and enhance the spatial data that was gathered from secondary sources of Al-Ain City. These observations were conducted in April and May of 2024, and thorough notes were taken on the state of the buildings, impromptu changes, and usage patterns in public areas. Concurrent with the field observations, on-site surveys were carried out to get other information that could not be obtained from satellite pictures or pre-existing maps.

Al-Ain City includes many villages that are considered rural areas, such as Al Amra and Al Salamat, that are not within the city's municipal boundary, as shown in Fig. 1, highlighted in red. Hence, the investigation targeted the municipal boundary of Al-Ain City. This study adopts a qualitative method to comprehensively analyze the physical expansion of Sha'abiat neighborhoods in Al-Ain City, UAE. The methodology utilized field observation to create physical mapping and understand the urban fabric of Sha'abiat. For the physical mapping, on-site surveys were conducted to validate the information from satellite images

Fig. 1 Distribution of Sha'abiat across Al-Ain City, UAE. (Source: Authors 2024)

and available maps, capturing additional details such as building conditions and public space usage.

4 Result

The spatial analysis of Sha'abiat neighborhoods in Al-Ain City reveals a complex evolution of housing typologies and urban forms over time. This section presents the findings from the spatial mapping, field observations, and the interpretation of satellite imagery to illustrate the morphological changes and challenges experienced in these neighborhoods. The goal was to capture how these neighborhoods have evolved from their inception to the present day. The results indicate the following typologies:

- Original Typology: The original Sha'abiat layout can still be seen in several neighborhoods. These areas, which show a planned growth phase, have a grid-like layout of single-family residences. The block size for Sha'abi houses typically ranges between 20 × 20 m to 30 × 30 m. However, the exact dimensions can vary depending on the specific development and location.
- Redesigned Typologies: On the other hand, a lot has changed in areas around these initial designs. Separated dwelling typologies have been developed through plot subdivisions and the conversion of residences into rental units. As a result, there is now a mix of single- and multi-family homes in irregular spatial patterns. Recent observations indicate that the neighborhood layouts have continued to evolve.

Figure 1 illustrates the distribution of Sha'abiat neighborhoods across Al-Ain City, highlighting their concentration around the city center and the variations in plot sizes over time. Based on the investigation and satellite maps, Fig. 2 shows 15 districts in Al-Ain City, including Sha'abiat. Those neighborhoods can be detected based on plot size compared to the recent developments. The spatial mapping and analysis reveal a clear trajectory of change, where the design of these neighborhoods featured a standardized grid pattern, and the block sizes ranged between 20 × 20 m and 30 × 30 m.

Moreover, several challenges related to the changes in urban form were highlighted during the observation. During the site visits, it was evident that many single-family houses had been converted into multi-rental units, often with minimal consideration for spatial planning or privacy. The fieldwork revealed a notable absence of open and communal spaces in the Sha'abiat neighborhoods. Most areas were found to have either limited or no designated spaces for com-

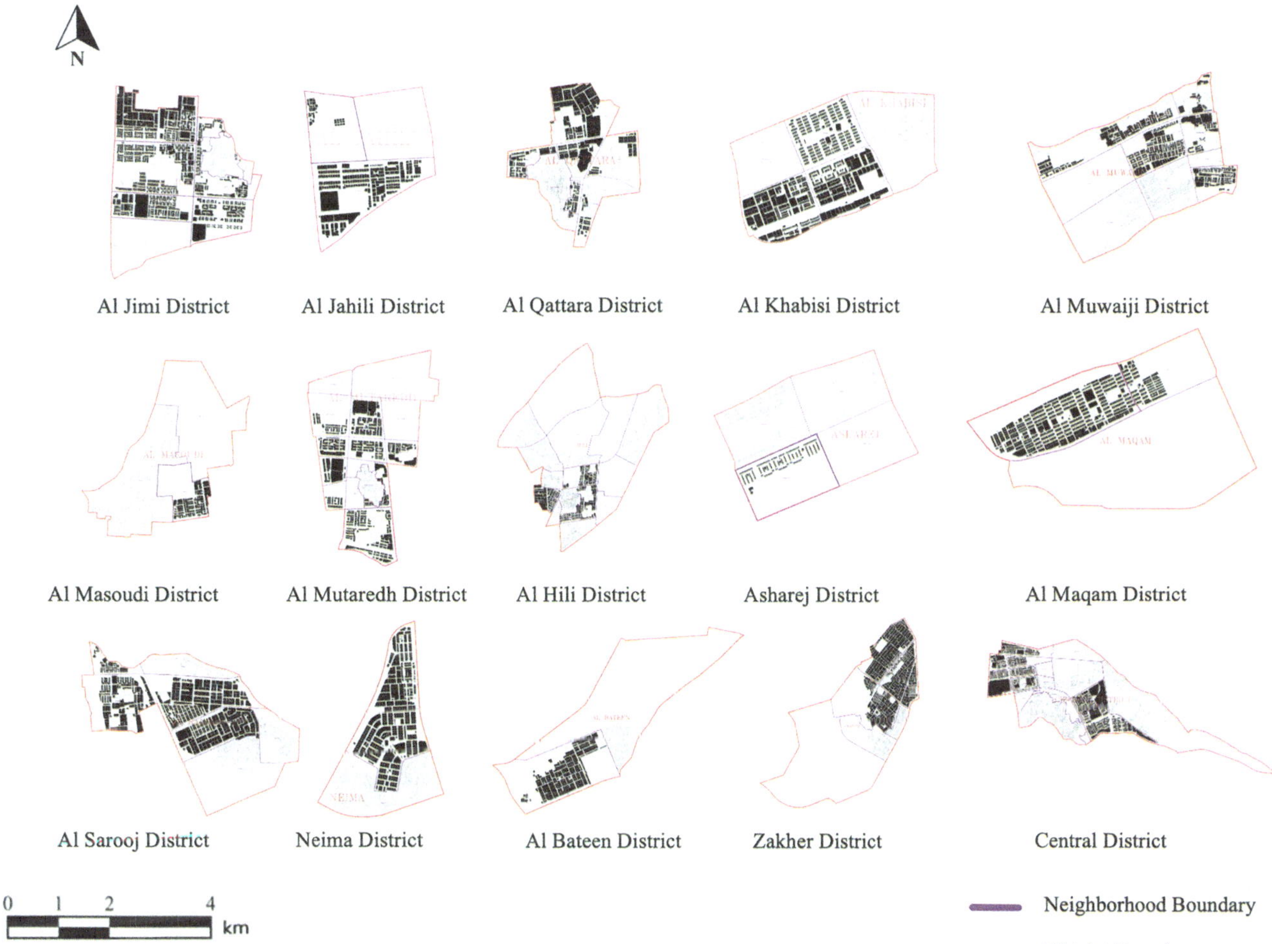

Fig. 2 The spatial layout and urban form of each Sha'abia in Al Ain Districts. (Source: Authors 2024)

Fig. 3 Samples of Sha'abi houses in Sarooj District, Al Ain. (Source: Authors 2024)

munal activities or green spaces. The following challenges were identified: *Overcrowding*: Increased population density, and overcrowding due to the division of plots and the conversion of single-family homes into multi-tenant units. This has led to reduced privacy, noise pollution, and strain on infrastructure. As shown in Fig. 3, one house has been divided into multiple units. *Lack of open spaces*: Limited open and communal spaces could affect the quality of life and opportunities for social interactions. Hence, one house can have a considerable number of people, and it was clear that parking lots are limited.

5 Discussion and Conclusion

The morphological evolution of Sha'abiat neighborhoods reflects broader urbanization trends in Al-Ain City. As the city has expanded and modernized, these neighborhoods have transitioned from their original, planned layouts to more complex, mixed-use environments. The spatial mapping and analysis underscore the importance of considering both historical contexts and contemporary demands when planning for sustainable urban development.

The findings of this study reveal a complex and evolving landscape within the Sha'abiat neighborhoods of Al-Ain City. The physical analysis provides a holistic understanding of the current state of these neighborhoods, highlighting both the preservation and transformation of their original typology. The transformation is shown in seven scenarios:

1. The owners are still in the house with minor modifications.
2. The owners lived in the house and rented part of it.
3. The house has been demolished and rebuilt as a villa.
4. The house has been demolished and rebuilt for commercial use (as an apartment or shop).
5. The house has been rented as a family house.
6. The house has been divided for workers' rent (the common case).
7. Abundant houses.

The current physical context of Sha'abiat neighborhoods in Al-Ain City is influenced by several dynamics that intertwine urban growth, socioeconomic changes, and cultural preservation. The key dynamics affecting the current physical context of Sha'abiat:

- Urbanization and expansion.
  Although there are many houses that have maintained their original typology, most Sha'abi houses have been rented, and have experienced substantial changes. These changes include structural modifications to the building, original plot divisions, and departures from the original spatial arrangement. Urban forms have become more fragmented because of the expansion of rental homes, many of which have been split into smaller apartments to accommodate several renters.
- Socioeconomic factors.
  This has led to a shift from the planned, coherent community structure, as well as an increase in density and reduced open areas. The segmentation and transformation of rental properties indicate broader socioeconomic phenomena, such as the need for reasonably priced housing and the commercialization of residential areas. Although these modifications address the different housing needs, they also threaten the areas' viability and livability.
- Dynamics of preservation and transformation.
  The dual phenomenon of preservation and transformation can coexist in Sha'abiat neighborhoods that have significant implications for urban planning and policy. Urban planners and policymakers need to find a balance between the needs of modern urbanization and the preservation of cultural heritage. This entails creating plans that balance the socioeconomic needs of an expanding urban population with the preservation of the artistic and architectural integrity of Sha'abiat communities. Creating conservation zones in Sha'abiat neighborhoods that have preserved their traditional typology is one such strategy. Regulations that limit changes in chosen zones and encourage the usage of conventional architectural types can help preserve these areas. Incentives for landowners to maintain and repair customary features can also help to promote cultural heritage conservation further.
- Infrastructure and public space.
  These interventions for transformed areas should prioritize public space enhancement, infrastructure improvement, and providing sufficient services to accommodate the increased density. The negative effects of fragmentation and overcrowding can be reduced with the support of policies that control the allocation of plots and the standard of rental dwellings.
- Resident perspectives and community dynamics.
  The findings emphasize the need for balanced urban planning approaches that address rapid urbanization challenges. Regulatory measures, incentives for preservation, and participatory planning processes are essential for creating sustainable and livable urban environments.

This paper, as mentioned, overviews the physical assessment of Sha'abiat and its sprawl all over Al-Ain as part of a broader study. The case study will be chosen for in-depth investigation. However, the mentioned strategies should be followed with the owners' perception of culture and the extent to which these houses could be a cultural asset. The interviews will highlight the diverse perceptions and lived experiences of residents in Sha'abiat neighborhoods. And explore the concerns either by keeping the houses or about the changing urban form and its impact on their quality of life.

This study will extend our knowledge of Sha'abiat neighborhoods as an urban housing model by merging physical investigation. The knowledge acquired will guide practice and policy, supporting the sustainable growth and regeneration of Sha'abiat communities and adding to the larger conversation on housing policy, urban form, and cultural heritage in rapidly urbanizing areas. It aims to inform urban planning and design solutions that are attentive to the physical and social features of Sha'abiat communities by combining physical mapping with qualitative interviews. The UAE's policies and practices may benefit from the lessons gained from these case studies, which can help Sha'abiat communities achieve sustainable growth and regeneration. By bridging the views of intersectionality between housing regulations, governance, and cultural heritage, the research seeks to uncover key variables affecting sustainable development in Sha'abiat neighborhoods.

This study's findings highlight the need for a strategic framework that addresses the challenges of overcrowding, lack of open spaces, and parking shortages while preserving the cultural heritage and original typologies of Sha'abiat neighborhoods. Such a framework could guide future urban planning efforts, ensuring that these neighborhoods are integrated effectively into Al-Ain's broader urban landscape. Finally, Planning actions in cities that are sensitive to the needs and preferences of residents requires an understanding of their perceptions. Urban development projects may be more successful and more widely accepted by including citizens in the decision-making process through participatory planning methods.

Acknowledgments This research was supported by the United Arab Emirates University Program for Advanced Research (UPAR) under Grant G00004166, cycle 2022–2026. The study was supervised by the Principal Investigator, Dr. Sahera Bleibleh.

References

Alexander C (1977) A pattern language: towns, buildings, construction. Oxford University Press

Broitman D, Koomen E (2015) Residential density change: densification and urban expansion. Comput Environ Urban Syst 54:32–46, ISSN 0198-9715,. https://doi.org/10.1016/j.compenvurbsys.2015.05.006

Burton E (2000) The compact city: just or just compact? A preliminary analysis. Urban Stud 37(11):1969–2006

Elsheshtawy Y (2004) Planning Middle Eastern cities: an urban kaleidoscope. Routledge

Elsheshtawy Y (2013) Dubai: behind an urban spectacle. Routledge

Elsheshtawy Y (2019) The Emirati Sha'bī House: on transformations, adaptation, and modernist imaginaries. Arab Humanit

Grant J (2002) Mixed use in theory and practice: Canadian experience with implementing a planning principle. J Am Plan Assoc 68:71–84. https://doi.org/10.1080/01944360208977192

Haughton G, Hunter C (2003) Sustainable cities. Routledge

Jenks M, Burton E, Williams K (1996) The compact city: a sustainable urban form? Routledge

Kropf K (2017) The handbook of urban morphology. Wiley-Blackwell. https://doi.org/10.51347/jum.v22i1.4827

Lynch K (1960) The image of the city. MIT Press

MacLean M (2017) Suburbanization, national space and place, and the geography of heritage in the UAE. J Arab Stud 7:157

Oliveira V (2016) Urban morphology. An introduction to the study of the physical form of cities. https://doi.org/10.1007/978-3-319-32083-0

Samuelsson K (2021) The topodiverse city: urban form for subjective well-being. Front Built Environ Sec Urban Sci 7. https://doi.org/10.3389/fbuil.2021.735221

Turner M, Singer R (2014) Urban resilience in climate change. In: Climate change as a threat to peace. PL Academic Research, Frankfurt, pp 63–81

Bibri S E, Krogstie J, Kärrholm M (2020) Compact city planning and development: Emerging practices and strategies for achieving the goals of sustainability,bı /Developments in the Built Environment, Volume 4, 100021, ISSN 2666 1659, https://doi.org/10.1016/j.dibe.2020.100021.

Bleibleh, S., & Hassan, B., (2025a) "Urban resilience: Sha'abiat's potential role in shaping the future of Al Ain." In Co. Ed. K. Mitchell, K., AlMulla, R., & AlSuwaidi, S. Future of Gulf Housing: Reflections and Propositions © 2025. SWALIF Publishing House. ISBN: 978-1-0687933-7-0.

Bleibleh, S., Hassan, B. (2025b) Placemaking, Spatial Behavior, and Community Interactions: Lessons on Social Housing and Sustainability. The Case of Al Ain City, UAE. In: Bagnato, V.P., et al. Sustainable Approaches to Environmental Design, Materials Science, and Engineering Technologies, Vol. 1. EDMSET 2024. Advances in Science, Technology & Innovation. Springer, Cham. https://doi.org/10.1007/978-3-031-76025-9_12

The Central Area of Ho Chi Minh City from the Perspective of Sustainable Urban Design

Le Ngoc Ha Mai, Nguyen Huu Khang Bui, and Ngoc Nhi Nguyen

Abstract

The central area of Ho Chi Minh City (HCMC) is an urban area that has gone through many historical ups and downs, with thousands of old buildings built before 1975 with high cultural, historical, artistic, and commercial values, which are still being used or reused. However, due to changing usage needs and economic development pressure, many problems have arisen with these core zones of the city, such as a lack of streets and urban utilities, lack of parks and public spaces, while new buildings have been built to replace or intersperse old buildings, breaking up the old urban complex. This study aims to assess the current status of these urban areas according to sustainable urban design principles. From there, we propose more sustainable solutions for this area, especially in organizing common spaces and building aesthetics. The research methods include (1) studying the history of the formation and development of the housing population in Zone 4 of the HCMC center, (2) scientific observation, measurement, and mapping the current status of the research area, and (3) evaluating the selected residential complex according to the United Nations MY Neighborhood matrix. The research results include drawings and diagrams of the current status of the residential area, the assessment matrix, and some proposals for public space and visual treatment to achieve harmony in the entire urban landscape. This topic is part of a more comprehensive study of the central area of HCMC, which will focus on sustainable and identity-rich urban development in the future.

Keywords

Sustainable urban design · Central area · Old buildings · Ho Chi Minh City

L. N. H. Mai (✉) · N. H. K. Bui · N. N. Nguyen
Van Lang University, Binh Loi Trung Ward, Ho Chi Minh City, Vietnam
e-mail: ha.mln@vlu.edu.vn; ha.mai@tum.de

1 Introduction

For many locals and tourists, the center of Ho Chi Minh City is part of District 1 and District 3, where administration, commercial, and service activities have long taken place. This area also clearly retains the grid plan from the French colonial period and many buildings that served previous political regimes, including public, housing, and religious categories.

The area was formerly known as Ben Nghe (Vương 1968). In the seventeenth century, when the Nguyen lord government in Dang Trong gradually established an administrative and military apparatus here, Ben Nghe gradually developed into an urban area from a pristine land by the river (Nguyen 2019). Ben Nghe port became popular, attracting foreign ships to trade in large numbers, and was a political and economic center of the Southern region, especially from the late eighteenth century to the first half of the nineteenth century. Before the French occupied Gia Dinh, the Ben Nghe area had about 40 villages with a population of up to 50,000 people (Tran et al. 2007). When the Gia Dinh citadel fell in 1859, most towns were destroyed, and the population was evacuated. Based on the Ben Nghe urban area, the French planned Saigon City (now the central area of Ho Chi Minh City) (Nguyen 2019). This plan turned a riverside urban area with most of the houses on stilts along the canals into a Western-style city called Saigon (Nam 2007).

Saigon was then the capital of the Indochina Federation during the period 1887–1901. Although known as the Pearl of the Far East *(La Perle de l'Extrême Orient)*, the architectural appearance of Saigon only really changed in the first decades of the twentieth century due to the investment of many large construction projects by the French, the Vietnamese, and the Chinese. From 1946 to 1975, Saigon was the capital of the Autonomous Republic of Cochinchina, the State of Vietnam, and the Republic of Vietnam. During this time, the city expanded, the population increased rapidly, and the architecture changed a lot due to the war. After

F. Alberti et al. (eds.), *Towards Sustainable Urban Regeneration: Utilizing Environmental, Cultural, and Social Assets*, Advances in Science, Technology & Innovation, https://doi.org/10.1007/978-3-032-02091-8_18

Vietnam unified in 1975, Saigon was renamed Ho Chi Minh City (since 1976), which remains to this day (Nguyen et al. 2020).

Up to now, Ho Chi Minh City has gone through nearly 50 years of construction and post-war recovery. To take advantage of available construction resources, the city government has re-planned the central areas and reused thousands of old buildings, villas, and tube houses in the central area for office, commercial, and residential purposes. However, these old central areas, placed in the context of a modern urban area and the need for sustainable development, have revealed many problems and contradictions. For example, the road network of the French colonial planning period has not been able to meet the needs of the ten million people of the city today. There is a severe lack of green parks in the urban core, which in turn leads to a lack of public spaces that can promote a healthy lifestyle for urban residents. Another problem is that under the pressure of economic development in recent decades, many old buildings have been demolished, replaced by new buildings, or arbitrarily renovated to meet new functions. Moreover, loose design and construction management have caused new buildings with entirely different functions, heights and architectural styles to spring up among the old buildings, creating an ugly appearance for the city's architectural face and destroying the inherent beauty of old urban areas.

Since 2012, the City People's Committee has approved the detailed urban construction planning project at a scale of 1/2000 (zone planning) of the existing Central Area of Ho Chi Minh City (930 ha) to manage the architecture in the HCMC's central area more effectively. Accordingly, the existing central area of HCMC is divided into five zones (Fig. 1). Depending on the characteristics of each zone and the connection of each zone with the surrounding areas, different development policies for each zone are proposed. However, the common ideal desire is still how to develop these zones in a sustainable direction, exploit the advantages of existing construction resources, build new ones in harmony, promote the economy, but still create a healthy living environment, and preserve the cultural identity and local urban architecture (Tp.HCM 2012).

This study aims to assess several neighborhoods in the central zones of Ho Chi Minh City according to sustainable urban design principles and in line with the development orientation of the city government for each zone. However, within the scope of a conference paper, the study will present assessments of a residential complex in Zone 4 (a low-rise zone). The United Nations MY Neighbourhood (The UN 2023; UN 2023), a comprehensive 'checklist' and guide on urban design principles will be used as a key guide to test the sustainability and development viability of the selected area. It examines four spatial dimensions-neighborhood, street, open public space, and building unit, to assess how each dimension aligns with the five sustainable city development goals of being compact, connected, inclusive, vibrant, and resilient. From there, find out the favorable features of this residential area and the existing problems, then propose solutions for solving them.

2 Methods

Research methods are used to answer the two research questions:

- How was the central area of Ho Chi Minh City formed and developed, and what is its significance to the city?
- What is the current status of the city's central zones from the perspective of sustainable urban design?

From these two research questions, corresponding scientific research methods will be used to answer them. Specifically, to answer question number one, the document research method will help provide comprehensive and in-depth knowledge about the history of the formation and development of the central area of Ho Chi Minh City. From there, we can gain a thorough understanding of the historical, cultural, architectural, and artistic values of this area.

Second, to be able to assess the current status of the central area of HCMC, we need to grasp the specific characteristics of the urban areas, roads, residential areas, public spaces, and buildings in the area. Scientific observation, measurement, drawing, and mapping of the current status of the study area will help to collect and map the information. Then, based on the United Nations MY Neighborhood matrix, the selected study area will be evaluated.

These methods are also general methods that cover a more comprehensive study of HCMC central zones. However, within the limits of the conference paper, these methods still reveal some disadvantages. Firstly, the area selected for assessment is limited to a residential complex in Zone 4 of the center of HCMC, not all five zones. The reason this residential area was chosen is because it gathers many old villas built in the early twentieth century with great historical and architectural values. Although it is only a small area equivalent to a separate residential area, it, on the other hand, includes many functions of living, working, service, and commerce, reflecting the vitality of Ho Chi Minh City while revealing many interesting stories about old and new architecture coexisting in an urban complex. However, some of the assessments in the matrix for the study area have not yet been identified by formal surveys or studies, such as the safety and welcomingness of public spaces. This will be addressed in a more comprehensive future study of Ho Chi Minh City's central area.

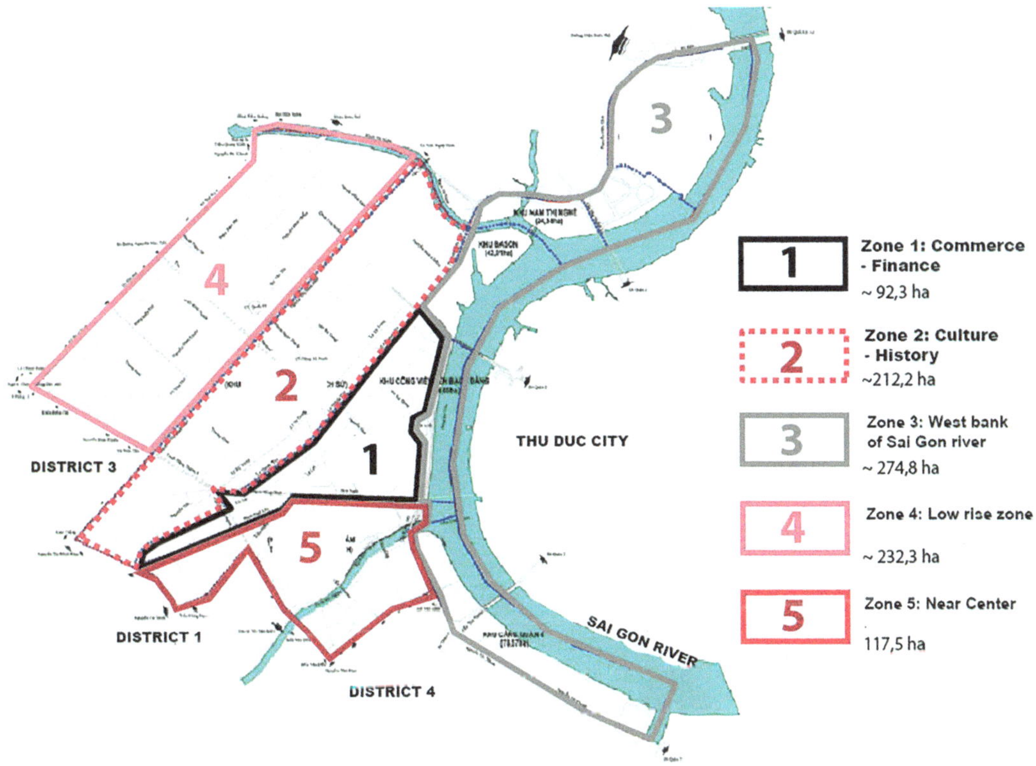

Fig. 1 Planning of five central zones of Ho Chi Minh City

3 Results and Discussion

3.1 The Research Area

The research area is located in Zone 4, which is in the central area of Ho Chi Minh City (Fig. 1). As mentioned above, this is a low-rise area with many French villas located in part of District 1 and District 3, with an area of about 232.3 ha. In the central planning of Ho Chi Minh City (6708/QD-UBND), this area functions as a low-rise residential, cultural, educational, and commercial area. For the streets located in the villa area, where many old villas are still preserved, it is necessary to strictly control the height and land use coefficient of the buildings. For high-rise buildings in Zone 4, it is necessary to control the setback and height of the building foundation according to current regulations, and there must be preferential policies on land use indicators to encourage the preservation of unique architectural landscape spaces with cultural, historical, and urban values in District 3. Especially in some areas, such as routes with many villas, canal landscape areas, or areas with urban railway stations, there must be preferential policies according to the TOD model (Transit Oriented Development—Development according to public transport orientation associated with land use).

The residential complex is bounded by four streets: Tran Quoc Thao, Vo Thi Sau, Nam Ky Khoi Nghia, and Dien Bien Phu (Fig. 2), in Vo Thi Sau Ward, District 3. This is part of the villa area for Westerners during the colonial period (Pugnaire 2023) and still retains the checkerboard street model planned by the French in the late nineteenth and early twentieth centuries (Bogle 1972). The villas here were built from 1920 to 1950 with a unique style, combining classical French architecture and indigenous architecture, called Indochina architecture (Nguyen 2020). According to HCMC's housing development program from 2021 to 2030 (Tp.HCM 2020b), old villas in this area have been listed and classified from type 1 to 3 according to the level of need for conservation for reuse (Tp.HCM 2020a). Specifically, there are 4 type-1 buildings, 19 type-2 buildings, and 9 type-3 buildings in this residential complex (Fig. 3).

From 1975 to now, in the process of development, the HCMC urban area has gradually expanded, serving from three million people to ten million people today. This land has gradually become a golden land, the administrative, commercial, and service center of the whole city. Therefore, most of the buildings here are reused for many different functions. Among them, there are many public buildings such as the City Children's Center, the Institute for Urban Development Research, the Vietnam Lawyers Association, the International Kidney and Dialysis Hospital, the National Assembly Guest House, the Office of the Youth Union of Central Agencies, the Center for Energy Conservation Research and Development, the Ho Chi Minh City Union of Science and Technology Associations, 03 kindergartens and many other private educational institutions (Fig. 4).

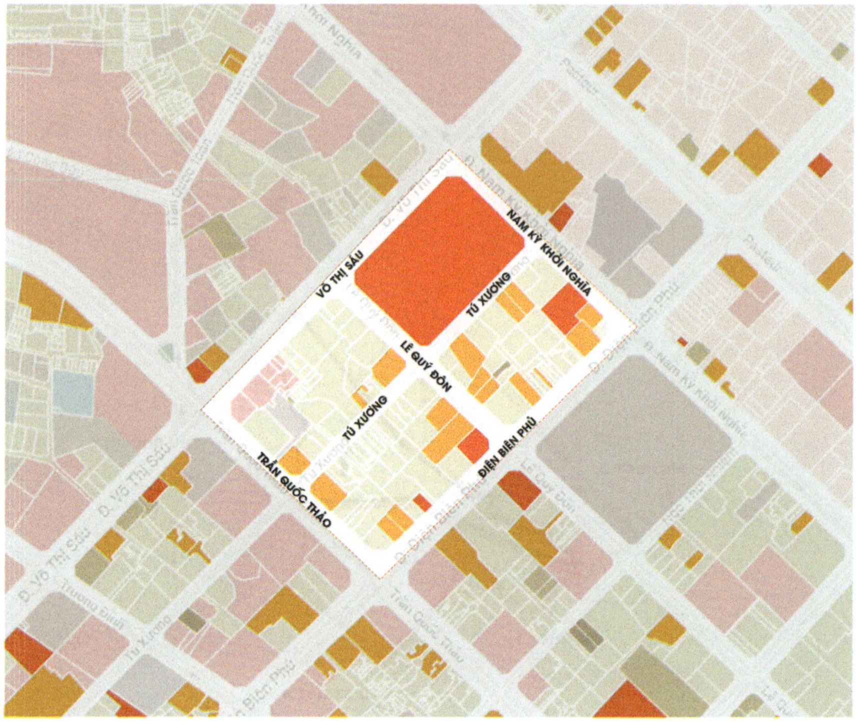

Fig. 2 The limited scope of research area

3.2 Assessment of the Current Status of the Residential Complex

Based on the UN's MY Neighbourhood matrix guidelines (The UN 2023), the four dimensions of the selected mixed-use residential area were scored as meeting or failing each of the five sustainable urban development goals: compact, connected, inclusive, vibrant, and resilient (Fig. 5).

3.2.1 Neighborhood

In the compact city target, 4/6 criteria are easily achieved. A survey of 40 buildings in the area found 48 cafes, restaurants, and shops, not to mention other functions such as offices, public administration, hotels, and other types of services. Therefore, it fully meets the 40–60% business floor area criteria of the compatibility of uses criterion. According to 2021 statistics of Ho Chi Minh City, Vo Thi Sau ward has an area of 2.2 km^2 and a population of 37,000 people, a density of about 168 people/ha. This has met the target of 150p/ha for a sustainable residential area.

Due to the restriction of new construction in the old villa area, the construction density is quite low, most of the green space around the City's Children Center and in the villa gardens are still preserved. Therefore, the criterion "preservation and integration of blue green infrastructure"' is still achieved, with the expectation that this green space will be maintained and expanded in the future.

In terms of transportation, all routes in the study area have buses, so residents can easily access public transport within a radius of only 100 m. However, both the proximity and walkability criteria are not achieved because the road system is not complete, one side of the sidewalk is limited, and there is no lane for cyclists at all.

Most of the criteria for the goal of a connected city are not achieved. For example, the street network is not really effective because the road area does not reach 30% (21%) and does not have enough 18 km of road per square kilometer (17.1 km). It is not really convenient for pedestrians and cyclists, and there is not a wide choice of routes or transport. One of the achieved criteria is a mixed-use development, where residents have many different activities in the residential area due to the interweaving of all functions in the same residential area, even the same building. Another plus point is that residents can walk from 5 to 15 min to bus stops and key services. However, the bicycle routes to these points are not smooth and unsafe because they have to share with other types of vehicles, such as buses, cars, and motorbikes.

For the inclusive city goal, most of the standards are achieved, such as mixed land use and a variety of functions in the neighborhood, providing adequate infrastructure and essential services such as water supply, drainage, waste collection, and full facilities such as schools, kindergartens, hospitals, etc. Although there are not many types of public transport accessing this area, the bus fare is supported by the city, so it is very cheap and accessible to all works and public spaces. There are no obstacles for people to access natural spaces in public spaces in the area, such as the city's children's center, which allows for the promotion of many open activities for children, teenagers, and everyone else.

Fig. 3 List of classified old villas in the study area

		Address	Classification time
Type 1: Preserve the architectural integrity of both the interior and exterior of the building	1	*226 Điện Biên Phủ, P.7, Q.3*	1st & 2nd (May, 2020)
	2	*167A Nam Kỳ Khởi Nghĩa, P.7, Q.3*	1st & 2nd (May, 2020)
	3	*169 Nam Kỳ Khởi Nghĩa, P.Võ Thị Sáu, Q.3*	9th (Apr 2022)
	4	*33 Lê Quý Đôn, P.7, Q.3*	1st & 2nd (May, 2020)
Type 2: Preserve the exterior architecture of the building, possibly repair the interior	1	*200 Điện Biên Phủ, P.7, Q.3*	1st & 2nd (May, 2020)
	2	*216/3 Điện Biên Phủ, P.7, Q.3*	1st & 2nd (May, 2020)
	3	*216/5 Điện Biên Phủ, P.7, Q.3*	1st & 2nd (May, 2020)
	4	*228 Điện Biên Phủ, Q.3*	4th (Apr, 2021)
	5	*228B Điện Biên Phủ, Q.3*	4th (Apr, 2021)
	6	*216B Điện Biên Phủ, P.Võ Thị Sáu, Q.3*	8th (Jan 2022)
	7	*62 Trần Quốc Thảo, P. Võ Thị Sáu, Q.3*	5th (May, 2021)
	8	*163 Nam Kỳ Khởi Nghĩa, P.7, Q.3*	1st & 2nd (May, 2020)
	9	*27 Tú Xương, P.7, Q.3*	1st & 2nd (May, 2020)
	10	*6 Tú Xương, P.Võ Thị Sáu, Q.3*	6th (Nov 2021)
	11	*6A Tú Xương, P.Võ Thị Sáu, Q.3*	6th (Nov 2021)
	12	*9A Tú Xương, P.Võ Thị Sáu, Q.3*	9th (Apr, 2022)
	13	*9B Tú Xương, P.Võ Thị Sáu, Q.3*	9th (Apr, 2022)
	14	*15 Tú Xương, P.Võ Thị Sáu, Q.3*	9th (Apr, 2022)
	15	*28 Lê Quý Đôn, P.7, Q.3*	1st & 2nd (May, 2020)
	16	*30 Lê Quý Đôn, P. Võ Thị Sáu, Q.3*	5th (May, 2021)
	17	*30bis Lê Quý Đôn, P. Võ Thị Sáu, Q.3*	5th (May, 2021)
	18	*35 Lê Quý Đôn, P. Võ Thị Sáu, Q.3*	5th (May, 2021)
	19	*37 Lê Quý Đôn, P.Võ Thị Sáu, Q.3*	9th (Apr, 2022)
Type 3: Can be repaired both inside and outside the building	1	*202 Điện Biên Phủ, P.7, Q.3*	1st & 2nd (May, 2020)
	2	*216D Điện Biên Phủ, P.Võ Thị Sáu, Q.3*	6th (Nov 2021)
	3	*204D Điện Biên Phủ, P.Võ Thị Sáu, Q.3*	8th (Jan 2022)
	4	*206 Điện Biên Phủ, P.Võ Thị Sáu, Q.3*	8th (Jan 2022)
	5	*208 Điện Biên Phủ, P.Võ Thị Sáu, Q.3*	8th (Jan 2022)
	6	*210 Điện Biên Phủ, P.Võ Thị Sáu, Q.3*	8th (Jan 2022)
	7	*212 Điện Biên Phủ, P.Võ Thị Sáu, Q.3*	8th (Jan 2022)
	8	*214 Điện Biên Phủ, P.Võ Thị Sáu, Q.3*	8th (Jan 2022)
	9	*169/4 Võ Thị Sáu, P.Võ Thị Sáu, Q.3*	7th (Jan 2022)

For the vibrant city goal, the neighborhood meets almost all the criteria. This was originally a villa area for French officers from the French colonial period, and the buildings continue to be used today, with the exterior architectural features not much repaired over time. Since 2016, when the city had a classification policy to manage and use these buildings, they have been preserved even better because most of these buildings are of types 1 and 2, meaning that the interior or both the interior and exterior of the building must be preserved. Therefore, they retain the charm of nostalgia and carry a unique local identity. At the same time, newly built service buildings that make use of the land surrounding the old buildings try to harmonize with the architectural style of the old houses and are often divided into smaller ones for easy business and rental. This meets the criteria of a diverse urban fabric and fine-grain and the appropriate high density of this goal.

In the final goal, a resilient city, we have some basic conditions to become a risk-responsive area, but there are no real moves to make the area truly resilient. Specifically, in addition to the advantages mentioned above, such as mixed land use and efficient density, the houses here provide safe, secure, and affordable housing, while the public buildings and spaces in the area will provide shelter in case of disasters such as floods, fires, or earthquakes. However, there are really no proactive designs to cope with climate change and natural disasters. The infrastructure system, urban equipment, and technical systems in the building are not equipped to minimize risks and improve the resilience of the area and the city after disasters.

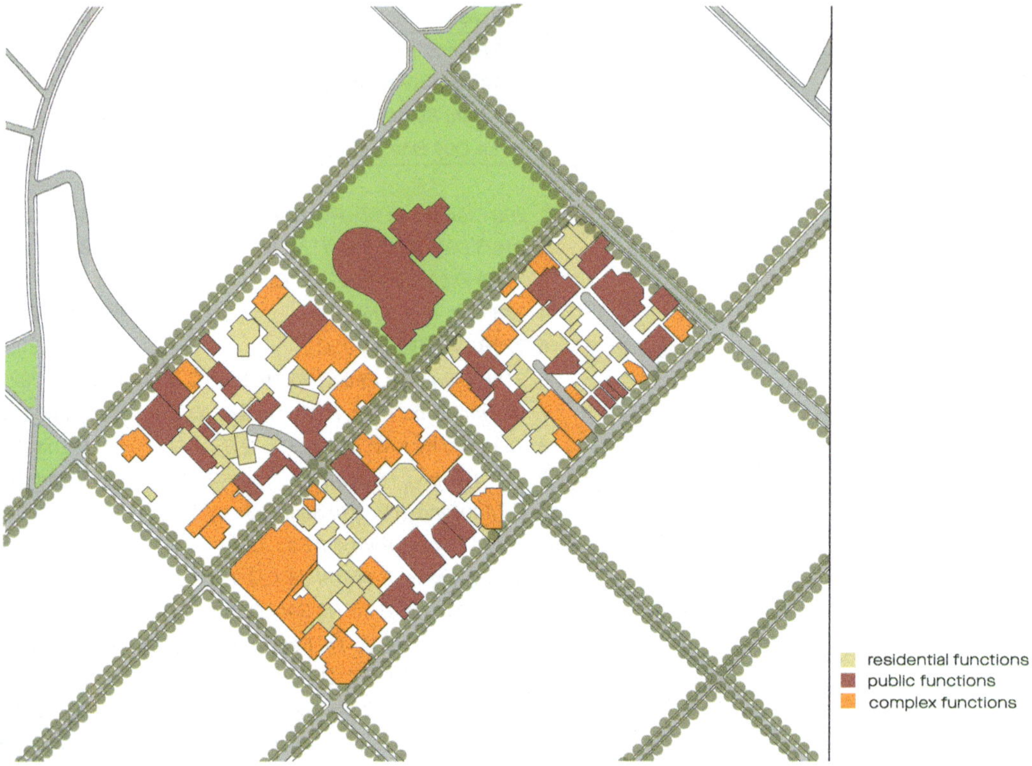

Fig. 4 Distribution of construction types

3.2.2 Streets

The biggest problem of this residential area is the incomplete traffic, so for the "Streets" dimension, the shortcomings will be dissected one by one to find solutions.

First, the sidewalk system for pedestrians is still lacking in both width and transparency, not up to standards for the disabled, and there is no dedicated lane for cyclists (Fig. 6). The traffic density on the street is always too high, and the non-compliance with traffic laws by too many motorbikes is very unsafe for cyclists. On the other hand, there are trees along the streets; however, the height is too high, and there are not many middle and low tree layers to create shade, reduce the temperature, beautify the street landscape, and create interesting resting and stopping points for pedestrians and cyclists. These factors do not encourage people to practice a healthy lifestyle.

Second, there is a lack of parking, and if there is, most of it is on the ground, which takes up valuable urban space, and there is no solution to help water seep into the city's underground water. That makes the subsidence and flooding of the city increasingly serious. There is a complete lack of bicycle parking. On Tu Xuong and Le Quy Don streets, it is easy to see spontaneous parking lots encroaching on the roadway and sidewalks for both motorbikes and cars, especially in front of or opposite large restaurants and cafes. This not only causes traffic congestion, hindering the smooth flow of traffic whether on the roadway or on the sidewalk but also makes it unsafe for road users, especially vulnerable people such as the elderly, children, and people with disabilities.

3.2.3 Open Public Space

For the "Open Public space" dimension, due to the urban expansion process in recent decades, the central areas of HCMC have gradually been surrounded by residents and become administrative, cultural, and commercial centers. Therefore, residential buildings in this area have gradually transformed into public and service functions, or integrated public, service, and functions in the same building. This flexibility inadvertently meets the need for easy and diverse access to public works and the types of services for the people. Looking at the map of the distribution of public works inside and outside the area (Fig. 4), it can be seen that schools, hospitals, committees, and commercial areas are all evenly distributed and within the accessible distance (400–450 m).

A big plus for this area is its vibrancy for stimulating the local economy. Small-scale and family-based economies have long been a characteristic of urban areas in Vietnam. This is also the attraction and cultural characteristic of HCMC. Moreover, taking advantage of old architecture also helps preserve heritage, honor urban architectural identity, and protect the environment by minimizing new construction.

However, the biggest drawback of not only this area but also of the whole of District 3 is that there is no park, although

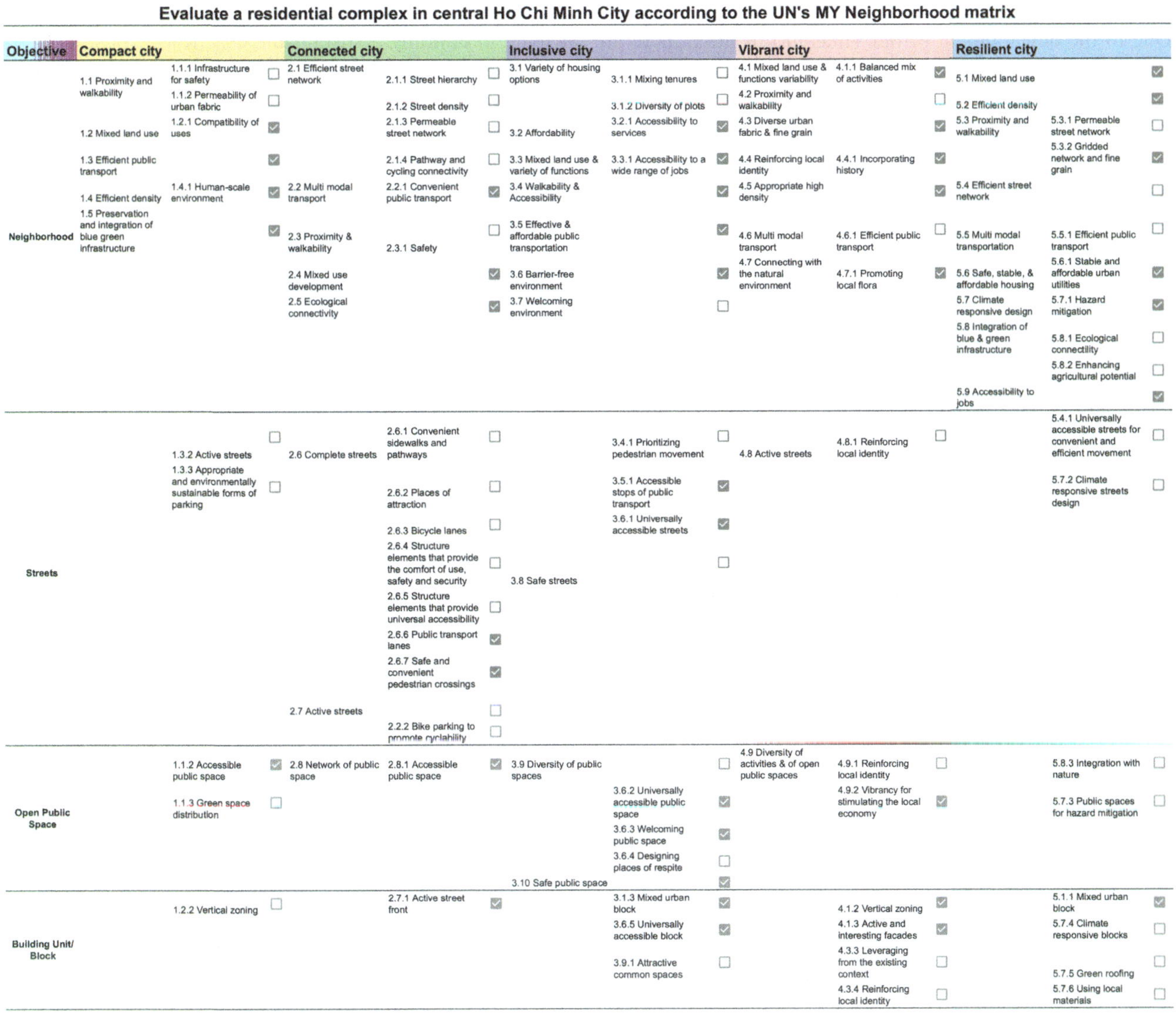

Fig. 5 Assessment matrix of the residential complex

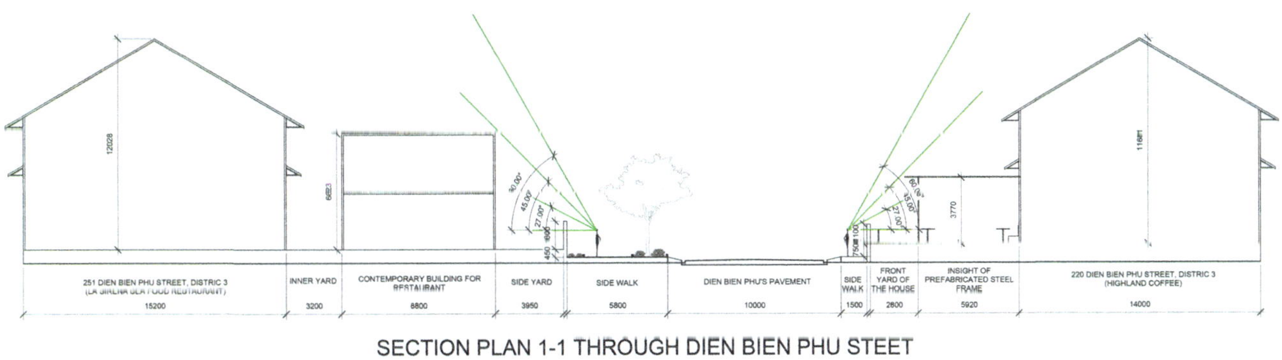

Fig. 6 Cross-section of Dien Bien Phu street

the density of trees is still quite good compared to the city's low average (0.55m²/person, statistics in 2022) because the green area is concentrated in the City's Children Center and scattered in the villas. However, it is clear that not having a park will significantly limit the spaces and activities for all people in society.

3.2.4 Blocks

Due to the development of old villas in this area, there are two opposite trends. One is the demolition of old buildings to build new modern buildings from 1990 to 2016, with heights and architectural forms completely opposite of the remaining old buildings (Fig. 7). The second trend is that after the policy of evaluating and classifying to preserve old villas in the central area of HCMC, many villas on Tu Xuong and Vo Van Tan streets were almost abandoned and seriously degraded. Both of these trends are harmful and cause huge losses to the economy as well as the heritage resources of the city.

The solution in recent years has been for residents to take advantage of the vacant land around these houses to build temporary shops with unsustainable structures. This creates diverse facades and creates economic value for the area. However, temporary structures without specific urban design regulations have led to chaos and are unsustainable, not environmentally friendly. More importantly, it obscures the main structures that should be accessible, used, and honored for their inherent historical and architectural value.

3.3 Proposals

With a hot climate all year round (average annual temperature of 28.2 °C), 2495 h of sunshine per year and frequent rain from May to November, streets need more shade trees and places to stop and rest, encouraging pedestrians and cyclists to practice healthy lifestyles. Street designs and solutions, including urban furniture, public art, bus stops, rest areas and local greenery, will enhance place recognition and enrich activities for locals and visitors. Not only streets but also public spaces are places that can integrate extremely well with nature, increase the ability to absorb rainwater, prevent flooding, avoid creating urban heat islands, and contribute to building a sustainable city. Roofs, balconies, gardens, and facades should be utilized to cover green areas, increase green areas, and integrate with nature.

Responding to climate change is also an idea that the surrounding area, in particular, and HCMC, in general, must pay attention to and must immediately implement sustainable solutions because HCMC is one of the cities with the highest risk of climate change in the world, especially the increasingly serious flooding situation. The city needs to consider using stable and affordable urban utilities such as water collection solutions, leveling asphalt and waterproofing concrete roads, and digging reservoirs in the natural space of the area to be able to collect water during heavy rains or high tides. Old buildings themselves have characteristics that can adapt to changes in climate and usage needs (Mai 2024). It is necessary to promote these existing characteristics while finding ways to renovate and upgrade old buildings in the central area to reduce emissions to the environment during operation.

For buildings, specifically old French villas, specific urban design solutions are needed to overcome the limitations they are facing. First, for houses surrounded by temporary structures used as business premises. There should be specific regulations on the height and density of temporary structures if they are placed in front of old structures to avoid obscuring the beauty of the villas. There should be specific guidelines on architectural styles and models for temporary structures to harmonize with old structures. Local materials that Vietnamese people often use in traditional construction, such as wood, coconut, cajuput, and unburnt bricks can be used for temporary construction spaces in front of and around old buildings, both to enhance local identity and protect the environment. Second, for villas that have not been encroached upon and obscured by temporary structures around them,

Fig. 7 Facades of Dien Bien Phu street

their problem is that the setback is too far, sometimes up to 50–60 m, making the facades of the houses discontinuous, reducing their commercial appeal and vitality. It is recommended to take advantage of the gaps between buildings to create public spaces or breaks, pauses in between, both enriching the activities of local people, attracting tourists, and encouraging pedestrians and cyclists.

4 Conclusions

The central area of HCMC has a long history of development with a system of streets and buildings built from the nineteenth century to the present. Sub-zones with different functions and development goals will have different missions and regulations on design and construction. Researching and evaluating these sub-zones helps to understand the history, development goals, urban morphology, and architectural characteristics of each zone. From there, there are management and usage plans, as well as the most suitable design proposals for each zone.

The residential complex selected for the study is located in Zone 4, a low-rise area, which is an old villa area built in the early twentieth century with a unique architecture that blends French and Indochinese architecture. Based on the UN's MY Neighbourhood matrix, this residential complex reveals the typical characteristics of a Vietnamese urban area.

The strengths of this residential area are its compactness, inclusiveness, and vibrancy. With the inherent strengths of an old residential area, this area has a natural appeal due to its nostalgia, along with many historical, cultural, and architectural features. It can easily be reused for many other functions, from administration, culture, education, health, to commerce and services. High population density, sustainable housing, easy access, mixed land use, and bustling commercial activities are its most prominent advantages.

However, the solution of utilizing vacant land around the villas to build temporary structures for business purposes is not sustainable. The temporary structures cover up and overshadow the true value of the old structures and, in many cases, disrupt the architectural appearance of the whole neighborhood.

The biggest problem of not only this residential area but also HCMC and urban areas in Vietnam in general is traffic. The lack of diversity in public transport and the lack of clear classification of roads for different road users have made the criteria of a connected city almost impossible to achieve. The "Streets" dimension is also the dimension that shows the most shortcomings. Therefore, developing adequate public transport and street planning oriented towards pedestrians and cyclists is HCMC's long-term and achievable goal.

Another issue is green space and the environment in the city center. The complete lack of parks and the concretization of the central area have created an urban heat island and aggravated flooding while limiting people's recreational and creative activities. The lack of trees and shady places along the roadside is also the reason why people rarely walk or cycle. Looking more broadly into the future, not only the research area and the central areas of Ho Chi Minh City, but the entire city must find ways to increase green coverage, increase green area per capita, and increase the variety of forms and types of trees in urban areas.

Finally, attention must be paid to the areas and buildings' resilience and responsiveness to climate change. Urban facilities and materials must be proactively researched and equipped to be water-permeable and recover as quickly as possible after natural disasters. Buildings must be inspected and maintained for efficient operation, avoiding large emissions into the environment while promoting their inherent responsiveness.

References

Bogle JE (1972) Dialectics of urban proposals for the Sai Gon Metropolitan Area.pdf

Mai LNH (2024) The adaptive reuse potential of the old residential buildings in Ho Chi Minh City

Nam S (2007) Di sản Sài Gòn. In: Sài Gòn xưa và nay. Trẻ, Xưa và Nay, pp 25–29

Nguyen TH (2019) Đô thị Sài Gòn—thành phố Hồ Chí Minh Khảo cổ học và bảo tồn di sản. Tổng hợp thành phố Hồ Chí Minh

Nguyen DH (2020) Kiến trúc đô thị và cảnh quan Sài Gòn Chợ Lớn xưa và nay. Văn hóa—Văn nghệ

Nguyen DH, Doling T, Vo CM (2020) Kiến trúc đô thị và cảnh quan Sài Gòn Chợ Lớn Xưa và nay. Văn hóa—Văn nghệ

Pugnaire M (2023) Quy hoạch vùng Sài Gòn—Chợ Lớn. In: Quy hoạch đô thị và kiến trúc ở Đông Dương. Mỹ Thuật, pp 139–148

The UN (2023). My Neighbourhood

Tp.HCM TỦ (2012) 6708_QD-UBND_165149.pdf

Tp.HCM TỦ (2020a) Quyết định 1550/QĐ-UBND về ban hành các danh mục biệt thự cũ trên địa bàn thành phố đã được phân loại (đợt 1 và đợt 2). https://qhkt.hochiminhcity.gov.vn/Uploads/Document/Legislation/QD1550_UB_2020_phanloai bietthu.pdf

Tp.HCM TỦ (2020b) Quyết định 4834/QĐ UBND về phê duyệt Đề án "Xây dựng Chương trình phát triển nhà ở thành phố Hồ Chí Minh giai đoạn 2021–2030." https://soxaydung.hochiminhcity.gov.vn/chi-tiet-tin-tuc/-/asset_publisher/6MeKi7djC3fc/content/uy-ban-nhan-dan-thanh-pho-ho-chi-minh-phe-duyet-e-an-xay-dung-chuong-trinh-phat-trien-nha-o-thanh-pho-ho-chi-minh-giai-oan-2021-2030-?category-related=205914&in

Tran G et al (2007) Sào Gòn Xưa và Nay. Trẻ, Xưa và Nay

UN T (2023) MY Neighbourhood digital matrix

Vương HS (1968) Sài Gòn năm xưa. Sống Mới

The Built Environment Policy Making: The Egyptian New Administrative Capital Model

Mahmoud Khalifa and Amany A. Khodair

Abstract

The public policies for state-building new cities have several goals, the most prominent of which are increasing the urban area, reducing population growth in city capitals, and facilitating urban development by creating an appropriate environment for human activity. The New Administrative Capital in Egypt aims to establish a new economic and administrative city, which is a modern capital that is compatible with technological development and with the requirements of the built environment that includes buildings, parks, or green spaces and their division into neighborhoods and cities, which can often include the supporting infrastructure, such as water supplies or power grids.

This paper seeks to answer the questions of the impact of public policy on the built environment and whether it has sustainable interests in the new administrative capital in Egypt. By examining Egypt's decision to establish a new administrative capital, the paper seeks to evaluate the Egyptian public policies of built environments with special emphasis on the efficacy of building a new administrative capital. The study is a qualitative approach to argue that Egypt's decision to establish a new administrative capital is an appropriate policy that positively impacts urban development and is an ideal model for built environments. The paper concludes that the decision to establish the "New Administrative Capital" is based on public policy regarding the built environment. Reinforced by decisions to consider the climatic environment and green spaces as new requirements for the planning and building controls in the Egyptian governorates.

Keywords
Public policy · Built environment · Sustainable development · New administrative capital · Egypt

1 Introduction

The significance of culture in achieving a sustainable built environment has become increasingly recognized, as the built environment plays a pivotal role in driving society toward sustainable development (Opoku 2015; UNESCO 2012). This is particularly critical for societies grappling with environmental challenges, such as climate change and the pressures of urban population growth. Addressing these challenges is essential for securing a better future. Cultural sustainability, along with other strategies, can enhance human life and provide meaningful legacies for future generations. To ensure future generations can meet their needs, society must rethink its current approach to resource exploitation. The need for public policies that shape modern lifestyles is urgent, as they have the power to change attitudes and foster sustainable development (World Economic Forum 2022). Public policies that support a sustainable built environment would help create vibrant cities and communities where people can live, work, and thrive, significantly contributing to social and economic prosperity.

One of Egypt's greatest challenges is escalating its Capital (Cairo) population at very high rates in a very short time. Until very recently, Cairo reached the maximum degree of population saturation and could not absorb more residents. This large population pressure—which was not matched by a simultaneous and proportional spatial expansion—resulted

M. Khalifa (✉)
College of Administrative Sciences, Applied Science University, East Al Ekir, Bahrain
e-mail: mahmoud.khalif@asu.edu.bh

A. A. Khodair
Faculty of Business Administration, Economics and Political Science, British University, Al Shorouk City, Egypt
e-mail: amany.khodair@bue.edu.eg

F. Alberti et al. (eds.), *Towards Sustainable Urban Regeneration: Utilizing Environmental, Cultural, and Social Assets*, Advances in Science, Technology & Innovation, https://doi.org/10.1007/978-3-032-02091-8_19

in major problems in the current built-up space, the valley, and the delta, including a significant increase in population density, erosion of fertile agricultural lands, and a low standard of living in its urban, environmental, economic, and social aspects.

To make the crucial transition from the current situation to the desired new situation on scientific grounds and with proper planning, sustainable built environment public policies were urgently needed. Thus, the General Authority for Urban Planning has prepared a comprehensive national strategic plan for urban development, intending to achieve the needed development and population expansion from now until 2052. During the past years, Egypt has taken proactive steps in urban development, which led to the decline of the phenomenon of slums, as well as the emergence of new cities, most notably the New Administrative Capital, which will be covered in this paper and the Environmental City of El Alamein (City-Eco) and much of the development of Egypt's regions, whether in the east, west, or north and south, for example, the development of a region The Suez Canal (as a model for regional development at the national level).

2 Methods of Research

This research adopts an institutional approach as its core theoretical framework. By viewing the study through an institutional lens, it explores the collaborative efforts of various entities, including the government, the New Communities Authority, civil society, and the private sector, alongside national, regional, and international bodies. These collaborations may occur formally or informally within the built environment. Moreover, this institutional perspective delves into how these interactions shape policy outcomes by analyzing how institutions operate, interact, and influence one another, as well as their impact on society.

Consequently, the study investigates how institutions shape societal perceptions of the significance of a sustainable built environment, with particular attention to the new administrative capital's role in advancing sustainable development. The research examines how strengthening partnerships through coordination can lead to effective participation in the planning, implementation, monitoring, and policy-making processes. This analysis will encompass the policies and objectives, the instruments used, the implementation strategies, and the overall impact on sustainable development.

New Administrative Capital in Egypt as a case study is a research approach that is used to generate a comprehensive understanding of a complex problem. One of the fundamental principles of scientific research is the need to examine an event or phenomenon from a deeper perspective. This is done through the use of a naturalistic design, which aims to control the variable that interests the authors. In general, Case studies are ideal for scholars who want to develop a deeper understanding of a certain subject. They allow them to explore the various characteristics and implications of the situation.

Data Collection The study employs a qualitative approach, primarily relying on the analysis of documents obtained from the relevant institution, the "New Communities Authority." In addition to primary data, the research integrates semi-structured interviews, supplemented by academic literature, articles, and reports as secondary sources. These diverse sources collectively inform the study's findings and analysis.

3 Literature Review

The concept and implementation of smart cities are relatively recent developments. While the idea of future smart cities can be traced back to ideal society narratives, such as "New Atlantis" (Cugurullo 2021), the term "smart cities" in its modern context first emerged in the 1990s. At that time, scholars began to emphasize innovation, globalization, and technology as key components of future urban planning (Gibson and Kozmetsky 1992). This led to the evolution of smart cities, which build on the foundations of earlier concepts like "Wired Cities" and "Intelligent Cities" by applying smart technology to manage urban complexities.

TWI Global defines a smart city as one that leverages ICT to enhance operational efficiency, facilitate information sharing with the public, and improve government services and citizen well-being (Burlacu 2022). Thus, smart cities are characterized by the innovative use of technology, combined with effective systems for monitoring, planning, and governance. Data sharing within smart cities goes beyond municipal boundaries, involving businesses, residents, and external stakeholders, potentially leading to greater insights and economic benefits (Paiho 2022). Early examples of cybernetic interventions in urban planning include Los Angeles' emergence of computational analysis techniques in the 1960s, leading to Singapore's formation of the National Computer Board in 1981 (Montes 2020).

Many scholars have tried to hypothesize about smart cities. They tried to develop different classifications and benchmarking methods to identify smart cities (Lai 2020; Praharaj 2018; Yin 2015; Angelidou 2014; Anthopoulos 2015). Yin (2015) assumed that the smart city must include four key perspectives: the application domain, technical infrastructure, data processing, and system integration. Furthermore, the application domains are classified into governance, citizens, businesses, and environment (Anthopoulos 2015). Compared benchmarking approaches according to six common dimensions: people, government, economy, mobility, environment, and living in three examples of smart city cases. Useful outcomes were deduced from this comparison.

Today, the use of smart technology has expanded to include many of the large cities in the world. This positively impacted the quality of life for residents in these cities. Based on economic and technological data, along with feedback from inhabitants, Zurich, Singapore, and Oslo were ranked globally as the top smart cities according to the 2021 Smart City Index (Bogdanov 2019).

Various studies have analyzed Egypt's New Administrative Capital in relation to comprehensive development efforts. Khalifa and Atwa (2021) emphasize the critical importance of public policies in the execution of Egypt's Comprehensive Development Plan. Notably, government policies focused on the Roads and Bridges Sector are seen as especially impactful, distinguishing them from other development projects. However, as noted by Serag (2017), the creation of the New Cairo Administrative Capital has generated significant controversy, with debates centering on whether such a project is truly necessary and adequate.

This issue was discussed in various planning disciplines. The first opinion is that the construction decision has been made and that the location of the new capital is the right one. The second opinion is that it is still a concept that should be explored. The third opinion, on the other hand, is against the whole idea of new capital. This study aimed to analyze the different lines of thinking (Abdelmoaty and Elsoad 2020). The goal of their study is to explore the various aspects of smart technology and its applications in the hospitality industry. Their study has been conducted in Egypt's new administrative capital, the study has looked into the different kinds of smart technology and how they can be utilized in the hospitality industry. It also analyzed the current usage of AI in the field, the results revealed that the hotels and airport in Egypt's new administrative capital do not currently offer new smart services to their guests, so they suggested the use of smart apps and services at airports to help passengers complete their airport procedures more easily. It also suggested the implementation of smart rooms and restaurants in hotels.

Ghalib et al. (2021) have presented a comparative analysis of the current and future administrative capitals of Kazakhstan and Egypt, despite the differences, the similarities between Egypt and Kazakhstan can be identified in various aspects of their economies, political and social structures, and governance, the study found that both of these capitals are important factors for nation-building and economic development, the authors concluded that the administrative capital of Egypt can follow the well-established steps of its counterpart in Kazakhstan. However, they noted that the country's new capital will also face various challenges and obstacles.

The concepts of smartness and city singularity were used as the starting point for one of the studies, and other elements, such as character, personality, and identity, were also considered. Abusaada and Rashed (2022) have utilized a descriptive-analytical approach for the study; their analysis suggests that common ground exists between urban life and everyday forms of existence. The authors identified 13 factors that influence the development of Egypt's new administrative capital city. They disregarded lifestyle factors since the project is still in its early stages, then proposed a framework for developing capital cities that will have a singularity across their urban forms, also, the study presents a framework for developing smart cities. It uses the four keys to enhance the urban structure of such places and claims that the smart city idea could lead to new directions in the design of capital cities all around the world.

Although the Egyptian smart city framework has gained widespread attention, its contribution to the development of the country's urban policies has been relatively little studied. Also, one of the study's objectives was to analyze the case study of Egypt's smart city policy and model. It utilized the pioneer and fourth-generation cities' narrative analysis to examine the policy's implications, it provided an overview of the Egyptian smart city system and its contribution to the country's innovation development. The case study was also analyzed using the administrative capital's interview with officials of the urban development company. The findings of the case study indicated that the NAC could serve as a successful model for Egypt's smart city system. The case study's smart city framework utilizes a comprehensive strategy to develop an urban system that can attract both innovation and knowledge (Ali 2022).

4 Conceptual Framework

4.1 Defining Built Environment

Since the 1950s, numerous researchers have delved into design theories and methods that emphasize the connection between architecture and the environment, considering factors such as climate, geography, hydrology, vegetation, and the habits of inhabitants. A wide array of studies has been conducted globally, investigating the relationship between architecture and urban design with the climate (Hillier 1976; Hawkes 2013; Murtagh 2016). These studies highlight that regional architectural features often emerge in response to specific climates, even when geological locations and cultural contexts differ (Elnabawi and Hamza 2020). Identifying early-stage urban design decisions that impact energy efficiency remains challenging (Saini 1980). Achieving environmentally responsible urban development requires balancing development needs with environmental considerations. It is crucial to meet the population's basic needs while enhancing living standards and ecosystems (Lee 1997).

Urban development ranks as the second-largest global industry, trailing only agriculture. The production of building materials demands vast amounts of energy and finite resources. The core objective of sustainable architecture is straightforward: to create appealing, comfortable, and affordable shelters that minimize environmental impact during both construction and use. In practice, environmental design incorporates principles of sustainability, environmental consciousness, and green, organic methodologies to craft design solutions that consider the site's characteristics, neighborhood context, microclimate, and local topography (Bianca 2000; Elnabawi and Saber 2022).

In engineering and social sciences, the term "urban environment" or "built environment" refers to the man-made surroundings that facilitate human activities, from individual buildings to entire cities (Sarkis et al. 2011). This has been described as "the space constructed by humans, where they live, work, and engage in daily activities." Urban ecology involves spaces created or modified to meet human housing and organizational needs. It encompasses architecture, urban planning, building technology, civil engineering, landscape architecture, and urban management during periods of growth and decline (Dearry 2004).

In recent years, the scope of research on the "urban environment" has expanded to include factors like access to healthy food, community gardens, mental and physical health, and infrastructure supporting walkability and cycling (Lavin and Higgins 2006). Fundamentally, the built environment encompasses all human-created spaces that facilitate human activity, from expansive urban areas to intimate personal environments (McClure and Bartuska 2007).

In this framework, the built environment comprises human and cultural influences (the motivations for construction), environmental aspects (natural and man-made settings), and technological elements (materials, energy, and financial resources, along with the methods and systems required to develop and maintain the built environment). Sustainable development in construction entails using design and building practices that strive to achieve a holistic balance across economic, social, and environmental aspects. The built environment goes beyond just constructs and systems; it also involves the interactions among people in local communities and their cultural experiences, to foster a sustainable environment that integrates social, economic, and cultural aspects of sustainability within the local economy (Beynon 2010).

Cultural sustainability should be a central focus when designing policies and executing development plans for the built environment. According to Moffatt and Kohler (2008), the built environment encompasses more than just physical structures, infrastructure, and transportation systems; it also includes the social fabric, cultural experiences, and human interactions. To foster sustainable communities, it is essential to thoroughly evaluate and align local and community needs—reflecting the cultural environment—with the objectives of built projects.

4.2 Copenhagen: A Model of Sustainable Urban Living

Copenhagen is regularly hailed as a model for sustainable urban dwelling, showcasing a built environment that prioritizes eco-friendliness, accessibility, and community engagement. The city's commitment to sustainability is evident in its large biking infrastructure, which inspires residents to use bikes as their primary mode of transportation. With over 390 km of devoted motorbike lanes, Copenhagen has successfully decreased traffic congestion and air pollutants, all at the same time as promoting a healthier way of life. Additionally, the combination of green areas, along with the King's Garden and the waterfront areas, creates inviting public nation-states wherein humans can gather, relax, and connect to nature (Peštová 2022).

The architectural landscape of Copenhagen blends modern-day design with historical charm, reflecting the city's wealthy cultural history. Notable initiatives, which include the revolutionary Copenhagen Opera House and the cutting-edge Ørestad district, spotlight the city's commitment to integrating present-day design with capability. Urban planning in Copenhagen emphasize blended-use traits that support vibrant neighborhoods, allowing citizens to live, paintings, and socialize within close proximity (Winter 2019). This approach no longer simply enhances the first class of existence but also fosters social interplay and inclusivity.

5 The National Strategic Plan for Urban Development in Egypt 2052

The primary objective of Egypt's National Strategic Plan for Urban Development 2052 is to expand the country's inhabited area from 7% to 14%. This expansion aims to create a range of development opportunities and establish new urban areas to accommodate population growth and meet the needs of residents. This goal has been pursued through key urban development initiatives, including the Suez Canal Corridor project and the new city of El Alamein, among others. These projects, which are planned and overseen at the national level, are also supported by a national road network project designed to promote urban development and improve access to the newly developed cities (National Strategic Plan for Urban Development—Vision 2052 2017).

5.1 Developmental Vision

Egyptian public policymakers envisioned Egypt as an advanced and competitive country at the global level. To achieve this goal, several criteria needed to be fulfilled; a balanced society adhering to the national identity and cultural heritage, as well as its religious and civilizational values, citizens enjoy the luxury of living and the existence of life within the framework of an integrated environmental system and a democratic climate that motivates sustainable development and community participation, actors.

In general, this strategy aims to double the Egyptian population from about 7% to about 14%, by implementing new cities, especially smart ones, thus eliminating the population density crisis, in Cairo and other urban cities. The large urban expansion, of course, means doubling the size of the networks of facilities and services provided to citizens, to accommodate the expected increase in the future, plans that have already begun to be implemented.

During the past 4 years, the coverage rate of drinking water service in rural areas was raised from 95% to 97%, network lengths increased by 10%, urban sanitation coverage increased from 79% to 94%, and in rural areas from 12% to 34%, which is more than double what has been implemented since the start of connecting the sewage service to rural areas. The total amount of drinking water produced at the state level is 10.5 billion m^3 annually, which is pumped through 2730 stations, as part of a huge investment plan to provide pure water to citizens and to safely dispose of sewage waste. (National Strategic Plan for Urban Development—Vision 2052 2017).

5.2 Main Pillars of the National Strategic Plan for Urban Development

This vision is based on three main pillars: economic efficiency (represented in the optimal use of the site and resources), social justice (population balance and poverty reduction), as well as dealing with natural and security risks, whether environmental, internal, or external. However, Economic efficacy was given priority, especially in agriculture. Giving priority to trying to achieve self-sufficiency and optimal use of land through: Developing a competitive agricultural sector, Exemplary exploitation of traditional water resources in the reclamation of new agricultural lands. And relying on the desalination of sea and brackish water to add new agricultural lands.

Additionally, attention was given to developing a highly competitive industry with a focus on export, developing integrated industrial complexes, and raising the quality of Egyptian production to international levels. Tourism is also considered an important industry for Egypt intending to increase the number of tourists to 53 million through; New areas for global tourism on the northern coast and the Red Sea, linking them to cultural tourism in the valley, developing the capabilities of workers in the tourism field, Raising the quality and diversity of the tourist product and activating marketing to increase the average length of stay for tourists. This is accompanied by the development of the infrastructure, especially the ports and the transport network. Egypt's logistics profile will play a significant role in the country's economic recovery.

Trade and Logistics Services are another industry that would help achieve the strategic goals. This necessitates a skilled labor force, improving transport and communications infrastructure, and providing a huge local market with the necessary areas for modern retail and logistics. Not to mention developing the communication and information technology in Egypt, which has recently witnessed a growth of 7–10%, in addition to telecommunications, which represents 5% of the national income, while the volume of electronic commerce represents almost 20%. This fact was reflected in the increasing number of professionals in ICT and inventions to reach 120,000 job opportunities (General Authority for Investment and Free Zones n.d.).

It is important to note that the success of the previously mentioned projects relies on social justice to ensure sustainability. Thus, the plan included launching regional development and restructuring the development system based on distributing population within the framework of intensive and focused regional development, focusing on developing poor areas in northern and central Upper Egypt as a priority. A framework that integrates with both existing communities and new cities involves adopting a strategy for expanding urban areas with small to medium-sized populations. This strategy encompasses diverse patterns and functions, including regional services, advanced infrastructure, agricultural support, and tourism development.

5.3 Egyptian Energy Policy

Egypt aims to produce 42% of its electricity from renewable sources by 2035. To reach this target, the country has introduced several key policies: initially, feed-in tariffs are used to attract investment, followed by competitive bidding processes to reduce Purchasing Power Agreement (PPA) costs. Additionally, various reforms have been implemented to ready the market for renewable energy and accommodate new market entrants.

The idea of reform was central to the enactment of Electricity Law No. 87 of 2015, which introduced a new legislative and regulatory framework aimed at overhauling the electricity market. This law was designed to boost competitiveness within the sector by allowing third parties to access and utilize the electricity grid for a fee. Additionally, it enhanced the role of the Electricity Regulatory Authority, establishing it as an independent public body tasked with overseeing, supervising, and coordinating stakeholders involved in electricity production, transmission, and distribution (Rogge et al. 2015).

During the 2014–2019 period, Egypt's approach to reaching its 42% renewable energy target predominantly focused on generation-side policies, such as feed-in tariffs and tenders. However, there were no incentives targeting the demand or consumption side to encourage end-users to choose renewable over fossil fuel-based energy. As a result, the capacity of fossil fuel-based power plants grew from approximately 32–53.5 GW, while renewable energy plants saw a more modest increase from 3.5 to 6 GW (EEHC 2015; NREA 2020).

6 The New Administrative Capital

As shown in Fig. 1, a 170,000-acre site was allocated for the creation of the "New Administrative Capital" to the east of Cairo. The aim is to establish the new capital as a prominent political, cultural, and economic hub within the Middle East and North Africa region. The city is envisioned to foster a thriving economic environment that supports sustainable development while preserving Cairo's unique historical and natural heritage. Its strategic location, close to the Suez Canal, regional highways, and major routes, positions it to attract a significant population. In the first phase, covering 10,500 acres, the city is expected to house approximately five million residents, along with 40,000–50,000 government employees relocating to the new administrative center, with plans to expand to accommodate 100,000 employees within 3 years (Staff 2021).

Seven key objectives have been outlined for the development of the New Administrative Capital: (a) creating a green city with 15 m^2 of green space per person; (b) establishing a sustainable city by incorporating sustainability principles in energy use and waste recycling, with 70% of building rooftops equipped with solar panels; (c) designing a pedestrian-friendly city, where 40% of the road network is designated for walking and cycling; (d) building a livable city with a housing distribution of 35% high-density, 50% medium-density, and 15% low-density, with 30% of the city's land allocated for residential purposes, including 1.5 million housing units; (e) ensuring accessibility through a comprehensive transportation network consisting of trains, metro, trams, trolleys, buses, and taxis; (f) developing a smart city where all services are delivered electronically, with full coverage by the global information network; and (g) positioning the city as a financial and business hub for the Greater Cairo and Suez Canal regions, with 30% of the city dedicated to labor and financial sectors.

As shown in Fig. 2, the significance of the "New Administrative Capital" project for Egypt stems from its potential to alleviate traffic congestion in the historic center of Cairo, also known as Khedive Cairo, by relocating government institutions, foreign embassies, and consulates. Additionally, the development of a predominantly residential city with a capacity of approximately seven million residents helps address the housing shortage. Furthermore, the new capital is envisioned as a vibrant economic and investment hub that will attract both domestic and international investments, creating thousands of job opportunities in the process.

7 Important Services Expected to Be Provided by the New Administrative Capital

7.1 Better Health Services

The capital will feature a medical district spanning 20,000 m^2, including three hospitals: a public hospital with 300 beds, a children's hospital, and a facility specializing in obstetrics and gynecology. These hospitals will be supported by a doctor training center and an advanced technical institute for nursing. In addition, a number of major private healthcare facilities are being developed, including the "Capital Care Complex," which will accommodate specialized medical centers, outpatient clinics, radiology and diagnostic centers, and rehabilitation facilities. The total investment in this project is expected to reach approximately 2 billion pounds. Furthermore, several private medical malls, such as "Capital Care Mall" and "Medical Capital Mall," will be established. These facilities will include medical clinics, radiology and diagnostic centers, and pharmacies, and will be strategically located near residential areas (El Yom n.d.).

7.2 Educational Services

Currently, the New Administrative Capital houses 50 international and language schools, including 4 Japanese institutions such as International Knowledge School, Al Bayan Educational School, Manar House School, and the British International School. These are complemented by government schools covering all educational levels. A notable educational facility being constructed by the Engineering Authority of the Armed Forces in collaboration with Misr Italia Real Estate Company will be located in the R2 district of the new capital.

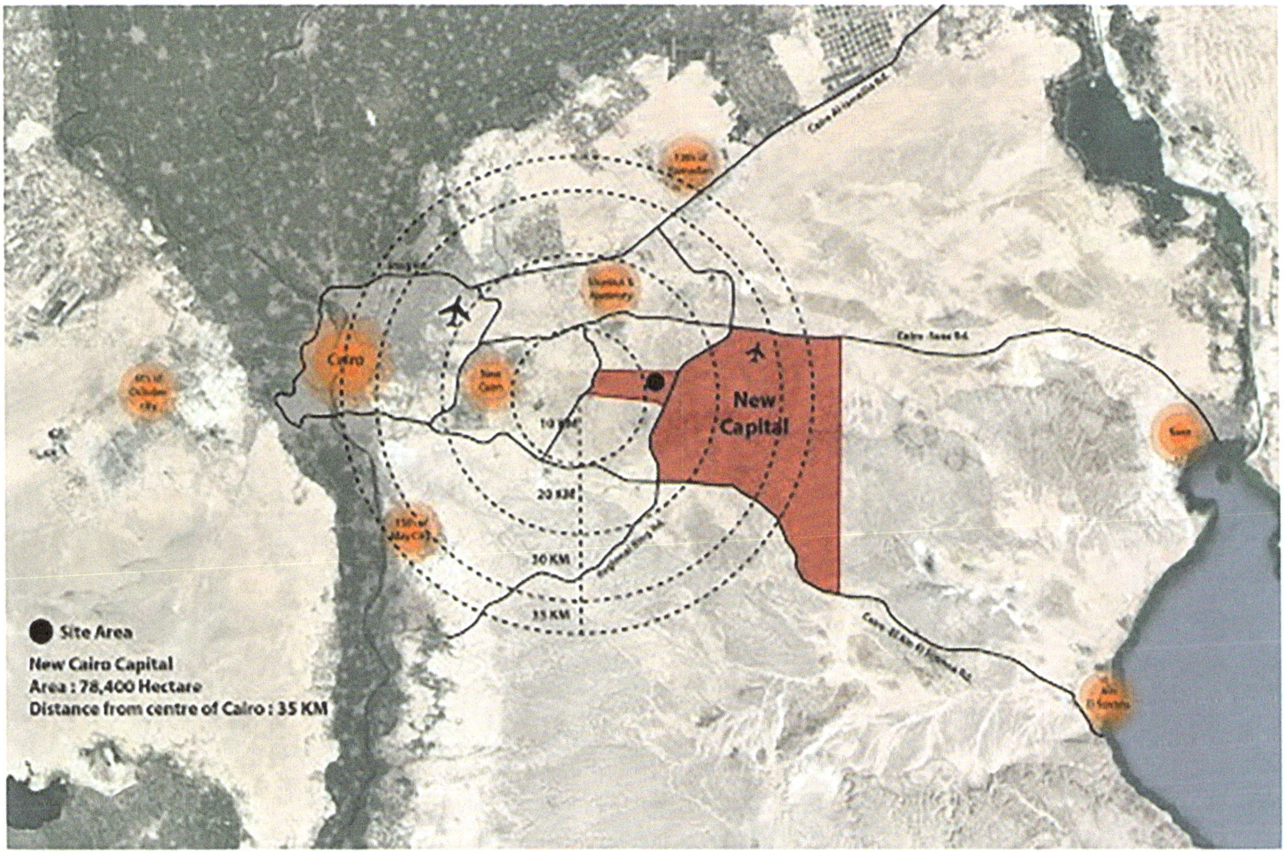

Fig. 1 The location of the new administrative capital. (Source: Capital Egypt: http://www.acud.eg/)

Regarding higher education, several foreign, international, and Egyptian universities are planned, with some already operational for their second academic year. The American International University will span 50 acres and accommodate 21,000 students. Coventry University, located in a prime area near Midtown Sky, is among these institutions. An area of approximately 450 acres has been set aside for the Japanese University, the Canadian College, and Sinai University. Additional institutions include the German International University of Applied Sciences, the Hungarian University, the Swedish University, the New Canadian University, and the University of Liverpool (Kandil n.d.).

7.3 Facilities and Infrastructure

In Fig. 3, as we have shown, the green belt, as well as the modern waste management systems for garbage disposal and recycling, will be implemented across all neighborhoods and residential buildings in the New Administrative Capital. While a dedicated water treatment plant is planned for the city, water is currently being supplied temporarily from the New Cairo and Badr stations.

The capital also has a private power station with a capacity of 4800 MW, completed in 2018, which meets the city's energy demands as well as those of surrounding areas. Built by the German company Siemens, the power station is among the largest in the world. It consists of four modules, each with two natural gas units and a steam unit powered by exhaust from the gas units to conserve fuel, along with two energy recovery boilers. Notably, this station is the first of its kind in Egypt and the Middle East, located in a mountainous area away from traditional water sources like rivers or seas, and operates using an air cooling system featuring 12 giant fans, a technology being used for the first time. The station includes eight gas units of the H-class model, the most advanced gas technology globally, with each unit capable of generating 600 MW of power (Minister n.d.).

So, to calculate the power generation capacity of a power station, we can use the following formula:

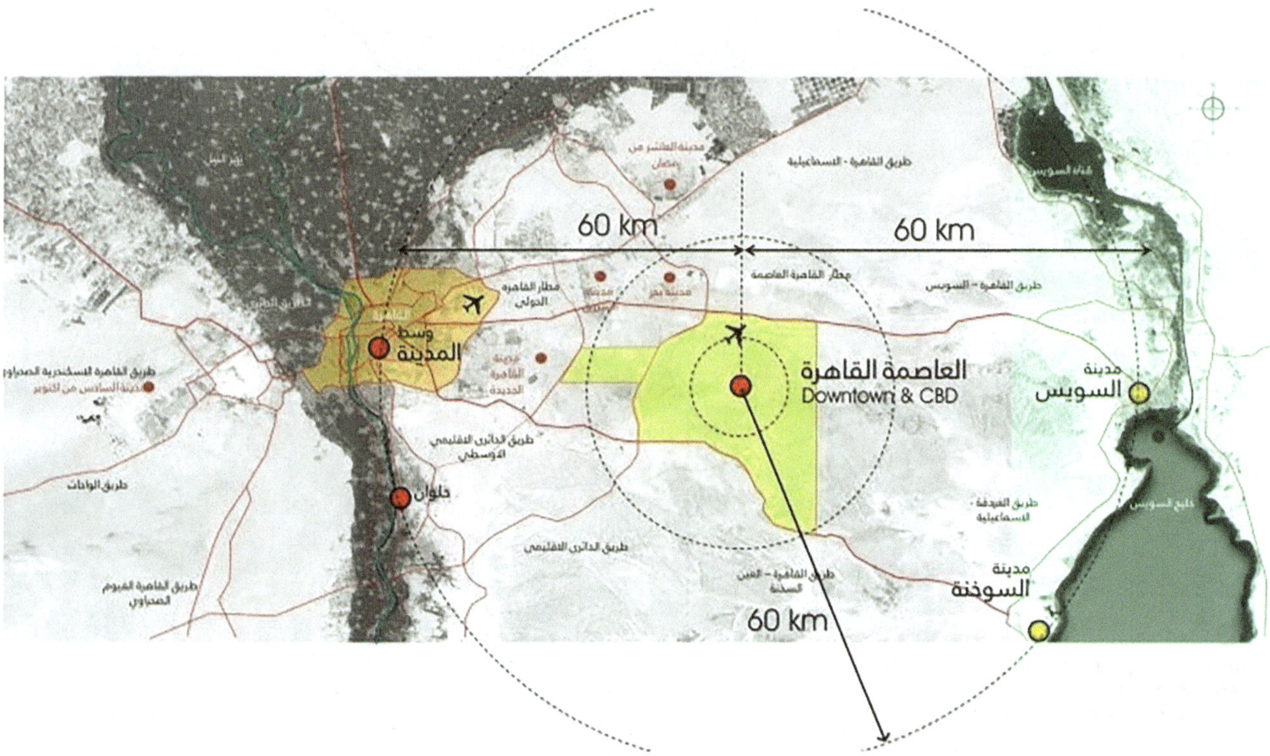

Fig. 2 The center location of the New Administrative Capital. (Source: Aqarmap: https://advice.aqarmap.com.eg/en/all-you-need-to-know-about-new-administrative-capital/)

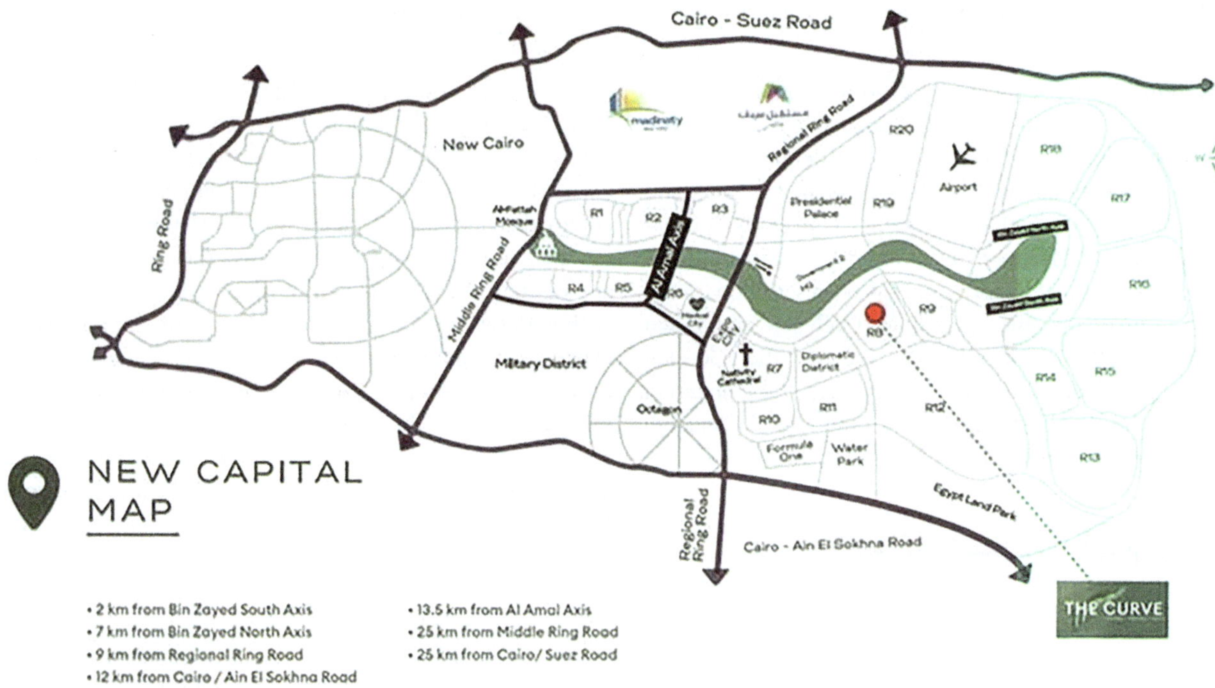

Fig. 3 The neighborhoods of the New Administrative Capital. (Source: Inland Properties Group: https://ipgegypt.com/en/projects/223-the-curve-new-capital)

$$\begin{aligned}\text{Power generation capacity} &= \text{Number of units}(\text{generators}) \\ &\quad \times \text{output per unit} \\ &= 8\,\text{Units} \times 600\,\text{MW} \\ &= 4800\,\text{MW}\end{aligned}$$

The New Administrative Capital is designed to be an ideal model for built environments as all its districts, such as the government district, the embassies and diplomatic quarter, and the business section in the New Administrative Capital, are all built according to the requirements of environmentally compatible buildings, which rationalize the consumption and sustainability of resources such as energy and water. Also, green areas are within the new global trend towards expanding green buildings to maintain the sustainability of resources and reflect the state's public policy to expand environmentally compatible buildings and sustainable cities.

The matter is asserted by decisions made by the Egyptian Cabinet on the built environment and establishing new cities and urban development, as well as the Supreme Council for Planning and Urban Development decisions on planning and building regulations and requirements for Egyptian cities, which included the necessity of assigning universities to review and prepare plans for proper building requirements for newly established cities.

Also, universities are assigned the responsibility of preparing a general theme/design for architectural fronts for buildings suitable to Egypt's climatic environment, and the distinctive characteristics of each city, the space lands suitable for construction will also be determined, through the map of climatic variables and the data available at the competent administrative authority, and made available to such universities or who is commissioned to prepare site validity certificates as per the planning and building requirements previously specified.

Therefore, we see that the decisions of the Council of Ministers and the National Strategic Plan for Urban Development in Egypt 2052 will depend on urban development in building new cities, on compatibility with the climatic environment and green areas, following the planning and structural controls and requirements in the Egyptian governorates.

8 Concluding Remarks

The establishment of Egypt's New Administrative Capital, formalized by Presidential Decree No. 57 of 2016, reflects a strategic alignment with built environment public policies. By designating the lands of the New Administrative Capital and the Sheikh Mohammed bin Zayed Urban Gathering as part of new urban communities, and involving key entities like the New Urban Communities Authority and the National Service Projects Authority, Egypt has demonstrated a forward-thinking approach to urban development. The incorporation of climate-conscious planning and green spaces into building regulations across governorates further solidifies Egypt's commitment to sustainable urban growth.

This case study highlights how the built environment plays a critical role in Egypt's economic, political, and social progress. The integration of individual buildings, transportation systems, urban landscapes, a new airport, a business district, and residential areas shows how physical infrastructure contributes to the nation's development. The government's approach to urban planning aligns with environmental and climate considerations, creating communities that foster cultural interactions and enhance societal well-being.

However, it is recommended that future policies place even greater emphasis on microclimatic conditions and context-specific socio-cultural factors. While the built environment encompasses infrastructure and transportation, it also extends to human interaction, cultural experiences, and the creation of green spaces within urban settings. These dimensions warrant further exploration and present opportunities for future research, particularly in crafting a more sustainable and resilient society.

References

Abdelmoaty AG, Elsoad AS (2020) Smart technology applications in tourism and hospitality industry of The New Administrative Capital, Egypt. J Assoc Arab Univ Tour Hosp 19(2):102–129

Abusaada EA, Rashed AR (2022) Exploring the singularity of smart cities in the New Administrative Capital City, Egypt. Ain Shams Eng J

Ali MA (2022) Smart city policy in developing countries: case study of the new administrative capital in Egypt. J Public Aff:1–16

Angelidou M (2014) Smart city policies: A spatial approach. Cities 41:S3–S11

Anthopoulos LG (2015) Comparing smart cities with different modeling approaches. In: Proceedings of the 24th international conference on world wide web, pp 525–528

Beynon D (2010) Defining cultural sustainability in multicultural built environments. Int J Environ Cult Econ Soc Sustain:255–266

Bianca S (2000) Urban form in the Arab world, past and present. Thames & Hudson, New York

Bogdanov OJ (2019) Scrutinizing the smart city index: a multivariate statistical approach. Zborn Rad Ekon Fak Rijeci 37(2):777–799

Burlacu MB (2022) Smart cities and transportation: reviewing the scientific character of the theories. Sustainability 14(13):8109

Cugurullo F (2021) Frankenstein urbanism: eco, smart and autonomous cities, artificial intelligence, and the end of the city. Taylor & Francis. ISBN 978-1-317-31362-5

Dearry A (2004) Impacts of our built environment on public health. Environ Health Perspect 112:A600–A601

EEHC (2015) Annual report 2014–2015. Egyptian Electricity Holding Company, Cairo

El Yom A-M (n.d.) Health minister announces medical city in new administrative capital. https://egyptindependent.com/health-minister-announces-medical-city-new-administrative-capital/. Accessed 8 Feb 2018

Elnabawi MH, Hamza N (2020) Outdoor thermal comfort: coupling microclimatic parameters with subjective thermal assessment to design urban performative spaces. Buildings 10(12):238

Elnabawi MH, Saber E (2022) Reducing carbon footprint and cooling demand in arid climates using an integrated hybrid ventilation and photovoltaic approach. Environment, Development and Sustainability 24(3):3396–3418

General Authority for Investment and Free Zones (n.d.). https://www.investinegypt.gov.eg/english/pages. Accessed 30 Nov 2022

Ghalib, El-Khorazaty H, Serag Y (2021) New capital cities as tools of development and nation-building: review of Astana and Egypt's new administrative capital city. Ain Shams Eng J 12(3):3405–3409

Gibson DV, Kozmetsky G (1992) The Technopolis phenomenon: smart cities, fast systems, global networks. Rowman & Littlefield Publishers

Hawkes D (2013) The environmental tradition: studies in the architecture of environment. Taylor & Francis, UK

Hillier BA (1976) Architecture as a discipline. J Archit Res:28–32

Kandil AM (n.d.) 50 Schools to be established in New Administrative Capital. https://www.egypttoday.com/Article/1/57702/50-schools-to-be-established-in-New-Administrative-Capital. Accessed 17 Sept 17 2018

Khalifa M, Atwa R (2021) New public policy and its impact on comprehensive development plan. Inform Sci Lett J 10(3):487–496

Lai CS (2020) A review of technical standards for smart cities. Clean Technol 2(3):290–310

Lavin T, Higgins C (2006) Health impacts of the built environment: a review. The Institute of Public Health in Ireland, Dublin

Lee NA (1997) Sustainable development in a developing world—integrating socio-economic appraisal and environmental assessment. Edward Elgar Publishing LTD, UK

McClure WR, Bartuska T (2007) The built environment: a collaborative inquiry into design and planning. In: Bartuska T (ed) The built environment: definition and scope, 2nd edn. Wiley, Oxford, pp 3–14

Minister EH (n.d.) Ahram online. https://english.ahram.org.eg/NewsContent/3/12/278625/-.aspx. Accessed 11 Oct 2017

Moffatt S, Kohler N (2008) Conceptualizing the built environment as a social-ecological system. Build Res Inf:248–268

Montes J (2020) A historical view of smart cities: definitions, features and tipping points. SSRN Electron J. https://doi.org/10.2139/ssrn.3637617. ISSN 1556-5068. S2CID 238125868

Murtagh NA (2016) The relationship between motivations of architectural designers and environmentally sustainable construction design. Constr Manag Econ 34(1):61–75

National Strategic Plan for Urban Development—Vision 2052 (2017) United Nations Development Programme

NREA (2020) Annual report 2020. New and Renewable Energy Authority, Cairo

Opoku A (2015) The role of culture in a sustainable built environment. In: Chiarini A (ed) Sustainable Operations Management: Advances in Strategy and Methodology, pp. 37–52. Springer, Cham

Paiho SE (2022) Opportunities of collected city data for smart cities. IET Smart Cities 4(4):275–291. https://doi.org/10.1049/smc2.12044. S2CID 253467923

Peštová V (2022) Sustainable urban development: case study Copenhagen. In: Proceedings from the EDAMBA 2021 conference, pp 390–399

Praharaj SH (2018) Evolving a locally appropriate indicator system for benchmarking sustainable smart cities in India. In: Sustainable development research in the Asia-Pacific region: education, cities, infrastructure and building

Rogge K, Breitschopf B, Mattes K, Cantner U, Graf H et al (2015) Green change: renewable energies, policy mix and innovation

Saini BS (1980) Building in hot dry climates. Wiley, New York

Sarkis J, Meade L, Presley A (2011) Sustainability in the built environment: Factors and a decision framework. In: Quaddus MA, Siddique MAB (eds) Handbook of Corporate Sustainability: Frameworks, Strategies and Tools, pp. 113–136. Edward Elgar Publishing, Cheltenham

Serag Y (2017) The new administrative capital of Egypt a critical review from the regional. In: 1st International conference on towards a better quality of life

Staff ET (2021) Egypt's road quality ranking advances thanks to national projects, Cairo. https://www.egypttoday.com/Article/1/105586/Egypt-s-New-Administrative-Capital-can-be-home-to-6. Accessed 6 Feb 2021

UNESCO (2012) Culture: A driver and an enabler of sustainable development. Thematic Think Piece, United Nations System Task Team on the Post-2015 UN Development Agenda. UNESCO, Paris

Winter AK (2019) The green city citizen: exploring the ambiguities of sustainable lifestyles in Copenhagen. Environ Policy Gov 29(1):14–22

World Economic Forum (2022) The Global Risks Report 2022. World Economic Forum, Geneva

Yin CX (2015) A literature survey on smart cities. Sci China Inf Sci 58(10):1–18

Urban Heritage and Cultural Identity: Methods and Applications for Conservation

Revival of Urban Heritage as an Engine of Social and Economic Recovery in Post-War Homs, Syria

Abeer Abdelal and Nadia Kandakji

Abstract

The war in Syria has mercilessly devastated every place it has touched, particularly targeting buildings of historical and architectural significance, resulting in abandoned neighborhoods, destroyed buildings, a torn, besieged, and impoverished society with little hope for lasting peace. However, amidst this despair, cultural heritage emerges as both a driving force and a symbol of resilience, embodying people's confidence in their identity, unity, and sense of belonging.

Considering the severely damaged old part of the Syrian city of Homs as a case study, this paper documents cultural heritage impact as a concept to apply creative solutions to confront post-war challenges, and revive destroyed historical parts of the city.

The paper examines the revival of a cohesive network of heritage symbols in the historical part of Homs. Besides the official role that was played by the city's authorities, this paper emphasizes the significance of spontaneous initiatives that were carried out by the city's residents to achieve this goal.

After the devastating years of war, buildings and public spaces that cater daily needs of residents played a crucial role in encouraging people and various activities to bring life back to their old city, proofing its inherent ability to overcome challenges.

Keywords

Cultural heritage · Recovery · Post-war city · Homs

1 Introduction

Culture is fundamental to a city's resilience, especially those that have endured extreme challenges such as wars or natural disasters. In the context of prolonged conflicts, like the ongoing war in Syria, the concept of urban resilience becomes particularly relevant. Urban resilience encompasses a city's ability to recover and thrive despite severe adversities, extending beyond physical reconstruction to include social, economic, and environmental dimensions. This resilience is often visible in the efforts to bring life back to the destroyed and abandoned parts of the city, through small community-driven projects and large-scale reconstruction initiatives. These initiatives collectively aim to restore social and economic vitality, providing a semblance of normalcy and hope amidst the desolation.

This paper examines how cultural heritage contributes to urban resilience in the post-conflict city of Homs in Syria, by focusing on these key aspects: Firstly, cultural heritage is a powerful tool facing social and economic challenges in the post-conflict city. It acts by strengthening the community's social fabric through shared history and belonging. Secondly, local communities play an essential role in cultural heritage revival. Residents' intimate knowledge of their city's history and traditions ensures that restoration efforts are authentic and reflective of the community's identity, leading to more effective and sustainable recovery. Thirdly, the paper explores the Integration of Urban Symbols and Social Behavior. Urban symbols such as buildings, public spaces, and landmarks are interconnected with community behaviors and practices. These dynamic interactions enhance the city's resilience and vitality.

Undoubtedly, the process of restoring cultural heritage is most effective and sustainable when it involves the active participation of local communities. Local residents possess a deep understanding of the cultural and historical significance of their city's landmarks, which enables them to be engaged in the preservation of their cultural heritage.

A. Abdelal (✉) · N. Kandakji
Al-Wataniya University, Hama, Syria
e-mail: abeer.abdalaal@wpu.edu.sy; Nadia.Kandakji@wpu.edu.sy

F. Alberti et al. (eds.), *Towards Sustainable Urban Regeneration: Utilizing Environmental, Cultural, and Social Assets*, Advances in Science, Technology & Innovation, https://doi.org/10.1007/978-3-032-02091-8_20

In Syria, the ongoing political crisis and severe economic sanctions have significantly impeded large-scale reconstruction efforts. This has left the authorities of the heavily destroyed city of Homs authorities and its citizens with limited resources to rebuild their city on a grand scale. Despite these formidable challenges, local communities in this city have shown remarkable resilience.

Research importance emerges from the necessity of documenting the recovery process, combined with the authors' personal connection to the Old City. By meticulously documenting observations and recovery activities, the authors tried to capture the determination of the residents and the transformative power of heritage in post-war recovery.

This paper aims to illustrate how concepts derived from local urban heritage can guide the social and economic rebuilding of post-war cities. It uses the old city of Homs as a case study to demonstrate how these heritage-based principles can be effectively applied in the recovery process. The paper identifies three distinct stages of recovery that Homs experienced, each with different key players and strategies: (1) Immediate response as phase one: offered a swift response from official bodies, international organizations, and local groups focused on immediate relief and stabilization. (2) Coordinated efforts as phase two: a more sustained effort with greater coordination between local and international actors, yet faced significant challenges due to limited resources and ongoing security concerns. (3) Community-led revival as phase three: substantial progress mainly driven by the individual initiatives of local Homs residents, utilizing their deep connection to the city's heritage and intimate knowledge of its history and cultural significance. This phase proved most successful in revitalizing the old city.

2 Research Methodology

To ensure a comprehensive understanding of the revival process in the old city of Homs, the research data for this paper were gathered from a diverse range of sources. Starting with interviews with official administration representatives, to gain insights into the policies and actions taken by local and national authorities to facilitate the city's recovery. Examine reports from international Organizations. In addition to discussions with residents, to uncover narratives of community-led initiatives that played a pivotal role in the city's rebuilding. Interviews were also done with representatives of non-governmental organizations and business owners, in addition to author's own observations, who are still reside in the city at the time of writing this paper.

3 Literature Review

In cities across the Arab region, the old city serves as a vital reflection of the community's cultural identity and history. These ancient urban centers are characterized by their unique architectural and urban morphology. Typically, these old cities feature a dense, interconnected urban fabric that creates a sense of enclosure and intimacy. This layout facilitates social interaction and fosters a strong community bond (Jabour et al. 2001). The ancient parts of contemporary Arab cities face a range of threats. These threats can be broadly categorized into natural and human factors (AlMalki 2004). The impact of war on urban heritage could be considered as one of the most effective human threats, especially in the "Global South" countries, including Afghanistan, Yemen, Iraq, Libya, and Syria.

In Syria, the civil war has severely damaged many historical buildings and sites. Similarly, in Iraq, the city of Mosul suffered extensive destruction, including the loss of significant cultural landmarks like the Great Mosque of al-Nuri and its iconic leaning minaret. Yemen's ongoing conflict has devastated the old city of Sana'a, a UNESCO World Heritage site, and Libya's civil strife has endangered many of its archaeological treasures.

Considering post-war recovery, it is not only about the rehabilitation of physical infrastructures but it also involves the restoration of social and cultural environments. This holistic approach is crucial for fostering a sustainable and resilient community (Adly et al. 2021). Effective recovery begins with the coordinated and integrated participation of all stakeholders, including society, government, NGOs, and architects. This collective effort ensures a comprehensive, sustainable, and fulfilling recovery process, also addressing the social, economic, and cultural needs of the affected communities (Krasny et al. 2014).

The post-war recovery process is heavily influenced by geographic location, environmental conditions, economic resources, and political stability. The relationship between these local factors and the success of post-war recovery efforts is complex, as each context presents its own unique challenges and opportunities (Salmo et al 2021). While international aid organizations can provide critical funding and support, the long-term success of recovery efforts ultimately depends on local capacity-building and community mobilization. A pertinent example of the importance of local involvement can be seen in Kosovo, where local NGOs and community groups were instrumental in rebuilding homes and public spaces after the conflict in the late 1990s (Earnest and Dickie 2012).

Another example of the management of urban cultural heritage revival strategies after wars, the case of Mosul city in Iraq, the war affected both tangible and intangible heritage components. Reconstruction of Mosul requires a tailored strategy, starting with the empowerment and activation of local authorities as key stakeholders and sources of knowledge. Local authorities possess an intimate understanding of the cultural significance and historical context of the area, making their involvement crucial for effective restoration efforts (Ali et al. 2002).

In the case of Homs city, several academics have recently researched the rebuilding of the old city in Homs. One notable study by Kassouha (2014) sought to develop a strategy for restoring the cultural identity of Homs during its reconstruction process. The research highlighted the mutual influence between the city's cultural identity and the post-war reconstruction process, emphasizing that urban identity is a key factor for a resilient recovery. She argued that successful recovery must be closely linked to economic development, the political landscape, and social cohesion (Kassouha 2014). In another paper discussing the role of local reconstruction planning in the Old City of Homs, researchers highlighted the significance of community participation in the recovery process. They found that the level of community involvement in cultural heritage recovery was limited and weak, primarily due to administrative and legislative barriers. They noted that bureaucratic hurdles and restrictive policies often prevented effective community engagement (AlZenbarakji and Atfeh 2016). They also pointed out that conflicts of interest among citizens themselves frequently hindered participation. These internal conflicts, whether stemming from differing priorities, economic interests, or social divisions, created obstacles to a unified and cooperative recovery effort. The research emphasized that for reconstruction planning to be successful, these issues need to be addressed, fostering an environment where community members can actively and collaboratively contribute to the preservation and revival of their cultural heritage (AlZenbarakji and Atfeh 2016).

As a fact, most of the existing literature discussing post-war Syrian cities often overlooks the opportunity to integrate the cultural dimension as a central component of the rebuilding process, instead focusing primarily on the broader aspects of physical and economic recovery.

4 Urban Heritage Agony in Syria

Syria, located in one of the oldest continuously inhabited regions of the world, is renowned for its rich cultural heritage. Almost every Syrian city boasts an old nucleus from which the modern city has expanded, such as Damascus, Aleppo, Hama, and Homs. These historical cores are often centrally located within each city, significantly influencing local social and economic performance. The old nuclei serve as vital hubs for cultural and economic activities. Unfortunately, the ongoing conflict in Syria has severely impacted these historical centers.

The Antiquities Law in Syria, established in 1963 (Legislative Decree 222 of 1963), designates architectural elements or groups of elements as having heritage value if they are more than 200 years old or possess historical, functional, or architectural significance. This legal framework aims to protect and preserve the rich cultural heritage of Syrian cities. However, the criteria for identifying heritage components have inadvertently created a legal loophole, leading to the loss of numerous architectural and urban symbols by the end of the last century. Focusing primarily on structures older than 200 years, the law overlooks many buildings and urban elements that, while younger, are still of considerable historical and cultural importance. This exclusion has resulted in the neglect and eventual demolition of numerous significant sites that contribute to the unique urban fabric and historical narrative of Syrian cities. Additionally, heritage conservation projects require substantial funding for maintenance, restoration, and protection of historical sites.

In pre-war times, many historical buildings and urban segments have suffered from neglect due to the limited financial resources available for their upkeep and preservation. In pre-war times, many historical buildings and urban segments have suffered from neglect due to the limited financial resources available for their upkeep and preservation. Later, the war in Syria led to widespread destruction of architecture and the displacement of people in several Syrian cities, including Aleppo, Homs, and Deir ez-Zor. Cultural heritage sites have been significantly damaged or destroyed, stripping these cities of their historical and cultural landmarks. Even cities with lesser levels of direct destruction, such as Latakia, have been indirectly affected by hosting waves of displaced families and experiencing the broader economic collapse across Syria.

5 Homs City

According to AlZehrawi (2003), one of the city's historians, Homs is distinguished by several significant heritage landmarks that contribute to its rich cultural and historical tapestry. Among these landmarks is the ruined castle, which stands as a testament to the city's ancient history and strategic importance (Fig. 1). The city was also famous of its seven gates, strategically placed on its surrounding wall. Despite the fact that none of these gates are still standing, they remain bustling centers of human activities, act as dynamic hubs of economic and social exchange, where the past and present intersect. The Heritage Market area (the Roofed Market) is another iconic feature, bustling with traders and shoppers and serving as a vital hub of economic and social activity.

Fig. 1 The strategic location of the historical ruined castle of Homs city. (Original source: Google Maps)

The Great Mosque situated at the entrance of the Heritage Market area is one of the city's most important religious sites. Additionally, Homs is also home to several historical churches that reflect the city's diverse religious landscape. These churches contain unique religious relics and have been integral to the spiritual and community life of the city's Christian population.

5.1 Historical Overview of Homs

Homs was mentioned as early as the second millennium BC, has endured numerous disturbances throughout history due to its strategic geographical location. Wars and floods occurring near the Orontes River, as well as devastating earthquakes in 1157 AD and 1170 AD, significantly impacted its development (AlZehrawi 2003).

The city's fortune began to change when the Ottomans took control of the region in 1616 AD. At that time, Homs experienced a period of social and economic stabilization. The Ottomans implemented administrative and infrastructural reforms that contributed to the city's growth and prosperity. This era of relative stability allowed Homs to flourish as a center of trade and culture, laying the groundwork for the rich heritage it is known for today (AlZehrawi & Al-Sibai 1992).

At the end of the eighteenth century, the area of Homs within the city walls was approximately 1.4 km^2, with a population of about 20,000. The city wall, which was constructed for protection against repeated attacks by Bedouins from the east, featured seven gates that controlled access and ensured security. The most important social and economic activities were concentrated in the north-western part of the city within the walls and in the immediate surrounding areas. This area became the focal point for the city's development, hosting markets, public squares, and key communal spaces. During this period, Homs was renowned for its textile products, which were highly valued and traded extensively. The city also acted as a significant exchange hub, facilitating trade between the products of the desert and the agricultural crops brought in from surrounding villages (Hussein Agha 2002), Fig. 2.

Most buildings in Homs featured facades made predominantly of bluish-black basalt stone, which contributed to the city's unique visual identity and linked the city's architecture to its natural geological surroundings.

The city was divided into eight distinct neighborhoods, each designed to protect the privacy of its residents according to their ethnic or religious affiliations (Hussein Agha 2002). For example, the Al-Hamidiya neighborhood was primarily inhabited by Christians and housed several historical churches, while the Bab Hood neighborhood, located in the western part of the old city, was predominantly Muslim. The southeastern neighborhood, named Bab-Drieb, formed a mixture of both Muslim and Christian residential zones. These neighborhoods provided a sense of community and belonging, and reflected the diverse cultural and religious tapestry of Homs (Hussein Agha 2002).

Albeit various origins of its communities, the city's main commercial and spiritual center was a place where everyone belonged, a vital network crystallized by time, the main arteries and squares intersected with the most important

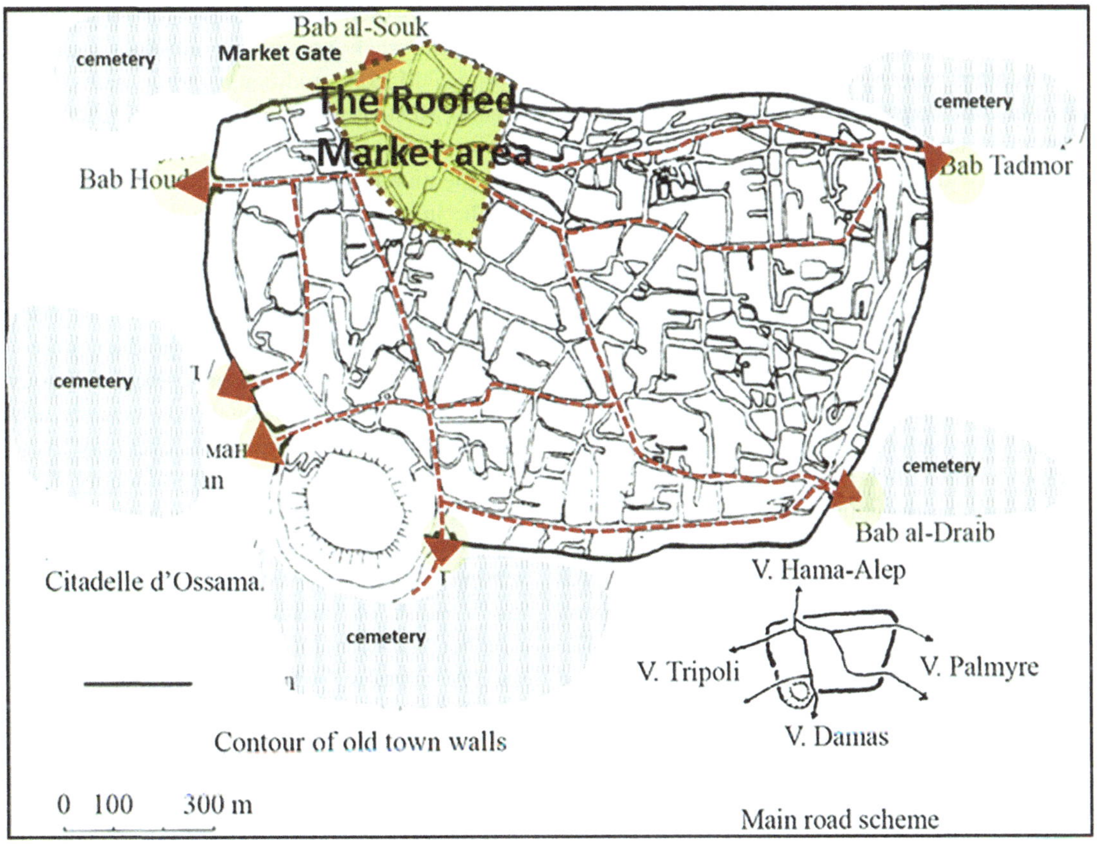

Fig. 2 The old city gates and commercial center in nineteenth century. (Original source: AlZehrawi 2003)

buildings and strongly linked with the socio-economic center of the city, the central market area (Belal and Shcherbina 2021). Around 1400 looms spread across the city, many of which were operated inside homes, which allowed for seamless interaction between residential and commercial areas (Hussein Agha 2002).

The late nineteenth century saw significant urban expansion, particularly in the commercial and administrative sectors. The main axis of the market extended towards the north, northwest, west, and southwest, creating new commercial and administrative areas and accommodating the development of modern residential neighborhoods (AlAhmad 1970).

At that time, Homs was renowned for its vibrant cultural and religious life, characterized by a rich tapestry of traditions practiced by its diverse communities. The city's Muslim residents, for instance, were known for engaging in various Sufi rituals, which played a significant role in their spiritual and communal life (Hussein Agha 2019). Additionally, the historical Church of Umm al-Zunar and the Cathedral of the Forty Martyrs served as focal points for important weekly and seasonal Christian prayers and celebrations. For entertainment and socializing, the city featured simple cafés attached to modest squares near each neighborhood's mosque. These cafés were popular evening gathering spots for men, who would meet to drink coffee, discuss daily matters, and listen to storytellers recount various tales (Asa'ad 1984). Outside the city walls, each family maintained its own cemetery, usually located near the city gates. Additionally, the city's outskirts were dotted with vineyards to the east and southeast, while the orchards to the west were considered the population's food basket. These orchards were not only crucial for supplying food but also served as popular spots for weekly picnics during the spring and summer, providing a respite from urban life and a place for families to enjoy nature (Hussein Agha 2019).

During the French Mandate (1925–1945), Homs experienced significant urban expansion, particularly in its city center. French planners aimed to modernize the urban landscape and introduced several key developments. One notable addition was the clock tower, which was strategically placed to emphasize the importance of the market square. This clock tower became a prominent landmark, symbolizing the blend of traditional and modern influences in the city. These developments revitalized the city's infrastructure and introduced a modern aesthetic and functionality that persisted well into the mid-twentieth century (Belal and Shcherbina 2021).

In the second half of the twentieth century, Homs underwent radical urban changes that caused significant damage to its urban fabric and cultural heritage sites. The push for modernization led to the construction of new street networks, commercial buildings, and residential complexes, often at the expense of historic neighborhoods and landmarks. Many traditional buildings and ancient structures were demolished to make way for new developments, resulting in the loss of valuable architectural heritage.

As the city expanded, insufficient attention was paid to integrating new developments with the existing historical context, leading to a fragmented and often incoherent urban landscape. As a result, many original residents of the Old City moved to settle in modern neighborhoods. The vacated homes in the Old City were often left to poor relatives or leased to families who had recently moved from the countryside in search of a better life (Harfoush et al. 2002). One of the most significant and disruptive changes occurred in the 1980s when the city's administration decided to demolish a complete neighborhood (Al-Arbaeen neighborhood) that was located next to the Heritage Market area. This decision aimed to annex the area to the expanding city center. The delicately woven urban fabric of Al-Arbaeen, characterized by its historical buildings and narrow streets, was replaced by large blocks designed to accommodate new official departments.

By the end of the twentieth century, the city administration in Homs issued new building regulations for the Old City, supervised by a specialized committee. The new regulations did not focus on enhancing permeability through the Old City; instead, the regulations maintained a fragmented approach, limiting the potential for economic and social revitalization (Harfoush et al. 2002).

Moreover, wide streets were constructed around the southern and eastern borders of the Old City. While intended to facilitate traffic flow and urban expansion, it instead contributed to the segregation of the Old City. This isolation exacerbated the deterioration of the area, as it became increasingly disconnected from the more vibrant and economically active parts of the city, as shown in Fig. 3.

5.2 Divergence in Heritage Value Perception

By the year 2010, a clear gap had emerged between what the official authorities and the local residents considered to be of heritage value in Homs. This created two distinct networks: formal and informal, each reflecting different aspects of the city's urban heritage.

Formal heritage network was defined by the official authorities and included elements that were legally recognized and protected under the Antiquities Law. This network typically focused on well-known historical and architectural landmarks such as ancient mosques and churches, the Heritage Market, traditional public baths, and the ruined Castle. Most of these elements were prioritized in official conservation and restoration efforts, often receiving funding and attention from both national and international bodies.

The informal heritage network was shaped by the local residents' perceptions and included elements that held cultural and social significance but were not officially recognized as heritage sites. This network comprised mainly of informal street markets, food product stores, craftsmen workshops, second-hand stores, local religious sites, and small scattered cafés. Though not formally acknowledged, they played a crucial role in the everyday lives of the residents and were integral to the city's identity. They are usually considered as a vital meeting place where people come not just to shop but also to socialize and conduct business, representing a unique intersection of rural and urban life, fostering a sense of interconnectedness and shared heritage. Another significant informal element of pre-war life in Homs was "Al-manzoul," which is a semi-private building used to host daily, seasonal, and family meetings for the male members of each family in Homs. It signifies the strict hierarchical social structure of Homs families, and is based on factors such as the family members' wealth, their education, and influence. Such space played a crucial role in maintaining family bonds and upholding social traditions.

6 The Impact of the Syrian War on Homs

6.1 The Old City: A Date with War 2011–2014

The Syrian war, which began in March 2011, brought widespread destruction and human suffering. Homs, in particular, became a major battlefield, with intense fighting lasting until 2017. The conflict had a profound impact on the city's population and infrastructure.

By 2014, it was estimated that around 587,000 residents of Homs had been displaced, out of a population of 800,000 in 2010 (UN Habitat 2014a). This massive displacement reflects the severity of the conflict and the extensive damage to homes, businesses, and public infrastructure. The displacement disrupted the social fabric of the city, as families were forced to flee to safer areas, both within Syria and abroad.

The war destroyed numerous buildings, including historical and cultural landmarks, severely affecting the city's urban heritage. The once vibrant neighborhoods, markets, and public spaces were left in ruins. Destruction and dis-

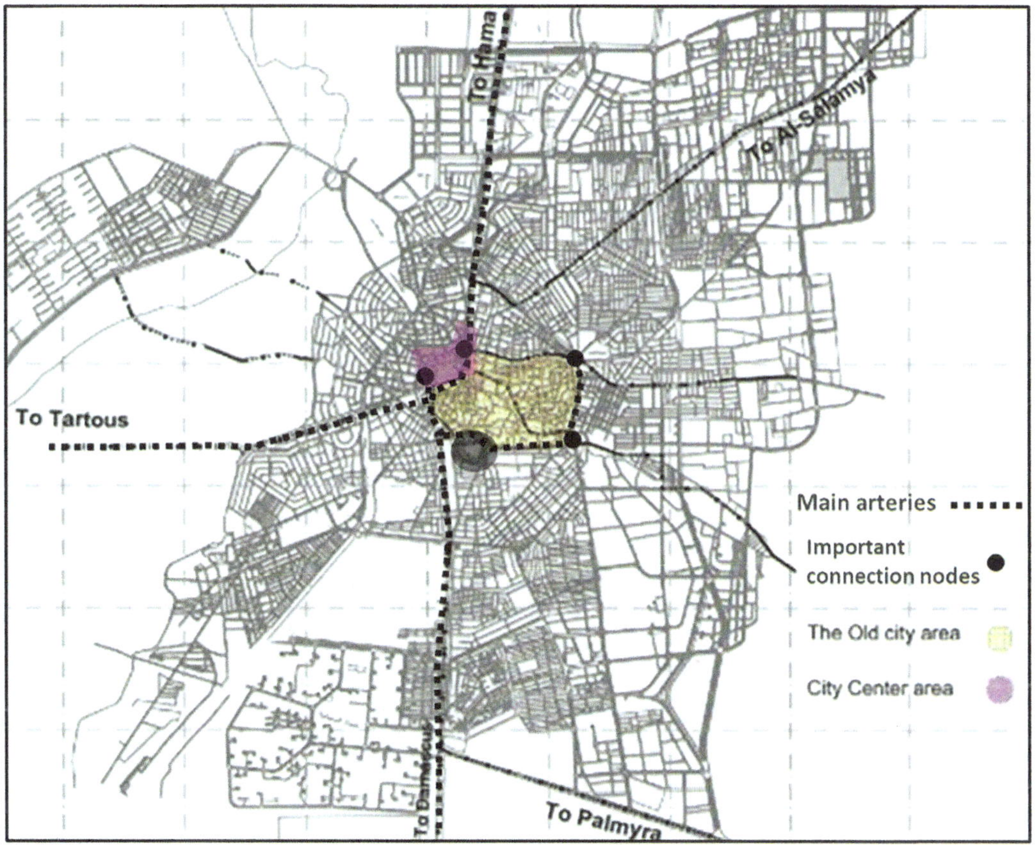

Fig. 3 The old city area and the city center of Homs. (Original source: Homs Municipality)

placement created immense challenges for the city's recovery and reconstruction efforts.

6.2 The Old City of Homs and the Syrian War

The Old City of Homs, with its interwoven fabric, poor permeability, and closed urban form, became a strategic stronghold for armed groups during the war. For 3 years, these features facilitated control of the area by these groups, leading to the abandonment of the Old City by most of its inhabitants. According to a UN-Habitat report from May 2014, large parts of the market were burned, and its shops were looted. Additionally, significant portions of the metal roof covering the market were demolished. During the battles, both minarets and church towers were repurposed as fighting positions, exacerbating the destruction of these historical and religious landmarks. The main streets of the Old City turned into perilous corridors, making movement extremely dangerous. In response, alternative paths were created by knocking holes in the walls of buildings. While these improvised routes allowed for safer movement, they also led to severe structural damage. Many buildings developed serious cracks, and some even collapsed entirely due to these makeshift modifications (UN Habitat 2014b).

In May 2014, the Old City of Homs was officially declared free of armed rebels, marking a significant turning point in the city's struggle (UN Habitat 2014b). Then by 2015, residents of Old Homs neighborhoods were permitted to return. Displaced residents returned to a city vastly different from the one they had left, with compromised infrastructure and limited access to basic services (water, electricity, and healthcare). The social fabric of the community was also frayed, with families scattered and neighborhoods depopulated.

7 Recovery: A Local Anecdote

Initial recovery efforts focused on ensuring the safety and habitability of the affected areas. Various agencies and organizations stepped in to provide material and in-kind assistance to support the return and restoration process. Building materials, tools, and other essential supplies were distributed, while some organizations offered financial aid or vouchers to help residents purchase necessities. Local and international NGOs played a crucial role in coordinating these efforts. This coordinated support, combined with the resilience of the residents, was vital in encouraging the return and the daunting task of rebuilding homes and lives.

Based on both official efforts of state institutions and international organizations, as well as individual or collective local initiatives, the Old City of Homs has undergone several stages of recovery following the end of the war. The follow-up of the recovery process revealed three distinct phases, each characterized by different agents and strategies. These observations, gathered by the authors, outline the progression of efforts to restore life in the Old City of Homs.

The first phase of recovery, which extended from 2014 to 2015, focused on immediate relief and the restoration of the city's most influential religious and heritage buildings. City authorities collaborated with international organizations, such as UN-Habitat, to secure funding and technical expertise for these restoration projects. The joint efforts ensured that the most critical heritage sites received immediate attention, which was seen as a crucial step in the broader recovery process. By focusing on religious buildings, the authorities aimed to reestablish key community hubs that would encourage the return of displaced residents and provide a foundation for further recovery efforts (UN Habitat 2014b). Driven by a strong emotional outburst rejoicing the end of the battles and hopeful for a speedy return, many original residents worked tirelessly to assist in debris removal.

Starting with religious and commercial Heritage Sites, the great Mosque and the Heritage Markets area (the Roofed Market) were restored as significant landmarks, aiming to support a sense of normalcy, local economy, and community. Historical Churches, mostly located in the northern neighborhoods of the Old City, were also identified for restoration. Residents primarily focused on restoring easily restorable houses that required less extensive repairs to enable a quicker return to habitability. Most residential returnees initially occupied the northern and eastern parts of the Old City, as these neighborhoods were the least affected by the conflict and the first areas where residents were allowed to return. Authorities recorded the return of only 16 families from 2014 until the end of 2015. Most of these families sought to protect what was left of their properties. These families' residents are primarily located near restored ancient churches and mosques, Fig. 4.

The restoration of religious buildings witnessed significant involvement from volunteers, many of whom were highly qualified builders eager to rehabilitate their city. "They believed that joining the restoration process of these buildings was a kind of blessing," said Mr. Haroun (the director of the Old Homs Project, representing the Islamic Charity Organization for Relief and Development, "Aoun" team from 2014 to 2017).

Among the most important international organizations that assisted and organized the process of return to the Old City of Homs were the UNHCR (United Nations High Commissioner for Refugees) and the UNDP (United Nations Development Program). All restoration operations, however, were carried out with the official approval of local authorities, including the Municipality and the Directorate of Antiquities.

In general, the UNHCR played a crucial role in supporting returnee families, including those returning to the Old City of Homs. Their efforts focused on the rehabilitation of homes and providing essential resources to ensure safety for returnees. According to the UNHCR Factsheet from November 2018, one significant initiative was the distribution of "doors and windows" packages. These packages offered a fast and economical solution to secure and restore the privacy of the damaged homes. The UNDP efforts also included the rehabilitation of the Heritage Market area, and supported the rehabilitation of many workshops, which were vital for revitalizing the local economy (UNDP 2017).

These coordinated efforts by the UNHCR and UNDP, in partnership with local authorities, helped to stabilize the area, restore essential infrastructure, and provide the necessary support for returnees to rebuild their lives.

The second phase of recovery lasted from 2016 to 2018. In this phase, the pace of recovery in the Old City of Homs significantly accelerated. It was characterized by the implementation of small service projects and community initiatives, which played a crucial role in encouraging families to return to their ruined neighborhoods.

Key Activities in this phase focused on supporting small service projects. These projects included the restoration and re-operation of public and private schools, daily necessity stores, and the reopening of various workshops and small stores.

According to an interview with a representative from "Aoun," the pace of such projects gradually increased during this period, leading to a substantial rise in the number of returning families. By 2018, approximately 2200 families had returned to the Old City, reducing their living costs compared to residing in other areas.

In this context, according to Mr. Haroun, in 2018, a community initiative in partnership with UNDP launched "Income Generating Equipment Project with UN-HABITAT." This project donated work equipment to 167 beneficiaries, leading to the establishment of about 63 craft workshops, most of which were located in the Old City area. The official city administration role was granting temporary or permanent permits for these workshops. This initiative helped to restore some of the Old City's original features. UNHCR supported returnee families to rehabilitate their houses (UNHCR 2018). On the other hand, financial contributions from expatriates also played a significant role in the restoration of some old buildings that were repurposed as luxurious restaurants, such as Agha Palace and Julia Restaurant, and

Fig. 4 Reinhabited partially destroyed buildings in the old city of Homs. (Source: Homs Emerges as Turning Point in Shaping Syria's Future—The New York Times)

two public baths, "Al-Osaiyati and Al-Othmany" traditional baths. These projects preserved some historical structures and enhanced the area's attractiveness and economic prospects.

The collective impact of these initiatives revitalized human activities in almost all neighborhoods of the Old City. The return of families, reopening of businesses and schools, and establishment of craft workshops fostered a sense of normalcy. This phase of recovery showcased the effectiveness of coordinated efforts between local authorities, international organizations, and the community.

Third phase of recovery, from 2019 to 2023, could be described as the community-led revival. During this period, the Old City of Homs experienced a significant revival of both its commercial activities and social fabric. This phase was marked by a resurgence of everyday basic functions, the return of traditional markets, and the reactivation of social spaces that contributed to the overall vitality of the area.

Key activities and observations concentrated on the revival of everyday basic functions. According to over 50 interviews with residents, workers, investors, and city authority representatives, it was noted that simple everyday functions, such as groceries, traditional food, and vegetable stores, returned to Old Homs, Fig. 5.

Additionally, informal markets that targeted low-income communities became active, featuring shops for used second-hand clothes, old furniture, upholstery, carpenter workshops, blacksmiths, perfumers, and local herbalists. These markets attracted frequent visitors from other parts of the city and neighboring villages, revitalizing commercial activities around the old city's zone, especially near the location of the north and west Old City entrance points.

It was noticeable that although most stores in the Heritage Market area had not reopened yet, but original owners of these stores refused to sell their properties despite facing financial difficulties. This determination to cling to their inherited shops underscored another key aspect of local cultural heritage.

Holding onto their properties and not giving them away to new owners helped in the current revival of the Roofed Heritage Market. This market, where families inherited their shops across generations, became a celebration of continuous family heritage and played a crucial role in the economic and social resurgence of the Old City, Fig. 6.

"It's a property that I've inherited from my ancestors, it preserves my dignity and my identity, I'll never sell it to a stranger" said in an interview with a store owner in this area.

One more traditional aspect was shown through preserving the "Al-Manzoul" concept. It narrated the continuous memory and persistent presence of Homs families across the city. This was particularly crucial as thousands of families in Syria had been separated since 2011.

Fig. 5 The revival of everyday basic needs for residents in the Old City's neighborhoods in 2022. (Source: Authors)

All mentioned contributions enhanced the social and economic revival in Homs. The resilience and determination of the residents, coupled with their strong sense of heritage, were instrumental in this recovery phase.

8 Results

Monitoring the three phases of recovery for the Old City of Homs revealed a dynamic series of activities and shifting responsibilities over time. In phase 1 (2014–2015), key players were basically official authorities and international organizations, focusing on restoring critical infrastructure and main heritage sites. While in phase 2, official activity stabilized, and local initiatives began to increase, driven by economic needs. Community initiatives fostered local economic activity through the establishment of some craft workshops. But in phase 3, official rehabilitation efforts declined versus a rise in the prominence of local residents' initiatives. Residents' initiatives to preserve and revive their heritage breathed life back into deserted streets and empty squares. These grassroots efforts were vital in sustaining the momentum of recovery. Residents also focused on reconnecting the Old City with surrounding areas, revitalizing old city entering points as movement nodes for people and goods. This reconnection was crucial for enhancing accessibility and economic interactions. These different practices and their impacts are listed in Table 1.

Fig. 6 Restored passages in the Roofed Heritage Market in 2023

Table 1 A summary of the recovery phases, the projects carried out, and the entities that contributed to these efforts

Temporal phase	Project description and initiated entity	Project location in the "old city" area	Project impact
First recovery phase 2014–2015	Official agencies: – Rubble removal – Restoring main religious buildings and the "heritage market"	Vicinity of the old city, some of its gates, and the "heritage market" area	– Improve movement around the old city – Invite returnees to reside in the area
	Informal entities (local's): – Contributing to non-hazardous rubble removal – Clean and beautify facades overlooking active paths	Old City neighborhoods the heritage market area	Re-opening of few daily services stores
Second recovery phase 2016–2018	Official agencies: – Further restoration of the "heritage market" – More rubble removal – Security enhancement – Rehabilitation of few schools	Main arteries in the "Old City" area	Reopening of more stores in the "heritage market," More inhabited dwellings around restored schools.
	Informal entities (local's): – Restoring partially damaged dwellings – Recovery of small craft workshops and local stores – Restoring local religious and educational buildings – Creating informal temporary markets	Main arteries, secondary & local streets in the "Old City" area The heritage market area	Lanes of street vendors occupied main axes, especially in the heritage market area, attracting new visitors Increase in the number of returnee families to the old city neighborhoods
Third recovery phase 2019–2023	Official agencies: – Monitoring commercial use permits, combating informal markets – Monitoring private restoration projects	The "heritage market "area and Main arteries	Control informal markets expansion in commercial zone in the "Old City"
	Informal entities (local's): – Connecting active spots all over the Old City zone – Various businesses are back in the heritage market area	Main arteries, secondary and local streets in the "Old City" area	Mutual impact between increasing local businesses and number of returnee families Partial revival of the heritage market

So, while official restoration projects provided necessary infrastructure and support, it was the local initiatives that truly brought life back to the Old City.

The map in Fig. 7 depicts the active areas dominated by local residents along vital routes in this part of the city. These areas have become hubs of human activity, driven by the dedication and resilience of the local community.

The process of reviving the Old City of Homs was executed gradually but steadily. The return of mosques' calls to prayer and churches' bells brought life back to scattered squares and streets associated with them, stimulating residential and service activities in their vicinity. This may explain why the first groups that returned were mainly composed of elderly people, under extremely difficult security and service conditions; they marked the beginning of a long recovery journey. At that time, the role of young people focused on voluntary work to live there.

Later, as services improved, especially with the reopening of schools, middle-aged families began to move back to live in the old city with their children. In fact, during the first phase, the restoration of the covered market did not succeed in convincing shop owners who had successfully opened businesses in other parts of the city to return back to work in the market. Instead, street vendors occupied this area during the first years following the start of restoration. Over time, these vendors who spread along the Old City's main streets

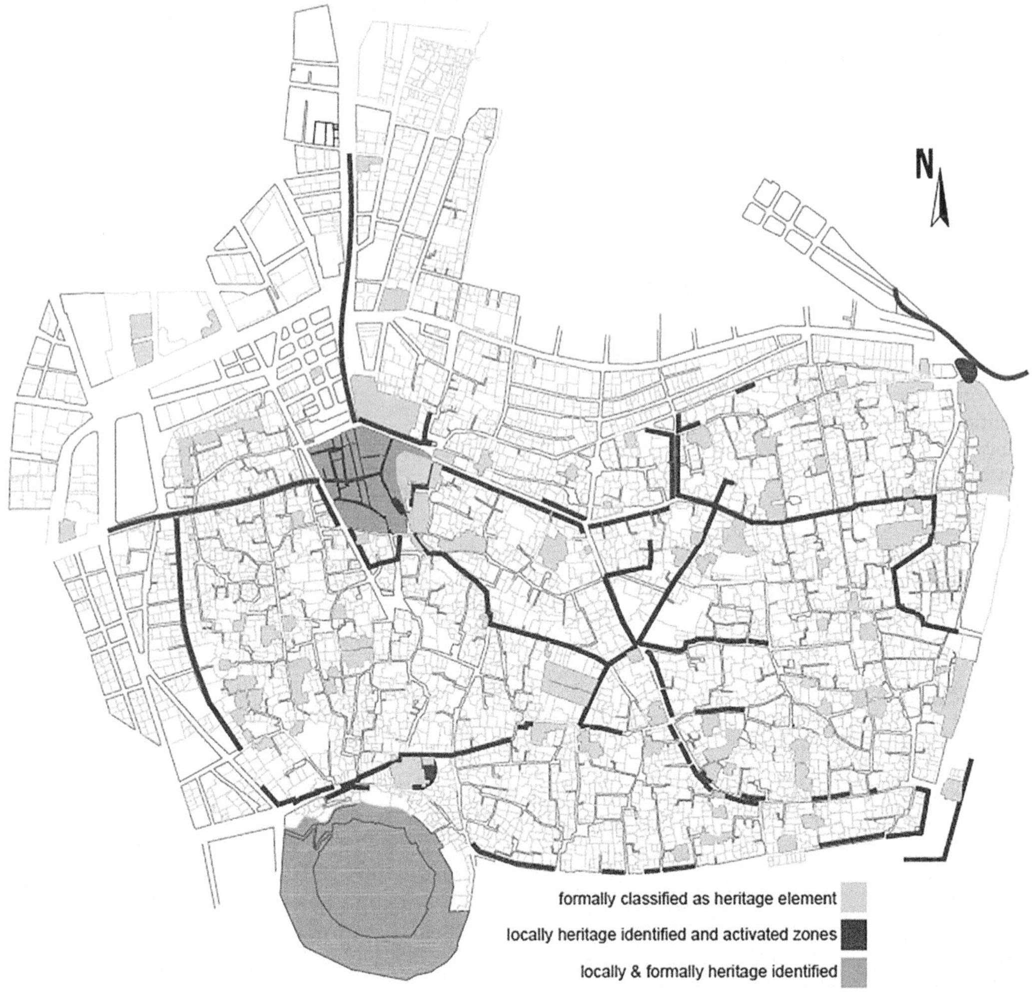

Fig. 7 Vital routes re-created by local residents. (Source: Authors)

were able to attract shoppers back to the whole area, revitalizing different markets and encouraging original business owners to reopen their stores. Additionally, the return of any skilled profession and the resumption of its activity was considered a victory and a new hope for those who knew the city before the war, as a local landmark that refreshes the memory of its residents and visitors.

The integrated network of architectural symbols (religious buildings, markets, etc.) and urban spaces (squares, streets, alleys) created a cohesive and vibrant environment that encouraged residents to return and rebuild their lives, re-establishing the Old City as a lively, interconnected community.

9 Conclusions

Post-war cities like Homs often grapple with social and economic fragility. While official authorities typically focus on security, damage assessment, infrastructure restoration, and rubble removal, comprehensive rebuilding plans often face significant financial and logistical challenges. In Homs, several years have passed since the major conflicts subsided, yet large-scale reconstruction projects in the most destroyed neighborhoods remain pending.

The old part of Homs is more than just a collection of historical buildings; it is a living reservoir of the experiences and resilience of successive generations. This area not only tells the story of the city's rich history and defines its identity but also embodies the community's enduring efforts to sustain life amid a harsh challenge such as war.

This paper demonstrates that the residents of the Old City possess a multi-layered knowledge of their cultural heritage, empowering them to revive both its tangible and intangible components. Their deep understanding of the city's history and traditions has enabled them to restore physical elements (squares, streets, buildings) and cultural practices (crafts, ancestral property preservation, family unity). Despite these efforts, challenges remain. Building facades are still broken, streets are marred by craters from shelling, and some stone walls are at risk of collapse or theft. These conditions pose ongoing risks to both the safety and the heritage of the Old City. Nevertheless, the residents' insistence, dedication, and resilience have laid the groundwork for a gradual yet determined recovery.

10 Limitations and Future Research

This paper acknowledges limitations in documenting recovery efforts and official reports due to their unavailability at the municipality or government levels. Future research could address these limitations by continually documenting the gradual alterations occurring within the Old City, comparing trends with the pre-war period. Additionally, comparative studies with other Syrian cities and international cases could reveal common patterns, unique challenges, and successful strategies for post-war recovery and heritage preservation.

References

Adly A, AlAraby M, Awad I (2021) Conflict by other means: postwar reconstruction in Arab states. Malcom H. Keer Carengie Middle East Centre. https://carnegie-mec.org/2021/02/05/conflict-by-other-means-postwar-reconstruction-in-arab-states-pub-83824

AlAhmad KA (1970) Landmarks and celebrities from Homs. Taha House for Printing and Publishing, Homs

Ali SH, Sherzad MF, Alomairi AH (2002) Managing strategies to revitalize urban cultural heritage after wars: the centre of the old city of Mosul as a case study. Buildings 1298(12)

AlMalki QF (2004) Urban and architectural heritage in the Arab world. AlWarra Institution, Amman

AlZehrawi N (2003) The population root of the city of Homs: from Kingdom to Kasbah, from the genesis to the end of 1918. Dar AlSalama Publisher, Homs

AlZehrawi N, Al-Sibai MO (1992) Homs: documentation study from 1840–1918. Dar AlSalama Publisher, Homs. https://ebook.univeyes.com/159437

AlZenbarakji L, Atfeh N (2016) Local planning concept and application towards reconstruction: a case study of the old city of Homs. A Thesis submitted in the Partial Fulfillment for the Requirement of the Degree of Master in Planning and Environment. Damascus. https://www.academia.edu/28845442/

Asa'ad MK (1984) The history of Homs: part 2, from 622 till 1977. Orthodox Archdiocese of Homs, Homs. https://ketabpedia.com/%D8%AA%D8%AD%D9%85%D9%8A%D9%84/%D8%AA%D8%A7%D8%B1%D9%8A%D8%AE-%D8%AD%D9%85%D8%B5-02/

Belal A, Shcherbina E (2021) Post-war planning for urban culture heritage recovery. E3S Web Conf 263(05054). https://doi.org/10.1051/e3sconf/202126305054. https://www.e3s-conferences.org/articles/e3sconf/abs/2021/39/e3sconf_form2021_05054/e3sconf_form2021_05054.html

Earnest J, Dickie C (2012) Post-conflict reconstruction—a case study in Kosovo: the complexity of planning and implementing infrastructure projects. In: PMI Research and Education Conference. https://ketabpedia.com/%D8%AA%D8%AD%D9%85%D9%8A%D9%84/%D8%AA%D8%A7%D8%B1%D9%8A%D8%AE-%D8%AD%D9%85%D8%B5-02/

Harfoush S, Khzam N, AlNajjar A (2002) Analysis of the old Homs police system in light of the international standards (Homs municipality). https://pdfcoffee.com/homs-building-code-pdf-free.html

Hussein Agha MG (2002) Homs City, the First Engineers under the Ottoman Caliphate: from 1315 to 1897. Approval of the Ministry of Information, Homs 73548-10/13/2002

Hussein Agha MG (2019) Al-quds Thursday in Homs City. Approval of the Ministry of Information, Homs 9/21/2019/117847

Jabour Z, Barakat H, AlJandali B, AlAssaf S (2001) City planning theories. Directorate of University Books and Publications, Homs

Kassouha S (2014) Towards a strategy for regaining cultural identity in the urban reconstruction of the war-ravaged city centre of Homs, Syria. A Thesis submitted in the Partial Fulfillment for the Requirement of the Degree of Master of Science in Integrated Urbanism and Sustainable Design. Ain Shams University, Egypt & University of Stutgart-Germany. https://iusd.asu.edu.eg/wp-content/uploads/2018/04/15-Kassouha.pdf

Krasny M, Russ A, Tidball K, Elmqvist T (2014) Civic ecology practices: participatory approaches to generating and measuring ecosystem services in cities. Ecosyst Serv 7:177–186. https://doi.org/10.1016/j.ecoser.2013.11.002

Legislative Decree 222 of 1963—Antiquities Law. http://www.parliament.gov.sy/arabic/index.php?node=5529&cat=9398

Salmo A, Scherbina EV, Alibrahim L (2021) Architectural and urban identity of Homs city. Monthly J Construct Architect 16(10)

UN Habitat (2014a) City profile Homs multi sector assessment. May 2014. https://www.alnap.org/system/files/content/resource/files/main/Homs%20RCP.pdf

UN Habitat (2014b) Neighborhood profile old city of Homs. June 2014. https://unhabitat.org/sites/default/files/download-manager-files/1421063714wpdm_Old%20city%20of%20Homs.pdf

UNDP (2017) Building resilience in response to the Syrian crisis. June 2017. https://www.undp.org/sites/g/files/zskgke326/files/migration/arabstates/2017 UNDP-Portfolio-all.pdf

UNHCR (2018) Syria Factsheet November 2018. https://reliefweb.int/report/syrian-arab-republic/unhcr-syria-factsheet-january-november-2018#

Revitalizing Urban Spaces: The Role of Minga and Community Engagement

Daniela Hidalgo Molina

Abstract

This study examines community participation tools for enhancing urban environments and fostering a sense of belonging, using the Luz y Vida neighborhood located at Monte Sinahí in Guayaquil-Ecuador as a case study. Rooted in the ancestral practice of the *Minga*, a form of collective work deeply embedded in Latin American Indigenous traditions, the project incorporated observation, participatory workshops, and community-led initiatives to improve urban aesthetics. Implemented in three phases, identifying cultural icons through workshops, conducting community cleanups, and painting facades, the initiative engaged local residents, children, and university students, strengthening collaboration and social cohesion. The findings highlight how grassroots efforts, guided by participatory design and the strategic use of local resources, can drive urban transformation while reinforcing community identity.

Keywords
Identity in the physical environment · Minga · Participatory design

1 Introduction

Urban development and sustainability increasingly depend on active community involvement in shaping the built environment. The Luz y Vida neighborhood in Guayaquil, Ecuador, exemplifies this approach by participatory design methodologies to enhance urban aesthetics and foster a sense of belonging. This article aims to present community participation tools that not only improve the physical environment but also strengthen social cohesion and collective identity.

Drawing on the rich tradition of the *Minga*, a Kichwa term meaning "collective work for community assistance", this study explores its relevance in contemporary urban transformation projects. Deeply rooted in Latin American Indigenous culture, the *Minga* embodies both solidarity and an effective mechanism for sustainable development. By combining participatory design principles with anthropological methodologies, the Monte Lindo project demonstrates how collaborative efforts can empower communities to take an active role in shaping their urban spaces.

This article examines the tools and processes used in this case study, including observation, participatory workshops, and collective action through *Mingas*, as fundamental components of community-driven urban interventions. Through a phased approach, encompassing exploration, cleanups, and aesthetic revitalization, the project underscores the transformative potential of grassroots participation in urban renewal.

2 Study Objective

The objective of this study is to explore the transformative potential of community-driven urban interventions, highlighting the role of the ancestral Minga tradition in fostering sustainable development, enhancing urban aesthetics, and strengthening social cohesion. Through a case study in Luz y Vida sector located in Monte Sinahí, Guayaquil, Ecuador, the article aims to demonstrate how participatory design approaches and collective efforts can address urban challenges while empowering communities.

D. H. Molina (✉)
"Improvement of Public Space, Through Placemaking as an Alternative for Transformation", Universidad Espíritu Santo-Ecuador, Samborondón, Ecuador
e-mail: mdhidalgo@uees.edu.ec

F. Alberti et al. (eds.), *Towards Sustainable Urban Regeneration: Utilizing Environmental, Cultural, and Social Assets*, Advances in Science, Technology & Innovation, https://doi.org/10.1007/978-3-032-02091-8_21

3 Literature Review

3.1 The Minga: An Ancestral Practice in Building Sustainable Communities

The Minga, a Kichwa term meaning "collective work for community assistance," is an ancestral practice deeply rooted in various Indigenous communities across Latin America. This tradition involves not only collaboration in physical tasks but also fosters solidarity and mutual support among neighbors.

In addition to its cultural value, the Minga has become a key tool for the management and development of human, economic, and natural resources within social organizations, cooperatives, and communities. Its impact on achieving common goals has been widely studied. For example, in the research by Vázquez Torres et al. (2018), its role in the maintenance campaign of San Roque, in Cuenca, Ecuador, is analyzed, highlighting its effectiveness in the protection of cultural heritage.

Furthermore, Cancino-Pérez (2020) emphasizes in his study on Minga and Volunteering that this practice consists of providing support in various tasks, where the compensation is not monetary but based on reciprocity. From an economic perspective, the Minga presents itself as an alternative that allows resources to be attracted and community ties to be strengthened.

An emblematic example of this dynamic is the Indigenous communities in Chimborazo, Ecuador, where the Minga plays an essential role in collaboration and the strengthening of local organizations. According to Bravo et al. (2023), social interaction within the community, regardless of age, balances aspects of daily life and motivates unity to achieve common objectives for collective benefit.

This urban image improvement project, using the *Minga* as a tool, is based on a practice common in the communities of the Ecuadorian Highlands. Although the term *Minga* is more widely known and used in the highlands, its meaning is also recognized on the coast. In this case, the *Minga* was applied to carry out community work in the Luz y Vida sector, located in Monte Sinahi, in Guayaquil, Ecuador.

This initiative aligns closely with participatory design methodologies, which draw heavily from anthropological approaches that identify and address local needs and social dynamics. Chambers (1994) traces the evolution of these methodologies from Rapid Rural Appraisal (RRA) in the 1970s to Participatory Rural Appraisal (PRA) in the 1980s, highlighting the shift from passive consultation to active community engagement. The application of the *Minga* in this context exemplifies this transition, as it empowers local stakeholders by transforming them from passive recipients to active agents in the planning and design process, fostering a sense of ownership and ensuring long-term sustainability.

Henry Sanoff introduces innovative participatory frameworks, such as *Design Games* and community workshops, which effectively incorporate public feedback into urban development (Sanoff 1979, 2000). His work aligns with more recent studies by John K.C. Liu and his team at the Building and Planning Research Foundation, who emphasize the critical role of community ownership in project sustainability (Delgado and Hidalgo 2022). Liu further underscores the value of direct engagement with residents, utilizing participatory tools to strengthen the connection between communities and their projects, thereby enhancing both cultural relevance and long-term impact.

4 Methodology

The article highlights approaches such as observation as a key process, participatory workshops with a focal group to identify elements representing identity, and Minga (community work) for improving the urban image of a given area. It specifically details the urban intervention conducted through Minga in the Monte Lindo neighborhood within the Monte Sinahí sector, using it as an example where participatory methods are essential for transforming urban aesthetics. Table 1 below presents a detailed list of tools utilized in this case study.

5 Case Study: Luz y Vida Community—Urban Image Transformation Through Community Participation

The Luz y Vida sector, located within Monte Sinahí (Fig. 1), lies beyond the urban boundary of Guayaquil. Over the past 20 years, Monte Sinahí has expanded significantly, covering 78,671 ha by 2022 (Hidalgo Molina et al. 2022). The existing buildings predominantly result from self-construction, which has gradually been consolidated and legalized over the past 10 years. The urban image enhancement project for the Monte Lindo sector was carried out in three distinct phases, as detailed in Table 2.

Table 1 Tools for participatory approaches to improve urban image

Tools	Key points	Relevance to architectural design	References
Observation as key process	Identifying points of interest to analyze how people interact with space	Observing user behavior within a space informs design decisions on spatial layout, functionality, and user experience	Alexander et al. (1977), Gehl and Svarre (2013), Hester (2006), Delgado and Hidalgo (2022)
Participatory workshops to identify elements representing identity	Participatory design, as a tool, draws heavily from anthropological methodologies aimed at identifying and addressing local needs and patterns	It helps identify community patterns, needs, and what people expect from a design	Chambers (1994), Sanoff (2000), Hester (1990), Delgado and Hidalgo (2022)
Minga (community work)	The Minga, a Kichwa term meaning "collective work for community assistance," represents a deep-rooted ancestral practice in Latin American indigenous communities	Minga emphasizes collaborative design, fostering community-driven architecture that reflects collective values and needs	Vázquez Torres et al. (2018), Cancino-Pérez (2020)

5.1 Phase 1: Exploration and Workshop to Identify Key Icons of the Luz y Vida Sector

The first phase began with an exploration of the area, establishing direct relationships with community leaders and identifying the primary site of communal activity, the local park. Serving as the central gathering space and decision-making hub, the park plays a pivotal role in community interactions.

A workshop involving 12 children employed a qualitative approach to explore the identity of Monte Lindo (Fig. 2). Utilizing a focus group methodology, participants identified key elements that symbolize their community, including fruit trees, corn, and mountains. This phase not only highlighted culturally significant icons but also fostered active engagement and a sense of ownership among participants.

5.2 Phase 2: Mingas for Community Cleanup and Urban Sidewalk Improvements

The second phase consisted of a series of Mingas (community workdays), organized by local leaders. Over the first 3 days, approximately 35 individuals of various ages participated in cleaning public spaces, removing trash, cutting overgrown vegetation, and addressing other environmental concerns (Figs. 3 and 4).

The fourth and fifth days focused on urban improvements, particularly the restoration of sidewalks. Participants worked collaboratively to fill potholes, addressing a critical issue for pedestrian safety and accessibility.

5.3 Phase 3: Painting Housing Facades

The third phase took place on the sixth and seventh days, involving ten students from the Universidad Espíritu Santo (UEES) Faculty of Architecture alongside children and their parents (Fig. 5). Together, they painted residential facades with designs inspired by the results of the children's workshop conducted in Phase 1. The designs aimed to reflect the unique identity of Monte Sinahí while transforming the urban environment into a vibrant and welcoming space.

Materials for the project were donated by private companies, and the community contributed additional resources through a bingo event organized prior to the workshop. This collaborative effort demonstrated how private and community resources can be leveraged effectively for meaningful urban transformation.

This project exemplifies community-driven urban intervention. Through the collective effort of Mingas, the Luz y Vida community actively participated in its transformation, demonstrating the power of collaboration and grassroots initiatives in fostering sustainable urban development. By engaging various stakeholders, including children, students, and community leaders, the project not only improved the physical environment but also strengthened social cohesion and community pride.

6 Conclusion/Discussion

This article highlights the critical role of community participation in urban transformation, demonstrating how participatory design approaches, rooted in the Minga tradition, can enhance the physical environment while fostering a sense of ownership and belonging among residents. The case study in

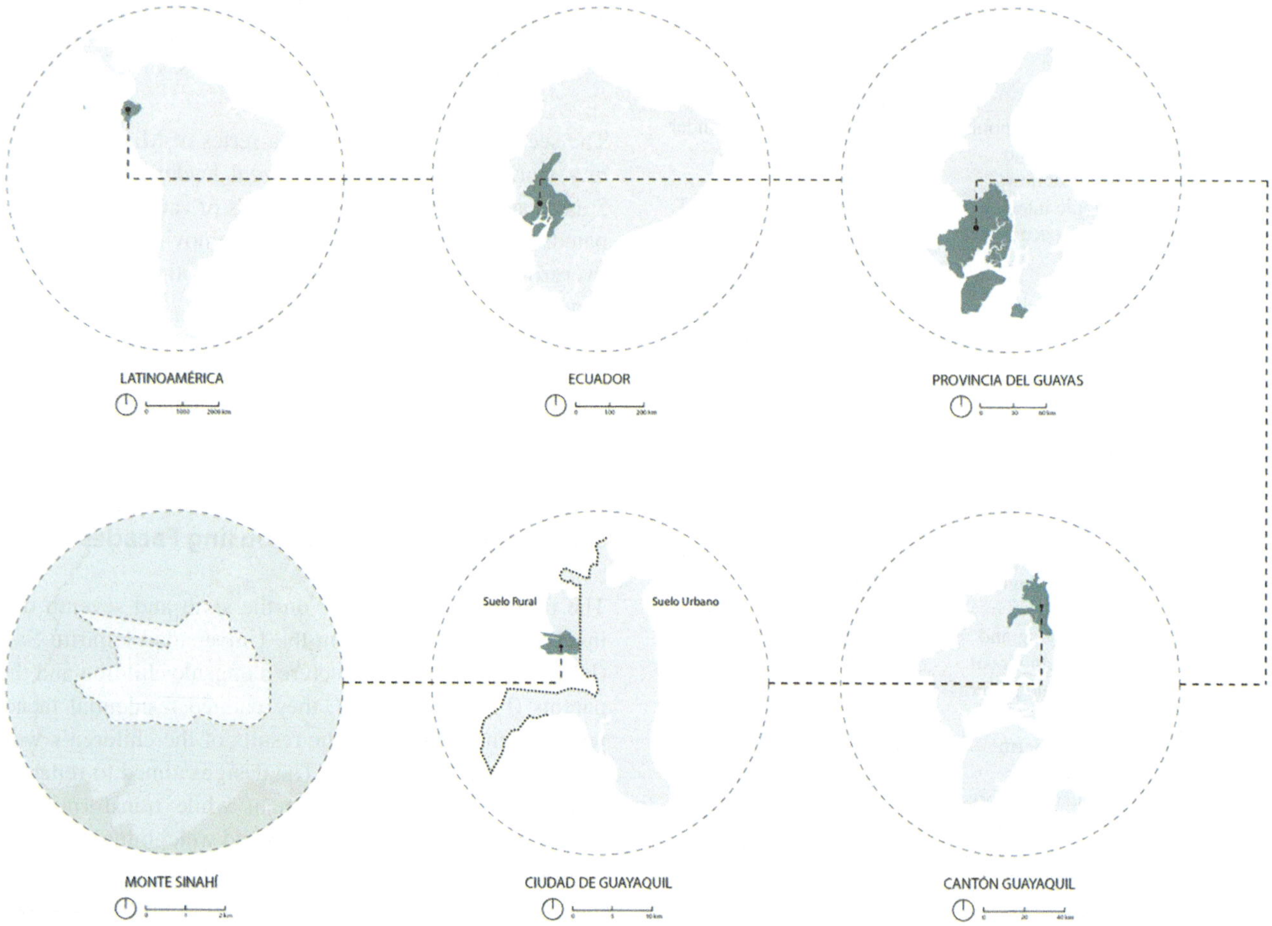

Fig. 1 Location map of the Monte Sinahí sector, Carla Orozco

Table 2 Phases of community urban interventions

Phases	Description
Phase 1	Exploration and workshop to identify key icons of the Luz y Vida sector: participants engaged in creative exercises to highlight elements defining the essence of the place
Phase 2	Mingas for community cleanup and urban sidewalk improvements: collaborative efforts to clean public spaces, repair sidewalks, and beautify the urban environment
Phase 3	Painting housing facades: revitalization of the community through the painting of residential facades, enhancing aesthetic appeal and fostering pride

Luz y Vida community in Monte Sinahí serves as a practical example of how collective action, driven by local communities, can significantly improve urban aesthetics and functionality. Through activities such as community cleanups, infrastructure improvements, and collaborative design, the residents not only addressed immediate urban challenges but also strengthened the social fabric of their community.

The integration of participatory methods into urban planning ensures that interventions are tailored to the specific needs and cultural identity of the community, fostering sustainable outcomes. This approach also empowers local stakeholders, turning them from passive recipients into active

Fig. 2 Workshop to identify key icons, 2023

Fig. 3 Minga for community cleanup, 2023

Fig. 4 Collective Community cleanup through Minga

agents in the development process, thereby ensuring long-term ownership and stewardship of the changes made.

Ultimately, the success of participatory processes like Minga is not only found in the physical transformation of spaces but also in the creation of a more resilient, inclusive, and cohesive community. When these efforts are supported by governmental policies and resources, the potential for widespread, lasting impact is maximized, promoting a more equitable and responsive urban development model. Therefore, community-driven approaches to urban design should be integrated into broader planning frameworks to ensure that cities evolve in a way that reflects and supports the needs and aspirations of their inhabitants.

Fig. 5 Painting housing facades, 2023

Acknowledgments I extend my gratitude to the Research Center of Universidad Espíritu Santo (UEES) for its support of this study, to the Dean of the Faculty of Architecture, and to the architecture students for their valuable contributions to the facade improvement workshop, and to the Monte Lindo community for their meaningful participation.

References

Alexander C, Ishikawa S, Silverstein M (1977) A pattern language: towns, buildings, construction. Oxford University Press, Oxford

Bravo JA, Murillo Pinos AS, Vaca Cárdenas A, Larrea Naranjo C (2023) La minga como práctica comunicativa que orienta la transformación social de los indígenas de Chimborazo, Ecuador. Commun Pap 12(26):23–37. https://communicationpapers.revistes.udg.edu/article/view/22921

Cancino-Pérez L (2020) Minga y voluntariado: Economía alternativa y trabajo precario en los asentamientos pro sustentabilidad en Chile. Psicoperspectivas 20(1) https://doi.org/10.5027/psicoperspectivas-Vol20-Issue1-fulltext-1955

Chambers R (1994) The origins and practice of participatory rural appraisal. World Dev 22(7):953–969

Delgado A, Hidalgo D (2022) Diseño participativo como herramienta para el desarrollo de proyectos en comunidades rurales del sector Samborondón. Casos: San José y Barranca Central. Editorial Universidad Espíritu Santo-Ecuador

Gehl J, Svarre B (2013) How to study public life. Island Press, Washington, DC

Hester RT (2006) Design for ecological democracy. MIT Press, Cambridge

Henry Sanoff (1979) Design games. Los Altos, CA: W. Kaufmann

Hidalgo Molina MD, Orozco Ganan CF, Velasco Caicedo C, Wong Chauvet NV, Jaramillo Loor JJ (2022) Atlas: Estudio espacial de los asentamientos informales: Caso Monte Sinahí. Universidad de Especialidades Espíritu Santo, Guayaquil

Sanoff H (2000) Community participation methods in design and planning. John Wiley & Sons, New York

Vázquez Torres L, Achig Balarezo MC, Cardoso Martínez F (2018) La minga como patrimonio intangible: Un estudio de caso en Cuenca, Ecuador. Revista de Estudios Culturales Latinoamericanos 8(2):57–72. https://dialnet.unirioja.es/servlet/articulo?codigo=6266263

The Medieval Aqueduct of Perugia: From Abandonment to Reuse as a Pedestrian Path

Pietro Matracchi, Mariano Sartore, and Maddalena Branchi

Abstract

The paper analyses the case study of the medieval aqueduct of Perugia and its reuse as a pedestrian path. The study concerns the terminal section of the ancient aqueduct that is identified with the arches that rise in the historic centre of the city in the area known as the 'Conca'. The remains of the aqueduct, which have survived the passage of time and are preserved in their new urban use, testify to the bond of interdependence between the water infrastructure and the city of Perugia, which has never failed over the course of seven centuries. The aqueduct, about 4 km long, dates back to the mid-thirteenth century and was built to carry water from Monte Pacciano to the Fontana Maggiore, in the heart of the historic centre of Perugia. The extraordinarily daring hydraulic work had a troubled life due to the frequent repairs that the municipality had to sustain and the effort it put into asking for opinions and solutions from various experts of the time. Due to ongoing functional and maintenance problems, the aqueduct was decommissioned in 1835 and replaced by a new one coming from Bagnara di Nocera Umbra and directed to the Monteripido reservoir.

Today, the arches of the Conca are a pedestrian hanging street, Via dell'Acquedotto, which connects the historic centre with the Conca area and with Borgo Sant'Angelo, an appendix of the medieval expansion of the city towards the north.

Despite the growth of the city, the aqueduct has been preserved as a pedestrian connection axis and still constitutes a persistent sign in the urban fabric of Perugia. Its importance as an element of identity for the city and its position constitute an interesting design opportunity to define future urban scenarios that could improve accessibility to the historic centre through the enhancement and expanded use of the medieval infrastructure.

P. Matracchi (✉) · M. Branchi
Department of Architecture, University of Florence, Florence, Italy
e-mail: pietro.matracchi@unifi.it, maddalena.branchi@unifi.it

M. Sartore
Department of Political Sciences, University of Perugia, Perugia, Italy
e-mail: mariano@urb.unipg.it

Keywords

Cultural heritage · Reuse · City identity · Accessibility · Conservation · Urban sustainable development

1 Introduction

Focusing on the final section of the aqueduct, represented by the arches in the 'Conca' area of the historic centre, the study highlights how this infrastructure remained a crucial element for the city for over seven centuries. Although it was decommissioned due to continuous maintenance issues and replaced by a new aqueduct, the ancient medieval aqueduct is now an elevated pedestrian path, called Via dell'Acquedotto, that connects the historic centre with Borgo Sant'Angelo. Despite urban expansion, the aqueduct continues to be an important pedestrian connection axis, with significant potential for future projects of urban sustainable development that improve accessibility to the historic centre, enhancing this ancient infrastructure (Chianella 2016).

F. Alberti et al. (eds.), *Towards Sustainable Urban Regeneration: Utilizing Environmental, Cultural, and Social Assets*, Advances in Science, Technology & Innovation, https://doi.org/10.1007/978-3-032-02091-8_22

1.1 The City of Perugia and Its Aqueducts: The Ancient Water Systems

The Etruscans left a lasting mark on the urban fabric, among which the mighty travertine walls stand out (Grohman 1981, pp. 9–10). With the arrival of the Romans, urbanization remained substantially confined within the Etruscan walls until the urban development of the twelfth century in the Middle Ages (Fiocca 1926; Radicioni et al. 2021). During these centuries, the city's water system was based on a capillary network of wells, used to capture spring water, and cisterns, used to collect rainwater. The water sources were distributed rationally, guaranteeing access to water resources for the entire population, exploiting the aquifers and local springs. Due to the discovery of wells and cisterns, the existence of an aqueduct in the classical age is considered unlikely but remains an unsolved problem (Grohman 1981, p. 16). The first city aqueduct dates back to the thirteenth century, functional to respond to the new needs of the growing population.

1.2 The Aqueduct in the 70s of the Thirteenth Century

During the Early Middle Ages, the urban expansion followed the morphology of the territory, characterized by deep valleys and steep slopes. In this context of urban development and demographic growth, in 1254, the General Council of the People of Perugia decided to capture and channel the waters of Monte Pacciano towards the new fountain designed for Piazza Maggiore (Silvestrelli 2005, pp. 113–119; Romalli 2006, pp. 317–318, 322–327).

Belforti (1785) wrote: 'Quindi è che gli antichi Magistrati nostri, intenti di proposito al pubblico bene, volendo provvedere al vantaggio di quella popolazione [di Perugia ndr], che andava ogni giorno crescendo, fin sul principio dell'anno 1254, determinarono in un Consiglio Generale del Popolo che da Monte Pacciano, lungo circa 3 miglia distante dalla città, si conducesse l'acqua della nuova fonte, che si era già ideata di dover fare nella Piazza Maggiore dinanzi alla Chiesa di San Lorenzo' [Our ancient magistrates, motivated by the public good and wishing to provide for the benefit of the population (of Perugia, ed.), which had been growing every day since the beginning of the year 1254, resolved in a General Council of the People that the water from the new spring should be brought from Monte Pacciano, over a distance of about 3 miles to the city, to the place already determined, in the Piazza Maggiore in front of the Church of San Lorenzo] (p. 400).

The task of verifying the feasibility of the project for the construction of a city aqueduct was entrusted to Friar Plenerio and Bonomo da Orte in the years 1254–1255 (Romalli 2006, p. 323). They outlined a route of about 4000 paces, which from Monte Pacciano passed through Sant'Orfeto and San Marco up to Perugia. Brother Plenerio also took care of the direction of the works, while the execution was supervised by master craftsman Bonomo from Orte. However, the works were interrupted for unknown reasons (De Felice 1995, p. 15). After various vicissitudes, in 1277, the interest in the construction of the work was rekindled and the direction was entrusted to Bevignate, of the Sylvestrine Congregation, with Boninsegna da Venezia, professor of hydrostatics from Venice (Romalli 2006, pp. 323, 329 note 34). The projects presented followed the original route (Monte Pacciano, S. Orfeto, S. Marco, Perugia) and introduced new elements, such as the vents to be built at S. Orfeto, above S. Marco and on a hill between S. Marco and S. Angelo, the lead pipes buried and installed in masonry boxes and the arches to overcome the morphological differences in the level of the land. Master Guido also suggested the construction of a larger cistern to collect rainwater as a reserve, while Master Coppo, having discovered further springs (the Faggeto spring), proposed to join them to those of Monte Pacciano. The Municipality General Council, after examining the four projects, commissioned Master Boninsegna from Venice, to carry out the necessary inspections and to draw up an accurate study. The work for the construction of the water infrastructure resumed under the guidance of Fra Bevignate, pupil of Boninsegna. The construction site was completed on 15 February 1280, when the water reached the acropolis of the city and fed the Fountain of Piazza Maggiore (1278–1280), decorated with marble panels sculpted by Nicola and Giovanni Pisano (De Felice 1995, pp. 16–17).

In the first medieval aqueduct of Perugia, the deepest depressions were overcome with arches that kept the pressure within manageable limits, avoiding sudden changes in the water flow. Despite the rationality with which the work was conceived, 10 years after the arrival of the water at the Fonte di Piazza, the first problems of malfunction began, mainly due to the accumulation of limestone in the water of the Barigiana main vein, which caused the occlusion of the pipes. Furthermore, the landslide terrain caused frequent interruptions in the water flow, as the subsidence of the ground damaged the pipes, which had to be replaced frequently. In 1293, the water completely stopped reaching the Fonte di Piazza due to the breakage of 32 pipes. By 1309, the aqueduct was in ruins, and, while the water was slow to return, the infrastructure was subject to numerous thefts. Magistrates and citizens attempted to raise funds to repair the aqueduct, appointing superintendents and custodians and promulgating laws against vandals. In 1317, the General Council of the People of Perugia charged Friar Vincenzo of the Order of Friars Minor with finding a way to bring water back to the Fontana Maggiore. The friar proposed abandoning the existing works and building a new aqueduct with a more direct route between Monte Pacciano and Perugia (De Felice 1995, p. 21).

1.3 The Aqueduct of 1322

The works for the construction of the second medieval aqueduct of the city, begun by the masters Ambrogio and Lorenzo under the direction of Paolo Barzi and Cola Gaboffi, were concluded in 1322 (Fig. 1). The new route of the Perugia aqueduct started from the 'conservetta', built downstream from the existing 'conserve delle Vene' where the waters of the main vein, the small fountain of the Barigiana and, if necessary, also those of the large conserve of Monte Pacciano were collected. From here began the new pressure pipeline that crossed the canal of Vajano on a first arch known as 'delle Piagge' and continued crossing the canal of Fornace (now the Rio torrent) on a second arch called 'del Monte Spinello'. After crossing the Carnevale canal, the pipeline went up the hill through a tunnel, crossed the Ponte d'Oddi canal on a further arch and continued for a stretch in another tunnel up to the convent of Santa Caterina Vecchia, near the city walls. The route continued under the monastery of Monteripido, crossed Porta Sant'Angelo and reached the reservoir of the convent of Sant'Agnese. From here, the distribution of water began, using the arches of the Conca to overcome the depression in the ground near the Etruscan walls, finally arriving at the Fontana Maggiore (De Felice 1995, p. 22). This second medieval aqueduct, although characterized by a spectacular design and beautiful arches, showed design defects from the beginning that would affect its functioning throughout its existence. The significant difference in height that characterized the route is among the most critical aspects: the water had to descend rapidly for 152 m over 1.5 km of the route to the Fornace ditch and then rise again by 141 m over another 1.5 km to the Monteripido convent, the highest point of the route. Considering further errors in the design of the second aqueduct, it can be deduced that the first medieval aqueduct was designed with greater rationality than the second, and that the early ruin of the first is attributable more to defects in execution than to errors in design (De Felice 1995, p. 23).

Over the course of five centuries, Perugia's second medieval aqueduct faced numerous difficulties that marked its history, but it remained an essential infrastructure for citizens until the end of the eighteenth century. In addition to defects in its construction, the historical chronicles report news of a continuous recurrence of landslides, breakages of entire sections of pipes and dismantling of masonry, excess limestone in some veins, but also thefts of lead pipes, illegal water withdrawals, damage caused by the owners of land bordering the buffer zone during agricultural work and destruction systematically carried out during wars or internal struggles (De

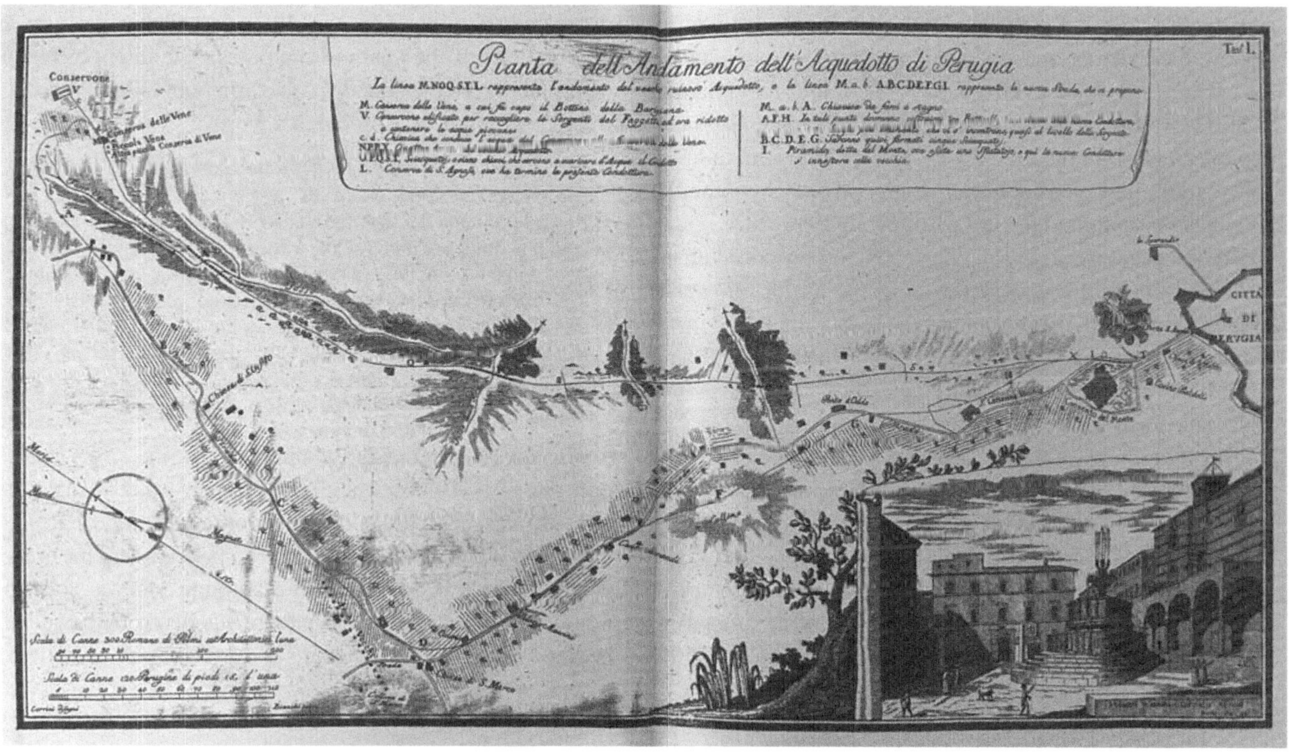

Fig. 1 Plan of the Perugia aqueduct (from De Felice 1995)

Felice 1995, p. 29). In 1796, a new redevelopment project for the aqueduct was recorded, but it was not followed up. At the end of the eighteenth century, the establishment of the Roman Republic and the fall of the Papal Government plunged the city into general disorder. The new republican government attempted to repair the damage to the infrastructure, but, in July 1799, groups of peasants and regular troops invaded and pillaged the territory of Perugia. Between 1795 and 1801, thefts and devastation further compromised the aqueduct, leaving the city's fountains without water.

2 Dynamics of Urban Transformation and Reconstructions Between the Nineteenth and Twentieth Centuries

The lack of functionality, until reaching the complete inactivity, is a recurring theme in the history of the aqueduct. This condition of inefficiency, at the beginning of the nineteenth century, led to new studies to identify alternative routes suitable for a new pipeline (Montanari 1950, p. 42). It was even hypothesized to move the Fontana Maggiore, a masterpiece of medieval art located in the square between the cathedral and the town hall, near the entrance to the Rocca Paolina, the most imposing fortification in Perugia built from the 1540s by orders of Pope Paolo III (Matracchi et al. 2023). The economic difficulties in which the Municipality of Perugia found itself limited the launch of new major projects. When the idea of definitively decommissioning the existing aqueduct had not yet been abandoned, the use of water collected from the roofs of the city buildings to feed the monumental fountain in the square had also been hypothesized (Montanari 1950, p. 42). In 1808, the municipality took into consideration some project hypotheses, which can be summarized in the following two:

- The proposal by architect Vincenzo Ciofi proposed abandoning the vein waters in favour of the use of rainwater. He identified the areas to build the reservoirs in the monasteries of Sant'Agnese, Monteripido and Santa Lucia and also calculated the total volumes of water necessary for the needs of the 'public fountains'.
- The solution by architect Giovanni Cerrini suggested the construction of new aqueduct, but with a path largely similar to that of the medieval one; it was also planned to reuse the existing lead pipes to create new, lighter ones. The availability of lead was a very critical aspect, since in the past there had been numerous thefts of this material, creating inevitable disruptions to the aqueduct. Thus, the possible use of iron pipes was also introduced as an innovative element (Montanari 1950, pp. 42–43).

The municipal administration commissioned the architect Andrea Vici to examine the proposals received. Ciofi's project was criticized for not having calculated the volume of water that would be lost, as well as for having neglected the possible deterioration of rainwater to be stored in the cisterns. Vici was in favour of the second solution which also favoured a lower water pressure, to the advantage of the tightness of the pipes, a central aspect for the efficiency of the aqueduct. But Vici's opinion sparked heated discussions, also through the publication of pamphlets and letters, among the supporters of the different proposed projects. Following the choice of Cerrini's project, in 1814, the contract was signed to start the works. However, years of disputes followed with the executing company, which did not supply pipes of suitable quality but made them with defective materials that did not pass the water pressure tests (Scaccia 1822). Until 1827, when the aqueduct was restored to its original function, the water once again reached the Fontana Maggiore (Montanari 1950, p. 46). An innovative aspect emerged with the initiative of the fountain-maker Germano Bartoccini, who in 1836 presented a proposal for the maintenance of the aqueduct to the Perugia magistrates, accompanied by detailed information on the entire structure (De Felice 1995; Chianella 2016, pp. 139, 141). This approach found an important precedent in the report by the Jesuit Giuseppe Boscovich, based on plans and profiles and with indications of some forms of degradation, such as limestone that obstructed the pipes and unstable parts of the structure, or malfunctions of the vents of the pipes. The intent to pursue the protection of the aqueduct also as a built cultural heritage is well expressed by Boscovich's words:

> '[…] sarebbe […] il lasciare perire le conserve, e gli archi, e i bottini, per ricavare il prezzo dei piombi, che buttare a terra in Roma le reliquie del Colosseo, che formano l'ammirazione dei popoli per ricavare il prezzo dei travertini […]', ['[...] this would be [...] like leaving the reservoirs, and the arches, and the drainage structures to decay in order to derive the value of the lead, in the same way as to destroy the relics of the Colosseum in Rome, that have been admired by people, in order to obtain the value of the travertines [...]'] (Belforti 1785, p. 464).

The aqueduct remained at the centre of the city's interest until the early 1930s, when a new urban water distribution network was inaugurated. Apart from the Monte Pacciano reservoirs, which are still used in lean periods, and the arches of the Conca transformed into a hanging road, a slow and progressive process of abandonment and decay of the ancient medieval aqueduct, which had served the city of Perugia for

Figs. 2 and 3 View of the buildings leaning against the aqueduct (P. Matracchi, July 2024)

centuries, began in this period (De Felice 1995, pp. 58–62). Serafino Siepi (1822, pp. 238, 261–262) describes the arches of the aqueduct as no longer used for the passage of water and at risk of imminent destruction. Those of the Conca section, so called because of the orography of the place characterized by a depression between two hills, are 22 arches. At the time, according to Siepi's testimony, they were already used as a pedestrian path by students and teachers to reach the University of Perugia. And precisely in order to make this use safer, the tops of the arches were equipped with parapets (Figs. 2, 3, and 4).

The studies and surveys carried out specifically have highlighted that, in the Conca, the surviving arches are 17, also considering the two at the ends which are documentable to a small extent (Fig. 5). In the remaining section, some arches are completely visible, and others are partially incorporated into houses. Above them, the pedestrian path extends without interruptions for over 200 m. The leanings of houses against the arches were early. In some of these buildings, in fact, stone slabs can be observed which bear the dates 1665, 1690 and 1745 engraved on them. These are datings which refer to a period in which the aqueduct was certainly still used in its original function. In these and other cases, the connection with the path on the aqueduct was therefore implemented only at a later time, as is also confirmed by some connections with the pedestrian path obtained with adaptations, via stairs. In the houses that have connections at the same height as the pedestrian path, when they were built, the height of the pedestrian path, which is variable, was evidently taken into account. We are precisely in the years in which Siepi complained about the risk of demolition of the arches, a time in which, at the same time, the precursory pedestrian function of the Conca section had been decided; other sections of the aqueduct arches survive in a serious state of decay (Romalli, p. 318, Fig. 2). It is not excluded that the conservation of the Conca aqueduct section had been paradoxically favoured by the incongruous leanings of buildings, which had made the arches intrinsic element to the constructions that partially incorporated them. In the sections outside the Conca, Siepi mentions over 40 arches that were later largely dismantled.

Fig. 4 View of Via dell'Acquedotto from Via Appia (P. Matracchi, July 2024)

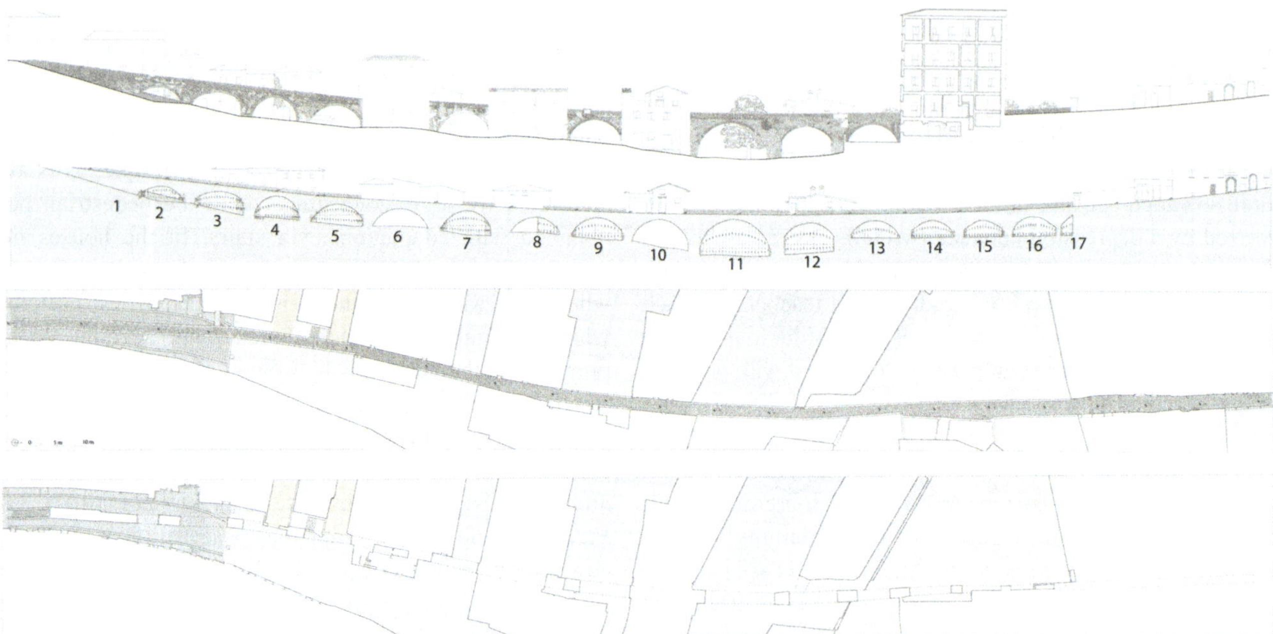

Fig. 5 Survey of the aqueduct, from the bottom: ground plan, level of the footpath, longitudinal section, elevation (original drawings scale 1:50 by A. Chianella, 2016)

3 Current State of Conservation

Today, much of the ancient aqueduct that once supplied Perugia with water is in a state of decay. The only exception is the Conservone Vecchio, still used as a cistern during periods of drought, while the majestic arches that survived outside the city walls are almost abandoned. The contribution focuses on the Conca arches, located near the historic centre of Perugia, which constitute the last part of the medieval aqueduct (Fig. 5). Currently, these arches have been transformed into an elevated pedestrian street, called Via dell'Acquedotto.

3.1 Urban Context of the Conca Area

The Conca arches are located near the historic centre, within the medieval walls. Apart from some archaeological finds dating back to Roman times, there is no historical information on the area until the mid-thirteenth century. The numerous engravings of the city made between the sixteenth and seventeenth centuries show the aqueduct almost devoid of lateral constructions and suggest that the buildings currently leaning against the infrastructure were largely built later. Comparing the current urban fabric with the historical cartography of the Gregorian land register of 1845, the main transformation dynamics of the last century can be identified (Figs. 6 and 7). This analysis highlights the demolitions and reconstructions that took place after 1845, such as the demolition of a building at the entrance to Via dell'Acquedotto, now a widening at the intersection with Via Ariodante Fabretti, and the expansion of some buildings adjacent to the aqueduct. From the consultation of nineteenth-century cartography, the aqueduct was already in a state similar to the current one. The urban fabric has progressively developed around and near the aqueduct itself, 'leaning' against it. Today the Conca presents itself as a densely urbanized part of the city, which unites within it both the historical buildings, the result of the medieval expansion of the fourteenth century, and most of the university structures of Perugia located in the more recent part of the area. This coexistence creates an evident dualism, with the medieval aqueduct which, as a suspended road, plays a crucial role in the city's connections (Figs. 8 and 9). It runs perpendicular to the main road of the area, connecting Via Fabretti with Via Appia and providing a direct pedestrian route between Borgo Sant'Angelo and the historic centre. The roads that branch off below the aqueduct are mainly intended for vehicular traffic. In the Conca area, apart from the parking lots in line along the access roads, there are no significant parking spaces, except for the parking lots reserved for university departments which, however, are unable to fully satisfy the

Fig. 6 Analysis of the transformations of the urban fabric from 1845 to 2000. In red, the demolitions. (elaboration by A. Chianella, 2016)

Fig. 7 Analysis of the transformations of the urban fabric from 1845 to 2000. In blue, the new constructions (elaboration by A. Chianella, 2016)

Fig. 8 Analysis of the intended uses of the Conca area. Residential buildings in light orange; care and accommodation facilities in dark orange (colleges, student residences, tourist and hotel facilities); school facilities in blue (from nursery schools to university buildings); religious building in pink (churches, convents, seminaries) (elaboration by A. Chianella, 2016)

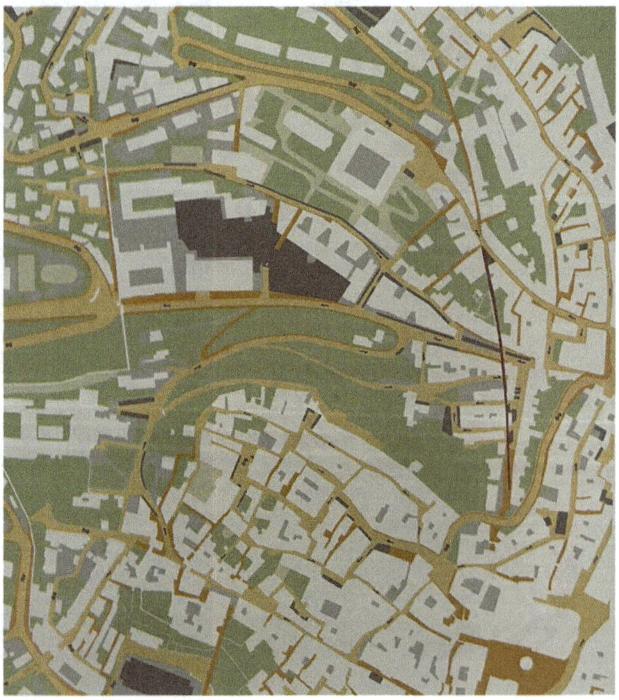

Fig. 9 Analysis of public mobility in the Conca area. Pedestrian areas are highlighted in orange, vehicular roads in yellow (elaboration by A. Chianella, 2016)

daily needs of users who frequent the area. As a result, car parking ends up invading the streets that cross or run alongside the aqueduct, compromising the view and accessibility of this important historical heritage.

3.2 Material Analysis and State of Conservation

For the material analysis and the state of conservation, the arches of the Conca were divided into different sections, including the initial embankment and the 17 arches, including those almost completely buried at the two ends (Fig. 5). The first arch towards the historic centre is visible for a very small part and remains incorporated into the nineteenth-century brick embankment. Arches 6, 8 and 10 are completely incorporated inside some houses, while the last four arches (13–17) are located in a private garden and cannot be visited.

The analysis of the structures and materials of the arches shows that extensive renovations have been carried out over time. The variations in materials highlight the original portions and the widespread reconstruction and restoration interventions. The original walls were made of sandstone blocks, replaced in many parts by stone and brick, used in particular for the renovation of edge arches. Sometimes the reconstruction also involved entire arches. One rebuilt entirely in brick, on the intrados, bears the date MDCCCXIX engraved on the mortar. Above the arches, the brick parapet, finished with a stone element, dates back to the nineteenth century (Fig. 10). It can also be observed that in some cases, to obtain the maximum width of the path on the aqueduct, the brick parapets were made by inserting brick supports (Fig. 11).

The analysis shows that many arcades have partially preserved the original sandstone ashlar archivolt (arch 1, arch 5, arch 7, arch 11 and arch 12), and others (arch 2, arch 3, arch 4 and arch 9) have extensive reconstructions with brick elements. Worthy of note is arch 7, at the confluence of two roads between Via Appia and Via del Cardellino. This arch is particularly interesting because it has not undergone partial or total reconstruction of the archivolt, preserving the sandstone ashlars even in the intrados. The sandstone masonry survives on the only visible abutment area, while the upper part of the masonry shows limestone elements and a brick parapet (Fig. 12).

Arch 11 is larger than the others, and below it, Via Santa Elisabetta leads to the lowest part of the Conca area. From the material analysis, it emerges that the predominant material in the lower part is sandstone; the archivolts are partially rebuilt in bricks. The highest part of the arch walls is mainly in limestone elements from Mount Subasio. On the parapet, we can see the presence of drainage channels, surrounded by

Fig. 10 Intrados of arch n. 9: date MDCCCXIX engraved in the mortar (P. Matracchi, July 2024)

Fig. 11 View of arch 3 of the aqueduct, at the crossroads between Via Appia and Via del Cardellino (P. Matracchi, July 2024)

Fig. 12 Detail of the aqueduct elevation (survey A. Chianella, 2016)

patches of mortar, which can be interpreted as elements added at a later time. The last visible arch is the twelfth. Also in this last case, the reconstruction of portions of the arch was carried out with sandstone blocks deriving from the original system. We can also identify the brick archivolts and parts composed of mixed brick and stone masonry as the result of reconstructions. The arch was also reinforced with the insertion of iron chains to increase the stability of the structure. From a conservative point of view, the degradation phenomena detected depend mainly on the exposure of the artefact to atmospheric agents. The action of rainwater run-off, and in particular of the drainage channels from the parapet, causes the proliferation of a black patina in the underlying arches that dirties especially the bricks on the facade. The sandstone ashlars are subject to the effects of exposure to atmospheric agents with erosion and exfoliation phenomena. The material deterioration of the artefact also derives from the action of biological patinas and higher plant organisms, which carry out a mechanical action inside the artefact through their roots. In most of the intrados of the arches, in addition to recurring phenomena also on the wall facings such as biological patina, lack, exfoliation and efflorescence, the phenomenon of cracks that develop parallel to the archivolts has been observed. This could be caused by the lack of connection between parts made at different times. In this regard, the insertion of transverse chains should be evaluated.

Finally, it is important to highlight the lack of valorization of this historic infrastructure work. Currently, there are no interventions or initiatives that highlight the historical importance of the artefact. The lack of information signs or other forms of signalling prevents visitors from recognizing the aqueduct as an architectural work of great importance for the history of the city. In particular, the widening at the intersection with Via Ariodante Fabretti, the starting point of the route towards the historic centre of Perugia, is in a state of evident degradation, invaded by parked cars (Fig. 13). This problem affects the entire area of the aqueduct, with the arches that in some places are used as a shelter for vehicles, compromising the visual integrity and accessibility of the work. A more careful and respectful management of this architectural heritage would be necessary to preserve its historical value and better integrate it into the contemporary urban context.

Fig. 13 View of the aqueduct from Via Appia towards the historic centre of Perugia (P. Matracchi, July 2024)

4 Conclusions

The Conca area in Perugia is an emblematic example of a condition of urban traffic congestion. The medieval structure of the neighbourhood and its contemporary development as the site of university buildings make traffic in some streets particularly intense, creating challenges for the usability and liveability of the area.

In this context, the medieval aqueduct offers a valuable alternative for pedestrian mobility in the city. A further improvement could come from the implementation of a lift which could be placed, for example, between arches 11 and 12. This would make the pedestrian route on the aqueduct more accessible, although we are aware that not everyone could benefit from it in the same way. In fact, the historic infrastructure does not offer the slope or the width necessary for a person in a wheelchair to pass independently.

The link between the city of Perugia, in particular the Conca area, and the pedestrian usability of the former medieval aqueduct is not new, but has its roots in the first quarter of the nineteenth century. Even then, the potential for a virtuous use of this important cultural heritage, capable of improving the quality of urban life, could be glimpsed. Today, the enhancement of this pedestrian path represents an important step towards a more sustainable city, where the past and the present dialogue to offer innovative solutions to the problems of contemporary urban life.

Authors' Contributions This paper is the result of a joint reflection by the authors. Section 1 was written by Pietro Matracchi and Mariano Sartore (Sects. 1.1 and 1.2) and Maddalena Branchi (Sect. 1.3); Sect. 2 was written by Pietro Matracchi; Sect. 3 was written by Maddalena Branchi; Sect. 4 was written by Pietro Matracchi, Mariano Sartore and Maddalena Branchi.

References

Belforti G (1785) Memorie istoriche sulla fonte di Piazza (Perugia, Biblioteca Augusta, ms. n. 1348)

Chianella A (2016) La città, l'acqua, i manufatti e la storia urbana. Ipotesi di restauro e rifunzionalizzazione per l'acquedotto medievale di Perugia [Master thesis in Ingegneria Edile-Architettura, Università degli Studi di Perugia]

De Felice A (1995) L'antico acquedotto della Fonte di Piazza di Perugia – dal 1254 al 1932. Editrice Volumnia

Fiocca L (1926) Perugia: sua origine, cinta delle mura e porte. Tipografia Guerriero Guerra

Grohman A (1981) Perugia in Le città nella storia d'Italia. Editori Laterza, Rome

Matracchi P, Angiovini G, Frattegiani C (2023) War conflicts and ideological conflicts. The 'Rocca Paolina' in Perugia between, dissimulations, rediscoveries and new usage opportunities. In: Amore R, Pascariello MI, Veropalumbo A (eds), City and war. Military defences, ruins permanences of urban memories and images, Part II, Picone R, Russo V (eds) Restoration and war. Federico II University Press

Montanari M (1950) La Fonte di Piazza e gli Acquedotti di Perugia. Donnini editore, Perugia

Radicioni F, Matracchi P, Stoppini A, Tosi G, Marconi L (2021) The Etruscan city gates of Perugia: geomatic techniques for the documentation and study of an urban history heritage. In: 9th ARQUEOLÓGICA 2.0 & 3rd GEORES, Valencia (Spain), 26–28 April 2021. https://doi.org/10.4995/Arqueologica9.2021.12058

Romalli G (2006) L'acquedotto medievale di Perugia e l'adduzione idrica nelle realtà comunali centroitaliane. In: Franchetti Pardo V (ed) Arnolfo di Cambio e la sua epoca. Costruire, scolpire, dipingere, decorare. Viella Editrice

Scaccia G (1822) Relazione sulla memoria dei Signori Deputati di Perugia intorno alle circostanze della nuova conduttura di acque per quella città, ed i provvedimenti, che si credono necessari, Perugia

Siepi S (1822) Descrizione topologico-istorica della città di Perugia, vol I. Tipografia Garbinesi e Santucci

Silvestrelli MR (2005) Acqua per la città. Lo spazio perduto della fontana di Arnolfo. In: Garibaldi V, Toscano B (eds) Arnolfo di Cambio: una rinascita nell'Umbria medievale. Silvana Editoriale, Cinisello Balsamo

From Cultural Preservation to Cultural Tourism: Examining the Heritagization of Rijal Almaa's Vernacular Architecture Within the Al Soudah Peaks Development

Saad Hanif

Abstract

Rijal Almaa, a heritage village in Al Soudah Peaks Development, has embraced a novel heritagization approach as the ambitious cultural project is set to recreate the vernacular architecture by blending modern and traditional elements. Therefore, the notion of "heritagization," which reframes heritage as a process rather than a fixed entity, has introduced fresh complexities that are both captivating and challenging. The research employs a case study research methodology to explore the role of heritagization in the evolution of Rijal Almaa's vernacular architecture within the Al Soudah Peaks Development. The research aims to analyze how vernacular architecture is being highlighted, transitioning from mere preservation efforts to becoming a key element of cultural tourism. Moreover, it examines the integration of the vernacular heritage into the broader framework of Saudi Arabia's Vision 2030, positioning it as a pivotal component of the Kingdom's ambitious cultural destination initiatives. Research concludes how vernacular heritage evolves under the process of heritagization and develops recommendations for transforming cultural heritage sites into cultural destinations while ensuring the balance between heritage preservation and cultural tourism. The findings and recommendations from this research will be applicable to other heritage sites currently undergoing heritagization for mass cultural tourism developments.

Keywords

Cultural preservation · Cultural tourism · Heritagization · Vernacular architecture · Rijal Almaa · Al Soudha Peaks Development · Commodification

S. Hanif (✉)
Faculty of Architecture, Eastern Mediterranean University, Famagusta, North Cyprus, Türkiye

Department of Architecture & Interior Design, College of Engineering, University of Bahrain, Sakhir, Bahrain
e-mail: m.saadhanif@emu.edu.tr, mhanif@uob.edu.bh

1 Introduction

Architecture, much like poems and rituals, embodies the culture, with its vernacular characteristics giving a distinct rationale toward its functions (Glassie 2000). Vernacular architecture is also shaped by the physical and cultural characteristics of a region and is passed down through generations via buildings, making it unique to each area despite common features and mutual influences with other regions. Such architecture belongs to nonindustrial or preindustrial societies and relies on local materials, tools, and techniques, eliminating the need for industrial production. Its elements are tailored to meet local needs, ensuring harmony with the surrounding environment (Senosiain 2003). Vernacular architecture remains largely unaffected by fashion trends and strives for perfection in a manner resistant to change or improvement (Rudofsky 1964). It evolves slowly through a trial-and-error approach, adapting over centuries primarily due to changing human needs. This architectural style is also designed to respond to the natural environment and the lifestyle demands of its users, earning it the label "architecture without architects." While individual builders create these structures, they ultimately reflect the values and culture of the society in which they are situated (Eyüce 2005). Thus, architecture shaped by its physical and cultural influences over time and preserved through local and regional techniques defines the term "vernacular architecture," highlighting that each region's architecture is distinct, though it shares common features, influences, and similarities with others (Ceylan 2022).

Vernacular architecture in numerous regions around the world, especially in areas experiencing rapid development due to globalization, has been subject to significant transformations over the years (Dayaratne 2020). Al Sayyad (2002) highlights this transformational character, arguing that cultural experiences are inherently hybrid, and vernacular forms are constantly evolving. According to him, hybrid cultural expressions are prevalent, especially among colonized cul-

F. Alberti et al. (eds.), *Towards Sustainable Urban Regeneration: Utilizing Environmental, Cultural, and Social Assets*, Advances in Science, Technology & Innovation, https://doi.org/10.1007/978-3-032-02091-8_23

tures, yet there are identifiable patterns of human adaptation within these hybrid constructions of traditionality that establish regional or subregional uniqueness. This characterization leads to depictions of vernacular traditions that are portrayed as static, incomplete, and frequently romanticized (Vellinga 2006, p. 83; Zwerger 2019). Vellinga (2006) critiques the characterization of vernacular traditions as static, incomplete, and often romanticized. Consequently, Vellinga proposes replacing "vernacular architectures" with "vernacular traditions," defining tradition as a conscious and creative adaptation of past experiences to current needs and circumstances. This approach allows for a more dynamic perception of vernacular architecture. Kopytoff (1986) similarly stressed the dynamic nature of vernacular architecture, before globalization, highlighting their continuous evolution in terms of use, function, and meaning (Zwerger 2019).

However, the significant increase in scholarly attention toward vernacular architecture has resulted in authors making simplifications, inaccurate statements, and erroneous conclusions, sparking global debate (Zwerger 2019). In recent decades, numerous global studies on vernacular architecture have aimed to extract valuable insights into traditional passive strategies and thermal comfort. Recently, the focus on designing more sustainable buildings to reduce environmental impact has gained significant public interest, particularly among architects and researchers (Altomonte 2008; DeKay and Brown 2014). This growing concern has led to an increased tendency to learn from the design strategies and methods of vernacular architecture, which are adeptly suited to local climates and regions (Na et al. 2022). Yet the complexities and breadth of issues related to the function and significance of vernacular architecture in the twenty-first century are considerable (Asquith and Vellinga 2006). Given the vast quantity and diversity of vernacular architecture worldwide, its presence in architectural education is surprisingly minimal. When academic attention is given to vernacular architecture, it typically focuses on historical documentation (Rapoport 2006). To advance the discussion on the significance of studying vernacular architecture in the present and into the twenty-first century, it should be viewed not only as an examination of traditions from the past but also as a source of new approaches, solutions, and advancements for future built environments.

Vernacular architecture faces significant material, structural, and constructive vulnerabilities due to various intrinsic and extrinsic factors. Extensive literature discusses the causes of degradation, including water in its various forms (Houben and Guillad 1989; Warren 1999; Keefe 2005; Bui et al. 2009; Cooke 2010; Illampas et al. 2012; Mileto and Vegas 2014, 2017). Ultimately, human activities play a crucial role in altering and adapting buildings to new uses and demands, impacting the preservation of vernacular architecture. Tourism and improper utilization also contribute significantly to damage to the integrity of the vernacular architecture (Mileto and Vegas 2022). The impact of tourism on vernacular architecture can be both beneficial and detrimental. On the positive side, tourism often encourages the preservation of historical sites and cultural heritage as these elements become central attractions for visitors. However, increased tourism can lead to the commercialization and commodification of vernacular architecture, causing it to lose its authentic evolution. The influx of tourists may also pressure local communities to modify or replace vernacular structures with modernized versions, disrupting the existing architectural integrity. Hence, appropriate measures are required when cultural tourism is initially planned for vernacular architecture and the peripheries.

Rijal village, situated in the Asir region and serving as the capital of the Rijal Almaa Province, is a compact vernacular settlement in the Kingdom of Saudi Arabia. It was historically a vital corridor connecting travelers from Yemen and the Levant to Makkah and Madinah, establishing it as an essential regional trade hub. The village is comprised of around 60 multistory palaces constructed from natural stone, clay, and wood. Over the years, The Saudi Commission for Tourism and Antiquities has initiated rehabilitation projects for the village, in collaboration with partners from both the public and private sectors (UNESCO). The newly planned urban development of the Rijal Almaa district within the Al Soudah Peaks Development project has adopted a novel heritagization approach in order to capitalize on the cultural tourism potential. Heritagization as described by Davallon (2019) is a process by which an item is recognized as *valuable for preservation as heritage*, setting it apart from items considered replaceable or tradable. This innovative initiative aims to recreate the area's vernacular architecture by integrating modern and traditional elements. Rijal Almaa, part of Saudi Vision 2030, is poised to become a prominent cultural tourism destination in the region. On a positive note, Vision 2030 is set to allocate substantial funds toward tourism infrastructure, which could involve the creation of entirely new cultural destinations while strengthening existing heritage preservation efforts (Alkhanbshi 2024).

This study employs a case study research methodology to explore the role of heritagization in the evolution of Rijal Almaa's vernacular architecture within the Al Soudah Peaks Development. Data were collected through an extensive review of academic literature, reports, published books, and online resources. Research aims to analyze how vernacular architecture is being highlighted, transitioning from mere preservation efforts to becoming a key element of cultural tourism. Moreover, it examines the integration of this vernacular heritage into the broader framework of Saudi Arabia's Vision 2030, positioning it as a pivotal component of the Kingdom's ambitious cultural preservation initiatives. This research concludes how vernacular heritage evolves under the process of heritagization and develops recommendations for transforming these sites into cultural tourism destinations while ensuring the balance between heritage preservation and cultural tourism.

2 Heritagization and Its Framework

Heritagization refers to the transformation of objects, places, and practices into cultural heritage, ascribed with values that essentially describe heritage as an evolving process. Heritagization is an intriguing process that involves the identification, preservation, and promotion of cultural heritage within a specific region or country. Although it offers significant economic benefits and enhances tourism, it must be approached with caution, as it carries the risk of over-commercialization, cultural commodification, and the potential erosion of heritage values. Walsh expressed concern about the transformation of spaces into heritage sites, viewing it as a process that turns authentic locations into tourist attractions by selectively referencing various historical images, often leading to the degradation or loss of the original sites (Walsh 1992). Building upon Walsh's ideas, Harrison described heritagization as the conversion of objects and locations from practical entities into items intended for display and exhibition (Harrison 2012), thus highlighting the concept of official heritage. Davallon (2019) conceptualizes heritagization as a process that frames heritage as both a valuable asset for preservation and a marketable commodity. Similarly, Di Giovine (2009) describes heritagization as a process deeply connected to the global development of tourism aimed at displaying the heritage to wider audience (Hanif and Riza 2024).

In recent decades, there has been a marked increase in what is recognized as heritagization within cultural heritage studies (Walsh 1992; Smith 2006; Harrison 2013). Harrison referred to this era as a "heritage boom," characterized by a proliferation of heritage. He also noted a widespread belief that once items, places, or traditions are officially recognized and classified as "heritage," they rarely undergo further change or transformation (Harrison 2013; Sjöholm 2016). For sites, objects, and practices to be acknowledged as heritage, they must be imagined, conceptualized, and narrated. Heritage, including intangible cultural heritage, must be enacted and materialized, especially within the framework of heritagization. This process involves converting abstract concepts into tangible forms that can be preserved, celebrated, and transmitted to future generations (Salemink 2021). Exploring the orchestration of heritagization process involves collecting, curating, exhibiting, serializing, and materializing the past across various locations, alongside the transformation of objects into heritage with altered statuses and meanings (Daher 2006; Gillot et al. 2013). Such transformation also incorporates creative strategies that can encompass both innovative methods of restoring, reinterpreting, or reconstructing heritage, as well as the integration of new functions within recently restored or rebuilt heritage structures (Gravari-Barbas et al. 2025).

The term "heritagization" can encompass several different aspects, including the designation of new heritage, the affirmation of recognized heritage, a reinterpretation of declared heritage, and the recreation of previously designated heritage (Sjöholm 2016). The emerging "heritagization" theoretical framework also refers to the process by which an object, location, or activity becomes part of its heritage (Hanif and Riza 2024). These items are elevated beyond mere functional objects when accompanied by meaning and/or cultural significance (Harrison 2012). Heritagization, as further elaborated by Skounti (2010), is a multifaceted process that involves the normative and institutional elevation of cultural and historical elements to the status of heritage, worthy of preservation, enhancement, and transmission. It involves both devaluation, through selection and delimitation, and revaluation, involving modifications and redefinitions to meet present-day needs rather than past configurations. This illustrates how heritagization is entangled in socioeconomic and ideological issues and is often characterized by heavy state control with limited citizen participation, as evidenced by certain contexts (Le Mentec and Zhang 2017). Consequently, this also challenges the existing or commonly accepted notions of heritagization where it is predominantly associated with cultural and historical heritage, emphasizing the recognition of places, practices, objects, histories, or ideas as significant inheritances from the past (Ashley 2014; Kajda 2019).

Hewison (1987) strongly criticizes this process, arguing that it compromises historical truth and accuracy in favor of commercial values, commodification, and popularization. He specifically critiques the proliferation of museums and related activities, asserting that instead of producing *genuine heritage*, society is creating a commodified version that lacks a clear definition but is eagerly marketed for profit. This perspective highlights the prioritization of commercial interests over authentic historical practice, emphasizing the need for a more nuanced understanding of heritage and its preservation in contemporary society (Mai Le 2021). This heritage-making process is often influenced by the broader political, cultural, social, and economic dynamics of modern society. It involves recognizing, preserving, and promoting cultural assets to sustain and celebrate a community's history, identity, and traditions. Zhu (2016), Su et al. (2019) demonstrated that heritagization operates within a structured top-down management framework. In this system, authorities determine which elements of local "authentic culture" are to be conserved and emphasized (Thouki 2022).

In the past decade, governments have actively employed the heritagization process to boost tourism and cultural consumption. The heritagization of specific historical regions leads to significant transformations in the urban landscape, such as the demolition of vernacular architecture and the displacement of original inhabitants. While some criticize these processes, others cooperate in hopes of participating in the cultural tourism industry and benefiting from it (Hanif 2024). Existing studies on heritagization can be broadly classified into two perspectives: one centered on Indigenous understandings of heritage and the other examining the social mechanisms involved in heritage formation. The concept of "heritagization" embodies at least two distinct epistemological viewpoints. Research, particularly in disciplines such as history, ethnology, and geography, often highlights the processes by which objects attain heritage status (Davallon 2019). Previous research has consistently highlighted the inadequately conceptualized and imprecise understanding of the dynamic and multilayered nature of heritagization. More comprehensive research is needed to fully grasp heritagization in the context of modern society, considering the rapidly changing relationship between cultural heritage and contemporary times. Therefore, further research is essential to clarify heritagization within today's world (Mai Le 2021; Hanif 2024). Since summarizing the effects of heritagization as a whole is challenging, more in-depth research is required to understand heritagization not only in modern society but also in various regions applying it, either consciously or unconsciously, to enhance their cultural heritage.

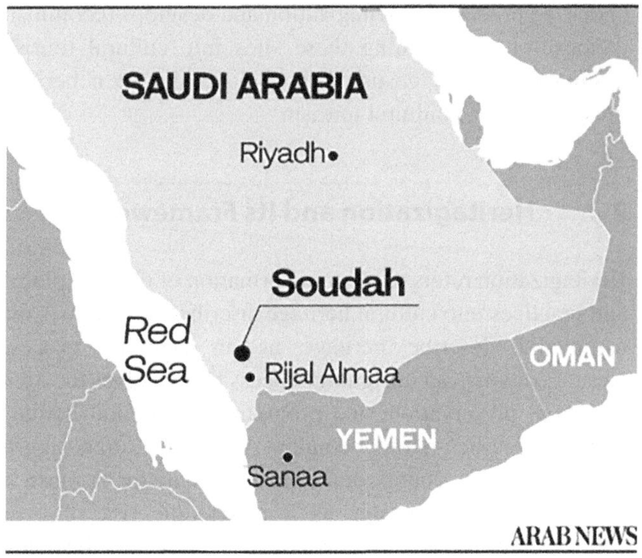

Fig. 1 Rijal Almaa is situated in the Asir region close to Al Soudah mountain. (Source: Arab news/BBC)

3 Case Study Area: Rijal Almaa

Rijal Almaa is a compact settlement featuring a distinctive urban layout, comprised of 60 multistory palaces constructed from stone, mud, and wood (Al-Garni 2011; Elassal et al. 2024). Situated in the Asir region, which boasts the highest mountain peaks in the Kingdom of Saudi Arabia, this area is renowned for its distinctive natural landscapes and the wide array of sports activities it offers. Al-Soudah Mountain stands as the tallest peak in the Kingdom, reaching an elevation of 3015 m (Fig. 1). The mountain is densely covered with juniper trees and is encircled by several agricultural villages, with Rijal Almaa being one of the most historic ones (Elassal et al. 2024). Rijal Almaa village is among the top heritage sites proposed for inclusion on the UNESCO World Heritage List. However, similar to other heritage sites in Saudi Arabia, Rijal Almaa remains relatively obscure, largely due to insufficient media coverage (Hoque et al. 2022).

Rijal Almaa, the vernacular village was originally constructed by the men of the region, who played a significant role in preserving the area's heritage, eventually gathering it within one of the fortresses of this remarkable village. The initial construction of the village did not take place at its current site but rather in Um Sedrah, located to the west of its present location. The village was once described as a lush area with running water and served as the capital of the Hilli Emirate during the reign of Governor Mousa Al-Kanani in 730 AH/1329 CE. It was referenced in 550 AH/1155 CE as a civic center and gained prominence through the dedication of its people to trade. This allowed it to become a commercial hub, receiving trade from the Red Sea ports of Al-Qunfudah, Al-Qahmh, and Jizan, as well as Indian trade through the port of Aden (UNESCO).

Hence, according to Criteria (V) and (IV), Rijal Almaa is an outstanding example of vernacular architecture and stands as an exemplary model of a deliberately preserved settlement that reflects a culture now at risk due to irreversible change. Developed during the tenth century AH/eighth century CE, the village is not only a significant representation of the Kingdom's cultural heritage but also a testimony to the vernacular lifestyle that largely disappeared after the agricultural revolution in the twentieth century. Additionally, Rijal Almaa exemplifies traditional human settlements that are seamlessly adapted to their natural environment and socio-economic needs.

3.1 Vernacular Architecture of Rijal Almaa

The vernacular architecture of Rijal Almaa, located in the mountainous Asir region near Al-Soudha, is a reflection of the area's cultural heritage and environmental adaptation. The village is known for its multistory stone houses, constructed from locally sourced flat stones. These structures, some rising as high as five to seven stories, are built close together, giving the village a fortress-like appearance. The thick stone walls provide natural insulation, keeping the interiors cool in summer and warm in winter, while small windows and doors minimize heat loss and protect the inhabitants from harsh weather. The buildings are arranged in a terraced layout, following the steep slopes of the mountains, which helps manage rainwater and prevent soil erosion. The use of wood and mud in the construction adds to the thermal efficiency and aesthetic appeal, with intricately carved doors and geometric patterns reflecting the cultural significance of the village.

Within Rijal Almaa, the structures referred to as "palaces" originally served as fortresses, playing a crucial defensive role during conflicts, such as the Ottoman invasion, due to the village's strategic position on a key trade route connecting Yemen with the Red Sea (Gingrich 2000). Although these fortresses have since been repurposed into residential buildings, their historical origins remain evident in their external appearance (Alsadi 2022). Communal spaces, such as courtyards and small plazas, play a vital role in the social life of the villagers. Despite modernization pressures, Rijal Almaa has retained much of its traditional character, with recent preservation efforts ensuring that the village continues to serve as both a living community and a historical site. This unique vernacular architecture highlights the balance between traditional stone and environmental adaptation in the Asir region of the Kingdom. The vernacular architecture of Rijal Almaa is not solely focused on aesthetics; it is also deeply rooted in social and humanitarian considerations (Fig. 2). The multistory buildings were designed to accommodate multiple families, reflecting the communal values of the local society (Visitsaudi.com 2024). Hence, Rijal Almaa exemplifies traditional human settlement and vernacular architecture, which is closely integrated with the ecological, social, and economic systems of the region. This adaptation is a key factor in the village's nomination for the UNESCO World Heritage List (UNESCO World Heritage Centre (n.d)).

3.2 Cultural Preservation of Rijal Almaa

The traditional village of Rijal Almaa, constructed in a way that harmoniously highlights its heritage elements and traditional characteristics, has also evolved naturally. Efforts were made over the years, to preserve the village's architectural techniques, to utilize local materials and forms, and to maintain its historical structure. The village's banana plantations, orchards, and gardens were also preserved, and the ecological balance was carefully upheld, even within the surrounding forestry environment, ensuring respect for historical authenticity. Consequently, Rijal Almaa is not only a preserved heritage village of traditional character but also a vibrant community where conservation efforts extend to farming practices (UNESCO). The villagers, supported by the General Authority for Tourism and National Heritage and the Rijal Almaa Heritage Village Municipality, played an active role in revitalizing and investing in the village. This was achieved through the establishment of a support fund, the creation of the Village People's Museum, and the development of an open-air theater (Al-Garni 2011; Elassal et al. 2024).

During the month of January 2018, the Saudi Commission for Tourism and National Heritage (SCTH) submitted the nomination file for the heritage village of Rijal Almaa in the Asir region to the UNESCO World Heritage Center. This initiative was part of a broader effort by the Commission to highlight the Kingdom's cultural heritage globally, preserving its diverse history, archaeological sites, and rich heritage. The SCTH emphasized that registering heritage sites serves to both illuminate Saudi Arabia's cultural contributions and rehabilitate these locations according to international standards.

The SCTH's initiative to include the vernacular village in the Heritage List is a key element of the Kingdom's Cultural Heritage Care program. This program encompasses various projects and initiatives aimed at preserving, enhancing, and showcasing national heritage sites. Preservation initiatives and revitalizing efforts were in compliance with UNESCO requirements at the historical village and were spearheaded by the Rijal Almaa Municipality, in collaboration with the SCTH, the Ministry of Islamic Guidance, and the Ministries of Transport and Environment. During a media visit, a visual presentation showcased the key preservation activities at Rijal Almaa, which included *"renovating the facades"* of several village buildings and enhancing the general counsel of the village (Arabnews 2018) (Fig. 3).

3.3 Cultural Tourism of Al Rijal Almaa

Cultural tourism endeavors are to strike a balance between economic advantages and the safeguarding of cultural heritage. It provides a unique travel experience, aiming to cultivate an appreciation for conserving historical values and enhancing cultural identity through cultural heritage, while fostering respect for both the natural and built environments (Rácz 1998). Boyd (2002) argues that the increasing trend of cultural consumption by tourists, along with the growing overlap between tourists and other types of visitors, creates difficulties in clearly defining cultural tourism or tourism driven by cultural interests. Boyd highlights that within tourism, the concept of "culture" has broadened to include not only cultural assets inherited across generations but also those that can be marketed as tourism products. This encompasses representing various ways of life across different

Fig. 2 Vernacular architecture of Rijal Almaa. (Source: Soudha. Sa)

Fig. 3 Rijal Almaa is a preserved historic village that has distinct vernacular architecture. (Source: Visitsaudi)

geographical regions, recounting the stories of people, their traditions and craftsmanship, beliefs, and other elements that have shaped their way of living (Boyd 2002; AlShaikh 2017).

In a similar vein, Rijal Alma is a significant example of vernacular architectural heritage and holds the potential to become a major cultural tourist destination, which could create various opportunities and contribute to the development of the local community (Hoque et al. 2022). Rijal Almaa has already attracted tourists from both within the Kingdom and around the world. As visitors navigate its narrow streets and intricately designed homes, they are invited to immerse themselves in the village's rich history. Central to this experience are the ancient mosque and a vibrant square, where cultural performances celebrate the village's deep religious and artistic heritage, enhancing its appeal as a tourist destination. At the heart of the village lies a comprehensive museum that showcases a diverse collection of artifacts, including traditional weaponry, farming tools, jewelry, and clothing, which together narrate the rich history of Rijal Almaa. The local market offers traditional garments and distinctive souvenirs, while the square serves as a venue for various cultural events, drawing visitors and fostering a strong sense of community. This illustrates the intangible heritage of the vernacular village and showcases the traditional lifestyle that has evolved in the village over the years.

Rijal Almaa stands as a key destination for tourism, essential for anyone seeking to explore the region's vernacular traditions. It is not only a historical village but also a dynamic vernacular settlement that has evolved over time, where history and modern hospitality seamlessly blend against the stunning backdrop of its landscapes (Fig. 4) (Discover Aseer). Subsequently, Rijal Almaa is being strategically integrated with its surrounding environment in the governorate to establish a cultural tourism corridor, fostering distinctive tourism experiences within the Kingdom's Vision 2030. This integration is anticipated to play a crucial role in advancing vernacular development. Additionally, the village's proximity to the city of Al-Abha and the Al-Soudah area positions Rijal Almaa as an interactive hub for tourists, which will contribute to the sustainable operation and long-term preservation of its cultural heritage.

3.4 Urban Evolution of Rijal Almaa in Al Soudah Peaks Development

Rijal Almaa has undergone several phases of development previously, including the creation of an open-air theater with a seating capacity of 1000 people. The village's market space, known for showcasing local products, has also been enhanced. Additionally, green spaces were expanded by approximately 7000 square meters, complemented by the installation of canopies and designated family gathering areas near the village entrance. The surrounding infrastructure has also been upgraded, including improved lighting along the highway leading to the village.

The Soudah Development Company (SDC), established by the Public Investment Fund (PIF), aims to transform Al Soudah and parts of Rijal Almaa in the Asir region into a premier cultural tourism destination as part of the Al Soudah Peaks Project. Al Soudah Peaks will make a substantial contribution to Saudi Arabia's tourism industry, positioning the Kingdom on the global tourism stage while emphasizing and honoring its vibrant cultural heritage. Visitors will have the chance to uncover the splendor of Rijal Almaa and other surrounding historical sites and immerse themselves in its cultural richness and the genuine hospitality of the Kingdom (Soudah Development).

Spanning 627 square kilometers, this area already benefits from a vibrant domestic tourism market, renowned for its expansive valleys, stunning hiking trails, and breathtaking mountain vistas surrounded by rain clouds, as well as its rich cultural heritage (Fig. 5). The destination will feature six distinct development zones: Tahlal, Sahab, Sabrah, Jareen, Rijal, and Red Rock (Fig. 6). Each zone will provide a variety of top-tier amenities, including hotels, luxurious mountain resorts, residential chalets, villas, exclusive mansion

Fig. 4 Rijal Almaa has evolved over the years to cater to the needs of increased visitor count. (Source: Discover Aseer)

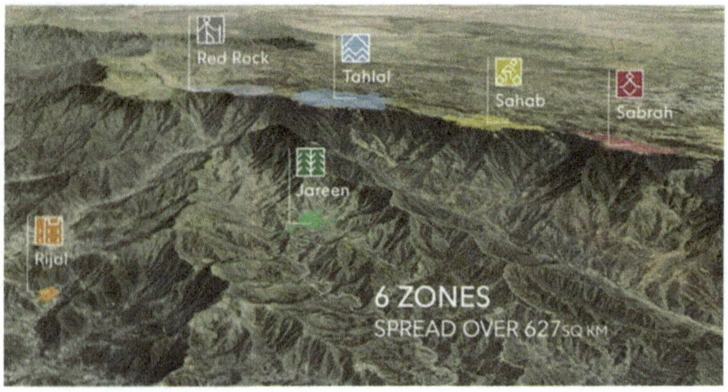

Fig. 6 Al Soudah Peak Development project is distributed in six zones. (Source: Soudha.sa)

Fig. 5 Development Project is set to replicate the vernacular architecture of Rijal Almaa. (Source: Soudha.sa)

plots, entertainment, and commercial venues, along with outdoor spaces dedicated to sports, adventure, wellness, and culture. Al Soudah Peaks Development aims to deliver 2700 hospitality units, 1336 residential properties, and 80,000 square meters of commercial space in Soudah Peaks by 2033. The master plan will be rolled out in three phases, with Phase 1 set to be completed by 2027, offering 940 hotel units, 391 residential properties, and 32,000 square meters of retail space. The village's development strategy focuses on showcasing its heritage and cultural significance, as well as revitalizing traditional crafts and industries. This initiative was supported by the local community, who generously contributed historical artifacts and heritage items to enrich the village museum (Elassal et al. 2024).

4 Heritagization Implementation in Rijal Almaa, Al Soudah Peaks Development

Rijal Almaa is evolving beyond its status as a vernacular village into a dynamic heritage-scape that brings history and tradition back to life. With a heritage spanning nearly seven centuries, this picturesque village is distinguished by its unique architecture, characterized by basalt stone structures

Fig. 7 Heritagization of Rijal Almaa's vernacular architecture. (Source: Supplied)

adorned with white quartz. The village is home to approximately 60 fortresses that tell stories of military strength and social hospitality. The Al Soudah Peaks Development project, centered on heritagization, aims to exhibit this rich history and enhance the infrastructure of Rijal Almaa, advancing entertainment, residential, and commercial sectors while improving access to the vernacular village's rich culture, natural environment, and heritage sites.

Drawing on Walsh's discussion of the conversion of spaces into heritage sites, where authentic locations are transformed into tourist attractions through selective referencing of historical imagery, often at the cost of the original site (Walsh 1992, p. 4), the historic village has undergone a similar heritagization process. This process highlights the village's rich cultural heritage, positioning it as a global tourist destination. This strategic investment and reaffirmation of Rijal Almaa's vernacular architecture align with the concept of heritagization highlighted by Sjöholm (2016), which encompasses various dimensions, including the designation of new heritage, the affirmation of existing heritage, the reinterpretation of recognized heritage, and the recreation of previously designated heritage.

The heritagization process in the case of Rijal Almaa merges economic benefits, tourism policies, and heritage preservation, encouraging a deep appreciation for historical values while reinforcing the significance of the vernacular architecture of the historic village (Figs. 7 and 8). It includes a wide array of activities, from boosting visitor numbers to strategically marketing cultural assets. The heritagization process, aimed at celebrating the preservation of cultural heritage and enhancing vernacular architecture, is integral to cultural tourism initiatives. These initiatives not only safeguard cultural heritage but also foster the creation of new cultural experiences, enriching the cultural landscape and leisure opportunities within a region. As a result, this process significantly contributes to elevating the appeal of heritage village as a cultural tourism destination.

Fig. 8 Heritagization of Rijal Almaa's to blend past and present. (Source: Supplied)

4.1 Heritagization Process in Rijal Almaa, Al Soudah Peaks Development

The heritagization process that occurred in the evolution of Rijal Almaa from a heritage village to a year-round tourist destination is multilayered. The heritagization process included recognizing, exhibiting, and promoting cultural heritage. First, the vernacular architecture of Rijal Almaa was recognized and reemphasized as a site having a significant cultural heritage. This led to the exhibition of the site having vernacular characteristics in terms of both tangible and intangible cultural heritage.

Second, this recognition not only reaffirmed the cultural significance of the historic village but also highlighted its previously overlooked potential as a prime cultural destination. Despite its rich heritage and architectural uniqueness, the village had not received the attention it deserved. The newfound focus on its cultural value has paved the way for its promotion as a key site for cultural tourism, showcasing the importance of preserving such historically significant locations.

Third, the designation of the place as a cultural tourism destination further highlighted its rich cultural heritage to the world through the development of further districts inspired by its vernacular architecture. This reframed the vernacular architecture within the frame of tourism potential by integrating modern facilities such as hotels, restaurants, and retail shops. Such designation also reframed the vernacular architecture toward the commodification spectrum, where the vernacular architecture is highlighted as not only heritage but also the intangible aspects of the village.

4.2 Heritagization of Vernacular Architecture in Al Soudha Peaks Development.

The integration of park areas, outdoor event spaces, boutique hotels, and authentic culinary experiences within the Rijal Almaa heritage village presents a significant opportunity to unlock the potential of cultural tourism of its vernacular architecture through the heritagization process. Vernacular architecture, in this context, emerged as a sanctuary for the human spirit, soul, and body, providing both refuge and shelter. This sentiment, expressed by the Andalusian judge Ibn Abdun, eloquently captures how our built environment reflects and embodies our spiritual values and beliefs (Bianca 2000). Dayaratne (2016), however, specifies that among such initiatives, the deployment of what is perceived as "an authentic vernacular character," generated by the uniqueness of culture, has always played a significant role.

Hence, the newly extended part which is to be known as the Rijal district will incorporate the same vernacular techniques of the village in constructing new buildings and the outdoor spaces. A similar stone technique is to be revived and reused in the expansion of the Rijal district as well as other zones of the project. This process of heritagization of the vernacular architecture is aimed to enhance the significance of stone construction and the relation it has with the topography of the region. The Rijal district will present opportunities to engage with authentic culture and heritage by blending the past, present, and future (Figs. 9 and 10). Set within the Rijal Almaa heritage village and surrounded by vernacular architecture, this destination will integrate green and outdoor event spaces to host a diverse range of year-round activities. The district will feature 80 boutique hotel rooms with elegant, heritage-inspired luxury interiors, along with boutique shops and authentic culinary experiences. Subsequently, the heritagization process is set to brand the tourism potential of the historic village's vernacular architecture. This initiative is to attract tourists not only locally and regionally but also internationally.

Fig. 9 New construction inspired by the existing vernacular architecture. (Source: Constructionweek.sa)

Fig. 10 Urban development in Al Soudha. (Source: Constructionweek.sa)

4.3 Leveraging Heritagization to Drive Vision 2030 for Al Soudha Peaks Development

Achieving the tourism goals set out in Vision 2030 will require a forward-looking strategy that capitalizes on emerging opportunities while continuing to foster infrastructure development and considering potential regulatory reforms by implementing a thoughtful mix of measure and strategic evaluation (Alkhanbshi 2024). Heritagization in this context has emerged as subtle yet an integral part of the vision. This is due to the dynamic nature of the approach that allows to identify, highlight, and showcase cultural heritage. Heritagization process in this case is applied to promote the cultural tourism by preserving and enhancing village's cultural heritage assets such as the vernacular architecture. In the Kingdom of Saudi Arabia, the heritagization of sites like Rijal Almaa aligns with Vision 2030's goals, which focus on economic diversification and boosting local tourism. This process blends heritage conservation with infrastructural development, creating new cultural experiences that enrich the region's offerings. By integrating modern tourism initiatives with historical preservation, heritagization helps turn heritage sites into vibrant cultural destinations that celebrate culture while contributing to sustainable tourism growth.

The Kingdom's ambitious Vision 2030 initiative seeks comprehensive growth across all sectors, including domestic tourism, as part of its wider strategy for economic diversification. Local tourism plays a pivotal role in this vision, as it supports efforts to diversify the national economy and create new employment and investment opportunities (Alkhanbshi 2024). Despite the tourism sector's rapid expansion, numerous opportunities remain for the regeneration of intangible aspects of the heritage including cultural events, marketing of cultural products, and film festivals, as well as business events (Monshi and Scott 2017). Such regeneration according to Wise and Jimura (2020) also takes into account social impacts, community dynamics, and the ways in which investments and developments shape human interactions with the heritage, which in turn generates a variety of long-term benefits and opportunities. However, leveraging heritagization within the new Rijal district of the Al Soudah Peaks project, as part of Saudi Arabia's Vision 2030, also presents the challenges associated with the heritagization process.

4.4 Challenges and Risks of Heritagization in Al Rijal Almaa of Al Soudah Peaks Development

Heritagization as a concept entails modifying the use of vernacular heritage to create favorable representations for political governance or tourism opportunities. According to Nilsson (2018), central to the idea of heritagization is the intentional manipulation of the heritage by authorities, often supported by their tourism sectors, to shape the presentation and interpretation of cultural heritage to reinforce a specific tourism agenda. This interpretation, being subjective and contingent on the viewer's perspective, grants tourism stakeholders and administrators the ability to harness cultural tourism in advancing the concept of heritagization. However, in this context, heritagization raises concerns. Critics fear that it may lose its inherent values and be commodified, serving merely as an economic tool for organized leisure aimed at tourism (Hayes and MacLeod 2007). Likewise, Bernbeck (2013) critiques heritagization as an inadequate trend in heritage conceptualization, arguing that it is often seen as a mechanism for tourism production, which neglects its wider implications and reduces heritage management to a simplistic strategy for tourism development (AlShaikh 2017). Gravari-Barbas and Jacquot (2019) highlights that heritagization, along with the processes of aestheticization and artification, can often conflict with areas that are primarily utilized as commercial and touristic spaces. This raises important concerns about sacrificing genuine vernacular heritage and traditions to accommodate fleeting tourists for the sake of commodification, ultimately neglecting those who advocate for a more conscientious and sustainable approach to cultural preservation (Spiess 2012).

In this context, the heritagization of Rijal Almaa requires a thoughtful and balanced approach, ensuring that the vernacular architecture of this historic village is not commodified solely for cultural tourism purposes. The development project must also consider the importance of allowing the vernacular architecture to evolve organically. A primary concern is the potential for excessive commodification, where the Nominated World Heritage Site may sacrifice its genuine cultural significance for a tourism-focused approach. Effective management is vital to prevent the spread of inauthentic or fake structures, which can diminish the historic character and integrity of these cultural sites for the sake of tourism (Boussaa 2003, 2014, 2015, 2018; Boussaa et al. 2023).

Heritagization, if not carefully managed, also has the potential to disrupt this natural evolution and accelerate development in a manner that marginalizes the local community. Many cultural preservation projects in Saudi Arabia encounter significant challenges due to the limited engagement of local communities, which jeopardizes their long-term sustainability. These heritage sites, identified as vital tourism assets within the framework of Vision 2030, are prioritized for urban rehabilitation and integration into the wider urban landscape (Al-Qahtani et al. 2023). Active community involvement is essential for the success and sustainability of such tourism initiatives. Engaging local communities in eco-tourism development supports sustainable practices, elevates the quality of tourism offerings, generates employment, and contributes to the preservation of cultural heritage (Altassan 2023). To achieve success, these efforts should draw inspiration from vernacular traditions and historical practices (Boussaa et al. 2021). Such efforts highlight the unique architectural style and organic layout that are prominently reflected in the interconnected vernacular architecture and the way people create a sense of belonging toward it (Alraouf 2014; Mohamed and Elghonaimy 2025). Consequently, heritagization encompasses both material and intangible aspects, influencing cultural practices of the local community as well as physical artifacts (Turgeon 2010; Gillot et al. 2013).

While the project aims to emulate the vernacular architecture of the historic village for the sake of turning it into a prime cultural destination, this approach increasingly aligns with the commodification and touristification of the village. Gravari-Barbas (2022) critically examines heritagization, challenging the predominant focus on locally oriented heritage while highlighting the significant influence of commodification on the dynamics and production of heritage. Saidi (2010), on the other hand, elaborates on this phenomenon by linking the term "touristification" with heritagization, which is frequently associated with the growth of cultural tourism or even the direct outcome of travelers' growing demand for it (Saidi 2010; Gillot et al. 2013). Heritagization, though intended to preserve and promote cultural heritage for economic and tourism benefits, raises critical concerns regarding the potential for over-commercialization to cater tourism needs and the subsequent erosion of cultural authenticity. This tension highlights the delicate balance that must be maintained between cultural preservation, cultural tourism, and economic development, particularly in projects driven by national strategies like Vision 2030.

Another critical concern is to limit or avoid the forces of gentrification which typically follow cultural tourism developments. Heritagization often triggers gentrification (Herzfeld 2006, 2009, 2010, 2016), which displaces the original residents and converts these areas into spaces for ethnic consumption (Comaroff and Comaroff 2009). Jimura (2021) elaborates on this by highlighting that the historic and contemporary cultural heritage frequently exists side by side. Some individuals perceive this scenario as a reflection of disunity and a lack of cohesion between historic and modern cultural heritage. This process tends to sanitize neighborhoods that were once considered impoverished or unsafe, while simultaneously driving out much of the original population due to gentrification pressures. When heritage sites are commercialized to promote cultural tourism, gentrification inevitably follows. In this regard, Rijal Almaa's vernacular architecture must strike a balance between the residential capacity of the local community and the emerging tourist accommodations. Another important consideration is the tendency for residents to voluntarily leave their homes, either for financial gain or due to the excessive commercialization of the heritage site. To address these challenges, there is a need for an effective urban plan devised in such a way that maintains the existing cultural preservation and prevents the commodification of the heritage for the sake of increased tourism in the future. This plan should involve close collaboration between researchers, stakeholders, and the local community to ensure a sustainable balance between cultural tourism development and cultural heritage preservation.

5 Recommendations for Future Heritagization Initiatives in Al Soudha

Based on the analysis of the research, the following recommendations are proposed for achieving a balanced heritagization of the vernacular architecture of Rijal Almaa within the Al Soudah Peaks Development. These recommendations are also applicable to other vernacular heritage villages and cultural heritage sites that are undergoing heritagization to enhance their cultural significance and tourism potential:

1. **Cultural and historical preservation**: Ensure that the unique cultural and historical elements of vernacular architecture are preserved and prioritized during the heritagization process, preventing over-commercialization and commodification solely for tourism purposes.
2. **Community engagement**: Actively involve local communities in decision-making and implementation, recognizing that their participation is key to the long-term sustainability of cultural tourism initiatives.
3. **Balanced development**: Implement a well-considered urban plan that integrates both residential and tourism spaces, while preserving the organic evolution of the village and preventing displacement caused by gentrification.

4. **Manage gentrification pressures**: The heritagization of Rijal Almaa must carefully consider the potential impacts of gentrification. Strategies should be implemented to prevent the displacement of original residents by ensuring affordable housing options and supporting the local community once the place becomes a cultural hotspot.
5. **Sustainable tourism practices**: Promote sustainable tourism practices that align with eco-tourism principles, ensuring that tourism activities do not negatively impact the environment or disrupt the cultural and social fabric of the heritage sites.
6. **Balance tourism and residential use**: To prevent over-commercialization, it is important to limit the scale of tourist accommodations within Rijal Almaa. Tourist infrastructure should be developed with a focus on the existing scale context of the settlement. Limiting the number of large-scale hotels and focusing on smaller, community-driven lodging options can help maintain this balance.
7. **Collaborative approach**: Foster collaboration among government bodies, heritage experts, developers, and local communities to ensure a balanced and holistic approach to heritagization that preserves the intrinsic values of heritage sites while enhancing their appeal as cultural tourism destinations.
8. **UNESCO World Heritage Designation**: In line with the heritagization process and reaffirming the site's historical and cultural significance, adopt a carefully planned approach that safeguards the potential designation or maintains the existing status as a tentative UNESCO World Heritage Site. This approach should ensure that development efforts do not compromise the integrity or eligibility of the site for inclusion on the World Heritage Site in the near future.
9. **Beware of commodification**: Since heritagization has the potential to impact people, places, and practices, exceeding the demand for commodification of tangible and intangible heritage becomes inevitable at some points. This calls for prior planning which can prevent the heritage from being exploited for mere economic benefits. Appropriate measures must be implemented to control the excessive marketing of the vernacular village and thus avoid the Venice syndrome.

Implementing these recommendations will help manage the heritagization process in a manner that protects the integrity of heritage sites and the authenticity of the vernacular architecture and vernacular traditions. This approach will guide the effective execution of heritagization, ensuring the preservation of cultural heritage while simultaneously promoting cultural tourism.

6 Conclusion

The development of the Al Soudah Peaks project as a cultural tourism destination in the Kingdom of Saudi Arabia is a key initiative, with the vernacular architecture of Rijal Almaa playing an integral role. Heritagization, the process of designating or reaffirming cultural heritage, offers a strategic approach to such urban evolution that seeks to blend modern techniques with traditional ones. In this context, heritagization emphasizes the importance of cultural heritage in shaping this ambitious transformation, celebrating the uniqueness, distinctiveness, and diversity of vernacular architecture that reflects the geographical distribution and distinctive characteristics of heritage sites.

However, the introduction of leisure activities and amenities for tourism must be carefully managed to avoid the commodification of the existing vernacular architecture of the village. The vernacular architecture of Rijal Almaa is economically valuable and can be leveraged in various cultural tourism and urban development sectors, but this must not come at the expense of degrading the authentic cultural heritage. Achieving the goals of Vision 2030 and the year-round tourism development strategy requires a balanced heritagization approach that safeguards the authentic cultural heritage while appropriately transforming it for cultural tourism initiatives.

This research has assessed vernacular heritage through the lens of heritagization, using Rijal Almaa as a case study, and has developed recommendations for transforming the heritage sites into tourism destinations while maintaining a balance between cultural preservation and cultural tourism. The insights and recommendations from this study are applicable to other heritage sites undergoing heritagization for mass cultural tourism. Moreover, the findings contribute to the Kingdom's Vision 2030 by highlighting the importance of effective heritagization practices for the Kingdom of Saudi Arabia's historical sites and aligning with the Ministry of Tourism's objectives of revitalizing cultural heritage for public benefit while supporting broader cultural and historical preservation goals.

Acknowledgments I wish to express my sincere gratitude to my thesis supervisor, Assoc. Prof. Dr. Muge Riza, for her unwavering guidance, expertise, and constant support throughout this research journey. Her insightful criticism and direction were instrumental in shaping this work. I would also like to thank Assoc. Prof. Dr. Boussaa Djamel and Prof. Dr. Islam Elghonaimy for their existing published literature that helped me better understand the current situation of the Gulf's cultural heritage. I also extend my thanks to the faculty of Eastern Mediterranean University for fostering an academic environment rich in resources. Additionally, I am grateful to my former instructors from the University of Bahrain for their continued support and invaluable academic advice.

References

Al-Garni M (2011) Rehabilitation and the redevelopment of the heritage town of Al Ghat as a model for the development of the heritage villages and towns in the kingdom. Record of the First Urban National Heritage Forum, 1st edn. Saudi Commission for Tourism and Antiquities Research

Alkhanbshi S (2024) Factors affecting Saudi internal tourism during achieving vision 2030. Effat College of Business. Effat University. https://www.researchgate.net/publication/386419985_FACTORS_AFFECTING_SAUDI_INTERNAL_TOURISM_DURING_ACHIEVING_VISION_2030y

Al-Qahtani A, Al-Takhifi K, Alabed A, Alzamil W (2023) The role of community participation in urban rehabilitation projects for heritage areas in Saudi Arabia a case study: Rijal Almaa Heritage Village. J Al-Azhar Univ Eng Sector 18:304–322. https://doi.org/10.21608/Auej.2023.283074. https://www.researchgate.net/publication/367541656_THE_ROLE_OF_COMMUNITY_PARTICIPATION_IN_URBAN_REHABILITATION_PROJECTS_FOR_HERITAGE_AREAS_IN_SAUDI_ARABIA_A_CASE_STUDY_RIJAL_ALMAA_HERITAGE_VILLAGE

Alraouf AA (2014) The rehabilitation of the Muharraq Historical Center, Bahrain: a critical narrative. In: Exell K, Rico T (eds) Cultural heritage in the Arabian peninsula, debates, discourses and practices. Routledge, London, pp 173–188. https://www.researchgate.net/publication/322330738_The_Rehabilitation_of_the_Muharraq_Historical_Center_-_Bahrain_A_Critical_Narrative

Alsadi A (2022) Al Qatt Al Asiri indigenous Saudi women's form of art Al Qatt Al Asiri indigenous Saudi women's form of art. Ansaq J Arts Lit Hum 9(2–4). https://journals.mejsp.com/public/assets/uploads/journals-researches/1668865346_2408293784.pdf

AlSayyad N (2002) Hybrid urbanism: on the identity discourse and the built environment. Praeger. https://www.academia.edu/85682771/Hybrid_Urbanism_On_the_Identity_Discourse_and_the_Built_Environment_edited_by_Nezar_Alsayyad

AlShaikh A (2017) The value of authenticity in heritagization: an exploratory case study on Dubai Historical District (DHD). Zayed University, United Arab Emirates, Dubai. https://www.zu.ac.ae/main/en/library-project-search/_projects/the%20value%20of%20authenticity%20in%20heritagization

Altassan A (2023) Sustainability of heritage villages through eco tourism investment (case study: Al-Khabra Village, Saudi Arabia). Sustainability 15(9):7172. https://doi.org/10.3390/su15097172

Altomonte S (2008) Climate change and architecture: mitigation and adaptation strategies for a sustainable development. J Sustain Dev 1(1):97–110. https://doi.org/10.5539/jsd.v1n1p97

Arabnews (2018) Rehabilitation of Saudi heritage village of Rijal Alma complete to join UNESCO's list. Arab News. https://www.arabnews.com/node/1411686/saudi-arabia

Ashley SLT (2014) Re-telling, re-cognition, re-stitution: Sikh Heritagization in Canada. Cultura: Int J Philos Cult Axiol 11:39–58. https://doi.org/10.5840/cultura201411214

Asquith L, Vellinga M (2006) Vernacular architecture in the twenty-first century: theory, education and practice. Taylor & Francis, London

Bernbeck R (2013) Heritage void and the void as heritage. Aust Archaeol 9:526–545. https://doi.org/10.1007/s11759-013-9245-0

Bianca S (2000) Urban form in the Arab world—past and present. vdf, Zurich. https://www.archnet.org/publications/10765

Boussaa D (2003) Dubai: the search for identity. In: Moser G, Pol E, Bernard Y, Bonnes M, Corraliza J, Giuliani V (eds) People places and sustainability. Hogrefe &Huber Publishers, Göttingen, pp 51–60. (PDF) Cities in the Gulf: Rapid Urban Development and the Search for Identity in a Global World. https://www.researchgate.net/publication/330481040_Cities_in_the_Gulf_Rapid_Urban_Development_and_the_Search_for_Identity_in_a_Global_World

Boussaa D (2014) Urban heritage and tourism in the Gulf: the case of Dubai in the UAE. J Tourism 2(4):174–181. https://www.researchgate.net/publication/309270185_Urban_Heritage_and_Tourism_in_the_Gulf_The_Case_of_Dubai_in_the_UAE

Boussaa D (2015) Urban regeneration, sustainability and urban heritage: a case study of Souk Waqif, Doha. J Urban Regener Renewal 8(4):389–400. https://doi.org/10.69554/GERC9962

Boussaa D (2018) 58. Cities in the Gulf: rapid urban development and the search for identity in a global world. In: E-planning and collaboration: concepts, methodologies, tools, and applications. Information Resources Management Association, pp 1230–1254. https://doi.org/10.4018/978-1-5225-0187-9.ch009

Boussaa D, Alattar D, Nafi S (2021) The search for identity in a global world: the case of Doha in Qatar. J Urban Regen Renew 14(3):295–311. https://hstalks.com/article/6154/the-search-for-identity-in-a-global-world-the-case/. Accessed 24 Dec 2023

Boussaa D, Boudiaf B, Awad J, Salameh M (2023) The resilient historic cities of Sharjah and Doha—urban regeneration and the search for identity in a global world. Future Cities Environ 9(1):17, 1–16. https://doi.org/10.5334/fce.199

Boyd S (2002) Cultural and heritage tourism in Canada: opportunities, principles and challenges. Tour Hosp Res 3(3):211–233. http://www.jstor.org/stable/23747289

Bui QB, Morel JC, Venkatarama Reddy BV, Ghayad W (2009) Durability of rammed earth walls exposed for 20 years to natural weathering. Build Environ 44(5):912–919. https://doi.org/10.1016/j.buildenv.2008.07.001

Ceylan S (2022) Vernacular architecture of Turkish eastern Black Sea region: a case study in Senoz (Büyükdere) Valley. Open House Int 47(3):513–532. https://doi.org/10.1108/OHI-08-2021-0174

Comaroff JL, Comaroff J (2009) Ethnicity, Inc. University of Chicago Press, Chicago

Cooke L (2010) Conservation approaches to earthen architecture in archaeological contexts. Archaeopress, Oxford

Daher RF (ed) (2006) Tourism in the Middle East: continuity, change and transformation. Tourism and cultural change, vol 9. Channel View Publications, Bristol

Davallon J (2019) 3. The game of heritagization. In: Roigé X, Frigolé J (eds) Constructing cultural and natural heritage: parks, museums and rural heritage. Institut Català de Recerca en Patrimoni Culural, Girona, pp 39–62. (IRPC Llibres, 4). https://www.researchgate.net/publication/331951762_The_Game_of_Heritagization_p_39-62_in_Constructing_Cultural_and_Natural_Heritage_Parks_Museums_and_Rural_Heritage_sous_la_direction_de_Xavier_Roige_Joan_Frigole_Girona_Institut_Catala_de_Recerca_en_Pa

Dayaratne R (2016) Re-dignifying vernacular for constructing national identity: elitism, grand traditions and cultural revival in Bahrain. A/Z ITU J Faculty Architect 13(1):127–138. https://doi.org/10.5505/itujfa.2016.43534

Dayaratne R (2020) Re-inventing culture and place searching for roots, identity and sense of belonging: winter camps of Bahrain. ISVS e-journal 7(4):64–76. https://isvshome.com/pdf/ISVS_7-4/P2.3-ISVS_e_journal_7.4-Ranjith-Paper.pdf

DeKay M, Brown GZ (2014) Sun, wind & light. architectural design strategies, 3rd edn. John Wiley & Sons, Hoboken, NJ

Di Giovine MA (2009) The heritage-scape: UNESCO, world heritage, and tourism. Lexington Books, Lanham

Elassal M, Sherif S, AlMaayn F (2024) Geoheritage as an important parameter for geotourist attractions: a case study in Asir Region, Saudi Arabia'. Kurd Stud 12(2):446–466. https://kurdishstudies.net/menu-script/index.php/KS/article/view/1821/1277

Eyüce A (2005) Geleneksel Yapılar ve Mekanlar (traditional buildings and places). Birsen Yayınevi, Istanbul

Gillot L, Maffi I, Trémon A-C (2013) "Heritage-scape" or "heritage-scapes"? Critical considerations on a concept/« Paysage patrimonial » ou « paysages patrimoniaux » ? : Réflexion sur l'usage d'un concept. Ethnologies 35(2):3–29. https://doi.org/10.7202/1026546ar

Gingrich A (2000) Trading autonomy for integration: some observations on twentieth century relations between the Rijāl Alma' tribe and the Kingdom of Saudi Arabia. Etud Rurales 155/156:75–91. http://www.jstor.org/stable/20122828

Glassie H (2000) Vernacular architecture. Google Books. Indiana University Press. https://books.google.com.tr/books?hl=en&lr=&id=ZBvYCwAAQBAJ&oi=fnd&pg=PP1&dq=vernacular+architecture&ots=pZtDCbEOy0&sig=zqra1mNRom2dEcNWcdCy1xw05w0&redir_esc=y#v=onepage&q=vernacular%20architecture&f=false

Gravari-Barbas M (2022) Tenements in New York and riands in Marrakesh. Mobilities and the new paradigm of heritagization. In: Cattan N, Faret L (eds) Hybrid mobilities, transgressive spatialities. Routledge, Milton Park, pp 15–36

Gravari-Barbas M, Jacquot S (2019) Mechanisms, actors and impacts of the touristification of a tourism periphery: the Saint-Ouen flea market, Paris. Int J Tour Cities 5(3):370–391. https://doi.org/10.1108/IJTC-11-2018-0093

Gravari-Barbas M, Guinand S, Lu Y (2025) "Creative" uses of heritage in the age of experiential tourism: the case of Tianjin's former international concessions. J Chin Architect Urban 3703. https://doi.org/10.36922/jcau.3703. https://www.researchgate.net/publication/389154717_Creative_uses_of_heritage_in_the_age_of_experiential_tourism_The_case_of_Tianjin's_former_international_concessions. Accessed 27 Feb 2025

Hanif S (2024) Reclaiming the lost cultural identity via heritagization: the Gulf states. MSc thesis, Faculty of Architecture, Eastern Mediterranean University, North Cyprus. https://doi.org/10.13140/RG.2.2.36710.13122. https://www.researchgate.net/publication/388156025_Reclaiming_the_Lost_Cultural_Identity_via_Heritagization_The_Gulf_States

Hanif S, Riza M (2024) Heritagization of historic sites: transformation of Al Diriyah from a ruin to a branded cultural capital. ISVS e-journal 11(12):29–48. https://doi.org/10.61275/ISVSej-2024-11-12-03

Harrison R (2012) Forgetting to remember, remembering to forget: late modern heritage practices, sustainability and the 'crisis' of accumulation of the past. Int J Herit Stud 19(6):579–595. https://doi.org/10.1080/13527258.2012.678371

Harrison R (2013) Heritage: critical approaches, 1st edn. Routledge, Milton Park

Hayes D, MacLeod N (2007) Packaging places: designing heritage trails using an experience economy perspective to maximize visitor engagement. J Vacat Mark 13(1):45–58. https://www.researchgate.net/publication/247764605_Packaging_places_Designing_heritage_trails_using_an_experience_economy_perspective_to_maximize_visitor_engagement

Herzfeld M (2006) Spatial cleansing: monumental vacuity and the idea of the west. J Mater Cult 11(1/2):127–149. https://doi.org/10.1177/1359183506063016

Herzfeld M (2009) Evicted from eternity: the restructuring of modern Rome. University of Chicago Press, Chicago

Herzfeld M (2010) Engagement, gentrification, and the neoliberal hijacking of history. Curr Anthropol 51(S2):S259–S267. https://doi.org/10.1086/653420

Herzfeld M (2016) Siege of the spirits: community and polity in Bangkok. University of Chicago Press, Chicago

Hewison R (1987) The heritage industry: Britain in a climate of decline. Methuen, London

Hoque N, Hannawi A, Ibrahim A (2022) Rijal Almaa and local tourism in Saudi Arabia. In: Mohamed M, Ibrahim A, Fekry M (eds) Cities of the future. Springer, Cham. https://doi.org/10.1007/978-3-031-15460-7_3

Houben H, Guillad H (1989) Traité de construction en terre. Parenthéses Ed., Marsella

Illampas R, Ioannou I, Charmpis DC (2012) Overview of the pathology, repair and strengthening of adobe structures. Int J Architect Heritage 7(2):165–188. https://doi.org/10.1080/15583058.2011.624254

Jimura T (2021) Cultural heritage and tourism in Japan, 1st edn. Routledge, Milton Park. https://doi.org/10.4324/9780429019173

Kajda K (2019) Heritage—public and expert discourse in the process of heritigization. Ex Novo J Archaeol 4:73–82. https://doi.org/10.32028/exnovo.v4i0.371

Keefe L (2005) Earth building: methods and materials, repair and conservation. Routledge, Milton Park

Kopytoff I (1986) The cultural biography of things: commoditization as process. In: Appadurai A (ed) The social life of things: commodities in cultural perspective. Cambridge University Press, Cambridge, pp 64–91. https://web-facstaff.sas.upenn.edu/~cavitch/pdf-library/Kopytoff_CulturalBiography.pdf

Le Mentec K, Zhang Q (2017) Heritagization of disaster ruins and ethnic culture in China: recovery plans after the 2008 Wenchuan earthquake. China Inf 31(3):349–370. https://doi.org/10.1177/0920203X17736508

Mai Le Q (2021) Tales of heritagization. Ph.D. thesis, Faculty of Philosophy, University of Bonn, Germany. https://bonndoc.ulb.uni-bonn.de/xmlui/handle/20.500.11811/9456

Mileto C, Vegas F (eds) (2014) La restauración de la tapia en la Península Ibérica. Criterios, técnicas, resultados y perspectivas. TC Cuadernos/Argumentum Ediçoes, Valencia/Lisboa

Mileto C, Vegas F (eds) (2017) Proyecto Coremans: Criterios de intervención en la arquitectura de tierra. Ministerio de Educación, Cultura y Deporte, Madrid

Mileto C, Vegas F (2022) Earthen architectural heritage in the international context: values, threats, conservation principles and strategies. J Cult Herit Manag Sustain Dev 12(2):192–205. https://doi.org/10.1108/JCHMSD-06-2021-0115

Mohamed AE, Elghonaimy IH (2025) Shaping tourism strategies through local perceptions: a case study of Muharraq. J Contemp Urban Aff 9(1):16–34. https://doi.org/10.25034/ijcua.2025.v9n1-2

Monshi E, Scott N (2017) 3. Developing event tourism in Saudi Arabia: opportunities and challenges. In: Almuhrzi H, Alriyami H, Scott N (eds) Tourism in the Arab world: an industry perspective. Channel View Publications, Bristol, pp 33–55. https://doi.org/10.21832/9781845416157-004

Na LTH, Park J-H, Jeon Y, Jung S (2022) Analysis of vernacular houses in southern Vietnam, and potential applications of the learned lessons to contemporary urban street houses. Open House Int 47(3):533–548. https://doi.org/10.1108/OHI-12-2021-0276

Nilsson PÅ (2018) Impact of cultural heritage on tourists. The heritagization process. Athens J Tour 5(1):35–54. https://www.athensjournals.gr/tourism/2018-5-1-3-Nilsson.pdf

Rácz J (1998) A kulturális örökség hasznosítása és menedzselése, kapcsolata a turizmussal Angliában egy tanulmányút tapasztalatai. Falu Város Régió 12(1):5–11

Rapoport A (2006) Vernacular design as a model system. In: Asquith L, Vellinga M (eds) Vernacular architecture in the twenty-first century: theory, education and practice. Taylor & Francis, London, pp 182–183. https://www.taylorfrancis.com/chapters/edit/10.4324/9780203003862-14/vernacular-design-model-system-amos-rapoport

Rudofsky B (1964) Architecture without architects, an introduction to nonpedigreed architecture. The Museum of Modern Art: Distributed by Doubleday, Garden City, NY. https://www.moma.org/documents/moma_catalogue_3459_300062280.pdf

Saidi H (2010) Tourisme culturel, cutlural tourism. Ethnologies, 32(2). http://www.ethnologies.ulaval.ca/archives/tourisme-culturel/

Salemink O (2021) Introduction: heritagizing Asian cities: space, memory, and vernacular heritage practices. Int J Herit Stud 27(8):769–776. https://doi.org/10.1080/13527258.2021.1890186

Senosiain J (2003) Bio-architecture. Routledge, Milton Park

Sjöholm J (2016) Heritagisation, re-heritagisation and de-heritagisation of built environments: the urban transformation of Kiruna, Sweden. Ph.D. thesis, Department of Civil, Environmental and Natural Resources Engineering, Division of Architecture and Water, Luleå University of Technology, Sweden. https://www.diva-portal.org/smash/get/diva2:999224/FULLTEXT01.pdf

Skounti A (2010) De la patrimonialisation: comment et quand les choses deviennent- elles des patrimoines? (About heritagization: how and when do things become heritage?). Hesperis-Tamuda 45:19–34. https://ahmedskounti.weebly.com/uploads/7/4/0/8/7408619/de_la_patrimonialisation_in_hespris-tamuda_2010.pdf

Smith L (2006) Uses of heritage. Routledge, London

Spiess A (2012) Reconciling economic growth, conservation and sustainable development. http://www.ndrd.org/Sustainable_Tourism_Development_in_the_GCC_-_General_Info___Abstracts.pdf

Su X, Song C, Sigley G (2019) The uses of reconstructing heritage in China: tourism, heritage authorization, and spatial transformation of the Shaolin Temple. Sustainability 11(2):411. https://doi.org/10.3390/su11020411

Thouki A (2022) Heritagization of religious sites: in search of visitor agency and the dialectics underlying heritage planning assemblages. Int J Herit Stud 28(9):1036–1065. https://doi.org/10.1080/13527258.2022.2122535

Turgeon L (2010) Du matériel à l'immatériel. Nouveaux défis, nouveaux enjeux. Ethnol Fr 40(3):389–399

UNESCO World Heritage Centre (n.d.) Rijal Alma [tentative lists]. UNESCO. https://whc.unesco.org/en/tentativelists/6030/. Accessed August 2024

Vellinga M (2006) 4. Engaging the future: vernacular architecture studies in the twenty-first century. In: Asquith L, Vellinga M (eds) Vernacular architecture in the 21st century. Taylor & Francis, Milton Park, pp 81–94. https://www.taylorfrancis.com/chapters/edit/10.4324/9780203003862-6/engaging-future-marcel-vellinga

Visitsaudi.com (2024) Rijal Almaa Village. https://www.visitsaudi.com/en/aseer/attractions/rijal-almaa-of-aseer

Walsh K (1992) The representation of the past: museums and heritage in the post-modern world. Routledge, London

Warren J (1999) An overview: artwork to earthwork. In: Warren J (ed) Conservation of earth structures. Butterworth-Heinemann, Oxford, pp 1–39. https://www.icomos.org/images/Doc_centre/BIBLIOGRAPHIES/Earthen-Architecture_09-2014.pdf

Wise N, Jimura T (2020) 1. Changing spaces in historical places. In: Wise N, Jimura T (eds) Tourism, cultural heritage and urban regeneration. The urban book series. Springer, Cham. https://doi.org/10.1007/978-3-030-41905-9_1

Zhu Y (2016) Heritage making in Lijiang: governance, reconstruction and Local Naxi Life. In: Brumann C, Berliner D (eds) World heritage on the ground: ethnographic perspectives, vol 1. Berghahn Books, Brooklyn, pp 78–96. https://doi.org/10.2307/j.ctvpj7hh2.7

Zwerger K (2019) Vernacular architecture: a term denoting and transporting diverse content. Built Herit 3:14–25. https://doi.org/10.1186/BF03545716

Knowledge and Valorisation of Sepulchral Cavities in the Neapolitan Underground

Federica Itri and Simona Scandurra

Abstract

The Neapolitan territory is characterised by a network of underground cavities consisting of tunnels and chambers destined for different uses. Many of them contain architectural elements of great historical and artistic value and traces of the eras and styles that have succeeded one another over time. Among these, of great interest are those created for sepulchral use. The first were concentrated mainly in the area north of the historic city. Later, from the third century until the late medieval, the Christian cult also resorted to the use of sepulchral cavities, the catacombs, assigning them a dual function, i.e. a cemetery area but also a place of devotion dedicated to the liturgy. Thus, *terresante* and *colatoi* found wide diffusion, especially below the churches of the ancient centre. The widespread diffusion of this architecture corresponds to a consolidated model. In this context, the paper proposes to employ digital information tools and, specifically, geographic information systems (GIS), to map the phenomenon in the Neapolitan territory, in order to understand the evolution of urban sepulchres and to support, through a knowledge system, new forms of valorisation.

Keywords

GIS · Terresante · Survey · Historical maps · Sepulchral architecture · Crypts · Churches

1 Introduction

Through geographical information systems (GIS), as is well known, it is possible to create dynamic maps based on georeferenced data that makes it possible to identify, analyse and visualise the characteristic and recurring elements of the territories studied (Carallo and Carolei 2021). Such systems make it possible to digitally represent physical space and, if appropriately informatised and interrogated, make it possible to explicate unusual narratives, clarifying and highlighting the stratified histories of places, making otherwise lost or forgotten knowledge accessible. In the context of cartography, for example, GIS becomes an irreplaceable tool for locating building manufacts at different scales of detail, allowing them to be related to one another and also to the other elements that compose the territory. This approach is fundamental especially if the intention is to work on the knowledge and valorisation of the historical building heritage, since the relationship between information data and graphic visualisations facilitates the collection, organisation and systematic reading of data from often very different sources and formats. In these cases, GIS can be used as a tool to facilitate not only the generation and sharing of knowledge, but also the systematisation and decoding of historical data, which is often not digitised or is difficult to interpret.

In the proposed research, GIS is used as a tool through which to produce a georeferenced and informatised mapping of the historical burial sites that fill the underground of the city of Naples.

Naples, in fact, with its rich millenary history and its highly stratified urban fabric, offers a particularly fascinating case study for the application of this research. The city, known for its deeply rooted cultural—pagan and religious—traditions, has developed over the centuries a complex system of urban burials reflecting on the one hand social and religious practices, and on the other hand responses to health and hygiene challenges in different historical periods. In this context, the proposed research focuses on a crucial historical

F. Itri (✉) · S. Scandurra
Department of Architecture, University of Naples Federico II, Naples, Italy
e-mail: federica.itri@unina.it; simona.scandurra@unina.it

F. Alberti et al. (eds.), *Towards Sustainable Urban Regeneration: Utilizing Environmental, Cultural, and Social Assets*, Advances in Science, Technology & Innovation, https://doi.org/10.1007/978-3-032-02091-8_24

period, circumscribed between the sixteenth and nineteenth centuries, when Naples faced urbanisation dynamics and changes such as to raise numerous doubts about the appropriateness of burial sites with respect to the healthiness of the air. This is, in fact, the period in which monitoring of the state of burials led to the production of a large number of documents, legal acts and technical reports concerning the state of burials in the city.

The study proposed here therefore began with the collection and analysis of bibliographic and archival documentation,[1] much of it previously unpublished. Through a laborious activity of systematisation of the documentary data, the research used digitisation as a pretext and means to enhance and make known the phenomena that have characterised the history of the transformations of the Neapolitan territory over time. Thus, the use of open data and, in particular, of GIS and Web GIS, which can be shared via the web, has made it possible to overcome the fragmentary nature of knowledge on the cultural and monumental heritage of the hypogea of the city of Naples, making historical-archival data accessible to all, which would otherwise have remained difficult to consult.

2 Case Study: Burials in Naples Between the Sixteenth and Nineteenth Centuries

The places and practices associated with the burial cult in the city of Naples have been transformed over time with the evolution of traditions that represent a veritable journey through centuries of the history of southern Italy, its popular culture and socio-religious changes. In particular, the numerous burial places of the dead found inside the ancient churches reveal a religious thought towards the deceased that is strongly ingrained in the population and has very deep roots, for which the cult of death has, over time, assumed great importance for the well-being of the human spirit. Beginning with the first Christian persecutions, communities of believers sought safe places for their dead and, in areas such as Naples that are characterised by easily workable tuffaceous foundations, they began to build proper underground cemeteries, leading to the birth of the 'catacombs' as a specific architectural typology destined for burials. The best known, in Naples, are the catacombs of San Gennaro and those of San Gaudioso which, as was typical, initially served as a burial place and, at the same time, as a place of worship.

Over time, even after the adoption of Christianity as the official religion of the Roman Empire, catacombs continued to develop, especially in southern Italy, until they gradually moved into churches, in an attempt to unite as much as possible the passing of the spirit with prayers and places of greater worship for believers. In the Middle Ages, in fact, the construction of hypogeal rooms began in correspondence with churches, at first only to preserve the relics of saints and to give burial to the upper classes or members of the clergy. Later, in the Renaissance, churches became enriched with private chapels and funerary monuments commissioned by the most politically influential families. These places became remarkably complex and richly decorated architectural structures, useful for burying the dead but also reflecting the power and earthly wealth of the noble families they belonged to.

Focusing on the city of Naples, it is important to emphasise that the organisation of funeral services has been entrusted to confraternities since the early Middle Ages. This entrustment is also confirmed in the seventeenth century, after the spread of the 'good death' and 'purgatory' cults introduced by the Council of Trent. During this period, the practice of double burial became widespread, with burial architecture designed to divide the treatment of the deceased into the phase of temporary burial and the subsequent phase of bone deposition (Carnevale 2008). Consequently, the hypogeal environments located under the churches take on forms and characteristics useful for facilitating the procedures and accompanying the entire process of decomposition of the bodies. In this regard, the most common types of hypogeal environments found under numerous churches in Naples and, more generally, in southern Italy are *colatoi* and *terresante*. The former consists of a masonry structure where the corpse was left to decompose (Fig. 1); *terresante*, on the other hand, are widespread mainly in Campania and represent a self-sufficient funeral system that solves decomposition and disposal at the same time (Fig. 2) (Fornaciari 2013).

When serious epidemics spread among the population in the seventeenth century, which also affected Naples, a large quantity of bones accumulated in a short period of time, so it was decided to purge the city's burial grounds and cemeteries to transport the remains to ancient caves,[2] as was the case

[1]The archival data consulted, kept in the State Archives of the city of Naples, consists of three folders containing police reports, documents and technical reports drawn up in the context of inspections of burial grounds carried out by the Ministry of the Interior–Superintendence of General Health between the eighteenth and nineteenth centuries. These documents provide detailed descriptions of burial practices within churches and, more generally, sacred spaces, as well as illustrating the different existing typologies. The richness and variety of these texts have made it possible to obtain a clear and precise picture of the burial practices active until the end of the nineteenth century and of the location of these spaces in the urban fabric.

ASN, Supremo Magistrato e Soprintendenza generale di Salute, b. 286/287/288. *'Travaglio sullo stato delle sepolture in questa capitale, eseguito dal Comitato Sanitario, ed espedienti progettati per rettificare il disordine'.*

[2]The Deputation of Public Health, set up in 1656 after a plague epidemic, was responsible for safeguarding burials and limiting the inspection of gravesites and the transport of bones to caves only to cases of real necessity.

Fig. 1 The *colatoi* of the Arciconfraternita dei Verdi dello Spirito Santo, Naples

Fig. 2 The *terrasanta* of the Chiesa di Santa Maria della Misericordia, Naples

with the well-known Cimitero delle Fontanelle. Only a few decades later, people began to consider it more hygienic to separate burial grounds from those frequented by people, and this led to the birth of the first extra-urban cemetery on the Poggioreale hill by Ferdinando Fuga and known as the Cimitero delle 366 fosse, inaugurated in 1763. Here, the burial process was reorganised, reducing the risk of miasmas and meeting the sanitary needs dictated by the Ospedale degli Incurabili. Despite this innovation, however, urban burials remained in use and the bones produced continued to be periodically transported to the caves of the Fontanelle Cemetery, worrying King Ferdinand IV who, in 1799, commissioned the physician Giuseppe Melchiorre Vairo to draw up a report clarifying the state of burials in the city. The report, in fact, highlights the sanitary problems especially near the *terresante*, as in these the bodies were buried too far above ground and then placed in niches above ground to complete the decomposition process. When the French occupied the capital of the kingdom in 1806, the Saint-Cloud Decree of 1806 attempted to separate the burial process from the ecclesiastical monopoly, removing its absolute management from the confraternities, parish priests and hospitals. The French imposed a new system of suburban cemeteries so that, while maintaining a link with religion, the city could sanitise itself.[3]

The Bourbon kingship also set out to preserve the health of the citizens by distancing them from all harmful and dangerous aspects of the cult of the dead. Through the Health Committee, it was decided to draw up, for each district, documentary tables on the state of burials at the time, focusing in particular on those linked to the numerous churches that already strongly characterised the urban fabric. These tables represent a document of fundamental importance for knowledge of the state of the sites, for the identification of underground cavities and for information on significant irregularities recognised at the time. They represent an initial attempt to constitute a true cadastre of the burial underground, at the time referred to as the '*Travaglio sullo stato delle sepolture di questa capitale*' (Survey of the state of burials in this capital city), as well as an initial instrument of urban regulation capable of reporting and punishing any abuses (Pezzini 2010). Meanwhile, in 1815, the architect Pietro Maresca was commissioned to build a monumental cemetery at Poggioreale.[4] The work continued and was interrupted several times over the years, until the urgent resumption dictated by the cholera epidemic of 1836. Nevertheless, the management of burials still remained a complex problem for the city of Naples for a long time. In fact, in 1889, an enquiry by the Directorate of Health revealed the enormous number of mass graves in operation here, as in many other Italian cities, despite the construction of the new cemeteries. Above all, the dead of the poorer classes continued to be interred in mass graves until the early twentieth century.

[3]The process of separating the world of the living from that of the dead was long and complex. In 1807, Interior Minister André François Miot proposed the construction of a cemetery in Fuorigrotta, which was then rejected. A few years later, Murat signed the decree for the construction of the new public cemetery at Poggioreale, on the opposite side of the one built by Fuga. The lay associations were dissolved and their assets incorporated into the factories of the churches. In this way, the government pursued the goal of abolishing the privileges of the confraternities, without compromising the functions offered to society.

[4]The cemetery was inaugurated in 1837, although the church was not yet completed.

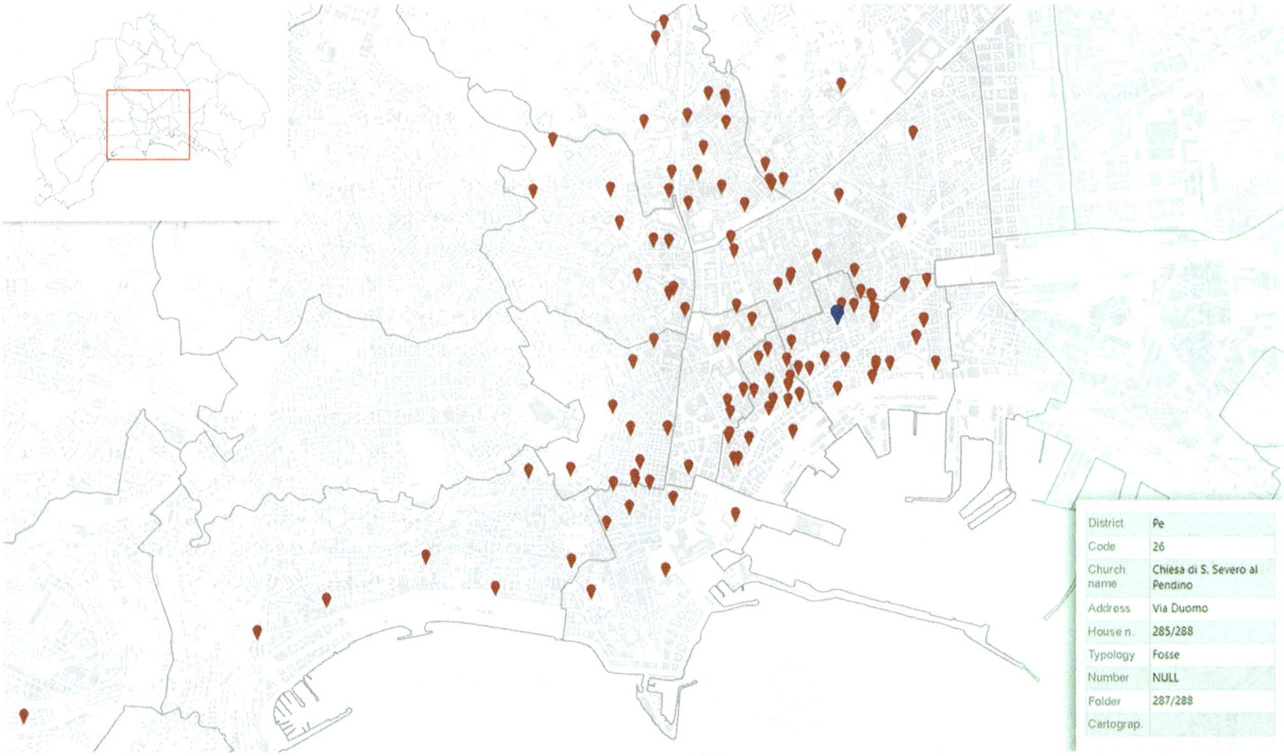

Fig. 3 Mapping of burials identified through historical records and reports in the period between 1760 and 1816

3 Methodological Approach: GIS as a Tool for Documenting and Exploiting the Phenomenon

From a methodological point of view, the research began with the collection, study, interpretation and synthesis of documents preserved in the State Archives of the city of Naples. This made it possible to identify churches and other sacred spaces characterised by the presence of burial sites.

Considering the enormous quantity of hypogea for sepulchral use found in the Neapolitan building fabric, an initial typological classification made it possible to divide the places into five categories, namely mass graves, gentilitial burials, *terresante, colatoi* and bone cemeteries (Guarino 2023).

Subsequently, the data was placed and organised on a current cartographic base of the city of Naples, directly in a GIS environment, constructing shapefiles consisting of georeferenced points representative of the different types of burial sites. In detail, one shapefile focuses on the geolocation of burials identified in police reports and technical reports referring to the historical period between 1760 and 1816.

For each of these burials, descriptive and technical-numerical data are documented and entered into the GIS database, as well as the name of the church, the district to which it belongs and the address. In addition, in order to make the process as transparent as possible, it was decided to report the source of information and, where necessary, the number of the archive folder from which each piece of data was retrieved. This first census made it possible to identify 116 burial places of various types referring to still existing churches (Fig. 3).

A second shapefile was organised to be much more specific. In this case, in fact, the reference is exclusively the '*Travaglio sullo stato delle sepolture di questa capitale*' compiled by the Health Committee from 1816. In this case, as many as 186 burial places were identified, corresponding partly to existing churches and partly to churches that no longer exist (Fig. 4). In fact, part of the work was dedicated to the study of the toponymic evolutions of churches (Alabiso et al. 2015), as well as the urban transformations of the city, in order to actually verify the persistent areas and those of which there is no longer any trace. For this part of the work, it was essential to compare the current urban fabric with that documented by the city's historical cartographies.

Of these, considering the date, the Duca di Noja's Map (1775) (Fig. 5a) and the cartography drawn up by the engineer Luigi Marchese[5] in 1813 (Fig. 5b–d) were crucial to

[5]With particular reference to the Portanuova, Porto and Vicaria neighbourhoods of the city of Naples, which over time have undergone numerous urban reorganisations that have led to the demolition of many religious buildings.

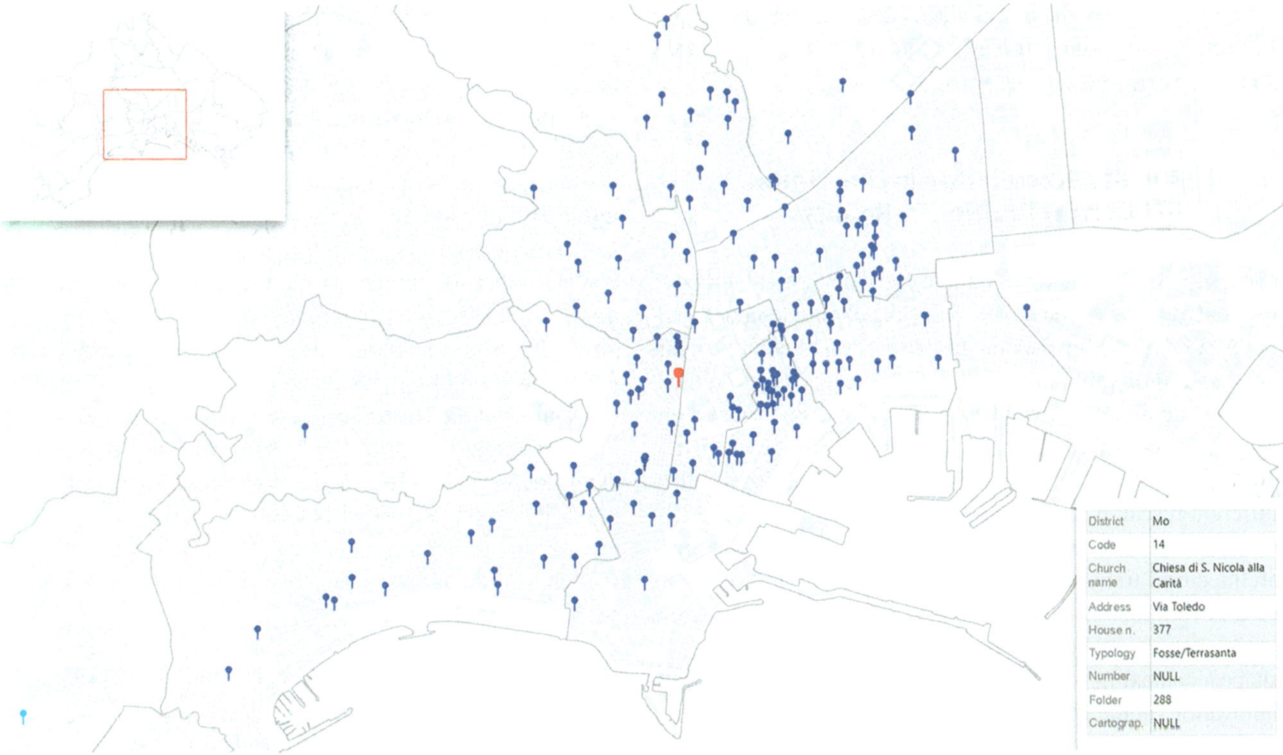

Fig. 4 Mapping of burials identified through the *'Travaglio sullo stato delle sepolture di questa capitale'* dating back to 1816

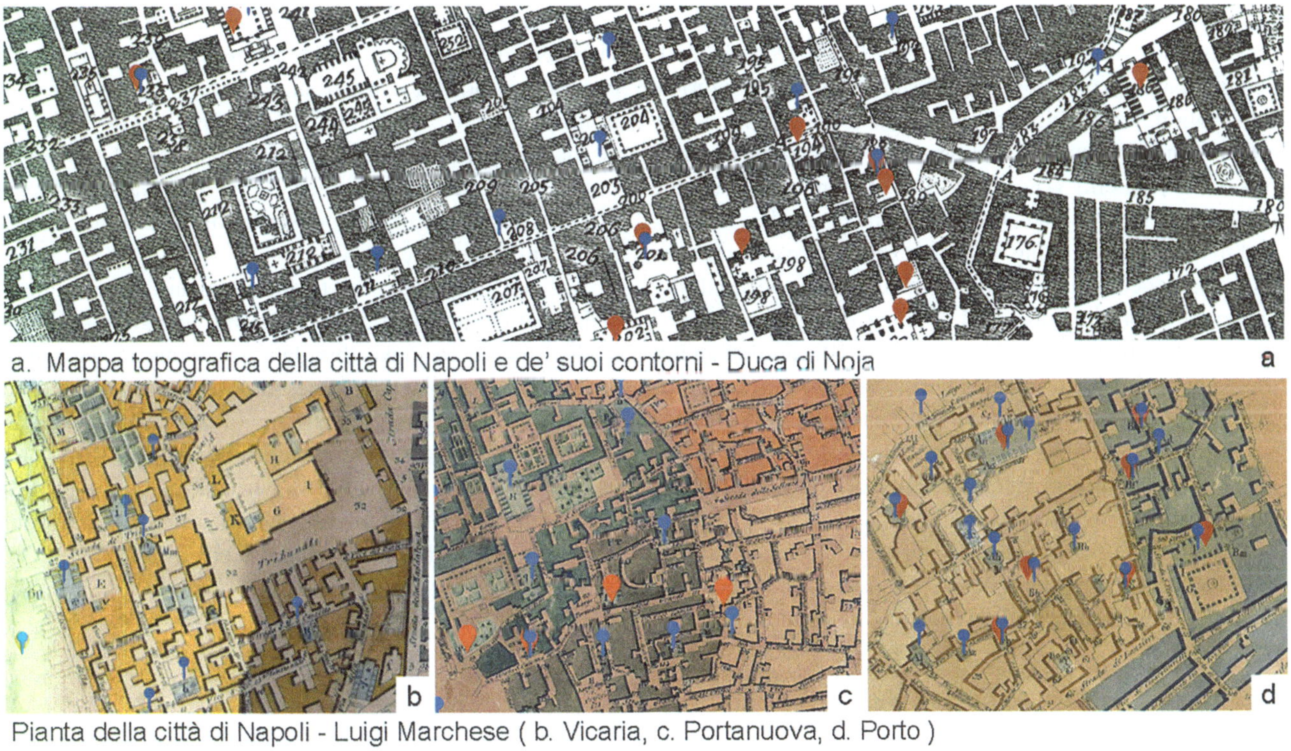

Fig. 5 Cartographies used to identify churches that no longer exist or have changed their names over time. In particular, *'Mappa topografica della città di Napoli e de' suoi contorni'* by Duca di Noja (**a**) and the *'Pianta della città di Napoli'* by the engineer Luigi Marchese (**b**, **c**, **d**)

accurately identify the burial places described in the archive documents, especially for the reference to the topography and toponymy of the churches at the time.

4 Results: Knowledge in the Digital to Support Fruition in Reality

The use of GIS made it possible to carry out critical analyses of the dynamics of the evolution of the phenomena detected in the study area and enabled the production of detailed analyses and in-depth evaluations of the artefacts examined, from an architectural as well as a cultural and historical point of view. The chosen approach (Bushmakina et al. 2017) demonstrated its flexibility and versatility, adapting itself to different research requirements as it went along. Indeed, the change in burial practices in Naples highlights not only architectural transformations, but also the cultural and religious evolutions that influenced the city. Brotherhoods, the establishment of new cemeteries and sanitary reforms helped shape a context in which tradition was often intertwined with innovation, in the context of a city in constant change.

The spread of these practices highlights the difficulties of managing post-mortem phases in a densely populated city like Naples, which was repeatedly afflicted by epidemics. During these periods, in fact, many underground cavities were reused as mass graves, as happened with the grain pits in Piazza Mercato.

This study offers a double reading of the burial phenomenon: a spatial perspective and a systematisation of historical information, preserving it from the risk of deterioration. The integration of historical cartographies and archival documents into the GIS was essential to accurately identify the location of churches and burial sites. The research is not limited to historical mapping, but also explores the current state of the hypogeum sites, verifying their current accessibility and state of preservation, and pointing out the *colatoi* and *terre sante*, which can now be visited for educational and/or tourist purposes. In its ongoing development, in fact, the survey intends to focus more on environments that are still accessible, offering itself as support for the construction of new knowledge itineraries, both digital and physical, recognising their enormous cultural value.

The use of GIS and Web GIS in research offers itself as a tool for the preservation and dissemination of knowledge, making historical data easily accessible to a wider public.

The maps produced are not simply static representations, but interactive tools that facilitate in-depth exploration. In fact, users can interact with the maps by selecting specific points to obtain detailed information or by performing more articulate queries useful for understanding the orographic, urban and social transformations of the city through the evolution of underground burials.

A crucial aspect is in fact the possibility of illustrating the evolution of burial practices and urban transformations over time, also thanks to the combination with historical cartographies. In fact, it is possible to compare the current urban fabric with that of the past, highlighting changes in the location and structure of burials, identifying areas of the city that have undergone significant transformations, such as the demolition of churches or the conversion of burial spaces into rooms intended for other uses.

5 Conclusions

Geographical information systems (GIS) have revolutionised the way we approach the study of land and its changes over time. Their ability to integrate georeferenced data with historical and cultural information makes them essential tools not only for urban planning and land management, but also for historical research and cultural heritage conservation.

The use of GIS in this study is not only a means to visualise geographical and historical data, but also becomes a tool for critical analysis to explore the evolutionary dynamics of the city. The mapping of ancient burials has allowed for a better understanding of the urban, social and religious transformations that marked Naples, offering new tools for the conservation and valorisation of the cultural heritage.

Although the result aims to document and preserve metric and historical information and current photographs (Fig. 6), it also intends to make this data accessible to the public, through the use of technologies such as open data and Web GIS. In this way, the project aims to contribute to the dissemination of historical knowledge and to raising awareness of the protection of a cultural heritage that is often little known, making information resources accessible that would otherwise remain hidden or inaccessible.

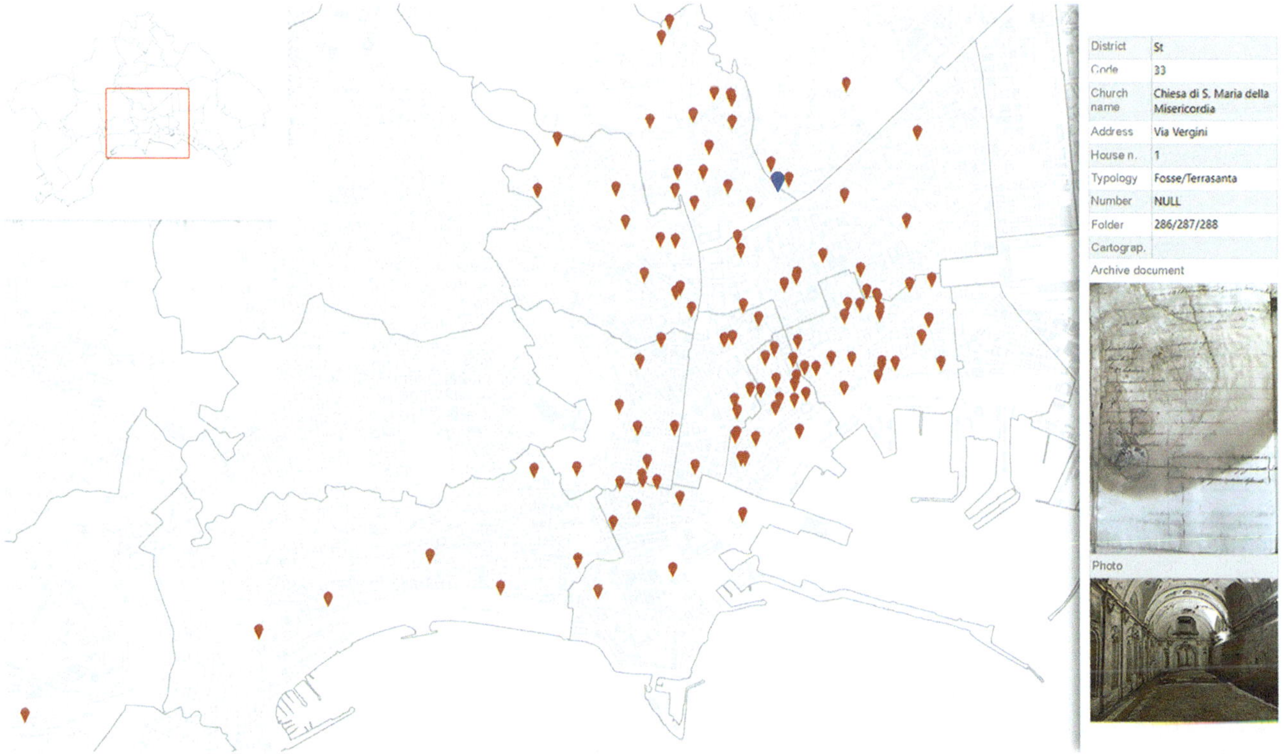

Fig. 6 Example of querying the GIS system that provides not only the possibility of knowing the information about the burial, but also the historical reference documentation. Moreover, in the case of still existing hypogea, it is also possible to display images of thems

References

Alabiso AC, Campi M, di Luggo A (2015) Il patrimonio architettonico ecclesiastico di Napoli. Artstudiopaparo, Napoli

Bushmakina YV, Balyberdina PA, Dmitrieva MK, Gogoleva MV (2017) The use of GIS for studying cultural heritage and historical urban landscape: the case of Perm and Usolie (Russia). Ge-conservaciòn 1(11):264–271

Carallo S, Carolei FA (2021) I Gis nella didattica a distanza. La formazione come strumento per la diffusione della cultura geografica. In: Conferenza ESRI Italia 2021. Digital week, 10–14 maggio 2021

Carnevale D (2008) La riforma delle esequie a Napoli nel decennio francese. Studi Stor 49(2):523–552

Fornaciari A (2013) Scheletrizzare o mummificare: pratiche e strutture per la sepoltura secondaria nell'Italia del Sud durante l'età moderna e contemporanea. Medicina nei secoli. Arte e scienza 25(1):205–238

Guarino M (2023) Practices of double burial and cult of the death in Naples. Anthropologie LXI(1):49–67

Pezzini F (2010) Disciplina della sepoltura nella Napoli del Settecento. Note di ricerca. Studi storici 51(1):155–208

Participatory Heritage: Reflection on the Elaboration of the Atlas of Territorial Heritage of Playa, La Habana, Cuba

Valentina D'Ippolito, Elena Tarsi, and Raffaele Paloscia

Abstract

The recognition and preservation of cultural heritage, both tangible and intangible, has been the subject of extensive international debate for a long time, but the issue still presents some challenges. This article is framed in the postcolonial approach and aims to contribute on how to overcome exogenous views in heritage recognition and preservation, especially in Global South contexts. By presenting the objectives, the methodology, results, and limitations of the elaboration of the Atlas of Playa, La Habana, Cuba, the authors propose a participatory and territorial approach to heritage recognition. The case of Cuba is relevant for three reasons: The reduction in available resources for the maintenance and conservation of buildings and infrastructure has exacerbated problems of decay and deterioration of cultural heritage more than in other contexts; the country, and particularly La Habana, has long been under significant pressure from foreign investors on multiple fronts. A possible future scenario of profound transformations is emerging with serious risks of losing part of the tangible and intangible heritage; Cuba's unique history and its relationship with memory make the case study paradigmatic.

Keywords

Tangible heritage · Intangible heritage · Participation · Conservation · International cooperation

V. D'Ippolito · E. Tarsi (✉) · R. Paloscia
Department of Architecture, University of Florence, Florence, Italy
e-mail: valentina.dippolito@unifi.it; elena.tarsi@unifi.it; raffaele.paloscia@unifi.it

1 Introduction

The recognition and preservation of cultural heritage, both tangible and intangible, is still, especially in some contexts of the Global South, a challenge that involves not only public administrations but also different stakeholders and local populations. Indeed, as explored by international literature on the subject, an inclusive approach (Roued-Cunliffe and Copeland 2017) can lead to a deeper and more comprehensive understanding of the heritage itself, allowing for a more accurate assessment of its qualities and conservation needs. Such an approach derives from a critical view of heritage studies and specifically from the postcolonial approach that, starting from the 2000s, has influenced many disciplines including urban studies.

This article is framed in the international debate on how to overcome exogenous views in heritage recognition and preservation, especially in Global South contexts. It aims to contribute to defining effective methodologies of cooperation between North and South that are horizontal and lead to respect for the peculiarities and differences of territories and the cultures that have shaped them.

The development of effective methodologies for engaging settled communities in enhancing local heritage has been one of the research objectives of the LabPSM—Laboratory City and Territory in the Global South at the Department of Architecture of the University of Florence for years. Thanks to research experiences and international cooperation (Paloscia et al. 2017, 2021; Paloscia and Tarsi 2012a, b; Tarsi 2012a, b, 2013, 2019) matured within the laboratory, community engagement strategies have been developed based on extensive literature on the subject (Colecchia 2019) and the many practices developed internationally (UN-Habitat 2012, 2023), but also on the specific approach of the territorialist school (Magnaghi 2000). Such an approach, developed in Italy, has been recognized internationally, starting from the translation of its manifesto in several languages (Magnaghi 2003, 2005a, b, 2011). It is based on a definition of territory

F. Alberti et al. (eds.), *Towards Sustainable Urban Regeneration: Utilizing Environmental, Cultural, and Social Assets*, Advances in Science, Technology & Innovation, https://doi.org/10.1007/978-3-032-02091-8_25

as the result of the coevolution between the natural system and the anthropic system, and of Territorial Heritage as the long-standing rules that have long defined the balance between these two systems. Territorial Heritage is considered the basis for any future sustainable project. The role of settled communities in recognizing such an heritage is the central focus around which projects have been organized. The territorialist approach has had an important influence on the innovation of Italian legislation related to spatial planning and on the structuring of plans at regional and municipal levels.

It is more difficult to recognize such heritage values that may have developed in territories subjected to centuries of colonization and exploitation of both resources and population (such as Latin America) without definitively disrupting every balance between humans and the environment. How is it possible, then, to transpose into these contexts the strategic vision that sees the territory as the basis for a future sustainable project and starts from the settled community?

In order to answer such research questions, the paper presents the methodology and the results of the most recent research, developed within the project "*¡Que no baje el telón!*" funded by the Ministry of Foreign Affairs of the Italian Government. It involves the development of the Atlas of Territorial Heritage of the Municipality of Playa, La Habana, Cuba, resulting from a strong community involvement in identifying and recognizing the tangible and intangible values of the territory. The Atlas constitutes an important basis not only for preserving memory but also as a "living" tool in the hands of the community to define future transformations centered on the idea of territorial heritage as an essential resource. The case of Cuba is relevant for three reasons: the reduction in available resources for the maintenance and conservation of buildings and infrastructure has exacerbated problems of decay and deterioration of cultural heritage more than in other contexts; the country, and particularly La Habana, has long been under significant pressure from foreign investors on multiple fronts. A possible future scenario of profound transformations is emerging with serious risks of losing part of the tangible and intangible heritage; Cuba's unique history and its relationship with memory make the case study paradigmatic.

The contribution is structured as follows: in Sect. 2, the framework of international literature on the postcolonial approach and the recognition of heritage by settled communities is outlined; in Sect. 3, the case study and the methodology adopted for the development of the Atlas, inspired by the territorialist approach, are presented; in Sect. 4, the authors discuss the results obtained and the limitations encountered, highlighting the peculiarities of the Cuban reality.

2 Heritage Recognition: The Postcolonial Approach and the Role of Participation

The concept of heritage has changed over time based on cultures, their values, and their relationship with the past and memory, assuming different meanings that reflect the understanding and value attributed to cultural and environmental resources. In ancient cultures, heritage was often linked to monuments and works of art of great importance, considered symbols of power and prestige. The focus was primarily on artworks, religious artifacts, and monumental buildings. Over time, the concept of heritage has expanded and enriched. During the nineteenth century in Europe, with the growing interest in the conservation of historical and architectural heritage, the first initiatives to protect monuments and places of cultural importance were born, a movement that led to the creation of agencies and organizations dedicated to the protection and preservation of heritage.

In late modernity, the process of patrimonialization has allowed the development of economies based on extracting capital from common cultural and landscape assets. The territories that progressively invested in these processes (Di Giovine 2009; Ryan and Silvanto 2009; Adie 2017) have on the one hand become richer (Bertacchini et al. 2024) but on the other inevitably suffered the negative impacts derived from them. There are many studies in the literature that analyze, for example, how the recognition given by UNESCO regarding the uniqueness of certain places or assets triggers a process of developing a tourism economy, often endangering the conservation of the assets themselves (Caust and Vecco 2017).

Moreover, there is another aspect that seems important to highlight: The concept of universal heritage introduced by UNESCO for the first time in 1972, linked to the characteristics of extraordinariness and authenticity, has effectively extended Western perspectives and values to all other cultures, denying their specific vision of heritage. Criticisms of this "colonial" approach led, in the 1990s, to a broader definition of World Heritage: Cultural landscapes were included, and greater importance was given to ancient traditions. This open attitude subsequently led, in 2003, to the inclusion of intangible heritage among the assets to be protected: oral traditions, music, theater, dance, rituals, festivals, and traditional crafts (Aikawa 2004).

This progressive trend to move beyond a Eurocentric approach to heritage studies has led to profound transformations in the practices and projects of recognizing and enhancing heritage (Chakrabarty 2000; Long 2000). Adopting a postcolonial approach to heritage studies means asking, "How do we organize and categorize knowledge? Who determines the selection and quality criteria of collections?

Who decides what is presented and represented?" (Cassián-Yde 2019). It therefore means adopting an inclusive approach that involves the participation of stakeholders and citizens.

The participation of residents and local organizations in the decision-making process regarding the recognition and conservation of cultural heritage offers numerous advantages by allowing the collection of a wide range of perspectives, experiences, and knowledge (Fortuna 2012). Firstly, an inclusive approach (Roued-Cunliffe and Copeland 2017) can lead to a deeper and more comprehensive understanding of the heritage itself, enabling a more accurate assessment of its qualities and conservation needs. Secondly, public participation fosters the development of a sense of belonging and shared responsibility toward cultural assets, which can contribute to preserving heritage for future generations and promoting its importance in society (Adell et al. 2015). At the same time, community participation creates stronger social bonds and greater cohesion (Friedmann 2004): Collaboration among individuals, groups, institutions, and organizations in safeguarding heritage can strengthen mutual understanding and respect for the diverse cultures present in society. Among the strategies to stimulate and strengthen community awareness of the richness of their territory, the Atlas is proposed as a cognitive tool in terms of resources and potential, as well as the enhancement of the identity of a place. It is useful both for exploring a vast territory in depth and for providing an analytical basis for future urban planning and transformation actions, all oriented toward considering territorial heritage as a key resource for sustainable development.

3 Case Study: The Atlas of Playa, La Habana, Cuba

3.1 The Project

The Atlas of Territorial Heritage of the Municipality of Playa is part of the Italy–Cuba international cooperation project *¡Que no baje el telón!* Conservation, management, and enhancement of the Faculty of Theatre Art, the product of a partnership between the Ministry of Culture (MINCULT) of the Republic of Cuba and the Italian Agency for Development Cooperation (AICS) of the Ministry of Foreign Affairs and International Cooperation (MAECI). This project, launched in 2018, has as its main objective the restoration and refunctionalization of the original location of one of the five schools that make up the Instituto Superior de Arte (ISA) complex: the *Facultad de Arte Teatral* (FAT) of the University of Arts, designed by architect Roberto Gottardi (Venice 1927—La Habana 2017).

The construction of the Atlas of the Territorial Heritage is an integral part of the project, as part of the various activities aimed at training technical personnel on restoration, consolidation, valorization, and management of cultural and territorial heritage, from the point of view that no architectural redevelopment can be truly effective if it is considered a punctual and discontinuous intervention; hence, the need to broaden our gaze to the surrounding area, to the relationships between the work and its context, and to the involvement of the population toward a continuous process of understanding and awareness of the entire heritage of the territory. The course started with an articulated fieldwork in La Habana from April 2022 to February 2023, continuing with the phase of graphic restitution of the collected and processed data carried out in Italy and concluding with the recently published volume (Paloscia et al. 2024) available both in paper and digital edition in open access.

The first stage, a training course on the valorization of territorial heritage, took place within the University of Arts, between April and May 2022, with the twofold objective of involving as many actors as possible among the local population and experts in the field and forming a multidisciplinary working group capable of managing all the stages of the subsequent process. The course, coordinated by Prof. Raffaele Paloscia and Elena Tarsi with Valentina D'Ippolito and Davide Perrotta, consisted of (a) an introductory part on the basic concepts of heritage, local identity, and good urban planning practices; (b) a phase of theoretical lectures on existing local experiences of land valorization; and (c) a final part of practical workshops specifically dedicated to the Playa Municipality, with on-site activities and surveys.

The training course attracted the attention of many students of heritage conservation and management, students of architecture, some experts in certain areas such as local development, as well as professionals from different disciplines such as architects, engineers, historians, and anthropologists, and representatives of some local institutions. An interesting dialogue was generated, enriched by different professional backgrounds on the topics discussed; each contribution contributed to building a knowledge base to then address the specific case of the Playa Municipality, its social and cultural context. In particular, the workshop phase, which focused on the Buenavista area and was devoted to the direct survey of the elements that make up the territorial heritage, constituted an invaluable contribution to the subsequent work, creating an essential methodological basis for the elaboration of the survey and cataloging data sheets used subsequently.

The course continued with the formation of a working group capable of coordinating and managing all the activities for the construction of the Atlas. Aware of the importance of the presence of different voices and competences, thanks to the relationships and connections created during the course, new actors were intercepted, including several local representatives, with the intention of involving as many actors as possible. The working group formed saw the constant participation of experts from the *Museo Municipal de Playa*, Italian specialists, and Playa's *historiador*, with the contribution of several students and professors from the *Universidad Tecnológica de La Habana José Antonio Echeverría* (CUJAE), the *Facultad de* conservación de la ISA, and representatives from the *Casa de cultura de Playa*, the *Consejo de administración municipal*, and the *Dirección de planificación física de Playa*. This multidisciplinary group of Cuban–Italian specialists met weekly from May 2022 to February 2023 at the Municipal Museum, carrying out various activities including: (a) research and study of existing documentation and the state of the art on the territorial heritage of the Playa Municipality (cartographic, iconographic, photographic, and historical material); (b) identification, location, and documentation of the most representative elements of the territorial heritage; (c) on-the-spot visits to collect qualitative and quantitative data, direct survey of the territory through photographic documentation and interviews; (d) processing of the data collected in the geographical information system (GIS), consisting of the data sheets and the archive of heritage values; and (e) open presentations of the work in progress to the community.

The bibliographical review of the existing documentation was aimed at learning about the history, practices, and transformations, in order to reconstruct a clear image of the Playa Municipality as a whole. Most of the necessary documentation was provided by the Museum, a fundamental point of reference for carrying out the research, thanks to the tireless work of its specialists, who were an active part of the whole process from the outset. The basic cartographic material was provided by the students of the CUJAE; the historical and photographic data were mainly from the archives of the *oficina del historiador de la Ciudad* and the *Casa de cultura de Playa*, while the iconographic material was provided by the Museum's specialists. Important references were the local development strategy of Playa and the cadastral database provided by the *Dirección de planificación física*.

The fieldwork consisted of daily inspections aimed at walking all the streets of the area in search of heritage elements that could be photographed and located on the map, confirming existing data and discovering new ones. An excellent starting point with respect to this was the list of Playa's buildings and sites registered in the Inventory of National Valuable Assets provided by the Office of Monuments and Historical Sites of the *Centro provincial de patrimonio cultural de la ciudad de La Habana*. To these, with a view not only to cataloging the heritage already known, but also to bringing out and making visible minor elements, were added those, so to speak, submerged, still not institutionally recognized and not inventoried but considered worthy of value and significance by experts and the community.

3.2 Methodology

The research has benefited from numerous experiences accumulated over time, which have provided suggestions and allowed the methodology to be updated, although it is difficult and slippery to establish a single, standard technique, since each place presents unique characteristics and specificities.

The *Atlante del Patrimonio Territoriale del Circondario Empolese Valdelsa* (Carta et al. 2005) is one of the first examples within a series of experimental applications in specific areas of the Italian context, in particular Tuscany, which have been a source of experience and a prelude to subsequent more extensive research works in Latin America such as the *Atlas del patrimonio local material e inmaterial de la ciudad de Leon, Nicaragua* (Paloscia et al. 2006), the *Atlas del patrimonio territorial de la Habana del Este* (Spitoni 2009; Paloscia 2010; Paloscia et al. 2021), and the *Progetto di cooperazione Caritalents a Barbados e Dominica* (Paloscia et al. 2017). In the Cuban context, the same methodology inspired the Atlas carried out in the municipality of *Centro Habana* (Rey Rodriguez and Rios Diaz 2010).

These projects adopt as a theoretical and methodological reference the territorialist vision, which in its original and complete formulation is a holistic approach to transformation that considers the territory as a living and complex organism, in continuous transformation, made up of various material and immaterial components, inevitably interconnected and traceable in three fundamental areas: the physical heritage, the socioeconomic heritage, and the heritage of new social practices. This theoretical–methodological framework takes concrete form in a series of immediately practicable objectives of the Atlas, modulated and articulated taking into account the diversity of the contexts, which can be summarized as follows: (a) to make explicit the tangible and intangible elements of the heritage of places in relation to the innovative social practices of their production, referring to the entire territory as a value (not only to monuments and environmental areas of greater manifest relevance); (b) denote and represent heritage values as a knowledge base for the definition of planning tools aimed at self-sustainable local development; (c) promote a use of the territory, including tourism, that is aware of its resources and qualities from multiple points of view such as historical, artistic, and environmental, linked to local material culture (gastronomy, productive knowledge, local handicrafts, etc.); and finally (d) to produce experimental information texts to be disseminated with educational/training and tourism value, as support for the identification of new eco-friendly activities that generate income for the local population. Through an interactive processing of resource data, using graphical, textual, and multimedia tools, the Atlas aims to represent the heritage of a given context, making its main features explicit, highlighting all the elements that constitute it, in order to make it accessible and to understand it in depth.

It is in this context that it is necessary to draw on postcolonial theories with respect to heritage and question how and why certain forms of knowledge are prioritized over others, how knowledge is organized and classified, how the history of places is represented, or who determines the criteria for the selection and value quality of elements (Collin-Delavaud et al. 2006). It is precisely the identification and representation of the elements that contribute to defining the value and identity of a place, which is why these steps are particularly delicate and could not do without the participation of the inhabitants, their presence, and their active involvement, since they are the true custodians of the values of the territories. Moreover, we must not forget that the value of heritage is not intrinsic, but is above all a social, cultural, and political creation of places (El Kadi et al. 2005). The Atlas is not only an important basis for preserving memory, but also a "living" tool available to the community, relevant to understand the present and valuable for the challenges of the future. With the intention of following the guidelines of the abovementioned territorialist methodology and adopting a postcolonial vision, it was essential to pursue the objectives of the survey: Create a space for multiple perspectives that show the different contexts that determine how we view objects or themes. This space was found in the multidisciplinary Italian–Cuban working group, which was immediately established as a community of dialogue in continuous learning. The challenge was to share views on different visions and cultural perspectives, work together to address historical injustices and persistent inequalities, and then promote the development of practices for recognition of territorial heritage.

In the view that the active and constant dialogue with local communities is fundamental, where it has been possible, especially for the immaterial section, reference people of the neighborhood have been involved and interviewed, thus collecting valuable direct testimonies. An important part was the organization of open meetings with the target community to present the progress of the work and discuss its contents. The in-progress elaboration of the Atlas was also presented at some schools, town councils, and events, in order to collect stimuli and suggestions and broaden participation.

3.3 Results

The tangible outcome of this work is the substantial volume (available in both print and digital formats) that consists of three main parts. The first part is an introduction that presents the objectives and methodology; the second part provides a historical and contextual overview of the Municipality of Playa; and the third part contains technical sheets. These technical sheets include all the territorial heritage elements encountered, organized by type, illustrated with photos and images, and precisely located on maps, therefore geolocated. The technical cataloging sheets of the territorial elements have been a crucial tool, forming a significant portion of the final Atlas. Essential for data collection during fieldwork, these sheets serve as a technical instrument for understanding, cataloging, and locating heritage. The entire territorial heritage is categorized into two main areas: tangible and intangible, within which five types are identified—urban, architectural, historical–archaeological, environmental, and intangible. A specific model of technical sheet was developed for each of these categories. The cataloging process and the preliminary development of the sheets gradually proved to be essential for gaining a clear understanding and vision of the heritage as a whole. The development of these sheets was not straightforward, as they are tools in continuous evolution. Their refinement was influenced by both the direct fieldwork experience and the opinions of Cuban experts, as well as the unique characteristics of the territory itself, with its originality and uniqueness. This process of refinement made the sheets more accurate and suitable for expressing the complexity of the local heritage. Once the sheets were revised, their graphic and textual representation made them a user-friendly tool for understanding each component and an effective means of communication, useful for spreading knowledge and awareness of the heritage value of the context.

Regarding the significance and utility of this Atlas, it is crucial to understand that the journey undertaken is still in its early stages and remains incomplete. First of all, this is because the detailed analysis presented in this volume covers only half of the Playa territory. Given the vastness and complexity of the municipality, this represents just the first volume, with the intention of applying the same approach and objectives to the yet unexplored parts of the territory. Secondly, the Atlas is conceived as a fundamental tool of knowledge that cannot be static or unchangeable. On the contrary, it must be a living, dynamic, and interactive instrument, intended for citizens and administrators, who are the primary users and key actors involved. They should continuously update, enhance, and refine all documentation related to heritage. Especially in the case of intangible heritage, many resources could emerge through progressive enrichment, involving an ever-increasing number of informed and engaged residents. The hope is that the Atlas will be able to adapt to the ongoing development and evolution of its core elements, responding to new needs through the integration of new features, additional data, or more advanced analysis and interoperability tools. To fulfill this role, it is essential that the Atlas be constantly updated through continuous research and territorial management efforts.

On a local level, the reconstruction of tangible and intangible elements that constitute specific identities has enabled direct engagement with local actors, such as schools, communities of practice, associations, informal groups, and individuals. These participants have often shown themselves to be engaged and enthusiastic about the work done and the recognition given to elements of their familiar environment. This positive outcome was made possible by involving the population in the process, collaborating with Cuban experts, and paying attention to lesser-known and seemingly secondary elements, understanding the value that communities also attribute to "hidden," minor, and less recognized heritage. Looking at a broader scale, at the municipal level, there has been difficulty in coherently connecting the heritage elements present in different locations and achieving a unified vision. This represents an organizational and managerial challenge that will need to be addressed in the future.

4 Discussion and Conclusion

In response to our research questions proposed in the introduction, we presented the methodology used for the elaboration of the Atlas of Playa. The results have allowed us to evaluate the efficacy of the process, underlying both the capacity for collaboration between the parties, demonstrating a mutual interest in adopting new approaches, and limitations and shortcomings revealed by the case study. The methodology for developing the Atlas, as we have seen, originated in the Italian context but evolved through Latin American experiences, which are strongly characterized by horizontal cooperation relationships. Cooperation with Cuba has been fundamental in this regard. The actors involved in the construction of the Atlas have indeed played a decisive role in defining the methodology itself as well as in cataloging the territory.

It is undeniable that the revolution in Cuba, by rejecting the values of Spanish and later American colonization, has managed to build heritage references based on the revolution itself that have deeply permeated the population. The awareness and sense of belonging of the Cuban population to their living places remain crucial resources for the defense and enhancement of cultural heritage. It has emerged that, probably due to the current historical, social, and political conditions in Cuba, there is a focus on intangible heritage and historical memory, while less attention is paid to the built and material heritage. In response to these needs, the Atlas has given ample space to the intangible section, which includes culture, memory, customs, identity, etc. In addition, focusing on lesser-known and less explored areas has had the double effect of gradually enriching the concept of urban heritage and involving more local people, who are enthusiastic about the recognition given to their area.

Cuba is undoubtedly a unique case study, both for its history and its present, but also for its uncertain future. The economic crisis that has hit the country at various times over the past decades, including the current one, has had significant repercussions on the deterioration of both the built and natural environments, contributing to a progressive decline in urban quality of life. The reduction of resources available for the maintenance and conservation of buildings and infrastructure has exacerbated the problems of degradation and deterioration. The country is increasingly subjected to exogenous pressure that seeks to penetrate the economic system, especially through the acquisition of real estate assets. On the other hand, there are growing endogenous pressures toward the development of a mass tourism economy. A possible future scenario of profound transformations is emerging, not necessarily positive, with serious risks of losing part of the cultural wealth, both tangible and intangible, that still exists. In this framework, more than in other contexts, the severe deterioration of the built heritage requires a cataloging effort that can serve as a bulwark against the risks of demolition it will face. The same applies to intangible heritage: The work of recognition and registration can make the substantial difference between a future of valorization or disappearance under the homogenizing pressure of tourism.

In this context, the work of collecting and cataloging heritage, carried out with the direct involvement of the population, becomes even more important. This participatory approach represents a bulwark against possible projects aimed solely at economic gain or uncontrolled gentrification processes, helping to ensure that decisions made regarding territorial transformation interventions closely reflect the values and priorities of the local community.

One of the main objectives of this work has been to strengthen the connection between local actors such as schools, resident communities, associations, informal groups, and individuals, and their territory; the Territorial Heritage Atlas constitutes a widely and easily accessible tool suitable for promoting its protection, enhancement, and management. From this tool, it is possible to extract specific components and develop further innovative tools for communication and knowledge transmission, useful both in education at various levels and in conscious and respectful tourism of local cultures and the environment. The Atlas could become a reference model, a pilot project replicable in other contexts, even within Greater Havana. It provides an analytical, interpretative, and communicative methodology to identify and enhance heritage, giving new meaning to the entire urban space, highlighting the uniqueness of each place, and contributing to the development of innovative and shared tools for the protection and management of the territory and the definition of feasible and sustainable urban, territorial, and landscape projects.

Promoting the development of decolonial heritage practices entails complex challenges and requires reflection on the future role of cultural institutions; this means transforming these institutions into learning communities that work together to address historical injustices and persistent inequalities. The work has been conducted with the aim of ensuring horizontality and respecting the specificities and differences of the territories and cultures that have shaped them, while maintaining an open and authentic dialogue with all members of the community and society. Cuban institutions involved shared valuable knowledge and information, while the Italian side contributed to the study and recognition, avoiding imposing external visions and leaving room for autonomy and independence. All the participants involved in the creation of the Atlas played a role in defining the methodology, cataloging the territory, and identifying heritage elements and values. The vision of the territory as a result of the coevolution between natural and anthropic systems, adapted to the Cuban context, characterized by significant historical and ideological events and a strong and incisive tropical natural environment, was partly absorbed by the inhabitants involved, who showed a strong interest in territorialist theory. This does not mean that a strategic vision that sees the territory as the basis for a sustainable future project has been incorporated, but the importance of the role of territorial heritage as a key element in safeguarding urban and cultural identity has emerged.

Acknowledgments The authors would like to thank all the participants in the elaboration of the Atlas of Playa, especially Jose Durand Galano, historian of the Playa Municipality; Belkis Diaz, Nydia Espinet Vasquez, and Maricela Marido Garcia, experts of the Playa Municipal Museum; and architect Gina Rey for their precious collaboration.

Attributions This paper is the result of joint reflection by the authors. The research project has been coordinated by Raffaele Paloscia. Sections 1 and 2 were written by Elena Tarsi, Sect. 3 by Valentina D'Ippolito, and Sect. 4 by all authors.

References

Adell N, Bendix RF, Bortolotto C, Tauschek M (2015) Between imagined communities of practice: participation, territory and the making of heritage. Göttingen University Press, Göttingen

Adie BA (2017) Franchising our heritage: the UNESCO World Heritage brand. Tour Manag Perspect 24:48–53. https://doi.org/10.1016/j.tmp.2017.07.002

Aikawa N (2004) An historical overview of the preparation of the UNESCO international convention for the safeguarding of the intangible cultural heritage. Mus Int 56(1–2):137–149. https://doi.org/10.1111/j.1350-0775.2004.00468.x

Bertacchini E, Revelli F, Zotti R (2024) The economic impact of UNESCO World Heritage: evidence from Italy. Reg Sci Urban Econ 105:103996. https://doi.org/10.1016/j.regsciurbeco.2024.103996

Carta M, Lucchesi F, Vannetiello D (2005) L'atlante del patrimonio territoriale del Circondario Empolese Valdelsa in Atti della IX conferenza nazionale ASITA, Catania, 2005

Cassián-Yde N (2019) Descolonizar las epistemologías urbanas: Saber experto y colectivos por el derecho a la ciudad, ¿quién puede decir "la verdad" sobre los problemas de la ciudad? J Lat Am Geogr 18:54–84

Caust J, Vecco M (2017) Is UNESCO World Heritage recognition a blessing or burden? Evidence from developing Asian countries. J Cult Herit 27:1–9. https://doi.org/10.1016/j.culher.2017.02.004

Chakrabarty D (2000) Provincializing Europe: postcolonial thought and historical difference. Princeton University Press, Princeton

Colecchia A (2019) Community heritage and heritage community. Participatory models of cultural and natural heritage management in some inner areas of the Abruzzo region. Il Capitale Culturale. Studies on the value of cultural heritage, (19), pp 129–164. https://doi.org/10.13138/2039-2362/1970

Collin-Delavaud A, Herschon JC, Orellana L (2006) Dossier: nouvelles formes d'intervention dans la gestion du territoire patrimonial urbain. Inf Geogr 22(70):6–10

Di Giovine MA (2009) The heritage-scape: UNESCO, world heritage, and tourism, Lexington Book

El Kadi G, Ouallet A, Couret D (2005) Inventer le patrimoine moderne dans les villes du Sud. Numéro spécial de la revue Autrepart, n°33

Fortuna C (2012) In praise of other views: the world of cities and the social sciences. Iberoamericana 12(45):137–115

Friedmann J (2004) Empowerment. Verso il potere di tutti. Qualevita, Pescara

Long DL (2000) Cultural heritage management in post-colonial polities: not the heritage of the other. Int J Herit Stud 6(4):317–322. https://doi.org/10.1080/13527250020017744

Magnaghi A (2000) Il progetto locale. Bollati Boringhieri, Torino

Magnaghi A (2003) Le projet local. Pierre Mardaga éditeur, Sprimont

Magnaghi A (2005a) The urban village. A charter for democracy and local self-sustainable development. Zed Books, London

Magnaghi A (ed) (2005b) La rappresentazione identitaria del territorio. Atlanti, codici, figure, paradigmi per il progetto locale. Alinea Editrice, Firenze

Magnaghi A (2011) El proyecto local. Hacia una conciencia del lugar. Edicions UPC, Barcelona

Paloscia R (2010) Guanabacoa. Hacia el Atlas del Patrimonio Territorial. In: Orefice P, Del Gobbo G, Presmanes GR, Benelli C (eds) Guanabacoa patrimonio cultural a valorar. Metodologías y buenas prácticas para la valorización del patrimonio material y immaterial. Edizioni via Laura, Firenze

Paloscia R, Tarsi E (2012a - a cura di) Capitale umano e patrimonio territoriale per il progetto locale: contributi di ricerca interdisciplinare in America Latina e Italia. Ed.it, Firenze

Paloscia R, Tarsi E (2012b - a cura di) Città e territori oltre il Nord, numero monografico di Contesti. Città, territori, progetti, n. 1/2012

Paloscia R et al (2006) Atlas del Patrimonio local, material e inmaterial de la Ciudad de Leon, Nicaragua. Consorcio La Toscana per Leon, Medina, Firenze

Paloscia R, Spellucci S, Morbidoni M (2017) Caribbean unveiled. The Caritalents youth and territory project in Barbados and Dominica. Didapress, Firenze

Paloscia R, Spellucci S, Spitoni L (2021) La Habana del Este. Atlas del patrimonio territorial. Editorial Universidad de Granada, Granada

Paloscia R, D'Ippolito V, Perrotta D (2024) La Habana - Playa. Atlas del Patrimonio Territorial. FUP, Firenze

Rey Rodriguez G, Rios Diaz M (2010) Valoración del Patrimonio Cultural. Atlas del Patrimonio Cultural de Centro Habana, Arquitectura y Urbanismo, a. XXXI, n.3

Roued-Cunliffe H, Copeland A (eds) (2017) Participatory heritage. Facet

Ryan J, Silvanto S (2009) The world heritage list: the making and management of a brand. Place Brand Public Dipl 5:290–300

Spitoni L (2009) La Habana del Este: territorio y arquitectura en la sugunda mitad del siglo XX. In: Basosi D, Lorini A (eds) Cuba in the world, the world in Cuba. Essay on history, politics and culture. FUP, Firenze

Tarsi E (2012a) Acqua Bene Comune. Piano partecipato di gestione delle acque e valorizzazione della risorsa idrica di Cajazeiras V, Salvador, Brasile. In: Paloscia R, Tarsi E (a cura di) Capitale umano e patrimonio territoriale per il progetto locale: contributi di ricerca interdisciplinare in America Latina e Italia. Ed.it, Firenze

Tarsi E (2012b) La città che vogliamo: laboratorio di progettazione partecipata con i ragazzi a Città del Guatemala. In: Paloscia R, Tarsi E (a cura di) Capitale umano e patrimonio territoriale per il progetto locale: contributi di ricerca interdisciplinare in America Latina e Italia. Ed.it, Firenze

Tarsi E (2013) Global challenges and local identity: integrated transformation of informal areas as a basis for social inclusion. In Perrone C (a cura di) Living landscape for living: policies, practices, images. Conference proceedings, Firenze, February-June 2012, Planum. J Urban II(27), pp 43–51

Tarsi E (2019) Patrimonio territoriale, comunità e territori fragili nel Global South. In: Scienze del Territorio, N.7/2019

UN-Habitat (2012) Young people, participation, and sustainable development in an urbanizing world

UN-Habitat (2023) Enabling meaningful public participation in spatial planning processes

Cosmopolitan Paradox: Dynamic Egalitarianism in the Redevelopment of the Cultural Center of the Philippines Complex

Chris Jehram M. Morallos and Leonido M. Gines Jr.

Abstract

The Cultural Center of the Philippines (CCP) Complex stands as a testament to Filipino cultural identity, yet it faces mounting challenges, including governance inefficiencies, commercialization, and physical decline. The nonimplementation of the 2003 master plan and the persistence of the edifice complex narrative have hindered its ability to evolve as an inclusive cultural space. This study critically examines how architectural monumentalism, centralized governance, and global–local tensions have shaped the complex's trajectory and explores strategies for its adaptive redevelopment.

Employing a mixed-methods approach—including archival research, stakeholder interviews, surveys, and spatial assessments—this research evaluates the CCP's architectural, cultural, and institutional landscapes. Findings reveal that monumentalism and state-driven narratives have marginalized participatory governance, limiting the complex's engagement with contemporary cultural dynamics. In response, the study introduces Symbiotic Pluralism, a framework integrating UNESCO's Historic Urban Landscape (HUL) principles, participatory governance, and sustainable urban strategies. By reconciling heritage conservation with inclusive modernization, the CCP can reassert its role as a living cultural institution that bridges its historical legacy and contemporary Filipino identity. This study underscores the imperative for responsive, community-driven policies to sustain the CCP as a dynamic, relevant, and accessible cultural hub.

Keywords

Cultural preservation · Egalitarianism · Urban redevelopment · Adaptive heritage management · Inclusive governance · Historic urban landscape · Symbiotic pluralism · Sustainable cultural landscapes

1 Introduction

The Cultural Center of the Philippines (CCP) Complex is an architectural and cultural landmark that has long symbolized Filipino heritage and identity. Designed to serve as the nation's premier venue for artistic and cultural expression, it has played a pivotal role in promoting Filipino arts and culture. However, despite its historical significance, the CCP Complex faces multiple challenges that threaten its relevance in an evolving society. Studies such as those by Espiritu (2015) highlight its importance as a national symbol while also revealing concerns about its preservation and the failure to implement the egalitarian principles outlined in the 2003 master plan by Planning Resources and Operations Systems, Inc. The lack of progress raises questions about governance efficiency and stakeholder collaboration, as emphasized by former CCP Chair Margie Moran (Caruncho 2018).

One major issue confronting the complex is the "Edifice Complex," a phenomenon described by Lico (2003) as the tendency to construct grandiose architectural monuments as symbols of power. This focus on monumentalism often overshadows the nuanced appreciation of culture within its contextual framework, leading to societal inequalities and stifling authentic cultural expressions (Horkheimer and Adorno 2017). While these grand structures were intended to

C. J. M. Morallos (✉)
Architecture Department, Far Eastern University, College of Engineering and Architecture, Technological Institute of the Philippines,
Quezon City, Philippines
e-mail: cmmorallos@up.edu.ph

L. M. Gines Jr.
History, Theory and Criticism (HTC) Studio Laboratory, University of the Philippines—College of Architecture, Quezon City, Philippines
e-mail: lmgines1@up.edu.ph

F. Alberti et al. (eds.), *Towards Sustainable Urban Regeneration: Utilizing Environmental, Cultural, and Social Assets*, Advances in Science, Technology & Innovation, https://doi.org/10.1007/978-3-032-02091-8_26

project national pride and modernity, they also reflected a top-down approach to cultural representation, where state-driven narratives dictated the architectural and artistic expressions that were deemed worthy of national significance.

This tension between monumentalism and cultural authenticity is further explored in the work of Ogura et al. in *Modern Architecture in the Philippines and the Quest for Filipino Style* (2002), which provides critical insights into the post-World War II period when Filipino architects sought to craft an architectural identity that resonated with the nation's cultural roots. This endeavor to blend indigenous cultural elements with contemporary design principles was not merely aesthetic but a profound expression of national pride and self-definition, as the country emerged from colonial influences and sought to establish its own voice in the global arena. The pursuit of a modern Filipino architectural style, especially during the Marcos era, was epitomized in monumental projects like the CCP. These efforts were part of a broader strategy to use architecture as a tool of nationalistic propaganda, showcasing the Philippines' cultural sophistication and modernity to the world. The integration of traditional Filipino aesthetics with modernist designs at the CCP and similar projects reflects a deliberate attempt to construct a visible symbol of the Philippines' progress and cultural heritage, which was intended to inspire national unity and bolster the regime's image both domestically and internationally. This architectural approach during the Marcoses' time highlights the complex interplay between cultural identity and political ambition, underpinning the monumentality of the era's architectural undertakings.

The cosmopolitan paradox underscores the inherent tension between global integration and local identity preservation, particularly in societies like the Philippines, where the convergence of international influences and indigenous traditions continuously shapes cultural and spatial development. This dynamic is especially evident in the evolution of public spaces, where modernization efforts often risk displacing historically and culturally significant elements. As cultural institutions navigate these complexities, the challenge lies in fostering a critical balance between embracing global perspectives and safeguarding local heritage. Rather than viewing this paradox as a conflict, it can be reframed as an opportunity to synthesize diverse cultural narratives, ensuring that heritage conservation coexists with contemporary urban and economic transformations. This approach not only reinforces cultural resilience but also establishes a model for sustainable and contextually grounded development, where tradition and innovation are not opposing forces but mutually reinforcing dimensions of cultural progress.

1.1 Research Question/Hypothesis

This study poses the question: How can egalitarian principles be effectively integrated into the redevelopment of the Cultural Center of the Philippines Complex to ensure its alignment with contemporary cultural and societal needs?

The hypothesis posits that a redevelopment strategy grounded in a dynamic approach to egalitarianism, informed by cultural literacy, can revitalize the CCP Complex, transforming it into a more inclusive and relevant cultural hub for modern Filipino society. The research aligns with the conference's theme on integrating art and culture within urban and architectural landscapes, reflecting broader challenges in urban redevelopment and cultural preservation.

2 Problem Statement and Study Aim

The Cultural Center of the Philippines (CCP) Complex, originally envisioned as a national cultural landmark and a democratic space for artistic expression, now faces critical challenges that undermine its relevance, sustainability, and cultural integrity. While conceived to promote Filipino identity and ensure broad access to cultural experiences, the CCP has deviated from this mission, struggling to adapt to evolving sociopolitical and urban landscapes. The failure to implement the 2003 business and master plan, which emphasized egalitarianism and inclusivity, reflects systemic governance issues and a broader misalignment between policy intentions and practical execution.

At the core of this issue lies the cosmopolitan paradox—the tension between global integration and local identity preservation (Gabriel and Veronis 2024). The CCP Complex exemplifies this conflict, as pressures from commercialization, governance stagnation, and urban development threaten its foundational role as a custodian of Filipino heritage. The complex suffers from the deterioration of infrastructure, the encroachment of profit-driven enterprises, and the loss of cultural focus—each of which has distanced it from its intended function. Additionally, limited community engagement, inadequate accessibility, and the absence of sustainable redevelopment strategies have compounded its decline, raising concerns about its long-term viability.

Addressing these challenges requires a holistic, multidimensional approach that reconciles cultural preservation with modernization, governance with participatory decision-making, and sustainability with heritage conservation. This study critically examines the structural, institutional, and sociocultural barriers that hinder the CCP's evolution and proposes an adaptive heritage management framework that ensures its continued relevance as both a cultural landmark and a dynamic, accessible, and resilient space for artistic and civic engagement.

2.1 Study Aim

This study aims to develop a strategic framework for the redevelopment of the Cultural Center of the Philippines (CCP) Complex, ensuring its continued relevance as both a custodian of Filipino heritage and a contemporary cultural hub. While originally envisioned as a democratic space for artistic and cultural expression, the CCP has struggled with governance inefficiencies, physical deterioration, commercialization, and obsolescence, all of which threaten its capacity to fulfill its foundational mission. Addressing these challenges requires a holistic approach that integrates heritage conservation, participatory governance, sustainable urban development, and inclusivity.

A critical focus of this study is the cosmopolitan paradox, which highlights the tension between globalization and local identity preservation. This paradox is particularly evident in the CCP's struggle to maintain cultural authenticity while adapting to contemporary socioeconomic and urban pressures. By analyzing the historical trajectory of the complex, assessing the impact of commercialization and physical decline, and identifying governance barriers, this study seeks to propose solutions that reinforce the CCP's cultural significance while enhancing its functionality and accessibility.

Central to this research is the need to redefine the CCP's role within modern Filipino society, ensuring that it remains a culturally vibrant, socially inclusive, and environmentally resilient institution. The study advocates for a balanced redevelopment strategy that prioritizes adaptive heritage management, fosters active community engagement, and integrates climate-conscious and economically sustainable practices. In doing so, it envisions the CCP not only as a repository of the nation's artistic and cultural legacy but also as a dynamic space for public engagement and cultural exchange in an increasingly interconnected world.

3 Review of Related Literature and Theoretical/Conceptual Framework

The redevelopment of the Cultural Center of the Philippines (CCP) Complex is deeply intertwined with the "cosmopolitan paradox," explored in the book Cosmopolitanism, which discusses the ethics of identity and cultural contamination, and is particularly evident in the Philippines, where the interplay between global influences and local traditions profoundly impacts daily life, societal issues, and the development of public spaces. Appiah adds a critical dimension to this discussion by framing cosmopolitanism not only as a philosophical ideal but also as a practical approach to empowering communities (2006). He explained that "cosmopolitanism is an empowering concept and one that forms the basis of mutual respect for oneself and for others." He concluded that "if people were to manage their own lives, which is what they are responsible for, then they need the powers to do so." This perspective underscores the importance of involving the community in decisions regarding the redevelopment of the CCP Complex, ensuring that these decisions foster local empowerment and reflect the true diversity of Filipino cultural identities.

The paradox becomes more complex when considering the community's response to these preservation efforts. As discussed by During (2011), the challenge of creating a unified representation of cultural heritage can lead to discomfort or disagreement among community members, potentially affecting the community negatively in both sociocultural and political domains. The pressure to conform to a unified, perhaps more universally appealing representation can weaken the cohesiveness within the community, prompting some members to disengage or even leave, representing a serious breach of their cultural rights (Babb 2012; Olwig 1999). By embracing the cosmopolitan paradox, we unlock the potential to meld global perspectives with local heritage, enriching our community's fabric. This union not only strengthens our identity but also propels us toward a future in which diversity and unity thrive side by side. It challenges us to find a balance where the CCP can be both a symbol of Philippine cultural pride and a dynamic participant in the global cultural landscape. This approach does not merely preserve the past; it actively engages with the present and anticipates the future, ensuring that the CCP remains a vibrant and relevant cultural institution.

Governance and collaboration within the CCP Complex have been persistent challenges, as highlighted by the failure to implement the 2003 master plan. The persistence of social exclusion mechanisms, particularly among marginalized groups, stems from the disconnect between policy formulation and local implementation. When policies fail to translate into meaningful actions, inequalities deepen, underscoring the need for governance reforms to foster inclusivity (UN-Habitat 2020). Mendoza et al. (2017) identify governance inefficiencies in the Philippines that undermine cultural institutions like the Cultural Center of the Philippines (CCP), where improved coordination could enhance public service effectiveness. Saldaen et al. (2021) similarly highlight the need for administrative reforms in local government units to strengthen cultural governance and institutional efficiency. Thompson (2008) further explores the complex interplay between political governance and cultural administration, revealing tensions between global aspirations and local heritage preservation. This disconnect is particularly evident in the CCP, where rigid administrative frameworks hinder its adaptability to contemporary cultural and societal needs (Herrera 2001). As a result, the CCP's role as a central hub for Filipino arts and heritage has been marginalized, emphasizing the need for more flexible and responsive governance structures that align its egalitarian vision with practical execution. Addressing these challenges is critical to fostering more inclusive, equitable, and sustainable cultural and urban development frameworks.

The concept of architectural monumentality, particularly during the Marcos regime, plays a significant role in shaping the CCP's identity. The analysis by Lico (2003) and Ogura et al. (2002) reveals how architecture was used not only as a form of cultural expression but also as a tool for political propaganda. The grandiose designs of the CCP, while symbolizing national pride, also reflect the complex interplay between cultural identity and political ambition. This historical context is crucial for understanding the challenges and opportunities in the CCP's redevelopment. The need to integrate local cultural elements into the architectural narrative is vital for ensuring that the complex resonates with contemporary Filipino society while respecting its historical significance.

In examining the monumental architecture of the Marcos regime, it is essential to contextualize these discussions within Clifford Geertz's perspective on culture. Geertz (1973) argues that culture functions as a system of symbols rather than a deterministic force, providing a framework within which social events, institutions, and behaviors can be meaningfully interpreted. This perspective offers a critical lens for analyzing the architectural strategies employed by the Marcos administration, particularly its use of built environments to project and institutionalize a national cultural identity. Architecture under the regime was not merely a means of construction but a deliberate instrument of power projection and societal influence, embedding political authority within the urban fabric. The Marcos-era architectural agenda aligned with Geertz's notion of culture as a semiotic system, where structures transcended their functional roles to serve as symbols of control and hegemony, reinforcing state narratives through the built environment.

Expanding on this theme, Espiritu (2015), in *The Marcos Romance and the Cultural Center of the Philippines: The Melodrama of a Therapeutic Cultural Policy*, examines the Cultural Center of the Philippines (CCP) as a key example of this architectural strategy. He highlights how the CCP was envisioned as a tool to cultivate a unifying national identity through the arts while simultaneously functioning as a mechanism of political legitimization. The regime's emphasis on promoting a "people's culture" and revaluing vernacular traditions reflects a populist approach to cultural policy that sought to merge authoritative symbolism with nationalistic sentiment. Flores (2017), in *Total Community Response: Performing the Avant-garde as a Democratic Gesture in Manila*, adds another dimension by exploring the CCP's dual role as both an artistic hub and a political stage. While fostering avant-garde artistic expression, the CCP also hosted state-sponsored performances, illustrating the complex intersection of cultural production, political narratives, and urban development during the Marcos era. These analyses collectively demonstrate how architecture functioned as a tool for sociopolitical control, embedding power within the built landscape while constructing a crafted cultural identity that reinforced both aesthetic and ideological objectives. This discussion deepens our understanding of the interplay between power, culture, and architecture, emphasizing how monumental spaces were strategically employed to consolidate political authority in a pivotal era of Philippine history.

However, as the CCP Complex evolves, it faces pressing challenges that extend beyond its historical and ideological legacy. The broader discourse on cultural governance and architectural preservation underscores the difficulties of maintaining monumental structures while adapting them to modern needs. The same mechanisms that once reinforced the CCP's role as a national cultural institution—state-driven architectural symbolism and centralized governance—now present obstacles to its sustainable transformation. Issues such as physical deterioration, commercial intrusions, loss of cultural focus, and governance inefficiencies have emerged as critical concerns. Addressing these requires a nuanced approach that balances cultural preservation with contemporary urban demands. By drawing on insights from global and local perspectives, this research aims to develop strategies that revitalize the CCP Complex as a dynamic cultural hub, ensuring that it remains a relevant and cherished landmark, deeply connected to the evolving cultural and social dynamics of the Philippines.

3.1 Theoretical/Conceptual Framework

The redevelopment of the Cultural Center of the Philippines (CCP) Complex is guided by a multidisciplinary theoretical framework that integrates Contextualism, Postmodernism, and Actor–Network Theory (ANT). This framework (refer to Fig. 1a), referred to as Symbiotic Pluralism, aims to create a space that is reflective of Filipino cultural identity, adaptive to contemporary challenges, and inclusive of diverse narratives.

Contextualism emphasizes the integration of architecture with its local environment, culture, and history. Theories by Norberg-Schulz (1980) and Smith and Relph (1978) stress the importance of "Genius Loci," or the spirit of the place, advocating for designs that enhance the cultural and historical narratives of their location. In the CCP Complex, this means incorporating traditional Filipino elements into modern structures to preserve cultural identity while meeting contemporary needs.

Postmodernism complements Contextualism by celebrating diversity and contradiction within architectural

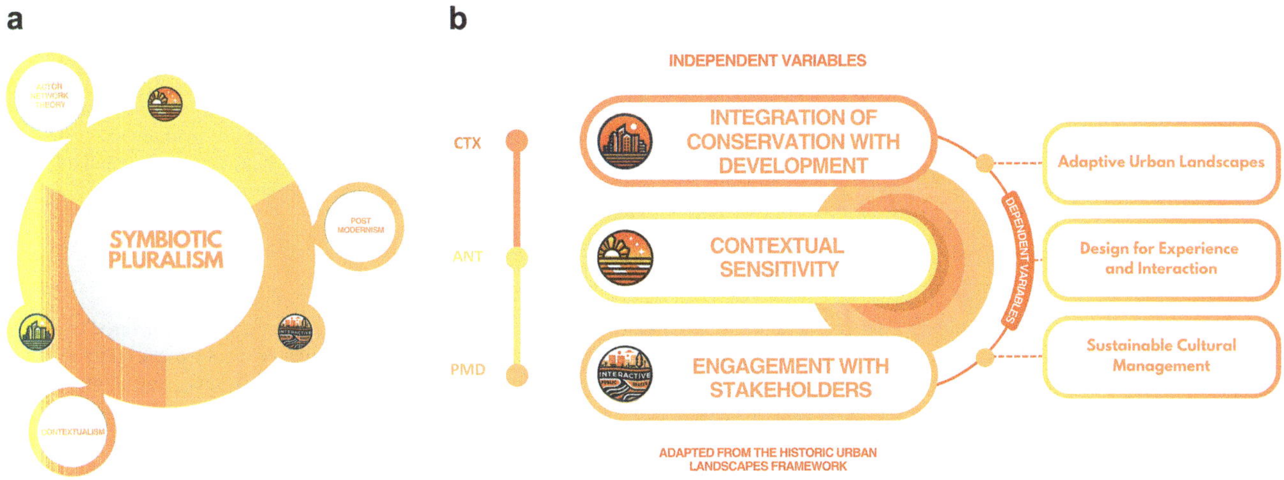

Fig. 1 (**a**) Theoretical Framework focuses on the manifestation of Symbiotic Pluralism through the exploration of Contextualism, Postmodernism and Actor Network Theory. (**b**) Conceptual Framework adapted from UNESCO's Historic Urban Landscapes

design. As outlined by Levin and Jencks (1978) and Venturi et al. (1977), this approach acknowledges the complex, layered nature of cultural identities, which is particularly relevant to the Philippines. The eclectic and inclusive spaces envisioned under Postmodernism allow for a creative amalgamation of historical and cultural elements, reflecting the multifaceted nature of Filipino society.

Actor–Network Theory (ANT) provides a framework for understanding the complex interactions between human and nonhuman elements within the CCP Complex. Law (1986) and Callon's (1984) ANT posits that spaces are not mere backdrops for human activity but active participants in cultural and social practices. This perspective encourages a holistic approach to redevelopment, where every element within the CCP—be it physical structures or cultural practices—plays a role in shaping the overall experience.

The concept of **Symbiotic Pluralism** integrates these theories to create a comprehensive approach to urban design that is deeply pluralistic and contextually grounded. It promotes the creation of environments that are not only aesthetically and functionally robust but also deeply embedded in the local cultural and historical context. This approach ensures that the CCP Complex becomes a vibrant, dynamic participant in the cultural life of the community, resonating with its cultural narratives while adapting to modern needs.

The **Conceptual Framework** (**see** Fig. 1b) derived from this theoretical foundation incorporates Adaptive Urban Landscapes, Cultural Identity and Space, and Interactive Public Spaces which are derived from the Historic Urban Landscapes framework of UNESCO (Brabec and Goetcheus 2014). These concepts emphasize the need for urban spaces that are responsive to societal changes, reflective of cultural identities, and foster active public engagement. The goal is to transform the CCP Complex into a space that not only preserves heritage but also enhances cultural vitality, ensuring it remains relevant and inclusive in the context of contemporary Filipino society.

4 Methodology

4.1 Research Design

This study employs a mixed-methods research design, integrating qualitative and quantitative approaches to explore the redevelopment of the Cultural Center of the Philippines (CCP) Complex through the Historic Urban Landscape (HUL) approach and Symbiotic Pluralism. The research is structured into three key phases: data collection, analysis, and framework development. This methodological framework (as tabulated below in Table 1) ensures a comprehensive assessment of the CCP Complex by integrating historical context, stakeholder perspectives, and experiential assessments, leading to an informed redevelopment strategy that upholds cultural integrity while promoting inclusivity and sustainability.

5 Key Findings and Trends

The Cultural Center of the Philippines (CCP) Complex stands as a pivotal institution for Filipino arts and culture, yet it faces significant governance, environmental, and inclusivity challenges. As cultural districts worldwide evolve toward sustainability and inclusivity, the CCP's redevelopment presents an opportunity to embrace a more participatory, climate-resilient, and community-driven approach. By critically

Table 1 Tabulated outline of the research design and the methodologies, and their relationship to the established subproblems and objectives

Section	Subproblem	Objective	Methodology
Data collection			
1. Archival research and document analysis	Subproblem 1	Determine the historical and cultural context of the CCP complex and assess the initial goals of the 2003 redevelopment plan	Conduct a comprehensive review of the 2003 business and master plan, historical records, and related documents. Analyze these sources to understand the initial goals and identify gaps in addressing contemporary needs
	Objective 1	Understand the original objectives of the CCP complex and the implications of the nonimplementation of the 2003 redevelopment plan	
2. Surveys and interviews	Subproblem 2	Analyze stakeholder perceptions regarding the current state of the CCP complex and the impact of commercial activities on its cultural mission	Administer surveys and conduct semistructured interviews with urban planners, architects, cultural heritage experts, government officials, and community representatives
	Objective 2	Assess current stakeholder perceptions of the CCP complex and how commercial activities influence its cultural mission	
3. Experiential landscape assessment and thick mapping	Subproblem 3	Evaluate how the physical and sensory aspects of the CCP complex affect its inclusivity and accessibility	Perform an experiential landscape assessment and thick mapping exercise. Assess the sensory experiences and emotional responses of users and map these experiences against the physical and cultural aspects of the CCP complex
	Objective 3	Identify areas of the CCP complex that are underutilized or inaccessible and evaluate user experiences	
Data analysis			
1. Thematic analysis	Subproblem 4	Identify key themes related to inclusivity, accessibility, and heritage preservation from qualitative data	Conduct thematic analysis of interview data and thick mapping results. Identify themes related to inclusivity and accessibility, drawing on gendered spaces and Latour's (2005) actor–network theory
	Objective 4	Analyze qualitative data to uncover themes related to the CCP complex's inclusivity and heritage preservation	
2. Quantitative analysis	Subproblem 5	Examine stakeholder perceptions of heritage value and modern amenities	Use statistical tools to analyze survey data, informed by Gehl (1980) on urban spaces and human-centered design. Examine correlations between perceived heritage value and desires for modern amenities
	Objective 5	Analyze quantitative survey data to measure stakeholder perceptions and identify correlations between heritage value and modern amenities	
3. Framework development	Subproblem 6	Develop a framework that integrates the HUL approach with principles of symbiotic pluralism for the redevelopment of the CCP complex	Synthesize findings from data analysis to develop a conceptual framework. Propose specific interventions to promote inclusivity, sustainability, and heritage conservation
	Objective 6	Propose a redevelopment framework that balances heritage conservation with modern inclusivity and sustainability	
4. Recommendations for accessibility and sustainability	Subproblem 7	Identify and recommend improvements for accessibility and sustainability in the CCP complex	Conduct an accessibility audit and review sustainability practices. Compare current redevelopment plans with best practices in accessibility and sustainability. Develop recommendations for improvements based on audit results and expert consultations
	Objective 7	Recommend enhancements for accessibility and sustainability based on audit findings and best practices	

examining governance structures, environmental sustainability, and cultural representation, this study positions the CCP within a broader discourse on urban cultural resilience. Drawing from international frameworks such as UNESCO's Historic Urban Landscape (HUL) and the UN's Sustainable Development Goals (SDGs), this paper explores how the CCP can serve as a model for cultural redevelopment in Southeast Asia.

5.1 Governance: Balancing Centralized and Participatory Models

Governance within the CCP Complex has long been a subject of contention. The existing centralized model (as reviewed from Chap. 1 of the 2003 Business and Master Plan), overseen primarily by the Planning and Architecture Review Committee (PARC), has led to bureaucratic stagnation and limited public engagement. Despite efforts to modernize its institutional structure, governance inefficiencies continue to hinder the full realization of previous master plans, such as the ambitious but largely unimplemented 2003 redevelopment blueprint. Survey results indicate that a sizable portion of stakeholders feel excluded from decision-making processes, reinforcing the need for a more participatory governance model.

Comparisons with other cultural districts, such as Singapore's Esplanade and Hong Kong's West Kowloon Cultural District, highlight the benefits of balancing institutional oversight with community-driven initiatives. These case studies suggest that the CCP could benefit from the establishment of advisory cultural councils, greater collaboration with local artists and urban planners, and a decentralized programming model that allows for more regional and grassroots participation. By shifting toward a governance model that integrates community input, the CCP can bridge the existing disconnect between institutional leadership and public expectations, fostering a greater sense of cultural ownership among Filipinos.

5.2 Sustainability and Urban Challenges

Environmental sustainability remains a critical concern in the redevelopment of the CCP Complex, particularly given its vulnerable coastal location and the increasing threats posed by climate change. The reclamation projects in Manila Bay have exacerbated issues related to flooding, rising sea levels, and the disruption of ecological systems. Observational assessments of the site highlight the discomfort experienced in open spaces, where the lack of shaded areas and green infrastructure significantly reduces visitor engagement.

Efforts to integrate sustainability into the CCP's redevelopment must go beyond superficial green interventions and instead embrace a holistic climate-responsive strategy. Inspiration can be drawn from waterfront cultural hubs such as Barangaroo Reserve in Sydney and Nordhavn in Copenhagen, where nature-based solutions have been embedded into urban design. In the context of the CCP, this could translate to the implementation of passive cooling strategies, the expansion of green roofs, and the integration of stormwater management systems. By aligning with UN-Habitat's New Urban Agenda (Amirtahmasebi et al. 2020), the CCP can reinforce its commitment to sustainable urban development while enhancing its role as an ecological and cultural landmark.

5.3 Inclusivity and Cultural Representation

The CCP Complex has historically been regarded as a bastion of elite and metropolitan artistic expression, often at the expense of grassroots and regional narratives. While it remains the country's premier venue for the performing arts, its programming has not always been reflective of the full spectrum of Philippine cultural identity. Survey results underscore the persistent challenge of accessibility—not only in terms of physical infrastructure but also in terms of socioeconomic inclusion. High ticket prices, limited transportation options, and the dominance of certain art forms contribute to an exclusionary cultural landscape.

To address these gaps, the redevelopment strategy must foreground inclusivity both in its spatial design and its artistic programming. One potential approach is the adoption of a more comprehensive framework based on the complex's current realities, wherein indigenous and regional art forms are given greater prominence through curated exhibitions, festival integrations, and traveling performances that extend beyond the confines of the CCP. Furthermore, improvements in universal design—such as tactile pathways, braille signage, and barrier-free architecture—can significantly enhance the physical accessibility of the complex. By fostering a cultural space that is truly representative of the country's diverse artistic heritage, the CCP can reaffirm its role as an institution that serves not just the elite but the broader Filipino public.

5.4 Understanding the Public

To gain an understanding of public sentiment and key themes related to the redevelopment of the Cultural Center of the Philippines (CCP) Complex, the researcher conducted a sensing survey as part of the preliminary data collection phase. This survey aimed to capture the public's views on various aspects of redevelopment, including cultural preservation, modernization, accessibility, and community engagement. By employing a mixed-methods approach, the study provided an overview of the community's priorities and concerns. The analysis of survey responses has allowed for the identification of significant trends and preferences, which will be instrumental in shaping a redevelopment strategy that resonates with the broader Filipino community.

5.4.1 Public Sentiment Analysis

Sentiment analysis was conducted using natural language processing (NLP) techniques on open-ended survey responses from the dataset. The analysis classified responses into three categories: Positive, Neutral, and Negative sentiments. Sentiment scores were calculated based on the polarity of words and phrases within each response. The overall sentiment was then classified based on the aggregated polarity score.

Findings

- **Positive sentiment**: 134 responses (47.9%) expressed optimism and support for the redevelopment, highlighting the importance of modernization while preserving cultural heritage.
- **Neutral sentiment**: 111 responses (39.6%) reflected a balanced or uncertain perspective, indicating a cautious attitude toward the proposed changes.
- **Negative sentiment**: 35 responses (12.5%) raised concerns, primarily focused on the risks of cultural dilution and the potential overshadowing of heritage by modernization.

5.4.2 Importance of Cultural and Historical Preservation

The importance of cultural and historical preservation was measured through a Likert scale survey question. Respondents rated their agreement on the importance of preserving the CCP Complex's heritage structures. The responses were aggregated, and the percentage of respondents who rated the preservation as "Important" or "Very Important" was calculated.

Findings

- **83.4%** of respondents prioritized heritage preservation, indicating a strong public sentiment toward maintaining the cultural and historical significance of the CCP Complex.
- This result underscores the community's desire to protect the identity of the CCP Complex while integrating modern elements.

5.4.3 Modernization and Accessibility

Similar to cultural preservation, modernization and accessibility were assessed using survey questions that asked respondents to rate the importance of various modern features, such as green spaces and public transportation improvements. The proportion of respondents who supported each modernization aspect was calculated and compared.

Findings

- **Green spaces**: 61.3% of respondents supported the inclusion of green spaces.
- **Public transportation**: 59.9% emphasized the need for improved public transportation.
- **Modern facilities**: 45.8% of respondents deemed the modernization of facilities for arts and culture important.

5.4.4 Community-centered Urban Planning

The preference for community-centered urban planning was evaluated through survey questions that gauged respondents' support for inclusive and community-focused development strategies. The percentage of respondents who advocated community-centered urban planning was calculated.

Findings

- **48.5%** of respondents supported development approaches that prioritize the needs of the local community and promote social cohesion, indicating a significant trend toward inclusive planning.

5.4.5 Reflecting on Filipino Cultural Identity

The study measured preferences for interactive public spaces and community engagement by asking respondents to rank the importance of various public space features. The average score was calculated and interpreted.

Findings

- The mean score for the importance of reflecting Filipino cultural identity was **4.40** on a 5-point scale, indicating a high level of importance placed on this aspect by the respondents.

5.4.6 Community Engagement and Interactive Public Spaces

Respondents rated the importance of reflecting Filipino cultural identity in the redevelopment efforts, using a Likert scale. The average score was calculated and interpreted.

Findings

- **Accessibility for all age groups**: 35.4% of respondents emphasized this as a top priority.
- **Open public spaces**: 29.6% of respondents highlighted the need for more open spaces.
- **Community events and workshops**: 18.2% supported these activities as essential features.
- **Interactive installations**: 17% expressed a desire for dynamic and engaging installations within the complex.

5.4.7 ANOVA and Cluster Analysis

Methodology

- **Analysis of variance (ANOVA)**: Conducted to determine if there were significant differences in opinions on cultural preservation, modernization, and accessibility across different demographic groups (e.g., age, income, region). F-statistics were computed for each aspect, comparing the means across demographic groups. Significant p-values indicated differences in opinions.
- **Cluster analysis**: Grouped respondents based on similar characteristics and preferences regarding the redevelopment. K-means clustering was used to identify distinct groups within the dataset, based on responses to key survey questions.

5.4.8 Computation

F-Statistic

$$F = \frac{\text{Mean Square Between}(\text{MSB})}{\text{Mean Square Between}(\text{MSW})}$$

where

$$\text{MSB} = \frac{\text{SSB}(\text{Sum of Squares Between})}{k-1}$$

$$\text{MSW} = \frac{\text{SSW}(\text{Sum of Squares Between})}{N-k}$$

- **Significance**: The computed F-statistic was compared to the critical F-value to determine if the differences were statistically significant.

Findings
- **Cultural preservation (by age)**: $F = 4.27$, $p = 0.012$, indicating significant differences across age groups.
- **Modernization (by income)**: $F = 3.58$, $p = 0.024$, showing significant differences across income levels.
- **Accessibility (by region)**: $F = 6.14$, $p = 0.004$, highlighting significant differences across geographic regions.

5.4.9 Cluster Analysis

Methodology: K-means clustering was used to group respondents based on their responses to key questions, identifying distinct segments within the population.

Computation

- **K-Means Algorithm**:

- $$\text{Distance} = \sqrt{\left((x_1-\mu_1)^2+(x_2-\mu_2)^2+\cdots+(x_n-\mu_n)^2\right)}$$

where x_i is an individual data point and μ_i is the centroid of the cluster.

The ANOVA analysis showed significant differences in opinions across demographic groups, particularly regarding cultural preservation, modernization, and accessibility. Cluster analysis identified three primary groups: Cultural Preservationists, Modernization Enthusiasts, and Community-Focused Respondents, each with distinct preferences that must be balanced in the redevelopment strategy (see Table 2).

This analysis provides a deep understanding of public sentiment and preferences regarding the redevelopment of the CCP Complex. The findings highlight the importance of balancing cultural preservation with modernization, ensuring accessibility, and fostering community engagement. Insights like these will guide the development of a strategic and inclusive redevelopment plan that aligns with the diverse needs of the Filipino community.

Table 2 Results from the ANOVA and cluster analyses conducted to understand the correlation of the themes with the respondents' sentiments

ANOVA results	*F*-statistic	*p*-value
Cultural preservation (by age)	4.27	0.012
Modernization (by income)	3.58	0.024
Accessibility (by region)	6.14	0.004
Cluster groups	**Characteristics**	
Cultural preservationists	Older respondents, prioritize heritage	
Modernization enthusiasts	Younger, higher income, support modernization	
Community-focused respondents	Value accessibility and community engagement	

5.5 Experiential Landscape Assessment: User Engagement and Spatial Performance

An assessment based on the book written by Thwaites and Simkins (2006) helped in evaluating CCP's spatial layout, revealing significant disparities in how different areas of the complex function. The Experiential Landscape Assessment of the CCP Complex is designed to provide a detailed, block-by-block evaluation of the public spaces within the complex, focusing on how visitors experience and interact with these spaces. The purpose is to understand how spatial design, activity variety, and visitor engagement influence the overall visitor experience, from functionality to comfort and social interaction. This research employs a multimethod approach combining observational fieldwork, visitor surveys, and environmental analysis to assess both objective (e.g., design elements, physical attributes) and subjective (e.g., comfort, visitor preferences) aspects of the complex's public spaces.

Data from ArcGIS-mapped surveys (refer to Table: Tabulated Discussion of Average Scores of all 10 Blocks against 13 Assessment Criteria for Experiential Landscape) indicate that spaces such as the Philippine International Convention Center (PICC) (Block 5) and Star City (Block 10) perform well in terms of activity variety and navigability, benefiting from a mix of commercial, entertainment, and cultural functions. In contrast, underutilized spaces like

Block 8.2 (Buendia Avenue Connection) and Block 9 (Magdalena Jalandoni) suffer from weak pedestrian infrastructure, lack of shade, and limited engagement opportunities (refer to Table: Experiential Landscape Assessment/ Urban Streetscape Assessment).

Design interventions must focus on optimizing these underperforming spaces to enhance visitor experience. Adaptive reuse of vacant areas, the introduction of interactive installations, and the improvement of wayfinding elements can collectively create a more dynamic and engaging environment. Moreover, a more seamless integration with Metro Manila's broader transit network would alleviate access difficulties, making the CCP a more inclusive cultural destination.

5.5.1 Lowest- and Highest-Rated Blocks

This section provides a detailed look at the highest- and lowest-rated blocks within the CCP Complex, offering insights into their strengths and weaknesses.

5.5.2 Highest-Rated Blocks (Fig. 2)

Block 5: PICC (Philippine International Convention Center)

- **Average score**: **4.03—highest-rated block** (based on all assessment criteria—refer to Table: Experiential Landscape Assessment/Urban Streetscape Assessment).
- **Strengths**:
  - **Cultural programming**: The PICC is consistently recognized for its capacity to host high-profile international events and cultural exhibitions, which contributes to its high rating for activity levels.
  - **Architectural significance**: The Brutalist architecture of PICC is iconic, creating a visually cohesive and impactful environment within the complex. The building's distinctive design enhances its cultural identity, and this contributes significantly to the overall rating for continuity and patterns in architecture.

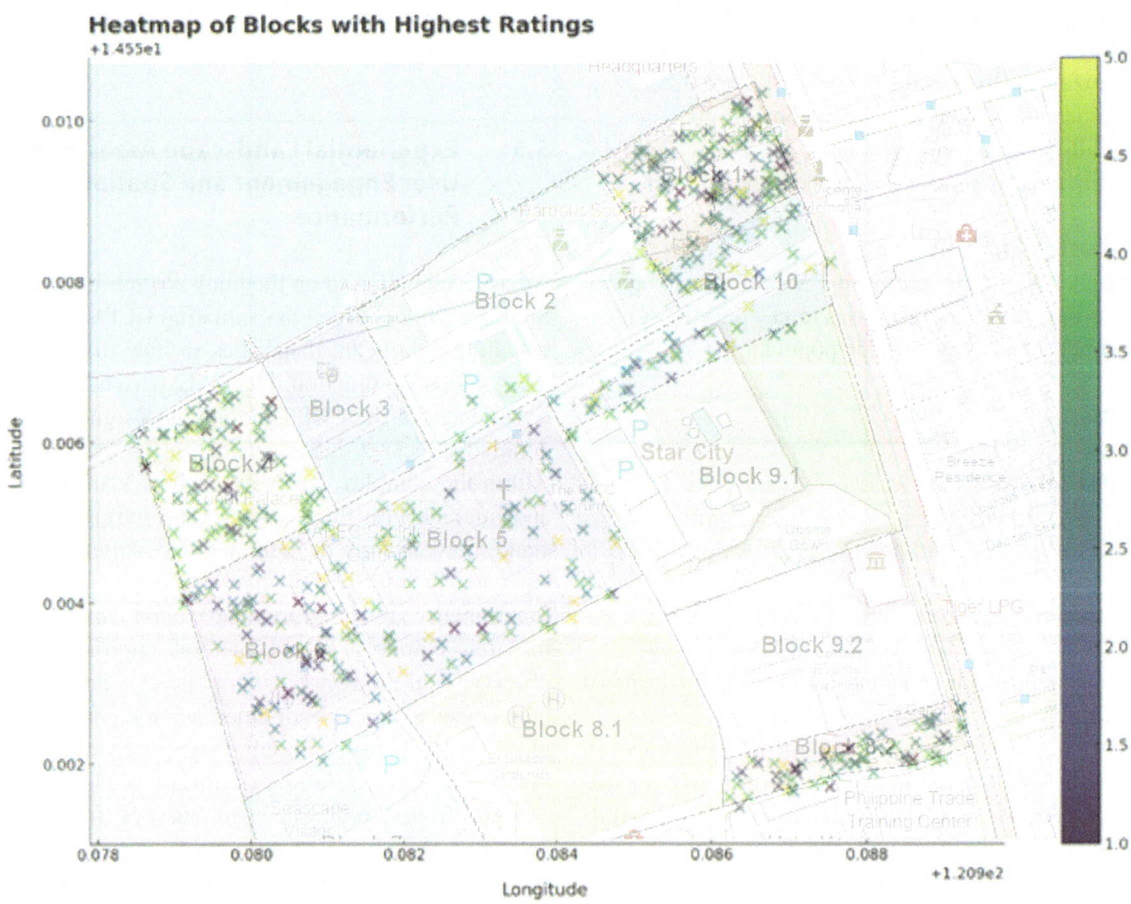

Fig. 2 Heat Map Projection of the highest rated blocks within the complex

  - **Ease of movement and clarity of navigation**: Block 5 has been praised for its well-designed circulation and clear signage, making navigation straightforward and reducing visitor confusion. The layout allows for easy transitions between spaces, whether for cultural events or casual visits.
  - **Privacy and security**: PICC offers clear boundaries between public and private spaces, making it a highly secure and private environment for eventgoers and performers alike.
- **Key Findings**:
  Block 5 is a cornerstone of the CCP Complex, effectively combining high cultural impact and well-executed architectural design. Its top rating in cultural programming reflects its capacity to host significant national and international events, while its architectural cohesion enhances the visual appeal of the complex as a whole. The ease of movement and security aspects ensure a positive experience for visitors, particularly during large events or exhibitions.

Block 10: Star City × Tanghalang Pambansa (Vicente Sotto Street)

Average score: **3.48—high activity levels** (based on criteria: variety in activities, completeness of activities, and interaction hotspots—refer to Table: Experiential Landscape Assessment/Urban Streetscape Assessment)

- **Strengths**:
  - **Commercial appeal**: Star City attracts large crowds due to its theme park offerings and entertainment value. This contributes to the block's high activity levels, which enhance its role as a vibrant destination.
  - **Variety of activities**: The variety of entertainment options in Star City (rides, games, and shows) stands out, making this block a dynamic and popular space. Visitors are drawn to the block's diverse activities, providing engagement for different age groups and interests.
  - **Interaction hotspots**: Star City functions as a prime interaction hotspot, encouraging both informal socialization and cultural experiences. This is bolstered by the park's lively atmosphere and entertainment programming.
  - **Sense of enclosure and security**: The enclosed layout of Star City helps provide a secure environment for visitors, while clearly defined spaces reduce security concerns.
- **Key Findings**:
  Block 10 is one of the most active and engaging blocks in the CCP Complex, thanks to Star City's high foot traffic. The variety of entertainment activities fosters a lively environment, making it a popular spot for visitors seeking both cultural engagement and leisure activities. While Star City excels in terms of activity variety, its lack of architectural cohesion with other blocks may limit its full integration into the cultural vision of the CCP Complex.

5.5.3 Lowest-Rated Blocks (Fig. 3)

Block 8.2: Buendia Avenue Connection

Average score: **2.95—lowest-rated for pedestrian infrastructure** (based on criteria: ease of movement, thresholds and changes, dynamics of corridors and segments, impact of climatic effects and continuity in patterns and architecture—refer to Table: Experiential Landscape Assessment/Urban Streetscape Assessment)

- **Challenges**:
  - **Lack of green spaces and shading**: Block 8.2 lacks the greenery and shaded areas that would provide visitors with environmental comfort, particularly under the intense sun of Metro Manila.
  - **Vehicular congestion**: The heavy traffic around Buendia Avenue creates significant obstacles for pedestrians, making it difficult to navigate safely or comfortably. The lack of dedicated pedestrian pathways and clear pedestrian zones exacerbate this issue.
  - **Underdeveloped pedestrian amenities**: Unlike other parts of the CCP Complex, Block 8.2 does not provide adequate rest areas, seating, or interactive elements that could encourage visitors to spend more time in the area.
  - **Visual appeal**: The commercial structures surrounding Buendia Avenue are uninspiring, and the block lacks cohesive design features that would integrate it into the overall aesthetic of the complex.
- **Key Findings**:
  Block 8.2 is currently one of the weakest performers in the CCP Complex, primarily due to its poor pedestrian infrastructure and lack of environmental comfort. The area's vehicular congestion and lack of green spaces make it less inviting for visitors, creating a barrier to the broader cultural goals of the complex. Improvements in pedestrian amenities, shading, and landscaping are crucial for transforming this block into a more engaging and comfortable space for visitors.

Block 9: Magdalena Jalandoni

Average score: **3.18—lack of cultural alignment** (based on criteria: variety of activities, ease of movement, thresholds and changes, and dynamics of corridors and segments—refer to Table: Experiential Landscape Assessment/Urban Streetscape Assessment)

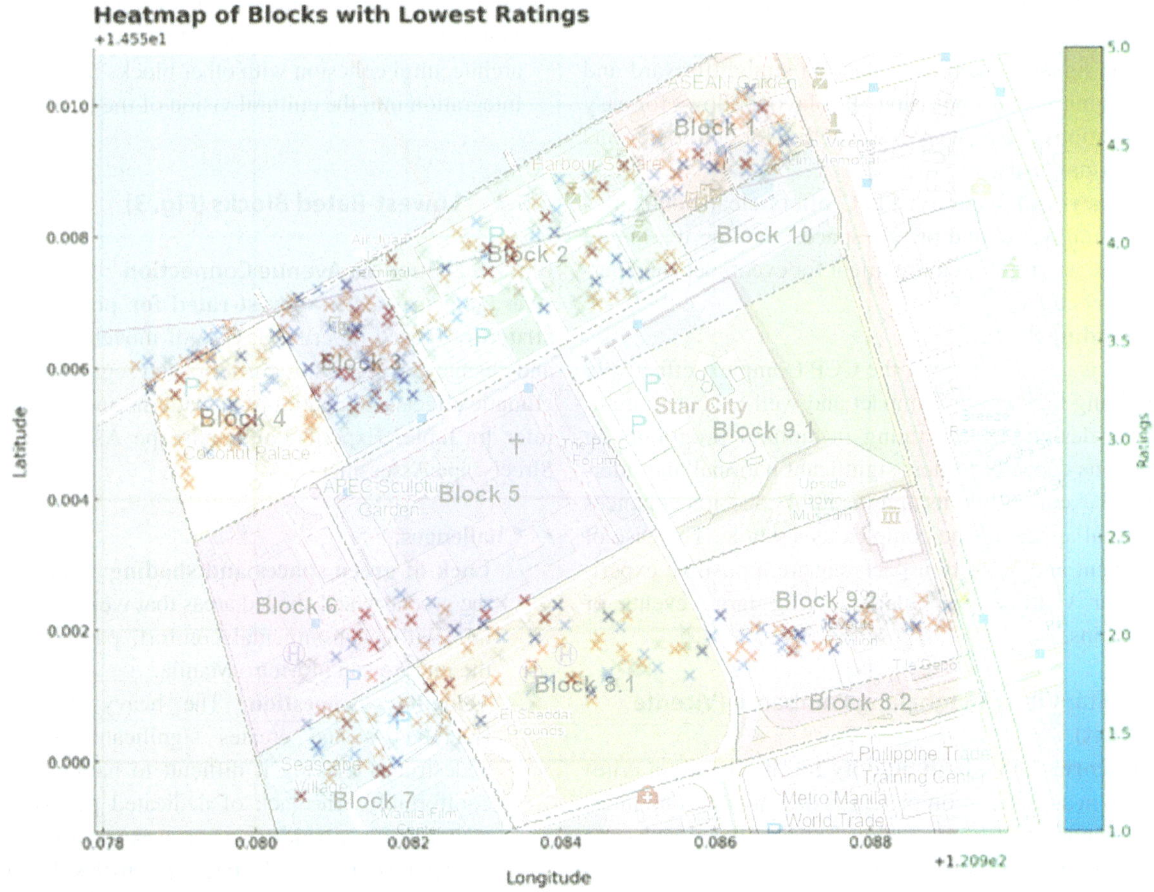

Fig. 3 Heat Map Projection of the lowest rated blocks in the complex

- **Challenges**:
  - **Disjointed pedestrian pathways**: The pedestrian experience in Block 9 is hampered by poorly defined and disconnected pathways, which make navigation difficult and less intuitive.
  - **Uninspiring commercial tenants**: The commercial tenants in Block 9, such as Wensha Spa and Pergola Events Center, are unrelated to the CCP's cultural mission, leading to a disconnect between the block's commercial focus and the cultural objectives of the complex.
  - **Lack of public–private boundaries**: The absence of clear public–private boundaries in Block 9 makes the space feel unwelcoming and less secure, especially for visitors who expect clear distinctions between private and public spaces in such a cultural venue.
  - **Underutilization**: Despite its location within the CCP Complex, Block 9 remains underused, and its lack of cultural programming or interactive spaces results in a lost opportunity for engaging visitors.
- **Key Findings**:
  Block 9 faces severe challenges related to commercial misalignment, poor pedestrian infrastructure, and a lack of cultural focus. These factors contribute to its underutilization and disconnection from the rest of the complex. The space could benefit from the introduction of culturally focused businesses, more defined pathways, and interactive cultural programming that would allow it to better serve the mission of the CCP Complex.

5.6 Integrating Governance, Sustainability, and Cultural Representation

The convergence of governance reform, sustainability integration, and inclusivity underscores the complex yet necessary transformation that the CCP Complex must undergo. From a governance perspective, a more participatory model is essential to bridging the disconnect between institutional leadership and public engagement. The adoption of UNESCO's HUL framework, coupled with collaborative urban planning efforts, would allow the CCP to evolve into a more dynamic cultural institution that is responsive to contemporary societal needs.

In terms of environmental sustainability, the CCP must position itself as a leader in climate-conscious cultural redevelopment. The lessons from global waterfront cultural hubs demonstrate that heritage conservation and ecological resilience can coexist, reinforcing the need for an integrated approach that prioritizes both environmental and cultural longevity. By embedding sustainable design principles into its redevelopment strategy, the CCP can redefine what it means to be a cultural landmark in a rapidly urbanizing and climate-vulnerable environment. Finally, the question of inclusivity remains paramount. The CCP's ability to remain relevant in the twenty-first century hinges on its capacity to embrace a more pluralistic and accessible vision of culture. From reimagining its programming to better reflect the diversity of Filipino artistic expression to ensuring that its physical spaces are welcoming to all, the redevelopment process must be guided by the principle that culture is not a privilege but a right.

6 Conclusion and Recommendations

The Cultural Center of the Philippines (CCP) Complex, conceived in the 1970s as a national symbol of Filipino culture, remains an iconic yet contested space within the country's cultural landscape. While it was envisioned as a beacon of Filipino identity and a custodian of cultural heritage, the CCP Complex has faced decades of challenges that undermine its mission, including governance inefficiencies, commercialization, environmental disruptions, and limited inclusivity. This study sought to bridge the gap between the CCP's foundational ideals and its present realities, uncovering critical issues and proposing a forward-looking framework to reestablish the CCP as an inclusive, vibrant, and sustainable cultural hub.

Through a comprehensive analysis of archival records, stakeholder perceptions, and the physical and sensory dynamics of the complex, this study explored the disconnect between the CCP's original mission and its current state. These findings underscored several persistent challenges:

- **Cultural preservation**: Cultural goals have often been deprioritized in favor of commercial interests, leading to a dilution of the CCP's role as a space for authentic Filipino expression.
- **Public accessibility**: The CCP Complex has become increasingly commercialized and physically disconnected from its community, exacerbated by insufficient pedestrian infrastructure and the adverse effects of Manila Bay reclamation projects.
- **Adaptability**: A lack of flexibility in architectural design and governance structures has hindered the CCP from evolving with urban and environmental changes.

Quantitative and qualitative analyses provided critical insights into how these challenges manifest. For instance, the block-by-block analysis revealed stark disparities in performance and utilization across the complex. Blocks such as the Philippine International Convention Center (PICC) and Star City performed well due to their integration of cultural programming and modern amenities. In contrast, underperforming areas like the CCP Open Grounds (Block 8.1) and Magdalena Jalandoni Avenue (Block 9), with mean scores of 3.15 and 3.19, respectively, highlighted issues such as lack of programming, underutilized spaces, and poor pedestrian infrastructure.

From a broader perspective, the public's overall satisfaction score for accessibility was 3.18, reflecting the barriers posed by inadequate public transportation and physical infrastructure. Sentiment and thematic analyses of stakeholder input further revealed deep concerns about the CCP's growing commercialization and its limited engagement with grassroots communities. These findings emphasized the need to realign the CCP's development with its mission, prioritizing accessibility, inclusivity, and cultural preservation.

The governance challenges identified in the research further exacerbate these issues. The centralized governance model of PARC (Planning and Architecture Review Committee) has led to fragmented oversight, where community representation and local cultural narratives have often been excluded from decision-making processes. Commercial interests have largely driven the development of the complex, overshadowing its cultural mission. Ineffective enforcement of design guidelines and lack of adaptive capacity have compounded these challenges, especially with the disruption caused by reclamation projects.

This study successfully redefines the Cultural Center of the Philippines (CCP) Complex as a dynamic, inclusive, and sustainable cultural institution by employing a multifaceted, evidence-based approach rooted in empirical data and theoretical frameworks. The findings reveal critical governance gaps, cultural misalignments, and spatial inefficiencies that have hindered the CCP's ability to fulfill its original mission. Through archival research, the study contextualized the CCP's foundational vision, demonstrating how the unfulfilled 2003 master plan has contributed to its governance challenges and deviation from egalitarian cultural policies.

The assessment of the complex's current state identified the adverse effects of commercialization, accessibility barriers, and declining cultural engagement, highlighting key areas for intervention. Analyzing the programmatic and spatial disconnect within the CCP emphasized the need for cultural revitalization strategies that integrate grassroots and regional narratives, ensuring broader representation and relevance. Additionally, the study addressed the threat of obsolescence, recommending adaptive reuse, modernization of outdated facilities, and spatial reconfiguration to enhance long-term functionality.

A critical dimension of the study focused on the lack of participatory governance and community representation, underscoring the necessity of inclusive decision-making frameworks that foster public trust and shared ownership. Furthermore, the study's accessibility and sustainability assessments informed strategies for improving transportation infrastructure, pedestrian access, and the integration of green practices, ensuring that the CCP evolves as a resilient and environmentally conscious cultural hub.

By synthesizing these findings, this research presents a comprehensive redevelopment framework that balances cultural preservation with contemporary urban demands, reinforcing the CCP's role as a premier national institution for Philippine arts and heritage while fostering engagement, inclusivity, and sustainability.

6.1 Egalitarianism and the Cosmopolitan Paradox

The study's theoretical foundation was deeply informed by the principles of egalitarianism and the cosmopolitan paradox:

- **Egalitarianism**: The analysis emphasized the need to democratize the CCP's governance, design, and programming to ensure equitable access and representation. This principle informed strategies to counteract the exclusivity of monumentalism, fostering an inclusive space that authentically represents Filipino culture across all demographics.
- **Cosmopolitan paradox**: The tension between the CCP's global aspirations and its local cultural responsibilities was addressed by proposing strategies that balance universal standards of modernity with the preservation of Filipino heritage. The framework acknowledges the CCP's potential to serve as a globally recognized cultural hub while remaining rooted in grassroots narratives and traditions.

6.2 A Vision for the CCP Complex

The CCP Complex stands at a critical juncture, with the opportunity to redefine its role as both a custodian of Filipino heritage and a dynamic platform for cultural exchange and education. By embedding egalitarian principles into its redevelopment and resolving the cosmopolitan paradox, the CCP can reconcile its monumental legacy with the diverse needs of contemporary Filipino society. This transformation not only positions the CCP as a beacon of cultural innovation and sustainability but also as a model for cultural institutions navigating the tensions between globalization and local identity.

As the Cultural Center of the Philippines (CCP) Complex navigates the challenges of the twenty-first century, its redevelopment must move beyond mere physical restoration and embrace a holistic transformation that integrates cultural preservation, social inclusivity, and environmental resilience. While the pressures of commercialization, governance inefficiencies, and ecological disruptions threaten its original vision, these challenges also present an opportunity to reimagine the CCP as a living cultural landscape—one that remains deeply rooted in Filipino heritage while engaging with contemporary urban realities.

This study underscores the necessity of a strategic, multidimensional approach that balances tradition with innovation, governance with participation, and conservation with adaptability. Ensuring the CCP's long-term relevance requires embedding inclusive decision-making structures, sustainable design interventions, and dynamic cultural programming into its redevelopment framework. By integrating these principles, the CCP can reclaim its status as the premier national hub for Philippine arts and identity, fostering a cultural ecosystem that is both historically grounded and future oriented. To achieve this vision, the following recommendations outline an adaptive heritage management framework centered on inclusivity, sustainability, and heritage conservation, providing concrete strategies for revitalizing the CCP Complex in a way that remains authentic, accessible, and enduring.

6.3 Recommendations

6.3.1 Toward an Inclusive, Sustainable, and Heritage-Driven Redevelopment of the CCP Complex

The redevelopment of the Cultural Center of the Philippines (CCP) Complex must address the evolving cultural, social, and environmental dynamics that challenge its role as a national cultural landmark. Anchored in the principles of inclusivity, heritage conservation, and sustainability, this study proposes a Dynamic Egalitarianism Framework that repositions the CCP as a living cultural institution—one that balances historical significance with contemporary societal needs. Applying an adaptive heritage management approach (as seen in Fig. 4), this framework is structured around four key elements:

1. **What**: Identifying the tangible and intangible heritage embedded in the CCP Complex.

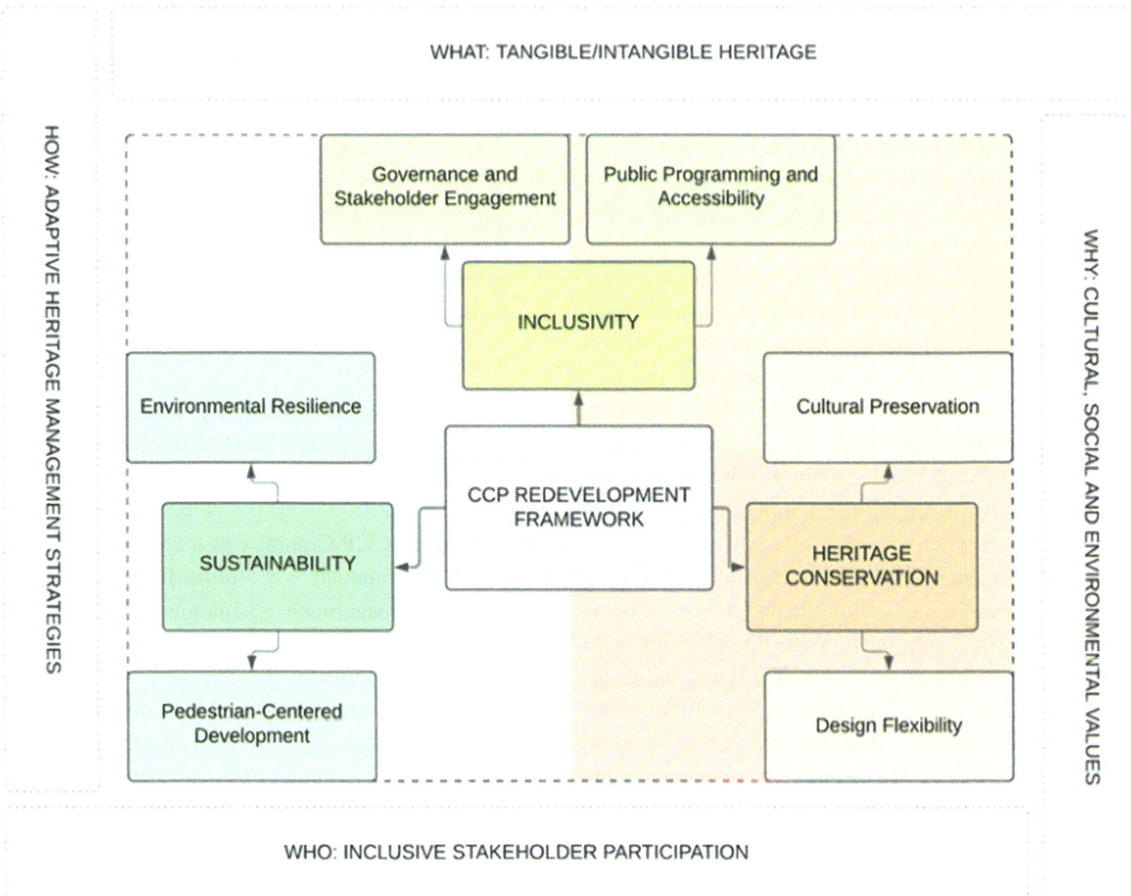

Fig. 4 Framework on promoting inclusivity, sustainability, and heritage conservation in the CCP complex

2. **Why**: Understanding the cultural, social, and environmental values driving redevelopment.
3. **Who**: Promoting inclusive stakeholder participation in decision-making.
4. **How**: Implementing adaptive strategies for sustainable heritage management.

6.4 What: Identifying the Heritage Components of the CCP Complex

The CCP Complex is more than an architectural ensemble—it embodies a tangible and intangible repository of Filipino artistic, historical, and cultural identity. Recognizing these components ensures that redevelopment efforts preserve, enhance, and contextualize their significance in contemporary urban life.

Tangible Heritage includes its iconic structures and spatial landscapes that define its physical identity. Key elements such as the Tanghalang Pambansa, Tanghalang Francisco Balagtas (Folk Arts Theatre), Manila Film Center, and CCP Open Grounds serve as cultural anchors, while the Promenade, Harbor Square, and Manila Bayfront contribute to its urban and ecological identity.

Intangible Heritage encompasses the rituals, performances, and artistic traditions that animate the CCP's spaces. Festivals such as Pasinaya, Cinemalaya, and the Virgin Labfest, alongside the practices of resident companies like Ballet Philippines and Tanghalang Pilipino, form an evolving cultural ecosystem. These cultural assets must be safeguarded and integrated into the CCP's adaptive reuse strategies to maintain its authentic connection to the Filipino identity.

6.5 Why: Cultural, Social, and Environmental Values Driving Redevelopment

The CCP Complex exists at the intersection of cultural heritage, social equity, and environmental responsibility. The redevelopment process must reconcile these three values to ensure its continued relevance in a rapidly changing urban landscape.

1. Cultural values: The CCP's redevelopment must reinforce its role as the premier platform for Philippine arts while adapting to contemporary cultural shifts. Ensuring the continued presence of regional and indigenous artistic expressions, integrating new media and interdisciplinary arts, and balancing global cultural narratives with local identity are critical to strengthening the Filipino artistic voice.
2. Social values: The exclusivity of state-driven monumentalism has distanced the CCP from the broader Filipino public. Prioritizing community engagement, accessibility, and participatory governance ensures that the CCP remains a public cultural space rather than an isolated enclave. Addressing barriers to participation, such as physical accessibility, socioeconomic limitations, and urban connectivity, is crucial for democratizing access to the arts.
3. Environmental values: The CCP's coastal location and connection to Manila Bay's ecological system necessitate a climate-responsive approach. With threats from reclamation projects, urban heat islands, and rising sea levels, integrating sustainability measures such as green infrastructure, water-sensitive design, and passive cooling strategies will enhance the complex's resilience while maintaining its environmental integrity.

6.6 Who: Inclusive Stakeholder Participation in Decision-Making

A heritage site's sustainability is contingent upon inclusive governance structures that incorporate diverse voices in shaping its future. The CCP's current governance model, which has historically been centralized and state-driven, must transition toward a multistakeholder participatory approach. The following key actors must be actively involved in the redevelopment process:

1. Government agencies: The CCP Board of Trustees, in collaboration with the National Commission for Culture and the Arts (NCCA), Department of Tourism (DOT), and Department of Environment and Natural Resources (DENR), must establish cohesive cultural policies that balance heritage conservation with urban development.
2. Cultural institutions and practitioners: Resident companies, artists, cultural workers, and scholars must be included in shaping programming decisions, spatial planning, and adaptive reuse projects.
3. Community and grassroots organizations: Local communities, independent art collectives, and heritage advocacy groups must have a voice in cultural policymaking, ensuring that the CCP remains reflective of broader Filipino cultural narratives rather than elite-driven agendas.
4. Private sector and global networks: Strategic partnerships with private investors, cultural sponsors, and international networks (such as UNESCO's Creative Cities Network) can diversify funding sources while reinforcing cultural diplomacy and global engagement.
5. General public: The integration of public consultations, citizen panels, and participatory design workshops will foster a sense of collective ownership, ensuring that redevelopment efforts align with the needs of its broader audience.

6.7 How: Implementing Adaptive Strategies for Sustainable Heritage Management

To sustain the CCP Complex as a cultural institution that is historically grounded yet future-forward, the following adaptive management strategies must guide its redevelopment:

1. Spatial adaptability and flexible programming: The CCP's physical spaces must be reconfigured for multiuse functions, allowing for hybrid cultural formats that accommodate contemporary art forms, digital exhibitions, and immersive experiences without compromising its historical integrity.
2. Sustainability-driven design: Architectural interventions must integrate climate-resilient infrastructure, including elevated walkways, renewable energy solutions, and biophilic design elements that mitigate environmental vulnerabilities.
3. Cultural and ecological reconnection: The CCP's coastal identity must be preserved through sustainable waterfront redevelopment, ensuring that public access to Manila Bay is restored while implementing nature-based solutions such as rain gardens, permeable pavements, and urban wetlands.
4. Participatory governance mechanisms: A Multistakeholder Cultural Council should be institutionalized to oversee the decision-making processes, ensuring that cultural workers, urban planners, and the public actively shape the CCP's future. Regular public engagement forums, citizen-led urban audits, and adaptive heritage policies must be embedded into governance structures.
5. Integrating digital and physical engagement: The CCP must embrace a hybrid cultural ecosystem where on-site performances, festivals, and exhibitions are complemented by virtual programming, expanding its audience reach while promoting global cultural exchange.

The Cultural Center of the Philippines Complex is not merely a physical landmark but a dynamic cultural landscape that must reconcile its monumental legacy with the evolving demands of contemporary society. Its redevelopment necessitates a strategic integration of inclusivity, sustainability, and heritage conservation within its governance and spatial frameworks. By transcending its historical constraints, the CCP can redefine itself as a resilient, participatory, and future-oriented cultural institution.

This underscores the importance of adaptive heritage management in securing the CCP's long-term relevance. Through participatory governance, climate-responsive urban strategies, and a culturally inclusive programming model, the CCP can reinforce its role as the epicenter of Philippine artistic and cultural identity. By embedding these principles into its redevelopment, the CCP Complex can serve as a model for cultural institutions navigating the intersection of heritage preservation and contemporary sociocultural transformations.

References

Amirtahmasebi R, Vuova Z, Fox EO (2020) The new urban agenda. United Nations Human Settlements Programme (UN-Habitat)

Appiah KA (2006) Cosmopolitanism: ethics in a world of strangers. W.W. Norton & Co, New York

Babb F (2012) Theorizing gender, race, and cultural tourism in Latin America. https://www.semanticscholar.org/paper/Theorizing-Gender%2C-Race%2C-and-%20Cultural-Tourism-in-Babb/dba5908f01a62a7a5ba17cfbd8dc3be23168a903

Brabec E, Goetcheus C (2014) The interrelationships of land, culture and heritage: the Gullah Geechee communities of the Southeastern United States. In: Conserving cultural landscapes. Routledge, Abingdon, pp 127–146. https://doi.org/10.4324/9781315813226-13

Callon M (1984) Some elements of a sociology of translation: domestication of the scallops and the fishermen of St Brieuc Bay. Sociol Rev 32(1 Suppl):196–233. https://doi.org/10.1111/j.1467-954x.1984.tb00113.x

Caruncho E (2018) New CCP chair Margie Moran Floirendo: the truth is in the arts. Lifestyle.Inquirer

During R (2011) Cultural heritage and identity politics (issue ISBN (Print) 9789461730756). https://edepot.wur.nl/181393

Espiritu T (2015) The Marcos romance and the cultural center of the Philippines: the melodrama of a therapeutic cultural policy. J Narrat Theory 45(1):141–162. https://doi.org/10.1353/jnt.2015.0003

Flores PD (2017) "Total community response": performing the avant-garde as a democratic gesture in Manila. Southeast Now Dir Contemp Modern Art Asia 1(1):13–38. https://doi.org/10.1353/sen.2017.0001

Gabriel C, Veronis L (2024) Cosmopolitan paradox? The labour market experiences of newcomer skilled workers. Cosmop Civ Soc 15(3):73–90. https://doi.org/10.5130/ccs.v15.i3.8777

Geertz C (1973) Thick description: towards an interpretive theory of culture. https://doi.org/10.4324/9780203931950-11.

Gehl J (1980) Life between buildings: using public space. https://ci.nii.ac.jp/ncid/BA8050894X

Habitat U (2020) Report of the 10th session of the WUF TEMPLATE TO REPORT ON THE OUTCOME DOCUMENT THE ABU DHABI DECLARED ACTIONS ASSEMBLIES AND ROUNDTABLES

Herrera MVT (2001) Creating a culture by governance: issues in managing culture. https://doaj.org/article/69a3e41886024c8aae80cb4a1955e212

Horkheimer M, Adorno TW (2017) The culture industry: enlightenment as mass deception. Routledge, Milton Park. https://doi.org/10.4324/9781315251196-23

Latour, B. (2005). Reassembling the social: An Introduction to Actor-Network-Theory https://doi.org/10.1093/oso/9780199256044.001.0001

Law EBJ (1986) Power, action, and belief: a new sociology of knowledge? Routledge, Milton Park. http://ci.nii.ac.jp/ncid/BA03757746

Levin R, Jencks CA (1978) The language of post-modern architecture. J Aesthet Art Crit 37(2):239. https://doi.org/10.2307/429857

Lico G (2003) Edifice complex: power, myth and Marcos state architecture. Ateneo De Manila University Press. http://ci.nii.ac.jp/ncid/BA68405289

Mendoza RU, Olfindo R, Poco LC (2017) Governance and institutions in the Philippines. SSRN Electron J. https://doi.org/10.2139/ssrn.2873942.

Norberg-Schulz C (1980) Genius loci: towards a phenomenology of architecture. https://ci.nii.ac.jp/ncid/BA00446954

Ogura N, Yap DLT, Tanoue K (2002) Modern architecture in the Philippines and the quest for Filipino Style. J Asian Architect Build Eng 1(2):233–238. https://doi.org/10.3130/jaabe.1.2_233

Olwig KF (1999) The burden of heritage: claiming a place for a West Indian culture. Am Ethnol 26(2):370–388. https://doi.org/10.1525/ae.1999.26.2.370

Saldaen FV, Celestino JC, Aspiras ED, Collado CP, Bautista RG (2021) Public management challenges in a local government unit in the Philippines: implications to the educational sector. Am J Educ Res 9(4):218–221. https://doi.org/10.12691/education-9-4-10

Smith CJ, Relph E (1978) Place and placelessness. Geogr Rev 68(1):116. https://doi.org/10.2307/213523

Thompson MR (2008) Philippine politics and governance. Philipp Polit Sci J 29(52):117–124. https://doi.org/10.1080/01154451.2008.9723510

Thwaites K, Simkins I (2006) Experiential landscape. https://doi.org/10.4324/9780203462096.

Venturi R, Scott Brown D, Izenour S (1977) Learning from Las Vegas: the forgotten symbolism of architectural form. MIT Press, Cambridge

The Paradoxical Heterogeneity of Al-Abdali's Imageability and the Dilemmatic Identity of Amman

Omar M. Amireh, Dania H. Al-Harasis, Zaineh S. Abu Omar, and Saja A. Zeadeh

Abstract

Recently, to catch up with the vast and rapid urban neoliberal developments that became widespread across the Middle East, Amman's authorities and developers initiated one of the most ambitious urban restructuring and spatial transformations of the city's image and urban topography. The Al-Abdali regeneration project, also known as the New Downtown Business District, is an integral aspect of this initiative. It is conceived as an avant-garde high-end project that would reorient Amman's socioeconomic compass and redirect its environmental spatial coordinates. In their effort to achieve and restore the lost livability of the area and to maximize its multifaceted purposes, the project designers focused on the complementarity, integration, compatibility, and vitality between the Abdali project and its direct surroundings and the old downtown. To assess if these goals have been attained in practice, a questionnaire survey and a comparative analysis were conducted as part of the present study. The aim was to clarify the role of visual attributes in transforming the city's identity and imageability. In this context, a distinction is made between the self or inner identity and the projected identity, allowing users' perceptions of Amman's old and new downtowns to be evaluated.

Keywords

Identity · Imageability · Urban

1 Introduction

As the largest redevelopment initiative in Amman, the Al-Abdali urban development project has had a profound impact on many aspects of urban life, yet its influence on the city's image and identity remains relatively underexplored. This foundational effort to create a new downtown includes multiple real estate subprojects and a central shopping area. To thoroughly investigate this new urban element, as part of this study, the project is decomposed into its subelements—namely blocks, buildings, open spaces, and streets. This strategy facilitates identification of the interconnections among these components as well as assessment of their roles in shaping the urban landscape of the new development. In evaluating the newly generated urban context, focus is given to spatial arrangements, visual appeal, usage, activities, and user experiences. This assessment takes into account comparisons with the older urban context, including both the surrounding areas and the original downtown, which holds historical significance as the heart of Amman's expansion.

To supplement these analyses, a survey was conducted, allowing users' cognition, perception, and behavioral, mental, and cognitive maps pertaining to the newly generated spatial and architectural identity of the old and the new downtowns to be evaluated. The ultimate aim of this study is to create a comprehensive urban formula that correlates the project with the surroundings, which is first applied to Amman's original identity and is later used to assess the project's capacity to change the urban scene of the whole city by linking its imageability with that of the old downtown.

O. M. Amireh (✉)
Prepared Under an Academic Sabbatical Leave, University of Jordan, Amman, Jordan
e-mail: amireh@ju.edu.jo

D. H. Al-Harasis
Welsh School of Architecture, Cardiff University, Cardiff, UK

Faculty of Engineering, Al-Balqa Applied University, Al-Salt, Jordan
e-mail: Al HarasisD@cardiff.ac.uk; dania.harasis@bau.edu.jo

Z. S. Abu Omar
Department of Architecture, School of Engineering, University of Jordan, Amman, Jordan

S. A. Zeadeh
Faculty of Engineering, Mutah University, Karak, Jordan

F. Alberti et al. (eds.), *Towards Sustainable Urban Regeneration: Utilizing Environmental, Cultural, and Social Assets*, Advances in Science, Technology & Innovation, https://doi.org/10.1007/978-3-032-02091-8_27

2 Background

2.1 The Dilemmatic Identity of Amman

Jordanian society is multicultural and diverse in its structure, exhibiting a varied identity (Zibin et al. 2023). Likewise, as the country's capital, Amman has a multicultural population, which adds to its architectural landscape's diversity (Atiyat 2022). Three historical processes have influenced Amman's identity: Jordan's conception of nationalism, the impact of the Palestinian crisis on Jordan, and the role of tribalism in forming national bonds (Smith 2019). Amman's rich legacy as an intersection of civilizations is reflected in its cultural diversity, which makes it challenging to define a cohesive identity in the face of globalization and cross-cultural exchange. Social dynamics exacerbate social integration challenges, which are also impacted by demographic changes and migration trends (Sharaf et al. 2023).

The first comprehensive plan of Amman, established in 1956, marked the inception of the Al-Abdali district. However, due to the plans published in 1978 and 1988, it began to take the shape it has today, with multiple skyscrapers and other developments in keeping with the metropolitan comprehensive plan for Greater Amman (Goussous and Qashmar 2019). In 2004, a proposal to designate a portion of Al-Abdali district for the development of mid-rise and high-rise buildings was made (Al Shawabkeh 2018; Zalloom 2018). As this was the first urban regeneration project in the country, some problems were inevitable, but the main concern pertained to the fact that it was located in close proximity to the historic old downtown. The project critics argued that the low-rise, clustered, local stone-covered buildings with stairs that connect Amman's horizontal landscape with its mountainous nature would lose their superiority and singularity due to the infiltration of unfamiliar urban structures.

There is also a concern that the growing emphasis on tall and modern mixed-use buildings as well as identical four-story residential apartments would jeopardize the city's ability to maintain its unique cultural, historical, and social identity. In sum, the cityscapes of Amman now comprise disparate and disconnected scenes.

2.2 The Projected and the Self or Inner Identity and the Urban Structure

Lynch (1960) introduced the concept of identity as a measurement of the qualities and uniqueness of the physical environment that, as he argued, could be traced in people's mental images of that environment. Lynch's concept was reidentified by Relph as an indication of both unique and common features of a particular environment (Järnefelt 1999). Accordingly, a greater understanding of Al-Abdali as a so-called "new downtown" and the old downtown of Amman can be gained by identifying their common as well as unique features. This is achieved by examining the buildings at both sites with respect to their self or inner identity, but also by making cross-site comparisons to arrive at their projected identity. This is highly relevant, given that the main aim of the Al-Abdali project is to give Amman's architectural identity a new value (Khirfan and Momani 2013). As noted by Qudah (2023), this strategy aligns with similar attempts made in other cities, whereby new architecture with modern shapes and positions would compete with the fast-growing or changing concepts of modern identities. Dahabreh (2020) further argues that, as Amman is going through a series of political, cultural, and economic transformations, having projects such as Al-Abdali promotes an expansive and multifaceted sense of identity.

According to Atiyat (2022), identity is defined as "the recognition of urban elements as distinct entities," structure reflects "the relationship between urban elements and the observer," and meaning is conveyed by "the relationship between urban elements and the emotional value for the observer." In other words, architecture represents the essential and the formal tool for reflecting the urban and city identity. Thus, it is not surprising that Melnik (2019) criticized the architectural style adopted in recent urban projects in the region, stating: "In exploring modern buildings in the Arab world, we can now see a lack of clear identity and blind adherence to Western methods, far from Arab civilization and society."

It is also noteworthy that, in his study of the neoliberal urban transformations, Daher (2013) described the Al-Abdali project as "the Vertex, the Heights, and the Lofts, as the occupants and residents will live a luxurious life and yearn for their own type of utopia and a privileged position that is, in most cases quite literally, above the rest of the city." Prompted by these concerns, Alhusban and Alhusban (2020) worked on a hybridization process that can be used to relocate (redefine) the city's identity. Their aim was to identify the most influential factors that shape the identity of Amman city from various perspectives. Hourani et al.'s (2014) earlier study was guided by the question: "Is the Abdali development not the epitome of market-driven urbanism, in which imperatives of local, regional, and global markets recenter cities upon such spaces of spectacle?" Their findings indicate that market-driven urbanism, or market-identity urbanism, plays a major role in marketing the emerging new identities. As Zalloom (2014) added social sustainability to that list, it is considered in this study to fully elucidate the impacts of urban regeneration resulting from the Al-Abdali project on the physical and sociocultural settings of Amman and its social sustainability. As evident from this brief review, although not all researchers agree on what effects the Al-Abdali new downtown will have on the urban identity of Amman, all agree that changes in the city's imageability and identity are inevitable.

2.3 The Paradoxical Heterogeneity of Al-Abdali and the Old Downtown Imageability

The Al-Abdali regeneration project, also known as the new downtown of Amman (Fig. 1), is the largest mixed-use development project to be constructed in Amman, with 384,000 square meters of land and a total built-up area of over two million square meters (Abdali 2004). It promotes a vision of being Amman's premium central business, tourism, medical, retail, residential, and entertainment district. The Al-Abdali megaproject was meant to create a modern, newly established downtown, dominating the cityscape (Zalloom 2014). As can be seen from Fig. 2, compared to its adjacent surroundings, Al-Abdali's urban morphology exemplifies both physical and social transformation. The physical transformation is achieved through geometric plot and street patterns, signifying a departure from the curved streets that smoothly follow the topography lines, as well as the materials used in building construction with a shift from organic patterns to those imposed, designed, and systematic in nature (Zalloom 2014).

These stark differences are made more apparent by the juxtaposition of Al-Abdali's new downtown with the old downtown. People in Amman are also affected by the recently launched project and are assumed to create distinct mental maps for the two spaces. Depending on how each space's components and features are perceived, people respond to them differently.

3 Research Design and Method

As the primary aim of this study is to elucidate users' reflections toward the old downtown of Amman and Al-Abdali as the new downtown, a survey was conducted focusing on three main aspects: (1) space, (2) uses and users, and (3) Amman's imageability. The questionnaire used for this purpose was intended to provide an in-depth assessment of the users' identity identification regarding the old downtown and Al-Abdali's new downtown. Therefore, its design was guided by the model shown in Fig. 3, corresponding to the inner self-identity of both downtowns and the projected identity of each on the other, as defined in Sect. 2.2.

As can be seen from Fig. 3, the proposed model is configured into a threefold vertical form, each corresponding to a set of questions pertaining to the aspects affecting each individual's perceptions of inner and projected identity. The horizontal dimension reflects the users' visual and sensual perception of the urban physical content (the space, the uses, and the users) and assesses their cognitive understanding of the projected contextual meanings associated with urban form and activities. Therefore, the data obtained through the survey can be used to convert subjective concepts—the uses (activities), the space or place (imageability), and the users (respondents)—into subjective (relative grading) and objective (absolute scale) factors.

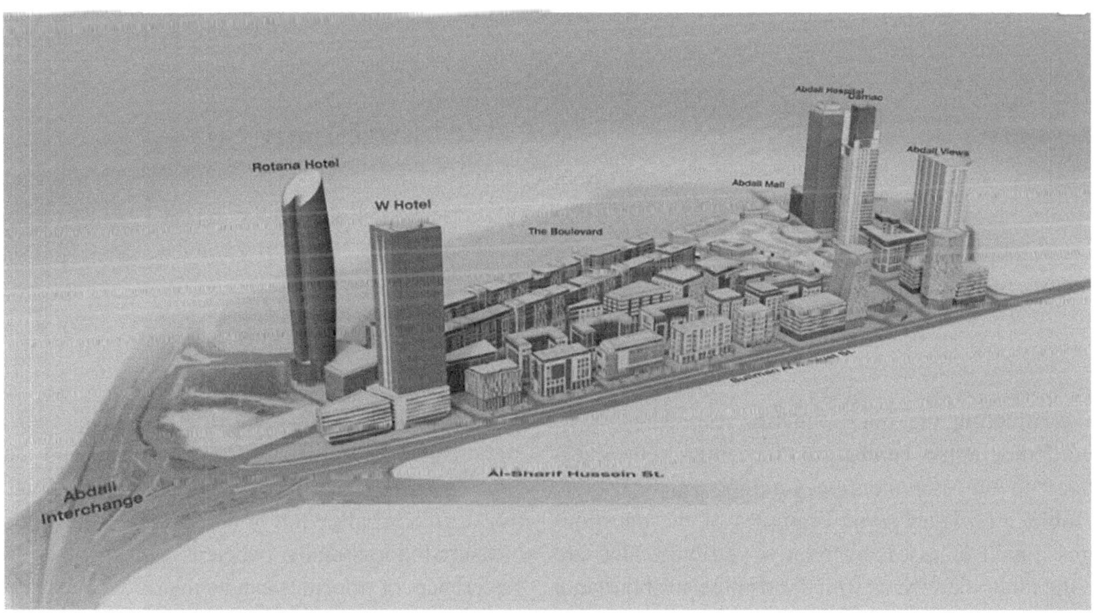

Fig. 1 A 3D model of the Al-Abdali regeneration project (as designed). https://www.theabdali.com/work (accessed 4 March 2024)

Fig. 2 Al-Abdali regeneration project with its surroundings (as projected). https://i257.photobucket.com/albums/hh216/yaz-maj/ABDALI-Sunrise.jpg (accessed 12 February 2024)

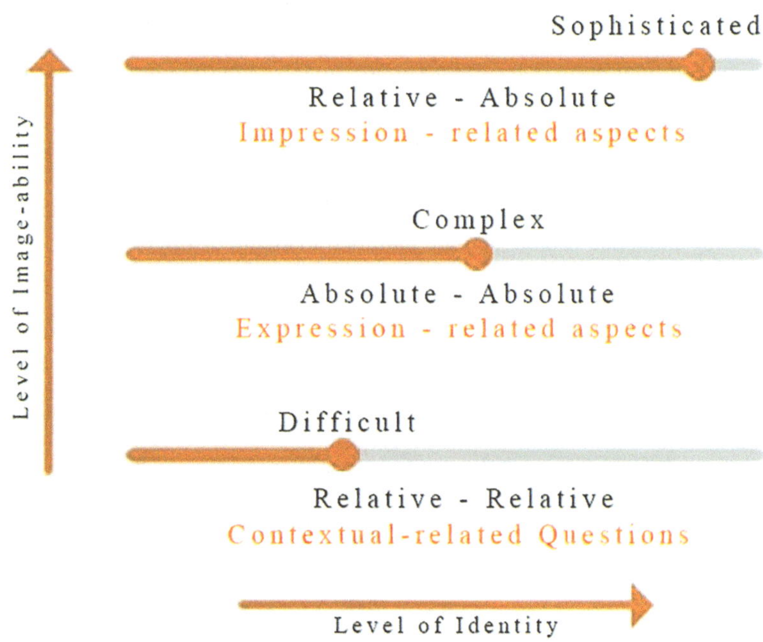

Fig. 3 Model adopted in Study 1

When completing the questionnaire, respondents provided their demographic data before proceeding to the questions related to the aforementioned relative and absolute aspects. Table 1 provides some examples of the questionnaire items aimed at assessing relative attributes that are related to the inner identity, as well as absolute attributes that reflect the projected identity.

As questions with contextual meanings and expressive significance are difficult to formulate and answer, the approach shown in Fig. 4 was adopted. Accordingly, the relative degree (of affection or friendliness, for example) can be measured horizontally, whereas the absolute importance, preference, or priorities can be measured vertically.

When these responses are plotted diagonally, it is possible to establish correlations between relative aspects and abso-

Table 1 Sample questionnaire items

Sample question assessing a relative attribute					
Describing the different characteristics of Al-Abdali's new downtown of Amman, which of the characteristics takes the highest value (5) and which takes the lowest value (1) in determining the identity of the place, knowing that this is a relative comparison, not an absolute one:					
	1 (the least)	2	3	4	5 (the most)
Cultural					
Heritage					
Historical					
Architectural					
Urban					
Sample question assessing an absolute attribute					
Evaluate each quality based on its relationship with the built environment of the old downtown of Amman, knowing that the highest value is 5 and the lowest is 1					
	1 (the least)	2	3	4	5 (the most)
Skyline					
Heights					
Form					
Transparency (solid and void)					
Scale (humanization)					

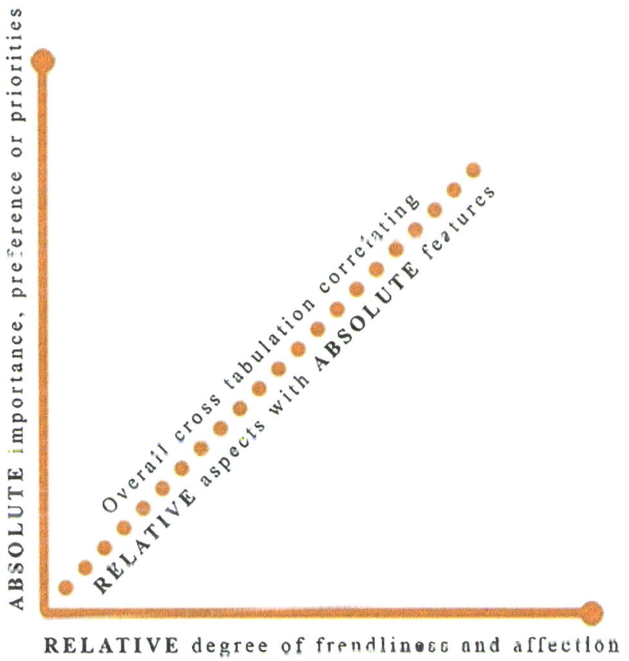

Fig. 4 Model adopted in Study 2

lute features. This comprehensive method with a twofold structure thus incorporates qualities and values with features and characteristics, tangible and intangible with content and context, and perception and cognition with expressions and impressions, allowing the constant aspects of identity to be identified. However, several questions target specific features, characteristics, elements, landscape, topography, materials, colors, etc., as the obtained responses are expected to reflect the variable aspects of identity. Finally, multifold questions with complex modules targeting mobility, wayfinding, compatibility, permeability, legibility, belonging, interaction, participation, and other collective conducts or activities are also included with the goal of identifying the conceptual aspects of identity.

The questionnaire comprising the items outlined above was subjected to a pilot study, whereby experts from the architectural field were asked to provide feedback on its comprehension, relevance, and other aspects that would affect user experience. Once the feedback was incorporated, the questionnaire was administered to 98 individuals, who represented a wide range of ages, genders, and backgrounds but were users of both the old downtown of Amman and the Al-Abdali's new downtown. Their responses were subsequently analyzed, compared, and discussed to form an understanding of the users' reflections on the transformation in Amman's identity.

4 Results and Discussion

Figure 5 depicts the paradoxical elements that define Al-Abdali's imageability and its impacts on Amman's identity. It indicates that, when both older and younger generations come to understand the physical characteristics of both downtowns, they foster a shared understanding of the city's identity. In contrast, when the older generation relies on their memories of the space, while the younger generation engages with it through their current cognition and perception, Amman's identity becomes paradoxically fractured. A third scenario occurs when both generations lack familiarity with the physical features of the downtown areas. In this case, the older generation draws on the physical attributes, while the younger generation focuses on the activities taking place, leading to a dilemma surrounding the city's identity.

4.1 Activities and Affordability in Terms of Purpose and Motive

When asked to rate the purpose of presence of various elements, most respondents opted for leisure and entertainment (which indicates the predefined function of space and correlated activities). While they also highly rated a wide choice of activities within Al-Abdali, most study participants found them unaffordable. This points to a paradox, as the services that are highly valued cannot actually be used by most Ammanis. As a result, there is a clear segregation in categories of users, which compromises the prospect of running a

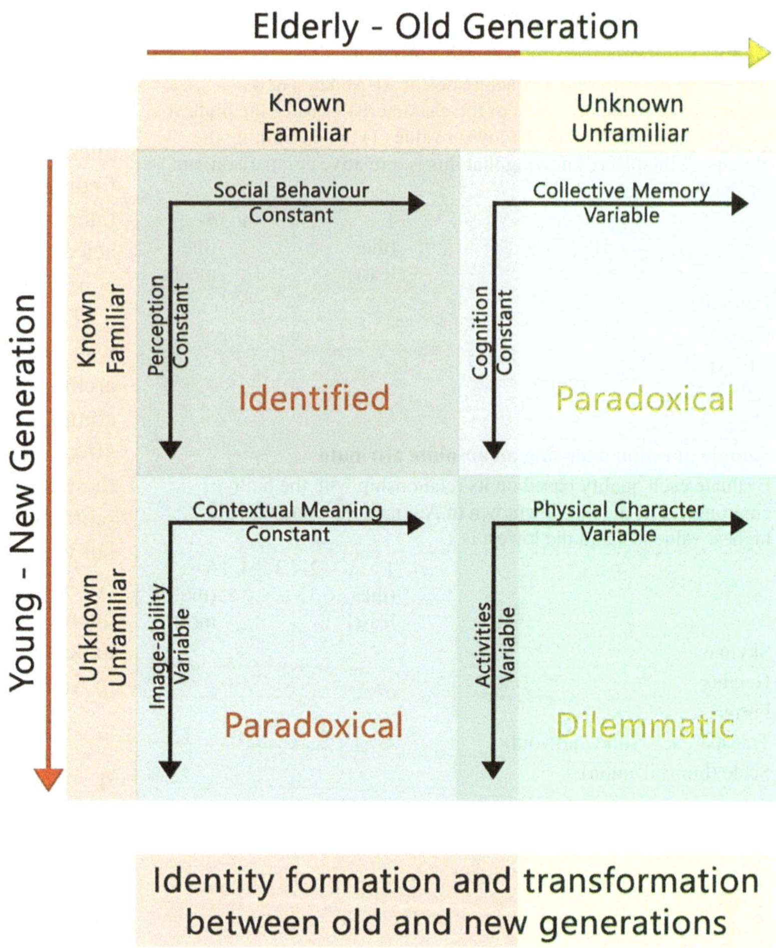

Fig. 5 Paradoxical elements that define Al-Abdali's imageability

business-oriented urban space due to the increased number of users with low contribution to the economic flow.

4.2 Activities and Affordability in Terms of the Collective Identity of Space

When the survey responses were analyzed to determine the perceived identity of space, it became apparent that most individuals associate it with its physical components. In particular, given the Al-Abdali's distinctive modern/futuristic character, emphasis is placed on indoor and outdoor public activities. However, most users prefer a connection between the outdoor and indoor spaces—both physically and visually.

As several questions pertained to Al-Nabulsi Street which forms the boundary of the Al-Abdali project, the survey included an image of the street to help the respondents recall their feelings and previous experiences more accurately. However, the participants showed a strong connection to the old downtown area and used it as a reference to help answer these questions. This preference indicates their strong connection to the identity of Amman, as evident in their appreciation for the details of the old downtown. When asked to compare the two sides of the street, they tended to use "comprehensive" to describe the range of activities in both cases, while opting for "contradicted" in reference to affordability.

This paradox can also be seen in the activities on offer (guided and spontaneous), which impacted the social categorization of space. In this case, Al-Nabulsi Street acts as a physical separation line for two contradicted typologies, yet it is still seen as comprehensive due to the type and quality of services provided on each side. This juxtaposition may suggest that the two sides are functionally working together, while still conveying a sense of contradiction.

4.3 Robustness Despite the Lack of Appropriateness

When rating the space's intangible features, participants prioritized robustness of values that contribute to the formation of its identity, while giving appropriateness the least priority.

Although these two values are usually perceived correlatively in the same context, this was not the case for Al-Abdali, as the increase in the appropriateness of space did not allow diverting activities to occur, causing more robustness. This surprising finding reflects the first paraxial aspect that prioritizes activities, which, regardless of affordability, result in greater robustness.

4.4 Space–Activity Relationships Impact Identity

Based on the survey responses, cultural quality does not seem to contribute to the commonality between the old and the new downtown. This is not surprising, given that Al-Abdali is criticized for its modernity, while the old downtown is seen as essential for preserving the cultural identity of Amman. This finding can be attributed to the fact that the old identity of Al-Abdali before neoliberal interventions had a cultural value, which was recalled from the participants' memory, while the cultural identity of the old downtown could be experienced in real time. As the use of space in Al-Abdali is activity-driven, it is evident that users have a memory of the identity of the broader context. As the context is larger than the enclosed space, it is predicted that the exclusion and the contrast will be felt, and the comparison will continuously be used to identify and feel the space. In this sense, past and present dilemmas occurring in users' minds will impact their perception and imageability of space.

5 Conclusion

5.1 Summary of Key Findings

In conclusion, we adopted a user-centric perspective to investigate whether Amman's society embraces and promotes transformative initiatives. We also examined whether these changes have reshaped the mental maps of Ammanis, substituting traditional conservative landmarks with neoliberal narratives.

Al-Abdali's new design has not only transformed the physical landscape but has also sparked a significant redefinition of Amman's identity. The survey conducted as part of this study yielded some unexpected findings, revealing how users perceive Al-Abdali's new downtown and its relationship to the previous downtown. The obtained results also highlighted how the paradoxical elements that characterize Al-Abdali's imageability influence Amman's identity, affecting its formation and transformation between older and newer generations. Participants indicated a continued connection to the old downtown, suggesting that their identity as Ammanis is still rooted in that area.

This research contributes to the current body of knowledge on the impact of urban projects. Its focus was on Al-Abdali's new design, which has been linked to feelings of segregation and social isolation, thus influencing the identity of the surrounding context. The main objective was to clarify the role of visual properties in defining, changing, and transforming the city's identity.

Interestingly, survey respondents, particularly those over 35, continue to reference the old downtown while navigating the new downtown. This observation, often centered on the physical characteristics of the spaces, underscores the contrasting nature of the two areas in terms of texture, materials, style, and the emotions they elicit. This contrast can evoke a sense of nostalgia and curiosity among users.

The study findings also indicate that the physical environment significantly impacts user experience. The visibility of the skyline and high-rise structures creates a unique sensory experience. As users find themselves more enclosed within such an environment, they often feel a sense of isolation from other parts of the city. This is particularly evident in Al-Abdali, where the height of some newly erected buildings created a distinct identity that starkly contrasts with old Amman.

In the old downtown area, physical structures still play a dominant role in defining spaces and shaping user experiences, whereas in Al-Abdali users tend to be more activity-driven. While the study participants appreciated the simplicity of the features in the new downtown, there was a need for higher sophistication in detail to convey a sense of value regarding those spaces. Such features were crucial in defining the spatial character, accessibility, imageability, and memory of the space. Nevertheless, all these aspects were rated more favorably in relation to the old downtown and its holistic identity.

Many study participants expressed feelings of segregation and social isolation in Al-Abdali, which significantly affected their experiences and impressions of the new spaces. This stance has influenced emerging trends in activities users partake within the area as well as their interactions with the space and one another. A prominent finding from this study is that users actively participate in shaping the image of their surroundings by observing and engaging with physical structures designed for various functions, choosing what to embrace or exclude.

5.2 Implications for the Field

Urban identity conceived in literature and arts differs from that comprehended in cultures and societies (Alazawi et al. 2015). These definitions also differ among users, depending on gender, profession, and social status. However, it is also important to note that some of the differences reported here may arise due to the difficulty in comprehending certain questions. To overcome these shortcomings, in future studies of this type, the questionnaire should also be piloted with members of the general public, not just experts in the field. Nonetheless, it can be concluded that physical features or imageability is less influential on people's mental map and space conception than the activities in which they may potentially engage. Furthermore, users rarely notice specific elements or detailed features, as they tend to concentrate on generic forms and overall spatial arrangements. In the end, retrieved memories of personal experiences play a significant role in forming people's conception of urban identity, suggesting that designers and planners should incorporate user perspectives when designing urban spaces and places. Focus should be given to individual and joint activities within these spaces, aiming to highlight culture and correlating activity, space, and the feeling of belonging (Amireh 2013).

5.3 Suggestions and Recommendations for Future Research

Further research on the imageability of Amman and the heterogeneity of Al-Abdali is certainly warranted to corroborate the findings obtained here. Moreover, once Phase 4 of the project is complete, it would be beneficial to replicate this investigation to determine if there are any changes in user perspectives.

References

Abdali (2004) Project factsheet: the new downtown of Amman. https://www.theabdali.com/. Accessed 15 Mar 2024

Al Shawabkeh RK (2018) New urban regeneration for the city of Amman using sustainable urban design principles: Al-Abdali area as a case study. Int J Appl Eng Res 13(19):14345–14358. https://doi.org/10.37622/ijaer/13.19.2018.14345-14358

Alazawi S, Amireh O, Abdullah M (2015) Hyper-hybrid restructuring of Amman spatial genealogy. Mitteilungen Klosterneuburg 65(8):166–176

Alhusban AA, Alhusban SA (2020) Re-locating the identity of Amman's city through the hybridization process. J Place Manag Dev 14(2):81–113. https://doi.org/10.1108/jpmd-07-2019-0066

Amireh O (2013) Designing architectural interactive design: users' experiences versus functions. Int J Archit Spat Environ Des 6:137–152

Atiyat D (2022) The image of Amman City: structure, identity, and meaning. Int J Sustain Dev Plan 17(6):1931–1935. https://doi.org/10.18280/ijsdp.170628

Dahabreh SM (2020) Trends in the contemporary architecture of Amman. Int J Eng Res Technol 13(7):1658–1670. https://doi.org/10.37624/ijert/13.7.2020.1658-1670

Daher RF (2013) Neoliberal urban transformations in the Arab city. Urban Environ 13(7):99–115. https://doi.org/10.7202/1027729ar

Goussous J, Qashmar D (2019) The dialectic dimensions of architectural identity in heritage conservation, the case of Amman City. In: Proceedings of the international conference on advanced research in social sciences, pp 36–46

Hourani NB, Hourani NB, Kanna A (2014) Neoliberal urbanism and the Arab uprisings: a view from Amman. J Urban Affairs 36(Suppl 2):650–662. https://doi.org/10.1111/juaf.12136

Järnefelt I (1999) Identity—identification. Nordic J Archit Res. https://core.ac.uk/download/pdf/235961203.pdf

Khirfan L, Momani B (2013) (Re)branding Amman: a "lived" city's values, image and identity. Place Brand Public Dipl 9(1):49–65. https://doi.org/10.1057/pb.2013.1

Lynch K (1960) The image of the city. MIT Press, Cambridge

Melnik V (2019) Urban identity of Amman (a dialogue between tradition and modernity). J Eng Archit 7(1):2334–2994. https://doi.org/10.15640/jea.v7n1a8

Qudah N (2023) 7 recent buildings redefining the architectural identity of Amman, Jordan. https://architizer.com/blog/inspiration/collections/7-public-buildings-redefining-the-architectural-identity-of-amman-jordan/

Sharaf F, Çinar H, Irmeili G (2023) Quality evaluation of public pedestrian spaces: the case of Abdali Development in Amman City, Jordan. Civ Eng Archit 11(5):2321–2335. https://doi.org/10.13189/cea.2023.110506

Smith H (2019) Ammani youth—paradoxical citizens on the margins [PhD dissertation, University of South Carolina – Scholar Commons]. https://scholarcommons.sc.edu/etd/5159/

Zalloom B (2014) Assessing the social sustainability of urban landscape. Case study of Abdali regeneration (the new downtown of Amman). Transformation of the urban character of Arab Cities since the late last century. https://www.researchgate.net/publication/309242305_Assessing_the_Social_Sustainability_of_Urban_Landscape_Case_study_of_Abdali_regeneration_The_new_downtown_of_Amman

Zalloom B (2018) Tracing the cityscape transformation under capitalism: the case study of Amman. J Urban Regen Renew 11(4):1–18

Zibin A, Jarrah M, Qaddoumi R (2023) Identity salience, change and stability in Amman identity salience, change and stability in Amman. Kervan Int J Afro-Asiatic Stud 27:319–351

Spatial Identities of Sarajevo's Residential Neighborhoods

Alma Hudović Kljuno and Leila Krivošić Dizdarević

Abstract

Since the establishment of the city in the in the fifteenth century, Sarajevo has been at the crossroads of many different civilizations, cultures, and religions. The unofficial name "European Jerusalem" is well deserved due to its rich and diverse historical heritage and heterogeneous confessionalism. The architectural and urban categorization of Sarajevo's neighborhoods is a reflection of diverse historical and cultural influences that left their mark on the very fabric of the city and have caused the creation of several spatial identities in the development of the city itself.

This study presents a chronological cross-section of the development of the city demonstrating the creation of its urban identities, from the Ottoman period in the fifteenth century to the present day, including the period of very turbulent early 1990s in which the country went through the war-induced destruction of the city. This article examines historical and contemporary spatial forms, typologies, and identities using examples from several case studies from different periods. The comparison should point out their advantages and disadvantages in terms of spatial organization and the quality of living. The graphical and textual supplements offer a concise overview explaining each of these very different urban matrices, which together create the architectural and urban heritage of the city of Sarajevo.

Keywords

Sarajevo · Heritage · Spatial forms · Urban identities · Residential neighborhoods

1 Introduction

The identity of the city is made up of a unique, inherited collection of assets, history, traits, and culture that differentiates it internally and externally, and as such, it is a complex system of functional values. The concept of city identity is studied from different points of view, which essentially defines its multidisciplinary character. Although identity is not easy to define, precisely because of its complexity, the concept itself enters the sphere of interest of various fields and is analyzed from the aspect of spatial phenomenon, architecture, urbanism, sociology, economy, and politics. Ultimately, as Cheshmehzangi (2015) states, urban identity can be defined as a socially constructed relationship between people and their environment, the environment and its elements, and elements with other elements—in other words, a set of multiple and mutual relationships between context and content.

The continuous transformations of the city, the way in which it changes and transforms, lead to the questioning of the concept of the identity of the city in the modern age. Dragićević Šešić (2009) argues in her text "Culture in the function of city development" about city identity as the starting point of the development strategy. We can simplify and say that city development has its roots in the city's identity and as such is closely related to the historical framework and other multidisciplinary processes that influence its development. The accelerated global development of cities has led to major changes in the way cities and urban spaces are created. While cities in Europe and the world are changing, creating betterment for their residents, the cities of the Balkan region and specifically Bosnia and Herzegovina, with their changes, create new identities with the loss of previous qualities.

A. Hudović Kljuno (✉) · L. Krivošić Dizdarević
Faculty of Engineering and Natural Sciences, Architecture Program, International University of Sarajevo, Sarajevo, Bosnia and Herzegovina
e-mail: ahudovic@ius.edu.ba; dleila@ius.edu.ba

F. Alberti et al. (eds.), *Towards Sustainable Urban Regeneration: Utilizing Environmental, Cultural, and Social Assets*, Advances in Science, Technology & Innovation, https://doi.org/10.1007/978-3-032-02091-8_28

1.1 Problem Statement

Sarajevo has a history of longitudinal development along the banks of the Miljacka River. Its spatial identities are clearly readable and distinguished from each other along this way. The last decade of the twentieth century led to radical changes at all levels of city and state life. The formation of a new, politically independent state and separation from the concept of socialist collective self-management create a basis for the development of a new city identity, as well as a departure and break with the traditions that the society of the socialist period, as well as earlier periods, had. The issue of national and cultural perception becomes the guiding idea that creates new policies for creating the identity of the state and the city, in its spatial and political domains. The new urban and architectural interventions influenced by neoliberal investments are creating new spatial identities which are not specific to this region or culture. The solutions are generic and common for many countries in the Western Balkans.

1.2 Study Aim

This study presents a chronological cross-section of the development of the city demonstrating the creation of its urban identities, from the Ottoman period in the fifteenth century to the present day, including the period of very turbulent early 1990s in which the country went through the war-induced destruction of the city. Several spatial entities from different historical periods are examined with the aim of showing specificum of each of these. The maps and photographs together with the textual explanation offer a concise overview that explains each of these very different urban matrices, which together create this linear city and ultimately are its architectural and urban heritage. The goal is also to point out their advantages and disadvantages in terms of spatial organization and the quality of living, comparing several case studies.

1.3 Methodology

Using methods of analysis and valorization literature review was an important starting point for understanding different approaches and thoughts about the identity of space and place. Investigation of historical and contemporary spatial forms, typologies, and identities helps in understanding the local context and background. The comparative analyses of certain spatial/urban entities in the city will provide a better understanding of the identities they create. The compilation of these very different urban matrices presented graphically will give insight into the architectural and urban heritage of the city of Sarajevo and its various spatial identities.

2 Literature Review

Even though the word *identity*, as a general term, was first recorded in 1560 (Dictionary 2024), the connection of man with the context, and thus the creation of identity, appeared much earlier. The modern age is marked in a peculiar way by the need for emphasizing identity and individuality, although these two concepts are different. Many theorists and philosophers of the nineteenth and twentieth centuries deal with the issue of identity in which they place man in the foreground, and specific fields of science deal with the issue of identity from different points of view, such as sociological, psychological, historical, spatial, political, cultural, and many others.

The spatial identity narrative has been mentioned in contemporary world literature since the 1990s, and until today, it creates the relationship between an individual and a place, that is, context. This kind of spatial analysis does not necessarily have borders. Sociologist Vukićević (2011) states in his studies on identity and integration in the Balkans that societies in the Balkans have a historically constituted identity.

In his book "Identity of Cities and City of Identity," Ali Cheshmehzangi (2020) explores the multilayered hybridity of urban identities tackling the questions of the notion of identity and human–environment relations, social behavior, regeneration of place, etc. He states that "Our cities are a true representation of our civilizations, history, and aspirations for the future. They age in an interesting process, they grow in their own paces, they decay with various reasons, and they may reborn and revive by new interventions."

Authors like Jones and Svejenova (2017) focus more on the impact of architectural interventions in creating city identity. In their text "Constructing City Identity through Architecture: A Multimodal Approach," they argue that the identity of a city is constructed over time through architecture, which involves three sign systems—material, visual, and rhetorical. These sign systems can help build the "uniqueness" of the city and make it distinct and recognizable among other cities.

The identity of the city as a specific topic of this research is a complex process of historical cultural tradition that depends on all the mentioned factors and is adapted to the place and context. The context becomes a sensitive intersection and the base on which new concepts of the city's contemporary identity are built and should be developed. In this aspect of research and writing, Idrizbegović-Zgonić and Čakarić (2016) are important because they question the conflicts between local and global narratives, using local narratives and thereby creating new principles.

The identity of the cities of Bosnia and Herzegovina is predominantly discussed in contemporary literature from the point of view of national identity, mostly in the post-socialist period, from the breakup of Yugoslavia until today. The city in this context usually does not occupy a primary place and is not a direct object of research, but it serves as a secondary space for political and other turmoil of individuals in a highly exploited and degraded transition period. The closest writings and theories about the identity of the city are the researches and writings of the historian of architecture, Amra Hadžimuhamedović (2012), which touch on the identity of the city and space through the treatment and valorization of the problem of the identity of the heritage in Bosnia and Herzegovina. She states that "Bosnian identity is based on the setting of culture as a three-dimensional network of beliefs and intentions, where every seemingly small change causes a process of active or potential change." Jasenka Čakarić's research on the topic of space as a relationship between the physical and the human is also significant in the field of the semiotics of space as well as urban transformations. She fundamentally examines the symbols and values of the city from the point of view of semiotics, but also the relationship of man toward them, thus speaking in an unusual way about the identity of the (constructed and nonconstructed) city (Čakarić 2017).

Although the history of today's Bosnia and Herzegovina should not be limited to the existence from the period of the Ottoman Empire until today, there is no doubt that this very period (mid-fifteenth century to the end of the nineteenth century) influenced the creation of the recognizable identity of Bosnian urban units. Professor Kamberović (2020) points out that the relationship of identity creation should be viewed in the context of time, which essentially implies that the Ottoman Empire, after the destruction of medieval Bosnia, brought with it established new social relations as well as a new culture that is still being inherited in Bosnia and Herzegovina today. Part of that heritage, in architecture, for example, Stari Most in Mostar or Sarajevo's Baščaršija, is truly part of the world's cultural heritage.

3 The Genesis of the City's Development

Today's Sarajevo was mainly built during three main eras of expansion: the first 140 years of Ottoman rule (until 1600), the peaceful years of Austro-Hungarian rule (from 1883 to 1914), and the formative period of socialist rule (from 1945 to the Olympic Games in 1984) (Donia 2006). The years following the aggression (1992–1995) and up to the present should be added to this, as neoliberal capital and international tendencies have replaced the city's building and urban development principles.

The first mention of Sarajevo as a town was in 1507 but its establishment dates back to 1461 when the first Ottoman governor of Bosnia, Isa-Beg Ishaković, formed a town by constructing several public buildings such as a mosque, a closed marketplace, a public bath, a lodging, a bridge, and the governor's palace unifying the already existing nearby villages. The city grew on the slopes of the surrounding hills with a commercial center in the valley, along the river Miljacka. This was a typical Ottoman town layout with separate living and commercial areas formed along the terrain configuration resulting in an organic urban structure. The principles of Ottoman urbanism were formed under the influence of religion (Islam) and artistic influences from the East (Bejtić 1953).

With the arrival of the Austro-Hungarians in 1878, Sarajevo experienced Western urbanization following urban planning matrices that shape urban centers according to the principles of a modern European city. This kind of planned development is accompanied by an abundance of architectural and construction projects that are often replicas and variations of European principles and achievements. The town was developed along the river in a west direction following the geometric street grid system and forming specific block construction with inner courtyards with maximum use of the building plot. Commercial and living functions were now under one roof. Buildings had multiple floors with rich ornamentations on the front facades. Streets became wider with pedestrian paths on both sides. New public buildings were introduced such as museums, theaters, but also public schools, hospitals, and factories. The city also got its first public park, railway station, communal infrastructure, water supply, sewerage, electrical installations, and electric tram in 1895 (Žuljić et al. 2015).

Due to its diverse demographic picture, Sarajevo has become a synonym for the coexistence of many cultures, nationalities, and religions and is colloquially known as the European Jerusalem. Although the beginnings of modernization came under Austro-Hungarian influence, the process continued even after the Second World War, during the Yugoslav period when the Marijin Dvor area became a place of the city's contemporary identity, an intersection between the past and the future. Besides residential buildings, this area was developed as a commercial and governmental quarter with *Unis* high-rise towers and a National Assembly building. Further to the west of the city along the main road axes, several residential neighborhoods, schools, and university buildings have been constructed.

In the period after 1945, Yugoslavia began with intense construction of factories and infrastructure. Large migrations of people from rural to urban/industrial areas caused a housing crisis. Residential complexes were built throughout the country during the socialist period, and the expansion of Sarajevo was oriented longitudinally from Marijin Dvor further west into the Sarajevo Field, which was mostly empty and uninhabited. During the 1970s and 1980s, the city experienced its urbanization peak. The settlements of the new Sarajevo were created as part of the postwar urban plan from 1963 in the spirit of social construction. Each settlement consisted of several residential buildings, service facilities, such as a school and a kindergarten, public transport stations, and a central public space, such as a park or a square.

During the aggression between 1992 and 1995, the whole country of Bosnia and Herzegovina including its capital Sarajevo experienced massive destruction and killing of civilians. In this period, the city was severely bombed from the surrounding hills and many buildings have been destroyed or severely damaged.

In the postwar period, many buildings, including residential buildings, were repaired and renovated. Due to heavy migrations and a great influx of people coming from the occupied territories of Bosnia and Herzegovina, the number of citizens increased in the postwar period so that new residential settlements and individual residential buildings were built supported by the neoliberal investment policies. The inherited practices of urban and architectural design from the socialist period are usually disregarded. The buildings have a minimal distance between each other, with sparse or no green area in between, and a lot of asphalted surfaces. Unfortunately, several such residential settlements have been constructed from the early 2000s onward creating low-quality living spaces.

After the dynamic changes that occurred in the 1990s and the socialist past, the city entered the phase of transition and capitalism, which essentially represents the total degradation of all traditional values due to the influence of neoliberal policies. In this regard, Jadžić (2011) characterizes the "neoliberal coup" that results in the destruction and deconstruction of the "social welfare state." The three-decade transition period, which, in a way, is still going on, has resulted in the permanent deformation of urban units.

4 Sarajevo's Spatial Identities

As mentioned earlier, topography and sociocultural influence greatly influence the identity of space. Certain areas can be distinct and recognizable through some natural features (river/channel, sea, and hills) and others through man-made features (ports, high-rise buildings, and great public areas like squares or parks). Every city has several microidentities that are specific to certain neighborhoods. The formation of the identity of the neighborhood is dependent on the typology of the buildings and their connection/distance to each other, their height, other available facilities and amenities such as commercial, educational, or recreational areas, but also on demographic factors (age, economic status of residents, etc.), on the safety of the area, its cleanness and maintenance and the intensity of the use of public spaces. The man-made features together with people who live and use the spaces within the particular neighborhood create together its very specific identity.

This chapter presents several very different spatial identities from various historical periods of urban development of Sarajevo pointing out their urban and architectural characteristics. The human factor and relationship between the residents and the built environment of the particular neighborhood are not taken into consideration for this analysis. All case studies are presented with maps and photographs.

4.1 Urban and Architectural Characteristics of the Ottoman Period Neighborhood

The organization of the city in the Ottoman period follows an organic approach to city development with clearly defined functions (residential areas on the hills and commercial areas in the valley, close to the river). The commercial area often had a public open space into which the network of roads flowed, while the residential areas—*mahallas*—were always outside this public zone in a separate, less trafficked area. The (residential) architecture adapted to the local morphology, fitting harmoniously into the context with its forms, clear building surfaces, and modest dimensions. The houses were mostly two floors high with pitched roofs and private gardens, separated from each other along the vertical access, which ultimately allowed a free view and access to the sun. The spatial organization of the *mahalla* follows the morphology of the terrain, and based on the philosophy of living and the tradition, the architecture is introverted (created from the outside toward the inner garden). Such an approach creates an organic network of streets forming a topology without the influence of geometrization (Fig. 1a, b).

4.2 Urban and Architectural Characteristics of the Austro-Hungarian Period Neighborhood

During the 40 years of Austro-Hungarian rule in Bosnia and Herzegovina, Sarajevo experienced the development of a modern city with an orthogonal street layout that formed multifunctional blocks with an inner courtyard. The buildings that formed a block were three to four floors high and covered with a pitched roof. Individual living (a house for one family) that was common during the Ottoman period was replaced by collective housing. The mixed-use buildings that formed a block regularly contained commercial facilities on the ground floor, while the upper floors had the function of housing with two to three apartments per floor. Residents of each building had access to the inner courtyard. During this period, the basic feature of urban planning was based on

Fig. 1 (**a**, **b**) Figure-ground plan of Vratnik neighborhood and photograph of Vratnik from 1938. (Sources: **a**—author, **b**—https://www.facebook.com/StareSlikeSarajeva/photos/pb.100070091262570.-2207520000/1624631560964094/?type=3&locale=hr_HR)

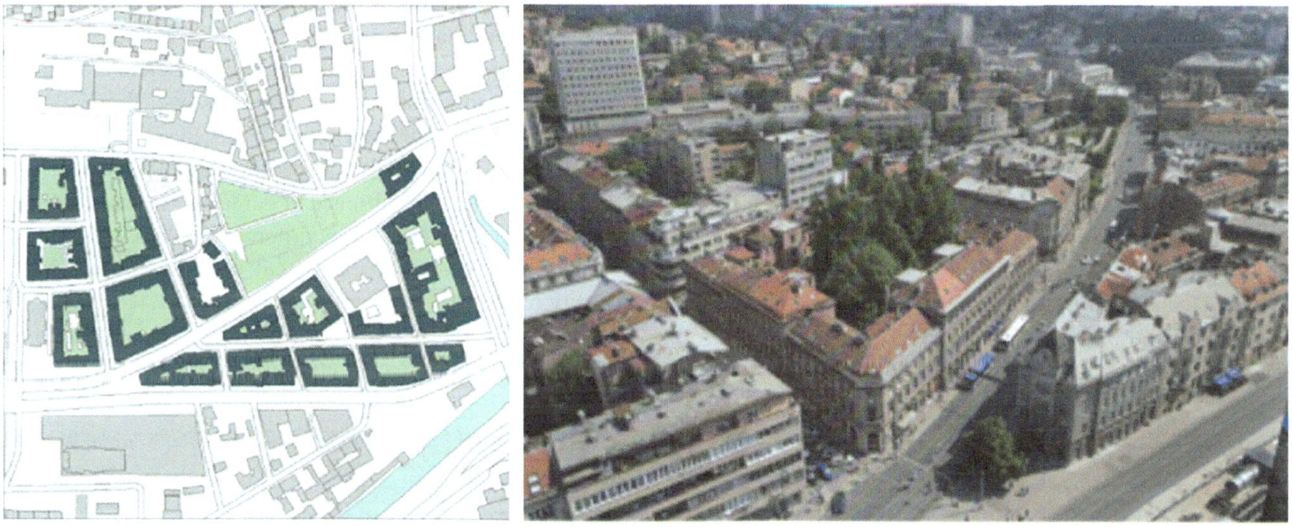

Fig. 2 (**a**, **b**) Figure-ground plan of Marijin Dvor neighborhood and photograph of Marijin Dvor from 2014. (Sources: **a**—author, **b**—Anida Krečo)

an orthogonal system of traffic roads, construction in blocks and rows. All of these principles were developed on the practical approach of the concept of development of a typical Central European city (Fig. 2a, b).

4.3 Urban and Architectural Characteristics of the Socialist Period Neighborhood

During the Yugoslav period, many residential complexes were built throughout the city following the established norms of urban and architectural planning. Great attention was paid to the ratio of built and unbuilt areas (concrete versus green areas), the height of the buildings and the distance between them, recreational areas, service areas, and public green spaces. Most of the residential neighborhoods included adjacent functions such as schools and kindergartens, commercial areas, healthcare centers, and easy access to public transportation. Residential buildings were minimum four floors high, covered with flat or pitched roofs with different architectural expressions and materialization. The diversity of residential building typologies forming different neighborhoods is scattered all over the city. There is a clear distinction between the early period of socialist emancipation and the examples from the late period, just before the 1992–1995 war. The early period was characterized by very strong block typologies arranged in a linear manner. Although with all the accompanying contents, the residential areas of this period, in their architectural scope and processing, are characterized by the simplicity of form and the lack of any ornamentation, in the spirit of socialist equality (Fig. 3a, b). In the late socialist period, as a combination of socialist realism and Le Corbusier's doctrines from the beginning of the last century, new settlements for urban workers were created. This period was marked by a more relaxed formation of volumes influenced by postmodernist ideas (Fig. 3c, d).

4.4 Urban and Architectural Characteristics of the Postwar/Transition Period Neighborhood

The postwar transition period was greatly (and still is) influenced by neoliberal investments and has resulted in the permanent deformation of urban units. Usual for the postwar construction is a partial approach to construction with the aim of achieving the highest possible profit. The buildings are the exclusive property of one person (the investor). The inherited rules of urban planning applied during the Yugoslav period are very often disregarded. The housing blocks often have a minimal distance from each other, creating problems with privacy and blocking the view and access to the sun (particularly on the lower floors). The empty space between the buildings is usually paved and serves as car parking with minimum or no greenery at all. The architecture is generic, and adjacent facilities like schools, kindergartens, and recreational areas are absent (Fig. 4a, b).

5 Discussion

The development of Sarajevo can be divided into several phases, in which the first phase of the city's urbanization began in the middle of the fifteenth century with the construction of the Gazi Isa bey Ishaković endowment buildings, which directed the further development of the city, with the concept of spatial composition and through certain new contents (Žuljić et al. 2015).

In general, the architecture of the Ottoman period is unobtrusive, respecting the unwritten rules of access to the sun and view for every house. "Bosnian architecture of the Ottoman period is characterized by horizontality, economy, and attachment to the human scale" (Bejtić 1953). Specific to this period that adds value to the residential architecture from this period is the emphasis on greenery. This unwritten design philosophy became the base for a conceptual design of a large residential development "Djuro Djaković/Ciglane" from the late Socialist period. The settlement follows the configuration of terrain and respects the inherited values from the Ottoman period by creating terraced residential units that line up vertically. Inspired by the characteristics of traditional house construction on the slope, the settlement creates a clear visual identity and bears the stamp of time in which it was built creating a specific mega-structure of continuous volumes. The absence of gardens is replaced by large open square terraces with incorporated bulky concrete flowerpots for extensive greenery that would contribute to the quality of living.

During the 40 years of Austro-Hungarian rule, buildings of historical importance were built in the city, which some

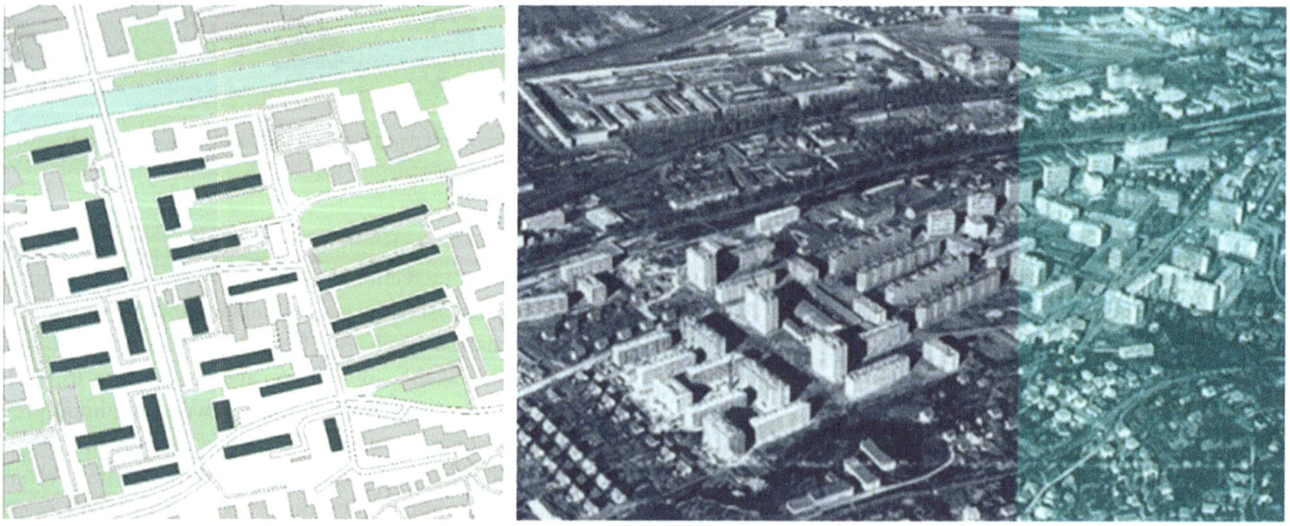

Fig. 3 (**a**, **b**) Figure-ground plan of Grbavica neighborhood and photograph of Grbavica from ca. 1960. (Sources: **a**—author, **b**—Ličina-Ramić, Urbanizacija Sarajeva (1971–1984) i njen utjecaj na društvene promjene, PhD dissertation, Filozofski fakultet Sarajevo, p. 37). (**c**, **d**) Figure-ground plan of Alipašino Polje neighborhood and photograph of Alipašino Polje from ca. 2012. (Sources: **c**—author, **d**—https://www.palelive.com/u-naselju-alipasino-polje-pocelo-probno-iskopavanje/)

Fig. 3 (continued)

theorists and historians of architecture consider unsuitable for the Sarajevo basin. Following this, Čelić (1954) states in his text "Monuments of Culture and Urbanism" that the capitalist era "erected buildings that not only aesthetically damaged the existing ensembles, but also irresponsibly deprived the environment of the numerous achievements of the earlier urban discipline: sun, greenery, vistas, and even regular air currents over the city." The orthogonal street layout with housing blocks of several floors was in total contrast to human-scale buildings constructed during the Ottoman period, thus forming a completely new (visual) identity of the city.

Planned housing construction, after the Second World War, is one of the fundamental determinants of a socialist city. The development of settlement is linked to the change in the economic and sociopolitical foundations of society, specifically as a response to solving the housing issue of a large number of people. In the context of this, we can say that the concept of a socialist city is essentially a concept of a humanized place that offers functions that meet the needs of modern man as a pillar of a socialist society. The paradigm of the new socialist city is a place that receives the rural population that is necessary to work in the industrial zones of the cities. This kind of city forms a new class of people, the proletariat, who become urban workers and are the essential foundations of the development of a socialist society (Korov 2012).

Unlike the former socialist general principles in city planning processes, where the ownership of building plots belonged to the state, which had the power not only over the land but also over all economically profitable ventures, in the post-socialist period, there is a restitution that excludes the state, in almost all processes, from planning to construction, and priority is given to private investors. The major problem in the postwar/transitional development of the city is modifications and revisions of existing urban plans that are subject to changes for the personal gain of the landowner. In addition to this, there is an issue of the lack of transparency in the construction process and noncompliance with the rules of the profession. The concepts of development of new settlements are far away from any inherited values of Ottoman or European spatial philosophies. This results in neighborhoods of low urban and architectural quality with repercussions on low-quality living.

6 Conclusion

Geographically, Sarajevo is at the crossroad between the East and the West, and therefore, historically, it was permanently under the influence of different political regimes that were executors of certain ideologies. Exactly, these ideologies have shaped the creation of the city and its various spatial identities.

During Ottoman rule, the focus was on a strict division between the public commercial space and the private residential space. The unwritten rules of designing a residential neighborhood highly respected the privacy of the house inhabitants. Under the influence of Islam, private life was inside the house and in the inner garden—hence a lack of public parks and gardens during this period.

Under the Austro-Hungarian rule, a new Western way of life was introduced. Life became more extroverted and happened in open public areas (streets, squares, and parks) that were available for everyone. The city was enlarged according to urban planning principles characteristic of Central European cities of the late nineteenth century respecting the geometric street scheme and closed housing blocks.

Fig. 4 (**a**, **b**) Figure-ground plan of Tibra neighborhood and photograph of Tibra from ca. 2020. (Sources: **a**—author, **b**—https://www.fokus.ba/vijesti/bih/kako-su-nastali-tibre-kolosalne-zgrade-na-otoci-i-dzinovski-sarajevo-tower/2201092/)

During the socialist Yugoslav period, the idea of brotherhood and unity was followed throughout all aspects of life. In terms of urbanism, this ideology was visible in the neighborhoods that were carefully planned respecting the distance between the buildings of mass collective housing and providing enough green public space, recreational areas, parking lots, access to public transportation, and other adjacent functions. The idea was to create high-quality living conditions for all people, regardless of their social position.

Today, the newly created residential neighborhoods do not in any way establish connections with the surroundings nor do they respect the existing urban context. This ultimately distorts the identity of space and underestimates the values of a traditional approach to building and spatial identity. The process of transformation of spatial characteristics does not acquire any elements from the previous identities and/or typologies, and this has resulted in the current *globalized* architectural and spatial formation of the city. It seems that the absence of clearly defined parameters for the development of the new neighborhoods might be the main reason for the current lack of spatial identity. A period of transition seems to be a fruitful ground for spatial manipulation and absolute detachment from inherited values. Architects and planners should return to building their ideals rather than maximizing the building on the site creating spatial conflicts without identity. In the context of this, the current research of spatial identity results in two ways, one as a recognition of the existing spatial challenges and the other as a basis for further guidelines of the development, with the appreciative and respectful approach.

References

Bejtić A (1953) Spomenici osmanlijske arhitekture u Bosni i Hercegovini. Prilozi za orijentalnu filologiju 3–4:229–297

Čakarić J (2017) Paradigm of the urban space semiotics. Facta Univ Ser Archit Civ Eng 15:167–178

Čelić D (1954) Spomenici kulture i urbanizma: neka iskustva i zapažanja. Naše starine: godišnjak Zavoda za zaštitu spomenika kulture i prirodnih rijetkosti 2:23–36

Cheshmehzangi A (2015) Urban identity as a global phenomenon: hybridity and contextualization of urban identities in the social environment. J Hum Behav Soc Environ 25(5):391–406

Cheshmehzangi A (2020) Identity of cities and city of identity. Springer, Singapore

Dictionary (2024). https://www.dictionary.com/browse/identity#word-history-and-origins. Accessed 6 Jun 2024

Donia RJ (2006) Sarajevo – biografija grada. Institut za Istoriju, Sarajevo

Dragićević Šešić M (2009) Kultura u funkciji razvoja grada. Kulturni kapitali integrativna kulturna politika. Kultura 122–123:20–40

Hadžimuhamedović A (2012) Raslojavanje bosanskog identiteta – kulturno pamćenje i njegova savremena interpretacija. Godišnjak Bošnjačke zajednice kulture Preporod, pp 235–254

Idrizbegović-Zgonić A, Čakarić J (2016) Universal Hyperlocals. New Arch Int J Contemp Architect 3(1):11–17

Jadžić M (2011) Prostor, urbanizam, politika u postsocijalističkom gradu. Jugolink- pregled post-jugoslovenskih istraživanja vol 1(1), pp 70–84

Jones C, Svejenova S (2017) Constructing city identity through architecture: a multimodal study of Barcelona and Boston. Res Sociol Organ 54B:1–42

Kamberović H (2020) Nasilje kao sudbina? Sarajevo na razmeđu carstava i država u 20. Stoljeću. IZMEĐU RATA I MIRA Sarajevo u prelomnim godinama 20. Stoljeća- Zbornik radova, UMHIS, pp 9–29

Korov G (2012) Zajednička ili zasebna? Paradigme u arhitekturi socijalističke Jugoslavije. In: Mrduljaš M, Kulić V (eds) Unfinished modernisations between utopia and pragmatism. Croatian Architects Association, Zagreb, pp 48–56

Vukićević S (2011) Fenomen identiteta i integracioni procesi na Balkanu, vol 1, pp 62–77. Institut za sociološka istraživanja -Filozofski fakultet Nikšić, Sociološka luča

Žuljić V, Čengić N, Čakarić J (2015) Sarajevo Metropola – model razvoja. Arhitektonski fakultet Univerziteta u Sarajevu

Exploring Urban Elements in a Postindustrial Destination: A Case Study of Bilbao

Jaramillo Estefanía, Alvarez Irantzu, Gandini Alessandra, Quesada Laura, and Garmendia Leire

Abstract

Over the past two decades, the utilization of visitor platforms such as TripAdvisor has emerged as a powerful tool, particularly in shaping perceptions of urban tourism in European destinations. These platforms often promote specific urban elements, resulting in a high influx of visitors to certain areas. The escalating tourist footfall has prompted concerns regarding the resilience of these elements under such pressure. This study explores the context in which urban tourism unfolds by analyzing main urban elements. Using the postindustrial city of Bilbao as a case study, it addresses two key research questions: (1) What are the physical and functional characteristics of the primary elements? and (2) Which specific area functions as a tourist magnet within the city of Bilbao?

To identify the physical and functional attributes of tourist landmarks, an examination of the must-see attractions highlighted on TripAdvisor is conducted. The research findings underscore the importance of establishing a spatial analysis, where the physical characteristics of landmarks and the proximity of hotspots play a pivotal role in shaping the tourism area and consumption patterns. In the context of evolving urban tourism, these insights offer valuable guidance for future research and policymaking, emphasizing the need to protect the Old Town as a vulnerable area from tourism's detrimental impact on the urban landscape and its residents' quality of life.

Keywords

Urban tourism · Primary elements · TripAdvisor · Bilbao · Spatial analysis

1 Introduction

Urban tourism, which generally refers to short trips or city breaks to urban centers or areas with high population density, is considered a particularly dynamic segment of tourism activity where cities are the key drivers of overall tourism volume growth (Namberger et al. 2019). The industry began to gain popularity in the 1990s when the European Union's freedom of movement rules and massive tax breaks for the aviation sector led to the liberalization of air transport (Amore et al. 2020; Bouchon and Rauscher 2019; Novy 2018). This, in turn, contributed to the continued growth and popularity of the low-cost airline boom, making travel more affordable and accessible than ever before (van der Borg 2022b).

As a consequence, the global demand for city experiences has soared by 50% in the last decade, and Europe stands at the forefront of this phenomenon capturing nearly 60% of all international city trips. Additionally, after the economic crisis of 2008, urban tourism emerged as a vital instrument for economic recovery and growth, gaining considerable traction as a strategic priority for cities (Koens et al. 2018). As the British newspaper The Guardian noted, tourism acted as a "lifejacket" for cities severely affected by the recession and city leaders intensified tourism promotion efforts, even amid fiscal austerity and public spending cut (Sequera and Nofre 2018).

J. Estefanía (✉) · Q. Laura · G. Leire
Mechanical Engineering Department, University of the Basque Country (UPV/EHU), Bilbao, Spain
e-mail: estefania.jaramillo@ehu.eus; laura.quesada@ehu.eus; leire.garmendia@ehu.eus

A. Irantzu
Graphic Design and Engineering Projects Department, University of the Basque Country (UPV/EHU), Bilbao, Spain
e-mail: irantzu.alvarez@ehu.eus

G. Alessandra
TECNALIA, Basque Research and Technology Alliance (BRTA), Parque Tecnológico de Bizkaia, Derio, Spain
e-mail: alessandra.gandini@tecnalia.com

F. Alberti et al. (eds.), *Towards Sustainable Urban Regeneration: Utilizing Environmental, Cultural, and Social Assets*, Advances in Science, Technology & Innovation, https://doi.org/10.1007/978-3-032-02091-8_29

Additionally, the continuous growth of the tourism industry, together with the emergence of the Internet, has positioned online platforms as a key source of information for travelers worldwide (Boivin and Tanguay 2019). In this context, TripAdvisor stands out as a particularly important tool, providing tourists with access to top attractions, user-shared experiences, and opportunities to discover new activities at various destinations (Nilsson 2020). Due to its ability to facilitate communication with large audiences across different locations and times, the website is recognized as the leading platform, hosting the largest travel community on the web (Miguéns et al. 2008). However, the continuous rise of travel together with easy access to the Internet involves the constant promotion of specific urban elements, resulting in a high influx of visitors in certain areas leading to a negative impact on the quality of life of residents.

This study examines the development of urban tourism by offering a theoretical framework for the primary elements that shape the urban fabric and identifying urban centers as the primary hubs of tourist activity. Using the postindustrial city of Bilbao as a case study, it addresses two key research questions: (1) What are the physical and functional characteristics of the primary elements? and (2) Which specific area functions as a tourist magnet within the city of Bilbao?

The article starts with the establishment of a theoretical framework considering the role of urban elements as tourism magnets, followed by a categorization of the primary elements; later, an approach of the tourist area as a scenario of tourism development is given. Next, the research methodology is explained, followed by the selection and analysis of Bilbao as a case study. Finally, conclusions and implications for further research are highlighted.

2 Theoretical Background

2.1 The Role of Urban Elements as Tourist Magnets

To comprehensively understand and analyze urban tourism as a transformative phenomenon, it is essential to acknowledge the significance of the context within which it unfolds (Araldi and Fusco 2017; Ashworth and Page 2011; Cocola-Gant 2018; van der Borg 2022b). In this sense, the urban landscape plays a fundamental role both in the creation of the tourist experience and in the setting in which both residents and visitors interact. As Boivin noted, tourism is inherently tied to the attributes and characteristics of places; therefore, its assessment cannot be isolated from the area, marking an intertwined relationship between tourism and the city (Boivin and Tanguay 2019).

In this manner, it is important to consider the urban landscape as a holistic concept that comprises a diverse set of layers, primarily distinguished by their tangible and intangible dimensions. The tangible realm includes a site's topography, geomorphology, hydrology, built environment, infrastructures, open spaces, land use patterns, and spatial organization. On the other hand, the intangible domain encompasses social and cultural practices, values, economic processes, and heritage, particularly concerning diversity and identity (UNESCO 2011).

Within the tangible dimension, the design, architecture, and aesthetic attributes of the built environment serve as powerful tourism magnets, significantly contributing to a destination's visual appeal and atmosphere (Moudon 1997). These inherent features play a crucial role in shaping tourists' perceptions, particularly attracting those drawn to the beauty of their surroundings (Freytag and Bauder 2018; Jansen-Verbeke 1986). In many European cities, the rich history embedded in the heritage enhances its esthetic values, functioning as a primary incentive for tourists' journeys (Ashworth and Page 2011). These elements not only attract visitors but also serve as primary objects of consumption, making them essential components of the tourism experience (Kádár 2013).

Ashworth comprehensive framework offers a robust analysis of urban tourism attractiveness, categorizing factors into primary and secondary elements (Ashworth and Page 2011). Primary elements are the core motivators that influence a visitor's decision to choose a destination, whereas secondary elements play a supportive role in enriching the overall tourist experience by satisfying tourists' basic needs. Given the objectives of this study, the analysis will focus on examining the primary elements.

2.2 The Primary Elements

The urban destination is defined principally for the existence of "primary elements" often located in the heart of the city and hold immense symbolic significance, identifying, differentiating, and shaping the character of the destination (Bouchon and Rauscher 2019; Jansen-Verbeke 1986). This symbolism, rooted in history, architecture, cultural heritage, and a unique urban environment, makes them the core attraction of urban destinations and the hub of tourist activity (Gospodini 2001; Sánchez-Montañés et al. 2023; Vinuesá 1995).

Primary elements encapsulate what makes a city distinctive and appealing, serving as the drivers for tourism engagement (Fig. 1). These elements were not necessarily created for tourism; instead, they serve as a leisure setting for both residents and visitors (Kádár 2013). Through a review of urban tourism literature in European destinations, a compre-

Fig. 1 Photography of urban elements in European destinations. (Source: Author)

hensive categorization is given, shedding light on the diverse elements that form part of the primary elements. This description recognizes that the appeal extends beyond the honeypot sites highly promoted on place marketing strategies, movies, or social media, encompassing different areas of the urban fabric where residents and visitors interact (Table 1).

Table 1 Reference examples (Source: Authors)

Urban elements	Reference examples	Image
Neighborhoods	Trastevere, Rome, Italy Alfama, Lisbon, Portugal Barrio Santa Cruz, Seville, Spain	Figure 1
Main landmark	Eiffel Tower, Paris, France Colosseum, Rome, Italy Sagrada Familia, Barcelona, Spain	Figure 2
Cultural heritage	Canal District, Amsterdam, Netherlands Old Town, Prague, Czech Republic Nyhavn, Copenhagen, Denmark	Figure 3
Museums/art galleries	Uffizi Gallery, Florence, Italy Louvre Museum, Paris, France Guggenheim Museum, Bilbao, Spain	Figure 4
Public space	St. Mark's Square, Venice, Italy Plaza Nueva, Bilbao, Spain Praça do Comercio, Lisbon, Portugal	Figure 5
Pedestrian areas	La Rambla, Barcelona, Spain Rialto Bridge, Venice, Italy Old Town, Bologna, Italy	Figure 6

2.2.1 Neighborhoods

Cities are characterized by constantly renewed diversity, offering a wide variety of activities to visitors (Boivin and Tanguay 2019). However, certain areas stand out as key points of interest. These zones are often traditional neighborhoods defined by physical or social boundaries, showcasing distinctive features that reflect the local culture, history, and lifestyle (Bouchon and Rauscher 2019; Jacobs 1992).

As Kadar argues, the freedom of exploration within these vibrant areas allows visitors to delve deeper into the authenticity of the place, offering a glimpse into the everyday life of residents, fostering a sense of community, and providing a more intimate connection (Kádár 2013). The charm lies not just in the sights but in the opportunity to engage with the local spirit, be it through conversations with shopkeepers, spontaneous encounters with street performers, or simply observing the rhythm of daily life.

2.2.2 Main Landmark

Landmarks, as major attractions, serve as iconic figures that embody significant historical, cultural, and architectural importance, anchoring themselves in the memories of both locals and tourists. They act as pillars of a city's identity, representing its cultural heritage and fostering a strong sense of place (ICOMOS 2008). Beyond attracting tourists, these elements also serve as vital touchstones for residents, shaping their understanding and appreciation of the city. Their role in shaping the mental maps of both residents and visitors makes them unique, irreplaceable assets that are remarkably recognizable and difficult to replicate (de la Calle-Vaquero et al. 2020; Noonan 2023).

These distinctive elements, not necessarily designed for tourism, are frequently promoted through travel guides, place branding strategies, films, and social media, establishing themselves as must-see sites for visitors. Their widespread popularity significantly enhances the destination's appeal as an urban attraction, influencing the distribution of tourism services around them and playing a key role as reliable indicators of emerging tourism facilities (Ignaccolo et al. 2023).

2.2.3 Cultural Heritage

Beyond iconic landmarks, visitors are often pleasantly surprised by the array of cultural assets that surround them (Bianco 2018). This attraction is typically exemplified by built heritage, where buildings of diverse architectural styles serve as living testimonies to a city's past, offering visitors a tangible connection to the history that reflects the cultural and historical evolution of the place (Alfonso 2003; Richards 2018).

On some occasions, they act as a rich backdrop surrounding and enriching the iconic landmark. In other scenarios, these constructions, statues, or memorials emerge as iconic elements, contributing to the overall allure of the destination. Many of these cultural treasures find recognition on the UNESCO World Heritage List, adding an extra layer of distinction to their historical and architectural value (Dunne et al. 2010). These physical and morphological features, particularly in historic cities, often act as powerful "pull factors" that attract visitors.

2.2.4 Museums and Art Galleries

The intrinsic connection to a specific location and time distinguishes museums and art galleries as truly unique attractions (Noonan 2023). Designed by renowned architects or iconic heritage buildings, these elements have become key players in global urban competition strategies (Richards 2018; van der Borg 2022b), playing a pivotal role in promoting cultural understanding, showcasing works, and creating a platform for cultural leisure appreciation of the aesthetic appreciation of iconic masterpieces.

Museums and art galleries also serve as vibrant centers of cultural exchange, enriching the tourism experience on multiple levels, which is why they have received so much attention in recent years as cultural amenities (Aall and Koens 2019). Likewise, museum attendance and tourism flows influence each other, giving rise to a powerful impact on tourism demand (Plaza 2000).

2.2.5 Public Space

Considering tourism is a mosaic of consumption experiences, where visitors not only engage with the principal attractions but also explore and interact with the surrounding environment, public spaces become essential for aesthetic appreciation and sociocultural experiences, serving as vital arenas for citizen interaction, social life, and community engagement—representing the lifeblood of a city (Álvarez-Herranz and Macedo-Ruíz 2021; Boivin and Tanguay 2019).

Originally enjoyed primarily by locals, these spaces now increasingly accommodate tourists, who have become a significant part of the user base (Nofre 2021); functioning as meeting points for diverse groups with various motivations and acting as sites of flows, connections, and even occasional conflict (Mordue 2017), public spaces play a critical role both as places of social interaction and as a place of consumption.

2.2.6 Pedestrian-Friendly Areas: Streets and Sidewalks

As visitors first explore a new urban destination, they often find themselves strolling and taking in their surroundings on sidewalks and streets to immerse themselves in the city's vibrant life (Freytag and Bauder 2018), sometimes to enjoy the urban environment, and at other times simply use them as routes for pedestrian movement between key tourist areas in the city center (De la Calle Vaquero 2019). This unique characteristic makes pedestrian-friendly areas distinct and essential elements in urban tourism, deserving individual attention as integral parts of the built environment (Araldi and Fusco 2017).

Acknowledging walking as a well-established tourist activity, urban destinations have increasingly cultivated areas with high walkability, blurring the lines between visitors and locals (Nientied 2021). The presence and distinctive features of these pedestrian-friendly spaces exert a direct impact on tourist flow and interactions between visitors and residents, ultimately shaping the overall visitor experience (Ignaccolo

Fig. 2 Right: High tourist traffic on the streets of Rome. Left: Concentration of tourists in Praça do Carmo, Lisbon. (Source: Author)

et al. 2023). In addition, as mere facilitators of movement, they offer valuable opportunities for social interaction and economic activity (Jacobs 1992; Levy 1999).

2.3 City Centers as the Tourist Area

Considering urban tourism notably affects only a small part of the physical space of destinations (Ashworth and Page 2011), city centers become the principal arena where the primary elements are placed, creating a readily identifiable perimeter for visitors. These areas are often located in the heart of the city, characterized by an immense symbolic significance, rooted in history, architecture, cultural heritage, and a unique urban environment, making them the core attraction of urban destinations and the hub of tourist activity (Sánchez Montañés et al. 2023, Vinuesá 1992).

Usually, these zones are recognized as traditional old neighborhoods with an important symbolic value, reflecting their inherent diversity and the various transformations over time (Lois 2020; Russo 2002; Sánchez-Montañés et al. 2023). Functionally, these spaces are vibrant areas characterized by a dynamic urban reality with functions encompassing residential areas, business districts, shopping areas, and even administrative centers, all contributing significantly to the richness of urban life (Gusman and Fernandes 2019).

Considering that tourists choose a limited number of attractions within a restricted area, typically found in city centers, these zones can be linked to the primary tourist area (Bugalski 2020; Vinuesá 1992). This zone possesses a continuous spatial concentration of primary elements and a predominantly high degree of walkability, making them vulnerable to high levels of tourism flow and the proliferation of tourism business (Fig. 2). Therefore, both the morphological characteristics that identify urban centers and their functionality pose new challenges associated with the impact that tourism generates in these areas.

3 Methodology

As previously discussed, primary elements within a destination's urban framework serve as critical focal points for visitor engagement. In this context, it is essential to analyze the intersection between the attraction promoted on tourist platforms and the primary elements established in the previous section. To achieve this, the following methodology is proposed.

First, a comprehensive review of TripAdvisor's "Top Attractions" in Bilbao is undertaken to categorize the promoted sites according to the primary elements framework. This step aims to provide insights into the physical and functional characteristics of these attractions, as well as their relationships and significance within the urban context. Both quantitative data, such as the number of reviews and visitor ratings, and qualitative factors, such as their ranking and user

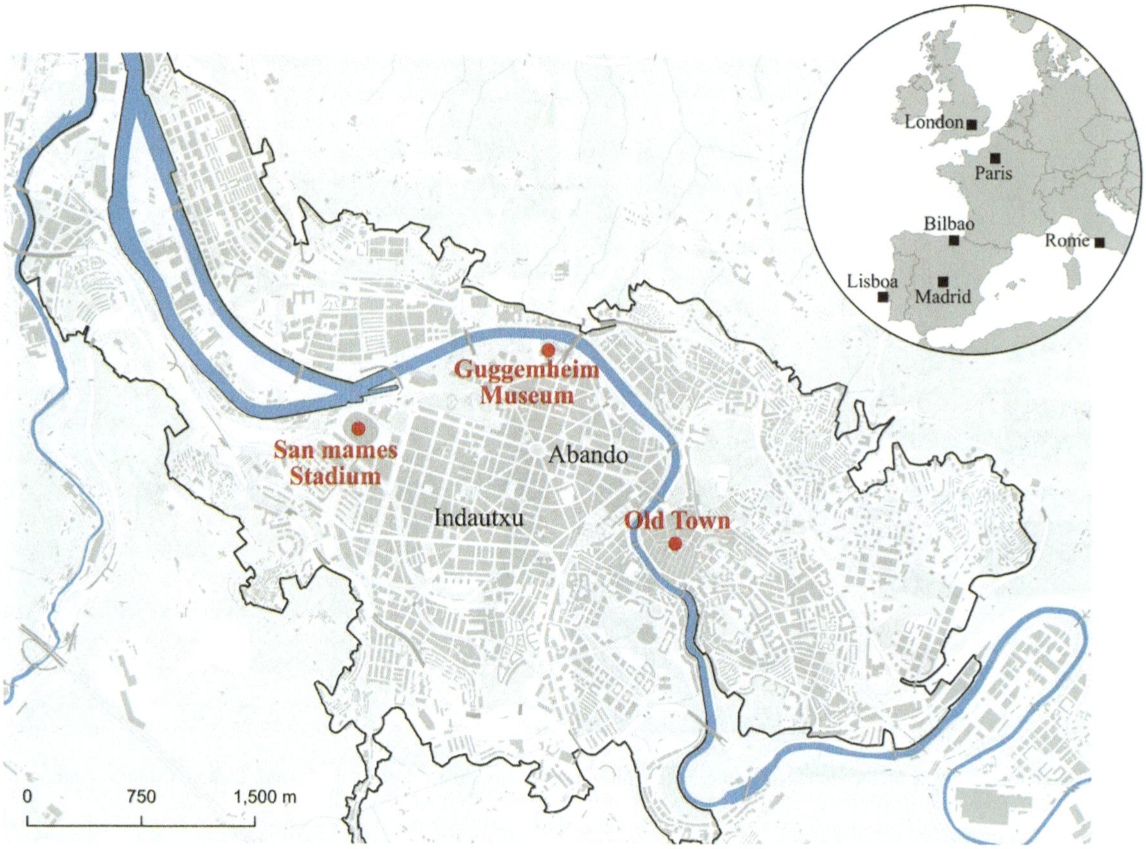

Fig. 3 Location of Bilbao. (Source: Author)

comments, will be considered to determine the relative importance of each element.

Second, a spatial analysis of these elements will be performed using geographic information systems (GISs). This analysis seeks to identify key variables related to the proximity and location of each component within the broader urban fabric. The spatial data will be used to visualize and evaluate the distribution and spatial characteristics of these elements, facilitating the identification of the primary tourist area within the city.

4 Bilbao as Case Study

4.1 Bilbao as a Tourist Destination

Since the late 1990s, urban centers have transitioned from production-based economies tied to the industrial era to consumption-led growth postindustrial districts (Novy 2018). In these areas, culture, leisure, entertainment, and tourism have taken a central role in urban regeneration and socioeconomic revitalization (Sequera and Nofre 2018), leading to a continuous renovation of space, activities, and image. Projects centered around waterfronts and old factories, or the creation of icons and well-known landmarks emerged, aim of promoting a new city image and offering activities and spectacles for both residents and tourists (Maitland and Smith 2009).

Bilbao, a medium-sized Spanish city, exemplifies this transformation (Fig. 3). After the decline of heavy industries (the mining of iron ore, shipbuilding, and the manufacture of steel), the Basque Government devised a plan to revitalize Bilbao with the broader goal of positioning the city as a European hub for service industries, modern technologies, and high-end tourism (Esteban 1996; Wouters 2022). The strategy aimed to reinvent its image through the introduction of revitalization initiatives fundamentally characterized by strong urban and architectural content, positioning it as a cultural hub and a tourist destination (Aguado-Moralejo et al. 2021; Alvarez 2010; Nieuwland and Lavanga 2021).

In this sense, the stakeholders commissioned the construction of a series of impressive buildings: a business center, a conference and performing arts center, a large transport interchange, a metro system, a new terminal for the city's airport, and new bridges over its river (Fig. 4). Each of these structures was designed by world-renowned architects (Martínez et al. 2024). However, it was the construction of one particular building—the Guggenheim Museum—that

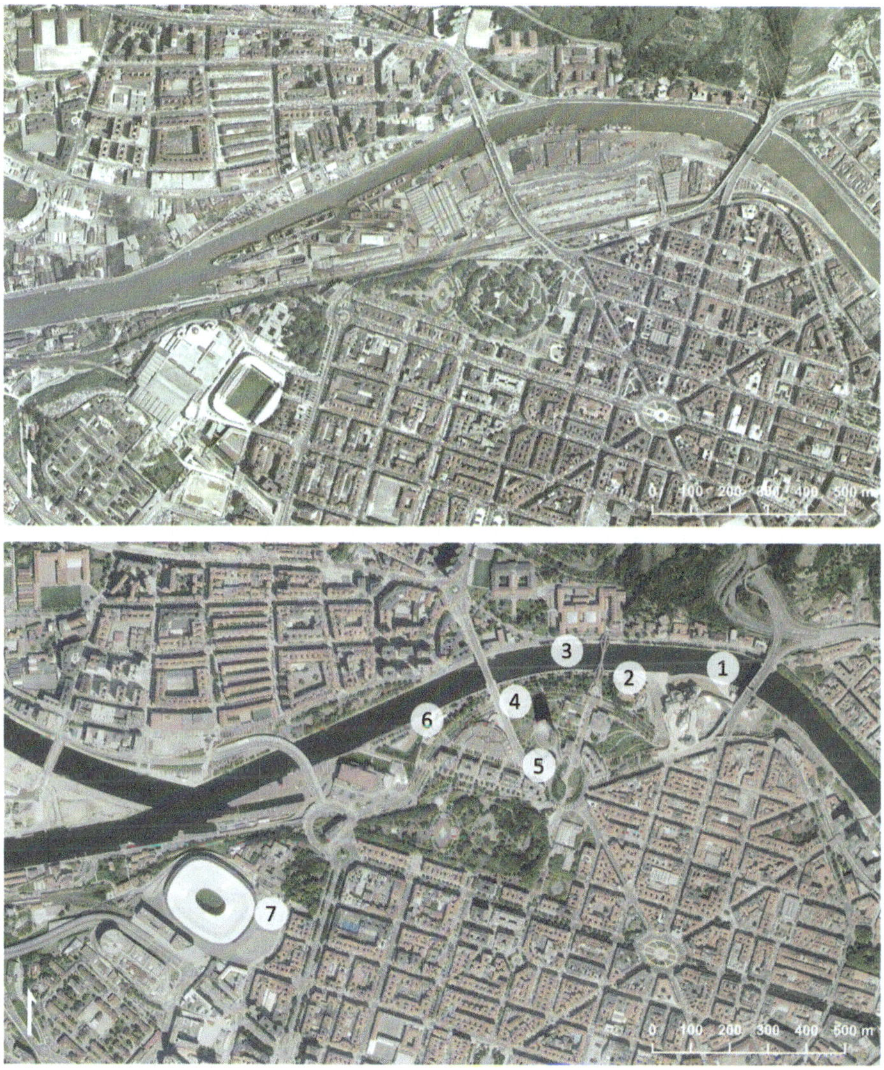

Fig. 4 Above: 1983 Bilbao orthophoto. (Source: Provincial Council of Bizkaia). Below: 2023 Bilbao orthophoto. (Source: Basque Government—geoEuskadi). 1. Guggenheim Museum, 2. University of Deusto Library, 3. Pedro Arrupe Walkway, 4. Iberdrola Tower, 5. Plaza de Euskadi, 6. Zubiarte Shopping Centre, and 7. San Mames Stadium. (Source: Author)

transformed the postindustrial Basque city into an internationally recognized tourist destination (Ignaccolo et al. 2023; Muschamp 1997; Plaza et al. 2009).

Prior to the construction of the Guggenheim Museum, the city had only a small art museum, which drew fewer than 100,000 visitors annually, mostly residents. The arrival of the Guggenheim Museum, however, significantly boosted the city's profile as a tourist destination. This phenomenon, known as the "Guggenheim effect" (Plaza et al. 2009), heightened the importance of urban leisure activities and transformed Bilbao into a new hub of entertainment. Between 1997 and 1999, visitors to the museum contributed to 58% of the tourism growth in the Basque Country (Noonan 2023). Today, Bilbao has emerged as a major tourist attraction in the Basque Country, welcoming 1,239,722 tourists in 2023—an 11.8% increase from the previous year—and is projected to experience an annual growth rate of 5% in visitor arrivals (Eustat 2024).

4.2 An Analysis of Bilbao's Primary Elements

The TripAdvisor platform for the city of Bilbao highlights a range of attractions, most of which are concentrated in the city's urban center. This list encompasses a variety of sites, from heritage buildings to modern architectural landmarks. Based on data regarding the number of reviews and the ranking of each attraction, the Guggenheim Museum and the Old Town emerge as the primary tourist draws, making them essential visits for travelers. They are followed in popularity by the Ribera Market, Plaza Nueva, the Museum of Fine Arts, and the Artxanda Funicular. Notably, the Old Town contains relevant attractions, such as the Arriaga Theater and the Cathedral of Santiago, suggesting that the reviews and ratings attributed to the area may also reflect visitors' experiences with these other elements located within its boundaries (Fig. 5).

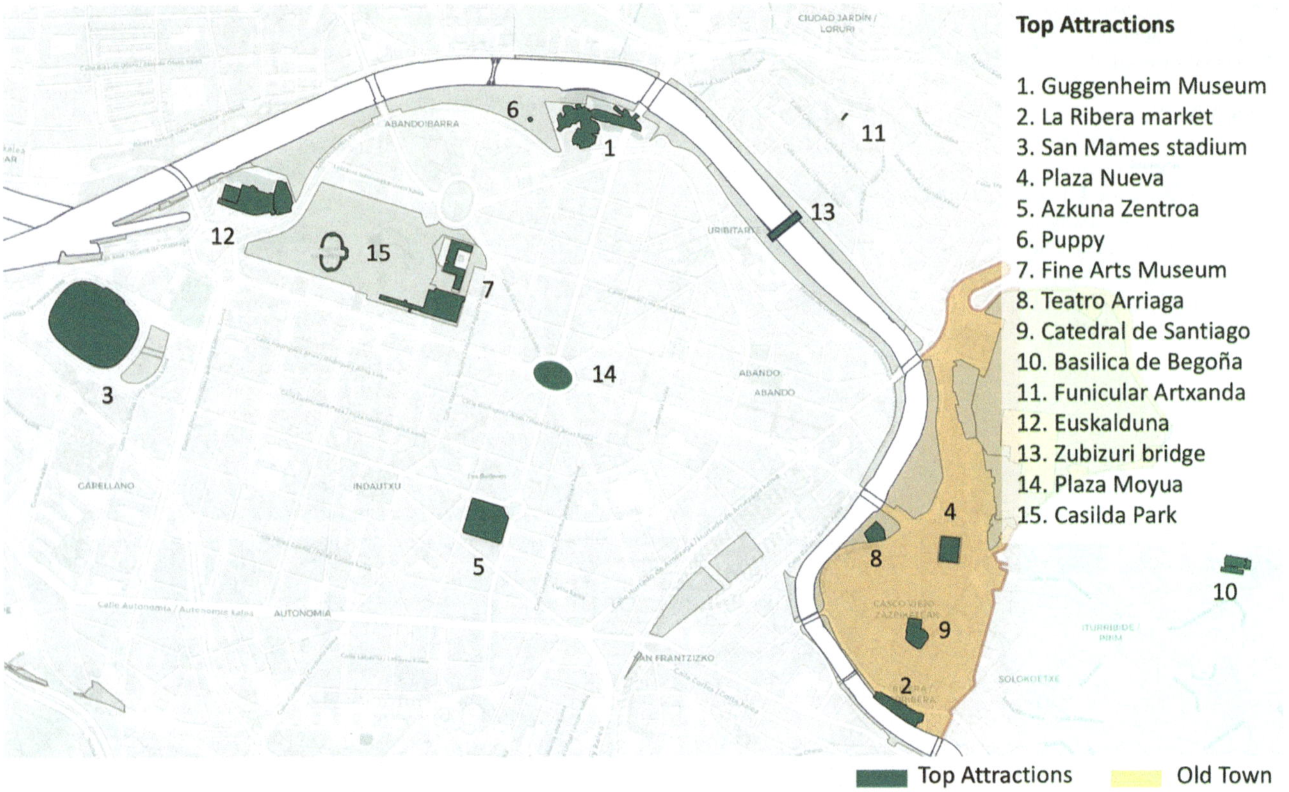

Fig. 5 Location of top tourist attractions in Bilbao. (Source: Author)

Additionally, the list shows a rich built environment, represented by emblematic buildings, cultural heritage, museums, parks, monuments, and public spaces, endowing the area with a distinctive visual appeal and a multifunctional character (Table 2). Similarly, each of these elements possesses a public or private nature, with some offering free access, which can influence a visitor's decision when choosing which areas to visit.

Regarding the relationship of the attractions with the primary elements, the Guggenheim Museum stands out as the primary landmark of the destination, attracting a significant number of visitors and shaping the city's image (Table 3). Similarly, there is a clear identification of the Old Town as a key area of interest due to its rich concentration of cultural heritage, allowing visitors to move around easily and at the same time observe its heritage and emblematic monuments.

In terms of public space, elements such as the La Ribera Market, Plaza Nueva, and the Azkuna Zentroa stand out; these areas, despite being a focus of interest for visitors, function as a meeting point for the locals. As part of the museum category, the Guggenheim and the Fine Arts museums are included; however, it is important to mention only visits inside the establishment are considered. For pedestrian areas, the most notable streets and walkways are located in the Old Town.

The Old Town serves as the area where some of the primary elements such as the Arriaga Theater, Plaza Nueva, Santiago Cathedral, and La Ribera Market are located. Given the surface area of 593,541 square meters, the proximity between landmarks is low, with a maximum walking distance of 7 min (500 m) between the Arriaga Theater and the La Ribera Market (Fig. 3). On the other hand, the Ensanche area corresponding to the neighborhoods of Abando and Indautxu encompasses the rest of the primary elements. This zone has a surface area of 1,789,800 m^2, approximately five times larger than the Casco Viejo, which translates into significantly greater distances between landmarks where the maximum walking distance is 30 min (2.2 km) between the San Mamés Stadium and the Zubizuri Bridge (Fig. 6).

4.3 Is Bilbao's Old Town a Tourist Magnet?

As mentioned above, the built environment and its primary elements shape its external image, attracting large tourist flows to specific areas. In this sense, Bilbao's Old Town stands out among other neighborhoods as a key area of interest, both for its distinctive physical characteristics and the functionality of its spaces. This area contains a rich cultural heritage, with key landmarks such as Santiago Cathedral, La Ribera Market, and Plaza Nueva forming a historically

Table 2 TripAdvisor list (Source: Authors)

No.	Attraction	Type	Reviews	Fee
1	Guggenheim Museum	Art Museums	18.192	Free/pay
2	Old Town	Neighborhoods/walking areas	5.405	Free
3	La Ribera market	Historic sites/flea and street markets	3.310	Free
4	San Mames Stadium	Other outdoor activities	650	Pay
5	Plaza Nueva	Points of interest and landmarks	3.794	Free
6	Azkuna Zentroa	Architectural buildings/civic centers	1.494	Free
7	Puppy	Monuments/statues	577	Free
8	Fine Arts Museum	Art museums	2.555	Pay
9	Casilda Park	Parks	640	Free
10	Funicular Artxanda	Rail services	3.809	Pay
11	Euskalduna	Architectural buildings/theaters	344	Pay
12	Zubizuri Bridge	Bridges	1.822	Free
13	Basilica de Begoña	Churches and cathedrals	518	Free
14	Teatro Arriaga	Theaters	288	Pay
15	Plaza Moyua	Points of interest and landmarks	786	Free
16	Catedral de Santiago	Churches and cathedrals	761	Pay

Table 3 Primary elements and top attraction (Source: Author)

Primary element	Attraction	Physical/functional characteristics
Main landmark	Guggenheim Museum	Architectural building and open space
Neighborhood	Old Town Viejo	Zone
Cultural heritage/ monuments	Teatro Arriaga, Catedral de Santiago, Basilica de Begoña, Puppy, and Euskalduna	Architectural building, monument, churches, and cathedrals
Public space/ infrastructure	La Ribera market, Plaza Nueva, Azkuna Zentroa, Zubizuri Bridge, Plaza Moyua, Funicular Artxanda, San Mames stadium, and Casilda Park	Architectural building, bridge, public transport, green area, open space, plaza, and park
Museums	Fine Arts Museum	Building
Pedestrian-friendly areas	Pedestrian streets in Casco Viejo	Walking areas

significant built environment. In addition, public spaces such as Plaza Nueva and the surrounding pedestrian streets serve as vibrant meeting points where locals and visitors interact naturally.

Considering the temporal and spatial constraints that visitors often encounter at a destination (Bertocchi and Visentin 2019), the proximity of the primary elements in Bilbao's Old Town acts as a significant "pull factor." The short distances between its key elements facilitate a compact yet immersive walking experience, allowing tourists to easily explore the area. Moreover, the pedestrian-friendly sidewalks, streets, and squares provide an authentic setting where visitors can engage with the city's daily life and interact with locals, contributing to a genuine experience of "real" and "authentic" urban culture. Public spaces such as Plaza Nueva serve as dynamic arenas where everyday urban activities and tourism seamlessly intersect, fostering a sense of coexistence between residents and visitors.

Given the continuous tourism growth Bilbao has experienced in recent years and the distinctive attributes that characterize the Old Town as one of its most iconic areas, it is crucial to assess tourism impact. Many in the field hold the view that the negative effects of tourism tend to concentrate in specific, highly recognizable areas near major landmarks (Alvarez 2010; Ashworth and Page 2011; García-Hernández et al. 2017). Therefore, the categorization and spatial analysis of the Old Town offer settings for identifying its vulnerability to the pressure of tourism growth in the area. Understanding these factors is a critical step in managing sustainable tourism and preserving the integrity of its urban elements.

4.4 The Recognition of the Tourist Area

It is acknowledged that each city destination possesses specific characteristics that offer a unique experience for visitors and influence urban dynamics and interpersonal interactions; however, the analysis of urban tourism shares a common component: its built environment, exemplified by the "primary elements," which act as a hotspot for both locals and visitors (Shoval 2018). Within this context, the proposed methodology can be applied to different urban destinations by utilizing spatial analysis to map key tourist attractions listed on visitor platforms such as TripAdvisor. This approach enables the initial identification of the tourist area and establishes a preliminary destination categorization based on the relationship between the size of the urban area and the tourist area.

Contemporary tourism trends increasingly seek a more authentic connection with the city, engaging with everyday

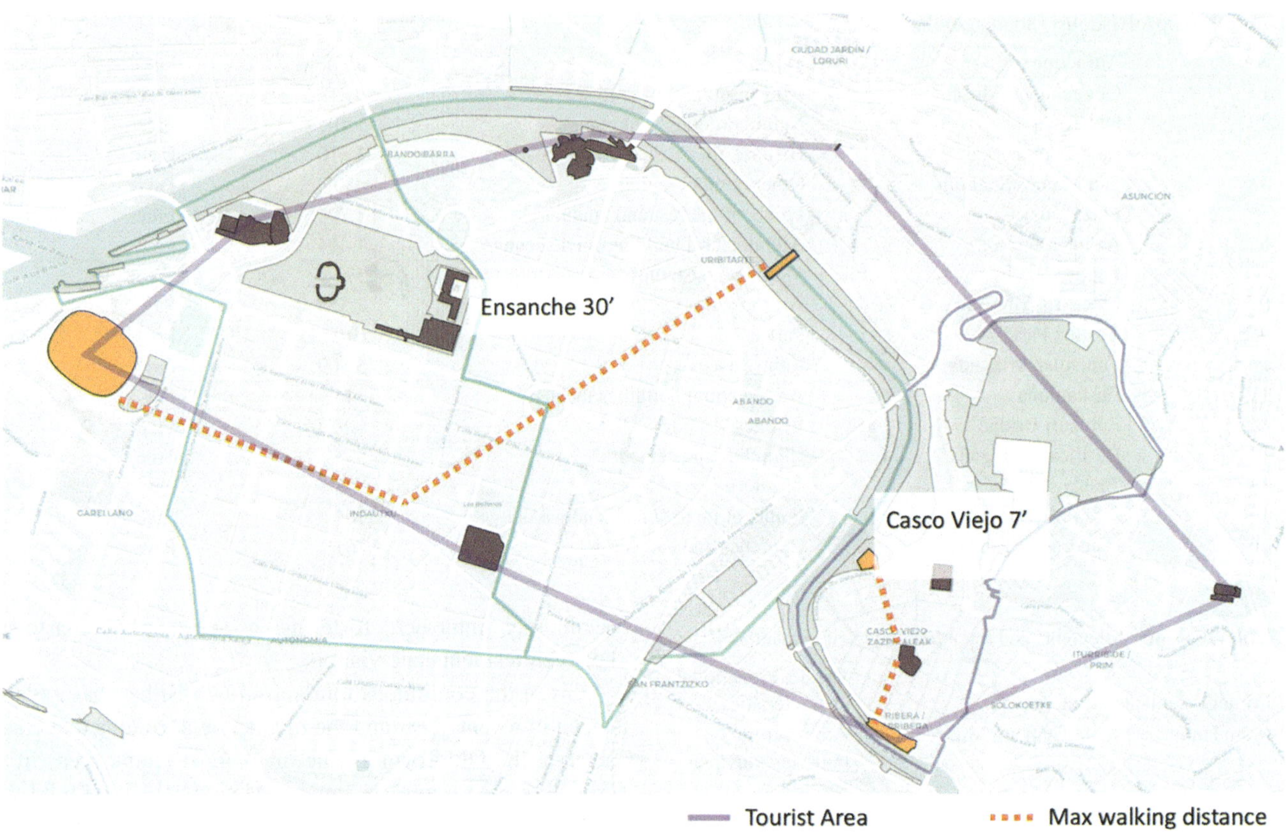

Fig. 6 Proximity between landmarks. (Source: Author)

life rather than merely traditional tourist attractions. Furthermore, tourism is expanding into areas of urban life that were previously not considered of interest (Bouchon and Rauscher 2019; Koens et al. 2018). In this context, the proposed methodology, alongside the recognition of the tourist area, contributes to understanding how these zones interact with surrounding neighborhoods, where cultural allure is often less historically rooted compared to traditional historic districts.

By mapping the main tourist attractions across different urban destinations, the spatial organization of the tourist area in relation to other urban elements can be analyzed. In metropolitan hubs such as Paris, London, and Lisbon, which feature larger urban areas, the primary elements are more widely dispersed. Consequently, variables related to mobility play a pivotal role in shaping tourism dynamics (Amore et al. 2020; Freytag and Bauder 2018). On the other hand, medium-sized cities—whether postindustrial or historically significant, such as Bilbao or Seville—exhibit a high concentration of tourist attractions in specific areas, where proximity significantly affects the dynamics of tourist influx (Barrero Rescalvo and Jover Báez 2020). Additionally, compact cities like Venice and Dubrovnik present unique challenges due to their constrained topography and urban compactness, which amplify the vulnerability of their urban fabric to tourism pressures (Ignaccolo et al. 2023; Stecker and Hartmann 2019).

In this sense, the study aims to bridge the existing gap between urban studies and tourism, where tourism often operates within a separate sphere, and urban planners only recognize tourism when it encroaches upon their exclusive territory (van der Borg 2022b). This dichotomy is encapsulated in Ashworth's observation that "those studying tourism neglected cities while those studying cities neglected tourism" (Ashworth and Page 2011).

5 Conclusions

The identification and evaluation of key urban elements, along with their relationship to prominent attractions featured on platforms like TripAdvisor, are fundamental to understanding tourism dynamics. These elements not only shape tourist experiences but also significantly impact the quality of life for residents. This study provides a detailed analysis of the interaction between tourism and the urban fabric, highlighting the need to consider the broader spatial contexts in which tourism activity occurs.

Through a spatial analysis of tourist elements within Bilbao's neighborhoods, the study reveals that the Old Town acts as a central hub, concentrating many primary attractions and functioning as a strategic zone where urban elements, visitors, and locals intersect. These findings emphasize the necessity of a comprehensive approach that prioritizes the development of frameworks and tools to effectively manage the complex challenges cities face as tourism pressures continue to increase.

While this research offers valuable insights into the physical and functional characteristics of Bilbao's main urban elements, it acknowledges certain limitations. Future studies are encouraged to utilize spatial analysis to map primary elements and identify tourist areas under constant pressure in other urban destinations. Additionally, it is recommended to delve into the supportive role of secondary elements in enhancing the overall tourist experience and addressing tourists' basic needs. Moreover, developing spatial indicators better to understand the interaction between tourism infrastructure and key assets would help differentiate the "tourism bubble" from broader urban contexts.

Tourism is an enduring phenomenon, making it essential to continually assess the environments where it unfolds to mitigate its impacts on specific areas. Without diligent management and forward-thinking strategic planning, destinations—especially historic city centers—are vulnerable to adverse physical and functional transformations. This, in turn, could erode a city's authenticity and negatively affect the quality of life of its residents. Addressing these concerns through careful planning will be crucial to ensuring that tourism contributes positively to the sustainable development of urban areas.

Acknowledgments The authors wish to acknowledge funding from the Investigo Program within the European Union's Recovery, Transformation, and Resilience Plan (Next Generation EU) and the sustainable and resilient built environment (SAREN) research group (IT1619-22, Basque Government).

References

Aall C, Koens K (2019) The discourse on sustainable urban tourism: the need for discussing more than overtourism. Sustainability 11:4228. https://doi.org/10.3390/su11154228

Aguado-Moralejo I, Echebarria C, Barrutia JM (2021) Airbnb: ¿economía colaborativa o nuevo modelo de negocio? Evidencia empírica reciente desde una ciudad de tamaño medio como Bilbao. Ingeo 97. https://doi.org/10.14198/INGEO.18910

Alfonso MJP (2003) El patrimonio cultural como opción turística. Horiz antropol 9:97–115. https://doi.org/10.1590/S0104-71832003000200006

Alvarez MD (2010) Creative cities and cultural spaces: new perspectives for city tourism. Int J Cult Tour Hosp Res 4:171–175. https://doi.org/10.1108/17506181011067565

Álvarez-Herranz A, Macedo-Ruíz E (2021) An evaluation of the three pillars of sustainability in cities with high Airbnb presence: a case study of the City of Madrid. Sustainability 13:3220. https://doi.org/10.3390/su13063220

Amore A, Falk M, Adie BA (2020) One visitor too many: assessing the degree of overtourism in established European urban destinations. Int J Tour Cities 6:117–137. https://doi.org/10.1108/IJTC-09-2019-0152

Araldi A, Fusco G (2017) Decomposing and recomposing urban fabric: the city from the pedestrian point of view. In: Gervasi O, Murgante B, Misra S, Borruso G, Torre CM, Rocha AMAC, Taniar D, Apduhan BO, Stankova E, Cuzzocrea A (eds) Computational science and its applications—ICCSA 2017, Lecture notes in computer science. Springer International Publishing, Cham, pp 365–376. https://doi.org/10.1007/978-3-319-62401-3_27

Ashworth G, Page SJ (2011) Urban tourism research: recent progress and current paradoxes. Tour Manag 32:1–15. https://doi.org/10.1016/j.tourman.2010.02.002

Barrero Rescalvo M, Jover Báez J (2020) Paisajes de la turistificación: una aproximación metodológica a través del caso de Sevilla. cuadgeo 60:13–34. https://doi.org/10.30827/cuadgeo.v60i1.13599

Bertocchi D, Visentin F (2019) "The overwhelmed city": physical and social over-capacities of global tourism in Venice. Sustainability 11:6937. https://doi.org/10.3390/su11246937

Bianco L (2018) Architecture, values and perception: between rhetoric and reality. Front Archit Res 7:92–99. https://doi.org/10.1016/j.foar.2017.11.003

Boivin M, Tanguay GA (2019) Analysis of the determinants of urban tourism attractiveness: the case of Québec City and Bordeaux. J Destin Mark Manag 11:67–79. https://doi.org/10.1016/j.jdmm.2018.11.002

Bouchon F, Rauscher M (2019) Cities and tourism, a love and hate story; towards a conceptual framework for urban overtourism management. Int J Tour Cities 5:598–619. https://doi.org/10.1108/IJTC-06-2019-0080

Bugalski Ł (2020) The undisrupted growth of the Airbnb phenomenon between 2014–2020. The touristification of European cities before the COVID-19 outbreak. Sustainability 12:9841. https://doi.org/10.3390/su12239841

Cocola-Gant A (2018) Tourism gentrification. In: Lees With Martin Phillips L (ed) Handbook of gentrification studies. Edward Elgar Publishing, Cheltenham. https://doi.org/10.4337/9781785361746.00028

De la Calle Vaquero M (2019) Turistificación de centros urbanos: clarificando el debate. BAGE. https://doi.org/10.21138/bage.2829

de la Calle-Vaquero M, García-Hernández M, Mendoza de Miguel S (2020) Urban planning regulations for tourism in the context of overtourism. Applications in historic centres. Sustainability 13:70. https://doi.org/10.3390/su13010070

Dunne G, Flanagan S, Buckley J (2010) Towards an understanding of international city break travel. Int. J. Tourism Res., 12:409–417. https://doi.org/10.1002/jtr.760

Esteban DM (1996) Luces y sombras del Bilbao del titanio

Eustat (2024) Turismo en Bilbao. Balance 2023. Bilbao Ekintza. Available at: https://www.bilbaoekintza.eus/documento/turismo-en-bilbao-balance-2023

Freytag T, Bauder M (2018) Bottom-up touristification and urban transformations in Paris. Tour Geogr 20:443–460. https://doi.org/10.1080/14616688.2018.1454504

García-Hernández M, De La Calle-Vaquero M, Yubero C (2017) Cultural heritage and urban tourism: Historic City Centres under pressure. Sustainability 9:1346. https://doi.org/10.3390/su9081346

Gospodini A (2001) Urban design, urban space morphology, urban tourism: an emerging new paradigm concerning their relationship. Eur Plan Stud 9:925–934. https://doi.org/10.1080/09654310120079841

Gusman C, Fernandes P (2019) Culture and tourism in Porto City Centre: conflicts and (im)possible solutions. Sustainability 11:5701. https://doi.org/10.3390/su11205701

ICOMOS (2008) Québec declaration on the preservation of the spirit of place

Ignaccolo C, Zheng Y, Williams S (2023) Tourism morphometrics in Venice: constructing a tourism services index (TSI) to unmask the spatial interplay between tourism and urban form. Cities 140:104369. https://doi.org/10.1016/j.cities.2023.104369

Jacobs J (1992) The death and life of great American cities. Vintage Books, New York

Jansen-Verbeke M (1986) Inner-city tourism: resources, tourists and promoters. Ann Tour Res 13:79–100. https://doi.org/10.1016/0160-7383(86)90058-7

Kádár B (2013) Differences in the spatial patterns of urban tourism in Vienna and Prague. Urbani izziv 24:96–111. https://doi.org/10.5379/urbani-izziv-en-2013-24-02-002

Koens K, Postma A, Papp B (2018) Is overtourism overused? Understanding the impact of tourism in a city context. Sustainability 10:4384. https://doi.org/10.3390/su10124384

Levy A (1999) Urban morphology and the problem of the modern urban fabric: some questions for research. jum 3:79–85. https://doi.org/10.51347/jum.v3i2.3885

Lois R (2020) Sustainability and visitor management in tourist historic cities. MDPI. https://doi.org/10.3390/books978-3-03936-245-5

Maitland R, Smith A (2009) Chapter 10. Tourism and the aesthetics of the built environment. In: Tribe J (ed) Philosophical issues in tourism. Multilingual Matters & Channel View Publications, pp 171–190. https://doi.org/10.21832/9781845410988-011

Martínez E, Rodríguez A, Altuzarra A, Álvarez I (2024) Remaking urban divides: shifting patterns of neighborhood differentiation in Bilbao, Spain. J Urban Aff 46:389–408. https://doi.org/10.1080/07352166.2022.2057319

Miguéns J, Baggio R, Costa C (2008) Social media and tourism destinations: TripAdvisor case study

Mordue T (2017) New urban tourism and new urban citizenship: researching the creation and management of postmodern urban public space. Int J Tour Cities 3:399–405. https://doi.org/10.1108/IJTC-04-2017-0025

Moudon V (1997) Urban morphology as an emerging interdisciplinary field. Urban Morphol 1(1):3–10

Muschamp H (1997) The miracle in Bilbao

Namberger P, Jackisch S, Schmude J, Karl M (2019) Overcrowding, overtourism and local level disturbance: how much can Munich handle? Tour Plan Dev 16:452–472. https://doi.org/10.1080/21568316.2019.1595706

Nientied P (2021) Rotterdam and the question of new urban tourism. Int J Tour Cities 7:344–360. https://doi.org/10.1108/IJTC-03-2020-0033

Nieuwland S, Lavanga M (2021) The consequences of being 'the Capital of Cool'. Creative entrepreneurs and the sustainable development of creative tourism in the urban context of Rotterdam. J Sustain Tour 29:926–943. https://doi.org/10.1080/09669582.2020.1841780

Nilsson JH (2020) Conceptualizing and contextualizing overtourism: the dynamics of accelerating urban tourism. Int J Tour Cities 6:657–671. https://doi.org/10.1108/IJTC-08-2019-0117

Nofre J (2021) The touristification of nightlife: some theoretical notes. Urban Geogr 42:1552–1561. https://doi.org/10.1080/02723638.2020.1855002

Noonan L (2023) The role of culture as a determinant of tourism demand: evidence from European cities. Int J Tour Cities 9:13–34. https://doi.org/10.1108/IJTC-07-2021-0154

Novy J (2018) 'Destination' Berlin revisited. From (new) tourism towards a pentagon of mobility and place consumption. Tour Geogr 20:418–442. https://doi.org/10.1080/14616688.2017.1357142

Plaza B (2000) Guggenheim museum's effectiveness to attract tourism. Ann Tour Res 27:1055–1058. https://doi.org/10.1016/S0160-7383(99)00131-0

Plaza B, Tironi M, Haarich SN (2009) Bilbao's art scene and the "Guggenheim effect" revisited. Eur Plan Stud 17:1711–1729. https://doi.org/10.1080/09654310903230806

Richards G (2018) Cultural tourism: a review of recent research and trends. J Hosp Tour Manag 36:12–21. https://doi.org/10.1016/j.jhtm.2018.03.005

Russo AP (2002) The "vicious circle" of tourism development in heritage cities. Ann Tour Res 29:165–182. https://doi.org/10.1016/S0160-7383(01)00029-9

Sánchez-Montañés B, Romero-Ojeda J-M, Castilla MV (2023) The impact of overtourism on architecture and urban space in historic cities: an understudied phenomenon. J Tour Anal Rev Análisis Turístico 30. https://doi.org/10.53596/jta.v30i1.439

Sequera J, Nofre J (2018) Shaken, not stirred: new debates on touristification and the limits of gentrification. City 22:843–855. https://doi.org/10.1080/13604813.2018.1548819

Shoval N (2018) Urban planning and tourism in European cities. Tour Geogr 20:371–376. https://doi.org/10.1080/14616688.2018.1457078

Stecker B, Hartmann R (2019) Case study: balancing the sustainability of tourism in city destinations—the case of Dubrovnik. In: Lund-Durlacher D, Dinica V, Reiser D, Fifka MS (eds) Corporate sustainability and responsibility in tourism, CSR, sustainability, ethics & governance. Springer International Publishing, Cham, pp 373–382. https://doi.org/10.1007/978-3-030-15624-4_22

UNESCO (2011) Recommendation on the historic URBAN landscape

van der Borg J (2022b) Urban tourism as a special type of cultural tourism. In: A research agenda for urban tourism. Edward Elgar Publishing, Cheltenham. https://doi.org/10.4337/9781789907407.00009

Vinuesá MAT (1992) Centro histórico, intervención urbanística y análisis urbano

Vinuesá MAT (1995) El turismo en las ciudades históricas

Wouters M (2022) Reinventing Bilbao, the story of the Bilbao effect. In: Piselli C, Altan H, Balaban O, Kremer P (eds) Innovating strategies and solutions for urban performance and regeneration, advances in science, technology & innovation. Springer International Publishing, Cham, pp 37–48. https://doi.org/10.1007/978-3-030-98187-7_4

Bridging Affordability and Resilience in African Housing: A Case Study of Housing Design Guidelines in Kenya

Garret Gantner

Abstract

Large-scale affordable housing programmes in rapidly developing cities of sub-Saharan Africa often rely on predetermined, best-practice standards, which are not always contextualized to local climates, urban conditions, or building practices. At the same time, the roll-out of such programmes represents opportunities to change standard practices in ways that improve environmental responsiveness and climate resilience. Methods to identify the realities of affordable housing on the ground in a specific local context are needed, while linking them to broadscale climate resilience principles and sustainable development objectives. This study uses secondary survey data conducted in three Kenyan cities and peripheral settlements, as well as primary architectural analysis of existing affordable housing models to generate and categorize architectural typologies. These typologies are then scrutinized for the ways their design, alteration, and implementation may contribute to urban sustainability and resilience, as well as low-carbon development, and make the goals of climate-resilient housing development in Africa more feasible. Through the proposed transformation of typologies and processes, the study finds that a ground-up reframing of building regulations and guidelines is needed not only to ease the transition towards more considered site and building design and material use but also to disincentivize standard, unsustainable industry practices.

Keywords

Affordable housing · Sustainable development · African architecture · Green building · Resilience

1 Introduction

For every six people on our planet, on average, one of them lives in an informal dwelling. One in five lives in 'inadequate' housing (ACAPS 2024). By the end of this decade, as many as three billion people may be in need of adequate and affordable housing (Agence Phare 2024).

The staggering scale of the global affordable housing problem brings about both a critical need to find means and methods of delivering better housing stock and an exigence to link them to holistic strategies of sustainable development. A 2024 study focused on the delivery of affordable housing found 'strong interaction' between the planning, design, and construction of affordable housing and 10 of the 17 Sustainable Development Goals, concluding that access to affordable and sustainable housing should be considered a pivotal influencer of sustainable development (CSTI 2023), something which the architectural, urban design and planning professions have, for the most part, yet to effectively leverage in communicating (and demonstrating) the value design may bring to this monumental task.

It is within this context of needs that the Kenyan government launched its affordable housing programme (AHP) in 2018, an ambitious framework now targeting the delivery of 200,000 affordable housing units annually. In 2023–2024, a project was launched to leverage the interaction between the design and construction of new housing stock with sweeping policy initiatives intended to promote sustainable development and address climate change, including the greening of the design and construction industries. As part of this, the development design standards and guidelines for affordable housing were undertaken.

1.1 Problem Statement

About 70% of Kenya's current housing stock consists of shacks, with an average size of just over 3 m by 3 m (Espelia

G. Gantner (✉)
University of the Witwatersrand, Johannesburg, South Africa
e-mail: garret.gantner@wits.ac.za

F. Alberti et al. (eds.), *Towards Sustainable Urban Regeneration: Utilizing Environmental, Cultural, and Social Assets*, Advances in Science, Technology & Innovation, https://doi.org/10.1007/978-3-032-02091-8_30

et al. 2024), a statistic which has triggered the roll-out of major policy initiatives intended to address the adequacy and affordability of housing. Yet in many cases, large-scale affordable housing programmes in the rapidly developing cities of sub-Saharan Africa rely on predetermined, best-practice standards, which are not always contextualized to local climates, urban conditions, or building practices. At the same time, the roll-out of such programmes, and responses to the affordable housing demand, represents opportunities to change standard practices in ways that improve environmental responsiveness and climate resilience.

Design guidelines are not universally applicable in all situations: nuances are always a part of the design process and part of an appropriate response. While it is impossible to codify recommendations for all possible scenarios, predictable variables may influence design criteria. Thus, methods to identify the realities of affordable housing on the ground in a specific local context are needed, while linking them to broad-scale climate resilience principles and sustainable development objectives.

1.2 Study Aim

One step (amongst many others) towards a non-superficial greening of the design and implementation of affordable housing in this context is the simplicity with which interested actors and project administrators can understand and evaluate the effectiveness of designed solutions.

This study investigates how the development of design guidelines for affordable housing can be made more applicable to ordinary industry players, reducing the boundaries between idealistic and realistic development scenarios. In doing so, it focuses on a ground-up reframing of design guidelines through the identification of typical building typologies, allowing flexible interventions and adaptations of the design and implementation of buildings, thereby easing the transition towards more considered site, and building design and material use, and disincentivizing standard, unsustainable industry practices.

2 Background and Literature Review

2.1 Development Trends

The scale and rapidity of current housing development trends, in which properties with lower-density housing types are often replaced with higher-density, high-rise (or at least mid-rise) flats, have had major negative impacts on infrastructure and urban systems: road networks, sewage systems, water and electrical supply, and stormwater management systems become severely strained (FSD Kenya, Bankable Frontier Associates, and Digital Divide Data 2015). The latter of these was recently laid bare by massive flooding, causing damage estimated to be over 4 billion KSh (35 million USD) **[Unsupported source type (DocumentFromInternetSite) for source ACA24.]**, an estimate likely to be on the conservative side, and symptom of developmental problems which are likely to be exacerbated without a tactical shift towards a systems-based (as opposed to object-based) approach to housing development. At the same time, the potential benefits of higher-density (access to non-motorized transport, mass transit, proximity to job centres, etc.) are not being fully exploited, due to a lack of site-, urban-, and system-scale considerations when developing new housing, despite the 'plethora of tools… at the disposal of the planners charged with the responsibility of shepherding a more sustainable future' (FSD Kenya, Bankable Frontier Associates, and Digital Divide Data 2015), including compact and mixed-use development, preservation of green and blue infrastructure, protection or restoration of functional ecosystems, and incorporation of carbon sinks in development plans, amongst others.

The inability of both public authorities and private sector developers to implement more sustainable development and affordable housing practices has multiple causes. The first, and perhaps most obvious, is the economic cost-benefit of new construction, and sale prices or rental rates targeting a population in which the lowest urban income-earners spend 48% of their income on food, and after other necessary expenses, can afford to spend only 8% of meagre incomes on housing (Makunda 2018).

Awareness of, and willingness to advocate for, more resilient and sustainable development practices would seem to be a logical and major contributing factor to implementation as well. Certainly, core knowledge of sustainability and resilient urbanism will have significant influences on the shaping of more sustainable development practices (Ministry of Environment and Natural Resources 2016); however, further research reveals that this is not as large a problem as might be initially assumed. Awareness levels within the built environment sector in Kenya have been found to be relatively high; despite this, uptake of green building in Kenya is low, exacerbated by uncertainty, both generally and in terms of availability of green products in the value chain and reliability of performance (Moko and Olima 2014). A preceding 2017 study (Mottelson 2023) also noted high awareness and showed that a majority of housing developers both intended and believed they had successfully implemented sustainable development projects. It should be noted, however, in the absence of a clear metric, concepts of sustainable development are largely self-defined, so these figures should be viewed with scepticism; the author acknowledges this by specifically calling for the development of green building guidelines.

2.2 The Regulatory Environment Around Affordable Housing and Sustainability

This brings us to the question of how such guidelines can be effectively deployed. There is no shortage of existing tools, policies, and frameworks already available. Polices abound: The Green Economy Strategy and Implementation Plan from the Ministry of Environment and Natural resources and the Vision 2030 plan from National Treasury both include the development and implementation of certification standards for green buildings as a goal (Mwau et al. 2020; Mwero and Aduda 2019), initiatives also alluded to in the National Housing Policy (National Treasury and Planning 2018); the Kenya Building Research Centre (KBRC) has been tasked with championing and coordinating the government's Green Building Agenda, which includes the development of green building standards; a draft Green and Resilient Building Regulations policy from the Department of the Environment and Forestry lists a set of standards which morphed into a type of localized green building rating system called GreenMark, itself an adaptation of the 2011 Green Africa Building Standard (Niko Green 2018). In parallel, the Kenya Green Bond Programme has been established to enable green investments from local banks and corporates in multiple economic sectors, including building and construction (Ochieng 2007). A contract has also been signed between the International Growth Centre (IGC) and the Green Bonds Programme to promote Excellence in Design for Greater Efficiencies (EDGE)—a green building rating tool—to industry partners. The city of Nairobi has developed the Nairobi Green Building Principles with the Kenya Green Building Society (Okyere and Asibey 2024).

While these efforts are resulting in a progressive increase of commercial green buildings in Kenya, there is little uptake on residential buildings (Mottelson 2023), and the type of building stock utilizing these frameworks leads to a perception that green building is expensive (Ouma Agayi and Karakayaci 2020). Price, lack of information, limited options in the market, uncertainty on returns, lack of interest, and fear of unknown materials and techniques are hampering the uptake of green building in the housing market (Moko and Olima 2014).

The difficultly in navigating the laundry list of policies, tools, and proposed requirements mentioned above ought to be obvious. Well-intentioned efforts to steer the building industry towards more sustainable practices have produced a slew of frameworks and policies meant to incentivize or require different practices and approaches, but these efforts do not inherently generate clear pathways towards better practices. Rather, they result in a confusing morass of documentation which is convenient and easy to ignore, sometimes necessarily so, if one actually wants to deliver projects as opposed to spending exorbitant amounts of time swimming through uncoordinated and sometimes contradictory policies and initiatives. The lack of clarity in these frameworks, poor usability of guidelines, and duplicate (and sometimes incompatible) policies by multiple public entities are, rather than steering the industry towards a greener value chain, inadvertently creating major impediments to the delivery of affordable housing (Republic of Kenya 2016).

Attempts to raise the standard of construction through tighter regulation risks making self-built or self-improved housing impermissible, thereby making it more difficult to (legally) achieve even modest upgrading. In rapidly urbanizing areas with public authorities often overwhelmed by the pace of development, greater flexibility is needed, not greater stringency (Shah et al. 2020). A set of recommendations by the Centre for Science and Technology Innovations (The Green Bonds Programme - Kenya n.d.), developed as a way of greening and updating Kenya's building code, goes further in that it makes specific design recommendations at the scale of building sites and various architectural elements typically regulated in a code. However, it does not address larger, urban scale considerations, and many of the recommendations would need to be turned into highly prescriptive requirements to be enforceable by code, which itself would leave little space for practices which require design flexibility better expressed in the form of guidelines. Greater flexibility in the implementation implies a greater importance for context-appropriate design.

2.3 The Need for a Classification of Context-Specific Architectural Housing Typologies

Many of the current developments in affordable housing suffer from critical sustainability failures at a building scale. High-density units often fail to adequately factor in spatial requirements and are frequently built with 100% site coverage, making passive daylighting and ventilation nearly impossible (UN Habitat 2022) and leading to low quality-of-life indicators and higher operational costs (FSD Kenya, Bankable Frontier Associates, and Digital Divide Data 2015). These (often extra-legal) developments of high-rise blocks or tenements result in unsafe buildings, poor services, a lack of open or green space, and high costs of retrofitting existing infrastructure (UN Stats 2024), and lead to higher operational costs (FSD Kenya, Bankable Frontier Associates, and Digital Divide Data 2015). At the same time, the affordable housing programme's flagship project in Nairobi—also built as high-rise flats in, likely an (unsuccessful) effort to capitalize on economies of scale—has attempted to increase the standard of quality (albeit without any evident emphasis on sustainable design), but has been plagued by slow uptake of the completed units due to an inability of tenants to raise the necessary funds to participate in a lottery of unit allocation (UNECE 2019), and revealed that construction costs per square meter ended up being significantly higher than other, smaller-scale housing typologies (Urban Shift n.d.).

Much existing research focuses either on urban informality or on tenement housing due to their commonality and problematic nature on many fronts. This research and analysis makes clear that these building types cannot solve the housing gaps on their own; a diversity of housing stock is needed, adaptable to various circumstances, contexts, and climates. Uniform prototypes will not provide a way forward as sufficiently as adaptable typologies. The question then becomes how to create clear frameworks for how the contribution of design towards sustainable development can be best implemented across a wide variety of housing types, focusing on core principles of adaptability to circumstances and conditions which promote sustainability. Frameworks are still needed, but clarity in how to apply them also needs to be achieved.

In attempting to provide an at least partial answer, the identification of housing typologies with clear adaptability guidelines for each housing type to achieve myriad objectives related primarily to passive design strategies can be developed.

3 Methodology

The study of investigated housing typologies targeting the bottom 40% (B40) of income earners in Kenya focused on building types already commonly used by private sector developers, as well possible building types implementable under the public-sector affordable housing programme (AHP) with the aim of achieving broader objectives of sustainable development.

The analysis and identification of building types came in two primary forms. First, an analysis of the financing mechanisms was reviewed, seeking out the degree of real affordability each has historically achieved and the physical characteristics of the resulting construction through those means. Put succinctly, this study sought to match *how* realization occurred with *what* was actually built, to establish 'base case' conditions which could then be categorized and used to develop applicable standards.

Second, general observation and research of market-based construction were done, particularly for housing units built with the B40 as target occupants. Implementation methods, both in publicly subsidized programmes and in market-driven development, were scrutinized to further identify the resulting building forms which have already proven viable, if not ideal.

A synthesis of the first and second methods allowed for defining characteristics to be identified, and relevant building types categorized, defined by a fixed set of criteria of physical building characteristics. Where characteristics are entirely or substantially similar, architectural forms were considered to be variations of a single typology. Where multiple characteristics differ, a separate typology was classified.

These physical characteristics include:

- *Quantity of units*: while this may vary widely within some categories, it was loosely defined as high numbers of units within a building (roughly >15), moderate numbers (roughly 5–15), or low numbers (≤5 units) within the building.
- *Number of stories*: defined as high-rise (>5 stories, thereby requiring a lift), mid-rise (3–5 stories), or low-rise (1–2 stories). Note that the stories here include the ground floor, so a two-storey building represents ground plus one.
- *Party walls* versus *open space surrounding the building*: building types which always (or nearly always) have one shared party wall with another building were deemed to be one category; building types which may or may not have any shared party walls another; and building types which do not have party walls but have some open space on all sides another.
- *Presence, length, and interior/exterior condition of horizontal circulation*: buildings may contain veritably no horizontal circulation (e.g. units accessed directly from the outside or from a stairwell) or may contain horizontal circulation as part of the building. When it is part of the building, it is treated separately based on whether it is typically interior versus typically exterior.
- *Access to units*: units may be accessed directly from the exterior (ground floor), directly from vertical circulation elements (stairwells), or from corridors (interior or exterior).
- *Access to the site*: whether buildings are typically accessed directly from the public realm (streets or pathways), or from enclosed site areas, was considered.
- *Incrementality options*: the degree of flexibility or adaptability for incremental building was also considered.

4 Identified Affordable Housing Typologies and Their Application

4.1 Typologies

Although the identification of typologies flowed sequentially from their categorization first to the extraction of defining characteristics only coming last, a clear presentation of relations between delivery mechanisms, building types, and design principles requires an overview of the characteristics to be given first, in order to establish familiarity of characteristics before linking them to associated design types and subsequently searching for methods to increase the incorporation of sustainable design principles. Developing a workflow in this way allows for the planning, formation of criteria, and design evaluation of a complete process, tying the delivery mechanisms of affordable housing to clear but flexible design criteria which may improve the 'base case', precipitating (if only slightly) a green shift in the planning, design, and construction processes.

The study identified eight different building types (Fig. 1), the identification and categorization of which refers to common existing typologies, especially in Kenyan urban centres. The absence of a typology does not necessarily indicate that it does not exist nor that it cannot be used as a launch point for design; it means only that it is less common in current practice. In some cases, the alteration or combination of multiple typologies may create hybrid forms which do not neatly belong in any one typological category. Where this occurs, professional judgment would be needed in applying relevant guidelines.

To ease the understanding and use of these typologies, a simplified, iconic graphic representation of each type was developed as a system, further allowing for illustrative examples of how a 'base case' can be manipulated and adjusted to improve performance against certain site conditions and passive design criteria. This can be further expanded to create particular design considerations which guide the user through important upfront design decisions.

4.2 Applying Type-Specific Strategies

The classification of typologies is, on its own, an academic exercise; the utility of it comes through the ability to identify prominent design considerations by building type and to facilitate design actions which improve the typical 'base case' scenario. A set of green building guidelines can therefore be developed not only as generic criteria, but with specific adjustments appropriate to each building type in mind. Further adjustments—not directly related to building type, but indirectly applicable to them—were developed based on other important factors, such as the density of the urban setting and climate zone being designed for. These adjustments are equally as important in establishing the parameters for design (which is the ultimate purpose of creating green building guidelines), but are beyond the scope of this study.

In this study, indicative criteria for adjusting the design of building types were done in a multitude of ways (though this is not intended to be a fully exhaustive list of possibilities): optimizing direct visual and physical access to green pockets (which itself is linked to the incorporation of carbon sinks at a macro-level and appropriate microclimates by climate zone at a site level), applying appropriate natural ventilation strategies (which itself is influenced by form and orientation by climate zone and by relevance of ventilation methods by climate zone), and the degree to which a building type allows for incrementality (itself linked to flexible design criteria). This list can be (and ought to be) expandable over time, creating a potential feedback loop which may continue to evolve the guidelines over time.

Graphic icons play a role in making the design goals and guidelines approachable and easily comprehensible, though it would be ill-advised to use them literally rather than indicatively. Myriad permutations of design inputs are possible, creating a flexibility often lacking in other green building standards. Refer to Fig. 2 below for a selected graphic sample of influences and adjustments, each of which would be accompanied by descriptive text in a complete set of guidelines.

5 Conclusions

If a massive public investment in new affordable housing construction does materialize, it represents an uncommon opportunity to reset dialogues and strategies around rapid urban development and recalibrate the design and construction industries towards a greater, greener, more resilient, and sustainable trajectory. Kenya's overarching climate change and green economic policies help bring this issue to the fore and make space for transformative change.

However, even in an environment where coordinated initiatives promoting sustainable development are present, and even if a regulatory framework is well-aligned with those initiatives, a green shift in the construction of affordable housing will not be self-propelled. Policies are not panacea; design application and implementation remain critical links in the sustainability chain. In many sub-Saharan African contexts, the influence of architectural and urban design on the success of major public initiatives is too often relegated to a second-tier consideration, a 'nice to have' financially out of reach due to the limited resources available to tackle a seemingly intractable problem at such an immense scale, value-engineered out of the process. Yet the data do not support this: housing costs in Kenya are driven predominantly by land, materials, and access to infrastructural services; design or professional fees are comparatively low (Wakhungu 2017).

What will propel this shift is, amongst other things, to disentangle a bewildering web of competing regulations which often fail to reference each other, collate existing resources, understand possible synergies between them, and synthesize these to generate realistic and contextually appropriate standards and frameworks. Prescribing conformance to a multitude of competing standards with limited contextual adaptability is counterproductive; space must be opened

Building Type	Definition	Graphic
Low-cost flats	Stand-alone multi-unit housing building of a large scale, $\geq$4 stories (typically more), with $\geq$4 units (typically more) on each floor, often built up to site boundaries on at least 2-3 sides. Units are typically arranged around a double-loaded corridor with access to units from the interior of the building. (NB: this building type is typically referred to as "tenements." This paper uses the term "low-cost-flats" in order to avoid the negative connotation associated with tenements.)	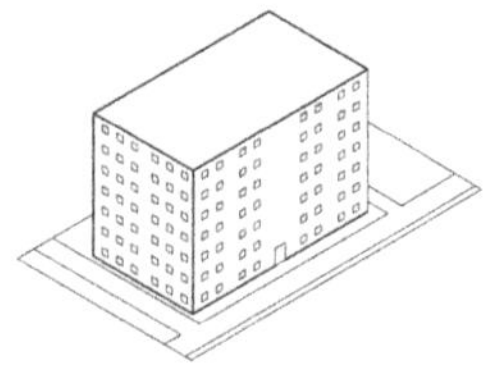
Infill building	Multi-unit housing building sharing two party walls with neighbouring buildings, typically on a narrow site, $\geq$2 stories, with more than one unit on a floor. Units are typically accessed from a stairwell inside the building.	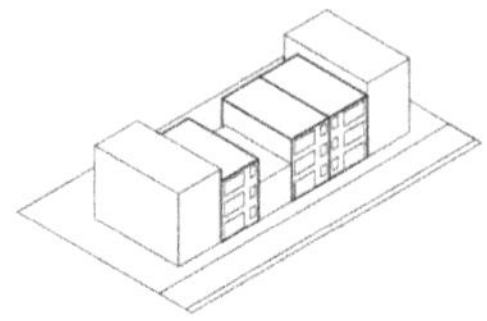
Freestanding block	Stand-alone, multi-unit housing building with open space on all sides (not sharing a party wall) or with a maximum of 1 party wall, $\geq$2 stories but typically <5 storeys, with $\geq$2 stories units per floor, typically more. Units are typically accessed from a stairwell inside the building, but without long corridors.	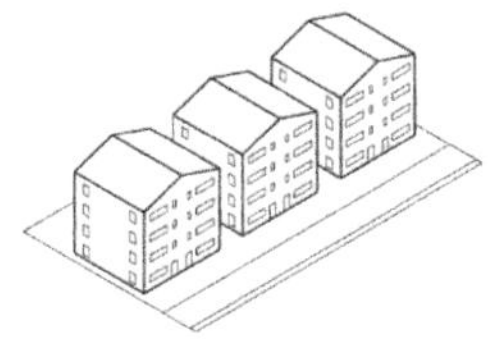
Linear block	Free-standing, multi-unit housing building with open space on all sides (no shared party walls), more than 1 but $\leq$4 stories, typically with access to units off exterior corridors or via multiple vertical cores within the building. Multiple units per floor, often back-to-back sharing a central, internal wall.	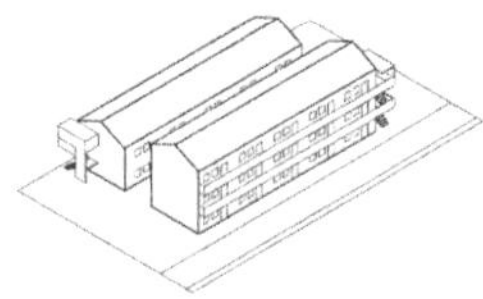
Rowhouse	2–3-unit housing building sharing two party walls with neighbouring buildings, typically on a small, narrow site, $\leq$3 stories, with either one unit per floor or units utilizing multiple floors. Units are accessed directly from the exterior, typically from a public street, or from access roads on larger sites.	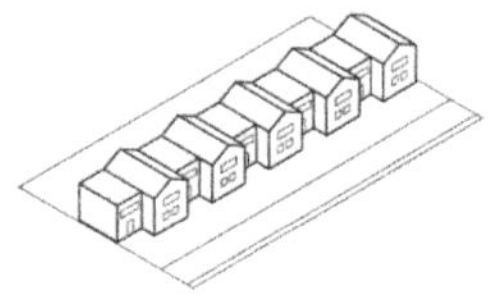
Swahili low-rise *ploti**	A lineup of individual units on a larger plot, $\leq$2 stories, with at least one-party wall against the back of another unit or (most frequently) a boundary wall, and typically minimal (if any) space between units, sharing a central, open, shared space/courtyard with a walkway. Units are typically accessed first by a single access point to the site, then directly from the exterior within the site. Though aesthetic resemblance may not be present, this type is modelled on a traditional Swahili communal house plan but is made of subdivided plots on a larger site. **"Ploti" is an adopted term coming from "plots" or "plotting"; it can be loosely translated to "sectional title in a larger estate"*	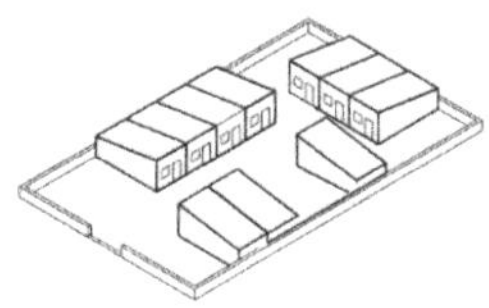
Bungalow	Small, free-standing, single-storey houses with open space on all sides, housing either one family unit or two units as a duplex. Units may be accessed either directly from streets or pathways, usually with some exterior space in between, or may be part of a cluster or grouped estate within a larger, protected site.	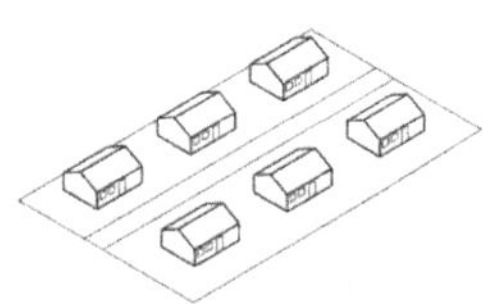
Improved shack	Very small single- or occasionally double-storey unit, typically a single room in the vicinity of 9 or 9.5m^2, with access to shared ablution facilities either as an addition to a larger unit or as a self-contained unit. Units may be accessed either directly from pathways (typically pedestrian paths in informal settlements) or streets, through semi-private spaces belonging to pre-existing compounds, or through pre-existing units as extensions.	

Fig. 1 Affordable housing types identified. Drawings by the author

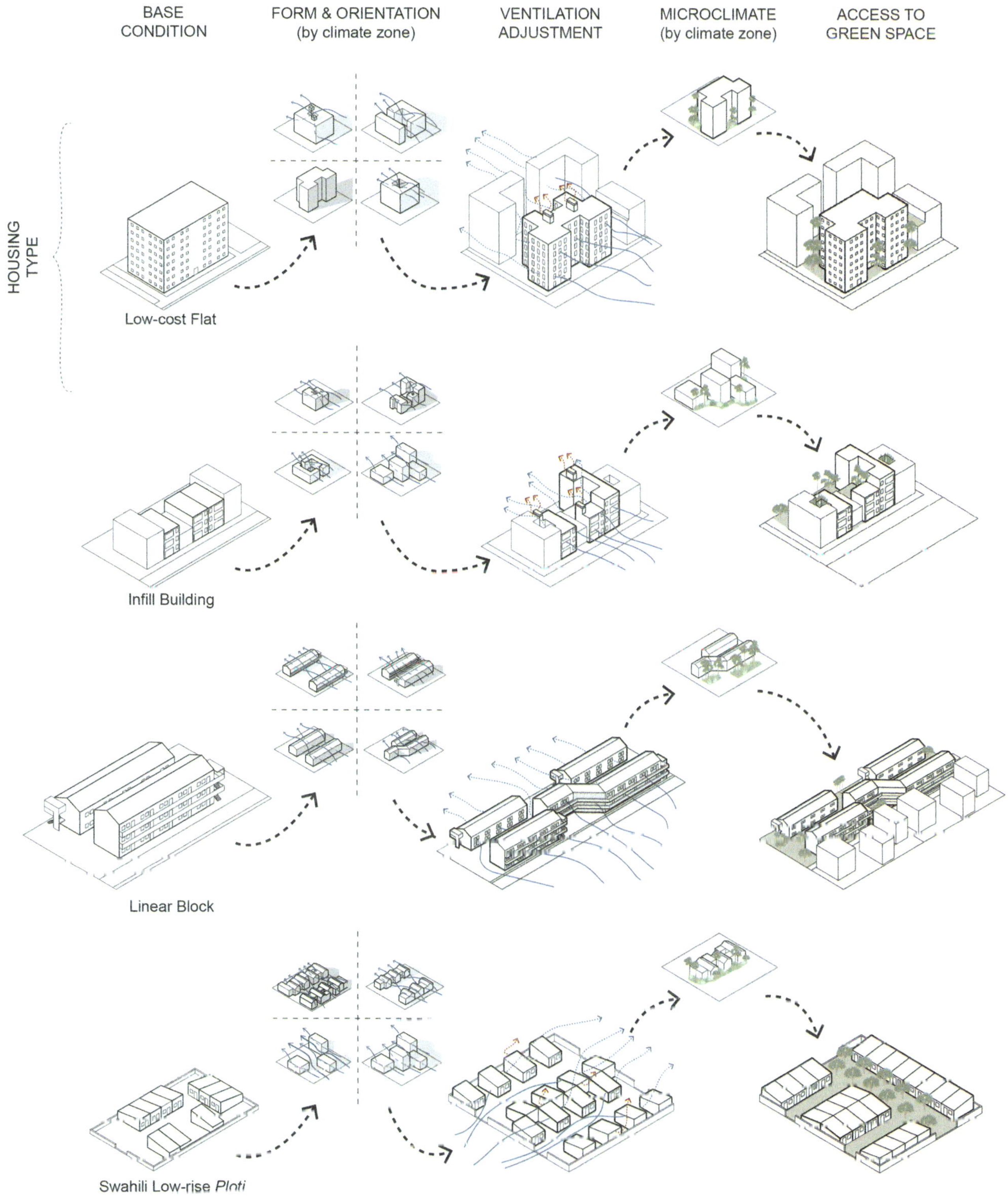

Fig. 2 Examples of adjustments to typologies to meet green building criteria. Drawings by the author

up for flexible but targeted design solutions which work across scales, targeting practical (rather than ideal) solutions which address social (World Bank 2019) and ecological needs. Workflows which consider match delivery mechanisms with typologies and implementation methods must be considered (Fig. 3).

At the same time, methodical support is needed to orient the course of design from project inception to completion,

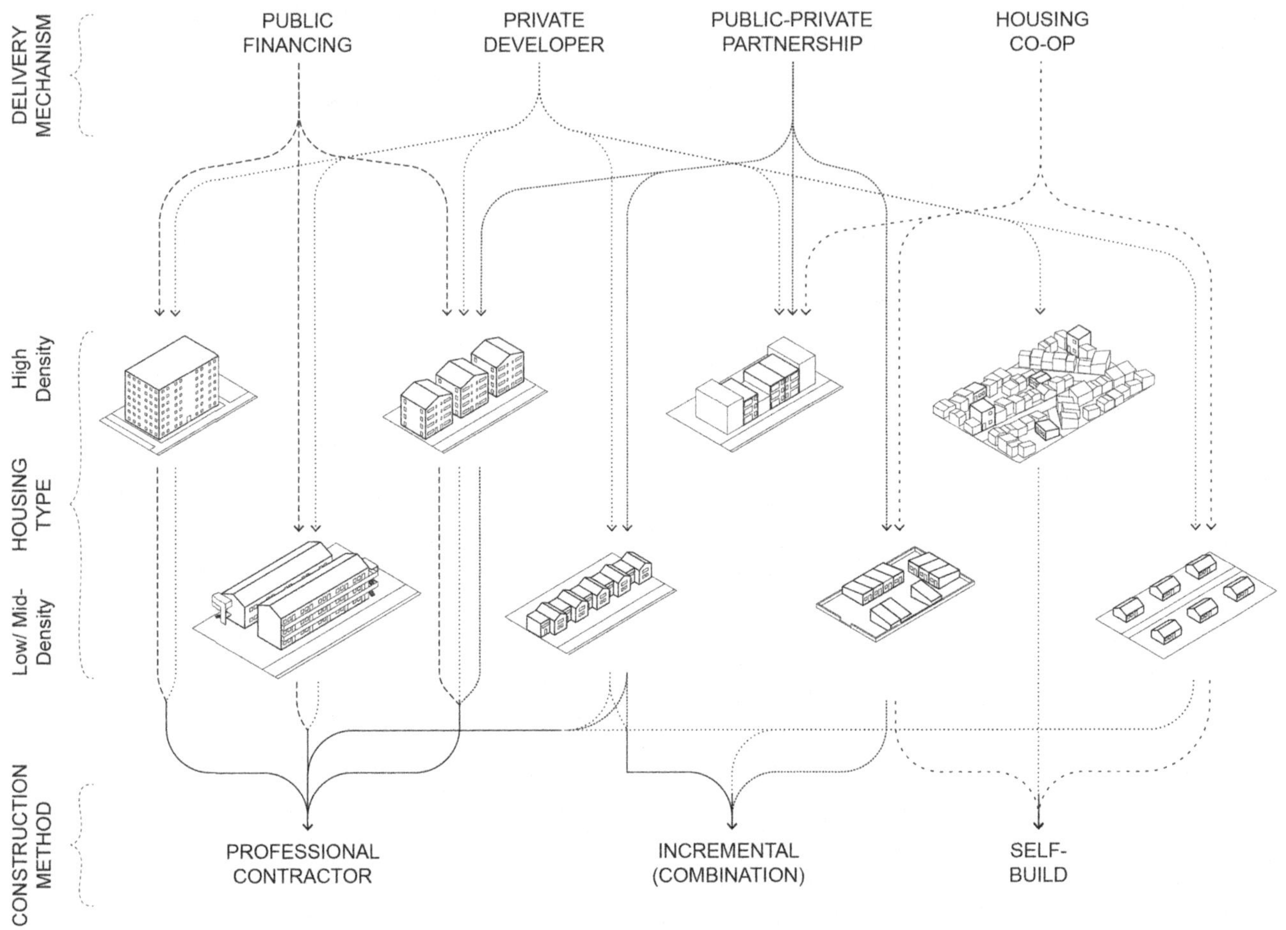

Fig. 3 Common finance mechanisms and construction types of housing typologies. Drawings by the author

engendering conscientious and resilient planning decisions, indicating flexible site, and building design targets appropriate for a multitude of building types which may be deftly applied. A ground-up and design-oriented reframing of the regulations and guidelines is thus an important—albeit limited—tool to help ease the transition towards more accountable design decisions which disincentivize standard, 'base case' industry practices. Similar approaches in other contexts will yield differing typologies and methods, but may contribute to a broader green shift within the growing design and construction industries in developing countries.

Acknowledgments The graphic imagery used here was produced by the author and Bantu Studio Design and Research and is used with permission. Contributions to the identification and naming of housing typologies identified were made by Moses Gathua and Baraka Mwau.

Competing Interests The author has no conflict of interests to declare that are relevant to the content of this chapter.

References

ACAPS (2024) Briefing note: Kenya floods. https://www.acaps.org/fileadmin/Data_Product/Main_media/20240514_ACAPS_Briefing_note_Kenya_Floods.pdf#page=1. Accessed from ACAPS

Agence Phare (2024) Housing for SDGs: building the case for housing as a cornerstone to achieve SDGs. Agence Française de Développement. https://www.afd.fr/en/ressources/housing-sdgs-full-report-access-housing-achievement-sustainable-development-goals. Accessed 1 Jul 2024

CSTI (2023) Scoping review: opportunities for increasing green infrastructure in Kenya's built environment. CSTI

Espelia, GRET, SDI Kenya, Bantu Studio (2024) Perspectives for the Kenyan affordable housing sector: approaching the market and evaluating recent initiatives. Agence Française de Développement

FSD Kenya, Bankable Frontier Associates, and Digital Divide Data (2015) Kenya Financial Diaries 2012–2013. FSD Kenya. https://www.fsdkenya.org/research-and-publications/datasets/kenya-financial-diaries-2012-2013-socio-economic-and-demographic-datasets/. Accessed 29 Jun 2024

Makunda CS (2018) Sustainable housing through sustainable planning practices: challenges and opportunities for formal housing provision in Nairobi, Kenya. In: Azeiteiro IU, Akerman M, Leal Filho W, Setti A, Brandli L (eds) Lifelong learning and education in healthy

and sustainable cities. Springer International Publishing, Cham, pp 539–549. https://doi.org/10.1007/978-3-319-69474-0_31
Ministry of Environment & Natural Resources (2016) Green economy strategy and implementation plan 2016–2030. Government of Kenya
Moko SK, Olima WH (2014) Determinants of house construction cost in Kenya: a case of Nairobi county. Int J Bus Commerce 4(1):19–36
Mottelson J (2023) On informal housing supply restrictions and livelihood in informal settlements: implications for sustainable development. Sustain Dev 31(5):3566–3578. https://doi.org/10.1002/sd.2611
Mwau B, Scerdlik A, Makau J (2020) Urban transformation and the politics of shelter: understanding Nairobi's housing markets. IIED
Mwero JN, Aduda VO (2019) End-users' perspective on green building: a study in Nairobi. Int J Sci Res Publ 9(4):649–669. https://doi.org/10.29322/IJSRP.9.04.2019.p8880
National Treasury and Planning (2018) Third medium term plan 2018–2022. Government of Kenya
Niko Green (2018) The GreenMark standard for green buildings. Green Africa Foundation
Ochieng CC (2007) A critical review of housing delivery in Nairobi: different actors—different socio-economic groups. Archnet-IJAR 1(3):140–152. https://doi.org/10.26687/archnet-ijar.v1i3.44
Okyere EA, Asibey MO (2024) Vacillant planning for creating green cities: the empirics of urban greenery in Kumasi, Ghana. Cities 152:105177. https://doi.org/10.1016/j.cities.2024.105177
Ouma Agayi C, Karakayaci Ö (2020) The role of changing housing policies in housing affordability and accessibility in developing countries: the case of Kenya. J Contemp Urban Aff 4(2):49–58. https://doi.org/10.25034/ijcua.2020.v4n2-5
Republic of Kenya (2016) National housing policy for Kenya. Republic of Kenya
Shah S, Aldridge S, Mugwe A, Njogoro I, Maina M (2020) Context and opportunities in the Kenyan affordable housing sector. Agence Française de Développement and Proparco
The Green Bonds Programme - Kenya (n.d.) The green bonds programme—Kenya. Green Finance. Green Growth. https://www.greenbondskenya.co.ke/. Accessed 12 Feb 2024
UN Habitat (2022) World cities report 2022: envisaging the future of cities. UN Habitat
UN Stats (2024) SDG goals: SDG 11. Department of Economic & Social Affairs, Statistics Division. https://unstats.un.org/sdgs/report/2019/goal-11/. Accessed from United Nations
UNECE (2019) Guidelines for the formalization of informal constructions. United Nations. https://unece.org/sites/default/files/2022-01/technical_guidelines_informal_settlements_e.pdf
Urban Shift (n.d.) Nairobi: Kenya: green building guidelines. C40 cities
Wakhungu J (2017) Adoption of green building concepts in residential developments in Nairobi City County. University of Nairobi
World Bank (2019) Managing risks for a safer built environment in Kenya. The World Bank Group

Design Approaches for Community Engagement and Social Sustainability

Investigating Universal Design Within the Framework of Social Sustainability in Northern Nicosia with Analyses and User Views

Ümran Duman and Buket Asilsoy

Abstract

In recent years, involving the universal design concept in urban spaces has gained importance with the increasing variety of users. Urban designs adopting the universal design (UD) concept with an aim of welcoming all individuals contribute to social sustainability (SS). Within this framework, first, based on the authors' previous work, a theoretical framework is redeveloped to clarify the connection between UD and sustainable urban settings in relation to urban design parameters (UDPs) and social sustainability. Adoption of the concept of UD for the urban spaces is also highly required for Northern Nicosia. Therefore, at the methodology part of the research, first, Marmara neighborhood, selected as the research area, has been evaluated. A qualitative assessment regarding the UD principles and a quantitative evaluation based on the Turkish Standards Institute (TSI) standards were utilized for the six urban items of the selected avenue. In light of the results at Yüzbaşı Tekin Yurdabak Avenue, the positive impact of the evaluated items on sustainable urban design parameters is insufficient and these items are mostly inappropriate regarding the measured international standards. Second, a survey was applied in three neighborhoods (Taşkınköy, Göçmenköy, and Marmara) with the participation of 150 randomly selected individuals in order to understand user satisfaction and views of the individuals about urban design parameters and universal design principles. The respondents relatively pointed out a noticeable disagreement about the presence of the principles of UD in urban spaces. Accordingly, it can be suggested that the urban areas in Northern Nicosia are neither convenient nor sufficient for welcoming the variety of all individuals. Therefore, the concept of UD should be adopted and the urban structure should be improved by applying the existing and required regulations in the city.

Keywords

Social sustainability · Universal design · Urban space · Analysis · Survey · Northern Nicosia

1 Introduction

According to the United Nations, the proportion of urban residents increased from 13% in 1900 to 29% in 1950, 50% in 2009, and 69% in 2050 (Shen et al. 2012). As such, since the latter half of the 20th, especially in developing countries, the world has witnessed its most rapid rate of urbanization (Chadchan and Shankar 2009). In 2008, more than half of the global population lived in urban areas for the first time and this proportion is expected to grow further (United Nations Population Fund 2007). The rapid urbanization that has been ongoing since the Industrial Revolution is a major contributor to the current environmental crisis we are facing as humanity (Yıldırım et al. 2020). As cities expand day by day, the diversity of their populations also grows. Many different types of people, including those with impairments, live in urban settings.

Within this framework, sustainable urbanism has emerged as a key concept for tackling the challenges that cities are facing. As a broad concept, sustainable urbanism involves disparate dimensions that can be defined within four main pillars as environmental, economic, social, and cultural. Since its initial definition more than three decades ago, sustainability has gained widespread acceptance as an essential framework for directing urban development and policy. It serves as the foundation for a substantial body of literature

Ü. Duman (✉)
Department of Architecture, Faculty of Architecture, Near East University, Nicosia, Cyprus
e-mail: umran.duman@neu.edu.tr

B. Asilsoy
Department of Landscape Architecture, Faculty of Agriculture, Near East University, Nicosia, Cyprus
e-mail: buket.asilsoy@neu.edu.tr

F. Alberti et al. (eds.), *Towards Sustainable Urban Regeneration: Utilizing Environmental, Cultural, and Social Assets*, Advances in Science, Technology & Innovation, https://doi.org/10.1007/978-3-032-02091-8_31

on urban design, architecture, and planning (Williams et al. 2000). By preserving natural resources, guaranteeing financial stability, and raising living standards, sustainability aims to create livable and healthy urban environments (Ranhagen and Groth 2012).

Social sustainability (SS) is a fundamental cornerstone of this concept. To achieve social sustainability, it is crucial to address aspects such as safety, human health, urban quality of life, stability, cultural identity, justice, accessibility, and the integration of disabled individuals into society. It can further be claimed that "social sustainability" has emerged as a distinct stand-alone topic (Turkington and Sangster 2006).

In addition, for boosting the social dimension of daily life, our cities and settlements require qualified urban areas (Arenghi 2020). As such, public spaces should not be neglected to be designed convenient for all users as they are among vital features of a city. A space shared by multiple groups of people, regardless of gender, color, age, or socioeconomic status, is referred to as an urban space (Lotfata and Ataöv 2020). Additionally, public spaces can act as centers for political, cultural, and economic activity and have the power to bring people together (Carmona et al. 2010; Jacobs 1961; Thompson 2002). In other words, public spaces share a significant role in urban planning and community socializing (Paşaoğulları and Doratlı 2004). Accordingly, designing urban spaces where individuals interact socially by the support of universal design (UD) concept contributes to social sustainability.

Furthermore, the existence of public spaces is crucial regarding urban planning and design. Urban spaces have to be planned properly for all users including various age groups and skills. For this reason, adopting the universal design (UD) concept is feasible and sensible as it aims to create inclusive designs for accommodating a diverse range of users in streets, squares, green spaces, and parks as common areas. It can be claimed that urban areas planned by taking account of the UD concept will accordingly contribute to SS.

In 1985, UD was defined as "products and environments created to be usable by all individuals, to the greatest extent possible, without the need for adaptation or specialized design" by Mace (Gossett et al. 2009). It emphasizes the importance of acknowledging and understanding that individuals will experience varying levels of ability throughout their lives (Pinna et al. 2020). To the greatest extent possible, this term seeks to design environments that are accessible, intelligible, and useful to all people. Promoting equal rights and opportunities for all users is its main objective (Lid 2014). It is significant to adopt the universal design concept in urban space designs which are used by everyone living in the city.

As the capital of Northern Cyprus, Northern Nicosia is the country's administrative center. Additionally, the city is known for its lively and varied social and economic activity. Besides, Nicosia hosts students from different age-groups in various private and public schools and universities. However, it cannot be claimed that urban spaces fully meet UD requirements in the city, despite the presence of regulations, legislations, and standards intended to ensure universal design principles.

Planning policies ought to be founded on accurate, scientific data regarding the intended public (Asilsoy and Oktay 2018). Therefore, in this study targeting the Northern Nicosia as the research context, two phases were involved as the research methodology. First, an assessment for a thorough analysis was made both quantitatively and qualitatively in Yüzbaşı Tekin Yurdabak Avenue of the Marmara neighborhood in terms of the Universal Design principles. Second, a questionnaire was applied in North Nicosia investigating the views and preferences of the users. Data were collected in order to evaluate residents' opinions about public spaces regarding sustainable urban design parameters (UDPs) and universal design principles.

Within this framework, the relevant literature is summarized in the next section. In the end of the next section, a theoretical framework was redeveloped for conceptualizing the link among UD, urban design parameters, and social sustainability for creating sustainable urban environments. Afterward, materials and methodology of the assessment made in the Marmara neighborhood and applied user survey about UD in the selected neighborhoods of Northern Nicosia are explained; findings and evaluation of the research are given. Finally, the last part involves a conclusion and recommendations.

2 Literature Review

Among all four pillars, social sustainability can be argued to be relatively neglected. Social issues have begun to be considered within the sustainability agenda only after the late 1990s (Monfaredzadeh and Krueger 2015). "Agenda 21" is one of the initial documents clearly highlighting the aspects of social issues and human development within the concept of sustainability (Harun et al. 2014).

Individuals with different characteristics are in social interaction in urban spaces used by everyone in the city. Urban spaces have to be used by different user groups in order to strengthen the social interaction. UD, a design method that aims to design for all, can be suggested as an efficient tool of urban space design for contributing SS. Social sustainability combines the organization of the social realm with the planning of the physical environment (Borowczyk 2018).

There are disparate studies seeking to define social sustainability. It can be argued that social sustainability encompasses both tangible and intangible measures as a broad concept. Dempsey et al. (2011) determined several nonphysical and physical aspects as a result of reviewing literature on social sustainability. Nonphysical factors include aspects such as social capital, mixed tenure, and community. In addition, urbanization, a pleasant public space, sufficient housing, local environmental quality and facilities, accessibility, neighborhood character, walkability, and sustainable urban planning are further physical factors (Stevenson 2020). Furthermore, social sustainability covers notions of social equity, inclusion, quality of life, poverty, participation, social justice, etc. (Manzi et al. 2010; Ketschau 2017; Pitarch-Garrido 2018).

As a wide term, there are physical and nonphysical components constructing the SS. In sum, concluding the existing literature about SS, it can be suggested that physical indicators of SS can be defined in two main headlines as sustainable urban forms and urban design parameters. Mixed land uses, density, compactness, sustainable transportation, and ecological design can be listed to be the items of sustainable urban forms. Furthermore, accessibility, connectivity, walkability, safety, adaptability, legibility, and comfort are the features of urban design parameters (UDPs) according to the related literature. In addition, human rights, quality of life, equity, security, poverty, and social justice can be summarized to be the nonphysical indicators of social sustainability.

Among physical indicators of SS, the UD has the potential to be a catalyzer for the social aspect of urban sustainability via directly strengthening sustainable UDPs in particular. The universal design (UD) concept as an inclusive approach involving a design philosophy focuses on creating designs that cater to the needs of diverse users within a single application. In the universal design framework, everyone is regarded as a user, regardless of age, physical attributes, developmental stage, or skill level. UD is not a passing trend or fad but a lasting design philosophy based on the understanding that the wide spectrum of human abilities is normal, not exceptional. UD addresses the barriers encountered by people with disabilities, older adults, children, and other groups that are often overlooked in the typical design process. UD offers advantages to all users.

In 1997, a multidisciplinary team of experts developed the fundamentals of universal design (UD) to define its scope and offer guidance. The concept gained recognition in the mid-1990s (Steinfeld and Maisel 2012). Together with a set of recommendations for each, these principles served as a useful tool for early adopters to understand universal design and they are still widely used today (Maisel and Ranahan 2017). In order to promote universal accessibility, the UD principles are explained in the context of the built environment (Yiing et al. 2013). These principles are described in Table 1 below.

From young children to frail older adults, the UD provides services for a broad range of users (Kristl et al. 2019). Universal design, utilized across various design disciplines, can be applied at multiple scales, from architectural and interior design to urban and landscape design. Universal design spans from product design to urban planning, integrating and interacting across different fields, in other words. Such that designing urban environments by the concept of UD will guarantee that the cities as a whole are suitable for all users. In sum, the study's conceptual framework is shown below in Fig. 1.

3 Research Methodology

3.1 Research Context

As an island in the Mediterranean, Cyprus's strategic location has made it a focal point for various civilizations throughout history. But in 1974, the island was divided into northern and southern parts with a buffer zone. İskele, Kyrenia, Güzelyurt, Nicosia, Famagusta, and Lefke are the six main districts of Northern Cyprus. Northern Nicosia as the capital of Northern Cyprus serves as the central district and covers approximately an area of 502 km^2. Nicosia district has boundaries defined by Kyrenia in the north, Famagusta in the east, South Nicosia in the south, and Güzelyurt in the west (Fig. 2).

Since the Hellenistic period, Cyprus has been recognized for its advanced urban development. Traditional Cypriot settlements are characterized by a cohesive design, featuring well-proportioned narrow streets, cul-de-sacs, and organi-

Table 1 Principles of UD (Imrie 2012)

Principle	Description
Equitable use	The design is useful and marketable to individuals with diverse abilities
Flexibility in use	The design accommodates a wide range of individual preferences and abilities
Simple and intuitive use	Use of the design is easy to understand, regardless of the user's experience, knowledge, language skills, or current concentration level
Perceptible information	The design communicates necessary information effectively to the user, regardless of ambient conditions or the user's sensory abilities
Tolerance for error	The design minimizes hazards and the adverse consequences of accidental or unintended actions
Low physical effort	The design can be used efficiently and comfortably and with a minimum of fatigue
Size and space for approach and use	Principle of appropriate size and space is provided for approach, reach, manipulation, and use regardless of user's body size, posture, or mobility

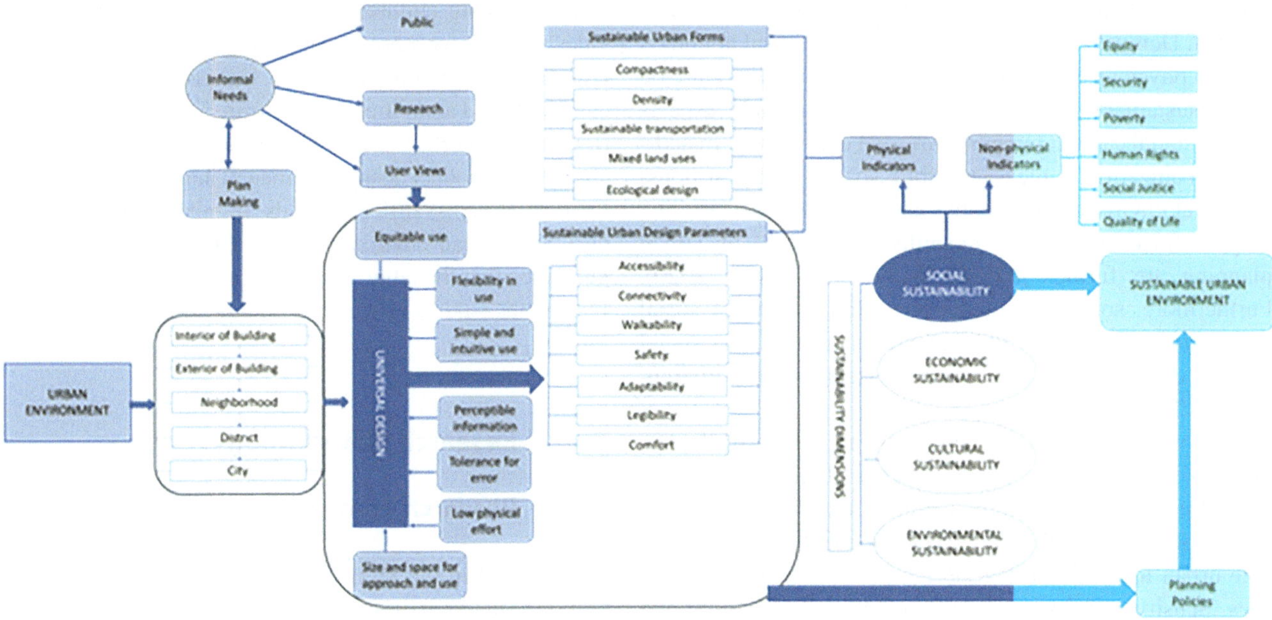

Fig. 1 Theoretical framework of the research. (Adopted from Duman and Asilsoy 2022)

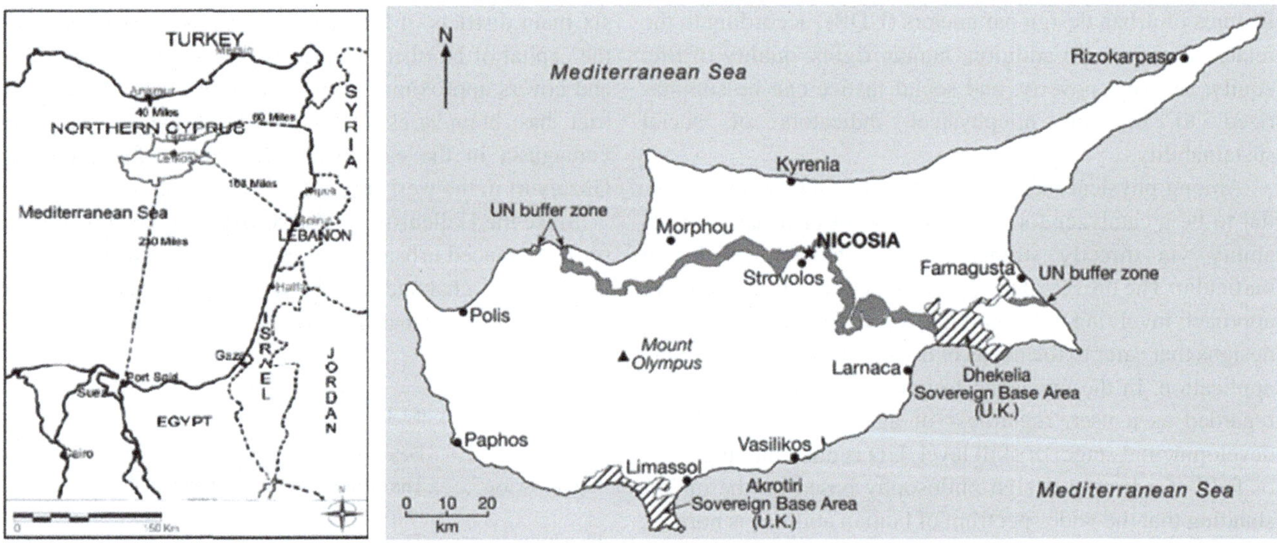

Fig. 2 Location of the island of Cyprus and Nicosia (Asilsoy and Oktay 2016; Oktay 2007)

cally designed public spaces where streets intersect (Oktay 2002). Nicosia was one of the cities representing such characteristics of the island within its built environments with the presence of the streets and squares, etc. However, in the meantime, with the effect of contemporary urban planning activities, it can be suggested that such a dynamic has been almost demolished.

Nowadays, as the capital of Cyprus, Nicosia has a relatively high population density and is the most densely populated city in North Cyprus. In the northern part of the city, there are 25 neighborhoods affiliated with Nicosia Turkish Municipality. According to the census conducted in 2006, while the total de-jure population of these neighborhoods was 56,146, it is seen as 61,378 in 2011 (State Planning Organization 2020).

Three neighborhoods—Taşkınköy, Göçmenköy, and Marmara—in Northern Nicosia were selected for the user survey. This area is significant as it encompasses key trade and business centers, as well as residential zones. To the north of the study area is Dr. Fazıl Küçük Boulevard, while Kemal Aksay Street runs along the eastern edge. Major arteries that connect to the city center are situated within and

Fig. 3 Taşkınköy, Gökmenköy, and Marmara neighborhoods (N. Nicosia City Guide, 2020)

Table 2 According to the last census, the population of Taşınköy, Göçmenköy, and Marmara (State Planning Organization 2020)

Neighborhood name	Population
Taşkınköy	3847
Göçmenköy	3003
Marmara	3081

around this region (Fig. 3). According to the latest census, the populations of these neighborhoods are listed in Table 2 (State Planning Organization 2020).

The neighborhoods feature a variety of urban spaces, including streets, parks, squares, avenues, and playgrounds. These areas are crucial for the social interaction of individuals having diverse characteristics such as physical and mental abilities, age, gender, and education level. Visual representations of the urban spaces in the selected neighborhoods are shown in Fig. 4.

However, when looking at the photographs seen in Fig. 4, it can be suggested that the public spaces have deficiencies both in quantity and quality. It can further be commented that a new agenda focusing on the redevelopment of these urban spaces including the implementation of universal design principles must come into effect with the help of both local and central government.

Marmara, chosen for the initial phase of the study, is a rapidly developing area in Northern Nicosia and is known for its high population density. The neighborhood features numerous urban spaces. In this study, Yüzbaşı Tekin Yurdabak, a particularly heavily utilized urban space within Marmara, was examined and analyzed (Fig. 5).

A two-lane road, Yüzbaşı Tekin Yurdabak Avenue that was selected for the research, was examined on-site. This avenue features two separate bus stops and parallel parking lots. Additionally, it includes various urban furnishings such as signs, lighting elements, and garbage cans placed at different points along the road. There are also three distinct pedestrian crossings (Fig. 6).

3.2 Research Design

The research design consists of two phases. Initially, the avenue, Yüzbaşı Tekin Yurdabak Avenue of the Marmara neighborhood, selected as the study area, was assessed. First phase involved a qualitative evaluation regarding the seven main universal design principles. In addition, the first phase also involved a quantitative assessment in relation to TSI standards, applied to six urban space elements within the selected avenue. Second, a questionnaire was applied investigating the views and preferences of the users.

3.2.1 Assessment

The present state of six randomly chosen items along the avenue was assessed based on their material, size, and location with respect to the principles of UD. Each item was photographed, and its dimensions were recorded. These urban elements were qualitatively examined. Additionally, in this phase of the research, the effects of these urban space items on various sustainable urban design parameters were analyzed.

Furthermore, in the quantitative part of the first phase of the research, the selected urban space items were evaluated based on their size characteristics regarding the Turkish Standards Institution's determined standards. Six items as parking lots, pedestrian crossings, garbage cans, sidewalks/walking paths, signs, and bus stops were assessed (Fig. 7).

3.2.2 The User Survey

In order to examine residents' perceptions about the UD principles in Nicosia urban spaces, this study was one of five sections that made up the questionnaire. In this research, only last part of the questionnaire with the demographic data as the first part was evaluated. Last section of questionnaire had seven items about universal design principles. A user

Fig. 4 Street, avenue, and park visuals from Taşkınköy, Göçmenköy, and Marmara (by Authors)

survey was conducted with participants during the first 2 weeks of September 2020. For the section assessing user opinions on various items, a Likert-type five-point scale was employed to capture responses, ranging from "strongly disagree" to "strongly agree." Responses were coded from 1 = strongly disagree to 5 = strongly agree. The measures of this study are listed below.

Demographic data: Information including factors such as gender and age was collected through six specific items.

User perceptions about UD principles: Data on user perspectives regarding universal design principles were gathered using seven specific items. Each item is used for one of the UD principles (Table 3).

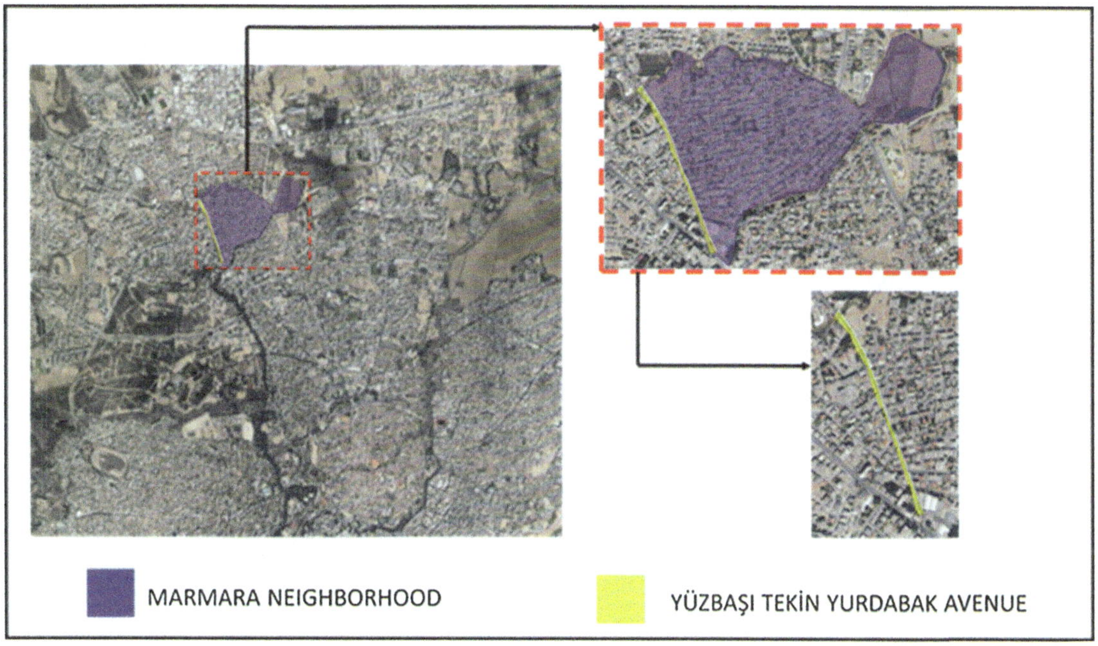

Fig. 5 Yüzbaşı Tekin Yurdabak Avenue of Marmara neighborhood (by Authors)

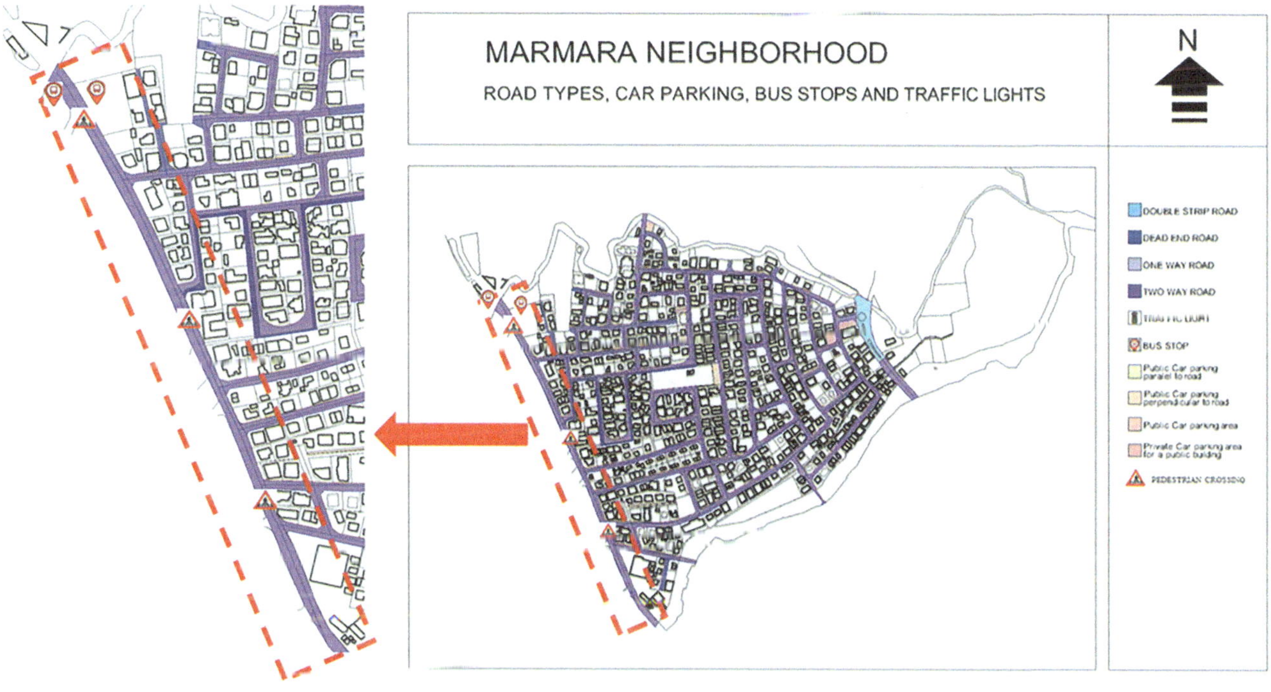

Fig. 6 The Yüzbaşı Tekin Yurdabak Avenue Analysis Details Map (by Authors)

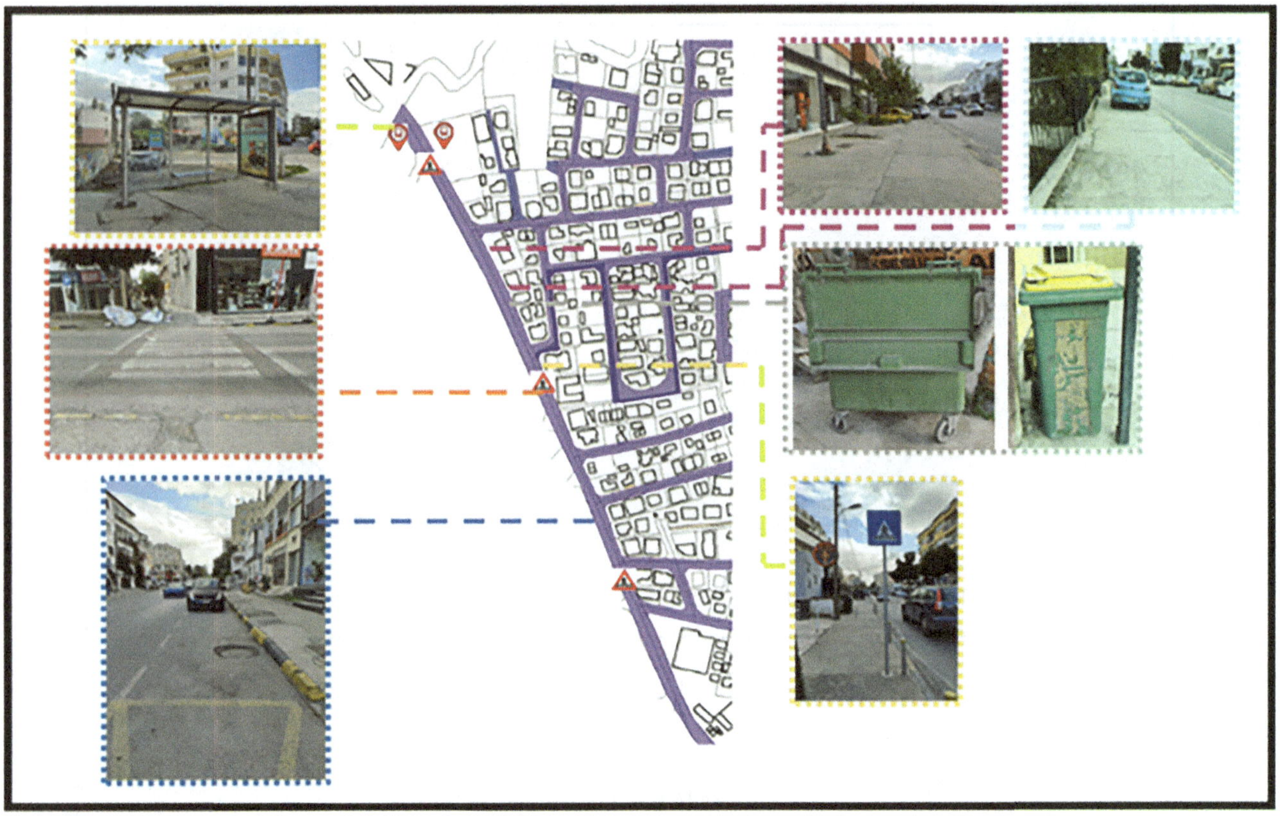

Fig. 7 The evaluated urban space items (by Authors)

Table 3 User survey items representing universal design principles in the last section (by Authors)

Equitable use	Item 1: "In the urban areas (streets, avenues, parks, squares, etc.) of the neighbourhood, everyone has the opportunity to move around under the same conditions"
Flexibility in use	Item 2: "Considering the diversity of users in the neighbourhood, there are elements with the same function designed with different features in urban spaces like street, avenue, park, square, etc. (such as garbage bins, book sharing points located at different heights)"
Simple and intuitive use	Item 3: "Urban spaces (streets, avenues, parks, squares etc.) in the neighbourhood can be easily found and used by everyone with the help of perceptible information"
Perceptible information	Item 4: "Information signs in urban areas of the neighbourhood (streets, avenues, parks, squares, etc.) can be understood by everyone"
Tolerance for error	Item 5: "Features (water items, urban furniture, etc.) found in urban areas of the neighbourhood (street, avenues, park, square, etc.) are designed to minimize the likelihood of accidents that may occur due to the carelessness and / or physical / mental abilities of the users (such as avoiding proximity by planting flowers around the water element)"
Low physical effort	Item 6: "Level differences in urban spaces (streets, avenues, parks, squares, etc.) of the neighbourhood can be overcome without exerting much physical effort by ramps achieving appropriate slopes"
Size and space for approach and use	Item 7: "Equipment (children's playgrounds, sports fields, benches, etc.) in urban areas (streets, avenues, parks, squares, etc.) in the neighbourhood are suitable for everyone's approach and use (they offer different sizes of options and / or adjustable features)"

4 Findings and Discussion

4.1 Findings of the Assessment

The findings of the qualitative evaluation regarding the seven main universal design principles are shown in Figs. 8, 9, 10, 11, 12, and 13. In addition, the quantitative assessment regarding the Turkish Standards Institution's determined standards involved in the first phase of the research which can be seen in Table 4. Based on the findings presented in Figs. 8, 9, 10, 11, 12, and 13 and Table 4, and as hypothesized in the study, it can be concluded that the existing UD

Urban Space Evaluation Form				
Neighborhood Name: MARMARA		Urban Spaces Name And Type: Yüzbaşı Tekin Yurdabak Avenue		
Evaluated Item	**Visual (S)**	**UD Principles**	**Evaluation**	**Sufficient (S)/ Insufficient (I) /Non(N)**
Garbage Cans		Equitable use	Garbage cans can be used by different profiles of users.	S
		Flexibility in use	They vary in material, size, and location. For this reason, it offers users opportunities for different needs.	S
		Simple and intuitive use	Everyone can understand the function of it.	S
		Perceptible information	It is made of material with a contrasting color to its surroundings, increasing its perceptibility.	S
		Tolerance for error	Users may have instant accidents because of the location.	I
		Low physical effort	Short individuals or children cannot reach easily.	I
		Size and space for approach and use	They are not suitable for the approach and use of children, short individuals, or wheelchair users.	I

	Sustainable Urban Design Parameters	Accessibility	Connectivity	Walkability	Safety	Adaptability	Legibility	Comfort
Score	✓ Positive × Negative ○ None	×	○	×	×	○	○	○

Fig. 8 Evaluation of garbage cans in Yüzbaşı Tekin Yurdabak Avenue of Marmara neighborhood (by Authors)

Evaluated Item	**Visual (S)**	**UD Principles**	**Evaluation**	**Sufficient (S)/Insufficient (I)/Non (N)**
Signs		Equitable use	It has no effect on the principle of equitable use.	N
		Flexibility in use	Visually impaired people could not understand the symbol.	I
		Simple and intuitive use	It can be understood by everyone without visually impaired individuals as the international symbol is used.	S
		Perceptible information	Although the international symbol is used, it is not possible for the visually impaired to understand.	I
		Tolerance for error	It is hazardous for persons who walk distractedly or have poor vision due to its position on the sidewalk.	I
		Low physical effort	It has no effect on the principle of low physical effort.	N
		Size and space for approach and use.	Its height is suitable for head recovery distance.	S

	Sustainable Urban Design Parameters	Accessibility	Connectivity	Walkability	Safety	Adaptability	Legibility	Comfort
Score	✓ Positive × Negative ○ None	○	○	×	×	○	○	×

Fig. 9 Evaluation of signs in Yüzbaşı Tekin Yurdabak Avenue of Marmara neighborhood (by Authors)

Evaluated Item	Visual (S)	UD Principles	Evaluation	Sufficient(S)/Insufficient (I) /Non(N)
Bus Stop		Equitable use	The seating component is in just one variety. (h:50cm. w: 122cm. d: 40 cm.)	I
		Flexibility in use	The level difference is not been passed with a pavement ramp to get to the bus stops.	I
		Simple and intuitive use	The stops cannot be observed from a distance because no necessary markers or guidance have been made.	I
		Perceptible information	There is no info board at the bus stops.	I
		Tolerance for error	Because transparent material is used, visually impaired persons are endangered.	I
		Low physical effort	Wheelchair users cannot reach bus stops without exerting physical effort. There isn't a ramp or textured surface.	I
		Size and space for approach and use	There is enough area for wheelchair users or parents with strollers to stand.	S

	Sustainable Urban Design Parameters	Accessibility	Connectivity	Walkability	Safety	Adaptability	Legibility	Comfort
Score	✓ Positive x Negative o None	✓	x	x	x	x	✓	x

Fig. 10 Evaluation of bus stop in Yüzbaşı Tekin Yurdabak Avenue of Marmara neighborhood (by Authors)

Evaluated Item	Visual (S)	UD Principles	Evaluation	Sufficient (S)/Insufficient (I) /Non(N)
Parking Areas		Equitable use	Not all individuals can use the parking spots next to the road equally. No parking space is available for those with disabilities.	I
		Flexibility in use	The arrangement of disabled parking spaces has not taken into account the various user kinds.	I
		Simple and intuitive use	The materials and designs used to construct the parking lots along the street make them difficult to recognize and comprehend.	I
		Perceptible information	On the street, there are not signs with information regarding parking times.	I
		Tolerance for error	The width of the parking space is not safe for vehicle landings.	I
		Low physical effort	A curb ramp is not taken into consideration for access from the parking area to the sidewalk.	I
		Size and space for approach and use	It does not adhere to the parking dimensions (250/500 cm) established for the vehicle, with the exception of the manoeuvring area.	I

	Sustainable Urban Design Parameters	Accessibility	Connectivity	Walkability	Safety	Adaptability	Legibility	Comfort
Score	✓ Positive x Negative o None	✓	o	o	x	x	o	o

Fig. 11 Evaluation of parking areas in Yüzbaşı Tekin Yurdabak Avenue of Marmara neighborhood (by Authors)

Evaluated Item	Visual (S)	UD Principles	Evaluation	Sufficient (S)/Insufficient (I) /Non(N)
Sidewalk/Walkway	 	Equitable use	Diverse user groups have been ignored. Both textured surfaces and sidewalk ramps have not been developed. Pavement surfaces have suffered partial destruction.	I
		Flexibility in use	The material size of the sidewalks along the street varies. Different users have not been considered, and designs, such as the pavement, have not been made with user needs in mind.	I
		Simple and intuitive use	Sidewalk areas are identified.	S
		Perceptible information	It has not been possible to identify the property area, pedestrian area, and safety lane individually.	I
		Tolerance for error	Various components, including lighting embedded in the pavement, cover the walking area and may contribute to accidents.	I
		Low physical effort	There aren't any ramps on the walkways to help maintain their continuity.	I
		Size and space for approach and use	Despite having the same user density, the sidewalk widths vary along the street.	I

	Sustainable Urban Design Parameters	Accessibility	Connectivity	Walkability	Safety	Adaptability	Legibility	Comfort
Score	✓ Positive × Negative ○ None	×	×	×	×	×	○	×

Fig. 12 Evaluation of sidewalk/walkway in Yüzbaşı Tekin Yurdabak Avenue of Marmara neighborhood (by Authors)

Evaluated Item	Visual (S)	UD Principles	Evaluation	Sufficient (S)/Insufficient (I) /Non(N)
Pedestrian Crossing		Equitable use	Due to deficiencies and errors in physical arrangements, it cannot be used equally by all individuals.	I
		Flexibility in use	User diversity is not taken into account.	I
		Simple and intuitive use	The caution lines painted on the ground make it visible from a certain distance. Additionally, there are ramps with speed limiters for cars that are close to the pedestrian crossing.	S
		Perceptible information	There are no guide way available for those who are blind.	I
		Tolerance for error	Due to its degraded terrain, it may result in accidents. Thankfully, there are ramps on both sides of the crossing to slow down drivers.	I
		Low physical effort	The intersection of the sidewalk and the pedestrian crossing is not on the same level, and no ramp is planned.	I
		Size and space for approach and use	In terms of size, it is appropriate for all individuals. (w = 400 cm)	S

	Sustainable Urban Design Parameters	Accessibility	Connectivity	Walkability	Safety	Adaptability	Legibility	Comfort
Score	✓ Positive × Negative ○ None	×	✓	✓	×	×	○	○

Fig. 13 Evaluation of pedestrian crossing in Yüzbaşı Tekin Yurdabak Avenue of Marmara neighborhood (by Authors)

Table 4 Assessment according to the TSI standards (by Authors)

Urban Space Items	Existing Dimension	TSI Standards	A/I
Garbage cans	Height: 86 cm Width: 48 cm	Height: 86 cm Width:48 cm	A
	Height: 128 cm Width: 122 cm	Height: 128 cm Width:122 cm	
Signs	Starting point height: 215 cm End point height: 278 cm	Starting point height: 105 cm End point height: 195 cm	I
Bus stops	Sitting element height: 55 cm	Sitting element height: 41–46 cm	I
	Thick, non-matte, colored, reflective strips: Not available	Thick, non-matte, colored, reflective strips: 100–140 cm	I
Parking lots	Width: 215 cm Length: Not limited	Width: 250 cm Length: 500 cm	I
Sidewalks/walkway	Width: 105–640 cm (variable)	Width: At least 150 cm	I
Pedestrian crossing	Width: 400 cm	Width: Min. 300 cm	A

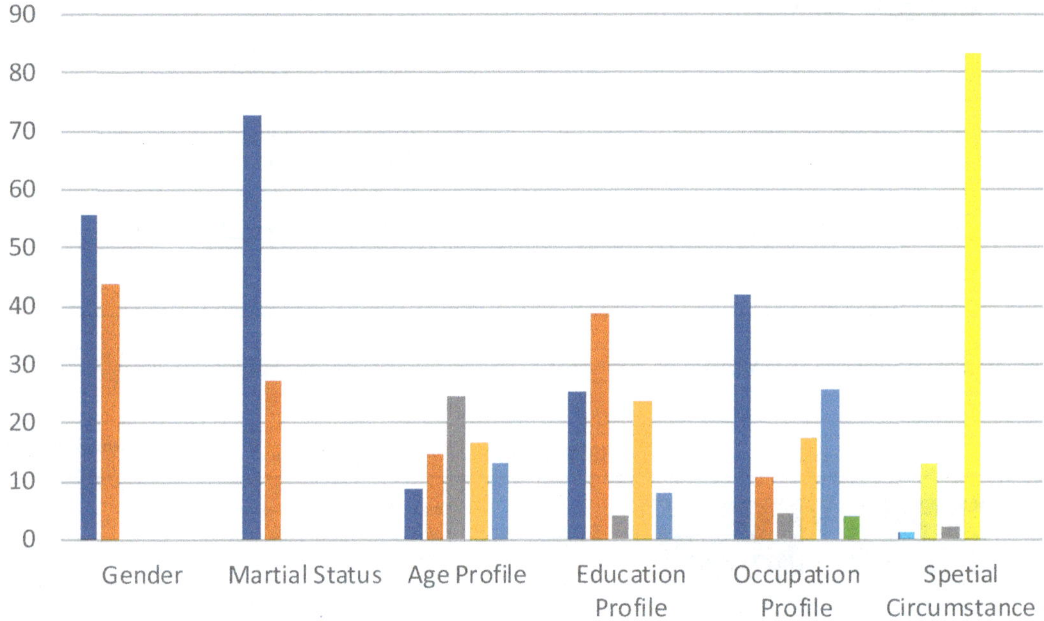

Fig. 14 Demographic data (%) (by Authors)

features on the avenue do not create an inclusive and high-quality environment. The qualitative evaluation indicates that many of the assessed urban space elements, such as signs and parking areas, largely fail to adhere to the principles of UD.

4.2 User Survey Findings

For the user survey, 150 residents aged 18 to 65 as a random sample were selected in three neighborhoods in Northern Nicosia. Given the similar population densities of these neighborhoods, an equal number of participants were chosen from each area. Respondents were randomly selected within each neighborhood and asked to complete a questionnaire. After data collection, the research results were analyzed using Statistical Package for the Social Sciences (SPSS).

4.2.1 Findings of the Demographic Data (Fig. 14)

Gender: Within the randomly selected 150 participants, there were 44.0% men and 56.0% women.

Marital status: Of the 150 participants, 27.3% were unmarried and 72.7% were married.

Age: The study included 24.6% of individuals who were in the 35–44 age range. 22.0% of the population was 65 years of age or older, and 16.7% was 45 to 54. Of those aged 25 to 34, 14.7% were and of those aged 55 to 64, 13.3%. The remaining 8.7% were in the age range of 18 to 24.

Education: Of the participants, 38.7% had graduated from high school, making up the majority. Of those, primary schooling accounted for 25.3%. 8.0% were Master's or PhD students, and 24.0% were four-year undergraduates. The remaining 4.0% were two-year undergraduates.

Occupation: 42.0% of participants were from the private sector, making up the majority. Out of them, 26.0% were retired. Among them were 4.6% students, 10.7% officers, and 12.7% housewives. Four percent were unemployed.

Special circumstance: When asked if they had any unusual circumstances, 83.4% of study participants chose the "other" option. 2.0% chose "senile," while 13.3% chose "chronic illness." One percent of the remainder chose "physical disability." The majority of individuals who chose the "other" option stated they had no unique circumstances.

4.2.2 Findings of the User Perceptions in Relation to Universal Design Principles

The user survey part was examined through reliability analysis as indicated in Table 5. The table presents alpha-reliability result for the seven-item scale. The scale's Cronbach's alpha score of 0.712 as the result points out that it has good reliability.

Regarding the findings of the item 1 "In the urban areas (streets, avenues, parks, squares, etc.) of the neighborhood, everyone has the opportunity to move around under the same conditions," 34.0% of the respondents replied "strongly disagree," 18.7% suggested "agree," 18.0% selected "unsure," and 17.3% replied "disagree." 12.0% as the rest elected "strongly agree."

Regarding the findings of the item 2 "Considering the diversity of users in the neighbourhood, there are elements with the same function designed with different features in urban spaces like street, avenue, park, square, etc.) (such as garbage bins, book sharing points located at different heights)," 62.0% of the respondents replied "strongly disagree," 17.3% replied "disagree," 8.0% of the respondents selected "agree" option, and 7.3% replied "unsure." 5.4% as the rest elected "strongly agree."

Regarding the findings of the item 3 "Urban spaces (streets, avenues, parks, squares etc.) in the neighbourhood can be easily found and used by everyone with the help of perceptible information," 42.0% of the participants replied "strongly disagree," 25.3% replied "agree," 16.7% of the participants selected "disagree," and 10.7% of them replied "strongly agree." The rest 5.3% elected "unsure" option.

Table 5 Reliability value of survey part (by Authors)

Cronbach's alpha	No. of items
0,712	7

Regarding the findings of the item 4 "Information signs in urban areas of the neighbourhood (streets, avenues, parks, squares, etc.) can be understood by everyone," 28.0% of the participants replied "strongly disagree," 28.0% replied "agree," 18.0% of the participants selected "disagree" option, and 15.3% of them replied "unsure." 10.7% as the rest elected "strongly agree."

Regarding the findings of the item 5 "Features (water items, urban furniture, etc.) found in urban areas of the neighbourhood (street, avenues, park, square, etc.) is designed to minimize the likelihood of accidents that may occur due to the carelessness and / or physical / mental abilities of the users (such as avoiding proximity by planting flowers around the water element.)," 37.3% suggested "strongly disagree," 20.7% of the respondents suggested "unsure," 18.7% replied "disagree," and 18.0% suggested "agree." 5.3% as the rest elected "strongly agree."

Regarding the findings of the item 6 "Level differences in urban spaces (streets, avenues, parks, squares, etc.) of the neighbourhood can be overcome without exerting much physical effort by ramps achieving appropriate slopes," 40.0% of the respondents answered "strongly disagree," 20.7% answered "agree," 15,3% answered "unsure," and 14.0% replied "disagree." 10.0% as the rest elected "strongly agree."

Regarding the results of the item 7 "Equipment (children's playgrounds, sports fields, benches, etc.) in urban areas (streets, avenues, parks, squares, etc.) in the neighbourhood is suitable for everyone's approach and use (they offer different sizes of options and / or adjustable features)," 26.7% of the respondents answered "unsure," 22.7% replied "strongly disagree," 19.3% replied "agree," and 16.7% of them replied "strongly agree." 14.6% as the rest elected "disagree" choice (Fig. 15).

4.3 Discussion

4.3.1 Discussion of the Assessment

The evaluation of the assessment across seven UD dimensions and in relation to TSI requirements and user survey analyses has led to the conclusion that urban spaces do not effectively create an inclusive environment for everyone. Based on the findings, it seems that applying universal design principles in urban spaces is crucial in order to create designs for all users.

In order to ensure sustainable urban spaces in the city of Northern Nicosia, aspects such as accessibility, walkability, and safety must be prioritized by enhancing the physical infrastructure throughout all urban areas. Additionally, it is required to increase the variety and quantity of urban furni-

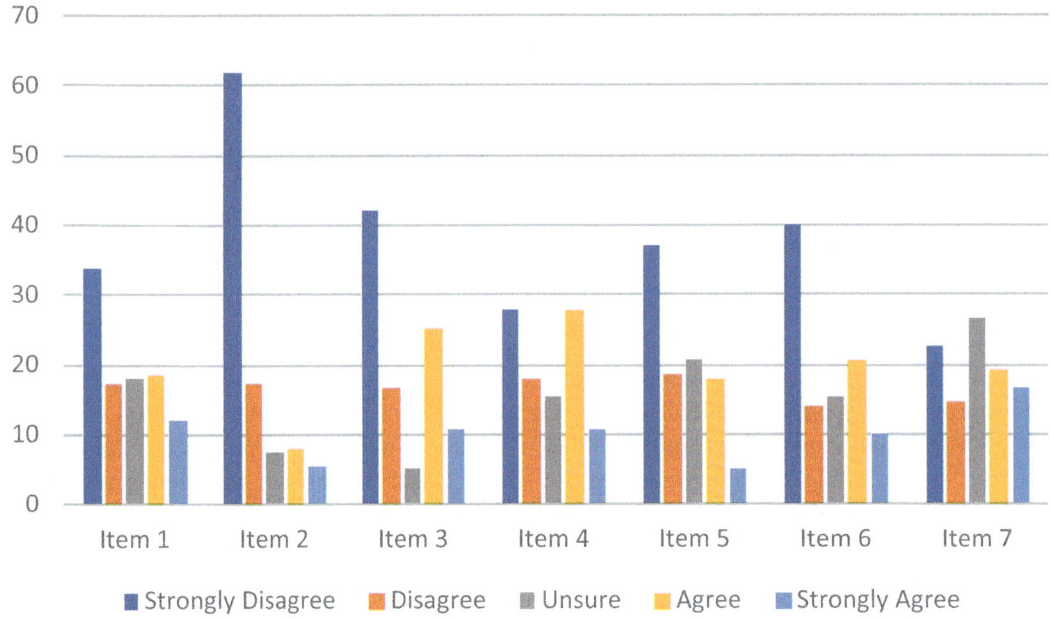

Fig. 15 The user views about universal design principles (%) (by Authors)

ture items in urban spaces and to adjust the existing items by optimizing their material, size, and placement.

This improvement should be guided by the adoption of UD principles. In this context, when implementing UD principles, current regulation (Article 19) under Chapter 96, along with the relevant TSI standards, can serve as key references. This regulation, Article 19 in Chapter 96, was specifically developed to ensure that roads and buildings are accessible for individuals with disabilities.

4.3.2 Discussion of the User Survey

Based on the findings, it can be claimed that the principles of UD are disregarded in the urban spaces of the study area. It can further be commented that principles of simple and intuitive use, flexible use, and the perceivable information all have low user satisfaction level. Such that, regarding the findings of the item "Considering the diversity of users in the neighbourhood, there are elements with the same function designed with different features in urban spaces like street, avenue, park, square, etc.)," representing the principle of flexible use, 62.0% of the respondents suggested "strongly disagree."

According to the results of the item "Urban spaces (streets, avenues, parks, squares etc.) in the neighbourhood can be easily found and used by everyone with the help of perceptible information" representing the principle of simple and intuitive use, 42.0% of the respondents replied "strongly disagree."

Furthermore, it is remarkable that for the item "Information signs in urban areas of the neighbourhood (streets, avenues, parks, squares, etc.) can be understood by everyone" representing the perceivable information principle, 28.0% of participants replied "strongly disagree" and one another 28.0% replied "agree." According to the rest of the findings, 18.0% of the respondents replied "disagree," 15.3% suggested "unsure," and 10.7% selected "strongly agree."

In brief, the findings indicate that adoption of the UD for the public spaces is also highly required for Northern Nicosia. However, as a rapidly urbanizing city, planning and design of urban spaces based on the sustainability principles involving the universal design concept have been widely undermined both in quantity and in quality. According to the survey findings applied in light of universal design principles in selected neighborhoods, it can be argued that the urban spaces are neither convenient nor suitable for welcoming the variety of all individuals in Northern Nicosia.

Despite the existence of various supporting frameworks such as standards, legislations, and master plans designed to ensure the implementation of UD principles, it cannot be claimed that urban spaces have successfully achieved the necessary inclusivity for all. Although incorporating standards for the adoption of universal design principles, the physical infrastructure in Northern Nicosia has not shown significant improvement. This conclusion is evident from the findings of the assessment and user survey analysis included. In order to solve these problems, the concept of UD should be adopted and the urban structure should be improved by applying the existing regulations.

5 Conclusion and Recommendations

In the last decades, sustainability in urban planning and design has been acknowledged as a key concept for combating the consequences of urbanization occurring within a global spectrum. Social sustainability is one of the main dimensions fulfilling the term of sustainability. One of the important elements that are frequently linked to social sustainability is public space. It can further be commented that public spaces have a vital meaning for all four main pillars.

In other words, in a period characterized by rapid urbanization, streets, squares, and parks as public spaces are becoming increasingly vital across both developed and developing countries. These urban areas have the potential to significantly improve the standard of living and advance social sustainability in modern cities by offering chances for equal involvement regardless of gender, age, nationality, or socioeconomic background (Rogers 2003; Lotfata and Ataöv 2020). Therefore, adopting universal design concept is vital for contributing to social sustainability when designing urban spaces.

The concept of UD is essential for ensuring that urban spaces are usable and accessible for everyone. This term is defined as "a process that enables and empowers a diverse population by improving human performance, health and wellness, and social participation" (Steinfeld and Maisel 2012). In the context of the built environment, UD's seven principles are established to support universal accessibility for all users (Yiing et al. 2013).

Within this framework, after the introduction, universal design is theoretically explained and its relation with social sustainability is conceptualized. Later in the methodology section of the study, a qualitative evaluation, based on the seven dimensions of UD and the quantitative assessment, aligned with TSI standards and European directives/regulations, was applied to six selected urban space elements along the avenue. The findings from Yüzbaşı Tekin Yurdabak Avenue indicate that urban spaces in Northern Nicosia do not fully provide a convenient environment for all individuals. This conclusion is further supported by the survey data gathered in this study as the second phase of the research. In sum, in all urban environments, the requirements of universal design must be prioritized to enhance comfort, adaptability, and flexibility. This approach is crucial for boosting social sustainability in cities (Kadir and Jamaludin 2013).

Beside universal design, there are other concepts like inclusive design (Newell et al. 2011; Clarkson and Coleman 2015), design for all (Persson et al. 2015; Bendixen and Benktzon 2015), etc., that share mostly common concerns to raise efficiency of urban environments within different scales. All these related terminologies require higher attention within urbanism circles in a scientific ground in order to gain opportunity to concentrate on the sustainability concept.

Acknowledgments The study was part of one of the authors' PhD thesis (Duman 2023) published online. The second author of the article is the supervisor of the study completed at Department of Architecture, Near East University.

References

Arenghi A (2020) Universal Design in Sustainable Urban Planning. In: Green planning for cities and communities. Springer, Cham, pp 119–138

Asilsoy B, Oktay D (2016) Understanding environmental worldview of Famagusta residents and its determinants through survey research. J Civil Eng Architect 10:544–552

Asilsoy B, Oktay D (2018) Exploring environmental behaviour as the major determinant of ecological citizenship. Sustain Cities Soc 39:765–771

Bendixen K, Benktzon M (2015) Design for all in Scandinavia–a strong concept. Appl Ergon 46:248–257

Borowczyk J (2018) Sustainable urban development: spatial analyses as novel tools for Planning a universally Designed City. Sustainability 10:1407

Carmona M, Tiesdell S, Heath T, Oc T (2010) Public places urban spaces, the dimensions of urban design, 2nd edn. Architectural Press, Oxford

Chadchan J, Shankar R (2009) Emerging urban development issues in the context of globalization. J ITPI (Institute of Town Planners, India) 6(2):78–85

Clarkson PJ, Coleman R (2015) History of inclusive Design in the UK. Appl Ergon 46:235–247

Dempsey N, Bramley G, Power S, Brown C (2011) The social dimension of sustainable development: defining urban social sustainability. Sustain Dev 19(5):289–300

Duman Ü (2023) Developing a framework of universal Design in the Context of sustainable urban Planning in northern Nicosia. Institute of Graduate Studies, Department of Architecture, Near East University, Nicosia, N. Cyprus

Duman Ü, Asilsoy B (2022) Developing an evidence-based framework of universal Design in the Context of sustainable urban planning in Northern Nicosia. Sustainability 14(20):13377

Gossett A, Gossett A, Mirza M, Barnds AK, Feidt D (2009) Beyond access: a case study on the intersection between accessibility, sustainability, and universal design. Disabil Rehabil Assist Technol 4(6):439–450. https://doi.org/10.3109/17483100903100301

Harun NZ, Zakariya K, Mansor M, Zakaria K (2014) Determining attributes of urban plazas for social sustainability. AMER Int Conf Qual Life 153:606–615

Imrie R (2012) Universalism, universal design and equitable access to the built environment. Disabil Rehabil 34(10):873–882

Jacobs J (1961) The death and life of great American cities. Vintage Books, New York

Kadir SA, Jamaludin M (2013) Universal design as a significant component for sustainable life and social development. Procedia Soc Behav Sci 85:179–190

Ketschau TJ (2017) Social sustainable development or sustainable social development-two sides of the same coin? The structure of social justice as a normative basis for the social dimension of sustainability. Int J Design Nat Ecodynam 12(3):338–347

Kristl Ž, Salaj AT, Roumboutsos A (2019) Sustainability and universal design aspects in heritage building refurbishment. Facilities 38:599

Lid IM (2014) Universal design and disability: an interdisciplinary perspective. Disabil Rehabil 36(16):1344–1349

Lotfata A, Ataöv A (2020) Urban streets and urban social sustainability: a case study on Bagdat street in Kadikoy, Istanbul. European Planning Studies 28(9):1735–1755

Maisel JL, Ranahan M (2017). Beyond accessibility to universal design. https://www.wbdg.org/design-objectives/accessible/beyond-accessibility-universal-design. Accessed 29 Nov 2020

Manzi T, Lucas K, Lloyd-Jones T (2010) Social sustainability in urban areas: communities, connectivity and the urban fabric. Earthscan/James & James, London

Monfaredzadeh T, Krueger R (2015) Investigating social factors of sustainability in a smart city. Procedia Eng 118:1112–1118. https://doi.org/10.1016/j.proeng.2015.08.452

Newell AF, Gregor P, Morgan M, Pullin G, Macaulay C (2011) User-sensitive inclusive design. Univ Access Inf Soc 10(3):235–243

Oktay D (2002) The quest for urban identity in the changing context of the city: Northern Cyprus. Cities 19(4):261–271

Oktay D (2007) An analysis and review of the divided city of Nicosia, Cyprus, and new perspectives. Geography 92(3):231–247

Paşaoğulları N, Doratlı N (2004) Measuring accessibility and utilization of public spaces in Famagusta. Cities 21(3):225–232

Persson H, Åhman H, Yngling AA, Gulliksen J (2015) Universal design, inclusive design, accessible design, design for all: different concepts—one goal? On the concept of accessibility—historical, methodological and philosophical aspects. Univ Access Inf Soc 14(4):505–526

Pinna F, Garau C, Maltinti F, Coni M (2020) Beyond architectural barriers: building a bridge between disability and universal Design. In: International conference on computational science and its applications. Springer, Cham, pp 706–721

Pitarch-Garrido MD (2018) Social sustainability in metropolitan areas: Accessibility and equity in the case of the metropolitan area of Valencia (Spain). Sustainability 10(2):371

Ranhagen U, Groth K (2012) A Symbio City approach. A Conceptual framework for sustainable urban development. SKL International, Stockholm. (ISBN 978-91-637-1467-2)

Rogers W (2003) The excellent city park system. What makes it great and how to get there. The Trust for Public Land Pub, Washington, DC

Shen L, Peng Y, Zhang X, Wu Y (2012) An alternative model for evaluating sustainable urbanization. Cities 29(1):32–39

State Planning Organization (2020) Yerel Yönetimler Raporu. https://www.devplan.org/belediyeler/Yerel-Yonetimler-2017-2019.pdf. Accessed 20 May 2021

Steinfeld and Maisel (2012) What is Universal Design? http://idea.ap.buffalo.edu/about/universal-design/. Accessed 29 Nov 2020

Stevenson N (2020) The contribution of community events to social sustainability in local neighbourhoods. J Sustain Tour:1–16

Thompson CW (2002) Urban open space in the 21st century. Landsc Urban Plan 60(2):59–72. https://doi.org/10.1016/S0169-2046(02)00059-2

Turkington R, Sangster K (2006) From housing to social mix: housing's contribution to social sustainability. Town Country Plann 75:184–185

United Nations Population Fund (2007) State of world population 2007: unleashing the potential of urban growth. UNFPA, New York

Williams K, Burton E, Jenks M (eds) (2000) Achieving sustainable urban form. Spon, London

Yiing CF, Yaacob NM, Hussein H (2013) Achieving sustainable development: accessibility of green buildings in Malaysia. Procedia Soc Behav Sci 101:120–129

Yıldırım S, Asilsoy B, Özden Ö (2020) Urban resident views about open green spaces: a study in Güzelyurt (Morphou), Cyprus. Euro J Sustain Dev 9(2):441–450

Community Engagement as an Innovative Urban Design Strategy: Conservation and Revitalization of Pantang Historic District, Guangzhou

Dingran Chen

Abstract

In the context of rapid urbanization, the conservation and revitalization of historic districts face significant challenges. This paper takes Pantang historic district in Guangzhou as a case study to explore the role of community engagement as an innovative urban design strategy for conserving and revitalizing historic districts. The Pantang historic district has undergone urban regeneration and is considered an important example of community engagement in China. Through methods such as on-site assessments, field visits, and literature and map surveys, this study systematically analyzes the current state of Pantang and the specific practices and outcomes of community engagement during its transformation.

The results indicate that community engagement significantly enhances the effectiveness of conservation measures, strengthens community cohesion, and increases residents' sense of belonging. Residents are highly satisfied with the environmental changes, especially the renewal of the physical environment, which has injected new life into the community and enhanced economic sustainability. Moreover, the regeneration process has significantly improved residents' happiness. The conclusion is that promoting community engagement as an innovative urban design strategy can achieve the sustainable development of historic districts, enriching the city's cultural diversity and historical treasures. The Pantang case demonstrates the great potential of community engagement in the conservation and revitalization of historic districts, offering crucial reference value for similar efforts in the context of rapid urbanization.

D. Chen (✉)
Politecnico di Torino, Turin, Italy
e-mail: dingran.chen@polito.it

Keywords
Community engagement · Urban design strategy · Historic district · Conservation · Revitalization · Sustainability

1 Introduction

With the acceleration of global urbanization, many historic districts are at risk of being neglected and demolished. This phenomenon is particularly pronounced in China. As a result of the rapid and unbalanced urban development prevalent in developing countries like China, historic districts are rapidly deteriorating (Li et al. 2021). Preserving these culturally and historically valuable areas has become a central focus of urban conservation efforts. Additionally, the planning and management of historic districts face the challenge of balancing the needs of residents with conservation issues. Community participation has gained attention as an innovative urban design strategy due to its societal benefits and applicability. For government, it serves as an effective open government tool (Polci and Pierantoni 2024). By fostering genuine communication, managers and designers can listen to and respond to public voices; meanwhile, the public can engage in the agenda-setting process (Ali et al. 2008). This enhances residents' sense of responsibility and belonging and improves communication between the public and officials.

In 2011, the United Nations Educational, Scientific and Cultural Organization (UNESCO) adopted the Recommendation on the Historic Urban Landscape (HUL). HUL aims to promote the management of historic urban areas, emphasizing the role of community participation. Additionally, the conservation and revitalization of historic districts align with the principles of the United Nations Sustainable Development Goals (SDGs), particularly SDG 11, which focuses on sustainable cities and communities. In this context, this study examines how community engage-

F. Alberti et al. (eds.), *Towards Sustainable Urban Regeneration: Utilizing Environmental, Cultural, and Social Assets*, Advances in Science, Technology & Innovation, https://doi.org/10.1007/978-3-032-02091-8_32

ment can serve as an innovative urban design strategy for urban conservation, using the Pantang historic district (Pantang) in Guangzhou as a case study. The main objective is to explore the role and effectiveness of community engagement in the conservation and revitalization of historic districts.

This study explores the social and cultural dimensions of Pantang through field observations and written materials. The current state of the community was analyzed via on-site assessment and subjective grading, supplemented by data from literature and map surveys. These methods highlighted the role of community engagement in Pantang's urban renewal process. As one of Guangzhou's important urban renewal projects, the experience of Pantang will provide valuable insights for the conservation and revitalization of other historic districts. Moreover, studying the role of community engagement in the conservation of historic districts has not only theoretical significance but also practical guidance.

2 Literature Review

2.1 Community Engagement

Community engagement is widely studied across disciplines such as sociology, economics, psychology, and urban studies (Esmaeilpoorarabi et al. 2020). Over the last 20 years, numerous countries have shifted from a top-down managerial approach to fostering institutional connections between government officials and citizens, a practice often referred to as "community engagement" (Head 2007). This methodology is crucial for governments, nongovernment institutions, and community groups, aiming to empower local stakeholders through meaningful participation (Tavares et al. 2024).

The most significant aspect of community engagement strategies is the interactive process with the public, which fosters cultural exchange and knowledge sharing. According to Kim et al. (2020), community engagement is an ongoing process rather than a one-time act, with participation being more important than specific planning strategies. However, this process is not instantaneous and is influenced by various social and cultural factors. Zhu (2015) found that residents' participation in community affairs is constrained by the social and natural environment of their neighborhoods.

In the context of urban design, community engagement is primarily considered from a social perspective. Gearin and Hurt (2024) noted that it can enhance community support for development, strengthening a community's sense of autonomy and influence.

2.2 Urban Design and Historic District Conservation

As one of the key approaches to urban conservation, urban design ensures that historic districts are preserved by integrating modern functionality with historical and cultural elements. Therefore, innovation in urban design strategies is crucial, and community engagement represents a significant innovation. Community participation in the regeneration process is essential. Polci and Pierantoni (2024) argued that active community involvement contributes to sustainable development, highlighting sociology's potential as a moral and political force in urban planning.

It is also crucial to address the physical aspects of historic districts—such as buildings, public spaces, and streets—as these elements shape local community memory. Hassen and Kaufman (2016) demonstrated through their research that street design is crucial for promoting community participation. Polat and Tümer Yildiz (2019) presented the "Participatory Urban Design Guidance" model, demonstrating its significance in historic area conservation through a case study of Bursa, Turkey. Their study integrates various community participation techniques into design guidelines, arguing that early community involvement helps local authorities overcome challenges.

The importance of community participation in urban design is increasingly recognized by scholars and practitioners. Community members, as direct stakeholders in historic districts, can enhance the feasibility and effectiveness of conservation measures through their participation. Literature shows that community engagement can increase residents' sense of belonging and community cohesion, which is significant for urban design and the preservation of historic districts.

Community engagement is an innovative urban design methodology that extends intervention temporality for deeper involvement, emphasizes inclusiveness and equality, and focuses on actionable outcomes. It empowers participants with evidence-based ideas and strategically forms stakeholder groups to implement urban improvements collaboratively (Tavares et al. 2024). In this context, community engagement plays a pivotal role in addressing the complexities of historic district conservation, thereby integrating social perspectives with urban conservation efforts. This approach ensures that the voices of local communities are heard and valued in shaping the sustainable future of historic neighborhoods.

In China, economic and political factors drive cities to enhance their competitiveness and attractiveness, often through redevelopment projects focused on history and culture. This strategy has elevated urban designers' roles in China's urban conservation practices (Xie and Heath 2017).

2.3 Study Site: Pantang Historic District

In China, the process of urbanization has increasingly focused on renovating historic urban districts to achieve sustainable development and economic growth (Li and Li 2024). As a critical component of Chinese cultural heritage, historic districts enjoy a profound historic connotation while helping preserve and present traditional Chinese culture and history. Conflicts between urban construction, residents' demands, and the conservation of historic districts are intensifying in China.

To mitigate tensions between reconstruction and cultural conservation, the government introduced the concept of "state-list famous historical and cultural cities" in 1982, leading to widespread urban regenerations in historic districts. Guangzhou (see Fig. 1), with a history of over two millennia, harbors numerous historic districts that testify to its significance as a trading hub along the Maritime Silk Road and serve as integral anchors in the contemporary urban fabric (Whitehand et al. 2011).

Pantang (see Fig. 2), with a history spanning over a thousand years, is adjacent to Liwan Lake Park, located south of Zhongshan Eighth Road and west of Pantang Road in the Liwan District of Guangzhou. It retains the ancient village layout and appearance, preserving the complete Qing Dynasty layout, texture, and modest style, making it highly representative of old Guangzhou's cultural landscape.

From 2015 to 2018, Guangzhou established the "Urban Renewal Bureau" and proposed the "Micro-renovation" method, initiating several projects including Pantang. Since then, Pantang has become one of Guangzhou's first urban renewal projects with the objective of "conservation." The first phase was in September 2016, focusing on transforming government-acquired public housing. In early 2017, the second phase began, improving public service facilities and planning public spaces. Designers employed "participatory planning and design," holding workshops to facilitate communication between designers and residents. This marked Guangzhou's first public participation "co-creation" workshop, fostering positive interaction and mutual support among residents, developers, and management departments.

In summary, within this context, the research gap can be identified. While there are many innovations in urban design for protecting historic districts, there is insufficient discussion on community engagement strategies, particularly empirical studies. There is also limited academic focus on how community engagement functions and its effectiveness in urban renewal, especially in China.

Fig. 1 Location of Guangzhou (source: Author)

Fig. 2 Location of Pantang historic district (source: Author)

3 Methodology

This paper adopts qualitative analysis to systematically analyze the conservation and revitalization process of Pantang through on-site assessments, field visits, and literature and map surveys.

3.1 On-Site Assessment

In the context of this paper, the on-site assessment focuses on evaluating the conservation benefits of urban design in Pantang through the three dimensions of sustainable development (social, economic, and environmental). These categories collectively provide a holistic framework for evaluating the functionality and impact of historic districts.

The first category, social sustainability, examines community integration into the urban design process and the resulting social benefits. It evaluates sense of cohesion, self-management, resident participation, well-being, and safety. High community engagement fosters ownership and responsibility, leading to sustainable social benefits and enhanced quality of life. The second category, economic sustainability, assesses the economic benefits and viability of the historic district. It evaluates the frequency of public space usage, impact of commercial functions of utilization, design innovation, quality and maintenance of public spaces, and visitor attraction, which are crucial for economic growth and for supporting local businesses. The third category, environmental sustainability, evaluates the accessibility, conservation and display of cultural and historical atmosphere, effectiveness of environmental measures, overall environmental quality, and transportation planning. Cultural conservation maintains historical identity and heritage, while environmental practices ensure livability and resilience.

Overall, the three categories of this ranking sheet are structured in a parallel manner, collectively covering all critical aspects necessary for a comprehensive evaluation of a historic district's functionality and impact. Each category focuses on different, yet interconnected, dimensions of sustainability.

3.2 Field Visit

Field visit is pivotal in studying Pantang, providing a nuanced understanding of community engagement. Fieldwork is conducted based on key urban design assessment principles (e.g., urban space and human behavior), with relevant photographs taken as supporting material. During the field investigation, many secondary sources are also referenced, including community charters, posted announcements, news reports, and planning documents, to analyze the discourse during the reconstruction process. This approach complements qualitative observations and provides concrete evidence to support the findings. Strict ethical guidelines are followed in conducting the fieldwork to ensure that observations are made in a respectful and unobtrusive manner to maintain the natural behavior of the residents.

Different areas within the historic district, including buildings, public spaces, and streets, are selected for this study to ensure a diverse and representative sample. Fieldwork is conducted in multiple phases to ensure a comprehensive understanding of the district's dynamics. The research process includes an initial reconnaissance to identify key sites and to understand the basic layout and functions of different areas within the district. Systematic observations are made at different times of the day and week to capture activity pattern changes. Public space usage, activity types, and human interactions are recorded, providing a wealth of data for an in-depth understanding of community life.

In addition to the textual records, there are extensive visual records to complement the observational data. Documenting the physical environment (buildings, public spaces, and streets) and community activities and changes over time, these visual data provide a powerful complement to qualitative observations, concrete evidence of changes and activities within the neighborhoods, as well as identifying problems.

3.3 Literature and Map Survey

This study systematically collects and analyzes documentary and cartographic data on the conservation of Pantang to gain a deeper understanding of its historical background, conservation policies, and planning design. The research design includes a detailed analysis of planning documents and renovation programs over the years to identify the specific implementation of community participation at different stages. Through textual and map analyses, the study aims to provide an in-depth understanding of the modes and effects of community engagement, offering comprehensive theoretical and practical insights.

4 Results

The field visit is categorized and focused on the physical environment (see Fig. 3). First, urban renewal has renovated and modernized most of the older houses in the neighborhood. Second, observing daily activities in the public space reveals the vitality and happiness of the residents. They often gather for activities like chatting, playing mahjong, playing shuttlecock, watching outdoor cinema, sunbathing, and so on. One of the images captures people praying in front of the Renwei Temple, demonstrating the cultural and spiritual life of the community. These activities not only enrich community life but also attract many tourists, making these public spaces "photo spots." Finally, street observation reveals management challenges due to outdated planning concepts. Traffic issues include vehicles or bicycles interfering with pedestrians and parking affecting road usability. Some areas have organized parking, while others have vehicles parked in the middle of the road. In addition, the area attracts tourists due to its proximity to Liwan Lake. During holidays, visitors mingle with residents in daily activities. Some tourists, unfamiliar with Cantonese culture, are curious about the local language, creating an interesting cultural attraction and integration. These activities reflect the vibrancy and variety of community life.

In addition to surveying and evaluating Pantang (as shown in Table 1), this study also focuses on the public areas of three important nodes within the local community (as shown in Fig. 4). These three nodes are important places where the local community holds events and showcases most of the community's cultural activities, and where residents are active in maintaining and managing their communities. They have developed their own community agreements, which are not legally binding but are voluntarily followed by the residents. This self-regulation reflects their commitment to maintaining the quality and harmony of their living environment.

Furthermore, the community has taken the initiative to start working toward solving the problems of the historic district. The overall good state of architectural conservation, great infrastructure, and high environmental quality of Pantang is attributed to the active participation of community members. As can be seen from Tang (2023) study, after urban renewal, most building functions have been changed from purely residential to semi-residential and semicommercial. The commercial functions have impacted the neighborhood, and environmental strategies and sustainability measures have been effectively implemented. Residents are more engaged, showing a sense of participation, resulting in a safe community environment. However, the utilization of public spaces is relatively low, and the design of public spaces is not very innovative. Neighborhoods have a high degree of connectivity to the city, as evidenced by the design of multiple entrances and exits (see Fig. 5); however, street management needs to be improved, as evidenced by the presence of haphazard parking in the photos.

In China, urban regeneration is usually a top-down process, with urban designers following specific planning policies and guidelines. However, the Pantang case represents a situation of greater community involvement. It was in Pantang that the first workshop between designers and residents in Guangzhou took place as part of an urban regeneration project. This pioneering approach drew much academic attention to this area.

Overall, the revitalization process of Pantang demonstrated the important role and active participation of community members. Although there is still room for improvement in the utilization of public space and the display of cultural ambience, the well-being of residents has been remarkably effective.

5 Discussion

5.1 Community Engagement as an Innovative Strategy for Urban Design

Community engagement as an urban design strategy involves bottom-up intervention in urban regeneration and conservation, reflecting a pro-people approach. This is facilitated through workshops between stakeholders and residents (Ye et al. 2024). The revitalization process of Pantang demonstrates the crucial role and active participation of community members, showcasing the innovative nature of community engagement in urban design.

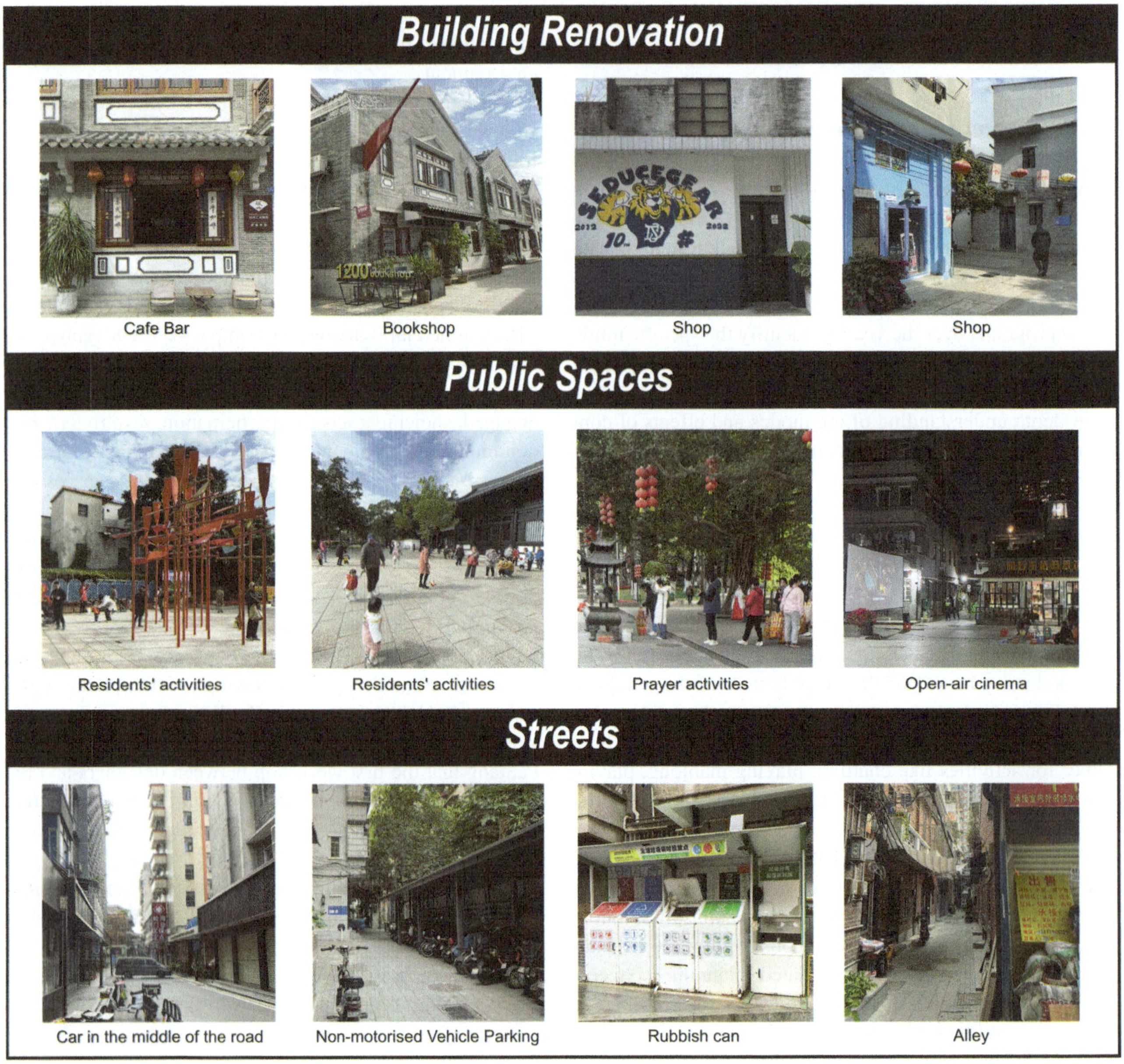

Fig. 3 Photograph of the built environment (source: Author)

5.2 Sustainable Development

Community engagement significantly enhanced the effectiveness and sustainability of conservation measures in Pantang. Residents are highly satisfied with environmental changes, the conversion of residences to commercial use, and the introduction of small businesses, which revitalized the community and enhanced economic sustainability. Wang et al. (2016) highlighted the challenges of community engagement in balancing public goals, participant freedom, and process control, and sustaining short-term impacts. In the long run, community engagement fosters knowledge and skill sharing, making conservation efforts more responsive to actual needs and minimizing negative external interventions.

5.3 Implications for Other Historic Districts

The experience of Pantang provides valuable lessons for conserving and revitalizing other historic districts, highlighting the transformative role of community engagement in urban design. Urban renewal significantly enhanced resident well-being, with public spaces frequently used for socializing, promoting a strong and positive community atmosphere, and attracting many visitors. Redesigned public spaces are

Table 1 On-site assessment of sustainability (source: Author)

Social sustainability	Rank
Community cohesion	◍
Community self-management	◍
Resident participation	◍
Resident Well-being	●
Safety	●
Economic sustainability	**Rank**
Frequency of public space usage	◍
Impact of commercial functions	◍
Innovative design of public spaces	○
Quality and maintenance of facilities and infrastructure	○
Visitor attraction	●
Environmental sustainability	**Rank**
Accessibility (connectivity)	◍
Conservation and display of cultural and historical atmosphere	◍
Effectiveness of environmental measures	●
Overall environmental quality	◍
Transportation planning	○

● High quality
◍ Medium quality
○ Low quality

multifunctional and central to community life. Zhu (2015) noted that shared neighborhood spaces foster passive and active contact, contributing to community engagement and collective activities. These spaces have both use and economic value, motivating residents to maintain community stability.

In conclusion, the decentralization of the conservation of Pantang to the local community underscores the importance of community engagement. Despite significant achievements, areas for improvement remain. The utilization rate of public spaces needs enhancement, and the conservation and display of cultural and historical ambience require improvement. Increasing the diversity and richness of community activities will attract more residents and tourists, enhancing the community's vitality and attractiveness. Additionally, balancing commercial development with the conservation issue is crucial. Future work should focus on optimizing public space use, conserving cultural ambience, and balancing commercial and community needs.

6 Conclusion

This study explores the role of community engagement as an innovative urban design strategy in conserving and revitalizing Pantang in Guangzhou. Through on-site assessments, field visits, and literature and map surveys, the study systematically analyzes the current situation and the specific prac-

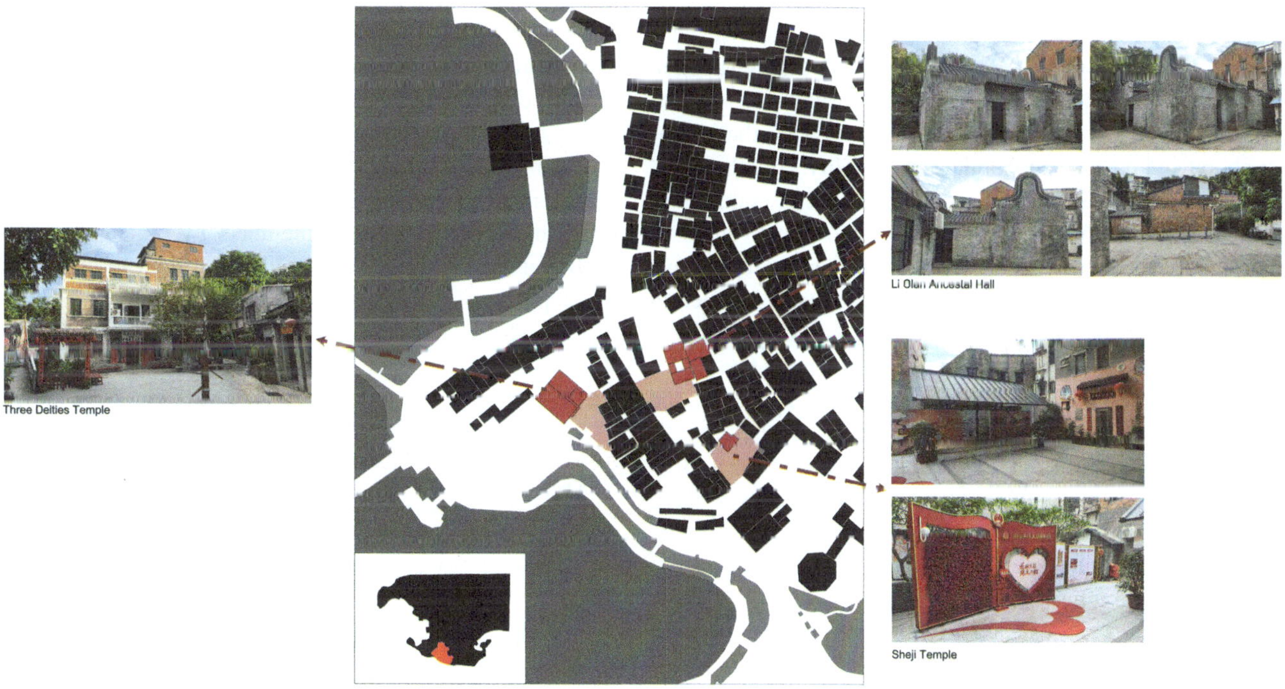

Fig. 4 Three important nodes within the local community (source: Author)

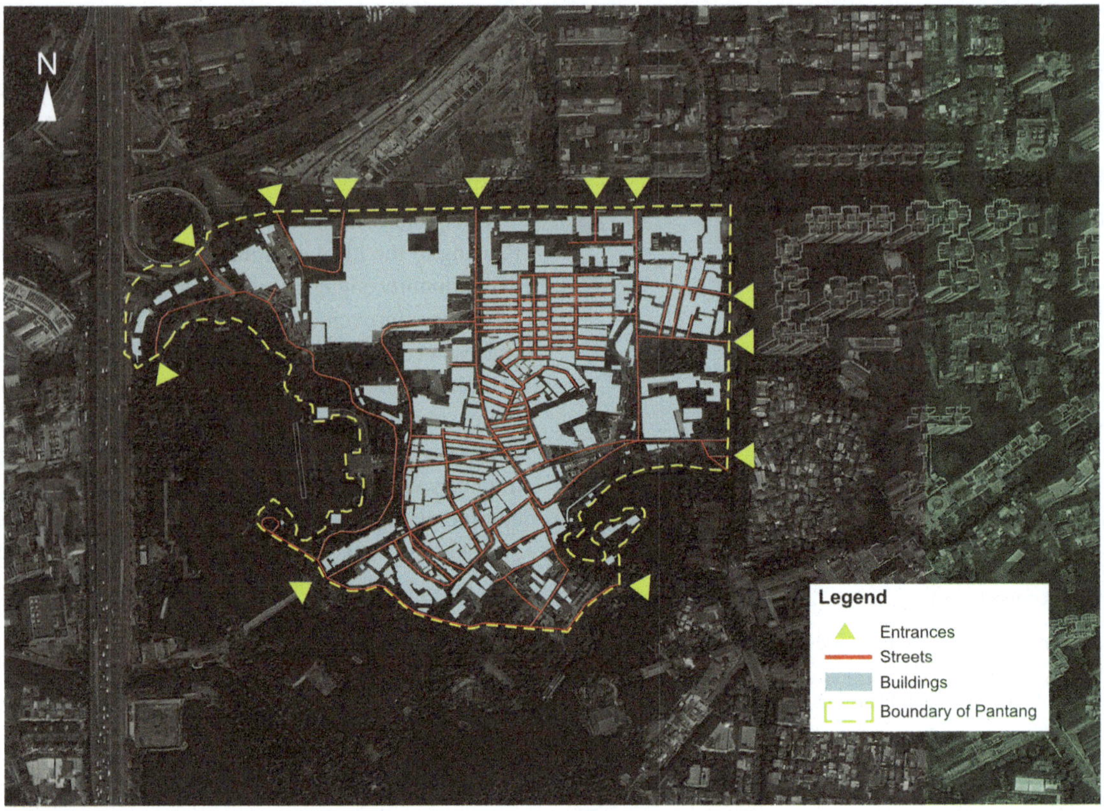

Fig. 5 Map of Pantang (source: Author)

tices and outcomes of community engagement in the district's transformation.

The findings reveal that community engagement significantly enhances the conservation effectiveness and sustainability of the historic district. Residents are highly satisfied with the environmental changes, particularly the renewal of the physical environment, which has revitalized the neighborhood and boosted economic sustainability. High levels of community engagement have strengthened community cohesion and responsibility, increased the effectiveness of conservation measures, made conservation more responsive to actual needs, and reduced negative impacts from external interventions.

While there is still room for enhancement in the use of public space and the display of cultural atmosphere, the overall enhancement of the environment and residents' wellbeing has been significant. Frequent use of public spaces fosters a strong community atmosphere and attracts a large number of tourists, bridging the neighborhood and the city. Redesigned public spaces have become central to community life, encouraging residents to actively maintain the community's stability and vitality. Despite high resident participation, the utilization rate of public spaces needs enhancement, indicating deficiencies in design and management.

In conclusion, the Pantang case demonstrates the significant potential of community engagement in the conservation and revitalization of historic districts and provides an important reference value for similar efforts in the context of rapid urbanization. The Pantang experience highlights the innovation of community engagement as an urban design tool, showcasing the positive impact of bottom-up community involvement on urban renewal and conservation. Future efforts should focus on enhancing cooperation and communication among community members by organizing workshops between designers and residents to achieve sustainable development of historic districts.

References

Ali R, Olden K, Xu S (2008) Community-based participatory research: a vehicle to promote public engagement for environmental health in China. Environ Health Perspect 116(10):1281–1284

Esmaeilpoorarabi N, Yigitcanlar T, Kamruzzaman M, Guaralda M (2020) How can an enhanced community engagement with innovation districts be established? Evidence from Sydney, Melbourne and Brisbane. Cities 96:102430. https://doi.org/10.1016/j.cities.2019.102430

Gearin E, Hurt CS (2024) Making space: a new way for community engagement in the urban planning process. Sustainability 16(5):2039. https://doi.org/10.3390/su16052039

Hassen N, Kaufman P (2016) Examining the role of urban street design in enhancing community engagement: a literature review. Health Place 41:119–132. https://doi.org/10.1016/j.healthplace.2016.08.005

Head BW (2007) Community engagement: participation on whose terms? Aust J Polit Sci 42(3):441–454. https://doi.org/10.1080/10361140701513570

Kim G, Newman G, Jiang B (2020) Urban regeneration: community engagement process for vacant land in declining cities. Cities 102:102730. https://doi.org/10.1016/j.cities.2020.102730

Li T, Li Q (2024) Virtual reality in historic Urban District renovation for enhancing social and environmental sustainability: a case of Tangzixiang in Anhui. Sustainability 16(7):2665. https://doi.org/10.3390/su16072665

Li M, Liu J, Lin Y, Xiao L, Zhou J (2021) Revitalizing historic districts: identifying built environment predictors for street vibrancy based on urban sensor data. Cities 117:103305. https://doi.org/10.1016/j.cities.2021.103305

Polat SİBEL, Tümer Yildiz HÖ (2019) Community engagement in developing urban design guidance for heritage sites: the case of Bursa, Turkey. Urbani Izziv 30(2). https://doi.org/10.5379/urbani-izziv-en-2019-30-02-001

Polci V, Pierantoni I (2024) Innovating the local plan through co-creation and the public sociology approach toward urban regeneration: an Italian case study. Sustainability 16(8):3160. https://doi.org/10.3390/su16083160

Tang CF (2023) Research on place-making from the perspective of healthy City-take Guangzhou Pantang Village as a case. Highl Sci Eng Technol 52:297–303. https://doi.org/10.54097/hset.v52i.9215

Tavares SG, Sellars D, Dupré K, Mews GH (2024) Implementation of the new urban agenda on a local level: an effective community engagement methodology for human-centred urban design. J Urbanism 17(1):24–46. https://doi.org/10.1080/17549175.2021.2021972

Wang W, Bryan-Kinns N, Ji T (2016) Using community engagement to drive co-creation in rural China. Int J Des 10:37

Whitehand JW, Gu K, Whitehand SM, Zhang J (2011) Urban morphology and conservation in China. Cities 28(2):171–185. https://doi.org/10.1016/j.cities.2010.12.001

Xie J, Heath T (2017) Conservation and revitalization of historic streets in China: Pingjiang street, Suzhou. J Urban Des 22(4):455–454. https://doi.org/10.1080/13574809.2016.1167587

Ye C, Chen W, Liu Y, He Q (2024) Institutionalisation of public participation in China's urban regeneration from the perspective of historical institutionalism: three-stage cases in Guangzhou. Polit Geogr 108:103036. https://doi.org/10.1016/j.polgeo.2023.103036

Zhu Y (2015) Toward community engagement: can the built environment help? Grassroots participation and communal space in Chinese urban communities. Habitat Int 46:44–53. https://doi.org/10.1016/j.habitatint.2014.10.013

Conflictual and Insurgent Planning: The Case of *Mondeggi Bene Comune*: Tuscany (Italy)

Naggila Taissa Silva Frota, Elena Tarsi, and Clarissa Figueiredo Sampaio Freitas

Abstract

The theory of insurgent planning (Miraftab 2009; Huq 2020) has been widely applied in the Global South, addressing experiences that challenge institutional frameworks and promote grassroots-driven transformations. More recently, conflictual planning (Vainer et al. 2013; Sánchez et al. 2016; Oliveira et al. 2020) has emerged as a framework to analyze planning processes shaped by urgency, tension, and the production of intermediary technical documents. Although the literature often focuses on the Global South, relevant cases in the Global North also contribute to these discussions, demonstrating the adaptability of both theories. This research combines a literature review and case study analysis to explore the overlap and distinctions between insurgent and conflictual planning within the Mondeggi Bene Comune (MBC) experience.

Insurgent planning focuses on community-based initiatives that create autonomous, "invented spaces" as alternatives to state-led or institutionalized planning. These spaces challenge hegemonic structures by fostering self-management, solidarity networks, and collective decision-making. As outlined by Miraftab (2009, 2012), insurgent planning emphasizes the imaginative and transformative potential of marginalized groups, transcending formal boundaries to envision alternative futures.

Conflictual planning, in contrast, is rooted in contexts of urgency and immediate conflict. It involves rapidly producing technical documents, such as recovery or development plans, to support claims and negotiate with institutional actors. This approach, developed in Brazil (Vainer 2003; Tanaka 2017), arises in response to crises or environmental disasters, where conflict acts as a catalyst for social organization and technical creativity.

The Mondeggi experience exemplifies both approaches. Insurgent planning is evident in creating autonomous spaces, such as the Public Assembly at the Faculty of Agronomy and the National Meeting on Common Goods, which strengthened solidarity networks and fostered collective action. Meanwhile, conflictual planning appears in the production of the *Dichiarazione di Gestione Civica di un Bene Comune*, a technical document advocating for a new land management model, and in the community's rapid mobilization to prevent the land's sale by the municipality of Florence.

Ultimately, the Mondeggi case demonstrates how grassroots movements in the Global North can simultaneously embody insurgent and conflictual planning practices, contributing valuable insights into alternative and community-driven approaches to urban and territorial management.

Keywords

Community mobilization, Grassroots movements · Urban sustainability · Agroecological practices · Urban transformation

N. T. S. Frota (✉)
Department of Architecture and Urbanism and Design, Federal University of Ceará, Fortaleza, Brazil

Department of Architecture, University of Florence, Florence, Italy
e-mail: naggilafrota@arquitetura.ufc.br

E. Tarsi
Department of Architecture, University of Florence, Florence, Italy

C. F. S. Freitas
Department of Architecture and Urbanism and Design, Federal University of Ceará, Fortaleza, Brazil
e-mail: clarissa@arquitetura.ufc.br

F. Alberti et al. (eds.), *Towards Sustainable Urban Regeneration: Utilizing Environmental, Cultural, and Social Assets*, Advances in Science, Technology & Innovation, https://doi.org/10.1007/978-3-032-02091-8_33

1 Introduction

Urban planning originated as a predominantly technical activity before the nineteenth century, rooted in Euclidean principles that largely ignored social complexities. It was not until the twentieth century that planning began integrating social sciences, evolving into a multidisciplinary and less fragmented discipline. During the 1950s and 1960s, it established itself as a scientific endeavor, primarily confined to institutional frameworks and conducted by state-affiliated professionals (Friedmann, 1987). In recent decades, urban planning has embraced emerging agendas, including socio-spatial inequalities and environmental challenges, fostering greater interdisciplinarity, expanding its scope of practice, and adapting its theoretical and practical paradigms to address dynamic social transformations (Marcuse et al. 2011).

Vasudevan and Novoa (2022) have addressed a long theoretical path from the beginning of the twentieth century to the present day, demonstrating how planning has undergone a paradigm shift that is beginning to include other actors, methods, and forms of knowledge in its development, especially in the context of Global South cities.

This group of cities has witnessed various urban planning experiments that question the *status quo* and that take place outside formal institutions (De Souza 2006; Huq and Shafique 2023; Rocco and Silvestre 2024; Yiftachel 2009). Some planning approaches have focused on giving greater visibility to these experiences, emphasizing their potential to challenge structural issues of territorial occupation and to improve the planning process, such as insurgent planning (Huq 2020; Miraftab 2009) and conflictual planning (Oliveira et al. 2020; Sánchez et al. 2016; Tanaka 2017; Vainer et al. 2013), which will be detailed in the next section.

Although the literature highlights numerous examples in the Global South, the Global North can also reveal relevant cases of insurgent and conflictual planning experiences. In New York, we can mention the Cooper Square neighborhood, which underwent a community planning process led by its residents to refute the proposal to build luxury developments put forward by the municipality (Angotti 2008). Another case was the Los Angeles River revitalization project, which underwent lengthy negotiations with the local community, led by the Friends of the Los Angeles River (FoLAR) group, to prevent the expulsion of the area's vulnerable population and ensure greater environmental justice (Yung et al. 2022). To expand studies on this type of experience, this research aims to relate the case of Mondeggi Bene Comune (MBC), located in the Metropolitan City of Firenze, in the Tuscany region of Italy, to the approaches of insurgent and conflictual planning, based on the identification of events that took place between 2009 and 2023.

In a nutshell, the object of study is a 170-hectare farm owned by the Città Metropolitana di Firenze which, after a period of underutilization and an attempt to sell it, underwent a process of occupation led by the Terre Bene Comune (TBC) group to occupy the idle area for sustainable agriculture ("Dossier Mondeggi," n.d.).

This article presents the initial results of the main author's recent doctoral research and is therefore the first approximation between the object of study and the theoretical concepts listed. To establish the theoretical framework, this research used the literature review method, considering authors from the Global North and South with consolidated and recent work in insurgent planning and conflictual planning. In addition, the authors carried out a documentary analysis and systematization of materials produced by different actors involved in the MBC experience. The selection of books, dissertations, and websites considered the diversity of authorship, such as the University of Florence and the *mondeggini* themselves (MBC residents), to address different points of view.

2 Urban Planning and Its Nuances

Urban planning is a tool undergoing constant transformation, increasingly shaped by actors outside the State. A notable milestone is Friedmann's (1987) pioneering proposal of radical planning, which emphasizes self-organized actions by communities as a challenge to the established hegemonic order. Later, Sandercock (1998) brought new contributions that broadened the perception of the limits of institutional planning and the tools for including different voices in the planning process.

Miraftab (2009) advances this discussion by describing the concept of insurgent planning as transgressive of time and space because it is historically anchored and has a transnational consciousness; counter-hegemonic because it destabilizes the imposed order and imaginative because it promotes new worlds that are necessary and possible. The contribution of insurgent planning (Miraftab 2012) lies in its emphasis on recognizing organized civil society as a key actor in the planning process. For Miraftab (2012), "*insurgent planning is a contested field of interaction among multiple actors including, but not confined to, professional planners, who determine the arenas of action to address the specific forms of oppression*" (p. 2).

More specifically, Miraftab (2012) identifies insurgent planning as the influence of the actions of groups excluded from the decision-making process of land management as an alternative to urban planning guided by the state and characterizes it as:

- Direct action undertaken by a marginalized group in opposition to an established order that does not favor the collectivity.
- The creation of autonomous, non-institutional discussion spaces called invented spaces.
- Claiming specific interests and values that have not been addressed by the state and hegemonic powers.

The concept of conflictual planning, on the other hand, was developed through reflections by Brazilian theorists on the challenges faced by urban planners, particularly in response to environmental disasters such as the collapse of ore tailings dams in Brazil (Vainer 2003a, b). This context required emergency measures to assist the population and robust social organizations to pressure the government for immediate responses. The Movimento dos Atingidos por Barragens (MAB—Movement of Those Affected by Dams) played a fundamental role in bringing together different groups and in long-term reflection on the issue (MAB 2022).

Another characteristic of conflictual planning is the presence of technical support groups, made up of either professionals (technical advisors) or students (university extension activities). These groups are dedicated to producing documents that act as instruments for claiming rights and often become facilitators of technical language. The *Plano de Recuperação e Desenvolvimento Econômico e Social das Comunidades Atingidas pelas Barragens de Itá e Machadinho* (Plan for the Recovery and Economic and Social Development of the Communities Affected by the Itá and Machadinho Dams) is a tangible example of a technical document developed autonomously in the context of conflict (Vainer 2003a, b). For this approach, the term conflict is based on the idea of a powerful and dynamic reality that can generate radical changes. Since then, this concept has been increasingly explored by Brazilian research groups, such as Núcleo Experimental de Planejamento Conflitual[1] (NEPLAC—Experimental Conflict Planning Center) and Laboratório Estado, Trabalho, Território e Natureza[2] (ETTERN—State, Labor, Territory and Nature Laboratory) which are affiliated with the Universidade Federal do Rio de Janeiro (UFRJ—Federal University of Rio de Janeiro). However, the concept is still in the process of construction.

More recently, new studies have been developed by NEPLAC and ETTERN (Vainer et al. 2013; Sánchez et al. 2016; Oliveira et al. 2020) about the case of Vila Autódromo, a precarious settlement threatened with eviction due to real estate interests in the context of the 2016 Olympics in Rio de Janeiro. As a way of remaining in their place of origin, the residents of Vila Autódromo, together with other actors such as the university, built the Popular Plan for Vila Autódromo (Autódromo 2012), which includes alternative solutions to those initially suggested by the City Council.

Nowadays, after the subject has matured further, conflictual planning can be understood as an approach to planning that deals especially with emergencies in which conflict is imminent (Tanaka 2017). In light of the analysis of the materials produced by these groups, we can say that Conflict Planning is characterized by:

- The technical support of academic or professional groups in the conception of urban planning and design alternatives.
- The emergency that requires intense social organization and the production of provisional technical pieces to support urgent demands.
- The conflict scenario as a catalyst for fostering creativity to generate solutions, actions, and possibilities.
- The circular knowledge that allows residents to move through different spheres of negotiation and power.
- The communicative actions that strengthen links with other groups and dispute narratives with the official media.

Ultimately, conflictual planning argues that only from the context of conflict is it possible to build collective actors who become political planners capable of promoting the right to self-management of the city (Oliveira et al. 2020).

Following this discussion, it is clear that although insurgent planning and conflictual planning are distinct concepts, the literature highlights several areas of convergence. Both frameworks challenge conventional institutional planning by proposing alternative and localized solutions grounded in community mobilization. They emphasize the role of collective action in reshaping urban planning. The main distinction lies in the context and response mechanisms. Conflictual planning is closely tied to situations of heightened social urgency and immediate threats, which necessitate the rapid production of technical documents and provisional solutions to address pressing demands. In contrast, insurgent planning focuses on the broader creation of "invented spaces" fostering transgressive, imaginative, and long-term transformations that challenge hegemonic systems. In other words, conflictual planning tends to operate within the constraints of immediate conflict, whereas insurgent planning emphasizes the construction of alternative futures beyond those conflicts.

3 Mondeggi in the Light of Insurgent and Conflictual Planning

One of the results obtained up to this point in the research is the temporal analysis of the main events that marked the MBC experience and which are close to the concepts dis-

[1] For further information, check the link: https://moradiaassessoria.org.br/grupos/nucleo-experimental-de-planejamento-conflitual-neplac.

[2] For further information, check the link: https://etternippur.wixsite.com/conflitual2021/apresenta%C3%A7%C3%A3o.

cussed above. The analysis starts by highlighting the historical occupation process of the area, which began in the tenth century as a large rural estate that housed the Santa Maria Nuova hospital (1488–1506), but which, for the most part, belonged to a sequence of noble families, with the della Gherardesca family being the uninterrupted owners for four centuries (1506–1938). In the 1960s, the estate was acquired by the Città Metropolitana di Firenze, and in 2001, it was taken over by the Mondeggi Lappeggi agricultural company, also owned by the province. In 2009, the farm company ceased its activities and Mondeggi went through a process of abandoning the land (Biggeri et al. 2023; Perrone 2016). The Città Metropolitana di Firenze, with the intention of paying off debts, put the land up for sale, backed by the Salva Italia Decree—Art. 66 of 2012 which states "*Le regioni, le province, i comuni, anche su richiesta dei soggetti interessati possono vendere o cedere in locazione, per le finalità e con le modalità di cui al comma 1, i beni di loro proprietà agricoli e a vocazione agricola*" (Salva Italia Decree 2012). This decree is part of the austerity measures introduced by Mario Monti's government (2011–2013), whose broader aim was to balance Italy's public accounts during an economic crisis which included, as in the Mondeggi example, the privatization of public farmland (Rosa 2013).

Within the parameters of conflict planning analysis, the proposal to sell the property is a major conflict that marks Mondeggi's experience. In 2013, the Terre Bene Comune (TBC) group, linked to the national Genuino Clandestino network, brought together families, students, farmers, and activists from the region and took the initiative to occupy the farm to prevent its sale, putting into practice another model of territorial appropriation. The *Genuino Clandestino* group is part of a national network of peasants who see agriculture from the point of view of sustainable development and encourage rural activity through smallholdings. At this point, we see the TBC group's confluence with the idea of conflictual planning, as the conflict over the use of the land is highlighted as a matter of urgency in an imminent commercialization process. To prevent the sale from taking place, radical measures were applied that were considered illegal from the point of view of the institutional apparatus.

The TBC group occupied the territory intending to promote "the participatory and autonomous management of public lands by local communities, within the framework of neo-rurality projects, favoring peasant, natural, community, social and subsistence farming projects, within a new relationship between town and country" ("Dossier Mondeggi," n.d.). This initial group was composed of around 40 individuals, including farmers, artists, students, and others. Mondeggi's experience included new agricultural technologies, planting indigenous species and using traditional techniques with the help of older generations. The early efforts were led by a small team of about five people, most of them women who had previous experience in their communities with social promotion, agriculture, education, and commercial activities (Perrone 2016).

However, this project to reoccupy unproductive land has grown and gone beyond sustainable cultivation. According to the ethnographic work carried out by Berg (2022), Mondeggi was considered by the *mondeggini* to be a laboratory for experimenting with new possibilities for cultivation, sociability, and the economy, in fact becoming a common good. At this point, we can see how close this initiative is to insurgent planning, based on its ability to imagine scenarios for the future that differ from the one initially put forward by the *Città Metropolitana di Firenze*.

As a way of building a path of recognition and dialog with those interested in consolidating the project and also outlining innovative methods of social and shared management, the TBC group renamed itself Mondeggi Bene Comune and collectively drafted the *Dichiarazione di Gestione Civica di un Bene Comune* (Declaration of Civic Management of a Common Good). It states that the model adopted by the *mondeggini* is based on building community life ("Dossier Mondeggi," n.d.). The drafting of this document can be seen as a conflictual planning action, as it creates an alternative territorial development policy that differs from the one imposed by the institutions. Another landmark moment in MBC's experience came with the implementation of the *Mondeggi Terreni Autogestiti* (MoTA—Mondeggi Land Self-managed) project, which further developed the idea of collective ownership by donating land. In both events, it is noticeable how the ability to imagine a new future is present and is put into practice with concrete actions symbolizing a rapprochement with insurgent planning practices. Ultimately, the community model of occupying agricultural property based on the common good promotes discussion about the social function of property and questions private property even more radically.

Other important moments in the MBC experience were the Public Assembly at the Faculty of Agronomy of the University of Florence and the National Meeting on Common Goods ("Dossier Mondeggi," n.d.). On these occasions, various issues were discussed collectively, such as the management of rural activities, the confrontation with the municipality, and the role of local communities in spreading peasant agriculture and food sovereignty. Both events can be considered "invented spaces" since they were initiatives proposed by the TBC to strengthen its solidarity networks through the exchange of experiences, raising awareness among actors outside MBC and winning over new supporters. It is important to note that in addition to MBC, there are similar experiences in Naples and Palermo that also function as a solidarity network. Berg (2022) recounts another moment when the *mondeggini* support another group that is undergoing a police intervention in order to reinforce the good relationship and thus guarantee the support of these support groups in other threatening circumstances.

Still in the temporal analysis, one of the most recent facts is that in 2021, the *Città Metropolitana di Firenze* was awarded funds from the *Piano Nazionale di Ripresa e Resilienza* (National Recovery and Resilience Plan) and earmarked part of it for the construction of the *Progetto di Rigenerazione della Tenuta di Mondeggi* (Project for the Regeneration of the Mondeggi Property). Various organizations are involved in this project, such as the University of Florence, the *Comune di Bagno a Ripoli*, and various professionals in the fields of architecture and engineering, among others (Biggeri et al. 2023). This institutional initiative signifies a change in the City Council's stance on how to deal with the occupation of Mondeggi, which is now more consolidated and organized. According to Berg (2022), the *mondeggini* questioned what would be the most appropriate stance, to collaborate with the project or to offer resistance, since accepting the project is a path to legalization and this could compromise the authenticity of MBC. At the same time, the *mondeggini* see the City Council's initiative as positive, because it means that the occupation has grown and consolidated to the point where it requires other measures to contain its progress, beyond just the previous attempts to sell it.

At first glance, the institutional intentions of the *Progetto di Rigenerazione* seem to be in line with the wishes of the TBC, including a mechanism for social participation, sustainable agricultural management techniques, and the generation of jobs and income. However, it is important to note that there is no questioning of the ownership model in the *Progetto* and no mention of the shared land management format proposed by the TBC, which remains public. The University of Florence recognizes the design process, which applies to situations that have some characteristics of informality and need greater flexibility to be regulated. These are generally very new situations that have no precedents, so there is no prior regulatory apparatus (Biggeri et al. 2023). This recognition highlights the mismatch between the MBC experience and the existing planning frameworks, which fail to accommodate the specific needs of such new and informal situations.

Furthermore, about the City Council's position, there is an attempt to reduce the importance and impact of the occupation, in the publication entitled *Mondeggi: Rigenerazione sociale, culturale e agricola per una Città Metropolitana sostenibile*, the text of the Mayor says "Dire 'Mondeggi' è stato a lungo evocare una sorta di sinonimo di ciò che appare 'abbandonato'. Non è così, non è vero." in the English version: "Saying 'Mondeggi' has long been a way of evoking a sort of synonym for what seems 'abandoned.' That is not the case, it is not true." (Biggeri et al. 2023, p. 11).

4 Final Considerations

Given this context, it is clear that the MBC experience both breaks with institutional planning by proposing a new model for occupying the territory and serves as a valuable example of insurgent and conflictual planning in the Global North. Mondeggi not only creates an autonomous community that questions the absolute right of private property but also builds an alternative path of reconnection between human beings and the environment.

As the next steps, this research intends to develop a more detailed timeline to visually highlight events in Mondeggi's history that relate to insurgent and conflictual planning. At this point in the research, as we seek a connection with contemporary challenges such as the climate crisis, new research questions present themselves: Does Mondeggi's experience represent a model of land occupation that can effectively influence the reduction of the effects of the climate crisis? Can its respectful way of cultivating the land using agroecological techniques and its call for a fairer form of access to productive land bring about a greater natural balance? In addition, the main author identifies gaps in the case study that warrant further investigation. For instance, a field visit is needed to determine: What is currently happening in Mondeggi? What is the condition of the residents? Has the agenda actually been transformative? Moreover, in theoretical terms, the research should advance the discussion of the European Green Deal platform to relate the proposals contained in this agenda to the MBC initiative. We believe that the MBC experience has a lot to contribute to a future in which planning takes into account different voices and different worlds.

References

Angotti, T. (2008). New York for Sale: Community Planning Confronts Global Real Estate. The MIT Press.

Autódromo, A.-A. de M. e P. da V. (2012) Plano Popular da Vila Autódromo—Plano de desenvolvimento urbano, econômico, social e cultural. https://comitepopulario.wordpress.com/wp-content/uploads/2012/08/planopopularvilaautodromo.pdf

Berg F (2022) No owners, no masters: a critical exploration of land access and community land stewardship at Mondeggi Bene Comune [Master's thesis, Lund University]. Lund University

Biggeri M, De Luca G, Ferrannini A, Pisano C (2023) Mondeggi: Rigenerazione sociale, culturale e agricola per una Città Metropolitana sostenibile, 1st edn. Firenze University Press, Florence. https://doi.org/10.36253/979-12-215-0195-7

De Souza ML (2006) Social movements as 'critical urban planning' agents. City 10(3):327–342. https://doi.org/10.1080/13604810600982347

Decreto Salva Italia (2012). https://www.normattiva.it/uri-res/N2Ls?urn:nir:stato:decreto.legge:2012;1~art66!vig=

Dossier Mondeggi (n.d.) Mondeggi Bene Comune. https://mondeggibenecomune.noblogs.org/. Accessed 12 July 2024

Friedmann J (1987) Planning in the public domain: From knowledge to action. Princeton University Press, Princeton

Huq E (2020) Seeing the *insurgent* in transformative planning practices. Plan Theory 19(4):371–391. https://doi.org/10.1177/1473095219901290

Huq E, Shafique T (2023) People move, policies don't: discursive partition against climate-impacted dwellers in urbanizing Bangladesh. Environ Urban 35(1):91–100. https://doi.org/10.1177/09562478221149863

MAB, M. de A. por B. (with Vainer, C., & Vieira, F. B.) (2022) Manual do Atingido

Marcuse P, Connolly J, Novy J, Olivo I, Potter C, Steil J (2011) Searching for the just city: Debates in urban theory and practice (First issued in paperback). Taylor & Francis Group, Routledge

Miraftab F (2009) Insurgent planning: situating radical planning in the global south. Plan Theory 8:32–50

Miraftab F (2012) Planning and Citizenship. Oxford University Press, Oxford

Oliveira F, Sánchez F, Vainer C (2020) Planejamento Conflitual na Cidade Olímpica: A experiência da Vila Autódromo. Contrapoder, Rio de Janeiro. https://contrapoder.net/wp-content/uploads/2020/09/Planejamento-Conflitual-na-Cidade-Olimpica-a-experiencia-da-Vila-Autodromo-Rio-de-Janeiro.pdf

Perrone C (2016) Grounds for future gendered urban agendas: policy patterns and practice implications. Town Plan Rev 87(5):589–606. https://doi.org/10.3828/tpr.2016.36

Rocco R, Silvestre G (2024) Insurgent planning practice. Agenda Publishing, London

Rosa OD (2013) Decreto Salva Italia: Nuove imposte per fronteggiare la crisi [Master's thesis, University Ca' Foscari Venezia] University Ca' Foscari Venezia. http://dspace.unive.it/handle/10579/3082?show=full

Sánchez F, Oliveira FLD, Monteiro PG (2016) Vila Autódromo in dispute: subjects, instruments and strategies to reinvent the space. Rev Bras Estud Urbanos Reg 3:408. https://doi.org/10.22296/2317-1529.2016v18n3p408

Sandercock L (1998) Making the invisible visible: a multicultural planning history. University of California Press, Oakland

Tanaka G (2017) Planejar para lutar e lutar para planejar: Possibilidades e limites do planejamento alternativo [Doctoral Thesis]. Federal University of Rio de Janeiro

Vainer C (2003a) O conceito de atingido: Uma revisão do debate e diretrizes.

Vainer C (2003b) O Plano de Recuperação e Desenvolvimento Econômico e Social das Comunidades Atingidas pelas Barragens de Itá e Machadinho. Uma experiência inovadora de extensão universitária e de planejamento. Cadernos IPPUR. https://ippur.com.br/wp-content/uploads/2016/05/CI_Ano_XVII_n1_jan-jul_2003.pdf

Vainer C, Bienenstein R, Tanaka G, Oliveira F, Lobino C (2013) O Plano Popular da Vila Autódromo: Uma experiência de planejamento conflitual. http://unuhospedagem.com.br/revista/rbeur/index.php/anais/article/view/4316

Vasudevan R, Novoa EM (2022) Pluriversal planning scholarship: embracing multiplicity and situated knowledges in community-based approaches. Plan Theory 21(1):77–100. https://doi.org/10.1177/14730952211000384

Yiftachel O (2009) Critical theory and 'gray space': mobilization of the colonized. City 13(2–3):246–263. https://doi.org/10.1080/13604810902982227

Yung B, Elias R, Hanuman S, Malik F, Greenberg J (2022) Urban displacement project (LA river redesign: a case study on displacement, gentrification, and housing justice in Los Angeles). https://www.urbandisplacement.org/wp-content/uploads/2022/03/LA_River.pdf

Shared Living with Local Connections: Creating a New Neighborhood Through the Participation of Young People

Yuno Tanaka

Abstract

The increasing rate of unmarried young people in Japan, coupled with the propensity toward spending more time as singles, has significantly influenced residential patterns. Traditionally, residential choices were determined by one's workplace; however, the growing prevalence of telework has expanded the range of options. Consequently, single young people have greater freedom of choice in housing, leading to a shift from long-term relationships with their neighborhoods to more short-term and tenuous ones. Considering the reduction in the number of local players due to population decline, it is necessary to consider how residents who do not intend to live permanently in a certain area can foster meaningful relationships with their neighborhoods while ensuring sustainable local management. This study posits that maintaining connection with the local community through daily interactions, engaging with intermediaries who facilitate relationship building within the local community, participating in local economic activities, and sharing intangible aspects such as the community's future vision and culture are important aspects of the residents' relationship with the local area. Therefore, the research focuses on complex-type shared residences with nonresidential facilities that fulfill the above conditions, to identify variations in complex-type shared residences and to understand the actual situation of facilities that have special characteristics. The novelty of this study lies in its focus on shared housing, where residents are generally more amenable to social interaction, and on the involvement of unmarried young people in local development.

Keywords

Shared housing · Local management · Short-term residence · Single person · Interview

1 Introduction

1.1 Backgrounds

Bauman (2000) describes the present day as "liquid modernity," highlighting the friability and brittleness of human bonds and networks in modern societies. In recent years, this trend has been confirmed in the relationships between people and their local areas in Japan. The rate of unmarried young people is increasing, with many spending more and more time as singles. While it used to be common for people to live within commuting distance of their workplaces, the spread of telework (Ministry of Internal Affairs and Communications 2023) due to the COVID-19 pandemic has led to more diverse residential choices. Consequently, single young people have more flexible housing choices, and the relationship with the local community in which they live is changing from a long-term one to a more short-term and tenuous one. Previously, local community development was generally conducted in a long-term approach by residents who lived in the community for a prolonged period and assumed that they would continue to live in the area in the future. However, the number of long-term, continuous community leaders is decreasing due to the declining population, an aging society with fewer children, and the increasing mobility of residents; therefore, it is necessary to reconsider how community development should strategies even for residents who do not intend to live permanently in a certain area so that they actively contribute to the community. Moreover, for new residents to be involved in the local community, it is essential to first consider how they can create contact with their local communities.

Y. Tanaka (✉)
School of Environment and Society, Institute of Science Tokyo, Tokyo, Japan
e-mail: tanaka.y.cp@m.titech.ac.jp

F. Alberti et al. (eds.), *Towards Sustainable Urban Regeneration: Utilizing Environmental, Cultural, and Social Assets*, Advances in Science, Technology & Innovation, https://doi.org/10.1007/978-3-032-02091-8_34

In this study, the following four points are considered important for residents to engage with the local community: having a place where they can interact with the local community through their daily lives (locational contact), having contact with mediators who promote building relationships within the local community (human contact), participating in local economic activities (economic contact), and sharing intangible aspects such as the future vision and culture of the local community (value contact). Therefore, this study will explore complex-type shared housing projects that include facilities other than residences that fulfill the above conditions, identify variations, and then conduct case studies to obtain an understanding of the actual situation. The novelty of this study lies in its focus on shared housing, which is assumed to have most residents who are not uncomfortable interacting with others, and on the involvement of singles in local community activities.

1.2 Literature Review

Ogawa and Nohara (2020) identify three primary types of local community development bases in Japan: self-governing, administrative, and private. Specifically, the self-governing type includes *chonaikai* (neighborhood associations), the administrative type includes *kominkan* (community centers) and *machidukuri-center* (community development centers), the private type includes area management offices and community cafes, and urban design centers have emerged in recent years as a type of collaboration among the three types. The challenges faced by private-type centers include a lack of manpower, insufficient financial resources, and aging staff (ibid). Koyano et al. (2016) surveyed community cafes as places for local residents to make connections and found that of the 78 cases, about half received subsidies or grants; however, the financial status was generally poor, with most community cafes operating at a deficit or break-even, with only 11.5% reporting profitability. This shows the difficulty of managing private community activity centers. In addition, participation in these local community development bases often requires a strong commitment to local engagement and frequent interaction, limiting casual involvement.

There are various ways to open private houses, such as the open house project (Open City Study Group 2017) which temporarily opens up a historical or characteristic house to the public, the house that is partially used as a small shop, and *sumibiraki* (Asada 2012), which opens a part of the house to the neighborhood as a common space. However, these projects generally are for detached owner-occupied houses, designed to foster long-term community relationships, and they are hardly an option for a single person living in rental housing.

Then, how do singles who do not intend to live permanently in the area live? The most common way is in rental housing units such as one-room apartments. However, these units typically offer minimal interaction with neighbors, leading to a lack of meaningful connections with the local community. Against this situation, shared houses have become a popular alternative for singles seeking social interaction. Although there is no clear unified definition of a shared house, the Ministry of Land, Infrastructure, Transport and Tourism (2017) defines it as a residence where several people who are not relatives live together in a single rental property. Shared houses have gained popularity, especially in urban areas. Despite some stagnation, presumably due to COVID-19, the number of shared house properties has continued to increase in recent years (Japan Share House Organization 2023), indicating that there is a certain demand for shared housing. Kubota (2014) categorizes shared houses in Japan into two main types: i) the self-managed type, in which people rent apartments or houses with their acquaintances and live together, and ii) the business type, in which a company acts as an intermediary between the owner and the residents to recruit and manage the property. Many business-type properties outsource part of the management of common spaces, such as the cleaning of common spaces by cleaning companies, and provide services to eliminate the causes of conflicts among residents. In recent years, the number of business-type share houses has been particularly increasing, and many of them appeal to residents with a variety of services and unique concepts, some of which emphasize local community interaction.

1.3 Methodology

To explore how single residents who do not intend to permanently reside in a certain area can engage with the local community, this study focuses on shared houses, where it is assumed that most of the residents are not uncomfortable with having relations with other people. The research aimed to identify shared houses throughout Japan that not only facilitate interactions within the house but also foster connections with the local community. Information was gathered through the property introduction webpage of Hituji Real Estate (2024), a major website for shared house properties in Japan. In addition to basic property information such as exterior and interior photographs, number of shared rooms, rent for available rooms, and distance from the nearest station, the property webpage also includes descriptions provided by the house suppliers. Properties that mentioned local community engagement in their descriptions were selected and classified based on factors such as location for interaction, type of people involved, and continuity of activities. Then, to gain a deeper understanding of the kind of involvement with the local community, case studies were conducted on two selected shared houses with permission for interviews. Based on the above research, I analyze and discuss how residents who do not intend to live permanently in a certain area can create connections with the local community and how they can continue and deepen their relationship with the local community afterward.

2 Classification of Shared Houses Characterized by Relationship with the Local Community

2.1 How to Connect with the Local Community

There are gradations in how residents interact with the local community. These interactions range from relatively minor engagements, such as shopping at local stores or greeting neighbors when they cross paths, to more substantial efforts like planning and organizing local community-related events independently. This section aims to identify the elements necessary for deeper involvement in the local community. As one of the relationships between housing and the local area, architect Rikien Yamamoto et al. (2013) proposed the concept of a "Local Republic." Yamamoto criticizes the current separation of housing and work and remarks that local communities need an economic basis. Besides, community designer Ryo Yamazaki (2016), in "The Origins of Community Design," takes up the imaginary socialists and points out the importance of handcraft work in local community development. Yamazaki argues that understanding the meaning of one's creations is crucial and criticizes the machine-like division of labor focused solely on efficiency. This suggests the importance of relatively small-scale, artisanal work being realized on a local scale, but in today's global society, it is not realistic for all people to be engaged in local-scale work. Therefore, it will be necessary to consider involvement through economic activities other than the main job (e.g., side jobs and hobbies). Furthermore, sociologist Eiji Oguma, in a conversation with Riken Yamamoto (Oguma and Yamamoto 2019), highlights the importance of sharing the same scenery and mention that invisible things such as the future and traditions create a community, and communities cannot be created by considering only the visible range. Based on the above insights, in this study, identifies key aspects of residents' interactions with the local community. In addition to having a base for daily life interactions (locational contact) and engaging with mediators who facilitate human relationships (human contact), the study considers participation in local economic activities beyond one's primary job (economic contact) and the sharing of intangible elements such as future visions and cultural values (value contact) as crucial. These dimensions are used to organize and analyze information on how residents can foster deeper, more meaningful connections with their local communities.

2.2 Results of Survey

For shared houses that are characterized by their involvement with the local community extracted from Hitsuji Real Estate webpages, the details of community engagement have been organized based on several dimensions: the location of interaction (locational contact), planner/visitor (human contact), participation fee (economic contact), and continuity (value contact). They were categorized as shown in Table 1. The locations where interaction takes place were categorized into three types: (1) a type in which a part of the shared house (the residence) is temporarily opened to the local community on the occasion of events; (2) a type attached to the same building as the shared house, but accessible from the outside; and (3) a type that encourages participation in events held in the local area, not on the site of the building in which the shared house is located. Specifically speaking, for example, (1) is a living room, vegetable garden, or courtyard shared by shared house residents, (2) is event space or a shop such as a café or gallery, and (3) is that the shared house manager supports shared house residents' participation in local events. Hosts are the people who operate the shop or host the event and can be categorized as shared house managers, shared house residents, acquaintances of residents, or outsiders who have no relationship with the shared houses. Guests ranged from closed events where only those who knew the shared house residents were invited, to open events that were widely advertised in the local community and open to anyone who wanted to attend. The continuity type was categorized into permanent facilities such as shops, regular events such as

Table 1 Classification of interactions with the local community (Source: authors' own research)

Location	Inside the shared house	Inside the building site but outside the shared house		Outside the building site
People (host)	Manager of the shared house	Residents	Acquaintances of residents	Outside person
People (guest)	Acquaintances only		Anyone	
Participation fee	Paid	Resident discount		Free of charge
Continuity	Permanent	Regular events		One-time event

weekly book readings, and one-time events such as concerts and art exhibitions. Some of the shared houses that hold art exhibitions limit the residents to artists and creators.

3 Case Studies

This section summarizes the results of interviews with stakeholders and field surveys of two case studies of shared housing projects that are actively working to engage with the local community. Interviews with stakeholders were conducted in July 2024 for approximately 1 h in a semi-structured format at each property, with a tour of the space used for local interaction.

3.1 Commune Tokiwa

Commune Tokiwa is a large-scale shared house in Saitama City, consisting of 55 units (19 units for families and 36 units for singles or two-person households), including 4 SOHO (small office, home office)-type units, as well as common spaces such as a studio, courtyard, and rooftop vegetable garden, and a shop. A 10-min walk from the nearest station, the project was completed in January 2020 under the concept of "people with dreams gathering, building a community, and living in harmony with the local area." The following is a summary of an interview with Mr. Funamoto, representative director of the owner.

3.1.1 Reason for Starting Commune Tokiwa

Since his youth, Mr. Funamoto has been conscious of the problem that rental housing in Japan is positioned as a temporary residence until one can purchase a detached house, and he felt that rental housing is too small and not affluent enough. He felt that a lifestyle change was needed in Japan's social structure, where workers work long hours at the workplace and go home only to sleep, and he aimed to create rental housing that would allow people to enjoy their lives. With the concept of valuing neighborhood relationships, he intentionally created variations in the size and layout of the residential units so that singles and families could get together in a mixed environment. The interiors are planned to be DIY-able, a rare feature in Japan, so that residents can personalize the interiors. Although the shared house is equipped with a variety of common spaces, the rent is about the same as that of ordinary rental apartments.

3.1.2 Common Spaces and Local Community Interaction

The building has a studio, courtyard, and rooftop vegetable garden as common spaces for shared house residents. There is also a handmade wooden stand at the entrance of the building for sharing vegetables from the vegetable garden, and a bulletin board where residents can freely post notices (Fig. 1). The common restroom on the first floor is wheelchair accessible and is designed to accommodate people with disabilities. The studio is equipped with a piano, screen, bookshelves, and game consoles and is free for use by shared house residents (Fig. 2). Reservations are required for occupied use, such as when holding events. The courtyard is a place where the building manager holds an annual summer festival and Christmas party, as well as other events planned by the residents. The rooftop vegetable garden has six plots for rent, and shared house residents who wish to use them can do so for 2500 yen per month. There is also a space for barbecues and other kinds of parties, as well as a water supply system. Although the events are open to shared house residents, one can invite their friends; so the events are not closed to the shared house community. Children living here become friends with other elementary school students from

Fig. 1 Wooden stand for sharing and resident bulletin board (Source: Photograph taken by Yuno Tanaka in 2024)

Fig. 2 Common studio for resident. (Source: Photograph taken by Yuno Tanaka in 2024)

the same school district, and sometimes, they play together with their families at the Commune Tokiwa.

3.1.3 Resident and Tenant Recruitment

When recruiting residents, Mr. Funamoto personally interviews prospective tenants to emphasize the importance of community interaction in the shared house. Once residents move in, some actively engage in planning events, while others may not be as social, but involvement is voluntary. Although the Commune Tokiwa has a community manager, she is not present to plan and operate events to treat residents as guests but rather to support resident-led events. The operator of the café in the retail zone is a nonprofit organization in Saitama Prefecture that provides employment support for people with disabilities because Mr. Funamoto wanted to support their activities. Although the building was completed in January 2020 and faced challenges due to the COVID-19 pandemic's first state of emergency in March, resulting in a delayed recruitment process, the housing units are now nearly fully occupied.

3.1.4 Social Relationships within the Shared House

Mr. Funamoto has been trying to provide employment for as many residents as possible. For example, he asked a resident web designer to create a website for his company's second shared house property. Some residents also offer work to each other, and he feels that a kind of business community has been growing. However, there are some difficulties in having households with various family structures living together, as their different lifestyles limit interactions with each other, and single residents sometimes complain to family households that their children's voices are bothersome.

3.1.5 Connections with Former Shared House Residents and Future Prospects

Although Mr. Funamoto hopes that residents will stay in the shared house for the long term, some have moved out due to factors such as job relocation. However, some residents continue to visit Commune Tokiwa events after moving out, because they are approached by current residents. Urawa-ku, where Commune Tokiwa is located, is a commuter town for students and professionals. In addition, many people move in from far away, and neighbors in the apartment houses do not know each other well. Under these circumstances, he believes that the people, a precious local resource, can be vitalized by helping the neighbors become acquainted with each other and making connections within the local community. If people can be vitalized, the local community will be activated, jobs will be created, and money will be circulated in the local area. He hopes that Commune Tokiwa will function as a nest to grow opportunities for this purpose. He believes that to live well, people need to have interpersonal relationships, and he would like to create relationships like weak ties (Granovetter 1973). Furthermore, his company opened a share house "Commune Urawa Ryoke" in 2023, which is a renovated old traditional house in a nearby area. There is a plum tree in the garden of Ryoke, and residents of Commune Tokiwa go there to help harvest the plum trees; in turn, the residents of Ryoke help as staff for events held at Commune Tokiwa. In addition, a new property is currently being purchased, and Mr. Funamoto hopes to include it in the exchange activities that transcend the boundaries of the property. He also hopes to expand their activities and connections to the local community, such as launching local businesses with the cooperation of the shared house residents. Some of the shared houses in Japan have specific concepts, such as houses for learning foreign languages, single mothers, or entrepreneurs. While such an approach is one way of doing business, if it is too specific, it will only attract people who are interested in that particular. Conversely, if there is no concept, it is challenging to identify the differences from other shared houses, making it difficult to attract people. He thinks that it is difficult to find the proper balance to attract a diverse range of people.

3.2 Kotona House

Kotona House is located in Kunitachi City, Tokyo, and combines a shared house and a common space open to the neighborhood called "Chanoma" (which means, in Japanese, a room in a house where family members eat, gather, etc.). Located in a local market street built in 1965, a five-minute walk from the station, the shared house has four rooms and opened in April 2015. The following is a summary of interviews with Mr. Suzuki, the owner and manager of Kotona House, and Mr. N, a resident of Kotona House. The content of the interview with Mr. N is noted as "Interview with Mr. N," and the rest of this section is based on the interview with Mr. Suzuki.

3.2.1 Reason for Starting Kotona House

The owner, Mr. Suzuki, previously worked in an area development project that utilized space under elevated railroad tracks. However, he felt limited in what he could do within the scope of his work for the local community and wanted to create a base of local activities in the area by himself. Therefore, he purchased a property, renovated it, and started Kotona House. At the time, he was an employee with limited time to invest in this venture. Fortunately, he met Ms. O, who was interested in managing a shared house, and the project started with Mr. Suzuki as the owner and Ms. O living in the shared house and managing it. Later, Ms. O started running a bookstore across the street from Kotona House; however, since she had a baby, Mr. Suzuki has been managing Kotona

House himself. When Kotona House was started, Mr. Suzuki had a strong desire to create a base in the local community and thought of the shared house more as a mechanism for obtaining financial support for the operation of the base. However, through the operation of the shared house, he has come to feel that the common space for the local community is realized only with the shared house residents, and he now prioritizes the shared house residents' lives when considering what to do next.

3.2.2 A Common Space and Local Community Interaction

To keep the management of the common space for the local community as concise as possible, the space is not rented to anyone but is only available to members (called Chanomers) or residents of Kotona House. The members pay a membership fee of 500 yen per month and can use the space for 500 yen per hour when hosting an event at the common space. Currently, regular events include children's inquiry-based learning class, English conversation class, and children's cafeteria, as well as regular monthly members' meetings. Irregular events include residents' meal meetings, coffee and conversation meetings, and sweets house building meetings (Figs. 3, 4 and 5). The children's cafeteria is popular, with 40–50 people participating each time, but there are some weekdays when there is no use of the common space at all, and he wants to increase the frequency of use. The lounge for shared house residents and the Chanoma space is completely separate, but the kitchen is shared by residents and Chanoma users. When kitchen usage times overlap, it is sometimes necessary to share the kitchen space. There are currently about 12–13 members of Chanoma, and although members come and go, about half of them are long-term members. Members' ways of involvement are varied: Some use Chanoma's space for their events and do not show up often to regular meetings and other opportunities for communication among members, while others do not organize their events but support the activities as members and show up to regular meetings to communicate with members. Profitability would improve if the common space were used as a classroom for lessons; however, if it were used only in that way, it would create a closed atmosphere that only a limited number of people could use. Therefore, Mr. Suzuki currently prioritizes having the neighborhood recognize that it is a place open to the local community, rather than focusing on earning money. The advantage of having both a shared house and a local community base is that it gives shared house residents a chance to connect with the local community. People who come to Chanoma are involved in the local community in a variety of ways and have diverse social networks that can be helpful to shared house residents when they need help.

Fig. 3 Event information on bulletin board at the entrance

Fig. 4 Common space "Chanoma"

Fig. 5 Signboard of children's cafeteria. (Source: Photographs taken by Yuno Tanaka in 2024)

3.2.3 Resident Recruitment

The manager makes final decisions about who will live in the shared house; even if there are empty rooms, he does not approach potential residents who want to focus on their private life because he considers that people who are not interested in connecting with the local community will have difficulty adjusting to the living environment of the Kotona House.

3.2.4 Social Relationships within the Shared House

(Interview with Mr. N) Resident Mr. N organizes a children's cafeteria named "Guruguru Shokudo." When he was looking for a shared house where he could be involved with the local community as a kind of "third place" (Oldenburg 1999) between his work and private life, he liked the location and the atmosphere of the area and moved in. Before opening the children's cafeteria, he held a board game party, a blindfolded movie party, and a *nagashi somen* (fine white noodles served flowing in a bamboo basket) party. The children's cafeteria was started because he wanted to use his time and energy for something and to create a place for children at a time when workplaces and schools were closed due to the COVID-19 pandemic. Now, in its seventh year of occupancy, he feels that through his activities, he has made bonds with people in the local community and that this has become something important to him.

3.2.5 Connections with Former Shared House Residents and Future Prospects

As for the involvement of shared house residents with the local community after moving out, Ms. O and a former resident help operate Guruguru Shokudo. Moreover, in addition to Ms. O's bookstore, new shops have recently opened on the local market street where the Kotona House is located. Mr. Suzuki believes that with new shops coming in and people living in the local market street, the atmosphere of the people will become more pleasant, and this will have a positive impact on the local community.

4 Findings and Discussion

The classification of shared houses confirms that there are several variations in the way shared houses interact with the local community. The following is a summary of the findings. While the location of interaction can be either inside or outside the shared house, it is necessary to ensure that people coming from outside do not invade shared house residents' privacy. Concurrently, it is necessary to provide opportunities for shared house residents to easily engage with local people in the context of their daily lives. For example, there could be opportunities for events, discounts on usage fees, and the creation of common spaces dedicated to specific hobbies such as vegetable gardening or art, in addition to the multipurpose common spaces.

In terms of human contact, the manager mainly acts as a mediator between the shared house residents and the local community when residents first move in, but diversity is observed in how the manager operates as a mediator. The manager may host events themselves, invite local events to be held in the common space, or support events hosted by the shared house residents. Once a community has grown within the shared house residents, it can be expanded through the friends and acquaintances of each resident. Before moving in, it is also important to confirm in advance whether the potential residents are comfortable interacting with others to avoid any misunderstandings later.

In terms of economic contacts, not only are shared house residents involved in the economic activities of the local community through paid events they organize in the common space but they have also expanded beyond the use of the common space, such as job offers among residents and a former resident of a shared house opening a bookstore in the same area. The interviewees expressed their visions for the future, such as creating jobs among shared house residents (commune Tokiwa) and increasing the number of people living in the local market street (Kotona House), which would propel the economy of the local community beyond the property itself.

Regarding value contacts, it is generally assumed that the more times a person is involved with the local community, the deeper their understanding of the local culture and future vision. Nonetheless, it is also important to have a way of involvement that is more casual and allows people to quit when they feel it is not a good fit. As in the case of Mr. N, a resident of Kotona House, who held various one-off events before organizing Guruguru Shokudo, or as in the case of the Commune Tokiwa's image of "nest" function, being a place where various ideas are born before starting long-term activities could help shared house residents get to know the local community.

Based on the above four aspects, this study summarized the variations in the way shared houses interact with the local community and their effectiveness. Compared to existing private local community development bases, the advantages of operating a shared house in combination with a common space that provides an interface with the local community is that rent income can be generated and management operations can be consolidated, eliminating the need to have a dedicated manager for the common space, thereby eliminating the shortage of financial and manpower resources. However, in the case study, the shared house does not focus on raising rent or earning income from fees for the use of the common space, and the common space is provided for reasons that emphasize social value rather than income and is not more profitable than a standard shared house. Therefore, it is desirable that the shared housing units be near full occupancy in order to keep the business running. At this point, the story of the resident who decided to move in because he liked the atmosphere of the surrounding area (Kotona House) and the human resources being mentioned as an attractiveness of the area (Commune Tokiwa) shows an awareness of the features of the area in each case. In other words, for outsiders, the uniqueness of the local area may connect to the originality of the shared house, thereby attracting people. The common space of the shared house will provide the opportunity for residents to get involved in local activities, and as some of them leave the shared house and set up a new local base, the atmosphere of the local community will become more active, which in turn will cause new potential residents who are attracted to the area to live in the share house. If such a cycle can be created, it would be expected to build a long-term relationship between the residents and the local community through the shared house.

Since this survey was conducted primarily focusing on the operation of places that interact with the local community, we do not have sufficient information on how residents think about having a relationship with the local community. However, the interview with Mr. N suggests that shared house residents' involvement with the local community not only contributes to local community development but can also play an important role as a third place for residents to live. Further research is needed to determine the meaning of local community interaction from the residents' perspective.

Acknowledgments I would like to express my sincere appreciation to Mr. Funamoto of Commune Tokiwa and Mr. Suzuki and Mr. N. of Kotona House for their kind cooperation in the interview. This work was supported by JSPS KAKENHI Grant Number JP 23 K13463.

References

Asada W (2012) Sumibiraki. Chikuma Shobo, Tokyo. (In Japanese)

Bauman Z (2000) Liquid modernity. Polity

Granovetter MS (1973) The strength of weak ties. Am J Sociol 78(6):1360–1380. http://www.jstor.org/stable/2776392

Hituji Real Estate (2024). https://www.hituji.jp/. Accessed 30 July 2024

Japan Share House Organization (2023) Share house market survey 2023. (In Japanese)

Koyano Y, Takehara A, Murota M (2016) Research on management conditions and categorical issues of community cafes. Report City Plann Instit Japan 15(4):337–340. https://doi.org/10.11361/reportscpij.15.4_337. (In Japanese)

Kubota H (2014) The current situation and significance of room-sharing/shared house in Japan. Sociol Rev Nihon Univ Sociol Soc 179:37–57. https://cir.nii.ac.jp/crid/1520853833334059904 (In Japanese)

Ministry of Internal Affairs and Communications (2023) Telecommunications usage trends survey

Ministry of Land, Infrastructure, Transport and Tourism (2017) Shared house guidebook. https://www.mlit.go.jp/common/001207549.pdf. Accessed 30 July 2024. In Japanese)

Ogawa A, Nohara T (2020) A study on functions and new roles of town development bases. Report City Plann Instit Japan 19(2):253–256. https://doi.org/10.11361/reportscpij.19.2_253. (In Japanese)

Oguma E, Yamamoto R (2019) Can architecture create communities? Economy, memory, and time. In: Urban beauty (No. 1). Sayusha, Tokyo. (In Japanese)

Oldenburg R (1999) The great good place: cafés, coffee shops, bookstores, bars, hair salons, and other hangouts at the heart of a community. Publishers Group West, Berkeley

Open City Study Group (2017) Machi wo hiraku gijutsu [methodology for opening the city]. Gakugei Shuppansha, Kyoto. (In Japanese)

Yamamoto R et al (2013) Chiiki shakai ken shugi. (revised and enlarged edition). LIXIL publishing. (In Japanese)

Yamazaki R (2016) The origin of community design in England. Ota Publishing, Tokyo. (In Japanese)

Normalizing Neglected Public Spaces: The Case of Portoviejo's Ground Transportation Node

Paúl Intriago and Jaire Cagigal

Abstract

Mobility, transport, and urban space are crucial in the design and development of inclusive cities. However, much research in transport tends to reduce the end user to a quantifiable entity whose decisions can be reduced to cost-benefit parameters. This leaves little room for perception and subjective experiences of the people who not only use transport services, but also experience and live in the environments they operate in.

This article seeks to address this deficiency by carrying out quantitative and qualitative research with end users of the multimodal Bus Terminal of Portoviejo, a medium-sized and rapidly growing city in Ecuador, Latin America. The study uses the Pedestrian Characterisation and Mobility Environment framework (PCEM), which examines urban space through the parameters of accessibility, security, comfort, and attraction for end users.

The results highlighted various deficiencies that are faced on a daily basis. However, the study's focus on perception shed light on the existence of an underlying problem: The deficiencies of the transport node had been internalized and were considered "normal" by those who were most affected by them. The impact of this finding is that participatory design processes face the risk of missing potential interventions that could be of a maximum benefit for end users, simply because they are unaware of the problems that they have to regularly face and that could be solved through appropriate urban design. It was by going beyond purely quantitative studies that this finding could be unearthed, whereby it is hoped that the methodology may be of use for urban planners in intermediate cities elsewhere in Ecuador and the Latin American region in general.

P. Intriago (✉) · J. Cagigal
Pontificia Universidad Católica del Ecuador SEDE MANABÍ (PUCEM), Portoviejo, Ecuador
e-mail: pintriago@pucesm.edu.ec; jcagigal302@pucesm.edu.ec

Keywords
Mobility · Public transport · Private vehicle · Informality · Bus terminal

1 Introduction

In order to examine how individuals interpret and relate to shared spaces within a city, it is necessary to delve deeper into their perceptions of public space. Lynch (1960) argued for several years that urban space is not only a physical setting, but also a network of meanings that influence the daily experiences of its inhabitants. Jacobs (1961) added that the quality of streets and public spaces determines not only the safety and comfort of pedestrians, but also the degree of vitality and social cohesion in cities. Years later, research by Appleyard and Lintell (1972) suggested that residents' behavior is influenced by their perceptions of the environment in terms of urban features. This view has been extended by authors such as Gehl (2010, 2011), who argued that the quality of urban space has a direct impact on well-being and social interaction.

Although this is not a new topic, several authors have emphasized for years the need to design spaces that encourage social interaction while meeting the needs of users. As an example, some authors like Arefi and Aelbrecht (2022) and Oktay (2023) argued that factors such as age, identity, and even beauty influence perceptions of urban space. Other authors, like Perelman and Marconi (2016), emphasize that there is a clear socioeconomic and demographic influence when assessing residents' perceptions of urban parks. In these studies, it is emphasized that the analysis of subjective perception allows not only to evaluate the functionality of spaces, but also to contribute to the design of more inclusive environments for city dwellers, even improving connectivity, quality of life, and mobility.

At the city scale, mobility plays an important role in the level of interaction and connectivity between city dwellers.

F. Alberti et al. (eds.), *Towards Sustainable Urban Regeneration: Utilizing Environmental, Cultural, and Social Assets*, Advances in Science, Technology & Innovation, https://doi.org/10.1007/978-3-032-02091-8_35

Recent studies have highlighted how even individual transportation experiences affect perceptions of the urban environment. Authors like Friman et al. (2020) found that the functionality, information, and comfort of public transport services significantly influence perceived accessibility and quality of the urban environment, and Olsson et al. (2021) observed that frequent travel by public transport is associated with lower levels of perceived accessibility in major city areas, suggesting that the frequency and quality of public transport services impact users' perceptions of accessibility and urban quality. The authors also mention that satisfaction with the transportation system, including related facilities, is strongly related to the perception of accessibility and connectivity of urban space. Ewing and Cervero (2010) emphasize that the design of public space, including the layout of streets and the quality of pedestrian spaces, can improve the perception of ease and safety of transportation. According to their research, facilities and the environment can encourage the use of active transport and even contribute to a positive perception of urban space.

In recent years, land transport terminals have emerged as crucial nodes in urban and regional dynamics, with significant impacts on mobility, the local economy, and urban planning. These infrastructures not only facilitate the movement of people and goods but also generate agglomeration effects that can transform the surrounding environment, as highlighted by the Victoria Transport Policy Institute (2021). Several studies have shown that the location and design of terminals influence the development of adjacent areas, promoting urban renewal and the revitalization of previously degraded areas (Spandou et al. 2010; Ližbetin and Stopka 2020). However, challenges related to congestion and environmental impacts also arise, highlighting the need for comprehensive planning that balances economic benefits with sustainability (Macharis et al. 2011; Babaei et al. 2023).

2 Justification

In the context of developing countries, land transport hubs are not only nerve centers of mobility but can also be epicenters of social exclusion and inequality. The dynamics of social exclusion around these hubs manifest themselves in several dimensions, starting with spatial segregation. Terminals, by attracting significant flows of people and economic activity, are usually located in central areas and have the capacity to generate or consolidate centralities (Crescenzi et al. 2016; Gómez 2016). But locations that are often neglected in terms of infrastructure and basic services for the surrounding communities accentuate social exclusion and limit opportunities for people living in the vicinity (Chikengezha and Thebe 2021). This situation contributes to an environment of inequality in which the most vulnerable populations, often located in nearby marginal neighborhoods, face problems such as lack of access to services, insecurity, and precarious living conditions, which affect the perception of public space by both residents and those who travel and perceive the area (Rodríguez Tabitas 2023).

According to Scholl et al. (2022), the planning of these infrastructures often lacks an inclusive perspective, prioritizing transport efficiency and economic benefits over the needs of local communities. This can result in the forced displacement of residents and the degradation of public spaces in neighboring areas. In countries from Latin America, research suggests that the creation of commercial and service zones around terminals tends to benefit mainly the most powerful economic actors, leaving low-income populations with limited access to the economic opportunities generated by these infrastructures (Inter-American Development Bank [IDB] 2016).

When the subjective perception of urban space and transportation hubs in Ecuador is studied, a complex reality influenced by the socioeconomic and political context of the country and each of its provinces emerges. According to Ballari et al. (2021) and Carrión Ordóñez and Godoy Chacha (2021), negative perceptions of urban space and transport in cities such as Cuenca are often associated with traffic congestion and the lack of efficient public transport systems. These problems contribute to the perception of dissatisfaction and stress among residents, who often face long travel times and lack of accessibility to services.

In cities such as Portoviejo, in the province of Manabí, Ecuador, the study of subjective perceptions of transportation related to mobility hubs is limited. The Portoviejo Bus Terminal, as a crucial point of interprovincial connectivity, handles a significant volume of passengers every day. According to recent data, the terminal receives approximately 2500 bus routes and 30,000 passengers every day, reflecting its importance in regional mobility (Ecuador Bus n.d.). However, it should be noted that there is a lack of updated and reliable data on mobility patterns, which demonstrates the low attention given to this node.

Portoviejo, as an intermediate city, has experienced significant population growth in recent years and is currently facing problems of traffic congestion and insufficient transportation infrastructure to meet the needs of its inhabitants. With 20% of the province's inhabitants, Portoviejo is currently the largest city in Manabí, with more than 320,000 inhabitants, compared with 238,000 inhabitants in 2001 who resided in rural and urban areas (Valencia 2022; INEC 2001). Some authors claim that it is necessary to work on quality transport systems and even green infrastructure (Cedeño and Zambrano 2024; Cárdenas Loor 2020). However, access to information is limited, and gathering information requires resources that not everyone has.

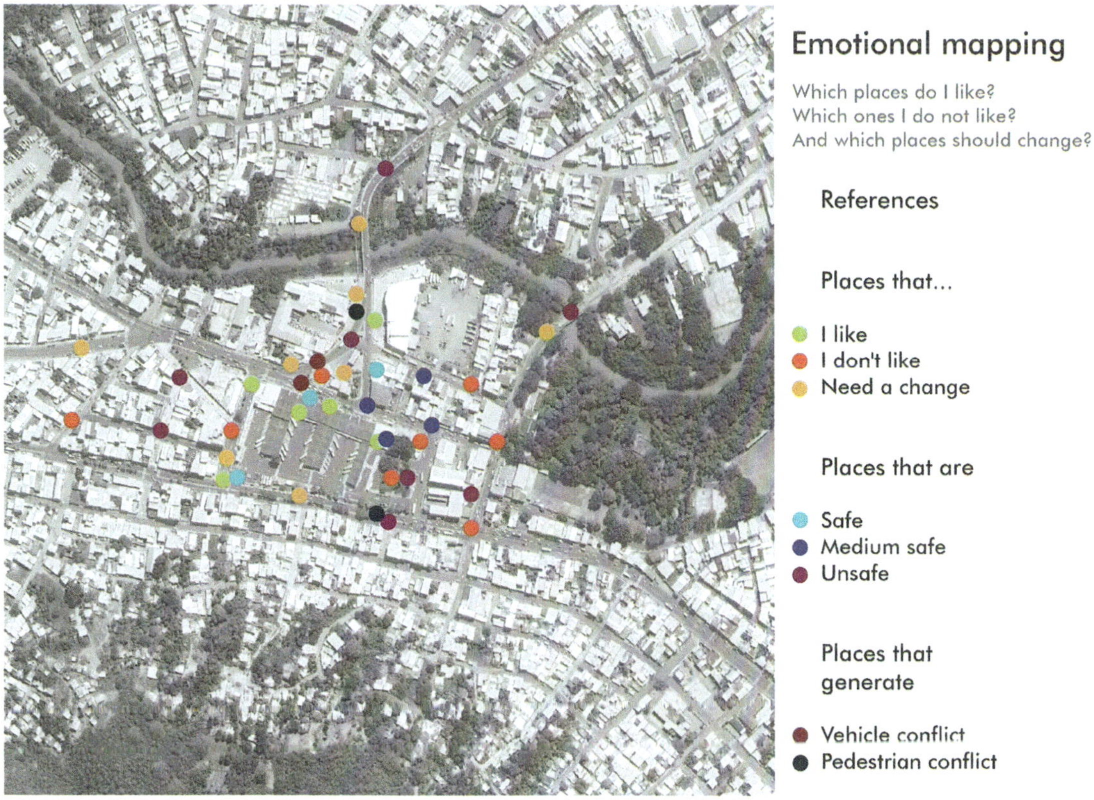

Fig. 1 Emotional mapping of the Bus Terminal in the city of Portoviejo, Ecuador, as part of the results of the Viva la Calle workshop, organized by the Inter-American Development Bank (IDB) and Superwien (adapted for this paper by Jaire Cagigal in 2025). (**Source:** Sixth semester students of the Architecture degree at PUCE Manabí (2024))

In this context, the sixth semester students of Architecture at PUCE Sede Manabí participated in the workshop Viva la Calle, organized by the Inter American Development Bank (IDB) and the urban planning office Superwien. The workshop was held during the month of February (2024), and one of the products of this workshop was the creation of emotional maps on the perception of the Bus Terminal in the city of Portoviejo and its surroundings. This methodology allows us to capture how people experience and value different spaces by assigning emotional attributes to specific locations on a map (Shukla and Pujara 2025), and the result is shown in Fig. 1.

3 Methodology

With this background, and with a little curiosity, the need arose to study the sector, establishing an intervention polygon, as seen in Fig. 2, to be evaluated within a methodology of study and analysis of public space. This polygon is made up of 13 streets that allow direct access to the Bus Terminal. For this purpose, we considered the characterization methodology developed by Talavera-García et al. (2014), based on Alfonos (2005), in their research "*Pedestrian quality as a method to evaluate urban mobility environments*." The use of this analysis methodology allows for a systematic approach to evaluate the pedestrian quality of an urban environment and to apply it in the case of the Portoviejo Bus Terminal, with the aim of obtaining data on the subjective perception of public space in this mobility node. With this premise, this study raises the research question: Is it possible to represent the quality of the pedestrian environment based on the user's perceptions in the Bus Terminal node of the city of Portoviejo?

The methodology of Talavera-García et al. (2014) allows us to characterize and evaluate the quality of urban space, establishing a frame of reference to improve urban planning and promote a more pedestrian-friendly environment. For the evaluation, a set of qualitative and quantitative indicators, that affect pedestrian mobility, are used. In this analysis, *accessibility* refers to the dimension of space available to pedestrians. *Safety* refers to the protection of pedestrians from vehicular traffic and the speed at which they travel. *Comfort* is related to factors such as urban trees and shade, noise, and the perception of the relationship between vertical and horizontal space. Finally, *complexity* includes the commercial complexity of the environment. To characterize and evaluate space, strategies such as on-site information gather-

Fig. 2 Intervention polygon in the surroundings of the Terminal Terrestre of the city of Portoviejo. (**Source:** Own elaboration (2024))

ing, satellite information, and qualitative information through surveys are used. For a correct application of the methodology, the use of formulas to obtain quantitative values and to standardize the results is included.

In our case study, we deviated from the methodology and chose to prioritize the use of surveys as the main tool to obtain information, which consisted of a 13-question survey applied to 32 people, considering the social and current context of the city and focusing on indicators as elements of analysis. To obtain data, we also included questions about the users' occupation and activities and about the frequency with which they travel the streets around the terminal. In the application of the surveys, users were asked to rate frequency, accessibility, safety, comfort, and attractiveness using a numerical scale with values between 1 and 5. Although the numerical scale of evaluation in the survey was set between 1 and 5, the results obtained ranged between numbers 2 and 3, so it was decided to divide the numerical results into intervals of 0.25. The structure of the survey made it possible to exchange opinions and open small spaces for discussion without becoming an interview. The methodology used is mixed, with a mostly qualitative approach that also provides quantifiable results. The information was collected both on site and online.

Due to constraints such as time, limited resources, and lack of security at the site, it was decided to use nonprobabilistic sampling, using surveys to measure the perception of public space. The nonprobabilistic sampling method is characterized by the selection of samples based on subjective and nonrandom criteria, which means that not all elements of the population have the same probability of being selected. This approach is particularly useful in exploratory research or in contexts where access to the population is limited, allowing researchers to obtain relevant data more quickly and cheaply (Etikan et al. 2016; Palinkas et al. 2015). Although it can introduce bias, its use in qualitative studies and initial data collection is crucial for generating hypotheses and subsequent more rigorous approaches (Creswell and Poth 2018; Vasileiou et al. 2018).

4 Results Analysis

The results were used to create maps of environmental quality based on the above categories. These numerical results are presented in maps as part of the results, and part of the qualitative information collected was taken into account for their analysis. The results show the importance of analyzing the node in different layers and in greater depth.

4.1 Population and Occupation

Figure 3 shows the results of the type of population around the Bus Terminal. It can be seen that almost half of the population travels through the sector by means of the intercantonal and interprovincial transport services (25.7%) and urban transport (22.9%) offered by the facility. On the one hand, the results highlight that 17.1% of users travel through the sector to make their purchases, while 14.3% are dedicated to work in the sector or to various types of commerce, and there is also a value of 5.7% of users who travel through the sector for tourist purposes. On the other hand, only 14.3% of the users are residents of the sector. Based on the above, it seems that most users in the sector are floating population. It is worth mentioning that the percentage of users related to transportation, whether interprovincial, intercantonal, urban, or even related to tourism, is the most representative, and this

Fig. 3 Surveyed population. (**Source:** Own elaboration (2024))

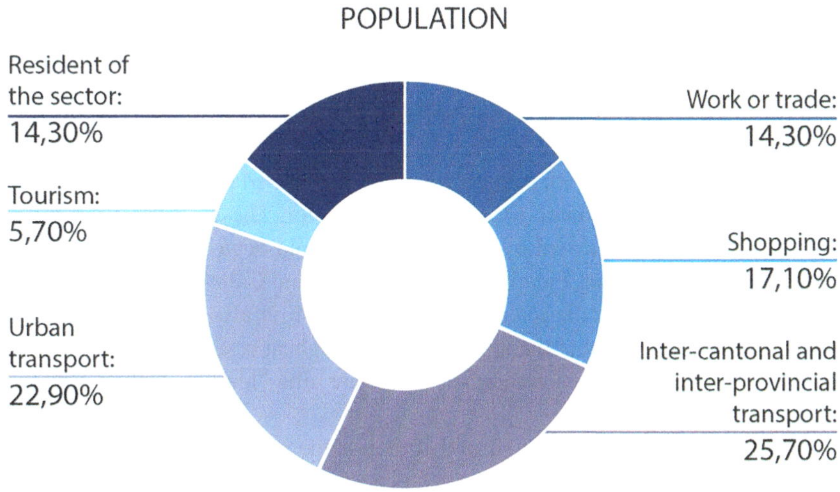

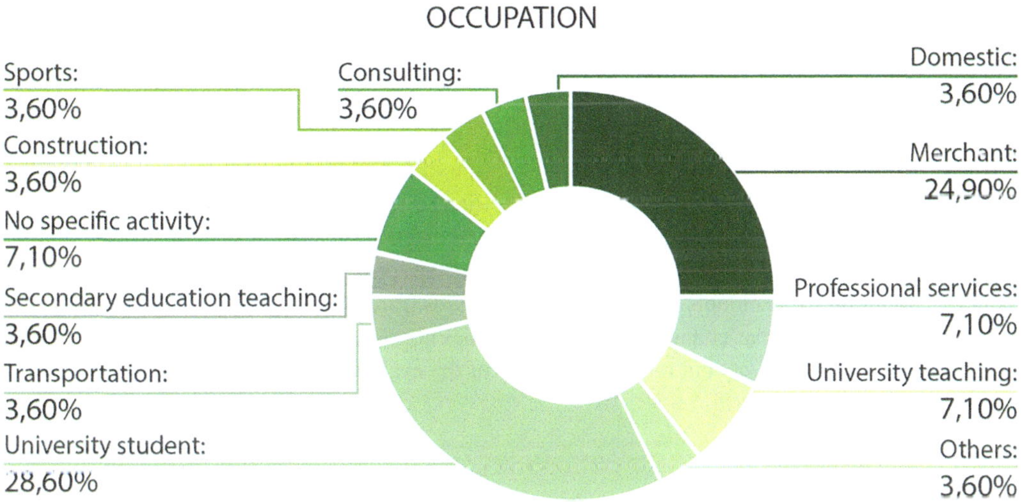

Fig. 4 Activities of the population surveyed. (**Source:** Own elaboration (2024))

allows us to affirm that this infrastructure is the main pillar of this urban environment in the city.

It should be noted that the majority of users are floating population. This causes the sector to reduce its activity outside working hours. The influx of people during working hours makes the users perceive the sector as a safer space. Some users interviewed claim that this movement makes the environment perceived as safe. They also add that this perception can change if the movement of people or flows are irregular. This was further explored, and some respondents expressed that long pauses by people outside the environment can make it feel unsafe.

Regarding the activities and occupations of the respondents, we can see in Fig. 4 that the percentage of merchants is among the highest (24.9%), just below the percentage of university students (28.6%). University students are the main users of transport and therefore travel the most, while merchants are a transient population who spend a good part of their time in the area. Personnel dedicated to teaching and professional services represent 14.2% (7.1% each). The rest of the professions occupy an equal percentage of 3.6%.

One of the aspects that was considered interesting is the percentage of users who carry out their daily activities in the area around commercial activities. This population became crucial at the time of the survey, since they were the users with the most knowledge of the environment and those who took advantage of the survey to comment on the reality of the Bus Terminal. These users were the ones most able to give an opinion on their perception of the site, including the number of residents. This allows us to understand that the residents prefer to make their passage through this node short, while the merchants are obliged by their work to inhabit and occupy the space.

4.2 Frequency Map

In this map, the areas most traveled by pedestrians were identified as a way of evaluating the streets about which there is greater knowledge. The frequency of trips is illustrated in Fig. 5, and it shows that the streets most traveled by the respondents are the main road axes of the sector. These axes are the most frequented by both pedestrians and occupants of vehicles (Av. del Ejército, Av. 15 de Abril, and Calle Pedro Gual). Although the results were expected, it is also important to highlight that the streets with the highest residential flow and low commercial flow are the least frequented.

Another aspect that stands out in this map is the number of corresponding values in the numerical classification interval. Throughout the data collection, we found that the greater the number of numbers in the rating intervals, the more knowledgeable the population is about the topic. For the respondents, mobility on these streets is common because of the size of the sidewalks, the proximity of services, and the safety that for some users means moving in a more crowded environment. This is striking because safety is also linked to the presence of vehicles, since walking on a street with constant traffic flow is not the same as walking on a street with no traffic flow. However, the conflicts that a vehicle can cause, such as noise, pollution, or accidents due to reckless driving, are not considered a factor that changes the perception of the environment in a negative way but rather makes it positive by feeling safe.

4.3 Accessibility Map

Figure 6 shows the map corresponding to the respondents' accessibility ratings. It is clear that the streets with the highest traffic flow are the best rated, despite the lack of adequate pedestrian infrastructure. For some respondents, the lack of pedestrian infrastructure is not the most important problem they face on the site, and therefore the simple fact of being able to walk along a street while feeling safe is enough.

In this map, it is also shown that Calle 5 de Junio (access to the Terminal Terrestre), shown in Fig. 7, is one of the shortest streets, but it is also the one with the lowest score and the one that is perceived the worst by the residents. The congestion at the entrance and exit of the Terminal Terrestre has an impact on the negative perception of the site. It is interesting to note that although this is the road with the lowest vehicle speed (where vehicles are constantly stopping), the stopped vehicle is perceived as a problem, unlike the moving vehicle, which is perceived more positively, as can be seen on the frequency map (Figs. 5 and 8).

Observing the environment, it is noted that at the intersection of 5 de Junio and 15 de Abril streets (Fig. 9), the pedestrian infrastructure is inadequate, but these streets are better perceived by users. In general, it can be observed that the lack of pedestrian infrastructure such as crossings, overpasses, ramps, or signage, and even the difference in level, does not represent a (conscious) obstacle for users at the time of the evaluation.

Fig. 5 Frequency with which users travel around the Bus Terminal in the city of Portoviejo. (**Source:** Own elaboration (2024))

Fig. 6 Accessibility map according to the evaluation of users who travel around the Terminal Terrestre of the city of Portoviejo. (**Source:** Own elaboration (2024))

Fig. 7 Calle 5 de Junio (access to the Terminal Terrestre of the city of Portoviejo). (**Source:** Paúl Intriago (2025))

As with the frequency map (Fig. 5), the range of intervals is wider. This confirms that respondents were able to generate broader responses due to their familiarity with the term being evaluated.

4.4 Safety Map

As can be seen in Fig. 8, the perception of safety is lower on the secondary streets surrounding the terminal, which do not have a high volume of pedestrian or vehicular traffic (Fausto Vélez and Venezuela streets). On the other hand, the highest-rated street is Miguel H. Alcívar Street.

Fig. 8 Safety map based on the assessment of users who travel around the Bus Terminal in the city of Portoviejo. (**Source:** Own elaboration (2024))

Fig. 9 Traffic flow around the Bus Terminal in the city of Portoviejo (Av. del Ejército). (**Source:** Paúl Intriago (2025))

It should be noted that safety, according to the evaluation methodology of Talavera-García et al. (2014), has to do with the relationship between the speed of vehicles on the streets and how it affects the perception of pedestrians walking on the sidewalks. However, around the Portoviejo Bus Terminal, the speed of vehicles is understood as a suggestion, since the law is not respected and the speed is always the highest that vehicles can reach on a street.

However, as seen in the frequency maps (Fig. 5) and accessibility maps (Fig. 6), vehicle speed is perceived as a factor that increases the perception of safety (also seen in Fig. 9). According to respondents, their understanding of the safety indicator is directly related to crime rates and social factors. When asked to give their assessment of road safety (based on the methodology), users chose to show their version of the concept of safety. This is seen as a barrier to understanding urban concepts for these users. Several respondents expressed that vehicle speed is of minimal concern when there is a latent phenomenon of instability and insecurity, not only in this urban environment, but on a national scale. They also feel that the discomfort that vehicles can cause is an everyday reality that normalizes the precariousness of urban space.

The number of values in the rating interval is considered low compared to the frequency or accessibility maps, reinforcing the idea that the greater the lack of knowledge about the concept being evaluated, the less variation there is between the data obtained.

4.5 Comfort Maps

4.5.1 Noise Comfort Map

Figure 10 shows the noise comfort rating map. The survey data reflect that the busiest streets are the ones that are perceived as the noisiest. This is obvious because they are the streets with the highest vehicle traffic and the highest commercial flows.

Regarding the rating interval, it can be observed that there are only two values. This means that, at a general level, the whole area of the Bus Terminal and its surroundings is considered noisy and perceived in a negative way. Although the users interviewed were familiar with the concept being evaluated, the results in this case were average. According to several respondents, the environment is simply noisy, which is why it is perceived negatively.

4.5.2 Comfort Map by Ratio

The comfort ratio refers to the relationship between the width of the street and the height of the boundary walls or the height of the buildings next to the sidewalk. The map in Fig. 11 shows respondents' ratings of the comfort ratio. The highest-rated streets are those that are wider, which is not surprising since the space is perceived as more open. According to the respondents, having "air flow" helps, but this also means accepting the pollution and noise generated by the high traffic flow in the area, as shown in Fig. 12.

Other users pointed out that these streets are better perceived for their width, because they allow a wider field of vision, where users can more easily detect dangers. The lowest rated streets are also the ones that are perceived as the least safe. According to some respondents, the number of closed walls and the narrowness of the pavement space negatively affect their perception.

4.5.3 Comfort Map by Presence of Trees

The comfort of the presence of trees is better perceived in streets where the shade provided by the vegetation is not accessible but visually strong. This information is interesting because, for some of the users interviewed, green spaces are perceived well when they are easily visible. This phenomenon can be seen in Fig. 13, where the presence of trees in the

Fig. 10 Noise comfort map according to the evaluation of users who travel around the Bus Terminal in the city of Portoviejo. (**Source:** Own elaboration (2024))

Fig. 11 Comfort map by ratio according to the evaluation of users who travel around the Bus Terminal in the city of Portoviejo. (**Source:** Own elaboration (2024))

Fig. 12 Comfort by ratio in the one of the highest-rated streets (Av. 15 de Abril) around the Bus Terminal in the city of Portoviejo. (**Source:** Paúl Intriago (2025))

park next to the Bus Terminal can be seen, while the surrounding streets do not have a positive perception of green spaces.

Some respondents elaborated on this observation, mentioning that the deterioration of public space makes the tree-lined area perceived negatively, but seeing green space and the presence of trees behind a boundary and well maintained are more pleasant.

Fig. 13 Comfort map by presence of trees according to the evaluation of users who travel around the Terminal Terrestre of the city of Portoviejo. (**Source:** Own elaboration (2024))

4.6 Complexity

As shown on the map in Fig. 14, complexity is highest on the busiest streets and along the main road axes, as shown in Fig. 15. The zoning in these areas allows for mixed uses, and because of the scale of the place, neighborhood-scale commerce, as well as commerce related to mobility (mechanics, workshops, and spare parts), dynamizes the sector.

Some respondents expressed the ease with which they can carry out activities around the terminal. However, other users expressed that this commercial area is a source of insecurity at the end of the working day, since the dynamism and movement in the sector stop. This leads to a negative perception at the end of the day.

5 Discussion

This section highlights some observations that go beyond the analysis of the results, related to daily life in the urban environment, urban design (or rather the lack of it), and the application of a methodology in this context.

The speed with which people live in the surroundings of this urban environment has led users to have a priority focus: daily survival. As can be seen in the analyses, many of the results, in addition to being represented on the maps, are accompanied by the opinions of several residents. The needs and deficiencies of daily life make it impossible for people to evaluate the reality of their environment on a numerical scale.

Based on latent needs, the users interviewed prefer to use the space provided to express an opinion that goes beyond a number. Many focus on the constant sensations that urban space evokes in them, while others make observations about its reality. This aspect also leads us to say that the respondents and inhabitants of the site are not guided by a structured evaluation process but rather prefer to focus on certain details to express the discomfort that urban space causes them. The perception of the quality of the space can be considered poor or even average. But in this process of expressing discomfort, there are no ideas of change or improvement. This is striking because many respondents expressed that they were complaining about a reality which they are accustomed to. The normalization of the deficiencies of the environment shows the importance of planning with the people, to show that quality design has an impact on the quality of life and that it is also a right.

The speed of the pace of life is not the only speed that exists in this urban environment. In the case of the Bus Terminal, we find that if the perception is relatively positive, it is due to the continuity of flows, both pedestrian and vehicular, and, in the latter case, regardless of the speed and the conditions that foster it. This adaptation is again the normalization of the precariousness of the environment by the inhabitants.

It was also noted that the lack of planning and socialization of urban processes in the sector has also created a barrier

Fig. 14 Complexity map according to the evaluation of users who travel around the Bus Terminal in the city of Portoviejo. (**Source:** Own elaboration (2024))

Fig. 15 Complexity around the Bus Terminal in the city of Portoviejo. (**Source:** Paúl Intriago (2025))

in the language used in urban planning, academia, and everyday life. The concepts are perceived differently by the respondents (although both studies were conducted in the same language, Spanish), which shows that the daily reality cannot be understood without living as a subject of that reality.

As far as the methodology used is concerned, the limitations of time and resources are a factor that makes it necessary to adapt it for its application. In the case of this methodology, the prioritization of qualitative aspects allowed us to approach the environment, and beyond the characterization, evaluation, and classification, it is used as a methodology to approach the problems of the urban environment in a structured way, allowing us to see the strongest aspects that we, as external researchers, do not always observe.

6 Conclusions

This research focused on answering the following research question: Is it possible to represent the quality of the pedestrian environment based on the user's perception in the Bus Terminal node of the city of Portoviejo? By applying the methodology of Talavera-García et al. (2014), modified to adapt to this context and local conditions, the results show that a representation based on perception is possible. It should be noted that more relevant information was found.

The maps generated reflect the perception of the surroundings of the Portoviejo Bus Terminal. The data, translated into numerical values, remain between 2 and 3, concluding that there is a perception that is not entirely positive. With this research, we have found that the representation of subjective perception is a tangible approximation tool to the problems of an urban environment, and that it goes beyond the process of individual observation and traditional cartographic studies. The study of the subjective perception of the urban environment in this case allows an approach to the reality of the city through the eyes of the user.

The study of these problems requires a deeper analysis and a methodology that goes beyond characterization and approach. However, in this research, the approach allows us to establish a basis for the selection of problems.

The possibility of interacting with the users allowed us to understand the reasons for the answers of the respondents and also to know that there is a generalized and shared dissatisfaction due to the lack of action by the State. For the users of the urban environment around the Bus Terminal in Portoviejo, the poor quality of the public space has been constant in recent years and is perceived as a permanent reality.

Finally, the impact of the research is showing how in order for urban planning to reach urban residents, the proposals need to be literally "brought to their front doors." It is important that people become familiar with the concepts that are explained in the design processes, or in this case, in the collection of information. In order to achieve this, it is necessary to move beyond traditional modes of planning, which are mainly desk based. The city is a right, and in the case of the inhabitants around the Portoviejo Bus Terminal, it is not fully fulfilled.

7 Further Research

We suggest continuing this research by taking into account the university student population, which represented a high percentage of respondents, despite the fact that the academic semester was not active in the city during the process. This will allow researchers to evaluate how the urban environment is perceived by a specific population.

In addition, it is suggested to carry out a detailed information gathering than the one used for this research. A similar methodology could be employed, but where each street is treated as a unique entity. This aspect is considered relevant because there is no publicly available or relevant information on similar data in the sector.

Acknowledgments Special thanks to the team of students from the Pontificia Universidad Católica del Ecuador: Sede Manabí: Sebastián Polanco, Ronal Muentes, Julexy Coaboy, and Maria Paula García for the creation of the emotional map. Special thanks to the team formed by Agustina Solis, Dayana Macías, and Fernando Moreira for helping us with the fieldwork and surveys. Also, special thanks to Michael Maks Davis for the language review and observations.

References

Appleyard D, Lintell M (1972) The environmental quality of City streets: a study of the effects of urban design on pedestrian movement. University of California Press, California

Arefi M, Aelbrecht P (2022) Urban identity, perception, and urban design. Urban Des Int 27:1–2. https://doi.org/10.1057/s41289-022-00179-9

Babaei A, Khedmati M, Reza M, Jokar A, Babaee Tirkolaee. E (2023) Sustainable transportation planning considering traffic congestion and uncertain conditions. Expert Syst Appl 227:119792. ISSN 0957-4174. https://doi.org/10.1016/j.eswa.2023.119792

Ballari D, Salgado-Castillo F, Hermida C, Orellana D, Hermida MA (2021) MODOS DE MOVILIDAD DE LOS NIÑOS Y NIÑAS EN EDAD ESCOLAR: EXPLORACIÓN DE LA INCIDENCIA DE FACTORES SOCIOECONÓMICOS, DE PERCEPCIÓN Y DE MESOESCALA URBANA UTILIZANDO RANDOM FOREST. Universidad-Verdad 2(79):44–58. https://doi.org/10.33324/uv.v2i79.381

Cárdenas Loor VS (2020) Arbolado urbano Como estrategia Para mejorar el confort térmico y la movilidad Urbana, frente al Cambio Climático Para la ciudad de Portoviejo, Provincia de Manabí-Ecuador [Tesina de especialización, FLACSO Ecuador]. Repositorio Digital FLACSO Ecuador

Carrión Ordóñez AA, Godoy Chacha JD (2021) Valoración del paisaje urbano circundante al tranvía Cuatro Ríos dentro de la zona del centro histórico de Cuenca (Ecuador). Revista de Investigación en Paisaje y Arquitectura 12:1–15. https://doi.org/10.18537/ripa.12.04

Cedeño JA, Zambrano ME (2024) Análisis del congestionamiento vehicular en diferentes intersecciones de la ciudad de Portoviejo, Manabí-Ecuador. Revista de Investigación en Matemáticas Aplicadas y Tecnología (RIEMAT) 3(2):45–60

Chikengezha T, Thebe V (2021) Living on the periphery and challenges of mobility: A tale of transport-induced social exclusion in Southlea Park, Harare, Zimbabwe. Urban Forum 33:267–279. https://doi.org/10.1007/s12132-021-09437-5

Crescenzi R, Di Cataldo M, Rodríguez-Pose A (2016) Government quality and the economic returns of transport infrastructure investment in European regions. J Reg Sci 56(4):555–582

Creswell JW, Poth CN (2018) Qualitative inquiry and research design: choosing among five approaches. Sage Publications, California

Ecuador Bus (n.d.) Terminal Terrestre de Porto Viejo. https://www.ecuadorbus.com.ec/terminal-terrestre-de-porto-viejo. Accessed 10 Apr 2025

Etikan I, Musa SA, Alkassim RS (2016) Comparison of convenience sampling and purposive sampling. Am J Theor Appl Stat 5(1):1–4

Ewing R, Cervero R (2010) Travel and the built environment: a meta-analysis. J Am Plan Assoc 85(1):35–45. https://doi.org/10.1080/01944361003766766
Friman M, Ettema D, Gärling T, Olsson LE (2020) Public transport quality, safety, and perceived accessibility. Sustainability 12(9):3563. https://www.mdpi.com/2071-1050/12/9/3563
Gehl J (2010) Cities for people. Island Press
Gehl J (2011) Life between buildings: using public space. Island Press
Gómez LA (2016) Terminales de Transporte, nodos de articulación entre la ciudad y la región. Universidad Nacional de Colombia. https://repositorio.unal.edu.co/bitstream/handle/unal/56469/74187475_2016.pdf
INEC (2001) Instituto Nacional de Estadística y Censo: CENSO 2001
Inter-American Development Bank (IDB) (2016) Transporte urbano y pobreza: Efectos de los sistemas de transporte rápido de autobuses apoyados por el BID sobre la movilidad y el acceso en Cali y Lima
Jacobs J (1961) The death and life of great American cities. Random House, Munich
Ližbetin J, Stopka O (2020) Application of specific mathematical methods in the context of revitalization of defunct intermodal transport terminal: A case study. Sustainability 12(6):2295. https://doi.org/10.3390/su12062295
Lynch K (1960) The image of the City. MIT Press, Cambridge
Macharis C, Caris A, Jourquin B, Pekin E (2011) A decision support framework for intermodal transport policy. Eur Transp Res Rev 3(3):167–178. https://doi.org/10.1007/s12544-011-0062-5
Oktay D (2023) Influences of urban design on perceived social attributes and quality of life: a comparative study in two English neighbourhoods. Urban Des Int 28:304–319. https://doi.org/10.1057/s41289-023-00218-z
Olsson LE, Friman M, Lättman K (2021) Accessibility barriers and perceived accessibility: implications for public transport. Urban Sci 5(3):63. https://doi.org/10.3390/urbansci5030063
Palinkas LA, Horwitz SM, Green C, Wisdom JP, Duan N, Hoagwood K (2015) Purposeful sampling for qualitative data collection and analysis in mixed method implementation research. Adm Policy Ment Health Ment Health Serv Res 42(5):533–544. https://arxiv.org/abs/2212.12007
Perelman P, Marconi M (2016) Percepción del verde urbano en parques de la ciudad de Buenos Aires. Multequina 25(1):13–22
Rodríguez Tabitas RF (2023) Influencia del ordenamiento territorial y la movilidad Urbana en la percepción de la satisfacción residencial: Caso de estudio: el norponiente de la zona metropolitana de Monterrey [Tesis doctoral, Universidad Autónoma de Nuevo León]. Repositorio Institucional UANL. http://eprints.uanl.mx/26045/1/1080312573.pdf
Scholl L, Fook A, Barahona Rebolledo JD, Rivas ME, Montes L, Montoya V, Pedraza L, Noboa N, Sandoval D, Lee S, Rodriguez Porcel M, Bocarejo JP, Vergel Tovar E, Urrego LF, Moreno JP, Bertucci JP, Oviedo D, Sabogal-Cardona O, Serebrisky T, Mojica C (2022) Transport for inclusive development: defining a path for Latin America and the Caribbean. https://doi.org/10.18235/0004335
Shukla S, Pujara T (2025) Emotion mapping methods in urban design and planning: a systematic search and review. Cities Health 9:1–20. https://doi.org/10.1080/23748834.2025.2461857
Spandou M, Garcia C, Macário R (2010) Urban revitalization and transport: local factors and driving forces from a stakeholders' view. In: En CITTA 3rd annual conference on planning research: Bringing City form Back into planning. Technical University of Lisbon
Talavera-García R, Soria-Lara JA, Valenzuela-Montes LM (2014) La calidad peatonal como método para evaluar entornos de movilidad urbana. Revista de Urbanismo y Movilidad 15(2):35–56
Valencia ME (2022) Portoviejo: Desarrollo económico, productivo y del comercio. FARO, Quito
Vasileiou K, Barnett J, Thorpe S, Young P (2018) Characterising and justifying sample size sufficiency in interview-based studies: systematic analysis of qualitative health research over a 15-year period. Soc Res Methodol 21(3):225–241
Victoria Transport Policy Institute (2021) Land use impacts on transport: how land use factors affect travel behavior. https://www.vtpi.org/landtravel.pdf

Enhancing Quality of Life for Older Adults in Rural Thailand: The Role of Community Residential Environments in Ageing-in-Place Strategies

Alisa Nutley

Abstract

As the global population ages, there is a growing need to support older adults in living independently and comfortably within their communities. This study examines the factors influencing the quality of life for older adults in rural Northern Thailand, focusing on the role of community residential environments in supporting ageing in place. Employing a mixed-methods approach, this research integrates quantitative surveys, qualitative interviews, and a detailed site survey to provide a comprehensive understanding of how physical, social, and environmental factors interact to affect well-being. The study involved 144 elderly residents, ensuring a robust and representative sample.

The findings highlight six key predictors of quality of life: healthcare access, social participation, infrastructure satisfaction, environmental quality, emotional support, and social support. Quantitative analysis reveals that active participation in community activities and robust social networks are the most significant predictors, underscoring the importance of social inclusion and engagement. Qualitative insights illustrate how community events foster a sense of belonging and how accessible infrastructure supports mobility and independence.

This study contributes to urban design by integrating health and well-being considerations into rural community planning. The ecological model of ageing and the concepts of 'Ageing in Place' and 'Active Ageing' provide the theoretical framework, emphasising the dynamic interaction between older adults and their environments. The results underscore the necessity of multidisciplinary approaches in urban design that prioritise health and social equity, particularly in under-researched rural contexts. Given the unique socio-economic and cultural dynamics of rural Thailand, this research offers actionable insights for policymakers and community planners. By enhancing social participation, improving infrastructure, and ensuring accessible healthcare, communities can significantly improve the quality of life for their ageing populations. The study's holistic approach offers a model adaptable to similar rural settings globally, contributing to the discourse on sustainable urban design and community well-being.

Keywords

Quality of life · Ageing in place · Rural Thailand · Community residential environments · Elderly well-being

1 Introduction

1.1 Context and Importance of the Study

The global demographic shift towards an ageing population presents unique challenges and opportunities for societies worldwide, particularly in rural communities in developing countries like Thailand. As the world's population ages, the proportion of older adults is expected to double by 2050, reaching nearly 2.1 billion (United Nations 2019). This demographic transition is accompanied by increasing concerns about the quality of life of older adults, especially those living in rural areas where resources and services are often limited (World Health Organization 2015).

In Thailand, the proportion of older adults is rapidly increasing, particularly in rural areas. This demographic shift necessitates a re-evaluation of community residential environments to support ageing-in-place strategies, which

A. Nutley (✉)
Faculty of Architecture, Chiang Mai University,
Chiang Mai, Thailand
e-mail: alisa.h@cmu.ac.th

F. Alberti et al. (eds.), *Towards Sustainable Urban Regeneration: Utilizing Environmental, Cultural, and Social Assets*,
Advances in Science, Technology & Innovation, https://doi.org/10.1007/978-3-032-02091-8_36

allow older adults to live independently and comfortably in their communities for as long as possible. Ageing in place has been recognised as a critical component of healthy ageing strategies, offering benefits such as continuity of care, familiarity with the environment, and sustained social connections (Wiles et al. 2012).

Rural areas in Thailand, such as the village of Baan Pong Nuea in the Hang Dong district of Chiang Mai province, provide unique contexts for studying the quality of life among older adults. These areas are characterised by traditional agricultural lifestyles, limited access to urban infrastructure, and strong community bonds. Understanding the specific needs and experiences of older adults in these settings is crucial for developing targeted interventions that support their well-being and independence (Knodel and Chayovan 2008).

1.2 Research Gap

Despite the growing recognition of the importance of ageing in place, there is limited research on how community residential environments specifically impact the quality of life for older adults in rural Thailand. Most existing studies focus on urban settings or developed countries, leaving a significant gap in the literature regarding rural and developing contexts. Moreover, while factors such as healthcare access, social participation, and environmental quality are known to influence the quality of life, their specific interactions and relative importance in rural Thai communities remain underexplored.

While previous research has extensively covered urban ageing dynamics, rural areas, especially in Southeast Asia, have received less attention. Urban studies have highlighted the importance of built environments, social support, and accessibility to services in enhancing the quality of life for older adults (Lager et al. 2015; Scharlach and Lehning 2013). However, rural communities differ significantly in their social structures, economic activities, and availability of services, which can uniquely influence the ageing process. In Thailand, studies have often focused on the socio-economic aspects of ageing, with limited empirical research on how rural community environments impact the quality of life of older adults (Phaswana-Mafuya et al. 2013).

By addressing these gaps, this study aims to provide a holistic understanding of how rural community environments in Northern Thailand influence the quality of life for older adults. This knowledge is essential for developing targeted interventions and policies that promote healthy and active ageing in rural areas.

1.3 Study Objectives

The primary objectives of this study are:

1. To identify and analyse the key factors influencing the quality of life for older adults in rural Northern Thailand.
2. To explore the specific roles of healthcare access, social participation, infrastructure satisfaction, environmental quality, emotional support, and social support in enhancing the well-being of the elderly.
3. To provide recommendations for policymakers and community planners on how to develop and enhance community residential environments to better support ageing-in-place strategies.

By achieving these objectives, this study aims to contribute valuable insights that can inform the design of age-friendly communities and improve the quality of life for older adults in rural settings. This knowledge is essential for developing targeted interventions and policies that promote healthy and active ageing in rural areas.

2 Literature Review

2.1 Theoretical Framework

This study is grounded in the ecological ageing model, originally proposed by Lawton and Nahemow (1973). The ecological model emphasises the dynamic interaction between older adults and their environments, suggesting that the ability of older adults to age in place depends on the fit between their capabilities and the demands of their environment. This model considers various environmental domains, including the physical, social, and policy environments, and how these domains support or hinder the ageing process. By applying this theoretical framework, the study aims to understand how community residential environments in rural Thailand impact the quality of life for older adults.

The concepts of 'Ageing in Place' and 'Active Ageing' further enrich this framework. 'Ageing in Place' refers to the ability of older adults to live in their own homes and communities safely, independently, and comfortably, regardless of age, income, or ability level (Wiles et al. 2012). 'Active Ageing' promotes the idea of all individuals participating in social, economic, cultural, spiritual, and civic activities regardless of age (World Health Organization 2002). These frameworks are used to evaluate how well rural communities in Northern Thailand support these ideals.

2.2 Review of Related Literature

Several key factors have been identified in the literature as critical to the quality of life for older adults. These factors include healthcare access, social participation, infrastructure, environmental quality, emotional support, and social networks.

1. **Healthcare access** is essential for managing chronic conditions and maintaining overall health. Scharlach and Lehning (2013) and others (Goins et al. 2005; Arcury et al. 2005) note that rural areas often face challenges in healthcare access due to transportation issues and a shortage of providers.
2. **Social participation** enhances life satisfaction and reduces loneliness, providing mental and physical health benefits (Victor et al. 2005; Holt-Lunstad et al. 2010; Berkman et al. 2000). Community involvement offers a sense of purpose and belonging, crucial for maintaining mental well-being.
3. **Infrastructure**, including transportation and public spaces, supports mobility and independence. Clarke and Gallagher (2013) and others (Rosenbloom 2009; Marottoli et al. 2000) emphasise that good infrastructure is vital for older adults' engagement in community life.
4. **Environmental quality** significantly impacts physical and mental health. Jackson (2003) and other studies (Evans 2003; Sugiyama et al. 2008) highlight that clean air, green spaces, and safe housing conditions are crucial for older adults' well-being.
5. **Emotional and social support** from strong social networks is vital for coping with the challenges of ageing. Wenger and Burholt (2004) and others (Uchino 2004; Cohen and Wills 1985) discuss the role of supportive relationships in mitigating isolation and promoting mental and physical health.

Studies such as those by Wahl et al. (2009) emphasise the need for adapting environments to meet the needs of the ageing population, enhancing accessibility, safety, and social services. Research by Scharlach and Lehning (2013) emphasised the role of community design in facilitating social inclusion and independence among older adults in urban settings. However, rural areas present unique challenges that affect the quality of life and independence of the elderly. These challenges include a lack of health and social services and increased social isolation, as noted by Keating and Phillips (2008) and Milbourne (2012). Furthermore, Wenger and Burholt (2004) and Victor et al. (2009) highlight the importance of social networks and community engagement in rural settings. They found that strong social ties significantly contribute to well-being and life satisfaction among older adults, mitigating the risks associated with social isolation.

3 Methodology

3.1 Study Design

This study employs a mixed-methods approach, combining quantitative surveys with qualitative interviews, and a site survey to provide a comprehensive understanding of the factors influencing the quality of life for older adults in rural Northern Thailand. This design allows for the integration of numerical data with rich, contextual insights and an examination of the physical characteristics of community residential environments.

3.2 Study Area and Participants

The study focuses on rural Northern Thailand, specifically in the village of Baan Pong Nuea, located in the Hang Dong district of Chiang Mai province. This area was selected due to its significant elderly population and unique rural community characteristics, which include traditional agricultural lifestyles, limited access to urban infrastructure, and strong community bonds typical of Thai rural settings.

The study utilised Taro Yamane's formula to determine a sample size of 144 elderly participants from Baan Pong Nuea, ensuring a 95% confidence level and a 5% margin of error. A purposive sampling method was employed to select individuals aged 60 and above, who had lived in the village for at least 1 year. This criterion was set to ensure that participants were familiar with the local environment and social structures. All selected participants were capable of providing informed consent.

3.3 Data Collection

Data were collected using a mixed-methods approach to capture both quantitative and qualitative insights into the quality of life of older adults.

1. **Surveys**: The World Health Organization Quality of Life Instrument–Older Adults Module (WHOQOL-OLD) instrument was used to assess the quality of life quantitatively, focusing on physical health, psychological well-being, social relationships, and environmental satisfaction. The survey consisted of 45 questions, each rated on a 5-point Likert scale from 1 (very dissatisfied) to 5 (very satisfied). Trained research assistants, fluent in the local dialect, administered the surveys face to face to ensure participants fully understood the questions and provided accurate responses.
2. **Interviews**: Semi-structured interviews were conducted using an interview guide developed based on the study's objectives and existing literature. The guide included

questions on daily routines, community engagement, accessibility to services, perceived safety, and environmental features. Interviews were held with all 144 participants, each lasting approximately 45 minutes, either in the participants' homes or community areas for comfort and convenience. The interviews were audio-recorded with participants' consent and later transcribed for analysis.

3. **Site survey for physical characteristics**: A detailed site survey was conducted to document the current layout and physical features of the community's residential environments. This survey encompassed mobility and infrastructure, including the village's main roads, secondary roads, and alleys, to assess accessibility and connectivity. It also identified key public and government buildings, such as healthcare facilities, educational centres, and community halls. Additionally, the survey documented the locations of local businesses and their contribution to the community's economy. Residential areas were mapped to understand living arrangements, and natural features like water sources, green spaces, and topography were identified to evaluate their impact on the living environment.

3.4 Data Analysis

1. **Quantitative data analysis:** Descriptive statistics, such as means, standard deviations, and frequency distributions, were calculated to provide an overview of the participants' demographic characteristics and quality of life scores. A multivariate regression analysis was conducted to identify the significant predictors of quality of life among older adults in rural Northern Thailand. The model included key community environment factors and demographic variables to isolate their effects on the overall quality of life scores.
2. **Qualitative data analysis:** The interview transcripts underwent thematic analysis to identify recurring themes and patterns related to ageing in place. This process involved coding the data, categorising the codes into themes, and interpreting these themes in the context of the study's objectives. To enhance the comprehensiveness of the findings, the qualitative results were triangulated with the quantitative data, providing a deeper understanding of how community environments impact the quality of life for older adults.
3. **Site survey analysis:** The site survey data were analysed alongside the qualitative and quantitative data to gain a comprehensive understanding of the physical characteristics of the community residential environments. This analysis encompassed various aspects, including mobility and infrastructure, public and government buildings, business buildings, residential buildings, and environmental features. The study assessed how the layout and condition of roads and alleys affected mobility and access within the village. It also evaluated the accessibility and distribution of key public and government buildings, such as healthcare facilities, educational centres, and community halls, and their impact on residents' quality of life. Additionally, the role of local businesses in providing necessary services and contributing to the community's economy was examined. The analysis further explored residential buildings, focusing on living arrangements and their influence on the well-being of older adults. Lastly, the impact of natural features, such as water sources, green spaces, and topography, on the living environment and quality of life was assessed.

3.5 Ethical Considerations

Ethical approval for the study was obtained from the Ethics Committee of Chiang Mai University (Certificate of Approval No. 198/66). All participants provided informed consent before participation and were assured of their right to withdraw from the study at any time without any consequences. Data privacy and participant confidentiality were rigorously maintained throughout the study. Identifiable information was anonymised in the data analysis and reporting to protect the participants' privacy.

3.6 Validity and Reliability

To ensure the validity and reliability of the findings, the survey instrument and interview guide were pilot tested with a small sample of elderly residents from the village, allowing for refinement of questions and procedures. Additionally, research assistants received training to administer the surveys and conduct interviews, ensuring consistency and accuracy in data collection.

4 Results

4.1 Descriptive Statistics

The study's participant group predominantly comprised individuals aged 60–79, representing 89.6% of the total sample. Within this age range, the distribution was further broken down into 63.9% in the early stage (60–64 years), 25.7% in the middle stage (65–74 years), and 10.4% in the late stage (75 years and above). Gender distribution showed a higher proportion of females (56.9%) compared to males (43.1%). Regarding marital status, the majority of participants (61.1%) were currently married. In contrast, 28.5% were divorced or widowed, and 9.7% were single. Educational attainment among participants revealed that most had a primary school education or lower (86.8%), with smaller proportions having secondary school or vocational certificates (4.2%) and higher education, vocational diplomas, or equivalents (2.1%).

Living arrangements varied, with a significant number of participants living with their children (70.8%), followed by those living with their spouse (52.8%), relatives (22.2%), and a smaller group living alone (12.5%). When examining personal disposable income, a substantial majority (88.9%) reported having sufficient income, while 5.6% had insufficient income with debt, and another 5.6% had insufficient income without debt. Health and care needs indicated that most participants (77.1%) required full-time care, 19.4% reported no care needs, and 2.8% needed part-time care. Furthermore, nearly all participants had a long-term residence in the community, with 99.3% having lived there for 10 years or more.

These descriptive statistics provide a comprehensive overview of the demographic characteristics, living conditions, and socio-economic status of the elderly participants in this rural Northern Thai community. The data illustrate older adults' diverse backgrounds and needs, highlighting the importance of tailored interventions to enhance their quality of life.

4.2 Inferential Statistics

1. **Quality of Life Scores**

The overall quality of life scores among participants were very high, averaging 4.90 on a 5-point scale, reflecting an excellent perception of life among the elderly in this rural community. Participants expressed particularly high satisfaction with social relationships, scoring 4.97, and community support, scoring 4.82.

2. **Significant Predictors of Quality of Life**

Healthcare access: Access to local healthcare services was a significant predictor ($\beta = -0.126, p = 0.058$). Participants with better access to healthcare facilities reported higher quality of life scores (see Table 1).

Community participation: Active participation in community activities was a significant predictor ($\beta = -0.363, p = 0.002$). This underscores the vital role of social engagement in enhancing life satisfaction, as indicated by the analysis summarised in Table 1.

Community mobility: Satisfaction with mobility within the community was a significant predictor ($\beta = 0.315$, $p = 0.006$). Participants who were satisfied with their mobility reported higher quality of life scores, as reflected in the results shown in Table 1.

Community infrastructure and elements: The factor of the infrastructure and community elements was also a significant predictor ($\beta = 0.359$, $p < 0.001$). Participants who were satisfied with community infrastructure, such as roads, public buildings, and recreational spaces, reported higher quality of life scores. The data in Table 1 clearly illustrate this relationship.

Environmental quality: Environmental factors, such as clean air and green spaces, were significant predictors ($\beta = -0.824$, $p = 0.002$). Participants who were satisfied with their environmental conditions had higher quality of life scores, as demonstrated in Table 1.

Emotional support: Emotional support from family and friends was a significant predictor ($\beta = -0.462$, $p = 0.002$). Participants who received emotional support reported better overall well-being. This finding is corroborated by the results shown in Table 1.

Social support: Strong social support networks were a significant predictor ($\beta = 0.383, p = 0.005$). Participants with robust social support systems reported higher quality of life scores, as evidenced by the data in Table 1.

4.3 Qualitative Results

1. **Key theme:** The qualitative data from the in-depth interviews provided rich insights into the lived experiences of older adults in Baan Pong Nuea. Thematic analysis identified several key themes that illustrate the factors influencing the quality of life and ageing-in place experiences in this rural community.

Table 1 Regression coefficient analysis, N = 144 (source: Nutley, 2024)

Predictor variable	Unstandardised coefficients (B)	Std. error	Standardised coefficients (β)	t	Sig. (*p*-value)
Healthcare access	−0.297	0.155	−0.126	−1.922	0.058
Community participation	−0.226	0.073	−0.363	−3.086	0.002
Community mobility	0.344	0.122	0.315	2.818	0.006
Infrastructure and elements	0.653	0.122	0.359	5.374	0.000
Environmental quality	−1.950	0.624	−0.824	−3.124	0.002
Emotional support	−0.439	0.138	−0.462	−3.185	0.002
Social support	0.503	0.174	0.383	2.894	0.005

Theme 1: Community Support and Social Interaction

Many participants highlighted the importance of community support networks and frequent social interactions in maintaining their quality of life. One participant (P24) noted, 'Being able to participate in community events makes me feel valued and connected'. Social activities, such as local festivals, religious gatherings, and community meetings, were frequently mentioned as crucial for fostering a sense of belonging and purpose. These events provide opportunities for socialising and strengthen the communal bonds that are integral to the participants' well-being. Another participant (P104) shared, 'Our village is like a big family. We look out for each other, and that's what makes life here so special'. This sentiment underscores the role of strong social networks in enhancing life satisfaction and mitigating feelings of loneliness and isolation among older adults. These findings are consistent with previous research that emphasises the importance of social interactions and community support in promoting well-being among older adults (Wiles et al. 2012; Victor et al. 2009).

Theme 2: Infrastructure and Accessibility

Accessibility issues were frequently mentioned, especially related to transportation and healthcare services. Participants expressed a desire for more accessible healthcare facilities within the community. One participant (P26) explained, 'Getting to the hospital is difficult, especially for those of us who don't have family nearby to help. We need more healthcare services closer to home'. Transportation infrastructure also emerged as a critical factor. 'Some parts of the roads here are in poor condition, making it difficult for us to access the field', mentioned another participant (P49). The need for well-maintained roads, accessible public transportation, and pedestrian-friendly pathways was emphasised as essential for ensuring mobility and independence for older adults. These concerns align with findings from Clarke and Gallagher (2013), who noted that infrastructure plays a crucial role in enabling older adults to remain active and engaged in their communities.

Theme 3: Safety and Security

Participants generally felt safe in their community, significantly contributing to their overall well-being. However, concerns were occasionally expressed about inadequate street lighting and pedestrian safety. One participant (P42) commented, 'I feel safe here, but I wish there were more street lights. It gets very dark at night, and it can be dangerous to walk around'. Safety concerns also extended to traffic conditions. 'There are many cars on the main road and not enough sidewalks. It's risky for older people to walk on the roads', noted another participant (P129). These safety issues highlight the need for improved infrastructure to ensure a secure environment for all residents. These findings are supported by research from Keating and Phillips (2008), which emphasises the impact of safety on the quality of life of older adults.

Theme 4: Environmental Quality

The natural environment of Baan Pong Nuea plays a significant role in the participants' quality of life. Many participants appreciated the clean air, green spaces, and overall tranquillity of their rural setting. One participant (P135) shared, 'Living close to nature is very calming. The fresh air and green surroundings make me feel healthy and peaceful'. However, there were also concerns about environmental maintenance. 'Sometimes the village gets neglected, and some areas are not well maintained. We need regular upkeep to keep our environment pleasant', suggested another participant (P21). The quality of the environment and its maintenance were identified as critical factors influencing physical and mental well-being. Moreover, there were concerns about air quality due to pollution. 'Sometimes the air quality is really bad because of pollution. It's worrying because we breathe this air every day', suggested another participant (P122). The importance of environmental quality in supporting the well-being of older adults is well documented in the literature (Jackson 2003; Milbourne 2012).

2. **Physical characteristics:** A comprehensive site survey thoroughly documented the physical characteristics of Baan Pong Nuea. The survey identified various elements crucial to the daily lives of elderly residents and provided insights into how these physical features support or hinder their quality of life.

The illustrations provided in Figure 1 offer a detailed overview of Baan Pong Nuea, focusing on various aspects such as mobility, business buildings, public or government buildings, residential buildings, and environmental features. Integrating this visual data with the data collected from the surveys and interviews offers a comprehensive understanding of the community's infrastructure and environment, directly impacting the quality of life of the elderly residents.

Mobility and infrastructure: Key aspects of mobility and infrastructure, as illustrated in Figure 1, emphasise the importance of main roads and Provincial Highway 1269. These roads are critical for connecting the village to nearby urban areas and facilitating access to healthcare services and other essential amenities. The accessibility and quality of these roads likely contribute significantly to the high satisfaction with public transportation and infrastructure noted in the survey ($\beta = 0.315$, $p = 0.006$; Table 1). Additionally, the village's secondary roads and alleys, depicted in the figure, are vital for daily movement and access to local facilities. Participant comments about the need for better road markings and lighting (Participants 49 and 129) reflect the importance of maintaining and

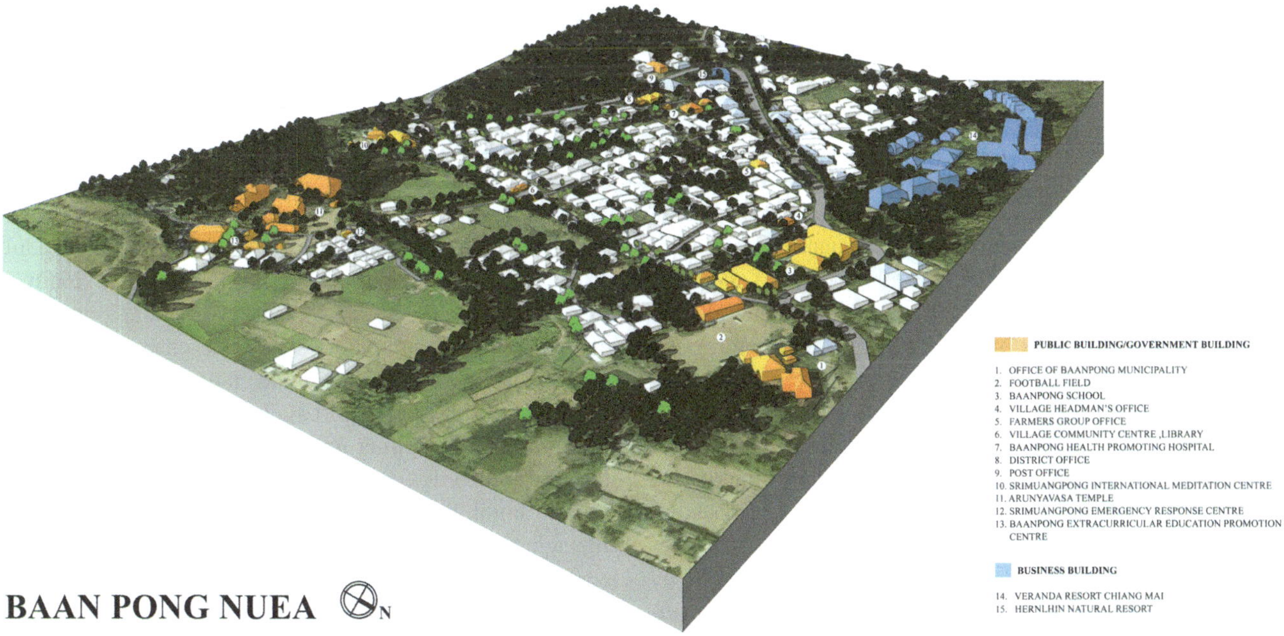

Fig. 1 Baan Pong Nuea's physical characteristics with the elements of the village (source: Nutley, 2024)

improving these pathways to enhance safety and mobility.

Public and government buildings: Several key public and government buildings are depicted in Fig. 1, such as the Baan Pong Health Promoting Hospital. This facility serves as a critical resource for the elderly, providing essential health services, a role underscored by the positive regression coefficient for healthcare access ($\beta = -0.126$, $p = 0.058$; Table 1). Educational and community centres, including Baan Pong School, the Village Community Centre, and the Baan Pong Extracurricular Education Promotion Centre, are also highlighted, serving as hubs for social interaction and community activities. The high participation in community activities ($\beta = -0.363$, $p = 0.002$; Table 1) and the strong emphasis on social support ($\beta = 0.383$, $p = 0.005$; Table 1) emphasise the importance of these spaces in fostering community engagement.

Business buildings: The presence of business establishments, detailed in Figure 1, includes small-scale businesses and resorts such as Veranda Resort Chiang Mai and Hernlhin Natural Resort. These businesses contribute to the local economy by providing employment opportunities and essential services. Their role in enhancing the community's social and economic vitality indirectly supports the well-being of elderly residents.

Residential buildings: Residential buildings, which form a major focus in Figure 1, depict the close-knit living arrangements within the community, where most participants (70.8%) live with family members. This set-up is vital for providing emotional and practical support, reflected in the high levels of emotional support ($\beta = -0.462$, $p = 0.002$; Table 1) and social support ($\beta = 0.383$, $p = 0.005$; Table 1) reported in the survey data. Such living arrangements foster strong social networks and robust support systems within the community.

Environmental features: The environmental features in Fig. 1, such as the Mae Tha Chang Canal, represent a crucial water source providing clean water, which is essential for health and well-being. This water source's reliability supports environmental quality, a significant predictor of quality of life ($\beta = -0.824$, $p = 0.002$; Table 1). Additionally, the area's natural topography, with its steep mountains and flat valleys, offers opportunities for physical activities and scenic beauty, enhancing residents' satisfaction with their living environment.

4.4 Integration of Quantitative and Qualitative Data

1. **Community support and social interaction:** The qualitative data highlighted the importance of community support networks and frequent social interactions, which aligns with the quantitative findings that indicated high satisfaction with social relationships and community support (average scores of 4.97 and 4.82, respectively). The regression analysis further confirmed that social support was a significant predictor of quality of life ($\beta = 0.383$, $p = 0.005$; Table 1). This suggests that strong social bonds and active community participation play crucial roles in enhancing the well-being of older adults, as both datasets

underscore the value of social engagement in this rural context. Previous studies have also shown that social networks and community engagement are crucial for older adults' mental and physical health (Wenger and Burholt 2004; Victor et al. 2009).

2. **Infrastructure and accessibility:** Participants' concerns about transportation and healthcare accessibility were mirrored in the quantitative data. Satisfaction with community mobility ($\beta = 0.315$, $p = 0.006$; Table 1) and community infrastructure ($\beta = 0.359$, $p < 0.001$; Table 1) were significant predictors of quality of life, highlighting the need for well-maintained roads, accessible public transportation, and local healthcare services. The qualitative accounts of difficulties in accessing healthcare and navigating poorly maintained roads reinforce the quantitative evidence that infrastructure improvements are essential for supporting the independence and mobility of older adults. These findings are consistent with research by Clarke and Gallagher (2013), which emphasises the critical role of infrastructure in enabling older adults to remain active and engaged in their communities, and by Davey (2006), who noted the importance of maintaining and adapting infrastructure to meet the needs of older homeowners.
3. **Safety and security:** The qualitative findings revealed that while participants generally felt safe, there were concerns about inadequate street lighting and road safety. This complements the quantitative data indicating the importance of environmental quality, including safety features, in influencing the quality of life ($\beta = -0.824$, $p = 0.002$; Table 1). The need for improved safety measures such as better street lighting and pedestrian-friendly pathways is clearly supported by both qualitative and quantitative data, emphasising the role of a secure environment in promoting the well-being of older adults. Research by Keating and Phillips (2008) and Jackson (2003) supports these findings, indicating that perceptions of safety are closely linked to mental well-being and physical activity levels among older adults.
4. **Environmental quality:** The natural environment's role in the participants' quality of life was evident in both datasets. Qualitative insights revealed an appreciation for clean air and green spaces, while concerns about air pollution highlighted environmental quality issues. Quantitative findings showed environmental quality as a significant predictor of quality of life ($\beta = -0.824$, $p = 0.002$; Table 1). This integration underscores the importance of maintaining and improving environmental conditions, such as air quality and green spaces, to enhance older adults' physical and mental well-being. These results align with the findings of Milbourne (2012) and Wahl et al. (2009), who highlight the importance of environmental quality in supporting the health and well-being of older adults.

5 Discussion

The findings from this study offer comprehensive insights into the factors influencing the quality of life for older adults in the rural community of Baan Pong Nuea, Northern Thailand. By integrating quantitative data, qualitative insights, and the community's physical characteristics, this research provides a holistic understanding of how various elements interplay to support ageing-in-place strategies.

Community support and social interaction: The high satisfaction scores for social relationships and community support highlight the crucial role of social networks in enhancing the well-being of older adults. The significant regression coefficient for social support ($\beta = 0.383$, $p = 0.005$; Table 1) underscores this point. These findings align with previous research indicating that strong social ties are vital for mental health and overall life satisfaction (Victor et al. 2009; Wenger and Burholt 2004). The qualitative data further enrich this understanding, with participants frequently mentioning the value of community events and social interactions in fostering a sense of belonging and reducing loneliness. Policies that promote community engagement and social participation are essential to maintain and enhance these social networks.

Infrastructure and accessibility: The importance of infrastructure and accessibility is evident from both the quantitative and qualitative data. Satisfaction with community mobility ($\beta = 0.315$, $p = 0.006$; Table 1) and infrastructure ($\beta = 0.359$, $p < 0.001$; Table 1) were significant predictors of quality of life. These findings are consistent with existing literature emphasising the role of a supportive built environment in enabling older adults to remain active and independent (Clarke and Gallagher 2013; Jackson 2003). The qualitative interviews revealed specific issues, such as the need for better road maintenance and more accessible healthcare facilities. These insights suggest that investments in infrastructure improvements, such as road repairs and the development of pedestrian-friendly pathways, are crucial for supporting the mobility and independence of older adults.

Safety and security: Safety and security emerged as critical factors influencing the quality of life, with concerns about inadequate street lighting and road safety frequently mentioned by participants. The significant impact of environmental quality on quality of life ($\beta = -0.824$, $p = 0.002$; Table 1) highlights the importance of a safe and secure environment. These findings are supported by previous research showing that perceptions of safety are closely linked to mental well-being and physical activity levels among older adults (Keating and Phillips 2008). Enhancing street lighting, improving road safety, and creating pedestrian-friendly environments are essential measures to ensure the safety and security of older residents.

Environmental quality: The role of environmental quality in enhancing the well-being of older adults is well documented in this study. Participants expressed appreciation for the clean air and green spaces, but concerns about air pollution were also noted. The quantitative data confirmed the significant impact of environmental quality on quality of life ($\beta = -0.824$, $p = 0.002$; Table 1). These findings align with existing literature highlighting the importance of environmental factors, such as air quality and green spaces, in promoting physical and mental health (Jackson 2003; Milbourne 2012). Efforts to maintain and improve environmental quality, such as reducing pollution and preserving green spaces, are crucial for supporting the well-being of older adults.

The integration of quantitative data, qualitative data, and the physical characteristics of Baan Pong Nuea offers a comprehensive understanding of the factors influencing the quality of life for older adults in this rural community. Quantitative analysis identifies key predictors of quality of life, including community support, infrastructure, healthcare access, and environmental quality. These findings are corroborated by qualitative insights, which highlight the importance of social interactions, accessible healthcare, and well-maintained infrastructure. The physical characteristics data provide context by detailing the locations and conditions of community resources and infrastructure, emphasising their crucial roles in daily life.

These findings reveal the complex interplay between social, infrastructural, and environmental factors in supporting the well-being of older adults. High satisfaction with social relationships and community support, alongside significant predictors such as community mobility and environmental quality, underscore the need for targeted interventions. Enhancing healthcare accessibility, improving road conditions, and maintaining environmental quality are critical areas for policymakers and community planners to address. Such interventions will create a supportive environment that fosters independence and well-being for the elderly population in rural settings, aligning with ageing-in-place strategies and ensuring a high quality of life.

6 Implications for Policy and Practice

The findings of this study have significant implications for policymakers and practitioners aiming to enhance the quality of life for older adults in rural communities. The integration of quantitative and qualitative data provides a comprehensive understanding of the key factors influencing well-being, which can inform the development of targeted interventions and policies. Here are the main implications for policy and practice:

1. **Enhancing social support networks**: Strong social support networks are critical for the well-being of older adults, as evidenced by their significant impact on quality of life scores ($\beta = 0.383$, $p = 0.005$; Table 1). Policies should focus on strengthening community ties and social networks through community engagement initiatives. Creating more opportunities for social interaction, such as community events, social clubs, and volunteer programmes, can significantly enhance the quality of life for older adults. Local governments and community organisations should prioritise funding and support for these initiatives to foster a sense of belonging and reduce social isolation.
2. **Improving accessibility**: Accessibility to essential services and infrastructure is a crucial determinant of the quality of life for older adults. The study highlights the importance of community mobility ($\beta = 0.315$, $p = 0.006$; Table 1) and community infrastructure and elements ($\beta = 0.359$, $p < 0.001$; Table 1) in supporting independence and well-being. Policymakers should prioritise the development of age-friendly infrastructure such as ramps, benches, well-maintained walking paths, and accessible public transportation. Ensuring that public spaces and transportation options are accessible to older adults can greatly improve their mobility and independence.
3. **Ensuring safety and security**: Implement improved safety measures, such as better street lighting and clearly marked roads. These changes can enhance the sense of security among older adults and encourage more outdoor activities.
4. **Promoting health and wellness programmes**: Healthcare access and support are significant predictors of quality of life ($\beta = -0.126$, $p = 0.058$; Table 1). Expanding and supporting community health initiatives that provide regular health check-ups, wellness programmes, and health education can help manage chronic conditions and promote healthy ageing. Policymakers should ensure that healthcare services are accessible and affordable to all older adults, particularly in rural areas where healthcare facilities may be limited. Mobile health clinics and telehealth services can also play a crucial role in reaching remote populations.
5. **Maintaining environmental quality**: Environmental quality, including clean air, green spaces, and well-maintained public areas, significantly contributes to the physical and mental well-being of older adults ($\beta = -0.824$, $p = 0.002$; Table 1). Policymakers should focus on preserving and enhancing environmental conditions in rural communities. This includes implementing policies that reduce pollution, protect natural resources, and promote sustainable land use practices. Investments

in green spaces and recreational areas can provide older adults with opportunities for physical activity and social interaction, further enhancing their quality of life.

By addressing these areas, communities can create environments that support and enrich the lives of their ageing populations.

7 Limitations and Future Research

This study has several limitations, including a sample size that may not represent all rural areas in Northern Thailand and a cross-sectional design that limits the ability to infer causality. Future research should consider longitudinal studies to better understand the long-term effects of community environments on quality of life.

8 Conclusion

This study explored the influence of community residential environments on the quality of life for older adults in a rural area in Northern Thailand. The findings highlight the critical role of social support, accessible infrastructure, safety, health programmes, and environmental quality in enhancing the well-being of older adults. Addressing these key areas can facilitate the development of targeted interventions and policies to support ageing populations in rural Thailand and similar contexts worldwide. These insights carry the potential to inform sustainable urban design and community well-being, contributing to the global discussion on ageing-in-place strategies.

Acknowledgements The author would like to express gratitude for being supported by the CMU Junior Research Fellowship Program at Chiang Mai University, Thailand, which provided essential funding. Heartfelt thanks are extended to the participants from the Baan Pong Nua community for their valuable contributions and willingness to share their experiences. Special appreciation is also given to the research assistants for their dedication and professionalism in data collection. Finally, the author wishes to thank the university's faculty and staff for their valuable guidance and assistance throughout the research process.

References

Arcury TA, Gesler WM, Preisser JS, Sherman J, Spencer J, Perin J (2005) The effects of geography and spatial behavior on health care utilization among the residents of a rural region. Health Serv Res 40(1):135–156

Berkman LF, Glass T, Brissette I, Seeman TE (2000) From social integration to health: Durkheim in the new millennium. Soc Sci Med 51(6):843–857

Clarke P, Gallagher NA (2013) Optimizing mobility in later life: the role of the urban built environment for older adults aging in place. J Urban Health 90(6):997–1009

Cohen S, Wills TA (1985) Stress, social support, and the buffering hypothesis. Psychol Bull 98(2):310–357

Davey JA (2006) "Ageing in place": the views of older homeowners on maintenance, renovation and adaptation. Soc Policy J N Z 27:128–141

Evans GW (2003) The built environment and mental health. J Urban Health 80(4):536–555

Goins RT, Williams KA, Carter MW, Spencer M, Solovieva T (2005) Perceived barriers to health care access among rural older adults: a qualitative study. J Rural Health 21(3):206–213

Holt-Lunstad J, Smith TB, Layton JB (2010) Social relationships and mortality risk: a meta-analytic review. PLoS Med 7(7):e1000316

Jackson RJ (2003) The impact of the built environment on health: an emerging field. Am J Public Health 93(9):1382–1384

Keating N, Phillips J (2008) A critical human ecology perspective on rural ageing. In: Keating N (ed) Rural ageing: a good place to grow old? Policy Press, Bristol, pp 1–10

Knodel J, Chayovan N (2008) Population ageing and the Well-being of older persons in Thailand. In: Population Studies Center Research Report. University of Michigan

Lager D, Van Hoven B, Meijering L (2015) Places that matter: place attachment and wellbeing of older Antillean migrants in The Netherlands. Eur Spat Res Policy 22(1):27–40

Lawton MP, Nahemow L (1973) Ecology and the aging process. In: Eisdorfer C, Lawton MP (eds) The psychology of adult development and aging. American Psychological Association, Washington, DC, pp 619–674

Marottoli RA, Mendes de Leon CF, Glass TA, Williams CS, Cooney LM Jr, Berkman LF (2000) Consequences of driving cessation: decreased out-of-home activity levels. J Gerontol Ser B Psychol Sci Soc Sci 55(6):S334–S340

Milbourne P (2012) Growing old in rural places. J Rural Stud 28(4):315–324

Phaswana-Mafuya N, Peltzer K, Chirinda W, Musekiwa A, Kose Z, Hoosain E, Davids A (2013) Self-rated health and associated factors among older south Africans: evidence from the study on global ageing and adult health. Glob Health Action 6(1):19880

Rosenbloom S (2009) Meeting transportation needs in an aging-friendly community. In: Cisneros H, Nyquist LM, Dyer-Chamberlain J, Hickie J (eds) Independent for life: homes and neighborhoods for an aging America. University of Texas Press, Austin, pp 129–157

Scharlach AE, Lehning AJ (2013) Ageing-friendly communities and social inclusion in The United States of America. Ageing Soc 33(1):110–136

Sugiyama T, Leslie E, Giles-Corti B, Owen N (2008) Associations of neighbourhood greenness with physical and mental health: do walking, social coherence and local social interaction explain the relationships? J Epidemiol Community Health 62(5):e9

Uchino BN (2004) Social support and physical health: understanding the health consequences of relationships. Yale University Press, New Haven

United Nations (2019) World population prospects 2019: highlights. Department of Economic and Social Affairs, Population Division

Victor CR, Scambler SJ, Bowling A, Bond J (2005) Loneliness in later life: a follow-up study. Aging Ment Health 9(3):257–265

Victor CR, Scambler SJ, Bowling A, Bond J (2009) The prevalence of, and risk factors for, loneliness in later life: a survey of older people in Great Britain. Ageing Soc 25(3):357–375

Wahl H-W, Fänge A, Oswald F, Gitlin LN, Iwarsson S (2009) The home environment and disability-related outcomes in ageing individuals: what is the empirical evidence? The Gerontologist 49(3):355–367

Wenger GC, Burholt V (2004) Changes in levels of social isolation and loneliness among older people in a rural area: a twenty-year longitudinal study. Canadian J Aging/La Revue canadienne du vieillissement 23(2):115–127

Wiles JL, Leibing A, Guberman N, Reeve J, Allen RES (2012) The meaning of "aging in place" to older people. The Gerontologist 52(3):357–366

World Health Organization (2015) World report on ageing and health. WHO, Geneva

World Health Organization (WHO) (2002) Active ageing: a policy framework. World Health Organization, Geneva

The manufacturer's authorised representative in the EU is Springer Nature Customer Service Centre GmbH, Europaplatz 3, 69115 Heidelberg, Germany. If you have any concerns regarding our products, please contact ProductSafety@springernature.com

Printed and bound by CPI Group (UK) Ltd, Croydon, CR0 4YY
21/09/2026
02219969-0003